ULTRA-REALISTIC

IMAGING

ADVANCED TECHNIQUES IN ANALOGUE AND DIGITAL COLOUR HOLOGRAPHY

ULTRA-REALISTIC IMAGING

ADVANCED TECHNIQUES IN ANALOGUE AND DIGITAL COLOUR HOLOGRAPHY

HANS BJELKHAGEN

DAVID BROTHERTON-RATCLIFFE

CRC Press
Taylor & Francis Group
Boca Raton London New York

CRC Press is an imprint of the
Taylor & Francis Group, an **informa** business

A TAYLOR & FRANCIS BOOK

CRC Press
Taylor & Francis Group
6000 Broken Sound Parkway NW, Suite 300
Boca Raton, FL 33487-2742

First issued in paperback 2020

© 2013 by Taylor & Francis Group, LLC
CRC Press is an imprint of Taylor & Francis Group, an Informa business

ISBN 13: 978-0-367-57643-1 (pbk)
ISBN 13: 978-1-4398-2799-4 (hbk)

Library of Congress Cataloging-in-Publication Data

Bjelkhagen, Hans I.
　Ultra-realistic imaging : advanced techniques in analogue and digital colour holography / Hans Bjelkhagen, David Brotherton-Ratcliffe.
　　pages cm
　　Includes bibliographical references and index.
　　ISBN 978-1-4398-2799-4 (hardback)
　　1. Holography. 2. Image processing--Digital techniques. I. Brotherton-Ratcliffe, David. II. Title.

TA1540.B54 2012
621.36'75--dc23

2012044176

**Visit the Taylor & Francis Web site at
http://www.taylorandfrancis.com**

**and the CRC Press Web site at
http://www.crcpress.com**

Contents

Foreword

It is my pleasure to write the Foreword to this book. I have known the authors, as well as their professional accomplishments in this field, for three decades.

Hans I. Bjelkhagen has devoted his entire professional life to holography. He is most well known for advancing the fields of silver-halide materials and their processing, interferometry, pulsed portraiture, Lippmann photography, and ultrarealistic color holography. He taught annual summer workshops at Lake Forest College from 1982 to 1997 that included courses on color holography, photochemistry, and pulsed portraiture.

As leader and organizer, Bjelkhagen cochaired with me the triennial International Symposium on Display Holography (ISDH) until I retired in 1997. In 2005, he revived and chaired a successful ISDH in the United Kingdom. With this encouragement, we held an ISDH in China in 2008, where he was a cochairman. He also helped organize the 2012 ISDH at the Massachusetts Institute of Technology. He is the current chairman of the annual Practical Holography Conference in California sponsored by the International Society for Optical Engineers (SPIE). Less well known is that he is a meticulous collector of all information pertaining to holography, maintains a detailed directory on holographers past and present, and collects holographic postage stamps that have been issued internationally.

David Brotherton-Ratcliffe is a polymath whose career spans both theoretical and experimental sciences including plasma physics, magnetohydrodynamics, nuclear fusion, laser engineering, diffractive optics, holography, and advanced systems design for the recording and production of both analogue and digital large-format holograms. His recent work in pulsed laser portraiture systems and computer-generated full-color, full-parallax holograms as presented in this book is truly state-of-the-art.

An abundance of books and articles on holography at the elementary level have already been published over the years since the mid-1960s. It is not necessary to repeat their contents here. This book is a condensation at the highest level of the authors' collective knowledge and experience in the area of display holography that includes history, theory, practice, and detailed designs of industrial systems. It includes over 500 photographs, diagrams, and design schematics for the recording and production of ultrarealistic holographic images.

Tung H. Jeong

Preface

Ultra-Realistic Imaging may be defined as any imaging technique that is able to record and reconstruct the visible electromagnetic light field scattered from a real-world object or scene with a resolution better than or equal to that of the unaided human eye. This book is devoted to a discussion of how the goal of ultra-realistic imaging may be attained through the application of the interferential methods of modern analogue and digital holography—and in particular through volume phase holography.

Holography was discovered more than 60 years ago. With the discovery of the laser and the off-axis technique in the early 1960s, it very quickly became obvious that the method constituted a unique imaging principle with enormous possibilities. However, the quality of the holographic image depended intrinsically on many things—such as the photosensitive material, the chemical processing scheme, the recording laser and not least, the illumination source with which the recorded hologram had to be viewed. This innate dependence of holography on critical associated technologies has been responsible for the waves of great enthusiasm for the subject being followed by periods of equally great disillusionment.

Holography has of course had its successes. The embossed security hologram and packaging industries are testament to this, but the role of holography as a technique of true ultra-realistic imaging is only just now becoming possible. This has been due to great progress in the key associated technologies of photosensitive materials, recording lasers, light sources, spatial light modulator technology and computers.

Digital techniques in holography are particularly exciting. Very large, high-definition full-colour holograms can now be written as a matrix of tiny elementary holograms, rather like a modern high-definition television screen is made up of pixels. One of the applications becoming possible with this technology is the creation of high virtual volume (HVV) displays. These are full-colour, full-parallax holograms that can replay 3D scenes having volumes from tens of cubic metres to cubic kilometres. Well-known holographers such as Nick Phillips and Paula Dawson took the first steps towards this type of display in the 1970s and 1980s using analogue laser transmission holography. Virtual volumes of tens of cubic metres were realised, but the images were noisy, monochromatic and had to be illuminated by bulky lasers. The new digital HVV holographic displays will have none of these deficiencies. In their ultimate incarnation, HVV displays can be used to create virtual holographic windows—these are ultra-realistic displays that seek to mimic a real window with a virtual 3D scene. As "space with a view" becomes an ever more costly commodity within the context of humanity's inextricable progression to increasingly high-density urban environments, holographic windows may well provide a valuable solution.

Such applications are of course simply light-years away from the conventional holography that most people are used to today. This radical transformation is only becoming possible now because of great progress in many associated fields. The information content of an HVV hologram is enormous—the required image processing has simply not been practical until recently. Likewise, the image data could not be written to a hologram at a sufficiently high resolution without the recent progress in HD spatial light modulator technology. Panchromatic photosensitive materials of sufficient resolution were not previously known. And compact narrowband laser and LED illumination technology was not available for the illumination of such displays.

This book is certainly not intended as a basic introduction to holography. There are already many excellent books fulfilling this need. Rather, it brings together a discussion of key methods that enable holography to be used as a technique of ultra-realistic imaging.

The book starts with a historical review of progress to date in holography. We felt that this was merited from a contextual point of view and we have also taken the liberty to document some of our personal work here. Chapter 2 is devoted to Lippmann photography. This 100-year-old interferential colour photographic technique is relatively unknown, but Lippmann photographs offer exceptional colour fidelity. When Lippmann photography was introduced by Gabriel Lippmann, the impression his type of colour images made on photographers of the day was clearly expressed by Edward Steichen who wrote to Stieglitz in 1908:

> Professor Lippmann has shown me slides of still-life subjects by projection, that were as perfect in colour as in an ordinary glass positive in the rendering of the image in monochrome. The rendering of white tones was astonishing, and a slide made by one of the Lumière brothers, at a time when they were trying to make the process commercially possible, a slide of a girl in a plaid dress on a brilliant sunlit lawn, was simply dazzling, and one would have to go to a good Renoir to find its equal in colour luminosity.

A discussion of CW recording lasers for holography is given in Chapter 3. This has been an area of great progress and there is every reason to expect that such progress will continue at an even greater rate into the future. Small DPSS and semiconductor lasers can now be integrated into portable full-colour holographic cameras. Such systems have potential for use in areas such as museum archives. Museums such as the British Museum in London and the Louvre in Paris have shown real interest here. Pulsed holography lasers are reviewed in Chapter 6. Recent progress in RGB pulsed laser design has been instrumental in achieving high-quality digital colour holograms; pulsed lasers constitute a crucial technology for HVV displays. Detailed optical designs are reviewed for many of the principal laser types with emphasis on attaining the parameters necessary for digital and analogue holography.

A full review of current photosensitive materials for colour holography is given in Chapter 4. Such materials are totally key to ultra-realistic holographic imaging. Some great materials are available today and there are indications that further progress will occur in the field. For example, processing-free photopolymers with index modulations approaching and even surpassing those observed in dichromated gelatin have been discovered. Modern methods of analogue holography are covered in Chapter 5; the latest work in this field has demonstrated the production of holograms that are almost indistinguishable from real objects. Work has underlined the importance of choosing the correct recording lasers here and progress in laser engineering has greatly helped this field.

Chapter 7 is devoted to the relatively new but extremely exciting field of digital holographic printing. Digital holographic printing is distinct from computer-generated holography (CGH). Here, we describe the detailed design of various types of digital holographic printers. We explain how ultra-realistic volume phase holograms may be printed as a matrix of elementary volume holograms using computer image data. Unlike CGH, each elementary volume hologram is created by an optical interference process. We discuss the generation of HVV displays and the design of HVV printers. The image processing algorithms required for the different types of digital holograms are developed in depth in Chapters 8 and 9. Chapter 8 introduces the mathematical and geometrical notation but is otherwise devoted mostly to horizontal parallax-only holograms. Chapter 9 deals with practical computational algorithms required for the full-parallax case.

3D image data acquisition systems must be used when digital holograms are to be printed of real-world objects or scenes. The most popular type of system is the *Holocam*—a camera on one or more motorised rails. These systems and the image processing algorithms required to convert the raw image data to the format required by digital printers are reviewed in Chapter 10. Other techniques such as structural light are also reviewed.

Chapters 11 and 12 are devoted to physical theory of the holographic grating and the hologram. Here, we develop various models from first principles. Paraxial and fully non-paraxial formulae are derived for image distortion, image blurring and chromatic aberration. Kogelnik's coupled wave theory is derived from first principles and expressions for diffractive efficiency are given. We also review N-coupled wave theory and discuss the question of diffuse holograms and polychromatic gratings. Of special interest is Chapter 12, which describes a new theory capable of treating the polychromatic grating as an infinity of parallel stacked mirrors. This theory lends a useful insight to the interpretation of Kogelnik's model—an issue which is not usually taken up in standard texts but which is nevertheless of some importance for a proper understanding of the process of holographic diffractive reflection and transmission.

Illumination sources are of fundamental importance to holography as they dictate how the holographic image replays. We give an up-to-date review of these sources in Chapter 13. Of particular importance are the new LED and laser diode sources. These devices are characterised by a much smaller value of étendue, a high power and a narrower bandwidth. They may be expected to improve, in a rather fundamental way, the displayed image properties of the polychromatic volume reflection hologram.

Finally, Chapter 14 contains a review of some of the most important applications of ultra-realistic holography. We include a section on scientific imaging where holographic microscopy, holography endoscopy and bubble chamber holography are discussed. Sections are also included on how digital holography can be used in advertising and display, urban planning, military mapping and architecture. Analogue holography is discussed in relation to its increasing interest from museums as a vehicle for both archival and travelling exhibitions. The book ends with a section on updateable and real-time digital full-colour holographic displays.

We have included a number of (mostly technical) appendices that should be of interest to workers in the field. Wherever possible in the book, we have tried to include enough detail so that the experienced reader may actually start using the techniques described. It is the authors' hope that this book will fill a gap that currently exists in the technical literature by providing a comprehensive treatment of holography and its key associated fields in the context of ultrahigh-fidelity full-colour imaging. By necessity, familiarity with a number of relatively advanced topics is assumed. Some of the chapters are completely non-mathematical; others, such as Chapters 8 through 10, despite the apparent complexity of expressions, only require a basic knowledge of mathematics. Chapters 11 and 12 probably require a slightly deeper knowledge, but nothing more than the mathematics learnt at the second or third year of a typical undergraduate course in mathematical physics. The book is, however, designed to be relatively modular and omission of the more mathematical chapters should not, in general, preclude a reading of the less mathematical ones.

Finally, we should point out that we have included within the book a number of photographs of holograms. In many previous books, it has always been possible to discriminate between a photograph of an actual object and a photograph of the hologram of the object. However, as the field progresses and the quality of images increases, this discrimination becomes increasingly difficult. Various hologram images we present here, particularly those recorded recently using the technique of full-colour analogue holography, may therefore appear to simply be photographs—which, of course, is not the case!

H. I. Bjelkhagen
D. Brotherton-Ratcliffe
London, UK

Acknowledgments

This book would not have materialised without the inspiration and encouragement of many people. Various people and institutions have also helped in providing material, which has made the book more interesting and more complete. First and foremost, we would both like to thank John Navas of Taylor & Francis, CRC Press for his encouragement and support in writing this book. Thanks also to Rachel Holt and Amber Donley of Taylor & Francis.

DBR: I would like to especially thank Nicholas Phillips, even though, sadly Nick is no longer with us, for the interesting and penetrating discussions I had with him, for his never-waning enthusiasm, and most importantly for his amazing inspiration: it was Nick's work that inspired me to take up holography in the first place. I would also like to say a special thank you to Stanislovas Zacharovas for his continual support with Geola over the years. I would like to thank Mikhail Grichine and Alexej Rodin for their part in introducing me to pulsed laser photonics and for many years of invaluable discussions; thanks also to Marcin Lesniewski for his vitally important input in the area of high numerical aperture Fourier transform objective design. I would like to thank Ramunas Bakanas for our long discussions on all aspects of laser physics and on the optics of digital holographic printers—and also for his unwavering commitment to Geola. I would like to thank Lev Isacenkov, without whom the Geola organisation would never have existed, and Eric Bosco and Lynne Hrynkiw, the founders of XYZ Imaging Inc., for their constant enthusiasm and great contribution to the practical realisation of direct-write digital holography. Thanks also to Yuri Sazonov and Olga Gradova, without whom Geola's work in digital holography would not have gotten very far, through lack of a suitable photosensitive film. Many thanks also to Paula Dawson, Juyong Lee and Jesper Munch for their support, first of Australian Holographics Pty. Ltd., and then of the Geola organisation. I should also mention Peter Hering from the Caesar Institute and Rob Taylor of Forth Dimension Holographics Inc. I would like to thank Andrej Nikolskij, Julius Pileckas and Jevgenij Kuchin at Geola, not only for their help in checking my formulae in Chapters 8 and 9, but also for their commitment over the years to Geola. Thanks also to John Fleming, who worked with me on the design project that led to Appendix 6, and to Chris Chatwin and Rupert Young for our fruitful collaborations at the University of Sussex. Finally, I would like to thank key members of the teams at Australian Holographics Pty. Ltd., LMC France Instruments SARL, Geola Technologies Ltd. (GT) and at the Centre for Laser Photonics (CLP) in North Wales—Geoff Fox (AH), Svetlana Karaganov (AH), Simon Edhouse (AH), Yasmin Farquhar (AH), Vytas Serelis (AH), Florian Vergnes (LMC and later manager of Geola UAB), Jerzy Lelusz (GT/CLP), Nataly Vidmer (GT/CLP), Lishen Shi (GT/CLP), Junhua Lu (GT) and John Tapsell (GT).

HB: I would like to thank Nils Abramson at the Royal Institute of Technology in Stockholm, who introduced me to holography in 1968 and who was responsible for my decision to select holography as a professional career. My Swedish colleagues, Per Skande and Johnny Gustafsson were involved in many of the commercial display holography projects carried out at Holovision AB in the 1970s. I would also like to express my gratitude to Tung H. Jeong, Lake Forest College, Illinois, for his collaboration in several projects including the *Lake Forest International Display Holography Symposia* with which I have been involved since 1982. Ed Wesly, from Chicago, has collaborated in many of my projects, both at Fermilab, Northwestern University, and holography business activities in Chicago. Thanks also to Max Epstein and Michel Marhic, Northwestern University, who were involved in many of the research projects carried out when I was working at the University in Evanston as well as being my colleagues at Holicon Corp. and the Light Wave hologram galleries in the United States. I am particularly thankful to Nicholas Phillips who invited me in 1997 to join him at the Centre for Modern Optics (CMO) at De Montfort University in Leicester, UK. Together, we worked on improving silver halide materials and processing methods, even after CMO moved to Wales. In particular, Nick had a lot of input into the European *SilverCross* emulsion project. Ardie Osanlou, my long-time CMO colleague, is gratefully acknowledged for working with me on many of our holography projects. Peter Crosby and Evangelos Mirlis, one of my PhD students, have both been particularly helpful in the emulsion and museum holography projects. In regard to my Lippmann photography research, I must acknowledge Darran Green's important contribution. Since the 1990s, one of my main interests in holography has been to develop ultra-realistic 3D images, which required recording high-quality colour holograms, in particular, of the Denisyuk single-beam reflection type. My very first such experiments were carried out in France together with Dalibor Vukičević, Strasbourg University, who devoted a lot of his time helping to record the first successful results. I had the honour to know and meet both Emmett Leith and Yuri Denisyuk many times. I am very thankful for the many valuable discussions I had with them and for the support they gave me in my research on holographic recording materials and display holography applications. I want to also mention Gennady and Svetlana Sobolev, Michael Shevtsov and Vladimir Markov, who for many years, have been very helpful, providing me with insights into ultrafine-grain silver halide emulsions. In addition, I would like to mention my Bulgarian colleague and friend, Ventseslav Sainov, who has worked with me on emulsions and colour holography projects. Last but not least, I would like to express my deep gratitude to Teresa Bjelkhagen, who provided me with translations of important Russian publications such as N. I. Kirillov's 1979 book on ultrahigh-resolution emulsions. She has also supported my work in holography from the beginning.

Authors

Dr Hans I. Bjelkhagen, Professor Emeritus of Interferential Imaging Sciences at Glyndŵr University, Centre for Modern Optics (CMO), located in North Wales, UK, was awarded his Doctoral Degree in 1978 by the Royal Institute of Technology in Stockholm, Sweden.

Over the last 15 years, Bjelkhagen has received much international recognition for his work in the field of colour holography and holographic recording materials. He has specialised in recording Denisyuk-type colour holograms. He has also researched and improved Lippmann photography over a period of many years.

In 1983, Bjelkhagen joined CERN in Geneva, Switzerland, where he was involved in the development of bubble chamber holography. A year later, he participated in an international team project, recording holograms in the 15-foot bubble chamber at Fermilab in Batavia, Illinois. Between 1985 and 1991, he was employed at Northwestern University, in Illinois, working on medical applications of holography.

In 1997, Bjelkhagen was invited by Professor Nick Phillips to join him at CMO at De Montfort University, Leicester, England. In 2004, CMO moved to the then newly established OpTIC in Wales.

In addition to scientific applications, Bjelkhagen is a well-known holographer who has recorded many holograms for 3D display purposes. From his early years in the field, he has been involved in large-format, high-quality display holography, using both pulsed and CW lasers. He has recorded many unique art objects, such as the Swedish *Coronation Crown of Erik XIV* (the crown dates back to 1561), and the Chinese *Flying Horse from Kansu* (from 100 AD). Bjelkhagen has worked with a number of famous artists, for example, Carl Fredrik Reuterswärd, creating holograms exhibited in many art museums and galleries around the world.

Bjelkhagen has also used pulsed holography to record a number of holographic portraits. In 1989, he recorded a portrait of the inventor of single-beam reflection holography, Yuri Denisyuk. The most famous person recorded by Bjelkhagen was President Ronald Reagan. His portrait was recorded on 24 May 1991. This was the first and, so far, the only holographic portrait recorded of an American President. A copy of this holographic portrait is held in The National Portrait Gallery of the Smithsonian Institution in Washington, DC.

Bjelkhagen has published more than 100 papers in refereed journals and conference proceedings, and holds 14 international patents. His most important academic contribution is a book on *Silver-Halide Recording Materials for Holography and Their Processing* published by Springer. He is a member of the Optical Society of America and is a fellow of the International Society for Optical Engineering (SPIE). He is the Chairman of SPIE's Photonics West Practical Holography Conference and SPIE's Holography Technical Group. He is an Accredited Senior Imaging Scientist and Fellow of the Royal Photographic Society (RPS) as well as Chairman of the RPS 3D Imaging & Holography Group. In 2001, he received the RPS Saxby Award for his work in holography, and in 2011, the Denisyuk Medal, from the D.S. Rozhdestvensky Optical Society, Russia.

Dr David Brotherton-Ratcliffe is the founder and scientific director of the well-known laser physics and holography organisation, Geola. He obtained a BSc (Hons) in Physics and Astrophysics from Queen Mary College, London University, in 1981. In 1984, while still at London University but now seconded to the United Kingdom Atomic Energy Authority at Culham Laboratories, he received a PhD for his work in nuclear fusion and magnetohydrodynamics. From 1985 to 1989, he continued to work as a theoretical physicist at the Flinders University of South Australia.

Brotherton-Ratcliffe first started to work in holography in 1982 during his doctoral studies, but it was not until 1989 that he founded Australian Holographics Pty. Ltd. and began working full-time in the fields of holography, optics and laser physics. During the 1990s, Australian Holographics became well known for its large-format display holograms, which were successfully marketed in Australia and throughout the Asia-Pacific region, often attracting significant media coverage.

In 1992, Brotherton-Ratcliffe founded the Geola organisation in Lithuania and began working seriously on high-energy pulsed laser technology. He founded associated companies in Australia, France and Romania, and travelled frequently between these countries for several years.

In 1999, Brotherton-Ratcliffe patented a key idea that proved to be highly influential: the printing of full-colour digital holograms, dot by dot, using RGB pulsed lasers. Over the years since 1999, Brotherton-Ratcliffe has come to be recognised as one of the leading workers in the expanding field of digital holographic printing.

Brotherton-Ratcliffe is the author of more than 14 patent families and over 60 publications in refereed journals and conference proceedings. He has published in the fields of plasma physics, magnetohydrodynamics, nuclear fusion, theoretical physics, laser physics, optics, holography and aerodynamics.

1

Ultra-Realistic Imaging and Its Historical Origin in Display Holography

1.1 Ultra-Realistic Imaging and Interferential Techniques

Modern high-definition flat-screen television displays offer a realism simply unthinkable just 10 years ago. Many such televisions now even offer the possibility of displaying high-quality three-dimensional (3D) images when used in conjunction with special glasses. From a certain perspective, today's television technology might well be regarded as falling within the category of "ultra-realistic imaging". However, if one examines the intrinsic information content of today's high-definition televisions, it is in fact far inferior to that of interferometric displays such as holograms or Lippmann photographs.

This book is about the technology of such interferometric systems and, most notably, about holography. Although the technologies underpinning the current display revolution are relatively mature, these technologies are nevertheless subject to certain well-defined limits. This is particularly true with 3D display. In contrast, interferometric displays, although they must still be regarded as being in their infancy, offer a clear potential to overcome some of the most important limits.

Interferometric displays have the innate and critical ability to reconstruct, to a high degree of precision, the original distribution of light emanating from a given scene. In the case of Lippmann photography, the frequency spectrum of the light can be reproduced with extraordinary precision. In the case of holography, it is the spatial structure of the image that is reproduced to an accuracy that can potentially reach a resolution of less than 1 μm. New techniques in colour holography provide the best of both worlds, offering the possibility of both excellent spectral and structural recording.

We can illustrate the huge difference between current non-interferometric displays and interferometric displays by comparing a modern 3D-enabled high-definition television with a full-colour digital hologram. The television display typically comprises several million pixels, each of which can be viewed from one point of view. The hologram, on the other hand, comprises roughly the same number of "holographic pixels", or hogels, but here, each hogel itself projects several million different images depending on the angle of view! The static information in the digital hologram is therefore a factor of some millions of times greater than that of the high definition static television display. This situation is even more marked for analogue holographic displays.

This incredible capacity to encode information, inherent to interferometric or holographic displays, leads to possible applications of this technology that would be extremely difficult to realise using non-interferometric principles. One such application is the holographic window. This is a 3D static or animated digital full-colour holographic display of extreme clarity and depth. It can be viewed from any distance and behaves just like a window—it could for example portray the view out of a New York skyscraper or the view out of a mountaintop restaurant. Such displays, when available at the correct price, are likely to become ubiquitous.

A physical window is something that we all take for granted. However, the window is very different from a high-definition television display. The light field surrounding a window is generally extremely complex—simply because the light traversing the window emanates from 3D objects, which are themselves complex. As a viewer approaches a normal television screen, the picture rapidly defocuses, and upon closer inspection, we see the pixels rather than a picture. In a 3D-enabled television, as we approach the screen, the perspective becomes completely erroneous. However, as we approach a window's surface, our eyes simply focus through the window to view, in perfect clarity, the objects behind it. We shall explore later on in this book the detailed physical requirements necessary for the construction of a full-colour holographic window. However, it is clear that the current techniques of digital holography known today are very close to the point of being able to realise such displays.

Large-format digital holographic windows are not the only application for ultra-realistic imaging as promised by interferometric techniques. Archival and the display of museum artefacts are, for example, growing problems that could be addressed by holography. High-definition full-colour holographic displays now have such a realism that it can be extremely difficult to tell whether the object in question is a hologram or indeed the real thing. Museums often cannot afford to display their collections due to insurance costs, and in addition, priceless items are all too often damaged when transported to exhibitions. Next-generation holographic technology promises to offer real solutions to these problems and to bring the realism of important and interesting museum collections to more people.

New techniques in analogue colour holography have recently been shown [1] to be capable of reproducing oil paintings with exceptional fidelity. Progress in this area promises to lead to a whole new industry of copying or reproducing paintings of significant value or cultural significance. Again, next-generation technologies are expected to produce such accurate reproductions that viewing the copy will be no different from viewing the original.

The basic idea behind what we mean by ultra-realistic imaging is then the reconstruction of totally realistic light distributions as viewed by the human eye. Eventually, we might hope to do this in real time, such as in true holographic television (see, for example, [2]), but the information content of such systems is so high that it is probable that many static applications of interferometric displays will precede this.

The interferometric science of holography is almost certainly set to become a key field in twenty-first-century display systems. In the remainder of this chapter, we shall review the historical development of display holography. Much of what has occurred in the past cannot be described as "ultra-realistic imaging" of course. However, in the 64 years since holography was invented, the field has made great progress. We find ourselves today at a stage in which holography can really start to seriously provide the promise of ultra-realistic imaging. This book is about the technological and scientific issues involved in transforming today's holographic technology into tomorrow's ultra-realistic display science. To set things into perspective, we shall start by retracing progress to date. Additional historical information is included in Appendix 1.

1.2 Before Holography

Long before holography came into being, there existed a colour photographic recording technique known as Lippmann photography, the roots of which can be traced back to the beginning of the twentieth century. During that period, Gabriel Lippmann began experimenting with colour photography in France. His work concerned optical wavefront reconstruction through the recording of standing waves in a volume medium. Lippmann's photographic recording technique was similar to the then unknown technique of holography; but it was not terribly effective as a viable commercial solution to colour photography at the time, because very long exposure times were required for practical use. This was due to the requirement of high resolving power, which could only be provided by materials of very low light sensitivity.

Lippmann was awarded the Nobel Prize in Physics for his invention in 1908. His technique was remarkable for its capability to record colour in a photograph but also for the way in which this was accomplished: this was a brand new idea—the recording of spectral information interferometrically in an ultrafine-resolution material. It was this idea of recording interference fringes throughout the depth of the emulsion that Denisyuk used in the former Union of Soviet Socialist Republics (USSR) when he introduced the technique of recording reflection holograms in the early 1960s. The technique of storing information in a recording medium as a physical interference structure constitutes the basis of the science of *interferential imaging*. Lippmann photography was the first interferential imaging technique discovered; this will be described in detail in Chapter 2.

Before the holographic theory was formulated as a coherent and comprehensive whole, a number of scientists (W. L. Bragg, H. Broersch and F. Zernike) considered the possibilities of x-ray microscopy for recreating an image from the diffraction pattern of a crystal lattice [3]. In 1920, the Polish physicist Mieczyslaw Wolfke wrote, "If an x-ray diffraction pattern of a crystal is illuminated with monochromatic light, a new diffraction pattern is created. This diffraction pattern is identical with the image of the object" [4]. However, it would take another 28 years until this idea was developed properly.

1.3 Early Holography

The holographic theory was presented to the world for the first time in 1948 by a Hungarian-born physicist and Nobel Prize winner, Dennis Gabor (1900–1979) [5–7]. At that time, Gabor (Figure 1.1) was trying to improve the resolution of the electron microscope by overcoming the spherical aberration of the lenses. He asked himself, "Would it not be possible to take first a bad picture, but one which contains the whole information, and correct it afterwards by a light-optical process?" [8]. He found that, by adding a coherent background as a phase reference, the original object wave was contained in the resulting interferogram, which he later called a *hologram*. Gabor concentrated on experimentally verifying his theory, using visible light of the best possible coherence that he could achieve for recording his holograms, which were of the in-line transmission type. He was forced to use filtered light from a mercury lamp, which he focussed through a pinhole. The resulting low intensity required a fast recording material. However, the demand on the material's resolving power was much more severe when the first off-axis transmission holograms were recorded in the early 1960s utilising laser light. This was even more pronounced for the first reflection holograms, which required an even higher resolving power.

Even if holography had already been invented in 1948, it was not until the early 1960s that the scientific community became properly aware of it. The reason for this was that the laser had only been invented in 1960 [9,10], and with this discovery, the quality of recordable holograms simply skyrocketed. Progress in holography immediately after the discovery of the laser occurred almost simultaneously in both the United States and in the former USSR. At the University of Michigan, Emmett Leith (1927–2005) and Juris Upatnieks (Figure 1.2) had been working on side-looking radar and communication theory. They applied the same principles to optical electromagnetic wavelengths and published their first articles on holography in the *Journal of the Optical Society of America* [11–13]. Their 1964 publication revealed the possibility of recording transmission holograms of 3D objects by introducing an off-axis reference beam [14]. This new technique was different from Gabor's technique because it separated the recorded image from the reference beam, making the holographic image substantially easier to view. A hologram

FIGURE 1.1 Gabor next to an off-axis transmission hologram portrait of himself. (Photo courtesy of the MIT MoH Collection.)

FIGURE 1.2   Upatnieks and Leith in the laboratory. (Photo courtesy of the MIT MoH Collection.)

of a model railroad engine (Figure 1.3) was put on display at the OSA spring meeting in Washington, D.C., in April 1964. At that time, Kodak produced the 649-F spectroscopic silver halide plates, which had a resolving power of 2000 line pairs per millimetre. This material was conveniently used by Leith and Upatnieks for their first laser-recorded transmission holograms, whose quality was remarkably good. The railroad engine hologram, illuminated by laser, displayed an extremely realistic 3D image and had a huge impact on the participants attending the meeting.

   In principle, when a laser transmission hologram is illuminated using the same laser light used for the recording, the recorded holographic image is essentially identical to the light field emanating from the laser-illuminated object itself. The displayed hologram and its associated publication initiated tremendous

FIGURE 1.3   Leith and Upatnieks' off-axis transmission hologram, which was displayed in April 1964 at the OSA Spring meeting. (Photo courtesy of the MIT MoH Collection.)

FIGURE 1.4 Yuri Denisyuk at the holographic museum in St. Petersburg next to an early, large single-beam reflection hologram.

worldwide interest in the new imaging technique of holography. The word *holography* was coined by Rogers; Gabor used the words *hologram* and *wavefront reconstruction* in his early articles [15].

Inspired by Lippmann's colour photographic technique, a different hologram-recording technique was invented in the former USSR by Yuri Denisyuk (1927–2006) [16,17] (Figure 1.4). For his first laser-recorded single-beam *reflection holograms*, he had to use a high-resolution emulsion made by him and Protas [18]. Like Gabor, both Leith and Denisyuk used mercury lamps for holographic recording experiments before they were able to use lasers.

1.4 Display Holography Milestones

As soon as the first laser-recorded holograms appeared, holography started to be promoted as a means of creating 3D images for display purposes. The interested reader is referred to a book by Johnston [19] on the detailed history of holography, including display holography. The reader new to holography may find it useful to also consult other books on holography to obtain information on the fundamental principles behind the different basic types of holograms and how they are recorded (see, for example, [20–27]). A recent book published in 2011 by Toal [28] gives a comprehensive discussion of most aspects of holography, including digital holograms and various modern applications, and is particularly recommended. In the following sections, we provide a brief historical review of the different types of hologram and how they were developed and promoted as a potential new 3D display medium. In addition to milestone achievements, some of the examples provided are also from the authors' own activities and projects in the field of display holography dating from the early 1970s until today.

1.4.1 Full-Parallax Transmission Holograms

The first type of hologram recording technique that could be used for display applications was the off-axis transmission hologram. Because only one laser wavelength was used for recording the first transmission holograms, only monochrome 3D images could be created. However, these off-axis transmission holograms provided remarkably realistic full-parallax 3D images. Lasers as well as yellow- or green-filtered mercury lamps were used for displaying the holograms. Lasers were used in public exhibitions until laser safety regulations were introduced. In 1972, two primary groups in the United States were developing laser standards; these resulted in the Z-136.1 (1973) standard. The American National Standards Institute created the concept of classifying lasers according to a scheme of graded risk of exposure and risk of injury involved. Later, laser safety regulations were introduced in Europe and other countries, mainly following the American regulations.

One of the main reasons why off-axis laser transmission holograms were able to provide such a realistic image was because the virtual image replayed in the same position that was occupied at the time of the recording by the object itself. This meant that the viewer's eyes focussed naturally on the exact location of the holographic image behind the plate. This was very different from the 3D display techniques in vogue at the time; with stereo photographs, one had to focus at the plane of the photograph independent of whether the image appeared behind or in front of this plane.

In addition to the accurate replay of the vertical and horizontal parallax, even the reflections contained in the first off-axis transmission holograms moved as in real life. The images were extremely sharp, and very soon, the analogy with looking through a window was coined. This analogy would become all the more appropriate as image quality and hologram size grew.

Lasers were found to produce the sharpest images on replay, but the phenomenon of speckle with lasers of too great a temporal coherence was disturbing. In the early days, filtered mercury lamps were therefore used; one could then no longer see the speckle, but the image was no longer as sharp as when using a laser, especially at larger image depths. The efficiency of the early transmission holograms was low and the noise high. Nonetheless, these early transmission holograms were received with surprise and appreciation; despite their imperfect images, these 3D "windows" constituted something simply never seen before.

1.4.1.1 Earliest Work at Conductron Corporation

In the beginning, most activities in display holography in the United States took place in and around Ann Arbor, Michigan. In the mid-1960s, the Conductron Corporation, which was owned by Keeve (Kip) Siegel, got involved in the field and started to develop and commercialise display holography. Gary Cochran and the staff at Conductron spent much time turning holography into a commercial 3D display technology. One of their first clients was General Motors, which ordered four large display holograms for the opening of the General Motors Building in New York. In 1966, Conductron was bought by McDonnell Douglas Electronics Co. in Missouri, but the research and production continued in Ann Arbor (not until 1970 was the operation moved to Missouri). The pulsed ruby laser, which had been developed at Conductron, was used for recording many large-format display holograms. Lawrence Siebert, who constructed the first pulsed laser at Conductron, recorded the first holographic portrait in 1967 (he was the subject) [29]. A picture of this first portrait hologram is shown in Figure 1.5.

The early pulsed hologram portraits produced very realistic images, but their monochrome appearance quickly led to them being regarded as "dead masks". One well-known McDonnell Douglas hologram was the 1972 *Hand in Jewel* hologram with a real image of a hand holding a bracelet. This image projected through Cartier's store window on Fifth Avenue in New York. The hologram was commissioned by Robert Schinella, who was marketing display holography applications at the time. Over the next few years, many display holograms were produced, some on square-metre plates. A well-known portrait hologram is the 50 cm × 60 cm hologram portrait of Dennis Gabor sitting behind a desk, which is shown in Figure 1.1. The hologram was recorded by Robert Rinehart at McDonnell Douglas to commemorate

FIGURE 1.5   Siebert in the first pulsed laser portrait. (Courtesy of the IEEE, 1968.)

the fact that Gabor was awarded the Nobel Prize in Physics in 1971. The hologram portrait was on display at the Royal Institute of Technology in Stockholm when Gabor received his Nobel Prize in December that year.

The first mass-produced hologram (500,000 copies) was a transmission hologram of chess pieces, which was included in the 1967 *World Book Encyclopaedia* and its *Science Year Annual*. A red filter was also provided to be used to illuminate the transmission film hologram with a white spotlight.

One of the first established artists to use the pulsed transmission hologram technique at Conductron was Bruce Nauman, who recorded a well-known series of holograms entitled *Making Faces* in 1968. His holograms were exhibited at the Castelli Gallery in New York. Another artist was Salvador Dalí, who made holograms with McDonnell Douglas and who also exhibited in New York at the Knoedler Gallery. Harriet Casdin-Silver (1925–2008) in the United States and Carl Fredrik Reuterswärd in Sweden are two other established artists who were attracted to lasers and holography, and created early holographic art pieces.

In 1973, McDonnell Douglas sadly closed its holography division because they had not achieved the marketing success they had hoped for. According to Kip Siegel's business philosophy, this may not have been a surprise. He had developed a reputation for believing that "What you have to do is to sell the *promise* of technology to investors", rather than a ready technology [30]. This idea was, and indeed continues to be, adopted by various display holography companies around the world; many start-up companies are still only able to sell the promise of technology to investors because even though holography has great potential, it is still currently a work in development. It is common that entrepreneurs promoting display holography and the investors who come into contact with the subject get very excited. Unfortunately, all too many underestimate the work required; there is a need to understand the technology and also its limitations as well as to develop a product that the market is willing to buy for a commercially profitable price.

1.4.1.2 Early Work in France

Soon after the world became aware of holography, universities and companies became involved in research and started to record display holograms. Early large-format transmission holograms were recorded in France at Viénot's Laboratoire d'Optique in Besançon. A very large and particularly impressive hologram was the 1 m × 1.5 m *Venus de Milo* transmission hologram recorded by Jean-Marc Fournier and Louis Tribillon, which is shown in Figure 1.6.

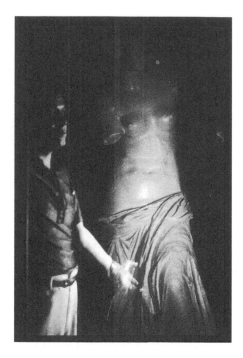

FIGURE 1.6   Jean-Marc Fournier and the 1 m × 1.5 m *Venus de Milo* transmission hologram. (Courtesy of J.-M. Fournier.)

1.4.1.3 Paula Dawson

Paula Dawson from Australia was one of the first artists to take advantage of the very large image depths offered by laser transmission holograms. She recorded her first art holograms in Besançon, France. In 1980, she produced a large-format laser transmission hologram that was 95 cm × 150 cm in size entitled *There's No Place like Home*. In 1989, back in Australia and working at Heytesbury Holography Bell Resources Ltd., Melbourne, she created another 95 cm × 150 cm hologram entitled *To Absent Friends*; this piece, which was of a life-size bar decorated for New Year's Eve was recorded using a large-frame continuous wave (CW) laser and portrayed the bar after closure. Figure 1.7 shows Paula Dawson and the bar installation in the holographic laboratory. More about Dawson's art holograms can be found in [31].

1.4.1.4 Nick Phillips and John Webster

Nicholas (Nick) Phillips (1933–2009) and John Webster, both working in the United Kingdom, were the first scientists to use pulsed lasers for recording commercial display holograms. Both worked on improving the pulsed ruby laser for its use in holographic recording. Based on the modifications they made and on their advice, the JK ruby laser, manufactured in Rugby, United Kingdom, became the laser of choice for the recording of high-quality pulsed holograms around the world. Many large-format holograms were recorded by Phillips, both with CW and pulsed lasers. Some of these holograms were exhibited at two major hologram exhibitions in London at the Royal Academy of Arts (*Light Fantastic I* in 1977 and *Light Fantastic II* in 1978). These exhibitions were responsible for bringing many people into the field of holography, including one of the authors (DBR). Phillips also taught at the Royal College of Art in London, where art students now had access to a pulsed ruby laser recording facility.

FIGURE 1.7   (a) Paula Dawson showing how large her holograms are and (b) the bar installation for the hologram entitled *To Absent Friends*. (Panel b courtesy of P. McLean, 1989.)

1.4.1.5 John Webster and Margaret Benyon

John Webster, working at the Central Electricity Generating Board in Southampton, needed extremely high-quality, high-resolution holograms for nuclear fuel inspection [32,33]. In addition, he also worked with artists producing display holograms. Notably, he recorded pulsed holograms together with the artist Margaret Benyon, MBE [34]. He has also recorded display holograms for museums, for example, in Italy, as well as using holographic techniques to perform interferometric inspection of art objects. Later, he continued to work on pulsed holography, acting as a consultant in Canada and in the United States.

1.4.1.6 Ralph Wuerker

Ralph Wuerker at TRW Systems, Redondo Beach, California, was also a pioneer in pulsed lasers, double-pulsed display holography, and interferometric inspections. He was involved in recording holograms under zero-gravity conditions on the space shuttle (SL-3 mission) [35], as well as recording pulsed holograms of museum artefacts as early as 1969 and later in 1972 in Italy [36,37]. Wuerker's ruby laser was flown to Venice, where he recorded holograms of Donatello's *John the Baptist* and Pisano's *Mother and Child*.

1.4.1.7 40 J Ruby Laser Facility at Fermilab

When one of the authors (HB) worked on bubble chamber holography at Fermilab, Batavia, Illinois, in the early 1980s, a 40-J pulsed ruby laser (made by JK-Lumonics) was in operation. This laser was the most powerful ruby laser made by JK Lasers (we shall have more to say about this high-resolution imaging application in Chapter 14). Edward (Ed) Wesly, who was a member of the bubble chamber holography team at Fermilab, used the same laser in 1986 to record a very large-depth pulsed laser transmission hologram entitled *Man on the Motorbike*. A photo of this 30 cm × 40 cm hologram is shown in Figure 1.8.

FIGURE 1.8 Ed Wesly on his motorbike (1986; 30 cm × 40 cm laser transmission hologram shot with the 40-J JK ruby laser at Fermilab).

1.4.1.8 Hologram of Swedish Crown

An example of an early, highly publicised display holography project in Sweden was the transmission hologram of the *Coronation Crown of Erik XIV* (from 1561). The 20 cm × 25 cm off-axis transmission hologram was recorded by one of the authors (HB) in the cellar under the Royal Castle in Stockholm. The crown was moved from the treasury down to a room in the cellar on the night of 20 August 1974. Figure 1.9 shows HB next to the hologram recording setup with the crown. The CW He–Ne laser and holographic equipment were arranged directly on the floor. Note the Styrofoam panels surrounding the setup, which prevented air motion during the recording. The hologram (Figure 1.10) was installed in a special display case, where it was illuminated by a mercury-arc lamp equipped with a yellow filter (Figure 1.11). It was produced for the International Stamp Exhibition *STOCKHOLMIA'74* in Sweden. The Swedish Post had issued a series of five stamps featuring the Royal Treasuries, including one of the

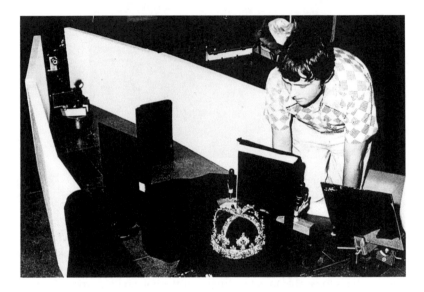

FIGURE 1.9   Recording of the *Crown* hologram in the cellar of the Royal Castle in Stockholm.

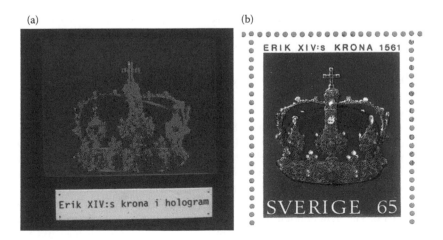

FIGURE 1.10   (a) The *Crown* hologram in the display case and (b) the Crown postage stamp.

FIGURE 1.11 Crown hologram display case with a mercury lamp at the bottom.

Crown. Because it was not possible to show the real crown at the stamp exhibition, the hologram was ordered by the Swedish Post Authority as the next best thing. The stamp show took place between 21 and 29 September 1974 in Älvsjö, south of Stockholm. The hologram was a main attraction at the show, and people waited in long lines to be able to get a glimpse of the holographic crown. Many people came to the stamp exhibition only to see the hologram.

1.4.1.9 Carl Fredrik Reuterswärd

The Swedish artist Carl Fredrik Reuterswärd, who is better known for his *Non-Violence Sculpture* outside the United Nations Building in New York, recorded many early off-axis transmission holograms at holographic laboratories in Stockholm [38]. On 20 January 1963, he announced somewhat eccentrically in the *New York Herald Tribune* that he was going to be closed for holidays during the period 1963 to 1972 to devote his time to lasers and holography (Figure 1.12). As it turned out, his holidays extended until the end of the 1970s.

Reuterswärd created many large-format art holograms—one example was a piece entitled *Gateaux Gabor*, 50 cm × 60 cm, which he recorded in 1978 together with one of the authors (HB) at Lasergruppen Holovision AB. The hologram, which portrayed a burning cake, was made to celebrate the 30th anniversary of Gabor's 1948 article on holography. This was a nice example of how an interference pattern in a hologram could be used by an artist. One could visualise the heat from the burning candles, which created a tremendous 3D "smoke" pattern above the birthday cake (made of wood and with white silicon sealant serving as the whipped cream). During the recording exposure, the 30 candles were actually burning (Figure 1.13). The light emitted from the burning candles (mainly in the yellow–red region of the spectrum) did not fog the green-sensitive Agfa plate, which was exposed using a 514.5-nm argon-ion laser.

FIGURE 1.12 The *New York Herald Tribune* advertisement.

FIGURE 1.13 (a) Carl Fredrik Reuterswärd lights the candles through the holographic plate holder and (b) the *Gateaux Gabor* or *Smoke without Fire* hologram.

1.4.1.10 Demise of Laser and Mercury Lamp Transmission Holograms

By the end of the 1970s, despite their amazing ability to portray large 3D scenes, creating the illusion of a window into another world, large off-axis transmission holograms had more or less completely disappeared due to the practical problems of displaying them using either mercury lamps or lasers. Indeed, laser safety regulations now prevented powerful CW lasers from being used in public exhibitions, and this severely limited progress in the field. Off-axis transmission holograms continued to be produced, however, as master holograms or intermediate holograms. These are often referred to as H_1 holograms; rather than being themselves the final product, they simply serve to make white light-viewable hologram copies. These copied holograms are named H_2 holograms and can be of either the transmission or the reflection type. However, the H_2 is most often an image-planed reflection hologram in which a limited-depth monochromatic 3D image appears both behind and in front of the holographic plate. Most pulsed holographic portraits are nowadays transferred to such reflection copies.

1.4.2 Pulsed Holographic Portraits

Holographic portraits recorded using pulsed lasers were offered by several holographers and companies after Siebert recorded the first portrait in 1968. One of the authors (HB) has been involved in recording pulsed portraits since 1971, first in Sweden at the Royal Institute of Technology and at Lasergruppen Holovision AB and later in the United States at Holicon Corporation, Evanston, Illinois [39,40]. Unlike the earliest pulsed transmission holograms, the final hologram portraits were always reflection copies that could be illuminated using simple halogen spotlights. The downside to this convenience was that the image depth was severely limited in comparison with the laser-illuminated transmission hologram. Carl Fredrik Reuterswärd also created pulsed art holograms in cooperation with HB in Sweden. One example was a hologram entitled *Cross Reference* in which Reuterswärd posed as Salvador Dalí. His moustache was shaped to form the letters *C* and *R*, which are the initials both of the work and of Reuterswärd's first and last names (Figures 1.14 and 1.15).

In the former USSR, Vadim Bryskin recorded pulsed portraits using solid-state lasers developed in St. Petersburg. Among the holographers involved in the early recording of holographic portraits in the United States were Peter and Ana Maria Nicholson, who worked at the Center for Experimental Holography, Brookhaven National Laboratory in Long Island. Later, they moved to the University of Hawaii, where many portraits were recorded. Peter Nicholson gained access to the pulsed ruby laser from the McDonnell Douglas laboratories; McDonnell Douglas had donated their laser to the Smithsonian upon the closure of their laboratories. After returning from Hawaii, Ana Maria Nicholson continued recording portraits, first at the Museum of Holography in New York and then at the Center for the Holographic Arts located in Long Island City, New York (Figure 1.16).

FIGURE 1.14   Salvador Dalí and Carl Fredrik Reuterswärd. (Courtesy of C. F. Reuterswärd.)

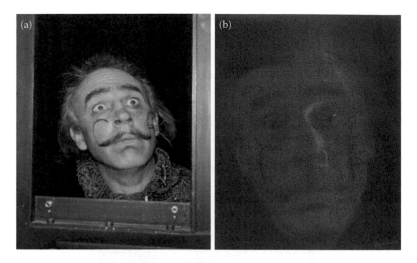

FIGURE 1.15 (a) Carl Fredrik Reuterswärd behind the plate holder. (b) *Cross Reference* hologram portrait.

FIGURE 1.16 Peter and Ana Maria Nicholson at the Long Island facility.

FIGURE 1.17 An example of combining a hologram and a gauche underpainting, carefully registered with the hologram entitled *Painted Margot* by Margaret Benyon.

Ron and Bernadette Olson operated a holographic studio in San Francisco, where they recorded many stunning pulsed portraits of people and animals using a neodymium YAG laser. Anaït Arutunoff Stephens (1922–1998) in California had her own in-house pulsed laser facility and created portraits and other art holograms.

As previously mentioned, Margaret Benyon, the first UK art holographer, active since the late 1960s, created many art pieces based on pulsed portraits. An example combining a hologram and a gauche underpainting, carefully registered with the hologram entitled *Painted Margot*, is shown in Figure 1.17 [41]. Regarding other portrait holographers from the United Kingdom, one should mention Edwina Orr and Martin Richardson, who both recorded portraits of famous people as well as creating pulsed art pieces. Martin Richardson's work has been documented in his book, *Space Bomb* [42]. Robert Munday is another holographer who started early on working at John Webster's laboratory and later with Nick Phillips at the Royal College of Art in London. In 1991, Munday started his own company, Spatial Imaging Ltd., where he recorded portraits of entertainment and sports celebrities. He has also pioneered the art of holographic stereography, a technique that he, together with Chris Levine, used to record the hologram portrait of Queen Elizabeth II.

There were also holographers in France who made early pulsed portraits. An interesting animated hologram portrait was recorded in 1978 at Laboratoire d'Optique in Besançon by Aebischer and Cargquille [43]. A pulsed portrait studio existed in Paris in the 1980s and 1990s operated by Anne-Marie Christakis. Pulsed portraits were also recorded by Yves Gentet, who later became an expert in colour holography. On the border between Germany and France, Paul Smigielski and his team at the French–German Research Institute Saint Louis and HOLO 3 (both located in Saint Louis) recorded pulsed holograms (of the display type, including portraits) and interferometric holograms [44]. Michel Grosmann and Patrick Meyruies at the Louis Pasteur University in Strasbourg have also been involved with display holography since the early days.

1.4.2.1 Use of Neodymium Lasers for Portraiture Applications

Most of the pulsed holograms prior to the mid-1990s were recorded using ruby lasers. However, the deep red light of ruby led to rather waxy-looking portraits. The problem was that the topmost layer of human skin was actually transparent to the ruby radiation. This meant that subjects had to be made-up heavily. The Olsons were one of a few to use neodymium YAG lasers as opposed to ruby lasers to circumvent this problem. One of the authors (DBR) worked with colleagues Mikhail Grichine and Alexey Rodin in Lithuania at the Geola Laboratories in the early 1990s to develop hybrid neodymium YLF/phosphate glass pulsed lasers suitable for holography [45–47]. Geola incorporated these pulsed lasers into semiautomated pulsed portraiture systems [48–54], which could also produce the H_2 copy. Many such systems were sold after the mid-1990s, and in fact, Geola continues to actively sell such systems today. Rob Taylor at Fourth Dimension Holographics, Inc. in the United States was one of the first clients for a Geola automated system. Taylor continues to produce holographic portraits today. The artist Juyong Lee in South Korea also currently operates a Geola system. Lee has made many large pulsed reflection holograms—for example, his 1999 *Dreaming History* series.

During the 1990s, as well as manufacturing pulsed lasers and automated pulsed laser holography systems, Geola also operated its own holographic portraiture studio in Vilnius, where holograms were produced of works by many artists, including the German artist H. R. Giger.

1.4.2.2 Reagan Portrait and Portraiture into the Future

Probably the most famous person to be recorded in a pulsed hologram portrait was President Ronald Reagan. One of the authors (HB) and a team from Holicon Corporation recorded the portrait (Figure 1.18) at the Brooks Institute of Photography in Santa Barbara, California, on 24 May 1991 [55].

Monochrome holographic portraits (often replaying as orange–yellow or green images) initially fascinated people, but their intrinsically waxy images and high prices always stifled sales. Even in the ex-Soviet Union, sales remain small today—Sergei Vorobyov continues to sell holographic portraits from his studios in Moscow, where he has done so for more than 25 years.

However, a new type of digital holographic portrait is available today. These holograms are fundamentally more appealing than the earlier pulsed portraits, and laser irradiation of the subject is not required. They can include limited animation, and the images from which they are made are derived using either modern structured light techniques or, in the case of horizontal parallax-only holograms, by a digital camera moving along a horizontal rail. We shall have a lot more to say about such modern digital techniques later on.

FIGURE 1.18 (a) The recording of the Ronald Reagan portrait—from left to right, Ronald Reagan, one of the authors (HB) and Ernest Brooks II; (b) the Ronald Reagan portrait.

1.4.3 Steven Benton's Rainbow Hologram

Among the articles on holography published toward the end of the 1960s, one particular publication is especially important. The article by Stephen Benton (1941–2003) described a white light-viewable transmission hologram—the so-called *Rainbow Hologram* or *Benton Hologram* [56]. Figure 1.19 shows Benton next to the 1977 *Crystal Beginnings*, a rainbow hologram recorded at Polaroid Corporation.

The rainbow hologram technique was an important improvement for recording transmission holograms because these holograms could be viewed using the white light from an ordinary halogen spotlight. The holograms were bright because they used the entire spectrum of the illuminating light and could represent much deeper images than the normal type of analogue reflection hologram. They were also extremely colourful, which again differentiated them from the intrinsically monochrome images of the laser transmission and white-light reflection holograms known before. The downside was that these new rainbow holograms did not encode the vertical parallax and exhibited a rather narrow viewing window in the vertical direction.

The rainbow technique is a two-step process in which the master H_1 is reduced to a horizontal section of the holographic plate (a slit with very little vertical extension) or to a narrow film strip. The transmission H_2 hologram is the final rainbow hologram that is generated from the strip master. When the rainbow hologram is illuminated with white light, a real image of the horizontal strip floats in front of the H_2 and acts to disperse the illuminating light into a rainbow spectrum. When the viewer's eye is coincident with the real image of the slit, a sharp image becomes visible. Depending on where the viewer's eyes are located in the vertical direction, the colour and magnification of the perceived image change.

1.4.3.1 Transmission Pseudo-Colour Techniques

It is also possible to generate rainbow holograms with mixed colours by using more than one slit. The *pseudo-colour transmission technique* is based on multiple white-light rainbow holograms in which different spectra are superimposed at the position of the observer's eyes. Often, these holograms are very impressive regarding both their colour composition and brightness. One problem though is that the colours and image magnification of any Benton-type hologram varies depending on the vertical position of the observer. The multicolour technique for rainbow holograms was introduced in 1977 by Tamura [57].

A slightly different type of transmission hologram, which produces a multicoloured image when illuminated by white light, was introduced by Hariharan [24]. Here, the hologram is made as a number of

FIGURE 1.19   Stephen Benton with his rainbow hologram *Crystal Beginnings*. (Photo courtesy of the Jeanne Benton MIT MoH Collection, 1980.)

superimposed component holograms formed from optical images of a subject. These images are formed either by an optical imaging system or by separate primary holograms of the subject. Each primary (slit) hologram is formed using coherent light of a given wavelength and is designed to be reconstructed upon viewing of the final transmission hologram at that particular wavelength. The use of a sandwich of two photosensitive media, each containing at least one component hologram, to form the product transmission hologram was also described by Hariharan. Grover and Tremblay [58] also demonstrated the possibilities of creating natural-colour rainbow holograms.

1.4.3.2 Large Rainbow Display Holograms and Embossed Holograms

Rainbow holograms, being transmission holograms, must be illuminated from behind. This makes them uniquely suitable for shop window displays where images can actually project out in front of the window. Large rainbow holograms recorded on silver halide materials are often used as promotional 3D displays. Rainbows are well suited to such applications because of their superior brightness and because of their ability to portray large depths without defocusing. The holograms can also be laminated to a mirror if standard front illumination is required. The rainbow technique became popular among artists early on and it was also quickly adopted as the holographic technique of choice for the mass production of document security holograms (Figure 1.20). The first company to use holograms on credit cards was MasterCard

FIGURE 1.20 Embossed mirror-backed rainbow credit card holograms.

FIGURE 1.21 Three *National Geographic* covers with embossed holograms. (a) November 1985, (b) March 1984 and (c) December 1988.

towards the end of 1982. The holograms were produced by American Bank Note Holographics, Inc. Later, VISA also introduced holograms on their credit cards.

The mass-produced rainbow holograms used for security applications were produced on nickel shims using relief patterns; these relief patterns were then embossed onto plastic materials. The manufacturing of document security holograms remains today the main commercial application of holography. Such holograms or similar types of optical variable devices (OVDs) are not only used on credit cards but also on banknotes, passports and legal documents. Mirror-backing is almost always applied to security holograms, which explains why these optical variable devices always have a mirror-like appearance.

Another early application of mass-produced embossed holograms was the inclusion of holograms on journal covers. For example, there were three *National Geographic* covers in the 1980s that featured such holograms, 11 million copies being produced for each issue (Figure 1.21).

1.4.3.3 US Rainbow Displays—Holographics North Inc.

John Perry of Holographics North in the United States has been a driving force in the application of large-format rainbow holograms for display applications for many decades [59]. The size of such displays has reached several metres in width but has usually been constrained to a metre in height by the availability of the photographic material. Perry has worked with artists as well as recording holograms for advertising companies and trade shows. Most often, such large-format rainbow holograms were recorded on film. One example of Perry's work is the Tonka Toys hologram, mastered by Holicon Corporation and transferred by John Perry. In 1987, both Hasbro (*Visionaries*) and Tonka Toys (*SuperNaturals*) produced toys with holograms attached. To promote this new line of toys, Tonka Toys ordered a large-format hologram to be exhibited in a New York Toy Show. The hologram is shown in Figure 1.22. By using the pseudo-colour technique described above, the text at the top of the hologram appeared in a different colour.

1.4.3.4 French Rainbow Displays—AP Holographie SARL

In France, AP Holographie, located in Metz and run by Jacques Bousigné, François Mazzero and Jean-François Moreau, was also well known for their large-format rainbow holograms. They

FIGURE 1.22 The Tonka Toys rainbow hologram.

FIGURE 1.23 *La Villette* rainbow hologram by AP Holographie.

produced many holograms for the 1983 Geneva Car Show. Another famous AP rainbow hologram was the square-metre *Parc des Folies à la Villette* produced in 1983 and is shown in Figure 1.23. The hologram was recorded from an architect's model and produced to promote La Villette Science Park in Paris.

1.4.3.5 Australian Rainbow Displays—Australian Holographics Pty. Ltd.

From 1989 to the mid-1990s, Australian Holographics Pty. Ltd., based in Adelaide, South Australia, produced some extremely large rainbow holograms. The company was founded by one of the authors (DBR) and was later (circa 1993) managed by Simon Edhouse. The chief holographer was Geoff Fox. Yasmin Farquhar ran the company's two popular holographic galleries in Adelaide and Hahndorf. The large analogue pseudo-colour rainbow holograms produced by Australian Holographics were made on a floating 25 tonne optical table (Figure 1.24) measuring 5 m × 6 m [60]. The recording laser was a Coherent Innova argon laser of 5 W, although large-frame krypton and argon-pumped dye lasers were later added for the production of metre-square reflection holograms.

One of the peculiarities of the Australian system was that the rainbow masters, which could be well over 2.3 m long, were all side-illuminated by a small spherical mirror and mounted in oil. Juyong Lee was one of the first artists to collaborate with Australian Holographics. Lee worked at the company's facilities in 1993 to produce a series of 10 pseudo-colour rainbow holograms measuring 2.2 m × 1.1 m, which formed part of the 1993 EXPO in South Korea. In 1994, Australian Holographics made a series of metre-square rainbow holograms for the Singapore Military, which were installed in the Singapore underground system; the installation was reported in newspapers and on Singapore national television like many of the company's higher profile projects. Paula Dawson was another artist to work at Australian Holographics, producing a hologram of a huge sculptured dome in 1995.

In 1994, Australian Holographics gained access to the large collection of artefacts owned by the museum of South Australia. The joint venture, the signing of which was again covered on national television and was even debated in the state parliament, led to the company producing some spectacular

FIGURE 1.24   Dmitry Konovalov (left), cofounder of General Optics Pty. Ltd., and Geoff Fox (right), chief holographer, sitting on the giant 25 tonne, 5 m × 6 m pneumatically suspended table at Australian Holographics.

large-format rainbow holograms of such items as the museum's T-rex skull (Figure 1.25) and its unique Thylacine exhibits. Some of these holograms, even though nearly 20 years old, still come onto the market today from time to time. When Australian Holographics closed in the late 1990s, its equipment was split between Adelaide University and the Geola Company in Lithuania.

1.4.3.6 Parameswaran Hariharan

Parameswaran Hariharan is an Indian-born scientist and a pioneer of holography in Australia. He worked at the Commonwealth Scientific and Industrial Research Organisation in Canberra. Hariharan's main research field was display holography and interferometry [24]. He introduced new processing techniques for holography and described how to produce pseudo-colour display holograms of both the transmission and reflection types, including sandwich techniques as well. He was the first to make a successful hologram in colour using three lasers and, for a time, collaborated with the artist Alexander in producing large holograms and surrealistic holographic stereograms.

1.4.3.7 Günther Dausmann

In Germany, Günther Dausmann has been involved in display holography since the beginning. His first company, Holtronic, offered both equipment and large-format CW and pulsed transmission and reflection holograms including stereograms. Later, he focussed on document security holograms, and in 2001, he produced the two-dimensional monochrome photopolymer hologram portraits for the German passport. His company, Holography Systems München GmbH, is now part of the successful document security company Hologram Industries in France, and his German operation has changed to Hologram Industries Research GmbH.

1.4.3.8 Dieter Jung and Rudie Berkhout

Among the established artists who have been attracted to rainbow holograms, one should mention Dieter Jung in Germany [61]. Jung created many large-format holograms—for example, the *Into the Rainbow*

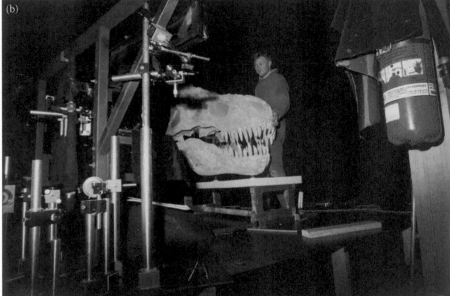

FIGURE 1.25 (a) Monique Haan preparing a T-rex skull from the South Australian Museum in 1994. (b) Holographer Mark Trinne mounting the heavy skull on the optical table at Australian Holographics. (c) The mounted skull illuminated by argon laser light.

FIGURE 1.26   The *4-D Landscape* hologram by Rudie Berkhout.

hologram series (one embossed version of *Into the Rainbow* is on the cover of the book made by Topac GmbH [61]). Rudie Berkhout (1946–2008) working in United States is also well known for his beautiful art rainbow holograms—an example is shown in Figure 1.26.

1.4.4 Circular Transmission Holograms

By surrounding an object with a film and adding a reference beam from above the object, it is possible to record a circular transmission hologram having a 360° image. This technique was first published in 1966 by Jeong et al. [62]. However, the technique has rarely been used to record display holograms. A large-diameter (120 cm) circular hologram of this type was recorded by Upatnieks and Embach in 1980 [63]. A special hologram viewer was constructed to rotate the hologram permitting it to be viewed from a stationary position. The project involved recording holograms of musical instruments and the hologram viewer had an associated audio tape to demonstrate the sound of the various instruments displayed.

1.4.5 Achromatic or Black-and-White Holograms

As a complement to rainbow and monochrome holograms, there has been interest in creating achromatic or black-and-white holographic images. With regard to rainbow holograms, Benton developed a technique to make an achromatic image hologram in 1978 [64]. His first such hologram was *Aphrodite*; later, he recorded a hologram of a mummy's skull, *Pum II*, which became part of a Smithsonian Institute programme to provide 3D copies of valuable artefacts in danger of decay or loss.

It should be mentioned that the emulsion thickness manipulation techniques for producing pseudo-colour reflection holograms can also be used to create black-and-white images. Orr and Trayner [65] produced impressive large-format black-and-white holograms using emulsion thickness manipulation in-between recordings. In this case, only two colours are necessary to create white. In the Commission Internationale de l'Eclairage (CIE) diagram, the colours should be chosen in such a way that a straight line drawn between them should pass right through the white region in the centre of the diagram. Obviously, there are many possibilities for choosing the primary colours that satisfy this condition. Orr and Trayner used yellow and blue, which they achieved by preswelling the emulsion with different concentrations of triethanolamine. The technique was used during the transfer process of the pulsed masters to the reflection copies. The 50 cm × 60 cm reflection hologram, *Kate McGougan & Stephen Jones Hat* (styled by Robin Beeche), made by Edwina Orr and David Trayner (Richmond Holographic Studios Ltd.) is an excellent example of a black-and-white hologram and is shown in Figure 1.27.

By producing an inverted rainbow spectrum using a diffraction grating, and using this to illuminate a transmission hologram, it is also possible to create achromatic images, a technique introduced by Kevin

FIGURE 1.27 *Kate McGougan & Stephen Jones Hat* black-and-white portrait hologram. (Courtesy of 1985 Edwina Orr and David Trayner. Richmond Holographic Studios.)

Bazargan in 1985 [66]. An important application of this technique is the holographic medical imaging application, which allows computerised tomography (CT) and magnetic resonance images (MRI) to be visualised as achromatic 3D holographic images. These multiplex holograms, developed and produced by Stephen Hart at Holorad LLC (former Voxel) Voxel in Salt Lake City, Utah, are still in use today. The image data is generated from a series of sequential slices of computerised tomographic or magnetic resonance imaging data [67]. A single piece of holographic film is then exposed multiple times to create the hologram by multiplexing these slices in depth. When the film is processed, the resulting hologram is referred to as a *Voxgram*. It must be examined on a special display unit called a *Voxbox*; it is this display device that produces the inverted rainbow spectrum required for illumination. Applications such as imaging brain tumours and related vasculature before surgery can be visualised in 3D and full parallax. A Voxgram hologram is shown in Figure 1.28.

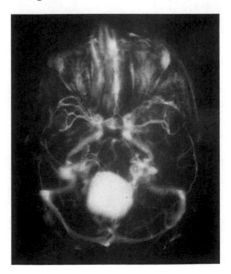

FIGURE 1.28 Achromatic Voxgram of a brain tumour.

1.4.6 Monochrome Reflection Holograms

In 1962, Denisyuk introduced the single-beam reflection recording technique; shortly after this period, such holograms started to be used for 3D displays. Because the technique was introduced in the former USSR, it is natural that such white-light holograms were mainly recorded in countries from the Soviet Union. Another reason was that the required ultrafine-grain silver halide emulsions already existed there. In Western countries, white-light holography was dominated by Benton's rainbow technique and this only changed when Agfa introduced their monochrome HD emulsions and when low-noise processing techniques had been developed. Nick Phillips (Figure 1.29) was able to record high-quality reflection

FIGURE 1.29 "Bridging the gap between Soviet and Western holography", Nick Phillips with a box of Slavich PFG-03 holographic plates.

FIGURE 1.30 The *Icon* Denisyuk hologram by Vanin.

holograms based on his low-noise processing on Agfa and, later, on Ilford materials [68–70]. By the late 1970s, high-quality reflection holograms on Agfa and Ilford emulsions were being used to record museum artefacts and also for advertising and product promotion projects.

Early Denisyuk reflection holograms from the former USSR were very impressive. This was due principally to the ultrafine-grain emulsions, which ensured that the diffraction efficiency and signal-to-noise ratio of the holograms were very high. Holograms of many artefacts from museums, such as the Hermitage in St. Petersburg were recorded and used at exhibitions. An example of a typical Denisyuk hologram is the *19th Century Icon* hologram produced by Valery Vanin in 1981 and reproduced here in Figure 1.30.

1.4.6.1 Sweden—Lasergruppen Holovision AB

A hologram of a *Gold Collar* is an early example of how a monochrome Denisyuk reflection hologram was used in a museum display in place of the real artefact. The 30 cm × 40 cm hologram was recorded in 1980 by Lasergruppen Holovision AB in Stockholm, Sweden. The three-ringed gold collar with filigree, granulation and carved figures was found in 1827 in Sweden and is from AD 400 to 550. Because of its high value, the collar is kept in the National Historical Museum in Stockholm. A local museum located near the place where the gold collar was found, *Falbygdens Museum*, was not allowed to exhibit the real gold collar, so they decided to exhibit a hologram of it instead. The collar hologram and its display case are shown in Figure 1.31. Today, modern colour holography can offer something even more realistic than this stunning yellow monochrome hologram from 1980.

Another early example of a large Denisyuk hologram produced by Lasergruppen Holovision AB is the 50 cm × 60 cm *Alfred Nobel Bust* plate ordered by the Swedish Nobel Foundation. The recording took place in 1981; the hologram recording setup in the laboratory is shown in Figure 1.32. Figure 1.33 depicts a photograph of the finished yellow–orange tuned hologram. This hologram, on display in the Nobel Foundation Building in Stockholm (Figure 1.34), has a label next to it stating that the hologram is based on five Nobel prizes in physics.

Another example by Lasergruppen Holovision AB is a 30 cm × 40 cm advertising hologram, the *Martini* hologram. This hologram was on display at Stockholm's International Arlanda Airport. Alcoholic beverages could not be advertised using photographic advertising, but advertising using holograms was still acceptable at that time. The hologram display cabinet at the airport is shown in Figure 1.35.

FIGURE 1.31 The *Gold Collar* hologram in the museum display.

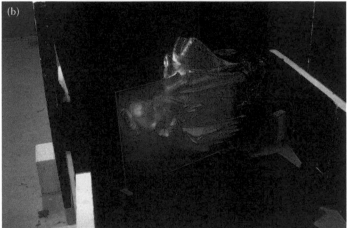

FIGURE 1.32 (a) Recording setup on the concrete floor showing the Alfred Nobel bust in the background and (b) the Nobel bust with the recording glass plate in front of it.

FIGURE 1.33 The *Alfred Nobel* hologram.

FIGURE 1.34 The hologram installed in the Nobel Foundation Building, Stockholm.

FIGURE 1.35 Hologram display cabinet at Arlanda Airport with the *Martini* hologram.

1.4.6.2 Mass Production of Reflection Holograms

A very early mass-produced reflection hologram was the *Marching Band Hologram*, a 10 cm × 12 cm silver halide Agfa film produced by McDonnell Douglas (Figure 1.36). It was included in two books, *Les Prix Nobel 1971* [71] and *Kosmos* 49:1972 as well as in McDonnell Douglas's own promotional materials.

A project to produce large quantities of reflection holograms was undertaken as a joint effort between Applied Holographics PLC and Ilford Ltd. in the mid-1980s [72,73]. The idea was to use a pulsed ruby laser to record the holograms in an automatic fast-copying process. The *Holocopier AHS1* used a 240-mm-wide Ilford roll film (Holofilm 250 PAR) and a small JK ruby laser in the copier. Automatic processing of the film was performed in a machine from Hope Industries. Using the Holocopier, it was possible to produce reflection copies from reflection masters tuned at the 694 nm wavelength. It was also possible to directly record Denisyuk film holograms in the machine. According to Simon Brown,

FIGURE 1.36   *The Marching Band* film hologram by McDonnell Douglas.

FIGURE 1.37   Tonka Toys figures with holograms.

it was possible to deliver high-quality copies in commercial quantities and at "realistic" prices. One of the first customers for the mass-produced holograms was Tonka Toys (Figure 1.37). Unfortunately, the toy hologram project turned out to be unique and eventually Applied Holographics gave up on display holography, turning to the more profitable document security market, trading as OpSEC both in the United Kingdom and in the United States.

In the mid-1990s, the Geola organisation in Lithuania used the concept of pulsed laser copying with its series of semiautomatic holographic portraiture systems based on neodymium glass lasers (initially the GP series, which was then followed by the more modern HS series). Geola offered versions of its systems with crystal amplifiers to achieve fast copy rates but few of these fast systems were sold.

1.4.7 Pseudo-Colour Reflection Holograms

Before true colour reflection holograms were developed, a pseudo-colour technique was used to create colour reflection holograms. Because it was difficult to produce such pseudo-colour reflection holograms on a large scale, they were produced mainly as artistic pieces in limited editions. The technique of creating different colours here was based on the fact that the colour in a reflection hologram is obtained by the diffraction of light from the recorded interference layers within the emulsion. The distance between these layers determines the colour. A certain distance between the interference fringes is generated during the recording of a reflection hologram depending on the laser wavelength used. However, this distance can be manipulated by various processing methods, with the result that different colours are obtained in the final hologram. Even using only a single-wavelength laser for the recording, such pseudo-colour reflection holograms were found to be able to display many different colours. Preswelling the emulsion before the recording resulted in shrinkage after processing, creating a colour of a shorter wavelength upon replay of the hologram compared to the colour of the laser light used. These methods were, however, very time consuming, and they were therefore mainly of interest for holographic art applications. The material most frequently used was the silver halide emulsion. The technique was introduced by Blyth in 1979 [74]. In an article by Hariharan [75], the pseudo-colour reflection process was described in greater detail. Normally the preswelling of the emulsion was performed by immersing the plates in various concentrations of triethanolamine solutions. Figure 1.38 shows an example of a pseudo-colour reflection hologram by John Kaufman [76]. Kaufman was one of the most experienced holographers who perfected this technique and who recorded many colourful art holograms. Other artists who produced many pseudo-colour holograms included Lon Moore and Larry Lieberman. At Holographic Images, Inc. in Florida, Lieberman developed a special step-rotation copying technique to produce large quantities of these types of holograms on film at more affordable prices.

For a long time, most reflection holograms were recorded on silver halide materials. This was definitely the case for the large-format holograms. However, other materials were also used. One material was the

FIGURE 1.38 *4-Color Rock Rotation* pseudo-colour reflection hologram by John Kaufman.

dichromated gelatin (DCG) emulsion, which was used from the very beginning for recording small reflection holograms—in particular, for pendants and jewellery items. Richard Rallison (1945–2010), Electric Umbrella and the International Dichromate Corporation, all located in Utah, produced large quantities of pendants for many years. An artist who has produced high-quality holographic jewellery and created beautiful art pieces recorded in DCG is August Muth (Figure 1.39) at Laserart Ltd., Santa Fe, New Mexico. Large and small DCG holograms were also produced in the Soviet Union. In 1987, Mike Medora established Raven Holographics Ltd. in the United Kingdom to record pseudo-colour DCG holograms. Richard Rallison was a partner in this company.

After DuPont and Polaroid started to manufacture photopolymers, this too became a material suitable for mass-produced monochrome reflection holograms. In 1987, a new monochrome photopolymer

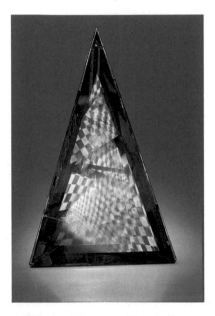

FIGURE 1.39   DCG sculpture *Internal Reflections* by August Muth.

FIGURE 1.40   *Mirage* photopolymer hologram.

material (DMP-128) was introduced by Polaroid Corp. [77]. Their display holograms were produced in-house for clients, and were marketed under the name *Mirage*. One example of such a hologram attached to a Polaroid ONYX camera packaging is shown in Figure 1.40.

1.4.8 Holographic Movie Films

There have been several experiments reported whose aim was to record 3D holographic movies, but this has been an extremely difficult task thus far. The Conductron Corporation had already performed some tests recording moving holographic images in as early as 1966. A 70-mm animated holographic film of a merry-go-round was made and viewed using a strobed laser source [78].

Gabor devoted a lot of time to designing a holographic projection screen for movie theatres [79]. The audience (in one plane or two) was covered by zones of vision having the width of the normal eye spacing, one for the right eye and one for the left with a blank space between the two pairs. The two eyes of the viewer had to see two different pictures: a stereoscopic pair. One of the movie projectors was used as the reference source and, for example, all the left viewing zones in the theatre as the object. The holographic screen, which was usually recorded as a reflection holographic optical element (HOE), became a very complicated optical system (when a picture was projected, it had to be seen only from all the left viewing zones). The recording process was repeated with the right projector as the reference and all the right viewing zones as the object beam. When viewing a 3D movie, the two projectors would then direct the left and right pictures to the two viewing zones in the theatre. Even if this worked in principle, it was found to be extremely difficult to create the large holographic projection screen with the required high-diffraction efficiency appropriate for a large audience. Gabor suggested that large screens should be made out of many smaller HOEs.

The most well-known experiments in the field of holographic movies were the ones performed by Victor Komar and his colleagues at the All-Union Cinema and Photographic Research Institute (NIKFI) in Moscow [80–82]. Figure 1.41 shows a photograph of a piece of 70-mm film strip with two different frames from recorded film scenes. The movies were rather short (only a few minutes) and it was only

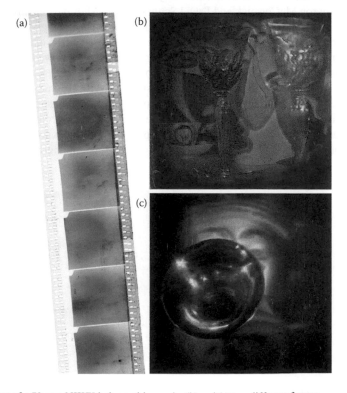

FIGURE 1.41   (a) Part of a 70-mm NIKFI holographic movie; (b) and (c) two different frames.

possible for four or five people to view them at the same time. The movies were projected on special HOE projection screens. Komar's first holographic colour film was produced in 1984 [83].

Paul Smigielski and colleagues at the Institute Saint Louis also made holographic movies [84]. Smigielski recorded short movies using a pulsed Nd:YAG laser—for example, a model train (*Holotrain* 1983) and a girl blowing soap bubbles (*Christiane et les Holobulles* 1985). The Australian artist Alexander worked with Smigielski at the Institute Saint Louis to make tests of holographic movies using a pulsed laser. Later in 1986, he created two movies: *Masks*, a 4-min movie, and *The Dream*, an 8-min movie. These movies were based on the multiplex technique and were recorded on a 25-cm-wide holographic film format [85].

1.4.9 Multiplex and Stereographic Holograms

The early multiplex technique produced display holograms from a combination of photographic or cinematographic recordings using special laser-copying techniques [86–88]. This made it possible to make holograms of people or objects as well as outdoor scenes without involving lasers in the first part of the process. One of the first holograms based on this method was a circular or semicircular hologram invented by Lloyd Cross [89]. This technique was similar to the rainbow hologram and was referred to as the *Integral* or *Multiplex* hologram; physically, this was a cylindrical 120° or 360° white light-viewable transmission hologram. The origination was done by recording the object or person on a rotating turntable using a conventional movie camera. Each movie frame was then converted to a vertical slit hologram on holographic film in a special printer. In this way, an animated holographic image could be recorded which appeared in the centre of the transparent cylindrical hologram when illuminated with a clear vertical-filament bulb positioned beneath the cylinder centre. Either the hologram rotated slowly, thus displaying a moving image, or the viewer walked around the stationary multiplex hologram to experience the animation. The best known multiplex hologram is Lloyd Cross' 120° hologram *Kiss II* from 1975. Three frames of this animated hologram are shown in Figure 1.42. Such holograms became popular in the mid-1970s and this situation lasted until the early 1980s. In addition to Cross, they were produced by Peter Claudius, Sharon McCormick and David Schmidt at the Multiplex Company in San Francisco, and by Hart Perry and Jason Sapan in New York. Jumpei Tsujiuchi, a display holography scientist in Japan, was involved in display and medical holography and also recorded many multiplex holograms. Eventually, the interest in multiplex holograms disappeared and they are nowadays very rarely produced.

Flat transmission stereograms have been envisaged by King et al. [90] since the 1960s but became popular only in the late 1980s; these holograms were simpler to record and enabled a greater variety of objects to be used as subjects. Like rainbow holograms, they could also be made as pseudo-colour holograms. Large-format transmission stereograms were made by several companies, but the ones made at Advanced Dimensional Displays in California are particularly well known and were reported by Newswanger and Outwater [91]. One Advanced Dimensional Displays hologram is shown in Figure 1.43.

FIGURE 1.42   Lloyd Cross' hologram *Kiss II*. (Courtesy of L. Cross).

FIGURE 1.43   Craig Newswanger and one of the authors (HB) next to a large Advanced Dimensional Displays stereogram.

FIGURE 1.44   The *Dizzy Gillespie* large transmission stereogram.

Mark Diamond in Florida also recorded large stereograms—for example, the animated hologram of *Dizzy Gillespie*, shown in Figure 1.44. The recording was made at Advanced Dimensional Displays together with Craig Newswanger.

By the early 1990s, most large stereograms had started to be recorded as reflection holograms, these types of holograms being more practical to display.

Walter Spierings and his company, the Dutch Holographic Company B.V., introduced the first full-colour reflection stereograms [92]. A particularly stunning example of this technology was Dutch Holographics' 1991 three-colour reflection hologram of Ricky Henderson, the US baseball star (Figure 1.45). Spierings named the technology multiple photo-generated holography (MPGH) because the single-parallax reflection hologram produced by the technique required several hundred photos to be taken with a 35 mm camera on a linear track.* The photos were illuminated by laser light and projected in sequence

---

* In the case of the Ricky Henderson hologram, the photographs were actually taken with a static camera, the subject being located on a rotating platform. We shall study different types of camera geometries in Chapter 10.

FIGURE 1.45   The first three-colour (multiple photo-generated holography) reflection holographic stereogram of US baseball star Ricky Henderson. (Courtesy of W. Spierings, 1991. Dutch Holographic Company B.V.)

onto a diffusion screen. Three separate H_1 transmission master holograms were produced using red, green and blue lasers. Each master was then transferred, one at a time, to a full-colour reflection H_2 on fine-grain silver halide emulsions. Spierings began working on the MPGH technique after being inspired by Steve Benton's work at MIT in the mid-1980s on computer-generated stereograms. At about the same time that Walter Spierings introduced his MPGH colour reflection holograms, another holographer, Rob Munday in the United Kingdom, was working on a similar (H_1/H_2) technique but with an important difference—this was the first of the digital stereogram systems. In 1991 Munday developed a system that he called the DI-HO (Digital Input–Holographic Output) to record H_1 holographic master stereograms and H_2 transfers using a modified high-resolution LCD screen. He used the Commodore Amiga 3000 Graphics Workstation with software that he had written himself named *Holomation*. The system could accept image input from sources such as a scanner or video as well as computer-generated artwork. The software also allowed for the planar separation of 2D graphic images which could then be combined with a 3D stereogram. Holograms up to a size 8" × 10" could be made with the system in both reflection and transmission format. Munday's first demonstration hologram was made in collaboration with MIT graduate Eric Krantz in 1992. It was a 7 ½" × 9" multi-colour H_2 reflection hologram of a *triceratops* dinosaur on a tiled surface (see Figure 1.46) made using multiple H_1's produced from digital data (the *triceratops* was a computer-generated 3D model). The H_1's were based on 70 slits, but Munday's technique allowed the number of slits to be increased to achieve better resolution if required.

Later in 1992 Munday also produced an embossed colour hologram called 'Z' using the above technique. The DI-HO system was used to record the stereogram frame by frame in three colour separations with each 'strip' colour separation contained on a single achromatic angled H_1 plate. The H_1 hologram was recorded on Agfa silver halide material using a Helium Cadmium laser. Munday then transferred the image to a surface relief photoresist H_2 before commissioning Applied Holographics plc to emboss the hologram. By the late 1990s, digital cameras and LCDs had largely replaced the old analogue techniques, giving birth to digital holographic printing.

FIGURE 1.46 *Triceratops* hologram (courtesy of R. Munday).

1.5 Digital and Analogue Full-Colour Holography

Conventional display holography more or less disappeared in the 1990s because the easy-to-make mono-chrome 3D image was not able to sustain its role in displays and advertising. It was realised that colour was simply a "must" in any serious holographic display application; but at the same time, early colour reflection displays such as Dutch Holographics' multiple photo-generated holograms were expensive, difficult to make and the quality was still somewhat lacking. Today, however, technology has appeared which holds definite promise of reversing this trend. Thanks to some crucial advances in areas such as holographic materials, lasers, electro-optics, image acquisition systems, computers and illumination technologies, full-colour holograms of both the digital and analogue type can now be produced with extraordinary realism. After a review of Lippmann photography in Chapter 2, we shall embark on a proper discussion of this new holography of ultra-realistic imaging.

REFERENCES

1. H. I. Bjelkhagen and D. Vukičević, "Color holography: a new technique for reproduction of paintings," in *Practical Holography XVI and Holographic Materials VIII*, S. A. Benton, S. H. Stevenson, and T. J. Trout, eds., Proc. SPIE **4659**, 83–90 (2002).
2. P.-A. Blanche, A. Bablumian, R. Voorakaranam, C. Christenson, W. Lin, T. Gu, D. Flores, P. Wang, W.-Y. Hsieh, M. Kathaperumal, B. Rachwal, O. Siddiqui, J. Thomas, R. A. Norwood, M. Yamamoto and N. Peyghambarian, "Holographic three-dimensional telepresence using large-area photorefractive polymer," *Nature*, **468**, 80–83, (4 Nov. 2010).
3. S. S. Shushurin, "On the history of holography," *Soviet Phys. USPEKHI* **14**, 655–657 (1972).
4. M. Wolfke, "Über die Möglichkeit der optischen Abildung von Molekulargittern," [Transl.: On the possibility of the optical recording of a molecular lattice]" *Physik. Zeitschr.* **21**, 495–497 (1920).
5. D. Gabor, "A new microscopic principle," *Nature* **161** (No.4098), 777–778 (1948).
6. D. Gabor, "Microscopy by reconstructed wave-fronts," *Proc. Roy. Soc. (London)* **A197**, 454–487 (1949).
7. D. Gabor, "Microscopy by reconstructed wave-fronts: II," *Proc. Phys. Soc. (London)* **64** (Pt. 6) No. 378 B, 449–469 (1951).
8. D. Gabor, "The outlook for holography," *Optik* **28**, 437–441 (1968).
9. T. H. Maiman, "Stimulated optical radiation in ruby," *Nature* **187,** 493–494 (6 Aug. 1960).
10. G. R. Gould, "The LASER, Light Amplification by Stimulated Emission of Radiation," in *The Ann Arbor Conference on Optical Pumping,* P. A. Franken and R. H. Sands eds, Proc. The University of Michigan, (15 June – 18 June, 1959), p. 128.

11. E. N. Leith and J. Upatnieks, "Reconstructed wavefronts and communication theory," *J. Opt. Soc. Am.* **52**, 1123–1130 (1962).

12. E. N. Leith and J. Upatnieks, "Wavefront reconstruction with continuous-tone transparencies," *J. Opt. Soc. Am.* **53**, 522A (1963).

13. E. N. Leith and J. Upatnieks, "Wavefront reconstruction with continuous-tone objects," *J. Opt. Soc. Am.* **53**, 1377–1381 (1963).

14. E. N. Leith and J. Upatnieks, "Wavefront reconstruction with diffused illumination and three-dimensional objects," *J. Opt. Soc. Am.* **54**, 1295–1301 (1964).

15. P. Kirkpatrick, "History of holography," in *Holography*, B. J. Thompson ed., Proc. SPIE **15**, 9–12 (1968).

16. Y. N. Denisyuk, *Dokl. Akad. Nauk.* (Academy of Sciences Reports, USSR) **144** (No. 6), 1275–1278. [Transl.: Photographic reconstruction of the optical properties of an object in its own scattered radiation field]. *Sov. Phys. Doklady* **7**, 543–545 (1962).

17. Y. N. Denisyuk, "On the reproduction of the optical properties of an object by the wave field of its scattered radiation," *Opt. Spectrosc. (USSR)* **14**, 279–284 (1963).

18. Y. N. Denisyuk and I. R. Protas, "Improved Lippmann photographic plates for recording stationary light waves," *Opt. Spectrosc. (USSR)* **14**, 381–383 (1963).

19. S. F. Johnston, *Holographic Visions - A History of New Science*, Oxford University Press, New York (2006).

20. R. J. Collier, C. B. Burckhardt and L.H. Lin, *Optical Holography,* Academic Press, New York (1971).

21. H. J. Caulfield (ed.), *Handbook of Optical Holography,* Academic Press, New York (1979).

22. Y. N. Denisyuk, *Fundamentals of Holography,* Mir Publishers, Moscow (1984).

23. R. R. A. Syms, *Practical Volume Holography*, Oxford Engineering Science Series **24**, Oxford University Press, New York (1990).

24. P. Hariharan, *Basics of Holography,* Cambridge University Press, Cambridge (2002).

25. G. Saxby, *Practical Holography*, Third Edition, IOP Publishing Ltd, London (2004).

26. G. K. Ackermann and J. Eichler, *Holography - A Practical Approach,* Wiley-VCH Verlag GmbH & Co. KGaA, Weinheim (2007).

27. S. A. Benton and V. M. Bove Jr., *Holographic Imaging*, Wiley-Interscience, John Wiley & Sons Inc., Hoboken, NJ (2008).

28. V. Toal, *Introduction to Holography,* CRC Press, Boca Raton, FL (2011).

29. L. D. Siebert, "Large-scene front lighted hologram of a human subject," in Proc. IEEE **56** (No.7) 1242–1243 (1968).

30. S. F. Johnston, *Holographic Visions—A History of New Science,* Oxford University Press, New York (2006), p. 163.

31. P. Dawson, *Virtual Encounters; Paula Dawson Holograms*, Macquarie University and Newcastle Region Art Gallery, Australia (2010).

32. J. Webster and B. Tozer, "Holographic post irradiation examination of advanced gas cooled reactor fuel elements," in Proc. *J. Brit. Nuclear Energy Soc.*, (1980).

33. B. Tozer and J. Webster, "Holography as a measuring tool," *J. Photogr. Sci.* **28**, 93–98 (1980).

34. M. Benyon and J. Webster, "Pulsed holographic art practice," in *Practical Holography*, T. H. Jeong and J. E. Ludman eds., Proc. SPIE **615**, 36–42 (1986).

35. R. F. Wuerker, L. E. Heflinger, J. V. Flannery and A. Kassel, "Holography on space shuttle," in *Recent Advances in Holography*, T.-C. Lee and P. N. Tamura eds., Proc. SPIE **215**, 76–84 (1980).

36. D. Westlake, R. F. Wuerker and J. F. Asmus, "The use of holography in the conservation, preservation and historical recording of art," in Society of Motion Picture & Television Engineers *SMPTE Journal*, **85**, 84–89 (1980).

37. R. F. Wuerker, "Holography of art objects," in *Recent Advances in Holography*, T.-C. Lee and P. N. Tamura eds., Proc. SPIE **215**, 167–174 (1980).

38. C. F. Reuterswärd, *25 Years in the Branch*, Benteli Verlag, Berne (1978).

39. H. I. Bjelkhagen, "Holographic portraits made by pulse lasers," *Leonardo* **25**, 443–448 (1992).

40. H. I. Bjelkhagen, "Holographic portraits," in *Holography, Commemorating the 90th Anniversary of the Birth of Dennis Gabor*, P. Greguss and T. H. Jeong eds., SPIE Institutes, Volume **IS 8**, 347–353 (1991).

41. M. Benyon, "Cosmetic series 1986-1987: a personal account," *Leonardo* **22**, 307–312 (1989).

42. M. Richardson, *Space Bomb – holograms & lenticulars: 1984-2004,* T.H.I.S. Limited, London (2004).

43. N. Aebischer and B. Cargquille, "White light holographic portraits (still or animated)," *Appl. Opt.* **17**, 3698–3700 (1978).

44. P. Smigielski, H. Fagot, A. Stimpfling and J. Schwab, "Holographie ultra-rapide," *Nouv. Rev. d'Optique appliqué* **2**, 205–207 (1971).

45. A. M. Rodin and D. Brotherton-Ratcliffe, "Compact 16 Joule phase-conjugated SLM Nd:YLF/Nd:Phosphate glass laser", in *The XIV Lithuanian-Byelorussian workshop on Lasers and Optical Nonlinearity*, Proc., Preila, Lithuania (1999) pp. 51–52.

46. A. S. Dementjev, A. M. Rodin, M. V. Grichine and D. Brotherton-Ratcliffe, "High-energy phase-conjugated Nd:Glass laser for the pulsed holography," *Spinduliuotes ir medziagos saveika, Konferencijos pranesimu medziaga*, Kaunas, Technologija, (1999) pp. 249–252.

47. M. V. Grichine, D. Brotherton-Ratcliffe and A. M. Rodin, "Design of a family of advanced Nd:YLF/Phosphate glass lasers for pulsed holography," in *Sixth Int'l Symposium on Display Holography*, T. H. Jeong ed., Proc. SPIE 3358, 194–202 (1998).

48. M. V. Grichine, D. Brotherton-Ratcliffe and G. R. Skokov, "An integrated pulsed-holography system for mastering and transferring onto AGFA or VR-P emulsions," in *Sixth Int'l Symposium on Display Holography*, T. H. Jeong ed., Proc. SPIE **3358**, 203–210 (1998).

49. A. M. Rodin, D. Brotherton-Ratcliffe and G. R. Skokov, "An automated system for the production of image-planed white-light-viewable holograms by pulsed laser," in *Practical Holography XIII*, S. A. Benton ed., Proc. SPIE **3637**, 141–147 (1999).

50. D. Brotherton-Ratcliffe, A. M. Rodin, and S. J. Zacharovas, "Evolution of automatic turn-key systems for the production of rainbow and reflection hologram runs by pulsed laser", in *Holography 2000*, T. H. Jeong and W. K. Sobotka eds., Proc. SPIE **4149**, 359–366 (2000).

51. A. M. Rodin, D. Brotherton-Ratcliffe and R. Rus, "Large-format automated pulsed holography camera System," in Optical Organic and Inorganic Materials, S. P. Asmontas and J. Gradauskas eds., Proc. SPIE **4415**, 39–43 (2001).

52. A. M. Rodin, D. Brotherton-Ratcliffe and R. Rus, "Large-format automated pulsed holography camera System," in *2nd International Conference Advanced Optical Materials and Devices,* Abstract Vilnius, Lithuania, 16–19 August, (2000) p. 60.

53. G. Gudaitus, S. J. Zacharovas, R. Bakanas, D. Brotherton-Ratcliffe, S. Hirsch, S. Frey, A. Thelen, L. Ladriere and P. Hering, "Portable holographic camera system HSF-MINI application for 3D measurement in medicine," in *2nd Int'l Conference "HOLOEXPO-2005, Science and Practice, Holography in Russia and Abroad,* Moscow (Russia) (2005) pp. 27–29.

54. R. Bakanas, G. A. Gudaitis, S. J. Zacharovas, D. Brotherton-Ratcliffe, S. Hirsch, S. Frey, A. Thelen, N. Ladrière and P. Hering, "Using a portable holographic camera in cosmetology," *J. Opt. Technol.* **73** 457–461 (2006).

55. H. I. Bjelkhagen, R. Deem, J. Landry, M. Marhic and F. Unterseher, "A holographic portrait of Ronald Reagan," in *Int'l Symposium on Display Holography*, T. H. Jeong, and H. I. Bjelkhagen eds., Proc. SPIE **1600**, 402–441 (1992).

56. S. A. Benton, "Hologram reconstructions with extended incoherent sources," *J. Opt. Soc. Am.* **59**, 1545–1546A (1969).

57. N. Tamura, "Pseudocolor encoding of holographic images using a single wavelength." *Appl. Opt.* **17**, 2532–2536 (1978).

58. P. Grover and R. Tremblay, "Multicolor wave front reconstruction in white light." *Appl. Opt.* **19**, 3044–3046 (1980).

59. J. F. W. Perry, "Design of large format commercial display holograms," in *Practical Holography III*, S. A. Benton ed., Proc. SPIE **1051**, 2–5, (1989).

60. D. Brotherton-Ratcliffe, "Large Format Holography," in *Sixth Int'l Symposium on Display Holography*, T. H. Jeong ed., Proc. SPIE **3358**, 368–379 (1998).

61. D. Jung, *Bilder · Zeichnungen · Hologramme,* Wienand Verlag, Cologne (1991).

62. T. H. Jeong, P. Rudolf and A. Luckett, "360° holography," *J. Opt. Soc. Am.* **54,** 1263–1264 (1966).

63. J. Upatnieks and J. T. Embach, "360-degree hologram displays," *Opt. Eng.* **19,** 696–704 (1980).

64. S. A. Benton, "Achromatic images from white light transmission holograms," *J. Opt. Soc. Am.* **68**, 1441A (1978).

65. E. Orr and D. Trayner, "Deep image reflection holograms in black and white and additional colours," in *Int'l Symposium on Display Holography*, T. H. Jeong ed., Vol. III, 379–388 (1989).

66. K. Bazargan, "A practical, portable system for white-light display of transmission holograms using dispersion compensation," in *Applications of Holography*, L. Huff ed., Proc SPIE **523**, 24–25 (1985).

67. S. J. Hart and M. N. Dalton, "Display holography for medical tomography," in *Practical Holography IV*, S. A. Benton ed., Proc. SPIE **1212**, 116–135 (1990).

68. N. J. Phillips and D. Porter, "An advance in the processing of holograms," *J. Phys. E: Sci. Instrum.* **9**, 631–634 (1976).

69. N. J. Phillips, A. A. Ward, R. Cullen and D. Porter, "Advances in holographic bleaches," *Phot. Sci. Eng.* **24**, 120–124 (1980).

70. N. J. Phillips, "Bridging the gap between Soviet and Western holography," in *Holography, Commemorating the 90th Anniversary of the Birth of Dennis Gabor*, P. Greguss and T. H. Jeong eds., SPIE Institute Volume **IS 8**, 206–214 (1991).

71. E. Ingelstam, *Les Prix Nobel en 1971*, The Nobel Foundation, Norstedt & Söner, Stockholm (1972).

72. S. J. S. Brown, "Automated holographic mass production," in *Practical Holography*, T. H. Jeong and J. E. Ludman eds., Proc. SPIE **615**, 46–49 (1986).

73. G. P. Wood, "New silver halide materials for mass production of holograms," in *Practical Holography*, T. H. Jeong and J. E. Ludman eds., Proc. SPIE **615**, 74–80 (1986).

74. J. Blyth, "Pseudoscopic moldmaking handy trick for Denisyuk holographers," *Holosphere* **8** (No.11), 5, (1979).

75. P. Hariharan, "Pseudocolour images with volume reflection holograms," *Opt. Commun.* **35**, 42–44 (1980).

76. J. A. Kaufman, "Previsualization and pseudo-color image plane reflection holograms," in Proc. *Int'l Symp. on Display Holography*, T. H. Jeong ed., Vol. **I.**, 195–207 (1983).

77. R. T. Ingwall and M. Troll, "Mechanism of hologram formation in DMP-128 photopolymer." *Opt. Eng.* **28**, 586–591 (1989).

78. S. F. Johnston, *Holographic Visions—A History of New Science*, Oxford University Press, New York (2006), p. 166.

79. E. Ingelstam, *Les Prix Nobel en 1971*. The Nobel Foundation, Norstedt & Söner, Stockholm (1972) pp. 169–201

80. V. G. Komar, "Progress in the holographic movie process in the USSR," in *Three-Dimensional Imaging*, S. A. Benton ed., Proc. SPIE **120**, 127–144 (1977).

81. V. G. Komar, "Holographic motion picture systems compatible with both stereoscopic and conventional systems," *Tekhnika Kino i Televideniya*. No.10, 3–12 (1978).

82. V. G. Komar and O. I. Ioshin, "Motion pictures and holography," *SMPTE Journal* **89**, 927–930 (1980).

83. V. G. Komar and O. B. Serov, "Works on the holographic cinematography in the USSR," in *Holography '89*, Yu. N. Denisyuk and T. H. Jeong eds., Proc. SPIE **1183**, 170–182 (1989).

84. P. Smigielski, H. Fagot and F. Albe, "Holographic cinematography and its applications," in *Int'l Conference on Holography*, D. Wang and J. Ke eds., Proc. SPIE **673**, 22–27 (1987).

85. P. Alexander, "Development of integral holographic motion pictures," in *Fifth Int'l Symposium on Display Holography*, T. H. Jeong ed., Proc. SPIE **2333**, 187–197 (1995).

86. R. V. Pole, "3-D imagery and holograms of objects illuminated in white light," *Appl. Phys. Lett.* **12**, 20–23 (1967).

87. D. J. Debitetto, "Holographic panoramic stereograms synthesized from white light recordings," *Appl. Opt.* **8**, 1740–1741 (1969).

88. J. T. McCrickerd and N. George, "Holographic stereograms from sequential component photographs," *Appl. Phys. Lett.* **12**, 10–12 (1968).

89. L. Cross, "The Multiplex technique for cylindrical holographic stereograms," in SPIE San Diego August Seminar (1977).[Presented but not published]

90. M. C. King, A. M. Noll and D. H. Berry, "A new approach to computer-generated holography," *Appl. Opt.* **9**, 471–475 (1969). *See also*: Ref.[1.87]

91. C. Newswanger and C. Outwater, "Large format holographic stereograms and their applications," in *Applications of Holography*, L. Huff ed., Proc. SPIE **523**, 26–32 (1985).

92. W. Spierings and E. van Nuland, "Calculating the right perspectives for multiple photo-generated holograms," in *Int'l Symposium on Display Holography*, T. H. Jeong ed., Proc. SPIE 1600, 96–108 (1992).

2

Lippmann Photography

2.1 Brief History of Interferential Colour Recordings

Lippmann photography is an interferential colour photographic recording technique that is capable of reproducing colour images with extraordinary precision. Today, more than a century after its invention, many wonderful, perfectly preserved colour images exist from the turn of the century in this unique medium. Colours are recorded in Lippmann photography in a natural way without the use of dyes.

The interest in recording images that are as realistic as possible commenced with the invention of photography. After the invention of black-and-white photography in the nineteenth century, much research was devoted to the possibility of recording natural colour images. It is important to mention this work in the context of today's quest for ultra-realistic imaging. The issue here is how to accurately capture the reflected light from scenes and objects. What is extremely interesting is that it was possible to record the full-colour spectrum of a scene using one of the very first colour photography techniques discovered.

Even before photography was invented, Johann Wolfgang von Goethe (1749–1832) published a book on light and colour (*Zur Farbenlehre*), in which light and colour recordings were discussed [1]. The reader may be surprised to find the poet Goethe involved in light and colour theory. Actually, one scientist who took Goethe's colour philosophy seriously was Thomas Johann Seebeck (1770–1831). Although the collaboration between them did not really advance colour theory, it resulted in some observations that Goethe included as an appendix in his book [1]. This can be regarded as a very early (circa 1810) contribution to colour photography. Seebeck conducted experiments in which solar spectra were projected onto paper impregnated with silver chloride. The recordings showed colours that were induced by corresponding colours in the solar spectrum. The philosophical explanation for this observation was that "light chose to impress itself on material objects in its own colours." It therefore made sense to describe such a process as a natural colour-recording technique. In the recording material used somewhat later in Lippmann colour photography, the colours of the object were actually recorded according to the Goethe–Seebeck description. Sadly, the recordings by Seebeck were not permanent; they disappeared quickly when exposed to light.

Sir John Herschel (1792–1871) made a systematic investigation of solar spectra recorded in silver chloride paper [2]. Herschel could record colours but could not find a method to fix the image. This type of photographic recording technique is often referred to as *heliochromy* (sun colouring) and the images as *heliochromes*. A major contribution was made by Alexandre Edmond Becquerel (1820–1891) [3–8]. Instead of using paper as the material substrate, Becquerel coated a silver chloride (Ag_2Cl) emulsion onto a polished silver plate (applying chlorination by a galvanic process). Employing such plates, the colours of the recorded solar spectrum were much brighter than any previously recorded spectra. In addition to spectra, Becquerel succeeded in recording images of objects. He recorded some coloured engravings and brightly dressed dolls, which required between 10 and 12 h of exposure in bright sunlight. Similar experiments with silver plates coated with silver chloride were carried out by Abel Niepce de Saint-Victor (1805–1870) [9–14] and Alphonse-Louis Poitevin (1819–1882) [15]. Much research went into making the heliochromes permanent but without success. However, we know that Niepce de Saint-Victor was able to improve the technique considerably and recorded many beautiful heliochromes.

In Westkill, New York, Reverend Levi L. Hill (1816–1865) claimed that he had been able to record and fix heliochromes (hillotypes). Hill's technique has been described in a book [16], which he published in 1856. In 1987, Joseph Boudreau [17] of Paier College of Art in Hamden, Connecticut, was able to record hillotypes according to the complicated procedure described by Hill. In Germany, Wilhelm Zenker (1829–1899) worked on the theory of colour photography and colour vision. His book on colour theory had the ancillary title *Photography in Natural Colours* [18]. He proposed an explanation for experiments that recorded spectral information in terms of standing waves of light, which formed within the light-sensitive layer and which were actually recorded. This structured layer, he maintained, would be able to selectively reflect different colours. Unaware of Zenker's theory, Lord Rayleigh (1842–1919) [19] suggested that "colour could be recorded as stationary luminous waves of nearly definite wavelength, the effect of which might be to impress upon the substance a periodic structure recurring at intervals equal to half the wavelength of the light." Lord Rayleigh maintained that the recording technique "produced just such a modification of the film as would cause it to reflect most copiously that particular kind of light." Lord Rayleigh never performed any experiments to verify his theory, but he mentioned, "I abstain at present from developing this suggestion, in the hope of soon finding an opportunity of making myself experimentally acquainted with the subject."

In 1889, in Strasbourg, Otto Wiener (1862–1927) investigated standing light waves and the polarisation of light. He was able to record such standing waves and thus prove that monochromatic light reflected off a mirror was recorded as a periodic interference pattern in an ultrahigh-resolution photographic emulsion [20]. In addition, he investigated the earlier experiments by Seebeck, Poitevin and Becquerel and explained the differences between them [21]. This work was done after Lippmann published his new technique of colour photography. The colours observed in the earlier experiments were obtained in two ways: by interference or by absorption and bleach-out. Zenker's theory was correct regarding Lippmann photography. However, it did not apply to the earlier colour recordings on chloride paper. There, a chemical bleaching process of pigments was behind the colours. The light-sensitive substances were bleached out only by those wavelengths of light that they absorbed, whereas they were not destroyed by light of their own colour.

In 1861, James Clerk Maxwell (1831–1879) demonstrated a colour photograph recorded according to the three-colour additive synthesis [22] that was based on the Young–Helmholtz three-colour theory of vision. An object (a tartan ribbon bow) was recorded on three separate black-and-white plates through three colour filters (red, green and blue). These plates were then projected separately through the same filters that were used for the recording. When the three projected images were registered with one another, the picture burst into colour.

2.2 Examples of Interferential Structures in Nature

It is interesting to mention that the principle of light reflection from periodic structures, upon which interferential colour-recording techniques is based, exists in nature. Opal, a sedimentary gem, gets its colours from tiny spheres of silica packed together. The sea mouse, *Aphrodita*, a marine worm, is named for its mouse-like appearance and behaviour. The spines, or *setae*, that emerge from the scaled back of the sea mouse are one of its unique features which can produce iridescence in a range of colours. Normally, these have a red sheen but when the light shines on them perpendicularly, they flush green and blue. These colours are believed to be a defence mechanism, giving warning to potential predators. The colours of moths, beetles and birds (peacock feathers) all possess light-reflecting structures creating

(a) (b) (c)

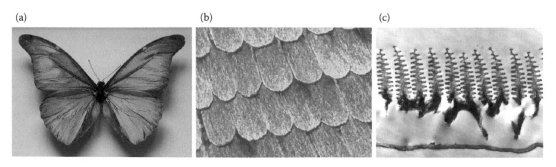

FIGURE 2.1   (a) *Papilio* butterfly, (b) wing scale and (c) wing structure. (Photos b and c courtesy of P. Vukusic, University of Exeter.)

iridescent bright colours. Perhaps, best known is the very bright coloured light reflected from some butterfly wings. The hue of this light changes with the viewing angle, and the colour appears highly saturated. The *Papilio achilles* butterfly of the *Morpho* type is shown in Figure 2.1. The reflection of the structured iridescent scales creates intense blue light. The colours are created by a tree-like structure of lamellae that are self-assembled from cuticles. Depending on the exact *Morpho* species, there are 6 to 10 branches on each side of the lamellae trunk with a density of 700 to 2000 lamellae per millimetre. More information about structural colours in nature can be found in several publications [23–31].

2.3 Lippmann Photography

Gabriel Jonas Lippmann (1845–1921) was born in Bonnevoic, Luxembourg, of French parents, on 16 August 1845. He entered the Normal School in 1867 and studied in Heidelberg, Germany, where he received the degree of Doctor of Philosophy in 1873. In 1875, he moved to Paris and became a professor of mathematical physics at the Sorbonne in 1883, a member of the Institute in 1886 and an Officer of the Legion of Honour in 1884. Previously, he had worked on thermodynamics, electricity and capillarity. At the Sorbonne, he was obliged to start teaching acoustics and optics, and it was in this way that he became interested in the theory of light and, in particular, colour theory. As early as 1886, he had developed a general theory of recording colours as standing waves in a light-sensitive emulsion. However, most of his time was devoted to perfecting a suitable recording emulsion for his experiments. This indicates that he had already developed the interference theory long before the result of the Wiener experiment [20] was published. Lippmann's work on what has come to be known as *interferential photography* or *interference colour photography* was published in several papers [32–35].

On 2 February 1891, Lippmann announced at the Académie de Sciences in Paris that he had succeeded in recording a true-colour spectrum [32]. In addition, the recording was permanent and could be observed in full daylight. A little more than one year later, on 25 April 1892, Lippmann gave a second presentation at the Académie de Sciences [33]. This time, he displayed four colour photographs of different objects: a stained glass window in several colours, a bunch of flags and a dish of oranges, a red poppy and a green parrot. Later, at a photographic exhibition in Paris, he displayed a landscape with a grey building surrounded with green foliage and blue sky. The size of these early photographs was 4 cm × 4 cm. Later, Lippmann used the format 6.5 cm × 9 cm for most of his colour photographs. He also recorded colour images in a dichromated gelatin emulsion [34], a material often used in holography today. Figure 2.2 shows two photographs by Lippmann.

Lippmann developed the first proper theory of the recording of monochromatic and polychromatic spectra [35]. He applied Fourier mathematics to optics, which was a new approach at that time. A bibliography on Lippmann and his inventions was published in 1911 by Lebon [36]. Although the new photographic colour-recording technique known as *Lippmann photography* was extremely interesting from a scientific point of view, it was not very effective for colour photography because the technique was complicated and the exposure times were too long for practical use. The difficulty in viewing the photographs was another contributing factor, in addition to the copying problem; this prevented Lippmann photography from becoming a practical photographic colour-recording method. Notwithstanding this, 100-year-old Lippmann photographs are extremely beautiful, and the fact that the colours are so well preserved in these photographs indicates something rather interesting about their archival properties.

The principle of Lippmann photography is illustrated in Figure 2.3. Owing to the demand for high resolving power in making Lippmann photographs, the material has to be a very fine-grain emulsion and consequently of very low sensitivity. The emulsion coating on the Lippmann plates must be brought into contact with a highly reflective surface, usually mercury. Light is thus reflected back into the emulsion and interferes with the incident light, producing a pattern of standing waves. These standing waves produce a very fine fringe pattern throughout the emulsion with a periodic spacing of $\lambda/(2n)$, which is recorded (here, λ is the wavelength of light in air, and n is the refractive index of the emulsion). The colour information is stored locally in this way. The larger the separation between the fringes, the longer the wavelength of the recorded image information. This picture is, however, only correct when fairly

FIGURE 2.2 Two examples of colour photographs recorded by Gabriel Lippmann: (a) Lippmann Autoportrait and (b) Sainte Maxime, Var, France, 1913. (Courtesy of Musée de l'Elysée Collection, Lausanne. Inventory nos. 9079 and 9985.)

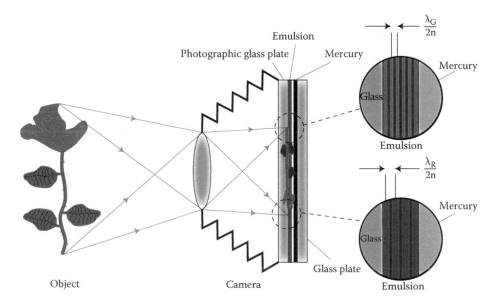

FIGURE 2.3 The principle of Lippmann photography.

monochromatic colours are recorded. A polychrome recording is actually rather more complex, but this was treated mathematically by Lippmann [35].

When the developed Lippmann photograph is viewed in white light, different parts of the recorded image produce different colours. This is due to the different spacings of the recorded fringes in the emulsion. The light is reflected from these fringes, creating precisely the colours that correspond to the original ones that had produced them during the recording. It is obvious that there must be a high demand on the resolving power of the recording material to record the fringes because the fringes are spaced at a distance of approximately half the wavelength of light. It is also clear that the processing of the photographic plates is critically important, as the separation between the fringes must not change because this would create erroneous colours on replay. In addition, one must find ways of obtaining high diffractive efficiency.

To observe the correct colours of a Lippmann photograph, the illumination and observation must be at normal incidence. When these angles change, the colour of the image changes. This change of colour with angle is called *iridescence* and is of the same type as found in peacock feathers and in mother of pearl. The image is recorded as a Bragg structure and is sensitive to the directions of illumination and observation. When a recorded Lippmann photograph is illuminated and observed perpendicularly, the correct colours are reproduced. However, when illuminated and viewed at oblique angles, the colours are tuned towards shorter wavelengths. Thus, red colours become orange, yellow or even green depending on the angles.

In the early 1890s, several scientists and researchers began to explore and further develop the new Lippmann photography technique. Among the most important contributions were those made by Auguste and Louis Lumière [37,38], Eduard Valenta [39], Richard Neuhauss [40], Herbert Ives [41] and Hans Lehmann [42]. There are several early books on Lippmann photography and colour photography [43–52]. An illustration of how a Lippmann photograph replays when illuminated in different ways is shown in Figure 2.4.

In 1908, Gabriel Lippmann (Figure 2.5) was awarded the Nobel Prize in physics for his invention. It was his idea of recording interference fringes throughout the depth of an emulsion that inspired Yuri Denisyuk in Russia in the early 1960s to introduce the technique of recording single-beam reflection holograms [53–55]. In recent years, there has been a revival of interest in Lippmann photography among scientists and holographers [56–83].

FIGURE 2.4 Lippmann photographic plate illuminated and observed in different ways: (a) negative reflected image seen when illumination and observation are not perpendicular, (b) positive red image seen in transmitted illumination caused by colloidal silver particles and (c) correct colour-reflected image seen when illumination and observation are perpendicular.

FIGURE 2.5   Gabriel Lippmann in his laboratory.

2.4 Theory of the Lippmann Process

2.4.1 Treatment of Monochromatic Recording

The transverse electric field of a monochromatic plane wave propagating in the $\hat{\mathbf{z}}$ direction and then reflected back on itself in the $-\hat{\mathbf{z}}$ direction can be expressed as

$$\mathbf{E}(z,t) = E_0 \cos 2\pi \left\{ \frac{nz}{\lambda} + \nu t \right\} \hat{\mathbf{x}} - E_0 \cos 2\pi \left\{ \frac{nz}{\lambda} - \nu t \right\} \hat{\mathbf{x}} \tag{2.1}$$

where λ is the wavelength in vacuum, ν is the frequency, n is the refractive index of the emulsion, E_0 is the peak electric field amplitude and $\hat{\mathbf{x}}$ the unit vector in the direction orthogonal to z. The interference between the two components to this light field creates a standing wave with the following electric field distribution

$$\mathbf{E}(z,t) = 2E_0 \sin \left\{ 2\pi \frac{nz}{\lambda} \right\} \sin \{2\pi \nu t\} \hat{\mathbf{x}} \tag{2.2}$$

We can then define the intensity I associated with this standing wave as the time-averaged value of the squared electric field:

$$I = \left\langle E^2 \right\rangle = 2E_0^2 \sin^2 \left\{ \frac{2\pi n z}{\lambda} \right\} \tag{2.3}$$

A factor of 1/2 comes in here from the integration of the time dependence of Equation 2.2 over one period. The distance between two intensity maxima is $d = \lambda/(2n)$. During the recording of the Lippmann photograph, these intensity maxima are recorded in the emulsion. For example, for $\lambda = 400$ nm (blue light), the distance between these maxima is $d = 200$ nm (in air).

2.4.2 Treatment of Monochromatic Replay

To view the recorded Lippmann photograph, a diffuse light source emitting a white-light continuous spectrum is required to illuminate the plate. This white light is reflected off the recorded interference fringes in the emulsion and, depending on the incident and reflected angles, the well-known Bragg condition will apply

$$\frac{m\lambda'}{n} = 2d\cos\alpha \qquad \forall \quad m = 1,2,3,...$$

(2.4)

For the recorded fringe distance d, the reflected wavelength λ' is then given by

$$\lambda' = \frac{2dn\cos\alpha}{m}$$

(2.5)

If illumination and observation are both perpendicular to the plate ($\cos\alpha = 1$), then exactly the same wavelength is reflected on replay as the wavelength used for the recording.

2.4.3 Treatment of Polychromatic Recording

To record a more complex wave front, such as a landscape scene in natural colours, Lippmann introduced Fourier mathematics to show how such a recording could be theoretically described [35].

To see how this works, we first make the observation that most objects do not reflect pure spectral colours but rather a continuous spectrum $f(\lambda)$. Therefore, if an object is illuminated by a white spectrum, such as sunlight, which we shall denote by $F(\lambda)$, then an observer of the object will observe an intensity proportional to the product $F(\lambda)f(\lambda)$ reflected from every point of the object. As such, we must replace the expression for the electric field in the Lippmann photograph at recording (Equation 2.1) by the Fourier integral

$$E_x(z,t) = E_0 \int_\lambda \sqrt{F(\lambda)f(\lambda)}\cos 2\pi\left\{\frac{nz}{\lambda} + vt\right\}d\lambda - E_0 \int_\lambda \sqrt{F(\lambda)f(\lambda)}\cos 2\pi\left\{\frac{nz}{\lambda} - vt\right\}d\lambda$$

(2.6)

The intensity at the emulsion surface may then be written as before as the time average of the squared electric field:

$$I = 2E_0^2 \int_\lambda F(\lambda)f(\lambda)\sin^2\left\{\frac{2\pi nz}{\lambda}\right\}d\lambda$$

(2.7)

If we now denote the spectral sensitivity of the recording material as $O(\lambda)$ and we make the further assumption that the Lippmann emulsion inherits upon exposure and processing an effective reflectivity $R(z)$ proportional to the time averaged squared electric field intensity present at recording, then we may write down the following expression for the reflectivity of the Lippmann photograph:

$$R(z) = 2E_0^2 \int_\lambda F(\lambda)f(\lambda)O(\lambda)\sin^2\left\{\frac{2\pi nz}{\lambda}\right\}d\lambda$$

(2.8)

Assuming that the recording material is isochromatic (the response is equal for all wavelengths within the spectrum), then the product $F(\lambda)O(\lambda)$ is a constant, which, with appropriate normalisation, can be taken to be unity, whereupon the integral in Equation 2.8 simplifies to

$$R(z) = 2E_0^2 \int_\lambda f(\lambda)\sin^2\left\{\frac{2\pi nz}{\lambda}\right\}d\lambda$$

$$= E_0^2\left\{\int_\lambda f(\lambda)d\lambda - \int_\lambda f(\lambda)\cos\left\{\frac{4\pi nz}{\lambda}\right\}d\lambda\right\}$$

(2.9)

The second integral is very similar to the Fourier transform of the reflected spectrum. Note, however, the factor 2 here.

2.4.4 Treatment of Polychromatic Replay

When the recorded Lippmann photograph is illuminated with a diffuse white spectrum (the same as that used for the recording), each individual volume element in the emulsion reflects the incident waves according to Equation 2.9. We may therefore write down an approximate expression for the reflected electric field due to reflection at point z

$$dE_x(z,\lambda') = -E_0\sqrt{F(\lambda)}R(z)\sin\left\{2\pi v't - \frac{4\pi n}{\lambda'}z\right\} \tag{2.10}$$

Note the phase term that is needed to account for the optical path length difference in the emulsion. Integrating all partial reflected waves, we obtain the total reflected field

$$
\begin{aligned}
E_x(\lambda') &= -\int_D E_0\sqrt{F(\lambda')}R(z)\sin\left\{4\pi v't - \frac{4\pi n}{\lambda'}z\right\}dz \\
&= -\int_D E_0\sqrt{F(\lambda')}R(z)\cos\left\{\frac{4\pi n}{\lambda'}z\right\}\sin\{2\pi v't\}dz \\
&\quad + \int_D E_0\sqrt{F(\lambda')}R(z)\sin\left\{\frac{4\pi n}{\lambda'}z\right\}\cos\{2\pi v't\}dz
\end{aligned}
\tag{2.11}
$$

where D is the thickness of the emulsion.

We may now use Equation 2.9 in Equation 2.11 to work out the reflected intensity as the time-averaged Poynting vector. In this calculation, we obtain several types of integral of which only one is non-zero due to simple orthogonality properties giving

$$I(\lambda') = \langle E_x(\lambda')H_y(\lambda')\rangle = \frac{nc\varepsilon_0}{2}F(\lambda)E_0^2\left\{\int_D R(z)\cos\left\{\frac{4\pi n}{\lambda'}z\right\}dz\right\}^2 \tag{2.12}$$

where ε_0 is the electric field constant and c is the speed of light is vacuum.

The integral here is the inverse Fourier transform of the second term in $R(z)$. Its square determines the emulsion's reflectivity to the wavelength λ'.

From Equations 2.9 and 2.12, we can see that at a given image point, the reconstructed colour spectrum is proportional to the product $F(\lambda)f^2(\lambda)$. This means that the reflected spectrum of the recorded object is replayed by a Lippmann photograph as the square of the original spectrum (of course, this is under the assumption that the induced reflectivity is strictly proportional to the time averaged squared recording electric field intensity). In the field of digital image processing, this is referred to as a gamma correction of two. In practice, this can be compensated for by the γ curve of the recording material. This effect and the phase shift caused by using mercury (an optical medium of higher refractive index than gelatin) as the Lippmann reflector will be discussed later when we describe the possibility of recording Lippmann photographs without a reflector and using only the Fresnel reflection of air in contact with the emulsion gelatin.

Finally, a very pertinent question is whether the time coherence of white light is sufficiently long to record standing waves in a Lippmann emulsion at all. The expression relating the coherence time τ_c to the spectral bandwidth Δv can be written as

$$\tau_c \Delta v \sim 1 \tag{2.13}$$

The coherence length L_c can likewise be defined as

$$L_c = \frac{c}{n}\tau_c \sim \frac{\lambda^2}{n\Delta\lambda} \tag{2.14}$$

Using monochromatic light from a laser, the coherence length can be very long. For white light (e.g., the sun's spectrum), the mean wavelength is approximately 550 nm and $\Delta\lambda \sim 300$ nm. This then equates to a coherence length of $L_c \sim 1$ µm! It is therefore remarkable that Lippmann photography works in practice and that it is capable of producing such wonderful and brilliant colour photographs.

2.5 Early Lippmann Emulsions

Let us look at how Lippmann plates were made and processed originally. The plates that Lippmann used were albumen emulsions containing potassium bromide. The plates were sensitised in a silver bath, washed, rinsed with cyanine solution and dried. The sensitivity was extremely low. Exposure times of 1 h or more were needed. When Auguste and Louis Lumière in Lyon, France introduced much finer-grained silver halide gelatin emulsions, these quickly became the recording material of choice for Lippmann photography [37,38]. Lippmann switched to the Lumière brothers' emulsions. These types of silver halide emulsions were much faster than the earlier albumen or collodion plates. Now the exposure time was only a few minutes rather than hours.

The most important contributions were made by Neuhauss and Lehmann; both were active in Germany. Neuhauss and Lehmann devoted a lot of time to the perfection of the Lippmann process. Many beautiful photographs were recorded by them, some of which have been preserved and are part of photographic collections in different museums around the world.

Almost all the research on Lippmann photography took place in Europe. Ives [41] is the only known scientist in the United States who was involved in the development of the technique. Most of the researchers of this time had their own way of preparing their silver halide emulsions. Later, at about the turn of the century, commercial Lippmann plates were produced by Kranseder & Cie A.G. in Munich and, after Kranseder's death, by Jahr's photographic company in Dresden. The recording procedure using the mercury plate holder was straightforward and more or less the same for all experimenters. As is the case in colour holography today, the recording material is the most important factor for recording high-quality Lippmann photographs as well as colour holograms. The preparation of the emulsion and its processing were absolutely critical to obtain good colour photographs. These emulsions were based on experience gained from making Lippmann photographs over several years.

2.5.1 Auguste (1862–1954) and Louis Lumière (1864–1948)

The Lumière brothers devoted several years to trying to commercialise the Lippmann technique. They were able to manufacture an ultrafine-grain panchromatic emulsion on which they recorded many beautiful colour photographs.

The Lumières' first successful emulsion was published in 1893 [37] and later modified by them [38] as presented in Table 2.1. The Lumière brothers recorded several colour interference photographs. They recorded the first colour portrait in the summer of 1893. It was a photograph of a woman, resting her head on her arm at a table with a green background of grapes and a glass of red wine on the table. The exposure time was 4 min in bright sunlight. Louis Lumière was the one who was mainly responsible for these photographs. At the International Photographic Congress in Geneva in August 1893, he presented

TABLE 2.1

The Lumière Emulsion

Solution A	Solution B
Gelatin (10 g)	Gelatin (10 g)
KBr (3.5 g)	$AgNO_3$ (5 g)
Water (200 mL)	Water (200 mL)

Note: Sensitisers, such as erythrosine, cyanine, anisolin and methyl violet were used.

several colour portrait photographs and landscapes. These images (3 cm × 5 cm) were projected on a 40 cm × 70 cm screen with a projector designed by the Lumière brothers.

2.5.2 Richard Neuhauss (1855–1915)

Neuhauss, who was a physician from Berlin, was one of the most experienced experimenters. He made many beautiful Lippmann photographs. Between 1894 and 1908, he produced about 2500 plates and performed experiments in his home in Gross-Lichterfelde outside Berlin. The test object was most often a stuffed parrot that was installed on his balcony, but also landscapes and portraits. His book contains important information on emulsion making [40]. Neuhauss stressed the importance of gelatin quality for making successful emulsions. For the plates, he used gelatin primarily from Lautenschläger in Berlin. His recommended emulsion is found in Table 2.2.

To prepare the emulsion, the gelatin is first dissolved in cold distilled water; this takes about 10 min. Then, the gelatin–water solutions A and C are heated until the gelatin is completely melted, after which solution C is cooled down to 35°C and solution A to 37°C. The book stresses the fact that the temperature of the solutions *must not exceed* 40°C. Solution B is then mixed with solution A under vigorous stirring. This mixture is then poured *drop by drop* into solution C while stirring. When finished, sensitisers are added to the mixture:

- erythrosine–alcohol solution (1:500)   1 mL
- cyanine–alcohol solution (1:500)   2 mL

TABLE 2.2

Neuhauss' Emulsion

Solution A	Solution B	Solution C
Gelatin (2.5 g)	$AgNO_3$ (1.5 g)	Gelatin (5 g)
Water (70 mL)	Water (5 mL)	KBr (1.25 g)
		Water (75 mL)

FIGURE 2.6 Reproduction of one of Richard Neuhauss' stuffed parrot photographs from 1899. (Courtesy of the Royal Photographic Society Collection, Bradford, UK.)

The emulsion (which should now appear completely transparent) is now filtered in the coating bottle, which constitutes a part of the coating process. The preheated plates (previously cleaned in a 50% nitric acid solution for 24 h) are coated as soon as possible after the solution is ready. The emulsion is coated by letting it float over the surface of the levelled plates until the entire plate is completely covered. The amount of emulsion prepared in this way will be sufficient for 8 to 10 plates measuring 9 × 12 cm. The plates are then quickly cooled by placing them on a levelled marble table. It is important that the plates do not dry out completely after the cooling because there is a danger of potassium nitrate crystallising within the emulsion before they can be washed to remove the unwanted salt. The plates are then rinsed and placed in a tray filled with water for 15 min, during which time the water bath must be changed once. After the wash, the plates are dried. Neuhauss recommends the use of a centrifuge for this purpose to avoid leaving drop marks on the finished plates. The emulsion must appear completely clear, otherwise the plates will not perform well. He also mentions another important point concerning the mixing of the emulsion, which is the importance of mixing and then coating the plates without delay. Neuhauss used a stuffed parrot for many of his emulsion and recording tests. One such recording that has survived is shown in Figure 2.6.

2.5.3 Hans Lehmann (1875–1917)

The following recipe (Table 2.3) from Lehmann was used for manufacturing commercial plates at Richard Jahr's dry plate factory in Dresden. Not until 1925, long after Lehmann's death in 1917, was his formula revealed by Jahr [84].

The gelatin is dissolved at 35°C and then filtered. Solution B is added to 80 mL of solution A. Potassium bromide (3.2 g) is then added to and dissolved in the remaining solution A. When the potassium bromide has been completely dissolved, the solution is slowly poured into the gelatin–silver nitrate solution. Note that adding the KBr solution to the $AgNO_3$ solution is actually opposite to the way that the other emulsions are mixed. All solutions must be kept at 35°C. After precipitation is completed, stirring is required for 3.5 min. Then, the following sensitising dye solutions (warmed to 30°C) are added:

- pinacyanol–alcohol solution (1:1000)   4 mL
- orthochrom T–alcohol solution (1:1000)   4 mL
- acridine orange–alcohol solution (1:500)   4 mL

This should be done in approximately 45 s. No further heating of the emulsion is allowed. Well-cleaned glass plates are then coated and cooled. A 10 min wash is needed to remove the potassium nitrate. Finally, the plates are dried in a horizontal position. The main secret behind this recipe is the particular sensitising dyes used and their combination. Colour photographs recorded in this emulsion produced the best colour-correct sensitivity and colour balance ever achieved in Lippmann photography (Figure 2.7). If you want to try one of the old recipes today, this formula would definitely be the first choice.

Lehmann's improved emulsion possessed approximately 10 times higher sensitivity compared with previous Lippmann emulsions. In addition, the emulsion allowed a good colour balance, which meant that white was easily obtained. The colour sensitivity maxima peaked at the following wavelengths: 635, 585, 509 and 475 nm. Lehmann also provided exposure recommendations for the new emulsion, which were applicable to landscape photographs in bright sunlight (Table 2.4).

TABLE 2.3

Lehmann's Emulsion

Solution A	Solution B
Gelatin (20 g)	$AgNO_3$ (4 g)
Water (390 mL)	Water (10 mL)

FIGURE 2.7   Reproduction of a *Still Life* 1908 Lippmann photograph by Hans Lehmann (Photo courtesy of Preus Photomuseum. Preus Photomuseum Collection, Norway.)

TABLE 2.4

Exposure Requirements for Lippmann Photographs using the Lehmann Emulsion

Aperture, F/number	F/3	F/3.5	F/4	F/4.5	F/5	F/6
Exposure time in seconds	6	8	11	14	17	25

For portraits, it was recommended that they not be made in direct sunlight but with a 20 s exposure time at F/3.5. Lehmann published articles on the practical aspects of recording Lippmann photographs and how to use the Zeiss equipment [85,86].

2.5.4 Edmond Rothé (1873–1942)

Edmond Rothé from Nancy University in France recorded Lippmann photographs without using the mercury reflector [87–91]. Instead, he used the light reflected at the gelatin–air interface only. Because of its simplicity, this technique is very interesting. The emulsion Rothé used for his experiments is given in Table 2.5. Rothé used the Lumière–Lippmann formula with minor modifications. We shall discuss the mercury-free, gelatin–air recording technique below.

Part A of Rothé's recipe is mixed at a high temperature. When cooled down to 40°C, the silver nitrate powder is added; this takes 1 to 2 min to dissolve under continuous agitation. Thereafter, the following sensitisers are added:

- cyanine–alcohol solution (1:500)   3 mL
- malachite green–alcohol solution (1:500)   2 mL
- glycin red–alcohol solution (1:500)   10 mL

The emulsion is then filtered. The plates should be coated, whirled and placed on a levelling table to set, then washed in running water for half an hour and dried. Both the demand for extremely high resolving power (very fine grains) and the colour sensitisation (isochromatism) of the Lippmann emulsion are

TABLE 2.5

Rothé's Emulsion

Part A	Part B
Gelatin (5 g)	$AgNO_3$ (0.75 g, fine powder)
KBr (0.53 g)	
Water (100 mL)	

equally important. Today, the dyes used for the sensitisation of silver halide emulsions are mainly the cyanine and isocyanine dyes.

2.6 Recording of Early Lippmann Photographs

2.6.1 Recording Equipment

Many of the professional photographic cameras at the end of the 1900s were cameras for glass-plate negatives. The main additional piece of equipment required for the Lippmann photographer was the mercury dark slide. At first, the slide had to be made up by the person who wanted to record a colour photograph. However, after some time, it was possible to obtain the equipment from camera manufacturers. Carl Zeiss Kamerawerke in Jena manufactured mercury plate holders, filters and viewing and projection apparatae [86] (Figure 2.8). Equipment was also made by other German companies: Stegemann, Braun

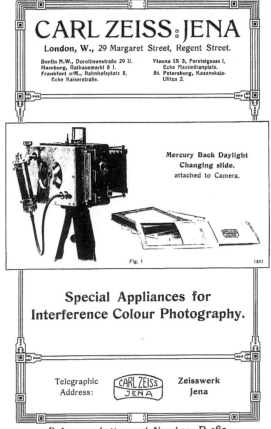

FIGURE 2.8 Zeiss interference colour photography catalogue.

and the plate manufacturer Kranseder & Cie. In England, dark slides were made by Watson & Sons and Penrose & Co.

Most Lippmann photographers used mercury for the required light reflection and phase locking. It worked rather well but involved both practical problems and safety concerns. Fog and streaks caused by mercury in contact with the silver halide emulsion was one such problem. A substitute for mercury was therefore of interest. As previously mentioned, Rothé recorded objects using the reflection off the gelatin–air interface only. Effectively, the reflection from this interface creates weak reflected light, which interferes with the (stronger) incoming light. Because lower modulation of the interference fringes was expected in this case, the efficiency of such an image was anticipated to be lower than when the mercury reflector was used. Notwithstanding this, Rothé reported good results and published several articles with details about his recording technique [87–91]. The main advantage was that an ordinary camera without special slides could be used and only Lippmann plates were needed for recording colour photographs. More about the difference between mercury–gelatin versus the air–gelatin interface later.

2.6.2 Processing Lippmann Silver Halide Emulsions

The processing of Lippmann colour photographs was more or less done in the same way by most experimenters. They used developers based on pyrogallol and ammonia, which were formulated to suit the particular emulsion. A surface developer will perform well because no image information is located deep inside the emulsion. In addition, the hardening effect on gelatin that pyrogallol provides is important. The Lumière developer was generally recommended. The recipe is given in Table 2.7. Development time was 1 to 3 min. Most often, the image was fixed by a 15 s immersion in a potassium cyanide bath (5%). However, the use of a safer sodium thiosulphate bath (15%–20%) was suggested by other Lippmann photographers. Rather later, fixing the developed image was not recommended because this would change the thickness of the emulsion and therefore change the colour of the image. The plate could be intensified by a bleaching and redeveloping process. Ives [41] tested other developers based on amidol and hydroquinone and used bleaching to create phase gratings which came out even more brilliant and with a narrower bandwidth when compared with those processed by pyrogallol. He was actually touching on modern processing methods that are now applied in holography.

2.6.3 Viewing Lippmann Photographs

There is one particular problem in viewing recorded Lippmann photographs. The reflection from the gelatin surface of the emulsion can cause colour distortion as a result of the phase shift in the gelatin–mercury interface during recording. Because the Lippmann photograph has to be viewed using perpendicular illumination, it is important to eliminate the specular surface reflection to see the image clearly. This can be accomplished by attaching a wedged glass plate (with an angle of ~10°) on top of the emulsion using an index-matching glue, most often Canada balsam ($n = 1.52$–1.54; Figure 2.9). The back of the photograph was usually painted black (Lippmann photographers often used asphaltum varnish mixed with machine oil for this purpose), covered with black paper and the edges sealed. Another possibility was to apply gum styrax ($n = 1.58$) which could be removed if desired after being slightly heated and the wedge plate reused for another Lippmann photograph. The refractive index of gum styrax is also a better match. Today, modern optical epoxy cements may be used as long as they do not affect the emulsion.

A good way to view a Lippmann photograph is by a small opening in a wall facing a brilliant white sky. If the observer stands with his back to the opening and holds the picture at arm's length, reflecting the sky, the image appears at its best. There were special viewing devices, such as the dioptric and the catoptric viewing apparatae, which facilitated the display of these beautiful colour photographs (Figure 2.10). The viewing difficulty inherent to Lippmann photographs resembles in some way today's difficulties in displaying and viewing holograms.

The photographs could also be viewed in an enlarged format by projecting the reflected image using a special projector. The image could not be projected like a modern slide however. Rather, the reflected image had to be projected by a projector of the aphengescope-type (Figure 2.11). Zeiss Works in Germany produced such viewing and projecting equipment for Lippmann photographs [86].

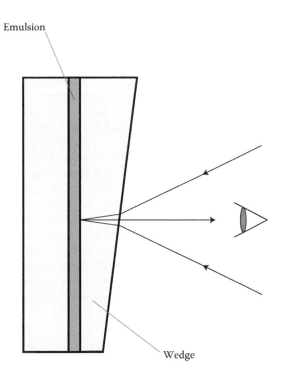

Emulsion

Wedge

FIGURE 2.9 Wedged glass prism (Wiener prism) attached to a Lippmann photograph.

(a)

(b) Binocular Dioptric Viewing Apparatus for
12 × 9 cm Plates (see Fig. 4) . . . . *Azzoppimmo* **225.—**
Stand for the above apparatus, rising from the
floor, to raise and lower (see Fig. 4) . . . . . *Azzoppirci* **30.—**

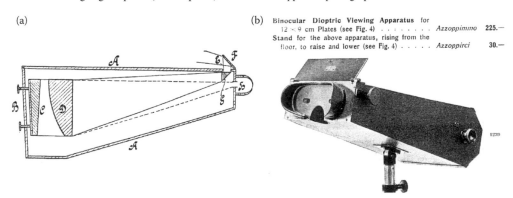

FIGURE 2.10 Zeiss viewing equipment for Lippmann photographs: (a) principle and (b) viewing apparatus.

(a)

(b) 20. Complete Optical Projection Apparatus for   Codewords   Marks
Interference Colour Photographs, comprising
the whole of the parts specified under heading
III, but **exclusive** of the projection lens . . . *Azzopperai* **638.—**

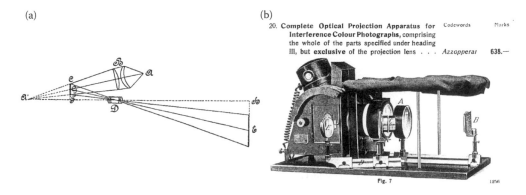

FIGURE 2.11 Zeiss projection equipment for Lippmann photographs: (a) principle and (b) projector.

2.6.4 Early Investigations of Lippmann Emulsions

The researchers working with Lippmann photography investigated the interference structure of the emulsion (Zenker's laminae) not only for the purpose of improving the colour-recording technique but also to prove Zenker's photochromic theory as applied to Lippmann photography. This theory, as previously mentioned, had convincingly been verified by Wiener's famous experiment on monochrome stationary light waves [20]. However, using optical microscopy to study the periodic silver grain structure in the emulsion pushed this technique to its very limits. Notwithstanding this, in 1897, Neuhauss succeeded in microscopically imaging the interference structure of red light recorded in a Lippmann emulsion [92] (Figure 2.12). He used a long wavelength recording in which the interference layers were most widely separated. In air, the distance between the interference layers was 330 to 380 nm, and in the emulsion, it was closer and depended on the refractive index of the emulsion. A fringe separation of 220 to 250 nm had to be microscopically resolvable to see the recorded structure in the emulsion. Neuhauss used different techniques for embedding the emulsion samples to be studied. He used paraffin, Canada balsam and glycerin, which slightly increased the separation of the layers in some cases. Flatau from the Anatomic Institute in Berlin helped him to prepare the thin emulsion layers (~2 μm in thickness) using a very fine microtome. The quality of the microscope lens was also considered to be very important, which is why the oil-immersion technique and a high-quality apochromatic lens (numerical aperture, 1.40) from Zeiss were used. Short-wavelength illumination also increased the resolution. Accordingly, Neuhauss used

FIGURE 2.12 Neuhauss' first microrecording of a Lippmann emulsion revealing the interference fringes. (Courtesy of R. Neuhauss, 1898.)

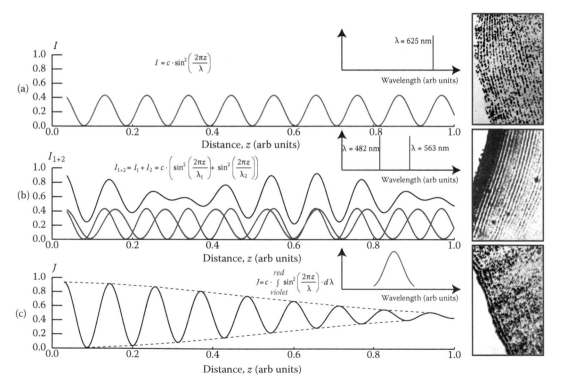

FIGURE 2.13 Lehmann's emulsion investigations.

sunlight passing through a dark blue dye to get a wavelength peaked at approximately 450 nm. Using these techniques, Neuhauss succeeded in recording photographs of the emulsion through a microscope having a linear magnification of approximately 4000×. The separation of interference layers was clearly visible and the distance between them could be measured. The results of the experiment were in good agreement with the theory. Thereafter, optical microscopy was often used by Lippmann photography scientists to study the influence of various recording and processing methods. Both Lehmann [42] and Cajal [93] made micrographs of Lippmann emulsions. Figure 2.13 shows the Lehman emulsion investigation.

The interest in Lippmann photography disappeared almost totally after 1907 when the Lumière brothers started to commercialise their autochrome plates for colour photography [94]. A much faster colour material was offered based on a different colour-recording principle with relative ease of handling. This became the first commercially successful colour photography technique and was used until the late 1930s when modern three-layer colour emulsions (with or without colour couplers) were introduced.

2.6.5 Museum Collections of Lippmann Photographs

Over a period of about 15 years, Lippmann photographs were recorded by a very limited number of scientists and photographers around the world. However, some of these photographs have been preserved and can be found in private and institutional collections. A rather large collection is in the Musée de l'Elysée Lausanne, Switzerland [95]. In France, Lippmann photographs are to be found at the following museums or institutes: Université Pierre et Marie Curie [96], Musée français de la Photographie [97], Société française de Photographie [98], all located in or near Paris, as well as in the Musée Nicéphore Niépce, Chalon-sur-Saône [99].

Preus Fotomuseum, Horten, Norway has some excellent photographs recorded by Neuhauss [100]. In England, the Royal Photographic Society has two Lippmann photographs. George Eastman House in Rochester, New York has several of Ives' photographs in their collection [101]. Museums of science and technology in a few countries may also have Lippmann photographs on display in their photographic exhibitions or in their collections.

2.7 Modern Lippmann Photography

Single-beam Denisyuk reflection holography [54,55] shows similarities with Lippmann photography. In both cases, an interference pattern is recorded in a high-resolution emulsion. For example, Kogelnik's theory [102] for the volume holographic grating is also applicable to Lippmann photography. The Bragg diffraction regime applies to both. The fundamental difference between a Lippmann photograph and a Denisyuk reflection hologram is that in the Lippmann case, there is no phase recording involved; the recorded interference structure is a result of the phase-locking of the light by the reflecting mirror. In holography, the phase information is actually recorded, being encoded as an interference pattern created between the light reflected from the object and a coherent reference beam. To some extent, a Lippmann photograph can be regarded as a reflection image-plane hologram recorded with light of very short temporal coherence. The reference wave is a diffuse complex wave front (the mirror image of the exit pupil of the recording lens).

The recording of monochromatic and polychromatic spectra has to be treated differently. The recording of monochromatic light in a Lippmann emulsion is easy to understand, and it is very similar to recording a volume reflection hologram. A broadband polychromatic spectrum such as a landscape image is very different. In this case, the recorded interference structure in the emulsion is located only very close to the surface of the emulsion that is in contact with the reflecting mirror. Thus, an emulsion thickness of only 1 to 2 μm is needed and actually preferred. A colour reflection hologram, on the other hand, is a result of the three colour process (red, green and blue) involving three monochrome recordings superimposed in the same emulsion. This can be done either by sequential or simultaneous exposure of the recording material when illuminated with the three primary laser wavelengths. In the case of the hologram, interference fringes are recorded throughout the whole depth of the emulsion. The thicker the emulsion is, then generally, the narrower the reconstruction band will be—which means that for the reconstruction of a colour hologram by a broadband white-light source, a compromise must be made between high image brightness and high colour saturation.

2.7.1 Modern Lippmann Emulsions

There was very little interest in making silver halide plates of the Lippmann type after interest in Lippmann photography disappeared. However, the need for such plates came back when holography started to become popular in the early 1960s. A full discussion of modern ultrafine-grain emulsions as used in holography, Lippmann photography and other high-resolution imaging applications will be presented in Chapter 4.

2.7.2 Recording and Processing of Lippmann Photographs Today

Recent progress in the development of colour holography has opened up new possibilities to investigate Lippmann photography. New and improved recording materials (silver halide and photopolymer) combined with special processing techniques developed for colour holograms may be expected to significantly improve Lippmann's interference photography. As mentioned in Section 2.3, a new interest in Lippmann's technology has been manifested by many recent publications [56–84]. Connes [61] published an excellent review of the history of standing light waves and colour interference photography. Nareid [62] and Fournier [64] both made contributions to the history of Lippmann photography. Nareid and Pedersen [63] have developed computer programs for modelling the Lippmann process. For monochromatic recording, Kogelnik's coupled wave theory [102] can be applied to Lippmann photography. For polychromatic recording, superposition of several frequencies in the recording light gives rise to aperiodic space gratings for which the Kogelnik theory cannot be applied. To treat such an aperiodic grating, Nareid and Pedersen [63] have, however, presented a treatment of wave propagation in a stratified medium where a first Born approximation is made. In addition, in Chapter 12, we shall discuss an alternative to Kogelnik's coupled wave theory in which a volume grating is treated as an infinite sum of infinitesimal discontinuities in the permittivity profile and the reflected signal wave is built up by summing the individual Fresnel reflections from such discontinuities in a consistent manner [103]. This model is able to naturally treat aperiodic (polychromatic) space gratings but at present has not been generalised to the diffuse images characteristic of Lippmann photographs.

Fournier and Burnett [66] describe colour rendition and archival properties of Lippmann photographs and compare Lippmann's technique with holography. They investigated old Lippmann photographs using modern electron microscopy. In addition, they recorded a monochromatic volume grating in DuPont photopolymer material using filtered light (520 nm, 10 nm bandwidth) from a slide projector. Marraud and Fournier [67] explained Lippmann's early introduction of Fourier mathematics in optics. A major contribution was made by Phillips et al. [69], comparing the theory of Lippmann photography and holography. Other important considerations, such as the influence of spatial coherence (camera lens aperture) and internal emulsion scattering were also treated in the article.

Rich and Dickerson [71] published results on recording Lippmann photographs on laboratory-made and commercial silver halide emulsions. Bjelkhagen [73] demonstrated the possibility of recording Lippmann photographs in DuPont panchromatic photopolymer materials. Bjelkhagen et al. [74] used the Slavich PFG-03C holographic emulsion to record Lippmann photographs using only the gelatin–air interface as the reflector. The history of Lippmann photography and modern recording techniques has been published by Bjelkhagen [75]. A potential modern application of Lippmann photography may appear in the document security field [76]. Optical variable devices (OVDs) such as holograms, are now common in this field. A Lippmann photograph can offer a new type of OVD, which belongs to the interference security image structures. It offers additional advantages over holograms for unique security documents.

2.7.3 Modern Lippmann Photographs Made with Different Modern Materials

One of the authors (HB) has spent considerable time recording modern Lippmann photographs on both photopolymer and silver halide materials. To record these photographs, the following camera was used:

- manufacturer: Eastman Kodak Co. (Folmer & Schwing Division)
- model: Auto Graflex Camera

FIGURE 2.14   Graflex 4 in. × 5 in. camera for Lippmann photography.

- format: 4 in. × 5 in.
- lens: Kodak Aero Ektar F/2.5, 178 mm

A photograph of the camera is shown in Figure 2.14. It could accommodate both sheet film and glass plates.

2.7.3.1 Photopolymer Recording Materials

The first experiments on Lippmann photography were carried out on panchromatic photopolymer materials. The colour holography photopolymer material HRF-700X from E.I. du Pont de Nemours & Co. was used [104–107]. Although less sensitive than the silver halide emulsions (which are also slow, according to modern photographic standards), it has the special advantages of easy handling and dry processing (only UV-curing and baking).

The recording of a DuPont Lippmann photograph is relatively simple. Before the recording took place, a reflecting mirror foil was laminated to the polymer film. This was a silver-sputtered (800 Å) polyester film produced by Courtaulds Performance Films [107] without the standard anti-dust oxide (InO) top layer. The mirror foil was laminated to the photopolymer material under safelight and then loaded into a conventional sheet film holder. The film cassette was then attached to the back of the camera, which had to be mounted on a tripod. In front of the camera lens, a colour correction filter was used to compensate for the low red sensitivity and the high green sensitivity of the photopolymer material. After the exposure was finished, the film holder was removed from the camera and opened in full daylight. The mirror foil was detached from the photopolymer film which was then exposed to direct sunlight (or strong white or UV light) for developing. DuPont recommends an exposure of approximately 100 mJ/cm^2 at 350 to 380 nm. The photograph must then be put into an oven at a temperature of 100°C for 2 h to increase the brightness of the image. A still life (a bowl of fruit) and a stuffed parrot were recorded using the technique described. Two 500 W halogen lights illuminated the scene at a distance of half a metre. The exposure time was between 5 and 10 min at aperture F/4. After processing, the back of the film was laminated to a black foil. For better viewing of the image, the front of the film was laminated to a wedged glass plate.

A modern Lippmann photograph using the DuPont photopolymer is shown in Figure 2.15; the spectrum of the Lippmann photograph is also presented in Figure 2.16. For conventional photographic reproduction, the sky was used as the illuminating field and a 35-mm camera with a macro lens was used with standard negative colour film (Kodak Royal Gold 100 ASA). No wedge glass plate was attached to the polymer film, as it was easy to view the image without the wedge. Then again, using a wedge plate may be expected to improve the contrast of the photograph. Note that a small vibration of the object probably occurred during the long exposure, which caused the image of the parrot to blur slightly.

FIGURE 2.15   Modern photopolymer Lippmann photograph produced by one of the authors (HB).

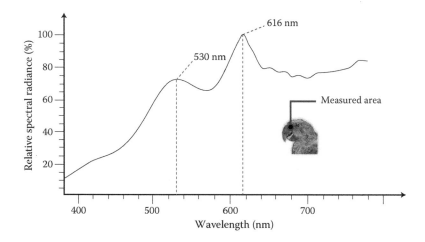

FIGURE 2.16   Spectrum recorded from Lippmann photograph of the parrot in Figure 2.15.

Figure 2.16 is interesting. The recording of a Lippmann photograph implies the recording of the entire reflected spectrum produced by the object rather than the common three-colour sampling used in most modern colour-recording techniques. This should be compared with the spectrograms produced by Fournier and Burnett [66] obtained from old Lippmann photographs.

2.7.3.2 Silver Halide Recording Materials

An easier solution than preparing one's own ultrahigh-resolution silver halide emulsions in the laboratory is to use the Slavich [108] commercial material designed for colour holography. Glass plates of a thickness of 1.5 mm and a size of 4 in. × 5 in. coated with the Slavich emulsion were used by HB to record Lippmann photographs without a mercury reflector (Figure 2.17).

Recording a Lippmann photograph onto ultrafine-grain panchromatic silver halide emulsion was done in the following way: the plate was inserted into a conventional dark slide with the emulsion side facing away from the camera lens. A modified graphic film pack adapter was used, inside of which black velvet was attached to reduce scattered light (Figure 2.18). The plate holder was inserted into the camera, which was mounted on a tripod. Because the plate was exposed without mercury, the exposure time had to be increased compared with a recording with a mercury reflector. After the exposure was finished, the plate holder was removed from the camera and the plate processed in a darkroom.

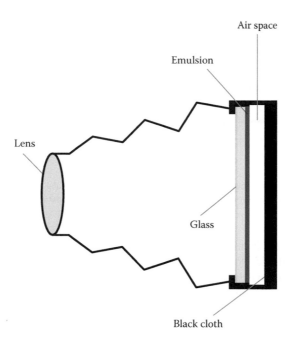

FIGURE 2.17 The principle of recording Lippmann photographs without mercury.

The reason why it is possible to obtain a Lippmann photograph without mercury can be explained in the following way. One must study the difference between reflections at the mercury surface and at the gelatin–air interface, as illustrated in Figure 2.19. A node is located at the mercury reflector (an optically thicker medium than gelatin), which is in contact with the gelatin surface. The phase shift there is π. On the contrary, an antinode is located at the gelatin-air surface when the reflection is obtained from a gelatin–air interface (an optically thinner medium than gelatin); in this case, because no phase shift occurs, a silver layer will be created at the emulsion surface after development. In the case of a mercury-gelatin interface, the first silver layer will be created at a distance from the interface of λ/4 inside the gelatin emulsion, whereas in the gelatin-air case, the second silver layer will be λ/4 closer to the gelatin-air interface surface compared with the same layer in the mercury-gelatin case. Because the coherence length of ordinary light is extremely short, this difference in distance from the gelatin-reflector surface is very important. The slightly increased modulation (caused by a higher degree of coherence) in the gelatin–air reflector case can somewhat compensate for the weaker reflection obtained. On the other

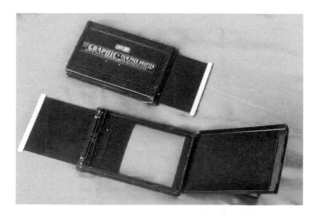

FIGURE 2.18 Modified Graphic Film Pack dark slide.

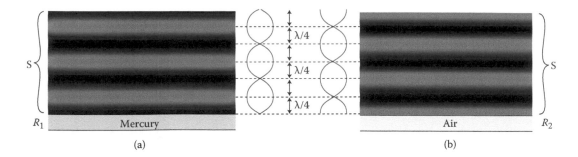

FIGURE 2.19 Light reflected at an optically thicker medium (mercury, R_1) and at an optically thinner medium (air, R_2). S is the gelatin emulsion. After development, the silver layers will be located at the antinodes.

hand, the exposure must be slightly increased to bring the recording up on the linear part of the Hurter–Driffield curve. During the development of the photograph, it is possible to also increase the interference fringe contrast using suitable developers. When using an air reflector, the surface reflection is in phase with the image. However, even in this case, it is recommended that a glass wedge be added to improve the image contrast.

The processing of the plates is critical. The Slavich emulsion is rather soft and it is important to harden the emulsion *before* the development takes place. Another fact is that the emulsion thickness is approximately 7 μm, which is much thicker than that necessary for Lippmann photography. The interference pattern is recorded only in a very thin volume at the top of the emulsion. This area has to be maintained intact after processing, another reason for the prehardening step. Emulsion shrinkage and other emulsion distortions caused by the developer must be avoided. The recommended bath to be used for this first processing step is listed in Table 2.6.

Among the old developers, the Lumière pyrogallol developer (Table 2.7) gave the best results and was used for all Lippmann photographs recorded by the author (HB). The development time was 90 s at approximately 18°C. The plates were not fixed to avoid shrinkage. After a 10 min wash, the plates were soaked in distilled water that contained a wetting agent. Finally, the photographs were slowly air-dried at room temperature.

TABLE 2.6

First Processing Bath for PFG03C

Distilled water	750 mL
Formaldehyde 37% (formalin)	10 mL (10.2 g)
Potassium bromide	2 g
Sodium carbonate (anhydrous)	5 g
Add distilled water to make	1 L
Processing time: 6 min	

TABLE 2.7

Lumière Developer

Solution A	Solution B
Pyrogallol (1 g)	Potassium bromide (30 g)
Distilled water (100 mL)	Distilled water (100 mL)

Note: Working solution: Mix 20 mL of Part A + 30 mL of Part B + 140 mL of distilled water. Add 10 mL of ammonia (s.w. 0.960) at 18°C just before using. Development time: 60 to 90 s.

Figure 2.20 shows a modern Lippmann photograph of one of the authors (HB) recorded on the PFG03C material using the simplified Lippmann technique without mercury. Figure 2.21 shows the recording setup. A colour correction filter was used in front of the camera lens to compensate for the fact that the Slavich material is not isochromatic. The aperture was F/4. The portrait was recorded on 2 July 1997 at 2:00 p.m. in direct sunlight. A diffusing screen was held above the subject's face to reduce the amount of glare off the face and a white screen was placed on the table to reflect additional soft light onto the face. The exposure time was 2 min.

After being processed, the back of plate was painted black. For better viewing of the image, a wedged glass plate (Wiener prism) was cemented to the emulsion side of the plate—as previously done for the old Lippmann photographs.

It should be mentioned that Darran Green, Bury St. Edmunds in England, has devoted more than 10 years to perfecting the Lippmann technique. He makes his own silver halide emulsion and has recorded many beautiful Lippmann photographs, two of which are shown in Figures 2.22 and 2.23.

FIGURE 2.20 Lippmann photographic portrait of one of the authors (HB).

FIGURE 2.21 Recording of the Lippmann portrait—bright sunlight at 2 p.m., softened with diffuser; exposure time, 2 min; aperture F/4.

FIGURE 2.22 *Opticks*, a modern Lippmann photograph by Darran Green, Bury St. Edmunds, UK.

FIGURE 2.23 *Rose Garden*, another modern Lippmann photograph by Darran Green, Bury St. Edmunds, UK.

2.8 Concluding Remarks

There can be no doubt that the colours of Lippmann photographs are especially realistic. These photographs show extremely beautiful colours of a brightness and vividness that make today's conventional colour photographs seem almost bland in comparison. In particular, the colour reproduction of human skin is simply amazing when recorded with the Lippmann technique. Modern colour photography is certainly not bad, but it cannot reproduce the type of colours obtained in Lippmann photographs, as the Lippmann photograph records and reproduces the entire spectral data.

Modern Lippmann photography may have limited applications in photography but may also very well appeal to artists and art photographers. Lippmann photographs contain no dyes or pigments and their archival stability is very high. In addition, the new photopolymer Lippmann photographs are not sensitive to humidity variations, meaning that their colours will not vary according to environmental changes. Furthermore, a Lippmann photograph is fundamentally difficult to copy, making it attractive for security applications. These features—unparalleled colour recording, great archival stability and resistance to copying—make Lippmann photography a unique photographic process.

Lippmann photography was the first interferometric imaging method to be discovered. It is capable of recording the spectral information of a two-dimensional scene to an exceptional precision. However, Lippmann photography led to another interferential method. This was the science of holography. Here, three-dimensional image data can be recorded to an exceptional precision—but at one wavelength only. The Lippmann photograph can, in some ways, be thought of as representing one extreme, with monochromatic holography as the other. In between these two extremes is polychromatic holography, in which the uncertainty in the recorded spectral data is roughly inversely proportional to the uncertainty in the geometrical data.

A discussion of Lippmann photography is therefore relevant in two ways. In itself, it constitutes a true technique of ultra-realistic imaging in the spectral domain and, as such, it may well have real applications. However, a discussion of Lippmann photography is also merited from the pedagogical and philosophical points of view due to its innate relationship to the wider field of holography.

REFERENCES

1. J. W. von Goethe, "Wirkung Farbige Beleuchtung," in *Zur Farbenlehre.* Vol. **II** - Supplement, Cottasche Buchhandlung, Tübingen (1810) pp. 703–724.

2. J. F. W. Herschel, "On the chemical action of rays of the solar spectrum on preparations of silver and other substances, both metallic and non-metallic, and on some photographic processes," *Phil. Trans.* **130**, 1–59 (1840).

3. A. E. Becquerel, "De l'image photographique colorée du spectre solaire," *C. R. Acad. Sci.* **26**, 181–183 (1848).

4. A. E. Becquerel, "De l'image photochromatique du spectre solaire, et des images colorées obtenues dans la chambre obscure," *C. R. Acad. Sci.* **27**, 483 (1848).

5. A. E. Becquerel, "De l'image photographique colorée du spectre solaire," *Annal. Chim. Phys. (Paris)* **22**, 451–459 (1848).

6. A. E. Becquerel, "De l'image photochromatique du spectre solaire et des images colorées obtenues à la chambre obscure," *Annal. Chim. Phys. (Paris)* **25**, 447–474 (1849).

7. A. E. Becquerel, "Nouvelles recherches sur les impressions colorées produites lors de l'action chimique de la lumière," *Annal. Chim. Phys. (Paris)* **42**, 81–106 (1854).

8. A. E. Becquerel, *La lumière ses causes et ses effets* 2, (Chap.4) (F. Didot Frères, Fils et CIE, Paris 1868) pp. 209–234.

9. C. F. A. Niepce de Saint-Victor, "Extrait d'un mémoire sur une relation existant entre la couleur de certaines flammes colorés, avec les images héliographiques colorées par la lumière," *C. R. Acad. Sci.* **32**, 834–841 (1851).

10. C. F. A. Niepce de Saint-Victor, "Second mémoire sur l'héliochromie," *C. R. Acad. Sci.* **34**, 215–218 (1852).

11. C. F. A. Niepce de Saint-Victor, "Troisième mémoire sur l'héliochromie," *C. R. Acad. Sci.* **35**, 694–698 (1852).

12. C. F A. Niepce de Saint-Victor, "Mémoire sur la gravure héliographique directement dans la chambre noire et sur quelques expériences scientifiques," *C. R. Acad. Sci.* **41**, 549–553 (1855).

13. C. F. A. Niepce de Saint-Victor, "Quatrième mémoire sur l'héliochromie," *C. R. Acad. Sci.* **54**, 281–284 (1862).

14. C. F. A. Niepce de Saint-Victor, "Cinquième mémoire sur l'héliochromie," *C. R. Acad. Sci.* **56**, 90–93 (1863).

15. L.-A. Poitevin, "Action simultanée de la lumière et des sels oxygénés sur le souschlorure d'argent violet; application à l'obtention par la photographie des couleurs naturelles sur papier," *C. R. Acad. Sci.* **61**, 1111–1113 (1865).

16. L. L. Hill, *A treatise on heliochromy; the production of pictures by means of light in natural colors,* Robinson & Caswell, New York (1856) [Reprinted by W. B. Becker, The Carnation Press 1972 State College, PA (1972)].

17. J. Boudreau, "Color daguerreotypes: hillotypes recreated," in *Pioneers of Photography* (SPSE, Springfield, VA. 1987) Chapter 18, pp. 189–199.

18. W. Zenker, *Lehrbuch der Photochromie (Photographie der natürlichen Farben)*, Zenker, Berlin (1868) [Reprinted by B. Schwalbe, Verlag Friedrich Vieweg und Sohn, Braunschweig (1900)].

19. R. S. Lord Rayleigh, "On the maintenance of vibrations by forces of double frequency, and on the propagation of waves through a medium endowed with a periodic structure," *Phil. Mag.* **24** (Ser. 5), 145–159 (1887).

20. O. Wiener, "Stehende Lichtwellen und die Schwingungsrichtung polarisirten Lichtes," *Ann. d. Physik Chemie* **40**, 203–243 (1890).

21. O. Wiener, "Farbenphotographie durch Körperfarben und mechanische Farbenanpassung in der Natur," *Ann. d. Physik Chemie* **55**, 225–281 (1895).

22. J. C. Maxwell, "Experiments on colour, as perceived by the eye, with remarks on colour-blindness," *Trans. Roy. Soc. Edinburgh* **21**, 275–299 (1855).

23. H. Ghiradella, "Light and color on the wing: structured colors in butterflies and moths," *Appl. Opt.* **30**, 3492–3500 (1991).

24. A. R. Parker, "515 million years of structural colour," *J. Opt. A: Pure Appl. Opt.* **2**, R15-R28 (2000).

25. S. E. Mann, I. N. Miacoulis and P. Y. Wong, "Spectral imaging, reflectivity measurements, and modeling of iridescent butterfly scale structures," *Opt. Eng.* **40**, 2061–2068 (2001).

26. P. Vukusic, R. Sambles, C. Lawrence and G. Wakely, "Sculpted-multilayer optical effects in two species of Papilo butterfly," *Appl. Opt.* **40**, 1116–1125 (2001).

27. C. Lawrence, P. Vukusic and R. Sambles, "Grazing-incidence iridescence from a butterfly wing," *Appl. Opt.* **41**, 437–441 (2002).

28. G. Tayeb, B. Gralak and S. Enoch, "Structural colors in nature," *Optics and Photonics News*, **14** (2) 38–43 (2003).

29. L. I. Rugani, "Structural and photonic properties of butterfly wings replicated," *Photonics Spectra* (No. 12) 18 (2006).

30. R. T. Lee and G. S. Smith, "Detailed electromagnetic simulation for the structural color of butterfly wings," *Opt. Eng.* **48**, 4177–4190 (2009).

31. H. D. Wolpert, "Optical filters in nature," *Optics and Photonics News* **20**, (2) 22–27 (2009).

32. G. Lippmann, "La photographie des couleurs," *C. R. Acad. Sci.* **112**, 274–275 (1891).

33. G. Lippmann, "La photographie des couleurs [deuxième note]," *C. R. Acad. Sci.* **114**, 961–962 (1892).

34. G. Lippmann, "Photographies colorées du spectre, sur albumine et sur gélatine bichromatées," *C. R. Acad. Sci.* **115**, 575 (1892).

35. G. Lippmann, "Sur la théorie de la photographie des couleurs simples et composées par la méthode interférentielle," *J. Physique* **3** (No.3), 97–107 (1894).

36. E. Lebon, *Savants du jour: Gabriel Lippmann. Biographie, bibliographie analytique des écrits,* Gauthier-Villars, Paris (1911).

37. A. Lumière and L. Lumière, "Sur les procédés pour la photographie des couleurs d'après la méthode de M. Lippmann," *Bull. Soc. franç. Phot.* (2 série) **9**, 249–251 (1893).

38. A. Lumière and L. Lumière, "Untersuchungen über die Herstellung einer lichtempfindlichen kornlosen Schict," *Eder's Jahrbuch der Photographie und Reproductionstechnik* **11**, 27–30 (1897).

39. E. Valenta, *Die Photographie in natürlichen Farben mit besonderer Berücksichtigung des Lippmannschen Verfahrens sowie jener Methoden, welche bei einmaliger Belichtung ein Bild in Farben liefern.* Zweite vermehrte und erweiterte Auflage. Encyklopädie der Photographie, Heft 2, W. Knapp Verlag, Halle a.S (1912).

40. R. Neuhauss, *Die Farbenphotographie nach Lippmann's Verfahren. Neue Untersuchungen und Ergebnisse.* Encyklopädie der Photographie, Heft 33, W. Knapp Verlag, Halle a.S (1898).

41. H. E. Ives, "An experimental study of the Lippmann color photograph," *Astrophysical J.* **27**, 325–352 (1908).

42. H. Lehmann, *Beiträge zur Theorie und Praxis Direkten Farbenphotographie mittels Stehender Lichtwellen nach Lippmanns Methode,* Trömer, Freiburg i.Br. (1906).

43. A. Berthier, *Manuel de Photochromie Interférentielle. Procédés de Reproduction directe des Couleurs,* Gauthier-Villars et Fils, Paris (1895).

44. F. Drouin, *La photographie des couleurs,* C. Mendel, Paris (1896).

45. R. Child Bayley, *Photography in colours,* Iliffe, Sons & Sturmey Ltd., London (1900).

46. E. Senior, "Lippmann's process of interference heliochromy," in *A Handbook of Photography in Colours*, Section III, Marion & Co., London (1900) pp. 316–343.

47. A. Berget, *Photographie directe des couleurs par la méthode interférentielle de M. Lippmann*, Gauthier-Villars, Paris (1901).

48. L. Tranchant, *La photographie des couleurs simplifiée*, H. Desforges, Paris (1903).

49. E. König, *Die Farben-Photographie*, Gustav Schmidt Verlag, Berlin (1904) [2: English translation by E. J. Wall, *Natural-color photography*, Dawbarn & Ward, Ltd. London (1906)].

50. B. Donath, *Die Grundlagen der Farbenphotographie*, Verlag Freidrich Vieweg und Sohn, Braunschweig (1906).

51. S. R. y Cajal, *La fotografía de los colores. Fundamentos científicos y reglas prácticas*, Líbros Clan A. Gráficas S. L., Spain (1912) [Reprinted : Colección Técnicas Artísticas, Madrid (1994)].

52. E. J. Wall, "Die Praxis der Farbenphotographie: Das Lippmann-Verfahren," in *Handbuch der wissenschaftlichen und angewandten Photographie*, ed. by A. Hay. Band VIII, Springer Verlag, Vienna (1929) pp. 219–230.

53. Y. N. Denisyuk, Dokl. Akad. Nauk. (Academy of Sciences Reports, USSR) **144** (No. 6), 1275–1278 (1962) [2:Transl.: Photographic reconstruction of the optical properties of an object in its own scattered radiation field]. *Sov. Phys. Doklady* **7**, 543–545 (1962).

54. Y. N. Denisyuk, "On the reproduction of the optical properties of an object by the wave field of its scattered radiation," *Opt. Spectrosc. (USSR)* **14**, 279–284 (1963).

55. Y. N. Denisyuk and I. R. Protas, "Improved Lippmann photographic plates for recording stationary light waves," *Opt. Spectrosc. (USSR)* **14**, 381–383 (1963).

56. K. F. Lindman, "Prüfung der Grundlagen der Theorie der Lippmannschen Farbenphotographie durch Versuche mit elektrischen Wellen," in *Acta Academiae Aboensis* **X** Sec.11., Åbo Akademi, Åbo (1937) pp. 1–62.

57. J.-L. Delcroix, "Utilisation des plaques Lippmann comme filtres," *Revue d'Optique.* **27**, 493–509 (1948).

58. H. Fleisher, P. Pengelly, J. Reynolgs, R. Schools and G. Sincerbox, "An optically accessed memory using the Lippmann process for information storage," in *Optical and Electro-Optical Information Processing* (MIT Press 1965) pp. 1–30.

59. R. E. Schwall and P. D. Zimmerman, "Lippmann color photography for the undergraduate laboratory," *Am. J. Phys.* **38**, 1345–1349 (1970).

60. L. Lindegren and D. Dravins, "Holography at the telescope—an interferometric method for recording stellar spectra in thick photographic emulsions," *Astron. Astrophys.* **67**, 241–255 (1978).

61. P. Connes, "Silver salts and standing waves: the history of interference color photography," *J. Optics (Paris)* **18**, 147–166 (1987).

62. H. Nareid, "A review of the Lippmann color process," *J. Photogr. Sci.* **36**, 140–147 (1988).

63. H. Nareid and H. M. Pedersen, "Modeling of the Lippmann color process," *J. Opt. Soc. Am. A* **8**, 257–265 (1991).

64. J.-M. Fournier, "Le photographie en couleur de type Lippmann: cent ans de physique et de technologie," *J. Optics (Paris)* **22**, 259–266 (1991).

65. J.-M. Fournier, "An investigation on Lippmann photographs: materials, processes, and color rendition," in *Practical Holography VII*, S.A. Benton ed., Proc. SPIE **2176**, 144–152 (1994).

66. J.-M. Fournier and P. L. Burnett, "Color rendition and archival properties of Lippmann photographs," *J. Imaging Sci. Technol.* **38**, 507–512 (1994).

67. A. Marraud and J.-M. Fournier, "Formation d'image accompagneé d'une analyse spectrale en champ complet: de J. Fourier à G. Lippmann," *Microsc. Microanal. Microstruct.* **8**, 37–39 (1997).

68. J.-M. Fournier, B. J. Alexander, P. L. Burnett and S. E. Stamper, "Recent developments in Lippmann photography," in *Sixth Int'l Symposium on Display Holography*, T. H. Jeong, H. I. Bjelkhagen eds., Proc. SPIE **3358**, 95–102 (1998).

69. N. J. Phillips, H. Heyworth and T. Hare, "On Lippmann's photography," *J. Photogr. Sci.* **32**, 158–169 (1984).

70. N. J. Phillips, "Links between photography and holography: the legacy of Gabriel Lippmann," in *Applications of Holography*, L. Huff ed., Proc. SPIE **523**, 313–318 (1985).

71. C. C. Rich and L. Dickerson, "Lippmann photographic process put to practice with available materials," in *Holographic Materials II*, T. J. Trout ed., Proc. SPIE **2688**, 88–95 (1996).

72. H. I. Bjelkhagen, "Silver halide emulsions for Lippmann photography and holography," in *Int'l Symposium on Display Holography*, T. H. Jeong and H. I. Bjelkhagen eds., Proc. SPIE **1600**, 44–59 (1992).

73. H. I. Bjelkhagen, "Lippmann photographs recorded in DuPont color photopolymer material," in *Practical Holography XI and Holographic Materials III*, S. A. Benton and T. J. Trout ed., Proc. SPIE **3011**, 358–366 (1997).

74. H. I. Bjelkhagen, T. H. Jeong and R. J. Ro, "Old and modern Lippmann photography," in *Sixth Int'l Symposium on Display Holography*, T. H. Jeong, H. I. Bjelkhagen eds., Proc. SPIE **3358**, 72–83 (1998).

75. H. I. Bjelkhagen, "Lippmann photography: reviving an early colour process," *History of Photography* **23**, (No. 3), 274–280 (1999).

76. H. I. Bjelkhagen, "A new optical security device based on one-hundred-year-old photographic technique," *Opt. Eng.* **38**, 55–61 (1999).

77. H. I. Bjelkhagen and S. De Souza, "Computer simulation of the Lippmann photographic process and recording experiments using holographic materials," in *Practical Holography XV and Holographic Materials VII*, S. A. Benton, S. H. Stevenson and T. J. Trout eds., Proc. SPIE **4296**, 300–311 (2001).

78. H. I. Bjelkhagen, "Lippmann photography: its history and recent development," *The PhotoHistorian, Journal of the Historical Group of the Royal Photographic Society*, APIS 2002 Special double edition, PH.141–142, (2003) pp. 11–19.

79. A. Labeyrie, J. P. Huignard and B. Loiseaux, "Optical data storage in microfibers," *Opt. Lett.* **23**, 301–303 (1998).

80. P. Hariharan, "Lippmann photography or Lippmann holography?" *J. Mod. Optics* **45**, 1759–1762 (1998).

81. W. R. Alschuler, "On the physical and visual state of 100 year old Lippmann color photographs," in *Sixth Int'l Symposium on Display Holography*, T. H. Jeong and H. I. Bjelkhagen eds., Proc. SPIE **3358**, 84–94 (1998).

82. W. R. Alschuler, "Lippmann photography and the glory of frozen light: Eternal photographic color real and false," in Proc. IEEE *31st Applied Imagery Pattern Recocnition Workshop* AIPR'02 (2002).

83. W. R. Alschuler, "Lippmann color process, the experience of amateurs in England," *The PhotoHistorian, Journal of the Historical Group of the Royal Photographic Society*, APIS 2002 Special double edition, PH.141–142, (2003) pp. 44–51.

84. R. Jahr, "Über die Herstellung sogenannter "kornloser" Platten für das Interferenz-Farben-Verfahren von Professor Lippmann nach Dr. Hans Lehmann," *Photographische Industrie* **23**, 1013–1014 (1925).

85. H. Lehmann, "Practical application of interference colour photography," *Brit. J. Phot.* (Col. Suppl.) (4 Nov. 1910) pp. 83–86, Practical application of interference colour photography. II. *Brit. J. Phot.* (Col. Suppl.) (2 Dec. 1910) pp. 92–95.

86. H. Lehmann, "Die Praxis der Interferenzfarbenphotographie, unter besonderer Berücksichtigung der in den Optischen Werkstätten von C. Zeiss in Jena konstruierten Specialapparaten," *Phot. Rundschau* **23**, 125–134 (1909) [English translation: "Practical application of interference colour photography," *Brit. J. Phot.* (Col. Suppl.) (4 Nov. 1910) pp. 83–86, "Practical application of interference colour photography II," *Brit. J. Phot.* (Col. Suppl.) (2 Dec. 1910) pp. 92–95.

87. E. Rothé, "Photographies en couleurs obtenues par la méthode interférentielle sans miroir de mercure," *C. R. Acad. Sci.* **139**, 565–567 (1904).

88. E. Rothé, "Photographies en couleurs obtenues par la méthode interférentielle sans miroir de mercure," *Bull. Soc. Franç. Phot.* (2 série) **20**, 548–549 (1904).

89. E. Rothé, "Sur la photographie interférentielle," *La Photographie des Couleurs* **1**, (No.2), 97–108 (1906).

90. E. Rothé, "Franges d'interférences produites par les photographies en couleurs," *C. R. Acad. Sci.* **147**, 43–45 (1908).

91. E. Rothé, "Variations des franges des photochromies du spectre," *C. R. Acad. Sci.* **147**, 190–192 (1908).

92. R. Neuhauss, "Nachweis der dünnen Zenkerschen Blättchen in den nach Lippmann's Verfahren aufgenommenen Farbildern," *Ann. d. Physik Chemie* **65**, 164–172 (1898).

93. S.R. y Cajal, "Chromomicrophotographie par la méthode interférentielle," *La Photographie des Couleurs* **2**, (No.7), 97–101 (1907).

94. A. Lumière and L. Lumière, "Sur une nouvelle méthode d'obtention de photographies en couleur," *C. R. Acad. Sci.* **138**, 1337–1338 (1904).

95. Musée de l'Elysée Lausanne, 18, avenue de l'Elysée, CH-1014 Lausanne, Switzerland.

96. Université Pierre et Marie Curie, L.D.M.C., Tour 22, 4 place Jussieu, F-75252 Paris, France.

97. Musée français de la Photographie, 78, rue de Paris, F-91570 Bièvres, France.

98. Société française de Photographie, 71, rue de Richelieu, F-75002 Paris, France.

99. Musée Nicéphore Niépce, 28, quai des Messageries, F-71100 Chalon-sur-Saône, France.

100. Preus Fotomuseum - Norweigan Photographic Museum, PB 254, N-3192 Horten, Norway.
101. International Museum of Photography, George Eastman House, 900 East Avenue, Rochester, NY 14607, USA.
102. H. Kogelnik, "Coupled wave theory for thick hologram gratings," *Bell Sys. Tech. J.* **48**, 2909–2947 (1969).
103. D. Brotherton-Ratcliffe, "A treatment of the general volume holographic grating as an array of parallel stacked mirrors," *J. Mod. Optics* **59**, 1113–1132 (2012).
104. W. J. Gambogi, W. K. Smothers, K. W. Steijn, S. H. Stevenson and A. M. Weber, "Color holography using DuPont holographic recording film," in *Holographic Materials*, J. Trout ed., Proc. SPIE **2405**, 62–73 (1995).
105. T. J. Trout, W. J. Gambogi and S. H. Stevenson, "Photopolymer materials for color holography," in *Applications of Optical Holography*, T. Honda ed., Proc. SPIE **2577**, 94–105 (1995).
106. K. W. Steijn, "Multicolor holographic recording in Dupont holographic recording film: determination of exposure conditions for color balance," in *Holographic Materials II*, J. Trout ed., Proc. SPIE **2688**, 123–134 (1996).
107. Courtaulds Performance Films, POB 5068, Martinville, VA 24115, USA.
108. SLAVICH Joint Stock Co., Micron Branch Co., 2 pl. Mendeleeva, 152140 Pereslavl-Zalessky, Russia.

3

Continuous Wave Lasers for Colour Holography

3.1 Introduction

In this chapter, we shall review the most common types of continuous wave (CW) lasers that can be used in colour holography. The range of commercially available, off-the-shelf CW lasers that are suitable for holographic applications has grown significantly in the last few years. There is now a relatively large choice of laser wavelengths available—many offered by the extremely reliable diode-pumped solid-state (DPSS) laser technology.

CW laser sources of interest to holography may be broadly divided into five categories. These are gas lasers, semiconductor lasers, DPSS lasers, dye lasers and diode-pumped fibre lasers. Historically, gas lasers were the most popular source. This is now changing with semiconductor and DPSS lasers becoming increasingly popular. High-coherence visible fibre lasers are also starting to become commercially available and we can expect this sector to expand in the future. Narrow-line liquid dye lasers continue to represent a very flexible but rather inconvenient laser source for holography. Narrow-line solid-state dye lasers are not yet sufficiently developed to offer a sensible solution for holographic applications, but this could change as considerable progress continues to be made in the sector.

Many digital and analogue colour holographic applications can require high-power CW lasers. In analogue holography, this is because the new colour materials, as we shall see in Chapter 4, have intrinsically low sensitivity and as such necessitate high exposure energies. Because a relatively low exposure time is always recommended in analogue holography due to vibration and movement concerns, this translates into a minimum power requirement for the laser. Of course, such minimum power will scale with the intended hologram size—meaning that extremely small holograms can still require only small laser powers. Nevertheless, as a general rule, a suitable laser source for a full-colour analogue holography setup will usually need to be greater than several hundred milliwatts.

High power may also be desirable when CW lasers are used to print digital holograms. Because a single digital hologram can be composed of more than a million component holograms, or "hogels", it is clear that one does not want to wait very long between exposures for the printing optics to settle to interferometric stability. As a result, one tries to use a very short exposure time—and this then leads directly to a requirement on the minimum laser power. The larger the power of the CW laser used, the faster one can print the hologram. Fringe-locking strategies can reduce the required power somewhat—but only to a certain point as, inevitably, the process of advancing optics from one hogel to the next induces vibrations over a wide spatial spectrum.

Digital holography and analogue holography impose differing requirements on the temporal coherence and power stability of CW laser sources. Because hogels are usually very small (typically 1 mm^2), one requires a coherence length of practically only a few centimetres, but because many such hogels must be printed, high power stability is vital. In contrast, analogue holography may require a coherence length of several metres if large depth scenes are to be recorded, but high power stability is not intrinsically necessary here, as exposure times may easily take into account a change in power.

Analogue colour holography usually requires the use of at least three laser frequencies, and these must be chosen carefully. In Chapter 5, we shall show how optimal wavelength sets can be defined from a colour model for three or more primary wavelengths. Digital holography has rather different constraints concerning the optimal wavelengths, as we shall see in Chapter 7. In Chapter 13, we shall discuss the illumination of the hologram. Here again, modern narrow-band illumination sources can require the recording lasers to be "matched" to the illumination system.

The choice of the laser to be used in colour holography can therefore be quite complex. First, it will depend on whether the hologram is to be digitally printed or whether an analogue recording is to be used. In the case of analogue reflection holography, it will then depend on the format of the hologram. If the hologram is to be replayed by a narrow-band source, then the recording lasers (usually) must also be matched to the illumination source. Finally, the laser wavelengths will, for both analogue and digital applications, have to be chosen with reference to a given colour model.

The question of whether to employ CW lasers or pulsed lasers for a given holographic application will depend on many issues. In digital holography, pulsed lasers confer a complete immunity to vibrational dimming of the hologram. In addition, small and relatively cheap commercial pulsed DPSS lasers will undoubtedly become available for this application in the future. In analogue holography, there are situations in which the use of pulsed lasers is obligatory—such as when the objects to be recorded are physically unstable or when the recording must be performed in a noisy environment. On the other hand, some of the best new materials are not sensitive to fast pulses. In this chapter, we shall restrict ourselves to a discussion of CW laser sources. Pulsed laser sources will be discussed in Chapter 6.

3.2 Gas Lasers

3.2.1 Helium–Neon Laser

The helium–neon (HeNe) laser [1] is a neutral atom gas discharge laser operating with a mixture of helium and neon gas in a glass tube typically having a length of approximately 15 to 100 cm (Figure 3.1). A voltage of approximately 1 kV is applied to the two electrodes, producing a small direct current; this maintains an electric glow discharge. The glass tube incorporates Brewster windows and two laser mirrors, forming a resonant laser cavity with a typical loss of less than 1%. The Brewster windows ensure a linearly polarised emission.

In the gas discharge, helium atoms are excited into a metastable state from which they efficiently transfer energy to neon atoms (having similar excitation energy) via collisions. Neon atoms have a number of energy levels below the pump level, giving rise to various laser transitions. The transition at 632.8 nm is the most common, but other transitions of interest to holography are 543.5 nm (green emission), 594 nm (yellow emission) and 612 nm (orange emission). The emission wavelength is selected using selective resonator mirrors, which introduce high losses at the wavelengths of competing transitions. Additionally, an intracavity etalon can improve performance significantly. Figure 3.2 shows a photograph of a Russian LGN-222 laser with etalon producing a high coherence emission at 632.8 nm of more than 70 mW. Commercially available lasers producing green and yellow emissions are available but only at low powers (typically several milliwatts). Typical tube lifetimes are more than 20,000 h.

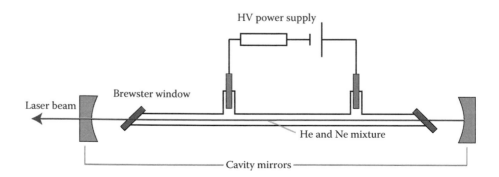

FIGURE 3.1   HeNe laser. An optional intracavity etalon can be used to improve the temporal coherence.

FIGURE 3.2 Picture of a Russian high-powered LGN-222 HeNe laser.

3.2.2 Argon Ion Laser

The argon laser [2] was invented in 1964 by William Bridges at Hughes Aircraft Corporation and is one of a family of ion lasers that use a noble gas as the active medium. Argon ion lasers are excellent sources of high-coherence green and blue light for use in colour holography (Figure 3.3). Optical powers from several watts to several tens of watts are common. The key component of this laser is the argon-filled tube, usually made from beryllium oxide ceramic. An intense electrical discharge between a hollow anode and a hollow cathode generates a high-density argon ion plasma. Unlike HeNe lasers, the energy level transitions that contribute to laser action come from ions. An optional solenoid wound around the tube can be used to generate an axial magnetic field that increases the output power by ensuring better plasma confinement. Typical CW plasma conditions in an argon laser discharge are characterised by current densities of 100 to 2000 Amps/cm^2, tube diameters of 1 to 10 mm, filling pressures of 0.1 to 1 torr and an axial magnetic field of the order of 1000 G.

Laser oscillation in the argon laser is restricted to a single line by an intracavity prism. High coherence is then attained by the use of an intracavity etalon. The laser is able to operate at a variety of wavelengths, the most useful of which for colour holography are (in order of decreasing power) the green line at 514.5 nm,* the blue–green lines at 488 nm (typically at 80% of the 514.5 nm power) and 496.5 nm (40% of 514.5 nm) and the blue emissions at 476.5 nm (30% of 514.5 nm) and 457.9 nm (15% of 514.5 nm). In addition to these lines, argon lasers also produce various weaker emissions at eight other wavelengths throughout the visible, ultraviolet and near-visible spectra, including 351.1, 363.8, 454.6, 465.8, 496.5, 501.7, 528.7 and 1092.3 nm.

A typical argon laser has a tube of approximately 1 m in length and can generate about 10 W of output power at 514.5 nm. The power consumption of such a laser is, however, several tens of kilowatts. The large amount of dissipated heat must be removed by water-cooling of the tube. Such water-cooling introduces significant vibration into the laser head; accordingly, an argon laser is best installed in a separate laser room and a beam tube used to deliver the beam to the holographic recording area.

3.2.3 Krypton Ion Laser

Krypton ion lasers are very similar to argon ion lasers. They provide a useful source of red radiation at 647.1 nm for colour holography. Generally, the emission at 647.1 nm is rather smaller than the 514.5 nm line for a given electrical power. For example, the single-frequency optical power available from a 2012 model Innova Sabre DBW25 argon laser manufactured by the company, Coherent, Inc., is approximately 6 W. The corresponding krypton laser generates 2.1 W at 647.1 nm.

In addition to the 647.1-nm line, krypton lasers can produce high powers in the violet at 413.1 nm (for example, the 2012 model Innova Sabre krypton laser from Coherent can produce an optical power at this

---

* An Innova Sabre argon laser, the largest commercially available laser from Coherent at the time of writing, produces 6 W of single frequency emission at 514.5 nm.

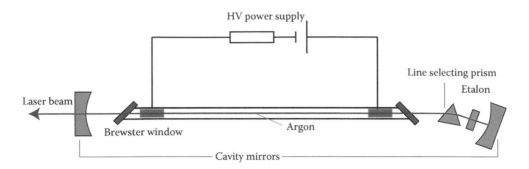

FIGURE 3.3   Basic schematic of an argon ion laser.

wavelength of 1.1 W). In addition to the two main lines, krypton lasers are able to generate weaker emissions at 406.7, 415.4, 468.0, 476.2, 482.5, 520.8, 530.9, 568.2 and 676.4 nm.

Like argon lasers, krypton lasers must be water-cooled and as such the laser is best installed in a separate laser room and a beam tube used to deliver the beam to the holographic recording area. Both argon and krypton laser tubes have a finite lifetime (usually somewhat greater than several thousand hours of operation). However, the tubes can break without warning if there is a problem with the cooling system or if, for example, the electricity supply fails inadvertently. The cost of the tubes in both these types of laser is a very large percentage of the total laser cost and, as such, manufacturers will often offer an automatic replacement service in return for a yearly service contract. The downside is that when one adds the annual service charge to the cost of the electricity required to run the laser, the bill can be very significant. Figure 5.25 in Chapter 5 shows a picture of two modern large-frame argon and krypton lasers.*

3.2.4 Helium–Cadmium Laser

The helium–cadmium (HeCd) laser [3] (see Figure 3.4) is one of a class of gas lasers using helium in conjunction with a metal that vaporises at a relatively low temperature. Both Penning and charge exchange collisions occur in the HeCd laser, and these two processes produce different laser transitions. The laser produces a blue-violet emission at 441.6 nm. Commercial (air-cooled) lasers are readily available with powers of up to approximately 130 mW. In addition to the 441.6-nm line, this laser may also produce a useful output at 325 nm. Typical coherence lengths available in commercial models today are usually no greater than 10 cm. The HeCd laser has been used extensively for recording holograms onto photoresists.

HeCd laser tubes are more complex than the tubes used in HeNe or ion lasers. They include a cadmium metal reservoir and a heater to control the cadmium vapour pressure. The vapour is propagated through the system through a process called cataphoresis. Outside the bore region of the tube, the cadmium vapour will coalesce on any cool surface. Cold traps and protective discharges are therefore required to prevent cadmium from being deposited onto the optical windows. A mechanism is also required to add helium to maintain an optimum pressure. Additionally, heaters to control tube temperature and various sensors inside the envelope for use in feedback control must be present. Typical tube lifetimes are approximately 5000 h. The power supplies are rather complex and must include power sequencing logic and multiple feedback loops.

---

* Mixed Argon-Krypton lasers exist, but it is not possible to individually adjust the output power of different wavelengths for simultaneous exposure of colour holograms.

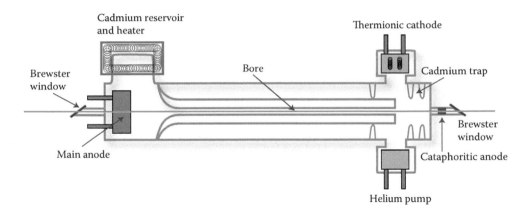

FIGURE 3.4   Basic tube design of a HeCd laser.

3.2.5 Other Gas Lasers

Various other gas lasers can, in principle, be used for holography. However, these lasers are certainly not popular. Some lend themselves to amateur fabrication, such as the helium–mercury laser and the helium–selenium laser. The high-power (>10 W) low-coherence 578.2-nm line of the copper vapour laser has been used as a very effective illumination source for large laser transmission holograms.

3.2.5.1 Helium–Mercury Laser

The helium–mercury (HeHg) laser is quite similar to the HeNe and Ar/Kr ion lasers in terms of operation. This laser operates through ion transitions of mercury. Another name for the HeHg laser is the "mercury vapour ion laser". It emits in the green line at 567 nm and produces a less powerful output in the red–orange line at 615 nm. The typical multiline output power for a 1-m tube may exceed 20 mW. This laser is not currently available as a commercial product but does lend itself to amateur construction [4,5].

3.2.5.2 Helium–Selenium Laser

This laser is very similar to the HeHg laser. It has been observed to generate up to 46 lines [5], the six most powerful of which are 497.6, 499.3, 506.9, 517.6, 522.8 and 530.5 nm. Up to 30 mW* is typically obtained from a 1-m tube with a 4-mm bore at 200 mA discharge current and 4.5 torr helium filling pressure. Like the HeHg laser, this laser is not currently available as a commercial product but lends itself to amateur construction.

3.2.5.3 Other Metal Vapour Ion Lasers

In addition to the ion lasers discussed in this section, there are several other ions that exhibit laser oscillation from levels excited by either Penning or charge exchange ionisation. Lasers constructed from these ions contain possible holographic sources, but in practice, these lasers have not been used to date for holographic applications. Thermally produced metal vapours of Zn, Te, As, Mg, Tn, Tl and Be all exhibit laser emission when excited in He and Ne discharges.

---

* The highest recorded power is 250 mW.

3.2.5.4 Copper and Gold Vapour Lasers

The copper vapour laser is the most useful of this class of neutral metal vapour lasers. The primary emission lines for this laser are 510 and 578 nm at powers up to and exceeding 100 W. The copper vapour laser is unusual with respect to its high power and high efficiency; its normal operation is at pulse repetition rates of several tens of kilohertz.

In neutral metal vapour lasers, a fast electric discharge directly excites the metal atoms. High repetition rates then permit high average power output. The copper vapour is generated from copper placed in the discharge tube, where it is heated to approximately 1500°C and produces a vapour of approximately 0.1 mb. Several millibars of neon are added as a buffer gas to prevent window contamination and a loss of copper. The overall wall-plug efficiency of this type of laser is approximately 1%, which is the highest for any visible gas laser. Copper chloride and copper bromide lasers offer an alternative technology, having a lower temperature discharge.

Gold can also be substituted for copper, producing a powerful laser emission at 627 nm. Neutral metal vapour lasers are not usually useful for recording holograms due to their typically multimodal structure and low coherence character. However, the 578-nm line of copper is an extremely effective source for illuminating laser transmission holograms.

3.3 Dye Lasers

3.3.1 Liquid Dye Lasers

Liquid dye lasers constitute a coherent source of laser radiation with a wide tuning range and have applications in various fields including holography. These lasers are particularly pertinent to colour holography as they provide a source of tunable radiation at virtually any wavelength across the visible spectrum.

A liquid dye laser consists of an organic dye mixed with a solvent. The mixture is either circulated through a dye cell or formed into a dye jet. An external laser, often an argon ion laser, is used as the pump source.* The dye solution must be circulated at high speed to circumvent triplet absorption and consequent beam extinction, and also to decrease the degradation of the dye.

The pump light excites the dye molecules into their singlet state where they emit stimulated radiation. In the singlet state, the dye molecules emit light via fluorescence; the dye is now transparent to the lasing wavelength. Within approximately a microsecond, however, the dye molecules change to their triplet state in which light is emitted via phosphorescence; here, the molecules absorb the lasing wavelength, making the dye opaque.

A ring laser design is often chosen for CW operation. Frequency-selective elements such as prisms, gratings and etalons serve to select a given wavelength and to ensure good coherence. Typical dyes include rhodamine, fluorescein, coumarin, stilbene, umbelliferone, tetracene and malachite green. Typical solvents used are water, glycol, ethanol, methanol, hexane, cyclohexane and cyclodextrin. The first narrow line width dye laser was demonstrated by Hänsch [6]. Subsequently, grazing incidence grating designs [7,8] and multiple prism grating configurations were reported.

Liquid dye lasers having good spatial and temporal coherence characteristics may easily be constructed in the laboratory. Critical elements such as the dye jets are commercially available. Many laser models are also available commercially. Using a large-frame argon laser as the pump, several watts of useful power may be attained across the visible spectrum from a liquid dye laser.

The great disadvantage of the liquid dye laser is the need for large volumes of organic solvents. There are also significant cleaning and servicing issues. For use in colour holography, the inconvenience of the liquid dye laser must be balanced against the great advantage offered by its wide choice of wavelengths at high power.

---

* Fast flashlamps may also be used.

3.3.2 Solid-State Dye Lasers

We should mention that solid-state dye lasers that use dye-doped organic matrices as the gain medium have been reported. The first solid-state dye lasers were reported in the late 1960s by Soffer and McFarland [9] and Peterson and Snavely [10] who demonstrated stimulated emission from polymeric matrices doped with organic dyes. In recent years, significant breakthroughs have been achieved in the development of practical, tunable solid-state dye lasers. Stable CW output power of several watts over some hours of operation in solid-state dye lasers has been reported [11]. In addition, high coherence operation has also been reported [12]. As such, solid-state dye lasers may represent an interesting future choice for colour holography.

3.4 Diode-Pumped Solid-State Lasers

DPSS lasers employ semiconductor laser diodes to excite ions within a doped laser crystal. Laser transition occurs from the excited ion state to a lower state. The most commonly used laser crystals in commercial DPSS laser systems today are doped with the rare earth metal neodymium. We shall discuss the properties of neodymium-doped materials at length in Chapter 6 in the context of pulsed laser sources.

DPSS lasers may be generally constructed using two forms of pumping:

- End-pumping
- Side-pumping

Here, to illustrate how DPSS lasers work, we will briefly describe the simple end-pumped system shown in Figure 3.5 [13]. This laser is built using two crystals of $Nd:YVO_4$ in series-connected double Z resonators. Each of the four ends of the two crystals is pumped by separate 18-W 808-nm laser diode bars. The radiation from these bars is collected into fibre bundles and focussed into the $Nd:YVO_4$ crystals through their polished ends. Approximately 54 W of actual laser pumping power is delivered to the two crystals. The laser then produces 35 W of output power at 1064 nm. The electrical power consumed by the pumping diodes is only 180 W.

DPSS lasers have significant advantages over other types of lasers. Their efficiency means that little waste energy is injected into the system and this means that these types of lasers are often much more stable. Diode stacks also typically have operational lifetimes of more than 10,000 h. This, together with the high wall-plug efficiency, translates into very economical operation. Many CW DPSS lasers using

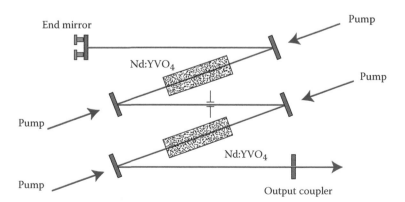

FIGURE 3.5 Simple DPSS laser built around two $Nd:YVO_4$ crystals in a Z cavity. Each crystal is pumped by separate diode bars through both its ends. The laser produces 35 W of optical power at 1064 nm for an absorbed pumping power of 54 W and a total electrical power consumption of 180 W.

neodymium-doped crystals have thus far been demonstrated. In addition, Ho-, Tm-, Yb- and Cr-doped crystals can be pumped using laser diodes.

3.4.1 Green Emission DPSS Lasers

The above example illustrates how DPSS laser technology can be used to construct extremely efficient and compact lasers. Single spatial and longitudinal mode operation can be ensured in DPSS lasers using an appropriate resonator design and frequency-selective elements. The 1064-nm line of Nd:YVO$_4$ can be doubled to produce a green emission at 532 nm by use of a non-linear frequency doubling crystal such as potassium dihydrogen phosphate (KTP). This may be internal or external to the resonator depending on the laser design. Q-switched operation may also be employed by the use of a passive Q-switch within the resonator. Today, many companies manufacture commercial laser systems operating at 532 nm. Perhaps the best known is the *Verdi* laser from Coherent.

3.4.1.1 *Verdi 532 nm Laser*

Figure 3.6 shows a basic optical scheme of the Verdi laser, which is manufactured by Coherent. A picture of the laser can be seen in Figure 5.25 in Chapter 5. The laser is capable of producing up to 18 W of highly stable single-frequency emission at 532 nm, with an estimated coherence length of more than 60 m. Maximum electrical power consumption is 1.3 kW. The Verdi is based on a single end-pumped Nd:YVO$_4$ crystal in a ring cavity.* Frequency doubling is internal via a lithium triborate (LBO) crystal. A Faraday rotator and waveplate ensure unidirectional lasing. This is an extremely good source for colour holography.

3.4.1.2 *Other Green DPSS Lasers*

The Nd:YVO$_4$^{4}F$_{3/2}$-^{4}I$_{11/2}$ transition has weaker Stark lines at 1074 and 1084 nm. Frequency-doubled emission at both of these lines (537 and 542 nm) has been reported in the literature. Commercial single-frequency lasers producing up to 50 mW at 542 nm were available at the time of writing. In addition to Nd:YVO$_4$, other neodymium crystals may be used in a DPSS laser. The most popular such lasers are based on either Nd:YAG (532 nm), Nd:YAP (539 nm) or Nd:YLF (523 and 527 nm). Yb:YAG can also produce a doubled emission at 515 nm (for example, Showa Optronics Company Ltd. produces a 50-mW single-frequency DPSS laser based on Yb:YAG at 515 nm).

3.4.2 Blue Emission DPSS Lasers

Nd:YAG has a transition at 946 nm, which can be doubled to 473 nm. Likewise, Nd:YVO$_4$ has a transition at 914 nm, which can be doubled to 457 nm. Examples of other less well-known materials are Nd:GSAG (943 nm) and Nd:YGG (935 nm). As early as 1997, more than 500 mW of single-frequency light at 473 nm was generated by a DPSS Nd:YAG ring cavity laser with frequency doubling using KNbO$_3$ [14]. Recently, Wang et al. have reported a highly stable single-frequency output power of 1 W at 473 nm from an end-pumped DPSS Nd:YAG laser using a ring resonator and intracavity frequency doubling with a periodically poled KTP (PPKTP) crystal [15]. Commercial, single-frequency DPSS lasers operating at 473 and 457 nm are available today. However, typical maximum output powers are currently limited to approximately 300 mW at 457 nm and 150 mW at 473 nm (for example, the 300 mW single-frequency BLSI model from CVI Melles Griot; Figure 3.7). Clearly, this can be expected to increase in the near future.

Neodymium-doped crystals have several transitions at approximately 1.3 μm. These may be tripled to produce a blue output using CW Q-switched operation. The most common include Nd:YVO$_4$ (447 nm), Nd:YAG (440 and 446 nm) and Nd:YLF (438 nm). The maximum output power currently available in

---

* A ring cavity is used to avoid the well-known "green problem", which can occur when intracavity frequency doubling is used with a conventional linear cavity.

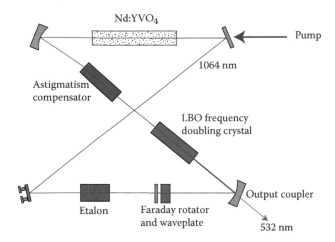

FIGURE 3.6 Simplified optical scheme of the DPSS *Verdi* laser from Coherent. The laser is capable of producing up to 18 W single frequency at 532 nm.

FIGURE 3.7 DPSS *BLS1* laser from Melles Griot, producing 300 mW of a single-frequency emission at 457 nm.

narrow-band commercial lasers is less than 50 mW. However, much higher outputs are certainly possible and again one can expect the maximum power offered by companies to increase in the coming years. We shall discuss similar lamp-pumped lasers in Chapter 6 in the context of pulsed operation.*

Yet another method to obtain a blue emission is sum–frequency mixing. Here, for example, an 808-nm diode is used to pump a Nd:YVO$_4$ crystal, which then emits at 914 nm. The 808-nm diode radiation is also used to pump a Nd:YLF crystal emitting at 1047 nm. Intracavity sum–frequency mixing at 914 and 1047 nm is then realised in an LBO crystal to produce an emission at 488 nm.

The Swedish company, Cobolt AB, use sum–frequency mixing of Nd:YVO$_4$ (914 nm) and Nd:YAG (1064 nm) in a PPKTP crystal to produce a single-frequency emission of 100 mW at 491.5 nm in their *Calypso* laser. The company also manufactures a version that produces a two-colour emission at 532 and 491.5 nm. A periodically poled crystal is the key component in such a multiline laser design. By placing multiple successive quasi-phase-matching gratings into a single crystal, it is possible to achieve multiple frequency conversion processes (in this case, doubling of 1064 to 532 nm and sum–frequency mixing of 914 and 1064 nm, giving 491.5 nm). Each grating has a different characteristic period along the direction of the propagation and is created in a lithographic process with subsequent exposure to a high electric

---

* The Lithuanian company, Geola, makes commercial lamp-pumped 50 Hz Nd:YAG Q-switched lasers giving average powers of more than 200 mJ at 440 nm. Nd:YLF and Nd:YAP lasers are also available for 438, 446 and 447 nm outputs.

field. Using two or more gratings in one crystal minimises the number of optical interfaces, reducing the complexity of the laser design and its production cost.

3.4.3 Red Emission DPSS Lasers

Red emission may be obtained by doubling one of the 1.3 μm lines in neodymium. Using a temperature-tuned PPKTP crystal, a record 1.3-W single-frequency red laser at 661 nm was achieved by intracavity second-harmonic generation in a $Nd:YLiF_4$ ring laser oscillating at the π-polarised transition ($\lambda \backsim$ 1321 nm) [16]. Intracavity second-harmonic generation of a diode-pumped $Nd:YLiF_4$ ring laser oscillating on the σ-polarised $^4F_{3/2}-^4I_{13/2}$ transition ($\lambda_\omega \backsim$ 1314 nm) with a temperature-tuned PPKTP crystal has also been reported, yielding up to 0.92 W of tunable ($\lambda_{2\omega} = 656-658$ nm) single-frequency output [17].

The Cobolt *Flamenco* is a commercial DPSS laser producing single-frequency emission of 400 mW at 660 nm (Figure 3.8). This laser is also based on frequency doubling of 1.3 μm in PPKTP. Using a type I critically phase-matched bismuth borate crystal, 620 mW of single-frequency emission at 671 nm has also been achieved using intracavity doubling of a π-polarised single end-pumped $Nd:YVO_4$ ring laser oscillating at the 1342-nm transition [18]. Various companies offer commercial CW DPSS Q-switched lasers using Nd:YAG, $Nd:YVO_4$ and Nd:YLF to generate up to 250-mW single-frequency emissions at 656.5, 660 or 671 nm.

3.4.4 Yellow Emission DPSS Lasers

Nd:YAG has a weak transition at 1123 nm. This can be doubled to produce an emission at 561 nm. In 2007, Zang et al. [19] reported single-frequency emission of up to 1.25 W at 1023 nm. Cobolt AB now produce a single-frequency laser (the *Jive*) capable of emitting up to 500 mW at the doubled wavelength (561 nm).

3.4.5 Orange Emission DPSS Lasers

Sum–frequency generation from the various infrared transitions of neodymium-doped crystals can produce emissions in the range 589 to 599 nm. For example, in 2008, Mimoun et al. [20] reported the generation of more than 800 mW of single-frequency emission at 589 nm by sum–frequency conversion of the 1319- and 1064-nm lines of Nd:YAG. However, few single-frequency orange DPSS sources were available at the time of writing.

3.4.6 Future DPSS Laser Technology

3.4.6.1 Praseodymium-Doped Lasers

Increasingly encouraging results have been reported recently concerning praseodymium (Pr^{3+})-doped crystal lasers. These lasers are extremely interesting due to their transitions in the visible spectrum. Unlike the Nd^{3+} ion, which has its transitions in the infrared and requires frequency conversion to

FIGURE 3.8   DPSS *Flamenco* laser from Cobolt AB, producing 400 mW of a single-frequency emission at 660 nm.

generate visible light, Pr^{3+} naturally lases at lines in the blue, green, orange and red. In addition, these transitions constitute four-level laser systems. In the past, however, a major difficulty with Pr^{3+} lasers has been how to pump the laser. The absorption transitions of Pr^{3+} are situated in the blue between 440 and 480 nm. The first experimental reports of Pr^{3+} lasers therefore used argon lasers [21,22], dye lasers [23] or an optically pumped semiconductor lasers [24].

Recent developments in high-power blue laser diodes emitting at approximately 445 nm have now enabled the construction of efficiently pumped Pr-doped lasers in compact setups. For example, Richter et al. [25,26] reported efficient CW lasing of praseodymium-doped $LiYF_4$ and $LiLuF_4$ crystals pumped either by an optically pumped semiconductor laser (at 479.5 nm) or a GaN laser diode (at 444 nm). Up to 600 mW (not single frequency) was obtained in the green at 523 nm, in the red at 640 nm and in the deep red at 720 nm. Additionally, more than 300 mW was obtained in the orange at 607 nm. Diode pumping of $Pr:KY_3F_{10}$, $Pr:YAlO_3$, $Pr:SrF_2$ and $Pr:SrAl_{12}O_{19}$ crystals have also been reported [27–31].

Although no commercial DPSS Pr^{3+} laser systems were available at the time of writing, one should expect their appearance within the next few years.

3.4.6.2 Microchip Lasers

Microchip lasers are miniature (typically submillimetre) monolithic solid-state lasers comprising a laser crystal with integral mirrors. Other optical elements, such as a Q-switch or a frequency conversion crystal can also be included in the monolithic cavity. These types of lasers are most usually diode pumped. Because of their monolithic nature and their very small size, they are usually extremely stable and alignment-free. Very often, they naturally produce single-frequency emissions. For example, Sotor et al. [32] recently reported stable generation of 160 mW of 532-nm single-frequency radiation from a $Nd:YVO_4$/KTP microchip laser. Work at the University of St Andrews in the late 1990s showed that single-frequency blue (33 mW at 473 nm) and red (10 mW at 671 nm) diode-pumped microchip lasers could also be produced using $Nd:YAG/KNbO_3$ and $Nd:YVO_4$/LBO, respectively [33]. Because of their small size, output power of DPSS microchip lasers is usually relatively small, but these lasers nevertheless constitute an interesting potential light source for colour holography.

3.4.6.3 Chromium Forsterite

The first reported lasing in a chromium-doped forsterite crystal was reported by Vladimir Petričević et al. in 1988 [34]. This vibronic laser has a broad range of transitions centred around 1.25 μm. Frequency doubling can potentially yield a useful red source, tunable over a wide range, although the thermal properties of forsterite are not good. Up to 1.1 W of emission has been reported at 1.2 μm [35], with pumping using a CW Nd:YAG laser. However, diode pumping of this material has proved difficult with very low efficiencies reported (only several milliwatts). We shall mention this laser system again in Chapter 6 when we talk about pulsed lasers. Narrow-band emission is possible when pumped with a Q-switched Nd:YAG laser—operation at 50 Hz with commercially available systems can deliver average powers of more than 50 mW at 627 nm.

3.4.6.4 Optical Parametric Oscillators

An optical parametric oscillator (OPO) is a parametric oscillator that oscillates at optical frequencies. It emits light not by stimulated emission, as in a true laser, but by a non-linear parametric process. Essentially, it converts an input laser beam (called the "pump") into two output waves of lower frequency using a non-linear optical process. The sum of the two output frequencies (called the *signal** wave and the *idler* wave) is equal to the input wave frequency. Generally, narrow-band OPOs require an additional narrow-band pump laser, making them complex systems.

---

* The wave with a higher frequency is called the signal.

Green-pumped OPOs are potentially attractive because of their capability to generate emissions ranging from the visible to the mid-infrared region within a single device. However, the available non-linear materials required for green pumping are limited by the absorption of visible wavelengths and associated thermal lensing. A different approach, which has had some success, is intracavity frequency mixing of the pump, signal and idler waves in an OPO cavity.

At the time of writing, there were no commercial single-frequency CW sources of visible radiation produced by OPOs.

3.5 Semiconductor Diode Lasers

3.5.1 Introduction

There has been great progress in semiconductor diode lasers in recent years. The concept of the DPSS laser first originated principally because laser diodes themselves did not have the output characteristics and mode quality required for use as a laser source in many applications. This equation has, however, now changed and diode lasers are themselves being used today as useful high-coherence sources for applications such as holography.

3.5.2 Operation and Construction

Like a light-emitting diode, a semiconductor laser diode consists of a "*p–n*" junction, which is usually deposited onto a crystal wafer. On the *p* side of the junction, the semiconductor is heavily doped with acceptor atoms. These are atoms with a missing electron in the valence band (a "hole"). On the *n* side of the junction, the dopant atoms each have an extra electron in their conduction band. When an electrical voltage is applied across the junction, electrons migrate from the *n*-doped semiconductor to the *p*-doped semiconductor, where they combine with holes producing laser radiation. A laser cavity is created by simply polishing the ends of the diode (Figure 3.9). Laser diodes differ from LEDs in that their depletion layers are rather thinner and a larger forward current is used. Additionally, much more care is needed in regulating the current than in an LED. A popular packaging format for low-power (≤1 W) diodes incorporates a pin diode in-line with the rear facet, allowing current control by monitoring of the back-emitted laser radiation (Figure 3.10).

The efficiency of semiconductor laser diodes is extremely high, sometimes reaching 80%. Tremendous progress is also being made with increasing the power and quality of the diodes. At the time of writing, Mitsubishi had just released a 638-nm multilateral mode laser diode (ML501P73) producing an unprecedented 1 W of emission at this wavelength (equivalent to a luminosity of 120 lumens) in a 5.6-mm-diameter capless package.

TABLE 3.1

Current Semiconductor Lasers in the Visible and Near Infrared Spectra

Wavelength Range of Emission (nm)	Type of Semiconductor Diode
430–550	GaN/AlGaN
447–480	ZnSSe
490–525	ZnCdSe
620–680	AlGaInP/AlGaAs
670–686	GaInP/GaAs
750–870	AlGaAs/GaAs
904	GaAs/GaAs
870–1100	InGaAs/GaAs
1100–1650	InGaAsP/InP

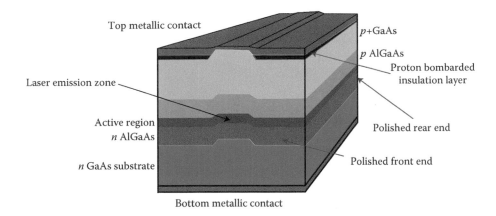

FIGURE 3.9   Typical structure of a double heterostructure edge-emitting laser diode. Dimensions of the active region are 200 µm in length, 2 to 10 µm in lateral width and 0.1 µm in transverse dimension.

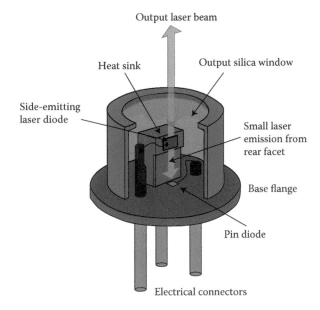

FIGURE 3.10   Typical mounting of a low-power laser diode (≤1 W) in a TO-CAN style package (base either 5.6 or 9 mm) incorporating monitor pin diode.

3.5.3 Mode Structure and Coherence

Lower-power laser diodes often produce single lateral and transverse mode outputs because only one mode will fit into the small active volume. The output from these diodes can also be highly coherent as the very short $p - n$ junction acts as a stable etalon with a high free spectral range. Generally, the smaller, less powerful diodes can have long coherence lengths and can quite often be used successfully for holographic applications if their temperature is kept reasonably constant. The problem here is that increasing the temperature will create mode hops. High-power laser diodes are, however, often not single mode in either space or time.

3.5.4 Operation at Single Longitudinal Mode

Various techniques are nowadays employed to ensure stable single longitudinal mode operation, such as etching a diffraction grating onto the top surface of the diode (distributed feedback laser) or using twin

FIGURE 3.11 Laser diode stabilised by a volume holographic grating producing a single-frequency output of 40 mW at 405 nm for embossed holographic applications.

FIGURE 3.12 *Cyan 488* laser from SpectraPhysics produces 150 mW of a single-frequency output at 488 nm from an external cavity-stabilised telecom diode. The laser head is just 12.5 cm long, and the unit consumes 6 W of electrical power.

semiconductor diode etalons (coupled cavity method). The US company Ondax uses a volume holographic element* to stabilise its laser diodes. It now offers a 640-nm diode laser producing up to 20 mW mW of highly coherent (>2 m coherence length) output power, which it advertises as a genuine "HeNe" replacement. Additionally, the company makes a full range of stabilised visible laser from 405 nm to 785 nm suitable for a wide range of holography applications (Figure 3.11). For example lasers, producing approximately 40 mW with good coherence at 405, 658, 685 and 690 nm—all in TO-CAN 5.6-mm packages were available at the time of writing. However, there have been reports in the scientific literature concerning much higher single-frequency powers. For example, more than 400-mW single frequency at 660 nm was obtained using an angled-grating structure in GaInP-AlInP [36].

Larger stabilised external cavities (often with piezo length control) can also be used very effectively to ensure very narrow band operation [37]. Again, the idea is to place the laser diode in a stabilised external cavity, dramatically improving the quality factor. The *Cyan 488* laser from SpectraPhysics is one such laser producing 150 mW of output at 488 nm (Figure 3.12). It is based on a highly reliable telecom-grade diode laser in an external cavity configuration with a proprietary wavelength-locking scheme that ensures stable single-frequency operation. The unit only consumes 6 W of electricity and is barely 12 cm long.

3.5.5 Amplification

Another commercial semiconductor diode laser producing high-power narrow band emission is the *DL RFA SHG pro* laser from Toptica. This laser produces no less than 2 W at 589 nm with a line width of less than 1 MHz. The laser is based on a tunable external cavity diode laser operating at 1178 nm. The output

---

* These elements are relatively thick normal-incidence quasi-lossless reflection holographic gratings of low permittivity modulation written in glass using UV lasers. They are employed as the output coupler of the diode laser cavity and, being extremely angle and wavelength selective, strongly favour operation at a single longitudinal mode.

of the external cavity diode laser is amplified within a polarisation-maintaining Raman fibre-amplifier, which preserves the spectrum of the seed laser. The amplifier output is frequency-doubled in a resonator to the target wavelength of 589 nm.

Amplification of seed emissions by a tapered diode in a master oscillator power amplifier configuration is also possible. Tapered diodes are available in the red and infrared frequencies and doubling can offer emissions at blue and green frequencies. Toptica produce commercial narrow-band devices using tapered amplification from 396 to 495 nm, with up to 400 mW of power at 459 nm (*SYST TA SHG pro*).

3.5.6 Summary

There is now a significant choice of commercial, high-coherence semiconductor lasers available at relatively low power. The powers on offer are generally increasing as diodes become more efficient and as manufacturers improve their technologies. At the time of writing, CW semiconductor lasers suitable for holography (coherence length >5 m)* were available at 405 nm (~40 mW), 440 nm (~30 mW), 488 nm (150 mW), 633 nm (~50 mW), 645 nm (~50 mW), 660 nm (~80 mW), 685 nm (40 mW) and 690 nm (~40 mW).

3.6 Fibre Lasers

A fibre laser is a laser in which the active gain medium is an optical fibre doped with rare earth elements such as neodymium, praseodymium, erbium, ytterbium, dysprosium or thulium. It is also possible to construct doped fibre amplifiers that can provide light amplification without lasing. Non-linear processes in the fibre, such as stimulated Raman scattering or four-wave mixing, can also provide laser gain. High-power single-frequency fibre lasers have been developed in the 1.5 μm region, where they are used for applications such as LIDAR. As of the time of writing, high-coherence commercial fibre lasers in the visible band suitable for holography were still relatively rare. However, this is likely to change. We have already mentioned the *DL RFA SHG pro* laser from Toptica, which uses a polarisation-maintaining Raman fibre-amplifier to achieve 2 W of single-frequency emission at 589 nm from an external cavity diode seed. The Canadian company, MPB Communications Inc., offers a series of visible fibre lasers emitting at 514, 560, 580, 592.5, 628 and 640 nm. However, the company currently only produces a single-frequency version for 514 nm. Maximum power is 300 mW. The US company IPG Photonics produces single-frequency 532-nm green fibre lasers (the GLR series) suitable for holography to a staggering 100 W (Figure 3.13). Menlo Systems, Inc. has recently released the *Orange-One SHG*, a single-frequency fibre laser with second harmonic generation producing more than 200 mW at 510 to 560 nm.

3.7 CW Laser Sources for Colour Holography Today

For low-power (<50 mW) holographic applications, there is currently a large choice of commercial CW laser sources right across the spectrum (Figure 3.14). This alone represents tremendous progress when compared with the situation 10 or 20 years ago. This really renders obsolete such popular lasers as the HeNe and HeCd lasers. However, as the required power increases towards the 150-mW level, the number of available wavelengths thins out. At more than 400 mW, there are no blue lasers to compete with the ion gas lasers, and at more than 500 mW, there are no red lasers. Efficient commercial lasers offering high coherence and high power are certainly available in the green (532 nm) and yellow (589 nm) 9 spectral regions. However, the blue region of the spectrum and the lower-wavelength reds are particularly poorly served. Published scientific research clearly demonstrates that the technology is known today to fill these gaps, but it may be a few more years yet before such research makes its way to the commercial marketplace.

---

* Excluding fibre amplification.

FIGURE 3.13   100-W single-frequency 532-nm green fibre laser (the *GLR-100* from IPG Photonics).

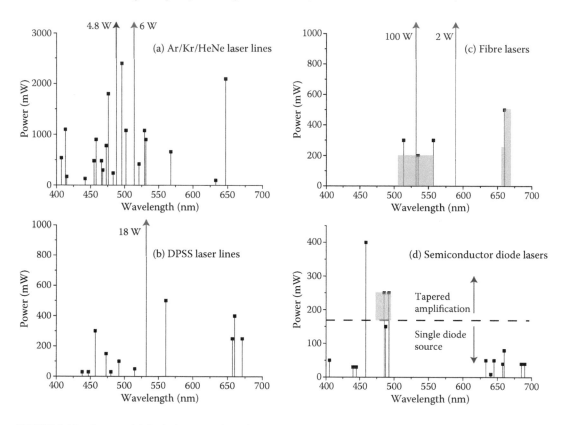

FIGURE 3.14   Commercial single-frequency laser frequencies available as of January 2012: (a) argon, krypton and HeNe lasers, (b) DPSS lasers, (c) fibre lasers and (d) semiconductor lasers.

REFERENCES

1. A. Javan, W. R. Bennett Jr. and D. R. Herriott, "Population inversion and continuous optical maser oscillation in a gas discharge containing a He–Ne mixture," *Phys. Rev. Lett.* **6**, 106–110 (1961).
2. W. B. Bridges, "Laser oscillation in singly ionized argon in the visible spectrum," *Appl. Phys. Lett.* **4**, 128 (1964); erratum: *Appl. Phys. Lett.* 5, 39 (1964).
3. G. R. Fowles and W. R Silfvast, "Laser action in the ionic spectra of Zn and Cd," *IEEE J. Quantum Elect.* qe-1 131 (1965).
4. D. C. Sinclair and W. E. Bell, *Gas Laser Technology*, Holt, Rinehart and Winston, New York (1969).
5. M. B. Klein and W. T. Silfvast, "New cw laser transitions in Se II," *Appl. Phys. Lett.* **18** 482–485 (1971).
6. T. W. Hansch, "Repetitively pulsed tunable dye laser for high resolution spectroscopy," *Appl. Opt.* **11**, 895–898 (1972).

7. I. Shoshan, N. N. Danon and U. P. Oppenheim, "Narrowband operation of a pulsed dye laser without intracavity beam expansion," *J. Appl. Phys.* **48**, 4495–4497 (1977).

8. M. G. Littman and H. J. Metcalf, "Spectrally narrow pulsed dye laser without beam expander," *Appl. Opt.* **17**, 2224–2227 (1978).

9. B. H. Soffer and B. B. McFarland, "Continuously tunable, narrow-band organic dye lasers," *Appl. Phys. Lett.* **10**, 266–267 (1967).

10. O. G. Peterson and B. B. Snavely, "Stimulated emission from flashlamp-excited organic dyes in poly-methyl methacrylate," *Appl. Phys. Lett.* **12**, 238–240 (1968).

11. A. Bank, D. Donskoy and V. Nechitailo, "High-average-power quasi-cw tunable polymer laser," in *UV and Visible Lasers and Laser Crystal Growth,* R. Scheps and M. R. Kokta eds., Proc. SPIE **2380**, 292–297 (1995).

12. F. J. Duarte, "Multiple-prism grating solid-state dye laser oscillator: optimized architecture," *Appl. Opt.* **38**, 6347–6349 (1999).

13. W. L. Nighan, Jr, N. Hodgson, E. Cheng and D. Dudley, "Quantum-limited 35 W TEM$_{00}$ Nd:YVO$_4$ laser," in *Lasers and Electro-Optics,* CLEO'99 Tech Digest, OSA (1999) p. 1.

14. H. Welling, "Single-frequency source in the blue spectral region," *Opt. Lett.* **22**, 1220–1222 (1997).

15. Y. Wang, J. Liu, Q. Liu, Y. Li and K. Zhang, "Stable continuous-wave single-frequency Nd:YAG blue laser at 473 nm considering the influence of the energy-transfer up-conversion," *Opt. Expr.* **18**, 12044–12051 (2010).

16. F. Camargo, T. Zanon-Willette, R. Sarrouf, T. Badr, N. U. Wetter and J. Zondy, "1.3 Watt single-frequency Nd:YLF/ppKTP red laser," in *Conference on Lasers and Electro-Optics/International Quantum Electronics Conference,* OSA Technical Digest (CD), paper CThZ7 (2009).

17. R. Sarrouf, V. Sousa, T. Badr, G. Xu and J. -J. Zondy, "Watt-level single-frequency tunable Nd:YLF/periodically poled KTiOPO$_4$ red laser," *Opt. Lett.* **32**, 2732–2734 (2007).

18. F. Camargo, T. Zanon-Willette, T. Badr, N. U. Wetter and J. -J. Zondy, "620 mW single-frequency Nd:YVO$_4$/BiB$_3$O$_6$ red laser," in *Lasers and Electro-Optics, Conference on Quantum Electronics and Laser Science,* CLEO/QELS (2009) p. 1–2.

19. E. Jun Zang, J. Ping Cao, Y. Li, T. Yang and D. Mei Hong, "Single-frequency 1.25 W monolithic lasers at 1123 nm," *Opt. Lett.* **32**, 250–252 (2007).

20. E. Mimoun, L. De Sarlo, J. -J. Zondy, J. Dalibard, and F. Gerbier, "Sum-frequency generation of 589 nm light with near-unit efficiency," *Opt. Expr.* **16**, 18684–18691 (2008).

21. R. Smart, J. Carter, A. Tropper, D. Hanna, S. Davey, S. Carter and D. Szebesta, "CW room temperature operation of praseodymium-doped fluorozirconate glass fibre lasers in the blue-green, green and red spectral regions," *Opt. Commun.* **86**, 333–340 (1991).

22. T. Sandrock, T. Danger, E. Heumann, G. Huber and B. H. Chai, "Efficient Continuous Wave-laser emission of Pr^{3+}-doped fluorides at room temperature," *Appl. Phys. B,* **58**, 149–151 (1994).

23. T. Danger, T. Sandrock, E. Heumann, G. Huber and B. H. Chai, "Pulsed laser action of Pr:GdLiF$_4$ at room temperature," *Appl. Phys. B-Photo* **57**, 239–241 (1993).

24. A. Richter, N. Pavel, E. Heumann, G. Huber, D. Parisi, A. Toncelli, M. Tonelli, A. Diening and W. Seelert, "Continuous-wave ultraviolet generation at 320 nm by intracavity frequency doubling of red-emitting praseodymium lasers," *Opt. Expr.* **14**, 3282–3287 (2006).

25. A. Richter, E. Heumann, E. Osiac, G. Huber, W. Seelert and A. Diening, "Diode pumping of a continuous-wave Pr^{3+}-doped LiYF$_4$ laser," *Opt. Lett.* **29**, 2638–2640 (2004).

26. A. Richter, E. Heumann, G. Huber, V. Ostroumov and W. Seelert, "Power scaling of semiconductor laser pumped praseodymium-lasers," *Opt. Expr.* **15**, 5172–5178 (2007).

27. P. Camy, J. Doualan, R. Moncorge, J. Bengoechea and U. Weichmann, "Diode-pumped Pr^{3+}:KY$_3$F$_{10}$ red laser," *Opt. Lett.* **32**, 1462–1464 (2007).

28. N. O. Hansen, A. R. Bellancourt, U. Weichmann and G. Huber, "Efficient green continuous-wave lasing of blue-diode-pumped solid-state lasers based on praseodymium-doped LiYF₄," *Appl. Opt.* **49**, 3864–3868 (2010).

29. M. Fibrich, H. Jelinkova, J. Šulc, K. Nejezchleb and V. Škoda, "Visible cw laser emission of GaN-diode pumped Pr:YAlO$_3$ crystal," *Appl. Phys. B,* **97**, 363–367 (2009).

30. T. Basiev, M. Konyushkin, D. Konyushkin, M. Doroshenko, G. Huber, F. Reichert, N. Hansen and M. Fechner, "First visible 639 nm SrF$_2$:Pr^{3+} ceramic laser," in *Lasers and Electro-Optics Europe (CLEO EUROPE/EQEC), 2011 Conference on and 12th European Quantum Electronics Conference, OSA Technical Digest (CD)* paper CA2-2 (2011) p. 1.

31. T. Calmano, J. Siebenmorgen, F. Reichert, M. Fechner, A.-G. Paschke, N.-O. Hansen, K. Petermann and G. Huber, "Crystalline Pr:SrAl$_{12}$O$_{19}$ waveguide laser in the visible spectral region," *Opt. Lett.* **36**, 4620–4622 (2011).

32. J. Z. Sotor, A. J. Arkadiusz and K. A. Abramski, "Single frequency monolithic solid state green laser as a potential source for vibrometry systems," *AIP Conf.* Proc. **1253**, 313–316 (2010).

33. R. Conroy, *Microchip lasers,* PhD Thesis, University of St Andrews, UK (1998).

34. V. Petričević, S. K. Gayen, R. R. Alfano, K. Yamagishi, H. Anzal and Y. Yamaguchi, "Laser action in chromium-doped forsterite," *Appl. Phys. Lett.* **52**, 1040–1042 (1988).

35. N. Zhavoronkov, A. Avtukh and V. Mikhailov, "Chromium-doped forsterite laser with 1.1 W of continuous-wave output power at room temperature," *Appl. Opt.* **36**, 8601–8605 (1997).

36. B. Pezeshki, M. Hagberg, M. Zelinski, S. D. DeMars, E. Kolev and R. J. Lang, "400-mW single-frequency 660-nm semiconductor laser," *IEEE Photonic Tech. L.* **11** (7), 791–793 (1999).

37. M. G. Boshier, D. Berkeland, E. A. Hinds and V. Sandoghdar, "External-cavity frequency-stabilization of visible and infrared semiconductor lasers for high resolution spectroscopy," *Opt. Commun.* **85**, 355–359 (1991).

4

Recording Materials for Colour Holography

4.1 Introduction

The recording materials used in modern polychromatic display holography are fundamental to achieving high image fidelity. It is perhaps not surprising that the earlier holographic recording materials used in holography are simply not suitable for recording full-colour holograms; polychromatic holograms put far higher demands on the recording material. Moreover, many of the earlier silver halide materials produced by commercial manufacturers such as Agfa and Kodak are no longer on the market. Additionally, these materials were not panchromatic, nor did they have sufficient resolving power to be used for recording colour holograms. The first commercial photopolymer materials did have sufficient resolving power, but many were only monochromatic. To record full-colour holograms, panchromatic materials with a resolving power of more than 10,000 line pairs/mm are required. In this chapter, we shall focus our attention on the materials required for recording ultra-realistic colour holograms and describe their characteristics and the demands put on them by this application.

The three main types of holographic recording materials that can be considered today are

- Silver halide materials
- Dichromated gelatin materials
- Photopolymer materials

4.1.1 Silver Halides

Silver halide recording materials are interesting for many reasons [1]. Historically, these emulsions were the first materials used for recording holograms; silver halide has, in the past, constituted the most important and most ubiquitous material employed in holography, particularly with respect to its numerous scientific and artistic applications. Most importantly, silver halide materials exhibit high sensitivity in comparison with many other alternative materials, they can be coated onto both film and glass and can cover very large formats. They can also be used to record both amplitude and phase holograms and are capable of a rather high resolving power.

Notwithstanding its advantages, silver halide does have various drawbacks: it is absorptive, has inherent noise and a limited linear response, it is irreversible and it needs wet processing. Phase holograms made with silver halide are also subject to printout problems. Agfa, Mortsel, Belgium; Ilford, Knutsford, UK; and Kodak, Rochester, NY all manufactured special holographic silver halide materials of a monochromatic form for a long time, but they no longer have any such materials on the market today. As a matter of fact, Ilford (now operating under the name Harman Technologies Ltd.) has started to manufacture holographic recording materials again, but these materials have too large a grain size for recording colour holograms. There are only very few commercial silver halide colour emulsion products on the market today. Among these are two Russian manufacturers and two European producers with rather limited production. However, it is difficult for such small commercial manufacturers to guarantee that each batch of material will possess exactly the same characteristics. As such, the serious worker in modern colour holography often needs to consider either sorting and calibrating commercial batches before use or, in many cases, actually producing the materials in-house.

4.1.2 Dichromated Gelatin

Dichromated gelatin (DCG) possesses a very high resolving power and a remarkable brightness due to a refractive index modulation of 0.08, which, until recently, was the largest known among holographic materials. Unfortunately, there are no commercial companies producing panchromatic DCG recording materials today, although the Russian company Slavich produces blue/green sensitive plates (PFG-04). Therefore, this is a material which one needs to make in-house if one wants to record colour holograms. In addition, until recently, the sensitivity of known panchromatic DCG materials was very low. Currently, the material is mainly used for producing holographic optical elements (HOEs)—for example, heads-up displays for aircraft. Recently, panchromatic DCG materials have been reported with greatly improved sensitivity. As such, DCG must be regarded as an interesting material for colour holography and in particular for digital techniques that write small elemental holograms, one at a time.

4.1.3 Photopolymers

Photopolymer materials can be used for recording phase holograms in which applications in the mass production of small (usually monochromatic) display holograms and optical elements currently constitute the main commercial interests. The sensitivity is not as high as with silver halide materials, but the advantages are a low light-scattering noise level as well as an innate suitability for the application of dry processing techniques. E. I. DuPont de Nemours & Co. Wilmington, DE, has been the main manufacturer of commercial photopolymer materials and has, for a long time, marketed these under the name of OmniDex. The DuPont material requires only a dry processing technique (exposure to UV light and heat treatment) to obtain a hologram. A new photopolymer material, Bayfol HX, suitable for colour holography, which requires even less postprocessing, has recently been introduced by Bayer Material Science AG in Germany.

One important application of colour photopolymer materials is document security. As a result, material manufacturers will often restrict sales to other users. At the time of writing, such a policy was in operation by both DuPont and Bayer, and with the best chemistries still under patent restriction, unlike silver halide and DCG, in-house fabrication of photopolymers cannot be regarded as a serious alternative for most workers. Although today's panchromatic photopolymer materials constitute an extremely interesting solution for colour display holography, from a purely practical point of view, there remains a sizeable question mark over whether these materials will be commercially applicable to display holography applications in the near-term. The only material freely available at the time of writing was the Polygrama DAROL photopolymer marketed by Lynx in Brazil.

4.2 Holographic Recording

Before a more detailed presentation of the materials suitable for recording ultra-realistic colour holograms is given, a general description is provided of how a holographic recording is made. In particular, we discuss how silver halide materials are able to store the interference patterns generated during the holographic recording process. What follows is therefore a short description of holographic recording theory with some definitions of common photographic properties that are of particular importance for recording colour holograms. For example, we shall explain how a latent image forms in ultrafine silver halide grains. The signal-to-noise ratio and diffraction efficiency in the holographic recording process will also be introduced. For more detailed information on the general principles governing the holographic recording process, and on silver halide materials in particular, the reader is referred to a book dedicated to this topic [1].

To record the entire light field scattered from an object, both the amplitude and phase of the electromagnetic waves involved must be stored in some way. There is, however, no material that can directly detect both the amplitude and the phase of a light wave. The practical solution is therefore to use the interferometric two-step process introduced by Gabor [2], in which phase information in an optical signal is converted to an implicit amplitude variation through mixing with a coherent reference; this amplitude signal is then recorded as a density or index variation within a photographic emulsion. This is the holographic process. The hologram it produces constitutes a micropattern created in a light-sensitive material as the result of the coherent interference between the signal (the object beam) and its reference (the reference beam). The time of exposure for a given material depends on the sensitivity of the material used as well as on the intensity of the interference pattern. Silver halide materials must be processed after exposure in a very specific way so that the recorded latent image of the interference pattern is successfully converted to local variations in optical density, refractive index or thickness of the recording layer.

A silver halide photographic recording material is based either on a single type or on a combination of types of silver halide crystals embedded in a gelatin layer. This is commonly referred to as a photographic or photosensitive emulsion. Actually, this photosensitive emulsion is not really an "emulsion" at all but rather a thin film of silver halide microcrystals dispersed in a colloid (gelatin). However, the term emulsion is commonly used in photography for this type of perpetual suspension. The emulsion is coated on a flexible or stable substrate material such as glass or plastic film.

There are three types of silver halides: silver chloride (AgCl), silver bromide (AgBr) and silver iodide (AgI). Silver chloride is used for low-sensitivity emulsions. Chloride/bromide emulsions have high light sensitivity. However, the bromide/iodide emulsions have even higher sensitivity. Silver iodide is never used alone but is used in a mixture with silver bromide; it normally constitutes 5% or less of such a mixture. Adding some silver iodide to fine-grained emulsions at low concentrations gives a higher sensitivity and contrast as compared with pure silver bromide emulsions of the same grain size. Silver halide crystals are cubical in shape, and in each crystal, a silver ion (Ag^+) is surrounded by six halide ions. The crystal normally possesses an excess of halide ions that originate from the emulsion manufacturing process. Silver halide grain sizes vary from approximately 10 nm for the ultrafine-grained Lippmann and colour holography emulsions to a few micrometres for high-sensitivity photographic emulsions (Table 4.1). Only a silver halide emulsion of the ultrafine grain type can be used for colour holography. For more detailed definitions and the mathematical theory of conventional photographic silver halide materials, the reader is referred to various scientific books on photography [3–6].

The resolving power of a photographic material is a measure of its ability to record fine detail. It can be defined as "the ability of a photographic material to maintain in its developed image the separate identity of parallel bars when their relative displacement is small." Normally, the resolving power of a photographic material is tested by using a resolution test chart. The highest number of lines per millimetre that can be resolved in the emulsion corresponds to the resolving power of the tested material. A line in this definition is a line with its adjoining space and corresponds to "line pairs" in electronic images. The resolving power of the holographic material is a critical feature that must be taken into account when defining its characteristics.

The resolution capability of an image reproduction process is normally described by the optical transfer function (OTF). For a given test input, the OTF is defined as the complex response (amplitude and phase) of the reproduced image for each spatial frequency, v. Usually, in practice, only the modulus of the OTF is quoted, this being known as the modulation transfer function (MTF). The MTF is a good aid in demonstrating the quality of a particular photographic emulsion as well as constituting a means for comparing different emulsions. Briefly, a test pattern containing a sinusoidal variation in illuminance combined with a slow continuous and linear variation in spatial frequency along one direction is recorded. The modulation M of the pattern in the test target can then be defined as

$$M = (H_{MAX} - H_{MIN})/(H_{MAX} + H_{MIN}) \qquad (4.1)$$

where H is the exposure* incident on the photographic material. When this pattern is recorded in the material, light scattering will take place in the emulsion, which will reduce the original contrast of

---

* See Section 4.2.1 for the definition of "exposure".

TABLE 4.1

Emulsion Grain Sizes

Type of Emulsion	Average Grain Diameter (nm)
Ultrafine-grain holographic emulsion	10–20
Fine-grain holographic emulsion	20–50
Fast holographic emulsion	50–100
Chlorobromide paper emulsion	200
Lithographic emulsion	200–350
Fine-grain photographic emulsion	350–700
Fast photographic emulsion	1000–2000
Fast medical x-ray emulsion	2500

the pattern. In this way, the modulation of the pattern will be decreased, in particular at high spatial frequencies. The effective recorded exposure modulation M' will then be given as

$$M' = \left(H'_{MAX} - H'_{MIN} \right) / \left(H'_{MAX} + H'_{MIN} \right) \tag{4.2}$$

where H' is the exposure within the emulsion.

The original modulation M is constant and accurately known; it is also independent of the spatial frequency. After the tested emulsion has been processed, the corresponding "exposed" modulation is obtained from observing the density variation. The ratio between the modulation M' in the emulsion and modulation M of the incident exposure is called the *modulation transfer factor*, also called the *response*

$$R = M/M' \tag{4.3}$$

If the response is plotted as a function of spatial frequency, this curve will then be the MTF of the material (Figure 4.1). The Fourier transform of the MTF is the line spread function of the emulsion, which indicates the width of a line image recorded in the emulsion.

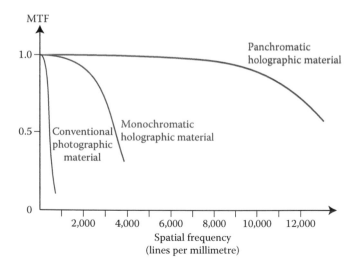

FIGURE 4.1   MTF for conventional photographic materials, for fine-grained holographic materials and for the ultrafine grain materials required for colour holography.

4.2.1 Sensitivity of Photographic and Holographic Materials

The time of exposure for a given material will depend on the sensitivity of the material used as well as on the intensity of the interference pattern. Some holographic materials must be processed after exposure in a specific way to obtain a hologram. The recorded intensity variations are converted during this processing step to local variations in optical density, to refractive index variations or to thickness variations of the recording layer.

The exposure of the recording material, H, is defined as the incident intensity, E, times the time of exposure, t. If the intensity is constant during the whole exposure time, which is usually the case, then

$$H = E\,t \tag{4.4}$$

Holographic materials are usually characterised using radiometric units. The radiometric equivalent of illuminance is irradiance. The unit of irradiance is Wm^{-2} and the exposure will then be expressed in Jm^{-2}. The sensitivity of a holographic emulsion is most often expressed in μJcm^{-2} or $mJcm^{-2}$. Knowing the sensitivity of the material used and having measured the irradiance at the position of the holographic plate, the exposure time can be found by dividing sensitivity by irradiance. Colour holographic materials must be sensitised in such a way that they are optimised for the laser wavelengths commonly used in holography.

4.3 Holographic Emulsions

The final quality of a holographic image will be a function of a number of factors such as the geometry and stability of the recording setup, the coherence of the laser light, the reference and object beam ratio, the type of hologram produced, the size of the object and its distance from the recording material and the recording material and the emulsion substrate used. It will also depend on the processing technique applied as well as the reconstruction conditions. We know that, if during the reconstruction of the hologram, the reference replay beam has identical characteristics to the recording reference beam, then no image aberration will occur.* This also applies when the reconstruction reference beam constitutes an exact conjugate of the original reference beam (time reversed). Theoretically, the holographic technique is the most perfect imaging technique in existence because both the amplitude and the phase of the light wave scattered from the object are recorded. However, in practice, the holographic image is subject to certain limitations imposed by the recording material. In fact, three main factors will determine the resolution of a holographic image: the recording wavelength, the numerical aperture and the properties of the recording material itself.

Of course, ideally, the ultimate resolution of a hologram should be independent of the properties of the recording material and should depend only on the wavelength that was used for the recording, on the size of the recorded area of the material (the aperture) and on the object distance. However, in practice, the limit on resolution may be set by the recording material. This is the case when the material cannot record spatial frequencies above a certain limit.

Additionally, even if we arrange for the replay reference beam to have identical characteristics as the recording beam, aberrations in the holographic image can be induced directly by the recording material. This is the case, for instance, when the refractive index of the recording material changes upon processing.

4.3.1 Demands on Recording Emulsion

A silver halide emulsion must comply with certain requirements to be suitable for the recording of colour holograms. The most important of these demands concerns the resolving power of the material. The recording material must be able to resolve the highest spatial frequencies of the interference pattern created by the maximal angle, θ, between the reference and the object beams in the recording setup

---

* The reader is referred to Chapter 11 for a discussion of the basic optical principles of holography.

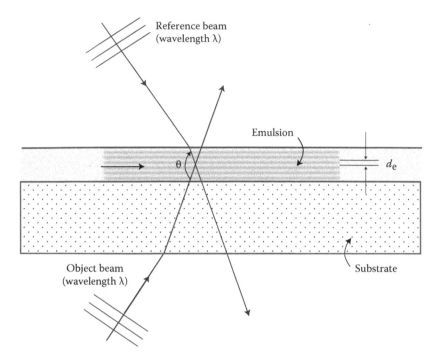

FIGURE 4.2   Demand on the resolving power from a material for recording a reflection hologram. The recording material must resolve the highest spatial frequencies of the interference pattern created by the maximal angle θ between the reference and the object beams in the recording setup.

(Figure 4.2). If λ is the wavelength of the laser light used for the recording of a hologram, then the closest separation d_a between the fringes in the interference pattern (in air) is given by

$$d_a = \frac{\lambda}{2\sin(\theta/2)} \tag{4.5}$$

In the recording layer, the fringe spacing d_e will depend on the refractive index n of the emulsion and is given by

$$d_e = \frac{\lambda}{2n\sin(\theta/2)} \tag{4.6}$$

We can use these formulae to find the resolving power needed to record a colour reflection hologram with blue laser light at λ = 440 nm. Assuming an emulsion having a refractive index of n = 1.62 and a maximal angle of 180° between the beams, Equation 4.6 shows that a minimum resolving power of 7360 lines/mm is required. This is the minimum resolving power needed to record the information. Close to its resolution limit, the material will exhibit a lower MTF and will thus make a low-quality hologram with poor fringe contrast and low signal-to-noise ratio. For high-quality colour holograms, the resolution limit of the material must be much higher than the minimum value obtained according to the above formula.

4.3.2 Resolution of Holographic Image

In holography, the resolution of the holographic image and the resolving power of the recording material are not directly related in the way they in photography. Equation 4.6 determines the minimum resolving power required to actually record a reflection hologram. This figure is not directly related to the resolution of the image recorded in or reconstructed by a hologram. If no lenses are involved in the holographic image formation process, the theoretical resolution of the image will be limited by diffraction

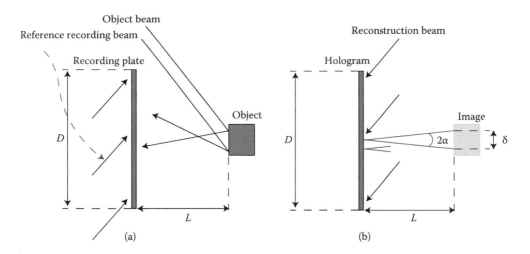

FIGURE 4.3 Diffraction-limited ideal resolution of the reconstructed image of a reflection hologram: (a) recording the hologram of an object and (b) replay of the recorded hologram showing that image detail is reconstructed with a characteristic length scale δ determined by L and D. This in turn is related to the angle of diffraction (2α).

and as such will be dependent on the area of the recording material, the recording laser wavelength and the distance between the recording material and the object (Figure 4.3). The image resolution δ is then

$$\delta \sim \frac{\lambda}{2\sin\alpha}$$
$$\sim \frac{\lambda L}{D} \quad \text{if} \quad L \gg D$$

(4.7)

where α is the angle indicated in Figure 4.3. Of course, if a limiting aperture is part of the system, diffraction will occur, causing the resolution to be slightly smaller than it would be if no aperture was present. For a circular aperture in incoherent light, the diffraction-limited resolution is given by

$$\delta \sim 0.61 \frac{\lambda}{\sin\alpha}$$
$$\sim 1.22 \frac{\lambda L}{D} \quad \text{if} \quad L \gg D$$

(4.8)

In a reflection hologram reconstructed using a broadband white-light source, the image resolution is mainly affected by the white-light source size, which has a much larger influence than the diffraction-limited resolution. More information about the illumination of colour reflection holograms and the influence on the image resolution and chromatic aberration using white-light sources will be provided in Chapter 11.

4.3.3 Image Resolution Determined by Recording Material

Theoretically, the resolution of the holographic image should be the true diffraction-limited resolution that can be obtained when the information is collected over an aperture equal to the size of the recording holographic plate. If the resolving power of the recording material is sufficient, the diffraction-limited resolution can be obtained under the assumption that the high-resolution recording material is also perfect in that the position of the recorded interference fringes will not be changed during the processing of the material. In practice, a stable support for the emulsion (such as a glass plate) is needed and the

(a)   (b)

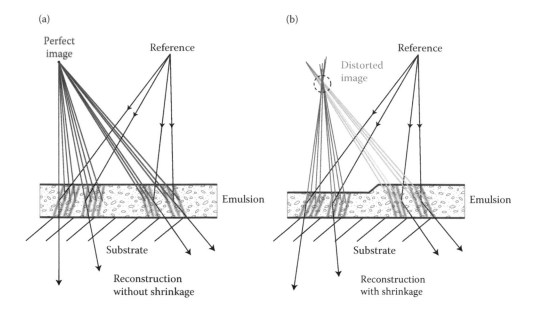

FIGURE 4.4 Illustration of how image distortion is introduced in an off-axis transmission hologram when the emulsion thickness changes. (a) shows the case with no change in thickness: here, a unique recording point produces image rays that coincide at the same unique point. (b) illustrates what happens when the emulsion shrinks (or expands). Now the image rays do not converge to a unique point.

processing methods applied must not affect the recorded fringe position in the emulsion, that is, no fixing. The most limiting factor controlling the resolution of a holographic image comes from distortions appearing in the emulsion. These aberrations, illustrated in Figure 4.4, are introduced by

1. Variations in the thickness of the recording medium before processing
2. Variations in the thickness of the recording medium during processing
3. Variation in the refractive index of the recording medium during processing
4. Deformation of the recording medium occurring between recording and reconstruction.

An ideal holographic recording plate should consist of a uniformly thick emulsion coated onto a perfectly flat plate of homogeneous glass having uniform thickness. In practice, the recording plate must be treated as one of the optical elements in the holographic system that can eventually affect the image quality.

4.4 Problems due to Short or Long Exposure

4.4.1 Pulsed Holography

RGB pulsed lasers are nowadays being used more and more frequently to record digital colour holograms. We shall discuss such lasers in depth in Chapter 6 and the digital hologram printers employing these lasers in Chapter 7. The main reason for the use of pulsed lasers as compared with CW lasers is that their short pulses confer complete immunity to environmental vibration at recording. For holographic applications, solid-state lasers today are operated mainly in the Q-switched regime, with associated pulse lengths of approximately 10 to 60 ns. From the photographic point of view, these exposure times can all be regarded as "short".

4.4.2 Reciprocity Failure

In Q-switched operation, laser energy is released during a very short time, producing a very high peak output power. For a 10 J pulse at 20 ns, the peak power is 500 MW. Such high power is certainly desirable when exposing holographic materials of relatively low sensitivity. However, short exposure times are associated with the problem of reciprocity failure or more precisely, the failure of the reciprocity law. In general, the exposure H of the photographic material is given by Equation 4.4. The reciprocity law was originally formulated by Bunsen and Roscoe [7]; it states that a given exposure H is independent of the two factors, E and t, separately. However, this is not true for extreme values of E and t, and this phenomenon can also affect hologram recordings at very long exposure times and low light levels (the Schwarzschild effect). In particular, using low-power CW lasers for recording colour holograms over long exposure times can also create problems for the ultrafine-grain emulsions.

Curves showing the reciprocity law (or its failure) are often plotted as $\log(Et)$ versus $\log(E)$ for a fixed optical density. A typical reciprocity law–failure curve is shown in Figure 4.5. Between any two points on a horizontal part of the curve, there is no reciprocity failure. However, if the two points of the curve are not on a horizontal part of the curve, then the reciprocity law does not hold between the corresponding exposure times. In reality, the exposure necessary for obtaining a certain density in the developed material is not constant but depends on the exposure time t. For very short exposures at high intensities E, as well as for very long exposures at low intensities, H has to be strongly increased to get the same density as the one required for the optimal values of E and t. These effects are called *high-intensity reciprocity failure* (HIRF) and *low-intensity reciprocity failure* (LIRF), respectively. The HIRF becomes of importance even for pulses much longer than those from a conventional Q-switched laser and the effect is roughly constant for times less than 10^{-5} s for conventional silver halide photographic materials.

To better understand these phenomena and how we might increase the sensitivity of an ultrafine grain emulsion, we must study the formation of the latent image. The silver halide crystal is an n-type photoconductor with a valence band of electrons and with a conduction band in which injected electrons are free to migrate throughout the crystal until trapped by a lattice defect. During the exposure of an emulsion, photons are absorbed by the crystals. When a photon of sufficient energy is absorbed, an electron from the silver halide crystal Ag^+X^- is promoted to the conduction band, leaving behind a positive hole that constitutes a free halogen atom:

$$Ag^+X^- + h\nu \rightarrow Ag^+X^0 + e^- \tag{4.9}$$

The silver ion will then attract the photogenerated electron to form a silver atom, the so-called *prespeck*

$$Ag^+ + e^- \rightarrow Ag^0 \tag{4.10}$$

The free electron is first trapped by a positively charged surface lattice defect. Once trapped, the electron will attract an interstitial silver ion, Ag^+ to the sensitivity site to form the silver atom prespeck, Ag^0.

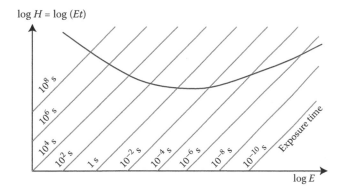

FIGURE 4.5 Typical reciprocity–law failure curve. Curves showing the reciprocity law (or its failure) are often plotted $\log(Et)$ versus $\log(E)$ for a fixed optical density. A material without reciprocity failure is a horisontal line for all exposure times.

Statistically, an isolated silver atom has an average lifetime of approximately 1 s. This lifetime can be calculated knowing the binding energy for the electron within the prespeck that has been experimentally measured to be approximately 0.7 eV. The Boltzmann statistical lifetime is then given by the standard formula

$$t = \tau e^{E_0/kT} \tag{4.11}$$

where $\tau = 10^{-12}$ s is the estimated electron collision period, E_0 is the binding energy, k is Boltzmann's constant (1.3805×10^{-23} *J*/K) and T the temperature in Kelvin. For $T = 300$ K and for $E_0 = 0.7$ eV, Equation 4.11 tells us that $t \sim 1$ s.

To create a sublatent image speck in the silver halide crystal where a diatomic silver molecule is formed by the process of nucleation, a second silver atom is required at the site of the first silver atom during its lifetime:

$$Ag + Ag^+ + e^- \rightarrow Ag_2 \tag{4.12}$$

A subspeck of two atoms is stable at room temperature ($E_0 = 1.74$ eV giving $t = 1.7 \times 10^{17}$ s). The sublatent image speck grows larger with further photon absorption, resulting in photogenerated electrons. The latent image is usually regarded as a collection of a few silver atoms at one site produced by the reduction of silver ions in the process of photolysis. Silver formed in this way is known as *photolytic silver*. A latent image of at least three to four silver atoms is needed for developability. Developability means the formation of a latent image that has the catalytic property of increasing the development rate of silver halide grains reduced to metallic silver by the reducing agent called a *developer*. For all the exposed grains, chemical development will then reduce the entire silver halide grains to metallic silver.

The chemical sensitisation of an emulsion is similar to the doping process of a semiconductor. The introduction of such impurities as sulphur, gold, or silver, alone or in combination, into the emulsion increases the grain's sensitivity (finishing). Chemical reduction is induced by raising the temperature. Depending on the impurities which have been introduced into the emulsion, the emulsion is called *sulphur sensitised, gold sensitised, sulphur plus gold sensitised*, or, at times, *reduction sensitised*. The sensitivity of a grain is defined as the reciprocal of the number of absorbed photons necessary to produce the developability of the grain. A highly sensitised grain requires fewer photons than a less sensitised grain to be developable. Grain size is, however, even more important for sensitivity: the larger the grain, the higher the sensitivity of the material. A typical large silver halide grain with the volume of 1 μm^3 (= 10^{-12} cm^3) contains about 2×10^{10} silver ions. In such a grain, just a few photons are expected to produce a stable latent image, which can later be used to trigger off the process of converting the entire grain to silver atoms. The overall amplification, from the quanta absorbed to the silver atoms produced, can be greater than 10^9 in this process. For a typical conventional holographic emulsion with the grain size of approximately 50 nm, the amount of silver ions in the grain is approximately 2.6×10^6, which translates to an amplification of about one million. However, note that this constitutes only approximately 1/1000 of the sensitivity of a conventional high-speed photographic film. The problem with ultrafine-grain colour holography emulsions is that it is impossible to make them very light sensitive.

Using pulsed lasers to record holograms with ultrafine-grain emulsions, one encounters the problem of HIRF. As the intensity increases, more absorbed photons are required per grain to produce the same density in the developed material as compared with exposures at lower intensity levels. HIRF is caused by the silver ions' motion and their concentration in the emulsion. The exposure time of about 1 s is sufficient for a mobile, interstitial silver ion to neutralise a trapped electron before another arrives. At high intensities, electrons are produced at such a rate that there is not enough time for the mobile silver ions to neutralise the trapped electrons. Because of electrostatic repulsion, the second electron is not trapped at the same site as the first electron. Recombination of holes and electrons may occur instead at an increasing rate or trapping of electrons may take place at some other site. In each case, the process of latent-image formation becomes inefficient.

It should be mentioned that LIRF can also affect the recording of large-format colour holograms on ultrafine grain emulsions using CW lasers. LIRF depends on the thermal stability of an isolated silver atom which, if not stabilised by combination with another silver atom within its lifetime (~2 s) will decompose into an electron and a silver ion again. This, of course, means that long exposure at a low light level becomes a very inefficient process in forming the latent image.

Another result of short exposure is the localisation of the latent image within the silver halide crystal. At exposure times longer than 10^{-2} s, the latent image is almost entirely localised on the surface of the silver halide crystals. At shorter exposure times, it is also formed inside the grains. Therefore, to develop the internal latent image specks as well, it is necessary to use the correct developing technique. A surface developer acts only on the latent image on the surface of the silver halide crystal. Internal development is performed on the internal latent image after the surface latent image has been bleached off. Total development is performed with an internal developer but without bleaching off the latent surface image. This type of development technique is definitely preferable for holograms exposed with a pulsed laser.

4.4.3 Holographic Reciprocity Failure

Early investigations on holographic reciprocity failure regarding fine-grain materials were performed by Vorzobova and Staselko [8,9]. They noted that there was a reduction in the γ value of holographic materials exposed with pulsed lasers. The diffraction efficiency for a Q-switched hologram decreased to 10% compared with that of a hologram recorded using a free-lasing mode. They concluded that the observed drop in diffraction efficiency of pulsed holograms was caused by a change in the optical characteristics of the photographic layer when the illumination time was reduced. In another Russian investigation, Benken and Staselko [10] studied the latent-image formation process in the Russian holographic materials LOI, PE-2 and IAE. They were particularly interested in the influence of the exposure time on the obtainable diffraction efficiencies. A dramatic difference in the diffraction efficiency of the PE-2 material was found when the material was exposed to a 20 ns pulse (Q-switched) as compared with a 300 µs pulse (free lasing): the longer pulse produced a 100 times higher diffraction efficiency. Using a special scattering technique, the authors could measure the diffraction efficiency of the dynamic latent-image grating as well as the efficiency of the static grating (the developed grating). The former was divided by the latter and the ratios compared. The ratio that was obtained was approximately 10 to 20 for free-lasing pulses, whereas for Q-switched pulses, it could sometimes be as high as 1000. This difference was due to some extent to latent-image fading, which will be discussed later in this chapter. Pantcheva et al. [11] discussed emulsion-manufacturing methods to reduce HIRF for silver halide materials for pulsed holography. In particular, HIRF can be reduced by:

- Creating hole traps by introducing reducing agents
- Increasing electron lifetime by introducing shallow traps for electrons using, for example, metal ions
- Formation of a few stable and active sensitivity-specks on the microcrystal's surface by chemical sensitisation.

The different methods listed above were tested using emulsions with a grain size of approximately 30 nm. The best reducing agent seems to have been ascorbic acid, which means that a pretreatment of the holographic emulsion in an ascorbic acid solution before exposure can improve pulsed hologram recordings. Introducing metal ions into the emulsion at the preparatory stage increases the lifetime of the latent-image speck. The best results were obtained using lead ions, but cadmium ions also gave fairly good results. Gold sensitisation using, for example, $HAuCl_4$ also reduces HIRF. This confirms similar results obtained for conventional photographic emulsions [12]. Johnson et al. [13] discussed a slightly different aspect of reciprocity failure. It concerned the decrease in diffraction efficiency of the reconstructed images in multiple-exposure holograms recorded with equal energy per exposure. The authors call this phenomenon *holographic reciprocity law failure*.

4.5 Increasing Sensitivity by Hypersensitisation and Latensification

A photographic material can be treated in different ways before exposure to increase its sensitivity. The technique of increasing the material's sensitivity is referred to as *hypersensitisation*. If the treatment is performed after exposure but before development, it is called *latensification* (latent image intensification). The total increase in the holographic material's sensitivity that can be obtained with the help of these methods depends on the material used, the manufacturing method, the ripening and finishing, etc.

4.5.1 Hypersensitisation

Here, we will only discuss triethanolamine (TEA) and water treatments. TEA is the technique recommended by the UK company Colour Holographic Ltd. for their colour emulsion before recording.

4.5.1.1 Water Solution of TEA

TEA [$(HOCH_2CH_2)_3N$] has been used extensively in holography to increase the sensitivity of recording materials [14–17]. The treatment of the material is performed in a bath with a TEA concentration of 0.7% to 2%, which provides an increased sensitivity factor of about 2. Higher concentrations (up to 10%) are recommended for materials intended for pulsed laser exposures according to Russian investigations. It is also recommended to use a bath at the temperature of 10°C to 15°C and not to dry the material in hot air. This method produces quite stable results, but it is advisable to expose the material soon after the treatment to keep the fog level low. Storing the material at low temperatures (−18°C) ensures better stability than storing it at room temperature. Kirillov [16] shows that the grain size in the emulsion is slightly reduced during TEA treatment, resulting in a holographic image of a higher quality. It should be mentioned that the TEA solution is also a swelling agent, which means that a TEA-treated emulsion of a recorded colour hologram may shrink after processing, which creates an erroneous colour replay.

4.5.1.2 Water Treatment

To avoid shrinkage caused by TEA hypersensitisation, it is possible to increase the sensitivity by simply treating the emulsion in water. Soaking the material in distilled water to which a few drops of a wetting agent have been added will remove excessive bromide and increase the concentration of silver ions, which in turn increases the material's sensitivity. The bath temperature should be 10°C to 12°C, and drying should take place in a low ambient temperature (13°C–15°C). The durability achieved by this method is low and the material must therefore be exposed directly after treatment. If this is not done, the fog level will increase.

4.5.2 Latensification

Latensification leads to the acceleration of the development process, giving an apparent speed increase at short development times. The methods used for latensification are very similar to the ones used for hypersensitisation. Primarily, only dry methods are of interest. Latensification using low-intensity light offers a possibility of true speed increase by actually using the LIRF mechanism. Postexposure can be made at a suitable wavelength depending on the spectral sensitivity of the material used. A very low light intensity must be used for a long time. The latensification exposure takes between 15 min and 2 h for normal photographic materials. The Geola organisation sometimes use light latensification for the Russian ultrafine emulsions. After the holographic image recording has been performed, an additional exposure of the material is performed using incandescent light. The incoherent light wavelength band used for latensification should be located in a part of the spectrum in which the material has low sensitivity. For example, an ordinary, safe-light lamp with a suitable dark colour filter can be used. The material is exposed at a distance from the lamp of one to two metres. The power density of between 2.0 and 5.0 µW/cm² at the film level is recommended depending on the emulsion's sensitivity. The exposure time can be varied between 10 min and 1 hour. At optimal conditions, an increased sensitivity of a factor of about 2.5 can be obtained.

4.5.3 Internal Latensification

The most important aspect of processing pulsed holograms is the process of internal latensification, which takes place when a certain type of developing agent is used to develop the recorded hologram. Electron injection methods using developing agents for latensification of the internal image have been reported by James [18]. Some developing agents, such as phenidone (1-phenyl-3-pyrazolidone) for example, can latensify the internal latent image. It has been suggested that latensification depends here on the initial formation of isolated silver atoms, which subsequently lose electrons to the conduction band of the crystal. Conduction electrons formed in this way act to build up latent subimage centres in the same way as the photoelectrons formed by the exposure act. If such an action occurs between the developer and the latent subcentres or the very small latent-image centres, as suggested, this action should also lead to latensification of the internal image. At any rate, this effect can be obtained by adding 1 to 2 g/L of phenidone to a metol-hydroquinone developer. The reason why developers containing phenidone work so well with holograms exposed with Q-switched pulsed lasers is that internal image latensification takes place in combination with the superadditive effect with another developing agent. Figure 4.6 illustrates the practical action of the Kodak D-19 developer with and without phenidone for a hologram recorded on the former Agfa 10E75 material exposed using a 13 ns ruby pulse at 694 nm [15]. The figure shows that sensitivity increases dramatically when using phenidone in the developer. Latensification taking place in the developer is, however, not directly recognised as a separate method for hologram treatment. Binfield et al. [19] found that the type of developer used also has an effect on holographic reciprocity when recording holograms exposed with CW lasers.

A special developer for pulsed holograms was formulated based on the earlier tests using phenidone. The developer SM-6 (Salim's mistake) was found more or less by a mistake. Salim, a student of one of the authors (HB), was asked to add 0.6 g of phenidone to a certain developer but instead added 6 g, which is a rather high concentration as compared with photographic developers based on phenidone. Nonetheless, after trying many other concentrations, 6 g turned out to be the optimal concentration for pulsed holograms when using fine-grain materials. The developer SM-6 is now widely used for printed digital colour holograms recorded with RGB pulsed lasers (Table 4.2).

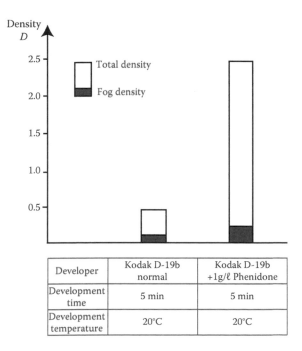

Developer	Kodak D-19b normal	Kodak D-19b +1g/ℓ Phenidone
Development time	5 min	5 min
Development temperature	20°C	20°C

FIGURE 4.6 Density obtained on Agfa 10E75 material for a fixed short exposure (13 ns pulse at $\lambda = 694$ nm) using the Kodak D-19b developer with and without phenidone.

TABLE 4.2

SM-6

Phenidone	6 g
Ascorbic acid	18 g
Sodium hydroxide	12 g
Sodium phosphate (dibasic)	28.4 g
Distilled water	1 L

Note: Developing time: 2–3 min at 20°C

4.6 Substrates for Holographic Emulsions

The material on which the emulsion is coated has a strong bearing on the final quality of the hologram. The best choice is often a glass plate as it is mechanically stable and optically inactive. Also, the light-scattering noise level in clear glass is very low. In many applications of holography, glass is actually the only possible support material. High-resolution imaging, hologram interferometry, HOEs and spatial filters are a few examples where a very stable emulsion support is important. In display holography, it is also often convenient to use glass plates, mainly because of the need for stability when using CW lasers. Producing master plates for hologram copying is another example suited to glass plates. Yet another example is the use of glass in the recording of expensive art holograms, where it is important to protect the emulsion well (if sealed with another glass plate after processing) against detrimental environmental effects (humidity, air pollution, etc.).

The use of film substrates has been growing steadily in recent years, especially in display holography for large-format colour holograms. In many cases, the use of film has many advantages as compared with that of glass (breakage, weight, cost, size, etc.). For example, for many display holography applications, film substrates are often sufficient and more economical than glass. Hologram copying in larger quantities is done mainly on film (sometimes the copies are laminated to a stable substrate after processing). The increased use of pulsed lasers has made hologram recording simpler when using film substrates. In Figure 4.7, the difference between holographic emulsions coated onto glass or film substrates is illustrated.

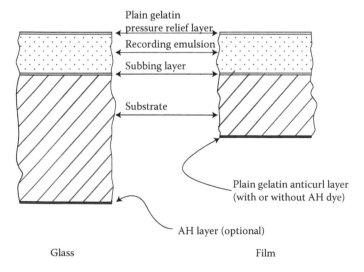

FIGURE 4.7 Holographic silver halide emulsion coated onto glass and film substrates, including additional layers used in holographic materials.

4.6.1 Glass Plates

Holographic glass plates are commonly made of soda-lime glass of high quality (free from graininess and molecular orientation) produced with the help of the flat-drawn sheet or the float process method. The refractive index (refractive dispersion) of glass varies depending on the light wavelength and is approximately 1.516 for $\lambda \sim 600$ nm. Good optical quality and high mechanical and thermal stability of glass are the main advantages when glass is used as a substrate for holograms. Young's modulus for glass is 70×10^9 Nm^{-2} and the thermal coefficient of expansion is only 8.1×10^{-6} °C^{-1}.

Glass thickness varies depending on the plate format and is typically between 1 and 6 mm. The emulsion coated onto untreated glass plates tends to peel off when dry, or frill off when it is wet. Therefore, a well-cleaned glass plate is often precoated with an extremely thin substratum of gelatin hardened with chrome alum—or sometimes with a layer of chrome alum solution alone—making the underlayer of the subsequently applied emulsion very hard. This process is referred to as *subbing*.

4.6.2 Film Substrates

Film substrates are of two main types: polyester (polyethylenterephthalate, PET) and a cellulose ester, commonly triacetate (cellulose triacetate) or acetate-butyrate. In addition to the light-sensitive emulsion coating (and the necessary subcoating), curl control and antihalation coating are often added here. The film may curl badly due to the variations in gelatin concentration caused by humidity. Therefore, a coating of pure gelatin is often applied to the back of the substrate to counteract the curl of the emulsion. If an absorbing dye is added to such gelatin coating it can serve as an antihalation layer at the same time. It should be mentioned that an antihalation cannot be used when a material is used for recording (Denisyuk) reflection holograms.

A film can also receive coating for static protection. The coating can be in the form of a layer containing matte particles that prevents close surface-to-surface contact and static electricity generation upon separation. The matting layer (coated on both sides of the substrate) also prevents individual film sheets from sticking together due to humidity variations during storage. In holographic materials these coatings are sources of noise explaining why the quality of a film hologram (especially on a polyester base) is not always as good as a hologram recorded on glass. As a matter of fact, a thin layer of pure gelatin is often coated over the light-sensitive emulsion (super coating). The reason for this is that if this layer is not applied, the emulsion grains affected by, for example, pressure marks could produce image defects during development.

Various important aspects must be considered when choosing a film base material for a given holographic application. The mechanical behaviour of the base material, with or without the emulsion, is strongly viscoelastic. Polyester is mechanically more stable (Young's modulus, 4.5×10^9 Nm^{-2}) than triacetate (Young's modulus, 3.8×10^9 Nm^{-2}) and it is also less sensitive to humidity. Because of the higher tensile strength, the polyester film can be made thinner than the triacetate film. On the other hand, this is birefringent and can cause many problems when recording reflection holograms (where the reference beam has to pass through the substrate). Polyester also has a higher inherent scattering level. During the manufacture of polyester, the polymer is biaxially oriented when it is drawn and tendered so that it has different refractive indices for each of the three orthogonal directions (α, β, γ). Polyester shows larger wavelength-dependent variations in refractive dispersion than triacetate does. For polarised light entering the polyester material at normal incidence, the refractive index for light at $\lambda \sim 600$ nm is $n_\gamma = 1.66$ and $n_\beta = 1.65$, respectively, depending on whether the electric field vector is oriented parallel or perpendicularly to the major axis. For polarised light propagating in the plane of the support, the effective refractive index is typically $n_\alpha = 1.5$. The refractive index for triacetate is approximately 1.48 at $\lambda \sim 600$ nm. This material is optically inactive and has a low inherent scattering level.

In general, polyester is recommended for transmission holograms in cases when a mechanically stable base is important, whereas triacetate is more suitable for reflection holograms, where birefringence causes severe problems if polyester is used. The inherent scattering levels in polyester may cause problems when recording colour holograms. The thickness of commercial film substrates for holography varies between 64 and 200 μm. Thinner film substrates apply to polyester materials only.

4.7 Commercial Recording Materials for Colour Holography

4.7.1 Manufacturing Companies

The market offers a very limited choice of silver halide recording materials for colour holography. In the past, Agfa-Gevaert, Ilford and Kodak manufactured materials for monochrome hologram recording. None of these companies have holographic materials on the market now. Recently, Harman Technologies (formerly Ilford) started limited production of a new holographic emulsion but with a silver halide grain size not suitable for colour holography. Kodak materials were used for the very first laser-produced holograms in the United States as well as for some early colour holography tests. The Kodak 649-F spectroscopic plate, often used in the early days, was actually panchromatic but had a rather low resolving power (only ~2000 lines/mm). In the following sections, materials from different manufacturers of colour holographic materials are presented. In general, the sensitivity of an emulsion depends on many factors, for example, the laser wavelength, the exposure time (reciprocity failure), the development (developer type, processing time, temperature, agitation, etc.) and storage conditions. Holographic sensitivity often varies from batch to batch, which is rare with conventional photographic materials. It is consequently recommended to make exposure and processing tests each time an important holographic recording is to be performed. It needs to be emphasised that, while it is nice to have high sensitivity for recording materials for colour holograms, it is not acceptable if it comes with an increased grain size. Unfortunately, it is much better to accept a longer exposure time (or to use a higher energy laser) and to obtain a scatter-free recording.

In Russia today, most of the manufacturing of holographic emulsions suitable for colour holography takes place in two companies, the Slavich Joint Stock Company [20] and Sfera-S AO [21]. Both are located in Pereslavl, a few hours' drive from Moscow. Slavich manufactures different types of silver halide materials for holographic purposes. The two main differences between these and previous western materials are the grain size and the silver content in the emulsion. The Russian emulsions have grain sizes as small as 10 nm, and the silver content is usually one-half (~0.25 g/cm^3) of the normal silver content present in western materials. Materials such as the old Soviet PE-2 and LOI-2 also possessed these characteristics. The LOI-2 materials were developed by Protas and are now manufactured under the name of PFG-02. The PE materials were developed by Kirillov and are of the highest quality—their grain sizes for the best materials do not exceed 10 nm. Panchromatic materials for colour holograms are now manufactured by Slavich under the name of PFG-03C.

Sfera-S was formed by Yuri Sazonov in 2004. Sazonov was earlier responsible for holographic plate manufacturing as director of the Micron plant at Slavich. In his new company, Sfera-S, which is located in a former Slavich building, he is now responsible for manufacturing a high-quality emulsion for colour holography which is available coated both onto glass plates as well as onto film. The main customer is the Geola organisation in Vilnius, Lithuania [22], where these films are used for the production of digital colour holograms (more about digital printing in Chapter 7). The Sfera-S emulsion is currently the highest-quality commercial panchromatic colour holography material available. It is sensitive to the short laser-pulse recording primarily utilised in the pulsed RGB digital holographic printers used by Geola. Both the Slavich and Sfera-S materials are available from Geola or from their international network of distributors.

Colour Holographic [23] is the primary manufacturer of ultrafine-grain holographic emulsions in the United Kingdom. Their emulsion is based on the material that Richard Birenheide manufactured in Germany. This type of emulsion (BB emulsion) was launched in 1996. Mike Medora of Colour Holographic acquired the rights to the BB products in 2001. Initially, these emulsions were monochromatic and later a panchromatic emulsion, the BBVPan, was introduced. Ulibarrena et al. [24,25] published reports concerning this new panchromatic ultrafine grain emulsion, which has a mean grain size of 20 nm. The shrinkage or swelling of the emulsion after the plate is processed is one concern because the emulsion requires TEA pretreatment to increase its sensitivity. In colour reflection holography, a change in emulsion thickness is directly related to the wavelength of reconstruction and so affects the final replay spectrum and the colour reproduction of the image. Ulibarrena is now working at Colour

TABLE 4.3

Commercial Silver Halide Holographic Recording Materials for Colour Holograms

Material	Spectral Sensitivity (nm)	Resolving Power (line pairs/mm)	Grain Size (nm)	Substrate
Colour Holographic				
RGB, BB-PAN	440–650	4,000	20–25	glass
Slavich				
Pan, PFG-03C	450–700	5,000	10	glass/film
Sfera-S Ltd.				
Pan, PFG-03CN	435–665	6,000	9	glass/film
Ultimate				
08-COLOR	460–650	10,000	8	glass/film

Holographics and is responsible for emulsion manufacturing, including the panchromatic emulsion, which is now called RGB BBPAN. The company's production facility is based in Maldon, Essex, UK.

The other European manufacturer of colour emulsion is Ultimate Holography in France. This emulsion is based on the work by Yves Gentet [26,27]. The company produces a panchromatic colour emulsion with an 8 nm grain size. However, the production capacity of such plates is limited. Ultrafine-grain monochromatic red- and green-sensitive emulsions are also manufactured. Table 4.3 lists current commercial silver halide materials suitable for colour holograms.

At the time of writing, both Colour Holographics and Ultimate could not guarantee the commercial supply of panchromatic silver halide holographic plates to all customers. Both companies cited quality control difficulties for the limited volumes involved.

4.8 SilverCross Emulsion Research Project

A recent two-year European research project (EC FP6 CRAFT project 005901) SilverCross [28] produced an ultrafine-grain panchromatic (isochromatic) emulsion (grain size of ~10 nm) intended for full-colour reflection holograms and HOEs. Holograms recorded with the SilverCross emulsion exhibit very little blue light scattering; extremely realistic looking bright three-dimensional colour holographic images have been recorded with the emulsion. The project was successful in demonstrating the feasibility of the materials' technology and the intention was to start manufacturing this type of emulsion coated on glass plates and eventually also on film if there was a market for such colour holograms. However, substantial funding was required for setting up a factory for such plate/film manufacturing and to date this has not been accomplished. Because there are currently so few colour hologram-recording facilities around the world, it is of course difficult to justify an investment in large-scale manufacturing of ultrafine-grain emulsions.

To understand the principles of making an ultrafine-grain silver halide emulsion suitable for colour holography in a small-scale operation, a short summary of the SilverCross project is provided here. The research project was carried out by the following European partners:

- The Centre for Modern Optics, NEWI, (now Glyndŵr University), UK
- THIS Ltd., UK
- CLOSPI-BAS, Bulgaria
- Université de Liège, Belgium
- Cristo Stojanoff, Germany
- Vivid Components Ltd., UK
- Geola UAB, Lithuania

TABLE 4.4

The SilverCross Emulsion

Making Step	Chemical	Mixing Method
Basic chemicals	**A:** Gelatin solution	440 mL of 0.57% at 35°C
	B: AgNO$_3$ solution	30 mL of 0.29 M at 19°C
	B1: Gelatin solution	0
	C: KBr and KI solution	15 mL of 0.59 M KBr$^+$ 15 mL of 0.029 M KI at 19°C
	C1: Gelatin solution	0
Precipitation		Jet B and C into A with stirring in <1 min
Preparation for freezing		Pour the precipitant into cells for freezing
Freezing		12–16 h at −20°C
Thawing		6–7 h at 20°C–24°C
Second freeze for storage		>12 h at −20°C
Coating preparation		Melt to 38°C
Dye addition		1.5 mL of 0.0026 M pinacyanol chloride, 1.5 mL of 0.0023 M quinaldine red
Coating		Apply 630 mL of emulsion/m^2
Drying		>12 h at 20°C–24°C
Packing		Place plates in light tight storage boxes and store between 0°C and 24°C
Sensitisation		Hypersensitise immediately before use, 1 min in 0.025 mL ascorbic acid sodium salt at 18°C ±1°C

Note: Recommended dyes: green, 2-(4-dimethylaminostyryl)-1-ethlquinolinium iodide (quinaldine red); red, 1,1′-diethyl-2′,2′-carbocyanine chloride (pinacyanol chloride).

The task was to devise a new nanoparticle, panchromatic, silver halide/gelatin emulsion suitable for making full-colour holographic recordings using primarily CW laser exposure. It was possible to make an emulsion with particles close to the target size of 5 to 10 nm. The SilverCross emulsions all have grain sizes of between 8 and 15 nm. The hypersensitised emulsion is very close to the target sensitivity of less than 2 mJ/cm^2. The emulsion (frozen) and the coated plates (refrigerated) have good conservation characteristics as well as coping with easy storage conditions. Thus far, only plates of dimensions up to 30 cm × 40 cm have been manufactured.

The SilverCross emulsion is an ultrafine-grain emulsion utilising the important technique of freeze-drying by Kirillov et al. [29]. One of the many difficulties to solve in producing an ultrafine-grain emulsion is crystal grain growth restriction, which Kirillov was able to achieve for recording reflection holograms. In essence, Kirillov's technique is the removal of the precipitation by-products by freezing and thawing the emulsion. The technique Kirillov used involved a number of freeze/thaw cycles plus showering of the emulsion with very cold water as well as additions of gelatin and other chemicals.

There are three main stages that take place during the fabrication of the SilverCross emulsion:

- The precipitation stage
- The washing stage
- The coating stage

Table 4.4 summarises the recommended manufacturing process.

4.8.1 Precipitation Stage

Four main ingredients are used for the holographic emulsion: silver nitrate (AgNO$_3$), potassium bromide (KBr), potassium iodide (KI) and photographic gelatin. The mixing of the chemicals is achieved by the method of "double jet precipitation" (Figure 4.8).

FIGURE 4.8 Production of the SilverCross silver halide emulsion in the laboratory. Double-jet mixing of the silver nitrate and potassium bromide solutions with the gelatin solution in the beaker. (a) Emulsion mixing equipment. (b) Simultaneous addition of the two solutions. (c) Solutions mixed at different levels in the beaker.

4.8.2 Washing Stage

The mixed silver halide gelatin solution is then placed in a container, where it is kept at a temperature of −25°C for 30 min. The next step involves cutting the emulsion into cubes and detaching it from the container. The container with the frozen, cut emulsion is then kept at the low temperature of −25°C for 24 h, after which it is placed in deionised water at a temperature of 8°C for washing. The washing stage involves three cycles of 5 min each with agitation (Figures 4.9 and 4.10).

The emulsion is turned into liquid form by placing it in warm water and by continuous stirring. A small amount of methanol is added to assist the melting. Once the emulsion is fully dissolved, spectral sensitisation and hardening takes place. The dyes and the hardener are added to the emulsion, one by one, with continuous stirring (400 rpm).

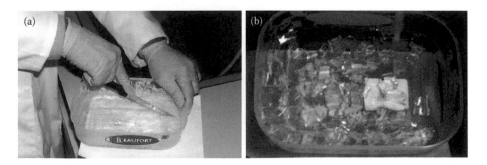

FIGURE 4.9 Frozen emulsion is cut into small cubes. (a) Cutting the frozen emulsion. (b) Frozen emulsion cubes.

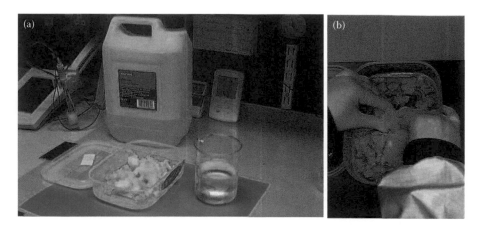

FIGURE 4.10   Frozen emulsion treated in 8°C deionised water. (a) Frozen emulsion and beaker with water. (b) Breaking up the frozen emulsion cubes.

4.8.3 Coating Stage

For making 10 cm × 12 cm test plates, manual coating can be performed. The coating is achieved using glass syringes. Eight millilitres of the emulsion is required for each test plate (Figure 4.11–4.13). The larger SilverCross plates were coated by Stojanoff at Holotech in Germany using a special high-quality glass plate coating apparatus.

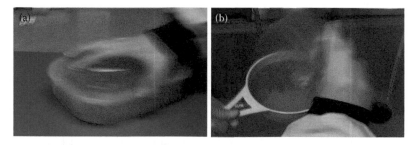

FIGURE 4.11   (a) Washing of the emulsion. (b) The by-products are removed using a sieve.

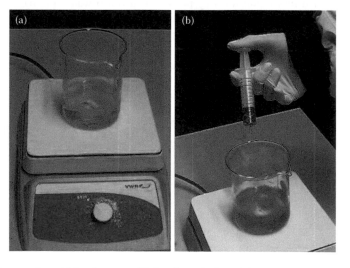

FIGURE 4.12   Hardener and sensitising dyes are added to the melted emulsion. (a) Beaker with the emulsion. (b) Dyes are added using a syringe.

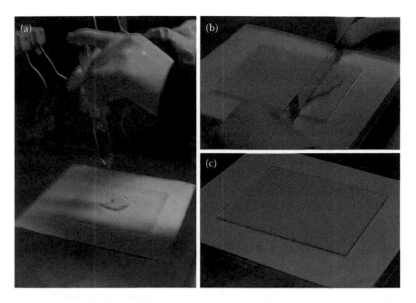

FIGURE 4.13 Manual coating of test plates. (a) Pouring emulsion on the glassplate. (b) Spreading the emulsion. (c) Coated plate.

4.9 Additional Silver Halide Materials for Holography

There are several other companies manufacturing silver halide materials for holography. These companies [30–32] have materials for applications other than colour holography—for example, products for recording transmission holograms and reflection holograms of the monochrome type. Microchrome [32] in California makes silver halide plates for the lithographic industry (similar to Agfa's former Millimask plates). These plates are used to image photomasks and patterns for microelectronics and holography use. The AGHD High Definition Plate is a blue–green sensitive emulsion coated on precision flat soda-lime glass. With an emulsion thickness of approximately 1.5 to 2 μm, it will resolve 2000 line pairs/mm. The other silver halide emulsion is the K1A Plate, which is on standard soda-lime glass that is extremely flat. The K180B high-resolution plate is an extremely high-contrast silver halide high-resolution plate that is used to image photomasks, patterns for microelectronics, holography and many other uses. This emulsion is coated on soda-lime glass that is extremely flat. It provides a more delineated "hard edge" and minimal background fog with good clear area density. These plates are primarily used to record transmission holograms from holographic companies making masters for embossed holograms. A high-resolution (2000 line pairs/mm) blue/green-sensitive film is also manufactured by the company.

Harman Technologies [31] manufactures the HOLO FX fine-grain holographic plates, which are coated with a fine grain holographic emulsion and are available in both red- and green-sensitive versions. With crystal sizes typically 30 to 40 nm, these plates are mainly intended for monochrome red and green holograms. The HOLO FX red-sensitive plates are sensitive from 600 to 694 nm, with a peak sensitivity at 660 nm. The HOLO FX green-sensitive plates are sensitive from 488 to 560 nm with a peak sensitivity at 532 nm.

ORWO FilmoTec GmbH [30] is a German company manufacturing ORWO holographic materials. The ORWO Holographic Film HF 53 is a silver halide recording material with highest resolution and low speed for holographic applications. It is green sensitised. The other green sensitive material is the HF 55 film, which is a material with lower resolution but higher speed for holographic applications. The holographic film HF 65 is a red-sensitised material. The ORWO film substrate is a clear triacetate of a thickness of 135 or 190 μm. The emulsion thickness is 6 μm.

Although the materials from these additional manufacturers are not suitable for full-colour reflection holograms, they can be used very successfully for either multicolour Benton-type transmission holograms or monochromatic laser transmission holograms.

4.10 DCG Materials

DCG is an excellent recording material for volume phase holograms and HOEs [33]. This grainless material has its highest sensitivity in the UV region but also extends into the blue and green parts of the spectrum. Recently, a panchromatic DCG material having useful sensitivity in the red, green and blue regions was reported. DCG is, however, still most often exposed with blue laser wavelengths. Depending on the processing parameters, diffraction efficiency and bandwidth can be controlled. It is easy to obtain high diffraction efficiency combined with a large signal-to-noise ratio. During the exposure of a DCG emulsion to UV or blue light, the hexavalent chromium ion (Cr^{6+}) is photoinduced to a trivalent chromium ion (Cr^{3+}), which causes cross-linking between neighbouring gelatin molecules. The areas exposed to light are hardened and become less soluble than the unexposed areas. Developing consists of a water wash, which removes the residual or unreacted chemical compounds. Dehydration of the swollen gelatin follows after the material has been immersed in isopropanol, which causes rapid shrinkage resulting in voids and cracks in the emulsion, thus creating a large refractive index modulation. The underlying mechanism is not completely understood because high modulation can also be caused by the binding of isopropanol molecules to chromium atoms at the cross-linked sites. Typical DCG material has a rather low sensitivity in the range of 100 mJ/cm^2.

There are very few commercial DCG recording materials on the market. Therefore, laboratories working on recording holograms and HOEs in DCG materials must usually prepare and coat their own emulsions. HOE manufacturers almost always prefer to produce their own emulsions. It does, however, take some time to learn how to make high-quality DCG plates.

4.10.1 Preparing Gelatin Plates

Often, it is convenient to use existing silver halide plates or film that can be easily fixed to remove the silver halide crystals. The diffraction efficiency of the final DCG hologram will depend on the initial hardness of the AgX plate used. The hardness can be decreased by soaking the plate in warm water, although too small a final hardness will lead to loss of the gelatin layer. The following steps should be employed (under standard lighting conditions):

- Soak in Kodak Rapid Fixer without hardener for 10 min at 20°C
- Soak in warm water at 40°C for 5 min
- Wash in running water at 20°C
- Soak in Kodak Rapid Fixer with 3.25% hardener for 10 min at 20°C
- Wash in running water at 20°C
- Dip in a bath with distilled water and photoflow at 20°C
- Drying

4.10.2 Sensitising the DCG Emulsion

The gelatin layer can be sensitised under dim red light conditions using the following steps. Note that some authors recommend omitting the drying step listed previously and proceeding directly to sensitisation:

- Soak the plate in a solution of ammonium dichromate (usually 5 g of ammonium dichromate in 100 mL of distilled water with some drops of wetting agent added) for 5 min
- Dry the plates on a hot plate or in an oven at 70°C
- Let the plates cool to room temperature for 1 h

Although the use of an AgX plate is the simplest technique, it is also possible to coat a 10-μm-thick emulsion directly onto an 8 in. × 10 in. glass plate using 1 g of ammonium dichromate mixed with 3 g of photographic grade gelatin dissolved in 25 mL of distilled water. The emulsion must be spin-coated onto the glass substrate or applied by the doctor blade-coating technique. Note that the sensitising solution and the sensitised emulsion should only be handled in red safelight illumination. The concentration of ammonium dichromate controls the exposure time.

4.10.3 Exposure

The required CW exposure can range from approximately 50 to 1000 mJ/cm^2 at 514.6 nm depending on the sensitisation (ammonium concentration and drying method) and the gelatin hardening.

4.10.4 Pulsed Laser Exposure of DCG

There are very few publications about DCG holograms made using pulsed lasers. The possibility of using a pulsed laser has been reported by Krylov et al. [34] and Blyth et al. [35]. However, this technique is not really recommended. Blyth succeeded in recording a weak holographic image using a 6 ns pulse emitted from a 532 nm Nd:YAG laser, requiring 12 mJ/cm^2. To obtain an acceptable image, 10 such pulses emitted over 1 s were needed. By turning off the Q-switch (using free-lasing operation), better holograms were produced. However, it would seem logical that to get sufficient cross-linking, a train of nanosecond pulses would be needed over a few microseconds. Nevertheless, this technique should not be completely discounted for applications in digital holography, where hogels of only fractions of a square millimetre are written using pulsed lasers. We shall take up this discussion again in Chapter 7, when we talk about the basic constraints placed on the pulse length in a digital pulsed-laser holographic printer. It is also worth mentioning that the SHSG processing method for silver halide holograms, which will be discussed in detail in Chapter 5, provides a method for recording pulsed laser holograms in a panchromatic silver halide material and then converting this material into a material with the same properties as DCG.

4.10.5 DCG Processing

Table 4.5 lists the recommended DCG processing. Depending on the required hologram/HOE characteristics, one has to carefully control the propanol/water mixtures as well as the temperature of the different baths. Cold baths produce better uniformity and lower noise. Warm baths can yield high index modulation but often with increased noise. The bandwidth can be controlled by the processing temperature and the ratio of propanol/water mixtures. The drying of the plates is critical. The relative humidity in the laboratory has to be low (<40%). Moisture will degrade DCG holograms that are hydroscopic. It is therefore necessary to cover the final hologram emulsion with a glass plate, which is cemented using an optical adhesive. Before sealing the plates, it is recommended to bake the plates for 2 h in a vacuum oven at 100°C temperature. To be able to record colour holograms in a DCG emulsion,

TABLE 4.5

DCG Processing

1. Wash in a 0.5% ammonium dichromate solution—5 min at 20°C
2. Wash in a Rapid Fixer solution—5 min at 20°C
3. Wash in running water
4. Wash under agitation in a 50:50 solution of isopropyl alcohol and water—3 min at 20°C
5. Wash under agitation in a 100% isopropyl alcohol bath—3 min at 20°C
6. Optional: dip plate in boiling 100% isopropyl alcohol bath
7. Drying

the spectral sensitivity of the material that is usually only sensitive in the UV and blue/green part of the spectrum must be modified by a dye. To obtain red sensitivity, it is common to use a methylene blue (MB) dye [36–42].

4.10.6 Panchromatic DCG Emulsions

The DCG material can be used for recording colour reflection holograms. It is possible to record high-quality bright-colour holograms using DCG. Kubota and Ose [43] recorded colour holograms in a glass plate using a fixed Kodak 649F plate, which was red-sensitised using the MB dye. Normally, the DCG emulsion swells during processing, something which should be avoided when recording colour holograms. The swelling causes a colour shift of the reconstructed image towards longer wavelengths for reflection holograms—for example, reconstructing at wavelengths of 520 to 530 nm when the hologram was recorded at the 488 nm wavelength. The swelling must be corrected without permitting the deterioration of the reconstructed image. One method of reducing the thickness is to bake the hologram at temperatures of approximately 150°C. Another possible method is to swell the gelatin layer before exposure with a suitable liquid, such as TEA, glycerin or ethylene glycol. Kubota and Ose used ethylene glycol, which was added into the sensitising solution. In addition, they baked the hologram at 150°C after processing. The thickness was reduced gradually by this method without affecting the diffraction efficiency. The recording wavelengths they used were 458, 514.5 and 633 nm. Their technique to obtain high-quality colour DCG holograms is listed in Table 4.6.

Later, Kubota [44] used a DCG plate to make a sandwich hologram in which the green and blue images were recorded in the DCG plate, and the red image in a silver halide plate (Agfa 8E75 HD). Because the red sensitivity of the DCG is rather low and the author wanted to record a large (8 in. × 10 in.) Denisyuk colour hologram using a 50 mW HeNe laser, he decided to make a sandwich. Employing the sandwich

TABLE 4.6

Production and Processing of MB-Sensitised DCG Plates

A. *Preparation of the Sensitised Gelatin Plate from a Kodak 649F Plate*

1. 15-min fix in Kodak Rapid Fixer with hardener
2. 15-min wash in running water (25°C)
3. 5 min in methyl alcohol
4. 5 min in clean methyl alcohol
5. 10-min soak in sensitising solution [$(NH_4)_2$-Cr_2O_7: 3%; MB: 6×10^{-4} mol/L; ethylene glycol: 3%]
6. Dry for more than 4 h in a desiccator with 28% aqueous ammonia solution and silica gel

B. *Exposure*

C. *Heating*

1. 24 h at room temperature, then 3 h at 60°C

D. *Development*

1. 30-s soak in mixture of one part Kodak fixer, two parts distilled water
2. 15-min wash in running water (25°C)
3. 3-min soak in hot water (47°C)
4. 3-min soak in mixture of 70% isopropyl alcohol and 30% distilled water (47°C)
5. 3-min soak in 100% isopropyl alcohol
6. 3-min dry in hot-air flow

E. *Baking*

1. Bake at 150°C until the colour shift is corrected

Source: From Kubota, T., and T. Ose. Lippmann color holograms recorded in methylene-blue-sensitised dichromated gelatine. *Opt. Lett.* 4, 289–291, 1979. With permission.

technique, one avoids the scattering problem* caused by sending blue laser light through a silver halide emulsion during recording. Since Kuboto's publication, there have been only a few articles on recording colour holograms in panchromatic DCG emulsions. However, these few articles clearly demonstrate that DCG must be regarded as an extremely useful material for full-colour holography.

Zhu et al. [45,46] recorded high-quality bright-colour holograms in a single-layer panchromatic DCG material, by using new types of multicolour photosensitisers and photochemical promoters for gelatin cross-linking. Using a red, green and blue laser, full-colour reflection holograms were successfully recorded at exposure levels of approximately 30 mJ/cm². An even lower exposure level of 15 mJ/cm² was observed in the case of a monochromatic grating of 80% diffractive efficiency recorded at 514.6 nm. The water-soluble dye rhodamine 6G (R6G) was used to achieve the green–blue photosensitivity. This dye has good compatibility with the red sensitiser MB and other chemical reagents in the photosensitive DCG layer. The laser wavelengths used for the colour hologram recording were

- 633 nm (He–Ne laser) or 647 nm (Kr-Ion laser)
- 514.5 (Ar–Ion laser) or 532 nm (CW Nd:YAG laser)
- 488 nm (Ar–Ion laser) or 442 nm (He–Cd laser).

The chemical reagent, potassium chromate, was used in the photosensitive layer as the cross-linking reagent. Under laser radiation (i.e., red laser for MB, green and blue lasers for R6G), the photochemical reactions of MB and R6G molecules are both photobleaching reactions. The photochemical reactions of the panchromatic gelatin material can be proposed as follows:

$$\text{dye} + hv \rightarrow \text{excited dye}$$
$$\text{excited dye} + e \rightarrow \text{leuco dye}$$
$$\text{leuco dye} + Cr^{6+} \rightarrow \text{dye} + Cr^{3+}$$
$$Cr^{3+} + \text{gelatin} \rightarrow C^{r3}+-\text{gelatin (cross-linking complex)}$$

When the dye (e.g., MB) is irradiated with a specific monochromatic laser line, its molecule absorbs a photon and passes to the excited state, whereupon the dye is reduced to its leuco (colourless) form by absorbing an electron from the surrounding medium. The leuco dye acts as an active reducing agent; it reacts with Cr^{6+} in the potassium chromate, and then Cr^{6+} is reduced to Cr^{3+}, whereas the dye then returns to its unexcited state. Cr^{3+} reacts with adjacent gelatin molecules to form a cross-linking complex and increases the refractive index of the exposed gelatin. Finally, a volume phase hologram based on the modulation of the refractive index is produced after the common DCG postprocessing procedures. To accelerate the photochemical reaction, a chemical reagent, 1,1,3,3-tetramethylguanidine (TMG), proposed by Blyth [47] is used in the photosensitive layer as a photochemical promoter. TMG has four methyl groups, which can donate additional electrons and the resonating structure of guanidine allows easier electron donation than other types of electron donors. The introduction of TMG as a strong electron donor can efficiently improve the photoreduction speed of the chromium ion and the sensitivity of the photosensitive system. Table 4.7 describes the production and postprocessing of the panchromatic DCG material.

Zhu et al. recorded both reflection gratings and full-colour Lippmann holograms in the new DCG material. The depth of the DCG layer was measured to be approximately 18 μm; they cite required exposures as 60 mJ/cm² at 633 nm, 30 mJ/cm² at 514.5 nm and 40 mJ/cm² at 442 nm. The reflection gratings clearly showed that the central playback wavelength was very close to the original recording laser wavelength, demonstrating that this new material can indeed replay the natural colours of an object. A 7 cm × 7 cm colour reflection hologram of a ceramic mask was recorded with the following wavelengths: 633, 514.5 and 488 nm; the exposures were performed sequentially, starting from the red component, followed by

---

* The grains in a standard red or green holographic emulsion will produce strong Rayleigh scattering when illuminated by blue light simply because the wavelength dependence of Rayleigh scattering scales as λ^{-4}.

TABLE 4.7

Preparation and Processing of Panchromatic Gelatin DCG Plates

---

A. Preparation of the Plates

1. Soak 2 g of French inert gelatin in 40 mL of deionised water at 25°C for 12 h
2. Heat the suspension in a thermostatic water bath to approximately 45°C, then maintain the temperature and stir the mixed solution slowly for 15 min
3. Add 2.5 mL of 0.5% potassium chromate solution at 45°C while stirring for 2 min
4. Add 0.6 mL of 25% TMG solution at 45°C while stirring for 2 min
5. Adjust the pH value to 9.18 with TMG or acetic acid solution (45°C, 2 min)
6. Add 0.3 mL of 0.4% MB solution to the suspension while stirring (45°C, 5 min)
7. Add 0.3 mL of 0.2% R6G solution to the suspension while stirring (45°C, 5 min)
8. Pipette out 8 mL of the mixed solution and spread it over an 8×24 cm^2 optical glass, keeping the coated plate horizontal in a dry location and in the dark

B. Postprocessing Procedure

1. Soak in Kodak F-5 Hardening Fixer solution at 25°C for 1 min
2. Wash holographic plate in running water at 25°C for 30 s
3. Swell holographic plate in warm water at 31°C for 1.5 min
4. Dehydrate in 60% isopropyl alcohol bath at 25°C for 1 min
5. Dehydrate in 90% isopropyl alcohol bath at 25°C for 1 min
6. Dehydrate in 100% isopropyl alcohol bath at 25°C for 2 min
7. Dry hologram rapidly with flowing hot air

Source: From Zhu, J. et al. True-color reflection holograms recorded in a single-layer panchromatic dichromated gelatin material. In *Holography, in Diffractive Optics, and Applications II*, edited by Sheng, Y., D. Hsu, C. Yu, and B. Lee. *Proceedings of SPIE* 5636, 245–253, 2005. With permission.

the green and blue components. The white-light replay of the DCG colour hologram demonstrated high colour saturation, low noise and high diffraction efficiency.

Artemjev et al. [48] have reported on how to record colour holograms using the Slavich PFG-04 DCG emulsion. In 1998, Wang et al. [49] explained how to avoid the problem of swelling of the DCG emulsion after processing colour holograms recorded in DCG emulsions. This problem has also been investigated by Kubota [50], who recommended that the reconstruction wavelength can be shifted to shorter wavelengths and controlled freely to a certain extent by using two different kinds of gelatin. No treatment is needed after the hologram is finished. More recently, Jiang et al. [51] have found a method for the wide-range quantitative adjustment of the playback wavelength of colour reflection holograms recorded in DCG. The main feature of this technique is to introduce a water-soluble organic reagent acrylamide into the DCG layer as a preswelling agent. Acrylamide will not react with the other chemical agents in the DCG emulsion. It will completely dissolve during the processing of the exposed DCG emulsion and result in a uniform emulsion shrinkage throughout the thickness of the emulsion. The shrinkage means that the image is reconstructed at shorter wavelengths after the dehydration process in isopropyl alcohol. By changing the concentration of preswelling, the final playback wavelength can be adjusted widely and quantitatively from red to green, and to blue, almost covering all the visible spectral range. A quantitative investigation to derive the relationship between the wavelength shift and the concentration of acrylamide added to the emulsion was presented. Because the problem of emulsion shrinkage after processing can make it difficult to record colour holograms in DCG materials, this new method was used to record high-quality colour holograms. By introducing new types of multicolour photosensitisers and photochemical promoters to conventional photo-cross-linking DCG systems, a single-layer panchromatic DCG material was fabricated in the laboratory. Its holographic recording characteristics, such as spectral response, photosensitivity to three primary colours and the angular and spectral selectivity of recorded volume holograms, were studied in detail. With this material, it was possible to obtain high diffraction efficiency (up to 85%) as well as high photosensitivity (as low as 20 mJ/cm^2) for three primary colour laser wavelengths. Colour reflection holograms with high colour saturation and brightness were recorded at the exposure level of 30 mJ/cm^2 using RGB lasers.

4.10.7 Commercial DCG Materials

At the time of writing, there were no manufacturers of panchromatic DCG materials. Below are listed several companies which supply blue/green-sensitive DCG.

> *Slavich* [20] in Russia is one manufacturer of presensitised dichromated plates for holography. The DCG emulsion is marked PFG-04. Plates up to a size of 30 by 40 cm can be ordered. These plates must be red-sensitised before recording colour holograms.
>
> *ORWO FilmoTec GmbH* [30] in Germany produces the ORWO Holographic Film GF 40DCG emulsion, which is only available on 125-μm triacetate film, 104 cm wide, in 10- and 30-m lengths. Emulsion thickness of 6 or 20 μm. Film sheets measuring 50 cm × 60 cm can also be ordered. Note that the film needs to be sensitised in dichromate solution as well as being red-sensitised before recording colour holograms. In fact, the GF 40DCG emulsion is actually a gelatin-coated film.
>
> *Holotec GmbH* [52] is a German-based company that offers custom-made presensitised DCG emulsions coated on both plates and film. The Holotec emulsion is based on the high-quality DCG materials developed by Stojanoff in Aachen. The company can supply large-format DCG glass plates (square metre size) or film (PET), presensitised—but not red sensitised.

4.11 Photopolymer Materials

Photopolymer materials have become popular for recording phase holograms and HOEs [53–61]. This is particularly true for the mass production of holograms because some photopolymer materials do not require any postexposure processing or, at worse, they only require dry processing techniques. Companies like Bell Laboratories, DuPont, Polaroid and Hughes produced photopolymer materials for recording holograms at a very early stage. A review of photopolymer materials was published by Lessard [62]. A SPIE Milestone publication was also produced by Bjelkhagen [63] covering a selection of articles on holographic recording materials for holography including photopolymers and DCG.

The main advantages of photopolymer materials for recording holograms are:

- Fast dry processing method
- High diffraction efficiency
- Relatively low material cost
- High stability of the recorded hologram

Its disadvantages are a rather low sensitivity and the short shelf life of the prepared material. Since the time Hughes started to make photopolymer materials, there have been a lot of different materials experimented with at various universities and research centres. In particular, there are several types of photopolymer systems that have been developed:

- Acrylamide-based systems
- Acrylate-based systems
- Polymethyl methacrylate systems
- Polyvinyl alcohol and poly-acrylic acid systems
- Polyvinyl carbazole systems

To increase the sensitivity, it is important to decrease the inhibition period of the system. The cause of the inhibition period, which results in a loss of sensitivity in the photopolymers, is the presence of oxygen in the sample. There are different ways of reducing the inhibition period by, for example, pre-exposure and including a second sensitising agent in the polymer material. Special self-developing materials exist which are intended for real-time hologram interferometry. Polaroid and DuPont are the main companies that have developed practical photopolymers for holographic applications. The Polaroid DMP-128 material was developed in the mid-1980s and, for many years, this material provided an excellent recording solution for HOEs and display holograms [64–66]. However, in 1998, Polaroid stopped producing holographic materials.

A photopolymer recording material, such as the DuPont material [67–69] consists of three parts:

- A photopolymerisable monomer
- An initiator system (initiates polymerisation upon exposure to light)
- A polymer (the binder)

To record a hologram in a photopolymer material, one starts with the exposure to the information-carrying interference pattern. This exposure polymerises a part of the monomer. Monomer concentration gradients, formed by variation in the amount of polymerisation due to the variation in exposure, give rise to the diffusion of monomer molecules from regions of high concentration to regions of lower concentration. The material is then exposed to regular light of uniform intensity until the remaining monomer is polymerised. A difference in the refractive index within the material is thereby obtained. The DuPont material requires only a dry processing technique (exposure to UV light and a heat treatment) to obtain a hologram.

The recording of a hologram on polymer is rather simple. The film has to be laminated to a piece of clean glass or attached to a glass plate using an index-matching liquid. Holograms can be recorded manually, but to produce large quantities of holograms, it is preferable to use a special machine. For hologram replication, laser line scanning techniques can provide the highest production rate. The photopolymer material needs an exposure of approximately 10 to 20 mJ/cm^2.

After exposure is finished, the DuPont film has to be exposed to strong white or UV light. DuPont recommends approximately 100 mJ/cm^2 exposure at 350 to 380 nm. After that, the hologram is put in an oven at a temperature of 120°C for 2 h to increase the brightness of the image. The process is suitable for machine processing using, for example, a baking scroll oven.

4.11.1 Recording Photopolymer Holograms with Pulsed Lasers

During the recording of an interference pattern in a photopolymer layer, a relatively slow diffusion process takes place; this makes use of the CW lasers most suitable for recording polymer holograms. However, there is a strong interest in using pulsed lasers to record holograms in photopolymer materials because these lasers confer a complete immunity to environmental vibration and stability problems. The question is how this can be accomplished. Weitzel et al. [70] and Mikhailov et al. [71] reported the recording of reflection holograms in DuPont HRF-800X071-20 photopolymer films. A pulsed laser with a pulse length of 25 ns was used to record holograms of a mirror. It was shown that the expected weak reflectance of the mirror holograms could be significantly increased by preillumination. Although pulsed preillumination enhanced only the reflectance, continuous incoherent preillumination significantly increased both the diffraction efficiency (which reached 80%) and the sensitivity (which increased 100 times), thus approaching the sensitivity of CW recording. The incoherent preillumination was provided by a broadband green-filtered light source that peaked at 500 nm. The results are comparable with those of hologram recordings obtained with CW exposure under the same processing conditions. The main reason for this reported behaviour of holograms recorded on DuPont photopolymer films is related to the kinetics of film components during exposure.

To better understand the behaviour of the photopolymer material under pulsed exposure, the reader is referred to the experimental results on polymerisation kinetics reported by Hoyle et al. [72]. A high level of polymerisation was observed in this article, even under a single high-intensity laser pulse. The rate of polymerisation depended on the type of polymerisable monomer as well as on the type of photoinitiator. In these experiments, the weak polymerisation ability was also significantly improved by accurate preillumination (prepolymerisation).

More details of recording pulsed holograms in DuPont photopolymers were published by Mikhailov et al. [73]. The preillumination technique should also apply to the recording of colour holograms using pulsed RGB lasers. This means that it might well be possible to record digital colour holograms using printers equipped with pulsed RGB lasers.

4.11.2 Panchromatic Photopolymers

Panchromatic photopolymer materials are suitable for recording colour holograms. This material is of interest for both recording one-off colour holograms, but most often, it is a suitable material for producing multiple copies from colour master holograms. DuPont introduced their panchromatic material in 1995 [74,75]. The new holographic photopolymer film was capable of producing high diffraction efficiency in colour volume holograms and HOEs. The properties of this panchromatic film allowed for a greater range of applications than had previously been considered feasible for colour holograms.

Since photopolymer materials became available, there have been several reports on using them for recording colour holograms [76–80]. In the earlier articles, the sandwich technique was used by combining two photopolymer sheets to create the colour hologram. Zhang et al. [76] recorded reflection gratings with red (647 nm), green (514 nm) and blue (488 nm) laser wavelengths using a quasipanchromatic photopolymer material prepared by the authors. The curves of percentage transmission versus wavelength and reconstructed wavelength versus exposure for the gratings were given. Using subtractive filters, colour reflection holograms were recorded employing red, green and blue laser wavelengths with simultaneous exposure.

Hubel and Klug [77] recorded colour holographic stereograms using multiple layers of the DuPont OmniDex photopolymer. Red, green and blue colour separations were reproduced at optimum replay wavelengths by exposing in blue and postswelling using monomer colour-tuning films. A theoretical analysis of the colour-reproduction was provided and the technique was compared with results using other materials. The signal-to-noise ratio, colour rendering and colour gamut area properties were shown to be comparable with those found when using DCG materials.

Kawabata et al. [78] introduced a new type of photopolymer system for recording reflection colour holograms. The photopolymer system using radical and cationic photopolymerisation controlled by the wavelength of light gave enhanced diffraction efficiencies and a balanced recording sensitivity (approximately 20–60 mJ/cm^2) in the blue to red region of the spectrum. In colour hologram recordings, diffraction efficiencies of approximately 60% were obtained when using a photopolymer film composed of different spectral-sensitive photopolymer layers.

In another article, Kawabata et al. [79] recorded colour reflection holograms on a specially prepared photopolymer material. Here, a suitable red–green sensitive dye and a blue–green sensitive dye were selected as the photosensitisers. Colour holograms recorded in a single-layered panchromatic photopolymer material containing these mixed dyes resulted in diffraction efficiencies of only 20% because of the multiple-exposure technique. The decrease in diffraction efficiency for the triple recording, as compared with a single wavelength recording, resulted from a change in the monomer form caused by the diffusion of monomers. To improve the recording properties, a photopolymer film that was composed of two parts was prepared: a red–green sensitive layer and a blue–green sensitive layer. A polyethylene film was positioned between the two polymer layers to prevent them from mixing. The dual recordings in each layer resulted in a colour reflection hologram with enhanced diffraction efficiency in comparison with that of the triple recording. In this case, they achieved a diffraction efficiency of 64% in red, 65% in green and 53% in blue. Here, the exposures of the red and green components were carried out simultaneously followed by the exposure of the blue component. It was found that simultaneous exposure gave an enhanced diffraction efficiency compared with separate exposures.

At the Dublin Institute of Technology, full-colour reflection holograms have been recorded by Meka et al. [80] using a new panchromatic acrylamide-based photopolymer layer. The recording laser wavelengths were 633, 532 and 473 nm. The reflection holograms, recorded using a combined single beam of RGB wavelengths, were spectrally characterised and compared with the recording wavelengths. An object having an additive colour diagram was recorded. The shrinkage effect of this new recording material on reconstructed wavelengths was also discussed.

4.11.3 Commercial Photopolymer Materials

Currently, the main manufacturer of holographic photopolymer materials is E.I. DuPont de Nemours & Co. DuPont manufactures the Omnidex 706 monochromatic film (blue/green sensitive) which is a polymer that only needs dry processing (as described in the previous section). The material has a coated compound polymer layer thickness of 20 µm. The photopolymer film is generally coated in a 12.5 in. width on a 14 in. wide Mylar polyester base which is 50.8 µm thick.

The panchromatic material, however, is not a fully commercial product. DuPont has restricted the distribution of this material due to concerns from its security customers; it is not on the general market—and with the exception of their in-house hologram production and a few selected customers, the material is not commercially available. Zebra Imaging Inc. in the United States records its holograms on the DuPont colour material. In Japan, Dai Nippon also uses the DuPont material. Dai Nippon has a domestic photopolymer product that it uses for its own security applications.

Bayer MaterialScience AG in Leverkusen, Germany [81] has developed a new panchromatic photopolymer material that was recently introduced to the market under the name Bayfol HX [82–84], although sales are restricted. The material has many advantages, such as long lifetime, stability, almost no shrinkage and no postprocessing (thermal or wet). It comes in rolls of up to 1.2 m wide. Besides creating fascinating optical effects, this material can also be used to make ID cards and other documents that are forgery-proof. The current holographic performance of the material is as follows:

- Refractive index modulation (δn), 0.03 in reflection holograms*
- Diffraction efficiency, 98%
- Colour sensitivity, 450 to 650 nm
- Suitable for both reflection and transmission holograms
- Sensitivity for reflection holograms, 100 µW/cm^2 to 50 mW/cm^2
- Photopolymer thickness, 10 to 25 µm
- Substrates PET or PC in roll format
- Environmentally stable (UV, heat and humidity)

Because the Bayer photopolymer film requires no subsequent chemical or thermal treatment, it is exceptionally suitable for cost-effective mass production of volume holograms. The Bayer photopolymer has real potential to become a leading material for light management within a variety of new technologies—for example, in improved three-dimensional digital and analogue holographic displays or for diffusers required in energy-efficient lighting technologies such as light-emitting diodes. However, like DuPont, Bayer has decided, at least for the moment, to restrict sales of its holographic materials to protect the value chain to the security industry.

Polygrama [85] in Brazil is another manufacturer of holographic photopolymer materials. Polygrama DAROL photopolymer is a dry film for holography, which is provided with adequate resolution, high contrast and low scatter. It is designed for use with blue–green lasers (488–535 nm) and red lasers (610–660 nm). It records a reflection hologram that must be thermally developed in a single process to deliver a hologram with high diffraction efficiency. It is totally moisture-resistant and stable. DAROL films are available as 20 cm × 10 cm sheets with a 30- to 40-µm-thick photopolymer onto optically clear polyester with laminated thin PET or HDPE cover. Colour sensitivity: 488 to 532 nm, 635 to 670 nm with a sensitivity of 5 to 30 mJ/cm^2, $\delta n \leq$ 0.100 on thermally developed reflection holograms. Lynx in Brazil is distributing the DAROL film [86].

Finally, it should be mentioned that, for 10 years, the Dai Nippon Printing Co., Ltd. has produced TRUE IMAGE colour reflection holograms and security labels [87,88]. Initially, these images were recorded on DuPont's panchromatic photopolymer material. More recently, Dai Nippon Printing, in conjunction with Nippon Paint Co., Ltd., has developed a new photopolymer material that is now used for manufacturing mass-produced volume reflection holograms. Nippon Paint is responsible for materials development, whereas Dai Nippon Printing is responsible for the development of the production technology.

---

* At the time of writing, the best results obtained by Bayer had increased δn in their experimental materials to 0.06.

REFERENCES

1. H. I. Bjelkhagen, *Silver Halide Recording Materials for Holography and Their Processing*, Springer Series in Optical Sciences, Vol. **66**, Springer-Verlag, Heidelberg, New York (1993).
2. D. Gabor, "A new microscopic principle," *Nature* **161** (No. 4098) 777–778 (1948).
3. C. E. K. Mees, *The Theory of The Photographic Process*, 2nd Edn. Macmillan, New York (1959).
4. C. B. Neblette, *Photography, its Materials and Processes*, 6th Edn. Van Nostrand, New York (1962).
5. M. Sturge (ed.), *Neblette's Handbook of Photography and Reprography*, 7th Edn. Van Nostrand Reinhold, New York (1977).
6. T. H. James (ed.), *The Theory of The Photographic Process*, 4th Edn. Macmillan, New York (1977).
7. R. Bunsen and H. Roscoe, "Photochemische Untersuchungen," *Ann. der Physik Chemie* **117**, 529–562 (1862).
8. N. D. Vorzobova and D.I. Staselko, "Exposure characteristics of high-resolution silver-halide photographic materials for recording three-dimensional holograms by means of a pulsed laser," *Sov. J. Opt. Technol.* **44**, 249–250 (1977).
9. N. D. Vorzobova and D. I. Staselko, "Diffraction efficiency of 3-D holograms recorded with short exposures," *Opt. Spectrosc. (USSR)* **45**, 90–93 (1978).
10. A. A. Benken and D. I. Staselko, "Light scattering in the formation of a latent image by pulsed laser radiation," *Sov. Phys. Tech. Phys.* **27**, 896–898 (1982).
11. M. Pantcheva, T. Petrova, N. Pangelova and A. Katsev, "Chemical sensitization of fine-grain silver halide emulsions for holographic recording," in *Holography '89. International Conference on Holography, Optical Recording and Processing of Information,* Y. N. Denisyuk and T. H. Jeong eds. Proc. SPIE **1183**, 128–130 (1990).
12. A. Hautot, "Reciprocity characteristics of silver bromide and silver chloride emulsions," *Photogr. Sci. Eng.* **4**, 254–256 (1960).
13. K. M. Johnson, L. Hesselink and J. W. Goodman, "Holographic reciprocity law failure," *Appl. Opt.* **23**, 218–227 (1984).
14. K. Biedermann, "Attempts to increase the holographic exposure index of photographic materials," *Appl. Opt.* **10**, 584–595 (1971).
15. H. I. Bjelkhagen, "Holographic recording materials and the possibility to increase their sensitivity," CERN, Geneva, Switzerland, EF-report 84-7 (1984).
16. N. I. Kirillov, [Transl.: *High Resolution Photographic Materials for Holography and Their Processing Methods* [in Russian)] NAUKA, Moscow (1979).
17. Yu. N. Denisyuk and I. R. Protas, "Improved Lippmann photographic plates for recording stationary light waves," *Opt. Spectrosc. (USSR)* **14**, 381–383 (1963).
18. T. H. James, "Electron injection by developing agents - Latensification of internal image," *Photogr. Sci. Eng.* **10**, 344–349 (1966).
19. P. Binfield, R. Galloway and J. Watson, "Reciprocity failure in continuous wave holography," *Appl. Opt.* **32**, 4337–4343 (1993).
20. SLAVICH Joint Stock Company, Russia; www.slavich.com (Sept. 2012).
21. Sfera–S Ltd, Russia; www.geola.com (Sept. 2012).
22. Geola, Lithuania; www.geola.com (Sept. 2012).
23. Colourholographic Ltd, UK; www.colourholographic.com
24. M. Ulibarrena, L. Caretero, R. Madrigal, S. Blaya, and A. Fimia, "Multiple band holographic reflection gratings recorded in new ultra-fine emulsion BBVPan," *Opt. Expr.* **11**, 3385–3392 (2003).
25. M. Ulibarrena, "A new panchromatic silver halide emulsion for recording color holograms," *Holographer. org*, 1–12 (Jan. 2004). www.holographer.org (Sept. 2012).
26. Y. Gentet and P. Gentet, ""Ultimate" emulsion and its applications: a laboratory-made silver halide emulsion of optimised quality for monochromatic pulsed and full color holography," in *HOLOGRAPHY 2000*, T. H. Jeong, and W. K. Sobotka eds., Proc. SPIE **4149**, 56–62 (2000).
27. The Ultimate Holography, France; www.ultimate-holography.com (Sept. 2012).
28. H. I. Bjelkhagen, P. G. Crosby, D. P. M. Green, E. Mirlis and N. J. Phillips, "Fabrication of ultra-fine-grain silver halide recording material for color holography," in *Practical Holography XXII: Materials and Applications*, H. I. Bjelkhagen and R. K. Kostuk eds., Proc. SPIE **6912**, 09-1–14 (2008).

29. N. V. Kirillov, N. V. Vasilieva and V. L. Zielikman, "Preparation of concentrated photographic emulsions by means of their successive freezing and thawing" [in Russian], *Zh. Nauchn. Prikl. Fotogr. Kinematogr.* **15**, 441–443 (1970).

30. ORWO FilmoTec GmbH, Germany; www.filmotec.de (Sept. 2012).

31. Harman Technologies, UK (former ILFORD); www.ilfordphoto.com/holofx/holofx.asp (Sept. 2012).

32. Microchrome Technology Inc. USA; www.microchrometechnology.com (Sept. 2012).

33. L. H. Lin, "Hologram formation in hardened dichromated gelatin films," *Appl. Opt.* **8**, 963–966 (1969).

34. V. N. Krylov, V. N. Sizov, G. A. Sobolev, S. B. Soboleva, D. I. Staselko and M. K. Shevtsov, "Study of nontanned dichromated-gelatin layers for hologram recording by cw and pulsed laser radiation," *Opt. Spectrosc. (USSR)* **69**, 115–116 (1990).

35. J. Blyth, C. R. Lowe and J. F. Pecora, "Improving the remarkable photosensitivity of dichromated gelatin for hologram recording in green laser light," in *Advances in Display Holography*. H. I. Bjelkhagen ed., Proc. *7th International Symposium on Display Holography*, River Valley Press. UK (2006), pp. 78–83.

36. A. Graube, "Holograms recorded with red light in dye sensitized dichromated gelatin," *Opt. Commun,* **8**, 251–253 (1973).

37. M. Akagi, "Spectral sensitization of dichromated gelatin,"*Photogr. Sci. Eng.* **18**, 248–250 (1974).

38. T. Kubota, T. Ose, M. Sasaki and K. Honda, "Hologram formation with red light in methylene blue sensitized dichromated gelatin," *Appl. Opt.* **15**, 556–558 (1976).

39. T. Kubota and T. Ose, "Methods of increasing the sensitivity of methylene blue sensitized dichromated gelatin," *Appl. Opt.* **18**, 2538–2539 (1979).

40. C. Solano, R. A. Lessard and P. C. Roberge, "Red sensitivity of dichromated gelatin films,"*Appl. Opt.* **24**, 1189–1192 (1985).

41. N. Capolla and R. A. Lessard, "Processing of holograms recorded in methylene blue sensitized gelatin," *Appl. Opt.* **27**, 3008–3012 (1988).

42. R. Changkakoti and S. V. Pappu, "Towards optimum diffraction efficency for methylene blue sensitized dichromated gelatin holograms," *Opt. Laser Technol.* **21**, 259–263 (1989).

43. T. Kubota and T. Ose, "Lippmann color holograms recorded in methylene-blue-sensitized dichromated gelatine," *Opt. Lett.* **4,** 289–291 (1979).

44. T. Kubota, "Recording of high quality color holograms," *Appl. Opt.* **25**, 4141–4145 (1986).

45. J. Zhu, Y, Zhang, G. Dong, Y. Guo and L. Guo, "Single-layer panchromatic dichromated gelatin material for Lippmann color holography," *Opt. Commun.* **241**, 17–21 (2004).

46. J. Zhu, J. Li, L. Chen, L.Wan and G. Dong, "True-color reflection holograms recorded in a single-layer panchromatic dichromated gelatin material," in *Diffractive Optics, and Applications II*, Y. Sheng, D. Hsu, C. Yu and B. Lee eds., Proc. SPIE **5636**, 245–253 (2005).

47. J. Blyth, "Methylene blue sensitized dichromated gelatin holograms: a new electron donor for their improved photosensitivity," *Appl. Opt.*, **30**, 1598–1602 (1991).

48. S. V. Artemjev, G.I. Koval, I. E. Obyknovennaja, A. S. Cherkasov and M. K. Shevtsov, "PFG-04 photographic plates based on the nonhardened dichromated gelatin for recording color reflection holograms," in *Three-Dimensional Holography: Science, Culture, Education,* T. H. Jeong and V. B. Markov eds., Proc. SPIE **1238**, 206–210 (1991).

49. J. Wang, R. Zheng and P. Liu, "A novel method to make Lippmann color holograms recorded in DCG," in *Holographic Displays and Optical Elements II,* D. Xu and J. H. Hong eds., Proc. SPIE **3559**, 30–34 (1998).

50. T. Kubota, "Control of the reconstruction wavelength of Lippmann holograms recorded in dichromated gelatin," *Appl. Opt.* **28**, 1845–1849 (1989).

51. M. Jiang, J. Zhu, Y. Hao, Y. Guo, and L. Guo, "Color holographic display based on single-layer and panchromatic dichromatic gelatin materials," in *Holography–Culture, Art, and Information Technology.* Proc. *8th International Symposium on Display Holography*, T. H. Jeong, F. Fan, H. I. Bjelkhagen and D. Wang eds., (2009) Tsinghua Univ. Press, China (2011), pp. 393–402.

52. Holotec GmbH, Germany; www.holotec.de (Sept. 2012).

53. W. S. Colburn and K. A. Haines, "Volume hologram formation in photopolymer materials," *Appl. Opt.* **10**, 1636–1641 (1971).

54. R. L. van Renesse, "Photopolymers in holography," *Opt. Laser Technol.* **4**, 24–27 (1972).

55. R. H. Wopschall and T. R. Pampalone, "Dry photopolymer film for recording holograms," *Appl. Opt.* **11**, 2096–2097 (1972).

56. B. L. Booth, "Photopolymer material for holography," *Appl. Opt.* **11**, 2994–2995 (1972).

57. E. S. Gyulnazarov, V. V. Obukhovskii and T. N. Smirnov, "Theory of holographic recording on a photo-polymerized material," *Opt. and Spectrosc. (USSR)* **67**, 109–111 (1990).

58. C. Carré and D.-J. Lougnot, "Photopolymers for holographic recording: from standard to self-processing materials," *J. Physique III, France* **3**, 1445–1460 (1993).

59. A. Fimia, N. López, F. Mateos, R. Sastre, J. Pineda and F. Amat-Guerri, "New photopolymer used as a holographic recording material," *Appl. Opt.* **32**, 3706–3707 (1993).

60. S. Martin, P. E. Leclère, Y. L. M. Renotte, V. Toal and Y. F. Lion, "Characterization of an acrylamide-based dry photopolymer holographic recording material," *Opt. Eng.* **33**, 3942–3946 (1994).

61. S. Bartkiewicz and A. Miniewicz, "Methylene blue sensitized poly(methyl methacrylate) matrix: a novel holographic material," *Appl. Opt.* **34**, 5175–5178 (1995).

62. R. A. Lessard ed., *Selected papers on Photopolymers, Physics, Chemistry, and Applications.* SPIE Milestone Series, **MS114**, Bellingham, WA, USA (1995).

63. H. I. Bjelkhagen ed., *Selected papers on Holographic Recording Materials.* SPIE Milestone Series, **MS130**, Bellingham, WA, USA (1996).

64. R. T. Ingwall and H. L. Fielding, "Hologram recording with a new photopolymer system," *Opt. Eng.* **24**, 808–811 (1985).

65. W. C. Hay and B. D. Guenther, "Characterization of Polaroid's DMP-128 holographic recording photo-polymer," in *Holographic Optics: Design and Applications*, I. Cindrich ed., Proc. SPIE **883**, 102–105, (1988).

66. R. T. Ingwall and M. Troll, "Mechanism of holograms formation in DMP-128 photopolymer," *Opt. Eng.* **28**, 586–591 (1989).

67. W. K. Smothers, B. M. Monroe, A. M. Weber and D. E. Keys, "Photopolymers for holography," in *Practical Holography IV*, S. A. Benton ed., Proc. SPIE **1212**, 20–29 (1990).

68. D. Tipton, M. Armstrong and S. Stevenson, "Improved process of reflection holography replication and heat processing," in *Practical Holography VII*. S. A. Benton ed., Proc. SPIE **2176**, 172–183, (1994).

69. S. H. Stevenson, M. L. Armstrong, P. J. O'Connor and D. F. Tipton, "Advances in photopolymer films for display holography," in *Fifth International Symposium on Display Holography*, T. H. Jeong ed., Proc. SPIE **2333**, 60–70 (1995).

70. K. T. Weitzel, U. P. Wild, V. N. Mikhailov and V. N. Krylov, "Hologram recording in DuPont photopoly-mer films by use of pulse exposure," *Opt. Lett.* **22**, 1899–1901 (1997)

71. V. N. Mikhailov, K. T. Weitzel, V. N. Krylov and U. P. Wild, "Pulse hologram recording in DuPont's photopolymer films," in *Practical Holography XI and Holographic Materials III*, S. A. Benton and T. J. Trout eds., Proc. SPIE **3011**, 200–202 (1997).

72. C. E. Hoyle, P. E. Sundell, M. Trapp, D. M. Kang, D. Sheng and R. Nagarajan, "Polymerization kinet-ics of mono- and multifunctional monomers initiated by high-intensity laser pulses: dependence of rate on peak-pulse intensity and chemical structure," in *Photopolymer Device Physics, Chemistry, and Applications II*, R. A. Lessard ed., Proc. SPIE **1559**, 202–213 (1991).

73. V. N. Mikhailov, V. N. Krylov, K. T. Vaitsel and U. P. Wild, "Recording of pulsed holograms in photo-polymer materials," *Opt. and Spectrosc. (USSR)* **84**, 589–596 (1998).

74. W. J. Gambogi, W. K. Smothers, K. W. Steijn, S. H. Stevenson and A. M. Weber, "Color holography using DuPont holographic recording film," in *Holographic Materials*, J. Trout ed., Proc. SPIE **2405**, 62–73 (1995).

75. T. J. Trout, W. J. Gambogi and S. H. Stevenson, "Photopolymer materials for color holography," in *Applications of Optical Holography*, T. Honda ed., Proc. SPIE **2577**, 94–105 (1995).

76. J. F. Zhang, C. R. Ma and H. Y. Lang, "Colour reflection holograms with photopolymer plates," in *Three-Dimensional Holography: Science, Culture, Education*, T. H. Jeong and V. B. Markov eds., Proc. SPIE **1238**, 306–310 (1991).

77. P. M. Hubel and M. A. Klug, "Color holography using multiple layers of DuPont photopolymer," in *Practical Holography VI*, S. A. Benton ed., Proc. SPIE **1667**, 215–224 (1992).

78. M. Kawabata, A. Sato, I. Sumiyoshi and T. Kubota, "Novel photopolymer system and its application to color hologram," in *Practical Holography VII: Imaging and Materials*, S. A. Benton ed., Proc. SPIE **1914**, 66–74 (1993).

79. M. Kawabata, A. Sato, I. Sumiyoshi and T. Kubota, "Photopolymer system and its application to a color hologram," *Appl. Opt.* **33**, 2152–2156 (1994).

80. C. Meka, R. Jallapuram, I. Nayedenova, S. Martin and V. Toal, "Acrylamide based photopolymer for multicolor recording," in *Holography-Culture, Art, and Information Technology. Proc. 8ᵗʰ International Symposium on Display Holography*, T.H. Jeong, F. Fan, H.I. Bjelkhagen, D. Wang eds. (2009) Tsinghua Univ. Press, China (2011). pp. 410–416.

81. Bayer MaterialScience AG, Germany; www.bayermaterialscience.com (Sept. 2012).

82. F.-K. Bruder, F. Deuber, T. Fäcke, R. Hagen, D. Hönel, D. Jurbergs, M. Kogure, T. Rölle and M.-S. Weiser, "Full-color self-processing holographic photopolymers with high sensitivity in red – the first class of instant holographic photopolymers," *J. Photopolym. Sci. Technol.* **22** (2), 257–260 (2009).

83. F.-K. Bruder, F. Deuber, T. Facke, R. Hagen, D. Honel, D. Jurbergs, M. Kogure, T. Rolle and M.-S. Weiser, "Reaction-diffusion model applied to high resolution Bayfol HX photopolymer," in *Practical Holography XXIIV: Materials and Applications*, H. I. Bjelkhagen and R. K. Kostuk eds., Proc. SPIE **7919**, 0I-1–15 (2010).

84. H. Berneth, F.-K. Bruder, T. Fäcke, R. Hagen, D. Hönel, D. Jurbergs, T. Rölle and M.-S. Weiser, "Holographic recording aspects of high-resolution Bayfol®HX photopolymer," in *Practical Holography XXV: Materials and Applications*, H. I. Bjelkhagen ed., Proc. SPIE **7957**, 0H-1–15 (2011).

85. Polygrama, Brazil; www.polygrama.com (Sept. 2012).

86. Lynx Comercio Importaco Ltd; www.lynx-us.com (Sept. 2012).

87. M. Watanabe, T. Matsuyama, D. Kodama and T. Hotta, "Mass-produced color graphic arts holograms," in *Practical Holography XIII*, S. A. Benton ed., Proc. SPIE **3637**, 204–212 (1999).

88. D. Kodama, M. Watanabe and K. Ueda, "Mastering process for color graphic arts holograms," in *Practical Holography XV and Holographic Materials VII*, S. A. Benton, S. H. Stevenson and T. J. Trout eds., Proc. SPIE **4296**, 196–205 (2001).

5

Analogue Colour Holography

5.1 Introduction

Colour holography is concerned with accurately capturing not only the three-dimensional (3D) shape information of an object but also the object's colour information. If the hologram produced is to be categorised as "ultra-realistic", the error in recording such shape and colour should be below the limits of human perception. Additionally, the field of view of such holograms should be as large as possible and both the vertical and horizontal parallax should be properly encoded. In this chapter, we shall discuss the problem of making holograms of real objects. In Chapters 7 through 10, we shall enlarge the discussion to computer-generated objects.

The most successful technique used to date to record ultra-realistic colour holographic images of real objects has been Denisyuk's reflection holography (Figure 5.1). Denisyuk's original holograms were of course monochromatic, but the technique can be extended by using three or more primary laser wavelengths, which are then used to create an achromatic laser beam. This "white" beam is used in place of the more usual monochromatic beam, characteristic of a conventional Denisyuk setup, and a Denisyuk colour hologram is recorded using a panchromatic photosensitive plate. The image replayed by a full-colour Denisyuk hologram constitutes a 1:1 scale representation of the object. As we shall see, the realism attainable today with the Denisyuk colour technique can properly be categorised as ultra-realistic because a viewer can really find it difficult to discriminate between the hologram and the real object. The most obvious disadvantage of the Denisyuk colour technique is that the final image appears completely behind the hologram surface. Although it is possible to copy a Denisyuk colour hologram to an H_2 hologram possessing a different plane, the price paid for this is a much smaller viewing angle. We shall return to this problem in Chapter 9 in which a solution is available using full-parallax digital techniques.

In general, Denisyuk colour holograms of real objects can be produced using either continuous wave (CW) or pulsed lasers. For the most part, CW lasers are used for their convenience—notably their larger available energies and the greater ease with which their beams can be spatially filtered and aligned. However, this is not a hard-and-fast rule and indeed some potential applications of colour holography, such as pulsed colour portraiture or the reproduction of interferometrically unstable objects, explicitly require the use of pulsed lasers.

Often, white light-viewable reflection holograms recorded with a single laser wavelength are referred to as *Lippmann holograms*. However, it would be more accurate to attribute the term Lippmann holography to Denisyuk colour holography because Lippmann photographs are, after all, intrinsically polychromatic. The Denisyuk colour holograms described in this chapter may therefore be properly termed Lippmann colour holograms.

In the following section, we shall present a brief history of full-colour analogue holography. What we actually mean by a *colour hologram* is an exact analogy with colour photography—in other words, a recorded holographic image with an accurate rendering of the object's colour. However, it needs to be pointed out that there are some colours that are impossible to record holographically. Holograms can only reproduce laser light that scatters off an object's surface without suffering a change in wavelength. However, in nature, we sometimes see colours that result from fluorescence. This process does not conserve wavelength and, as such, there is no coherence between the illuminating and scattered light. For example, some dyed and plastic objects achieve their bright saturated colours precisely by fluorescence. Using the techniques of photographic image acquisition (Chapter 10) and digital holographic printing (Chapters 7–9), one can nonetheless get around this limitation.

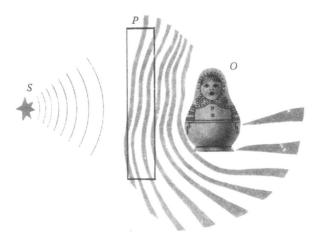

FIGURE 5.1 Denisyuk single-beam reflection hologram recording setup. *S* is the expanded laser beam illuminating the object *O* through the recording plate *P* and also acting as the reference beam. (From Denisyuk, *Fundamentals of Holography*, 1984. With the permission of Mir Publishers, Moscow.)

5.2 Origins in Monochromatic Holography

The field of display holography started in 1964 when Leith and Upatnieks [1] presented the possibility of recording transmission holograms of 3D objects by introducing the off-axis reference beam technique. Their hologram of a model railroad engine was on display at the OSA spring meeting in Washington, D.C. in April 1964 (Figure 1.3). This laser-illuminated hologram displayed a very realistic-looking 3D image; it had a huge impact on the participants at the meeting.

In principle, when the object is illuminated in a dark room with the same laser wavelength as that used for an off-axis transmission recording, it is not possible to see any difference between the recorded holographic image on display and the laser-illuminated object itself. In both cases, however, the viewer experiences speckle patterns covering the image or the object, as illustrated in Figure 5.2, which shows a painted egg and its associated laser transmission hologram. The size of the speckles depends on the observing aperture; the smaller the aperture diameter, the larger the speckles. It should be noted here

FIGURE 5.2 (a) A painted egg is illuminated with a green laser and (b) its holographic image is reconstructed from an off-axis transmission hologram with the same green laser. In both cases, the viewer sees a speckle pattern, which is caused by the viewer's eye aperture (iris) or, in the case of a photograph, the camera aperture. It is not possible to determine which one is the laser-illuminated object and which one is the image.

that speckle is a potential problem for ultra-realistic holographic imaging, particularly in high-light conditions when the aperture of the human eye is small. We shall see in Chapter 11 that one of the main effects that limit the perceivable depth in a hologram is the temporal coherence of the illuminating light source. However, too great a temporal coherence means that we start to see speckle degrade the image. This is the same problem encountered in the design of RGB lasers for projection televisions. However, the more wavelengths used to record a hologram, the less the speckle effect. For three laser wavelengths, it turns out that an illumination source with a bandwidth of approximately 2 nm is fairly ideal—at this bandwidth, both speckle noise and chromatic blurring fall below human sensitivities for normal ambient lighting conditions. This means that, from a fundamental point of view, it is easier to attain an ultra-realistic colour holographic image than a monochromatic one!

Another potential problem is human perception of colour. Humans easily compensate for different types of illumination. For example, a white surface in daylight also appears white in artificial illumination but is actually yellow. This means that there are some intrinsic problems with how exactly to record colour holographic images. The best solution is to record a hologram for a defined illuminating light source. However, one must realise that this has limitations: if the hologram and its designed light source are placed in strong ambient illumination, there is always the danger that the human eye will then perceive the holographic image incorrectly. Of course, this is exactly the same problem as a properly colour-calibrated high-definition television display being placed in high ambient illumination—here, the human observer will make erroneous colour corrections based on the presence of the strong ambient light source giving the impression of a bad image.

The displayed toy train hologram at the 1964 OSA spring meeting and Leith and Upatniek's article generated a tremendous interest in holography. As only one laser wavelength was used to record and display the hologram, the 3D image was, of course, monochromatic. However, multicolour wave front reconstruction was introduced in their article; the authors mentioned that it should be possible to illuminate a scene with coherent light of three primary colours (red, green and blue; RGB). Transmission holograms were the main topic of the article and therefore the three reference beams proposed for recording such a colour hologram had to come from three different directions to avoid cross-talk when the recorded hologram was viewed. The hologram would then comprise three incoherently superimposed R, G and B holograms, showing a full-colour 3D scene upon reconstruction. Leith and Upatnieks also pointed out that holography is related to the photographic Lippmann colour process described in Chapter 2. It is most likely that this off-axis transmission colour recording technique will appear in the future. Even if the recording setup (with CW RGB or pulsed RGB lasers) is rather complex, the display of such a hologram is more feasible thanks to small, cheap solid-state lasers or LEDs currently on the market. So far, no such colour hologram has been recorded according to the authors' knowledge, but it is only a question of time until we will see these displays.

Another type of transmission hologram that can be used to produce holograms that display images in different colours is the well-known rainbow or Benton hologram. This is an early technique that, in particular, appealed to artists who created beautiful pseudo-colour holograms. A true-colour variant of the rainbow hologram is also possible to attain using the holographic stereogram technique (using multiple sets of colour-separated photographs) and more recently using the direct write digital holography (DWDH) and master write digital holography (MWDH) techniques. As we shall see in Chapter 11, however, a true-colour image is only available at a certain vertical position in front of the hologram. As soon as the viewer moves up or down in front of the hologram, the colour of the image changes. Strictly speaking, this means that such holograms cannot be included in the category of ultra-realistic colour displays. Nevertheless, such holograms can produce stunning large depth scenes. Using direct write digital holography techniques with hogel sizes of as small as 250 μm, Stanislovas Zacharovas at the Geola organisation is producing wonderful full-colour images for applications such as postcards, first as silver halide masters and then as embossed shims for stamping (more about this application in Chapter 14).

It should also be mentioned that there is a reflection version of the pseudo-colour hologram that artists have used over many years to make beautiful multicoloured reflection holograms. Using this technique, it is possible to make reflection holograms using a single-wavelength laser and multiple exposures; at each exposure, the emulsion thickness is chemically changed and the objects being recorded are repainted. The final holographic image shows a multicoloured object; however, the colours are essentially fake.

5.3 History of True Full-Colour Holography

As already mentioned, the first principles for recording colour holograms were established a long time ago by Leith and Upatnieks [1]. Mandel [2] pointed out that it might be possible to record colour holograms directly using a polychromatic laser source and an off-axis setup. Lohmann [3] included polarisation as an extension of the suggested technique. These first methods concerned mainly transmission holograms recorded with three different laser wavelengths with substantially different reference beam directions to avoid cross-talk; we shall see in Chapter 11 that from fundamental considerations, the transmission hologram actually possesses a better reference beam angle discrimination than the corresponding reflection hologram. The recorded hologram must be reconstructed using the same three laser wavelengths from the corresponding reference directions. Colour holograms can be made this way, but at that time, the complicated and expensive reconstruction setup prevented this technique from becoming popular. The first transmission colour hologram was made by Pennington and Lin [4]. They used the 15-µm-thick Kodak 649-F emulsion with a spectral bandwidth of approximately 10 nm. This narrow bandwidth essentially eliminated cross-talk between the two colours (633 and 488 nm) at the reconstruction. In general, the cross-talk problem was solved by Collier and Pennington [5], who used spatial multiplexing and coded reference beams, which made it possible to record colour holograms in thin media.

The technique by which colour reflection holograms could be made is rather obvious because white light can be used for viewing such holograms. The problem here is the severe demand on the recording materials (Chapter 4). The lack of any suitable material in the early days resulted in early colour reflection holograms being of rather poor quality. Lin et al. [6] made the first two-colour reflection hologram that could be reconstructed in white light. They recorded a reflection hologram of a colour transparency illuminated with two wavelengths (633 and 488 nm; Figure 5.3). The material used here was the panchromatic spectroscopic Kodak 649-F plate, which was processed without fixing to avoid emulsion shrinkage.

Very few improvements in colour holography were made during the 1960s and practical progress did not occur until much later. Nevertheless, some important articles on colour holography were published during this period [7–11]. In 1979, a high-quality colour reflection hologram was recorded by Kubota and Ose [12]. They avoided the problem of using coarse-grain commercial holographic silver halide materials. Instead, they used a panchromatic dichromated gelatin (DCG) plate in which a much higher-quality colour hologram could be recorded (Figure 5.4). However, it was not until 1986, when Kubota [13] recorded a

FIGURE 5.3   First two-colour reflection hologram, recorded in 1965 by Lin et al. [6].

FIGURE 5.4 Example of one of Kubota's colour holograms recorded in a DCG emulsion.

20 cm × 25 cm sandwich colour hologram, that it was possible to demonstrate the true potential of colour display holography and the possibility of attaining ultra-realistic 3D images. This was one of the first really good Denisyuk colour reflection holograms recorded and is reproduced in Figure 5.5. Kubota used a DCG plate for the recording of the green (515 nm) and the blue (488 nm) components, and an Agfa 8E75 plate for the red (633 nm) component of the image. Because the DCG plate is completely transparent to red light, the silver halide plate (containing the red image) was mounted behind the DCG plate in relation to the observer.

In the former USSR, where finer-grain silver halide emulsions existed, some early colour holograms were recorded. The problem with these holograms was the brownish stain in the emulsion, which did not render good colour images. The processing technique used in the former USSR was mostly based on colloidal or pyrogallol-based developers. These techniques work very well when recording monochrome (red or green) holograms. However, a developed and bleached colour hologram has to be absolutely clear to obtain a colour-correct image. This is the reason why the early DCG colour holograms have much better colour rendering than early silver halide colour holograms.

The sandwich technique to record colour reflection holograms was also used by Sobolev and Serov [14]. For some time, the sandwich technique became the primary method of recording improved quality colour reflection holograms. Smaev et al. [15] and Sainov et al. [16] used this technique as well.

A 1983 review of various transmission and reflection techniques for colour holography can be found in a publication by Hariharan [17]. Regarding reflection colour holography, an extensive contribution was

FIGURE 5.5 Kubota's 1986 Japanese doll sandwich hologram *Dojo*.

made in the early 1990s by Hubel and Solymar [18]. In their publication, they gave a quantitative and exact definition (according to their opinion) of what a colour hologram was. They maintained that "A holographic technique is said to reproduce 'true' colours if the average vector length of a standard set of coloured surfaces is less than 0.015 chromaticity coordinate units, and the gamut area obtained by these surfaces is within 40% of the reference gamut." In addition "the average vector length and gamut area should both be computed using a suitable white-light standard reference illuminant." More about this definition and the subject of colour rendering will be discussed below. Hubel and Ward [19] and Hubel [20] recorded colour reflection holograms in Ilford emulsions. However, due to the emulsion's 40-nm grain size, the recorded holograms suffered from blue-light scattering noise. The sandwich technique was used by Hubel and Ward, combining the green/blue and red Ilford emulsions (SP 672T for blue and green and SP 673T for red) with the recording laser wavelengths at 458, 528 and 647 nm.

5.3.1 Silver Halide-Sensitised Gelatin Technique

An interesting processing technique for silver halide colour holograms is the silver halide-sensitised gelatin (SHSG) technique, which is used to convert a silver halide-recorded hologram into a DCG type hologram. The advantage of the SHSG hologram over the DCG type is the high sensitivity of the panchromatic silver halide emulsion. A SHSG-processed hologram does not suffer from the printout associated with bleached holograms. SHSG processing was first introduced by Pennington et al. [21]. There are many publications on this technique, although most of them describe the processing of the old type of western monochrome holographic silver halide emulsions. This technique works best for the ultrafine-grain emulsions. During the recording of a colour hologram, any light scattering in the silver halide emulsion will be recorded as well. This means that even if one converts a silver halide hologram into a DCG type hologram (by removing the silver halide grains) the hologram will still show the recorded scattered light. Only the ultrafine-grain silver halide emulsions can therefore be used for successful SHSG processing of recorded full-colour holograms.

The SHSG technique of generating reflection holograms exposes the silver halide emulsion and then processes it in such a way that local tanning occurs within the emulsion. Then the material is fixed to remove all the silver halide grains from the emulsion, leaving only the gelatin. The last step of the processing is to dehydrate the material in a hydrophilic solvent similar to the usual DCG processing. This method for recording colour holograms was first developed by Usanov and Shevtsov [22–24], who introduced the SHSG method, which is based on the formation of a microcavity structure. The Russian technique can be explained in the following way. The gelatin in a photographic emulsion is adsorbed on the silver halide grains. In fact, only part of the gelatin molecules are adsorbed. The molecular chains are also linked within the gelatin mass of the emulsion. The thickness of the adsorbed layer in a dry emulsion is 2.5 to 4 nm. Each silver halide grain is surrounded by gelatin molecules linked at different points by active groups that are able to form complex compounds with the silver grains produced during development. The Russian method is based on the hypothesis that these adsorbed layers are less active and will be more difficult to harden than the surrounding gelatin mass. Variations in hardening between exposed and unexposed areas will therefore occur. After the silver and silver halide grains are removed from the emulsion, and the hologram is dehydrated, microcavities remain causing refractive index variation. One important point here is that the material needs additional hardening before the fixing step. After this, it is dehydrated with graded isopropanol solutions. Because a fixing step is applied in SHSG processing, which means that material is removed from the emulsion, the preferred method for obtaining reflection HOEs is the rehalogenating method.

There are different techniques for obtaining reflection phase silver halide holograms as explained in Figure 5.6. Emulsion shrinkage must be avoided when recording colour holograms. Conventional rehalogenating bleaching (a) is not recommended in this case. The fixation-free rehalogenating bleaching process (b) is the preferred technique for all types of hologram processing with regard to bleached colour reflection holograms. Because the SHSG processing requires fixation, which normally results in emulsion shrinkage, one has to try to avoid this by extensive hardening of the emulsion before fixation. The reversal (solvent) bleaching technique (c), which results in emulsion shrinkage, is not suitable for colour holograms. This bleaching process was popular for recording monochrome reflection holograms with red lasers, which resulted in orange/yellow images.

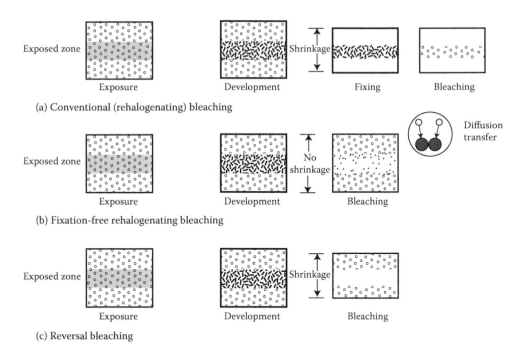

(a) Conventional (rehalogenating) bleaching

(b) Fixation-free rehalogenating bleaching

(c) Reversal bleaching

FIGURE 5.6 Different bleaching techniques for obtaining phase holograms with silver halide emulsions.

To solve the problem of SHSG emulsion shrinkage, selective hardening of the emulsion after bleaching and general emulsion hardening are the most important parts of the SHSG processing. After the two-step bleach and selective hardening process is completed, additional surface emulsion hardening is required. This can be performed by drying the emulsion in ethanol and then exposing the dry emulsion to formaldehyde vapour in a heated, sealed container, according to the method recommended by Usanov and Shevtsov. In this case, the hardening process can be performed in 15 to 30 min. Extensive hardening of the emulsion surface occurs during the dry-hardening step. The hard emulsion surface allows a high vapour pressure to build up inside the microcavities during the final dehydration step. Under this pressure, the microvoids expand and isopropanol is replaced by air, thus providing high-refractive index variations in the emulsion without any significant emulsion shrinkage. An example of a small Russian SHSG colour hologram is shown in Figure 5.7.

FIGURE 5.7 Russian SHSG-processed colour hologram.

In the early 1990s, an extensive research project to understand and further develop the SHSG processing technique was undertaken by one of the authors (HB) and his coworkers at the Centre for Modern Optics at De Montfort University, Leicester, United Kingdom. SHSG constitutes an important technique for processing colour reflection holograms and HOEs. By better understanding the gelatin hardening process and its role in SHSG processing, a recommended SHSG process was established for the ultra-fine-grain Slavich PFG-03C emulsion (Table 5.1). The effect of selective hardening was investigated to establish the optimum hardening condition. A chrome hardener was added to a modified rehalogenating bleach solution, and a warm-water bath was introduced for efficient gelatin cross-linking to occur; this was followed by an additional surface hardening in a sealed chamber with glutaraldehyde vapour which was found to work better than formaldehyde vapour. The dried hologram was placed in the chamber for a certain time; a slow-acting fixing solution was applied and then followed by conventional DCG

TABLE 5.1

Main SHSG Processing Steps

Processing Step	Time (minutes)
1. Prehardening in a formaldehyde solution	6
2. Develop in Agfa G282c developer at 22°C (diluted 1+5)	3
3. Bleach in the PBU-metol SHSG bleach at 22°C (diluted 1+1)	15
4. Treat in warm deionised water solution at 60°C (in safelight)	10
5. Dehydration in:	
50% water/50% industrial methylated spirits	3
100% industrial methylated spirits	3
6. Dry in oven at 45°C	5
7. Harden in chamber with glutaraldehyde vapour	25
8. Fix in SHSG fixing solution	2
9. Wash and dehydrate in	
50% water/50% isopropyl alcohol	10
100% isopropyl alcohol at 20°C	10
100% isopropyl alcohol at 70°C	2
10. Dry in oven at 45°C	
The following bath is used for the first prehardening step:	
Formaldehyde 37% (formalin)	10 mL (10.2 g)
Potassium bromide	2 g
Sodium carbonate (anhydrous)	5 g
Deionised water	1 L
The SHSG fix is mixed in the following way:	
Ammonium thiosulphate (anhydrous)	10 g
Sodium sulphate (anhydrous)	20 g
Deionised water	1 L
The SHSG bleach was based on the rehalogenating PBU-metol bleach, and the modified version is mixed in the following way (stock solution):	
Cupric bromide	1 g
Potassium persulphate	10 g
Citric acid	50 g
Potassium bromide	20 g
Borax	30 g
Deionised water	1 L
Add 1 g metol (*p*-methylaminophenol sulphate) after the other constituents are mixed.	

Note: To make this bleach for operations at a pH ~ 5, borax sodium tetraborate decahydrate was added to obtain the optimal condition for SHSG processing. A hardening compound is needed as well. Cr^{3+} ions were introduced into this bleach by adding 2% chromium (III) potassium sulphate.

isopropanol dehydration. Using this process, high-efficiency HOEs were obtained. The results of the CMO research project were published by Kim et al. [25–27] (Figure 5.8). Note that it is possible to obtain high diffraction efficiency (>95%) with no emulsion shrinkage after processing.

To investigate the microstructure of SHSG-processed emulsions, scanning electron microscopy of the Slavich PFG-03C plates was performed. To show the importance of sufficient emulsion surface hardening in a correctly hardened plate, microvoids located along the interference fringe pattern, and propagating through the emulsion, are shown in Figure 5.9. At a magnification of 80,000× (Figure 5.10), where a cross-section of 1 μm of the emulsion is depicted, microvoids are clearly visible. The size of the voids is approximately 100 nm.

The SHSG processing technique is rather time-consuming and complicated, and as such is mainly recommended for obtaining colour hologram masters or HOEs for contact copying in photopolymer film materials.

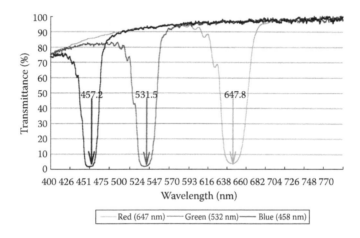

FIGURE 5.8 Recorded red, green and blue mirrors in SHSG-processed Slavich PFG-03C plates.

FIGURE 5.9 Microvoids in this SHSG-processed PFG-03C emulsion are visible and located along the interference fringes within the 7-μm emulsion cross-section (magnification, ×6000). (From Kim, J.K. et al. *Appl. Opt.* 41, 1522–1533, 2002. With the permission of the Optical Society of America.)

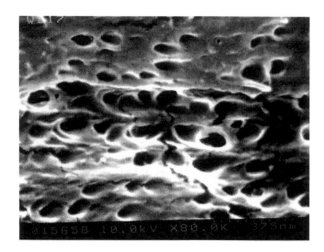

FIGURE 5.10   Scanning electron microscopy investigation of the SHSG-processed emulsion at 80,000× magnification. Microvoids (100 nm) are visible; 1-μm emulsion cross-section. (From Kim, J.K. et al. *Appl. Opt.* 41, 1522–1533, 2002. With the permission of the Optical Society of America.)

5.3.2 Colour Holograms in Single-Layer Silver Halide Emulsions

The sandwich technique served to demonstrate the potential of colour holography, but it was not really a simple technique for recording colour holograms. DCG emulsions are also difficult to use because of the necessity of preparing the panchromatic emulsion, which then requires a long exposure for the production of anything but the smallest colour holograms. In Russia, however, there was a different type of experimental silver halide emulsion for the recording of colour holograms.

In the mid-1990s, special panchromatic plates of the highest possible resolving power were prepared by Slavich for one of the authors (HB). These plates were prepared by Sergey Polyakov and his Slavich emulsion team and quickly hand-delivered by Henryk Kasprzak after they were finished. The first colour holograms on these plates were recorded in the laboratory by HB together with Dalibor Vukičević at the Louis Pasteur University in Strasbourg, France. The possibility of recording colour holograms in a single-layer ultrafine-grain panchromatic silver halide emulsion was published by Bjelkhagen and Vukičević [28] and later by Bjelkhagen et al. [29]. Their very first recorded colour holograms are shown in Figures 5.11 and 5.12a. These full-colour Denisyuk holograms were recorded with the following laser wavelengths: 633, 532 and 488 nm. The holograms were recorded with sequential exposures—first the blue exposure, then the green and finally the red exposure. Using sliding mirrors in the Denisyuk setup, the mixed RGB laser-light illumination of the object was not possible to observe. The red, green and blue beams were adjusted one at a time. The exposure time for each of the colour exposures was determined according to the sensitivity of the corresponding colour. Although the plates were panchromatic, they were not isochromatic, which translates to a different exposure for each colour. Therefore, the first time the mixed colours could be seen was after wet processing and after the recorded holograms had dried. The quality and colour rendering in these very first colour holograms were surprisingly good, thanks mostly to the ultrafine grains and the very high quality of the specially prepared Slavich emulsion. As already mentioned in Chapter 4, the key to success in recording analogue colour holograms is having access to suitable recording materials. The Denisyuk recording setup is straightforward and, apart from having access to stable output lasers with long coherence lengths and a vibration-free environment, there are no other requirements.

By about 2000, the Gentet brothers [30] in France, who had teamed up with Shevtsov from St. Petersburg, demonstrated high-quality colour reflection holograms recorded in the Ultimate ultrafine-grain silver halide emulsion manufactured in their laboratory. Their holograms were presented at the Holographic Millennium Conference in Austria. The most well-known Gentet holograms are the 30 cm × 40 cm holograms of butterflies and a toy fireman, which are reproduced here in Figure 5.13.

FIGURE 5.11 CIE test target hologram—this was the first hologram recorded in the special panchromatic Slavich emulsion.

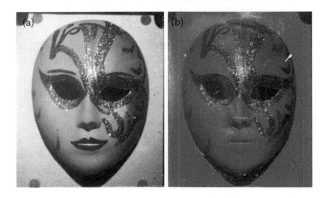

FIGURE 5.12 (a) Colour hologram of a ceramic mask with blue background; (b) monochrome hologram of the same mask.

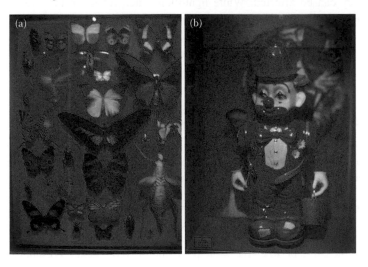

FIGURE 5.13 Two 30 cm × 40 cm Denisyuk colour reflection holograms produced by Yves Gentet and exhibited at the Holographic Millennium Conference in Austria.

The first Gentet holograms had a rather low blue component and the selected objects were mainly red or green. For example, the butterflies are mounted on a red background rather than a white one. Avoiding too much of the blue component means almost no blue light scattering in the emulsion, which results in extremely high contrast colour holograms. The problem with this approach is that white is reproduced more of a yellow colour. More recent holograms by Gentet have included blue objects and blue backgrounds.

5.4 Colour Recording in Holography

Methods for recording colour transmission holograms are based mainly on the geometry of the recording setup and are less affected by the material used for the recording. However, the original idea by Leith and Upatnieks [1] of using three different RGB reference directions to record an off-axis three-colour transmission hologram is now feasible. To display such a hologram today, one can use small, inexpensive lasers or LEDs. However, as we have already mentioned, it seems that there is no publication yet about the practical application of this colour transmission hologram technique.

Colour reflection holography presents few problems with regard to the geometry of the recording setup, but the final result is highly dependent on the recording material used and the processing techniques applied. The single-beam Denisyuk recording scheme has produced the best results so far and is the only one that can provide both 180° horizontal and vertical parallax. In addition to silver halide and DCG plates, photopolymer recording materials have also been used for recording colour reflection holograms.

There are at least five fundamental problems associated with the recording of colour reflection holograms in silver halide emulsions, which are the most convenient materials for large-format colour holograms because of their high sensitivity:

- Scattering occurring in the blue part of the spectrum during recording of a colour hologram requires a panchromatic ultrafine-grain silver halide material.
- Multiple exposures of a single emulsion may affect the diffraction efficiency of each individual recording [31,32]. Here, there is a difference between how many holograms can be recorded in an emulsion using the same laser wavelength and holograms recorded with different wavelengths, which is the case for colour holograms.
- Depending on the bandwidth of the light source used for the illumination of a colour hologram, the efficiency can be affected. White light-illuminated reflection holograms also normally show an increased bandwidth upon reconstruction, thus affecting the colour rendition.
- Shrinkage of the emulsion can often take place during processing, causing a wavelength shift.
- A set of laser recording wavelengths must be chosen to obtain the best possible colour rendition of the object.

5.4.1 Colour Theory and Colour Measurements

The problem of choosing the optimal primary laser wavelengths for colour holography is illustrated in the 1931 CIE (*Commission Internationale de l'Eclairage*) chromaticity diagram (Figure 5.14) and in the 1976 CIE version (Figure 5.15). The diagrams are useful devices for predicting the colours that can be matched by additively mixing a set of primary colours. Spectral colours are located along the horseshoe-shaped curve in the diagram. All visible colours are represented by points situated inside the diagram. Fully saturated colours are located along the periphery of the curve. The straight line joining the extremities of the curve (extreme red and blue) is the locus of purple. White is located in the centre of the area. The colours along the curve between the white point and the red region represent colours generated by a black body radiator, such as a hot filament at a certain temperature (K). By mixing different spectral colours, all possible colours can be synthesised.

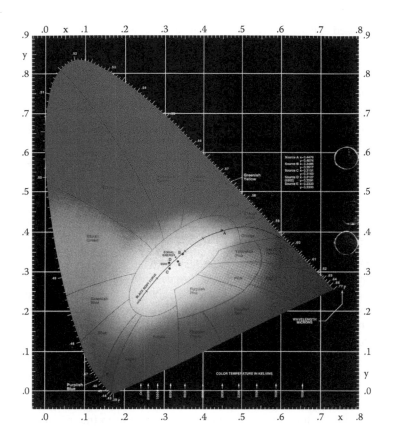

FIGURE 5.14 1931 CIE chromaticity diagram.

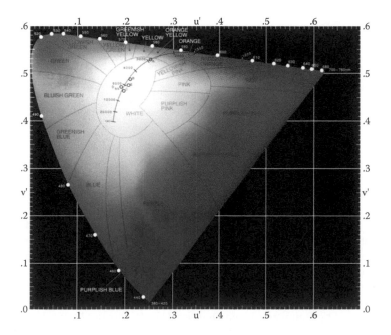

FIGURE 5.15 1976 CIE chromaticity diagram.

Some definitions used in discussions of colour can be useful to know:

Hue is that attribute of visual sensation that has given rise to colour names, for example, blue, green, yellow, and red.

Saturation, which is also referred to as the purity of colour, can be understood as the degree of purity in a given colour.

Brightness (or luminosity) is the attribute of a visual sensation according to which a given area appears to emit, or transmit, or reflect more or less light.

Lightness (or relative brightness) is the achromatic (colourless) continuum that goes from white, through gray, to black.

Luminance factor is the ratio of the luminance of a reflecting or transmitting surface, viewed from a given direction, to that of a perfect diffuser receiving the same illumination.

Chromaticity coordinates are the ratios of each of the three tristimulus values of a sample colour to the sum of the tristimulus values.

Colorimetry is the numerical expression of colours.

Primary colours are three colours of constant chromaticity used to specify an unknown colour by such amounts that are required in an additive mixture to match the unknown colour. Any three colours can serve as primary colours provided none of them can be matched by additive mixture of the other two.

Tristimulus values are the amounts of the primaries required to establish a match with a sample. This is done either by the addition to the sample of all three primaries, or of only one primary to the sample to match any pair of primaries, or alternatively by the addition of any pair of primaries to the sample to match the remaining primary.

Chromaticity coordinates for any given colour are computed from the tristimulus values *X, Y, Z* as follows:

$$x = \frac{X}{X+Y+Z} = \frac{\text{red}}{\text{red}+\text{green}+\text{blue}} \tag{5.1}$$

$$y = \frac{Y}{X+Y+Z} = \frac{\text{green}}{\text{red}+\text{green}+\text{blue}} \tag{5.2}$$

$$z = \frac{Z}{X+Y+Z} = \frac{\text{blue}}{\text{red}+\text{green}+\text{blue}} \tag{5.3}$$

Chromaticity is defined as a point in a two-dimensional rectangular coordinate space, with *x* and *y* denoting ordinate and abscissa, respectively. Two of the three primaries (corresponding to *X* and *Y*) are selected in such a way that their luminance factors are zero. Accordingly, the luminance factor of any colour is given directly by its *Y*-tristimulus value. The coordinate *z* is normally not plotted because $z = 1 - (x + y)$. Colour is approximately specified in the CIE system by *x*, *y* and *Y*.

After the three different primary spectral colours have been selected, a triangle is made by joining the three points corresponding to the spectral colours in the diagram. The colours within the area covered by the triangle correspond to all the colours that can be produced by an appropriate mixture of the chosen spectral colours. It may seem that the main aim in choosing the recording wavelengths for colour holograms is to cover the maximum area of the chromaticity diagram. However, there are many other considerations to be taken into account when choosing the recording laser wavelengths for colour holograms as discussed by Buimistryuk and Dmitriev [33], Bazargan [34,35], Kubota and Nishimura [36,37] and Hubel [20]. It is useful to be aware of Wintringham's [38] gamut of surface colours. Surface colours refer to colours of natural and man-made objects. Normally, these colours are of low saturation.

Such colours are also found in many of the objects considered for display colour holography. Pointer [39] has extended the gamut of surface colours to include some of the highly saturated fabric dyes that were introduced after Wintringham's publication. In early colour hologram tests, the recording wavelengths were 476.5, 514.5 and 632.8 nm, mainly because these wavelengths were available in common CW lasers of that time. These wavelengths cover the Wintringham data sufficiently well. However, as will be explained in later sections, using very narrow band laser wavelengths, it is not sufficient to select three laser wavelengths within the Wintringham area of the diagram. Another important factor to consider is the reflectivity of the object at the selected laser wavelengths. Thornton [40] has shown that the reflectivity of an object at three wavelength bands, peaked at 450, 540 and 610 nm, has a very high bearing on colour reconstruction in conventional colour imaging. In this case, it is assumed that the primary RGB colours are rather broadband. The luminosity of the colour image is affected by the drop in luminous efficiency with very short or very long recording wavelengths. According to Hubel's [20] colour rendering analysis, the wavelengths required to maximise the gamut area are 456, 532 and 624 nm. Kubota and Nishimura [36,37] approached the wavelength problem from a slightly different point of view. They calculated the optimal trio of wavelengths based on a reconstructing light source at 3400 K, a 6-μm-thick emulsion with a refractive index of 1.63 and an angle of 30° between the object and the reference beam. Kubota and Nishimura obtained the following wavelengths: 466.0, 540.9 and 606.6 nm. Bazargan [35] found the ideal wavelengths to be 450, 540 and 610 nm. In the next section, the wavelength selection problem will be described and its influence on colour rendering discussed.

A practical factor to consider is the availability of wavelengths in common CW lasers. In the early days of colour holography, the main CW lasers were argon ion, krypton ion, helium–neon and helium–cadmium lasers. Today, as we discussed in Chapter 3, there are a variety of solid-state lasers having many different wavelengths within the visible electromagnetic spectrum. With regard to pulsed RGB lasers, the reader is referred to Chapter 6.

5.4.2 Selection of Laser Wavelengths

Choosing the correct recording laser wavelengths is a key issue where accurate colour rendition is a primary concern. So far, most colour holograms have been recorded using three RGB primary laser wavelengths, resulting in rather good colour rendition. However, the colours recorded are not identical to the original colours and colour desaturation may sometimes constitute a problem.

As a starting point to understanding the selection of laser wavelengths, the minimum requirement is three RGB laser wavelengths. This follows from the tristimulus theory of colour vision, which implies that any colour can, for the human observer, be matched as a linear superposition of three primaries. The tristimulus values of an object define the colour appearance of the object as illuminated by a certain light source and for an average human observer:

$$X = \int_\lambda \bar{x}(\lambda)S(\lambda)E(\lambda)d\lambda \tag{5.4}$$

$$Y = \int_\lambda \bar{y}(\lambda)S(\lambda)E(\lambda)d\lambda \tag{5.5}$$

$$Z = \int_\lambda \bar{z}(\lambda)S(\lambda)E(\lambda)d\lambda \tag{5.6}$$

In the above equations, $\bar{x}$, $\bar{y}$ and $\bar{z}$ represent the colour-matching functions of the average observer, $E(\lambda)$ represents the power output of the illuminant over the visible spectrum and $S(\lambda)$ is the spectral reflectance curve of the object. Each colour has a different spectral curve. The nature of the illuminant in colour holography plays a most important role regarding colour reproduction. The reason is the fundamental difference between white light produced by narrow-band monochromatic laser wavelengths and the broadband light of a common illuminant such as daylight. Due to the narrow-band response of the laser illumination sources, the tristimulus values of a hologram are given by:

$$X^h = \sum_{i=1}^{N} E(\lambda_i)S(\lambda_i)\bar{x}_i(\lambda_i) \tag{5.7}$$

$$Y^h = \sum_{i=1}^{N} E(\lambda_i)S(\lambda_i)\bar{y}_i(\lambda_i) \tag{5.8}$$

$$Z^h = \sum_{i=1}^{N} E(\lambda_i)S(\lambda_i)\bar{z}_i(\lambda_i) \tag{5.9}$$

where i counts the laser wavelengths from 1 to N that are used during the recording of the hologram. There is a fundamental difference between the tristimulus values of the object given by Equations 5.4 through 5.6, and the tristimulus values of the hologram given by Equations 5.7 through 5.9. For the calculations of the tristimulus values of the object, all the components of the spectral curve are taken into account. For the hologram though, the only information within the spectral curves that is preserved is that located at the points corresponding to the recording wavelengths. It is apparent from Equations 5.7 through 5.9 that the monochromatic laser light introduces a sampling of the spectral properties of the object. Undersampling can lead to significant differences between the tristimulus values of the hologram and the tristimulus values of the object and hence an overall difference in colour. To demonstrate the effect of undersampling in colour holography, an example given by Peercy and Hesselink [41] is employed. In Figure 5.16, we plot the spectral reflectance curves of two different objects: A and B. Object A has a grey colour and B has a bluish–purple colour. At the wavelengths 477, 514 and 633 nm, both objects have the same value for spectral reflectance. Assuming a holographic recording of objects A and B with these laser wavelengths, the holographic images of both objects will appear to have the *same* colour because the hologram of the scene preserves only the surface-reflectance sampling of the wavelength information. The colour reproduction problem in this example is caused by under-sampling in the wavelength domain, which leads to aliasing.

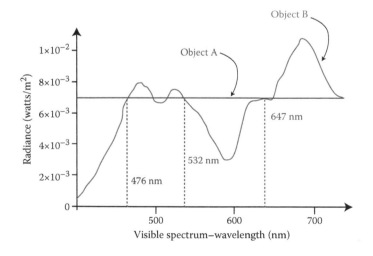

FIGURE 5.16 Aliasing due to undersampling of two objects A and B. (Modified from Peercy, M. S., and L. Hesselink. *Applied Optics* 33, 6811–6817, 1994).

It is important to ensure that a sufficient number of laser wavelengths are employed to avoid under-sampling, but it is also critical to define the minimum number of laser wavelengths to produce a hologram that demonstrates a visibly acceptable error in colour rendering. Increasing the number of recording laser wavelengths will improve colour rendition, but, at the same time, it will considerably increase the complexity and cost of the recording setup.

There have been theoretical investigations carried out which have studied the minimum number of laser wavelengths needed to give an error in colour rendition that is small enough to be undetectable by an observer. Peercy and Hesselink [41] as well as Kubota et al. [42] obtained results which indicated that more than three laser wavelengths are needed to reduce the colour error.

Laser wavelength selection is very important to be able to record holograms with the best possible colour rendition. A computer simulation study was performed at the Centre for Modern Optics, De Montfort University. Mirlis et al. [43] and Bjelkhagen and Mirlis [44] presented the results of these computer simulations based on Equations 5.4 through 5.9. Error values were calculated for different numbers of laser wavelengths, taking into account all possible combinations of wavelengths between 400 and 700 nm. In this way, sets of optimal wavelengths were defined. It was also found that once above seven laser wavelengths, further improvement in colour rendition was minimal; four or five can be considered the optimum number for practical high-quality colour holography (Figure 5.17). In Table 5.2, the average colour rendering error for three to seven optimal laser wavelengths is listed. The Macbeth ColorChecker target (Figure 5.18) was used to illustrate the improved colour rendering. In Figures 5.19 through 5.22, the error for each Macbeth colour is illustrated, as calculated in the computer simulations. Figure 5.19 shows the average colour error for holograms recorded with the laser wavelengths 476, 532 and 647 nm.

If, instead of the actual three wavelengths used for recording of the colour holograms in our laboratory, the three optimal laser wavelengths defined by the computer simulation were used (Table 5.2), the colour error decreases as shown in Figure 5.20. It is obvious that to produce the best colour reproduction, one should try and find lasers that emit as close as possible to these optimal wavelengths. In Figures 5.21 and 5.22, the colour error is plotted for the optimal sets of four and five wavelengths, respectively.

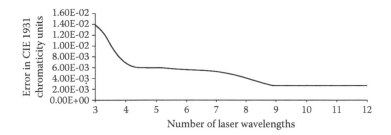

FIGURE 5.17 Total average colour error versus the number of laser wavelengths which are used in the recording of a colour hologram.

TABLE 5.2

Total Average Colour Error for Three to Seven Optimal Wavelengths

Number of Wavelengths	Optimal Laser Wavelengths (nm)	Colour Error
3	466, 545, 610	0.0137
4	459, 518, 571, 620	0.0064
5	452, 504, 549, 595, 643	0.0059
6	451, 496, 544, 590, 645, 655	0.0040
7	445, 482, 522, 560, 599, 645, 655	0.0026

FIGURE 5.18   Macbeth ColorChecker target.

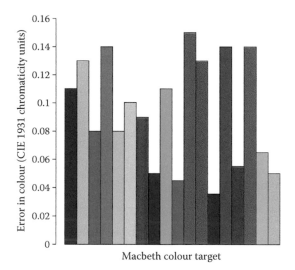

FIGURE 5.19   Computer simulation bar graph that displays the error for each Macbeth colour patch recorded in a hologram with three actual laser wavelengths at 476, 532 and 647 nm. (From Bjelkhagen, H. I., and E. Mirlis. *Applied Optics* 47, A123–A133, 2008. With permission of the Optical Society of America.)

By comparing the error graphs with each other, it is apparent that colour reproduction will be enhanced dramatically by employing one of the optimal sets. Optimal wavelength sets obtained in earlier publications are set out in Table 5.3.

We should underline that the optimal wavelength sets are a product of simulation and there is no guarantee that they correspond to easily existing or commercially available lasers suitable for holography. More work needs to be done in this area to define the best possible set of wavelengths for $n = 3$, 4 and 5 among current commercial laser emissions. Of course, high-power tunable lasers such as dye lasers

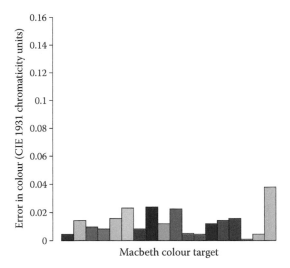

FIGURE 5.20 Computer simulation bar graph that displays the error for each Macbeth colour patch recorded in a hologram with three optimal laser wavelengths at 466, 545 and 610 nm. (From Bjelkhagen, H. I., and E. Mirlis. *Applied Optics* 47, A123–A133, 2008. With permission of the Optical Society of America.)

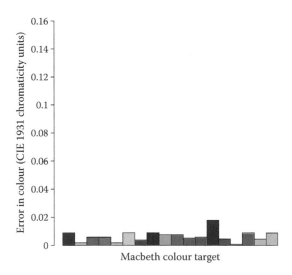

FIGURE 5.21 Computer simulation bar graph that displays the error for each Macbeth colour patch recorded in a hologram with four optimal laser wavelengths at 459, 518, 571 and 620 nm. (From Bjelkhagen, H. I., and E. Mirlis. *Applied Optics* 47, A123–A133, 2008. With permission of the Optical Society of America.)

might be used to match exactly the optimum set, but the technology becomes rather messy. Solid-state tunable lasers exist, but currently, the larger temporal coherence required for display holography can usually only be attained at the price of a very significant reduction in power.

We have already pointed out that the problem of speckle improves in a colour hologram and this has a beneficial influence on image resolution: there is an averaging of the speckle effect between different wavelengths when recording a colour hologram—this averaging effect is more pronounced when the number of recording laser wavelengths is increased [45]. Another improvement implicit to colour holography is lower susceptibility to moiré patterns, which may appear on the surface of the glass plate when

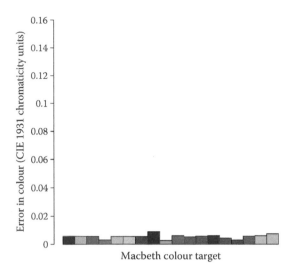

FIGURE 5.22 Computer simulation bar graph that displays the error for each Macbeth colour patch recorded in a hologram with five optimal laser wavelengths at 452, 504, 549, 595 and 643 nm. (From Bjelkhagen, H. I., and E. Mirlis. *Applied Optics* 47, A123–A133, 2008. With permission of the Optical Society of America.)

TABLE 5.3

Optimal Laser Wavelengths According to Different Investigations

Publication	Three Wavelengths	Four Wavelengths	Five Wavelengths
Mirlis et al. [43]	466, 545, 610	459, 518, 571, 620	452, 504, 549, 595, 643
Thornton [40]	475, 550, 625	460, 520, 580, 640	450, 500, 550, 600, 650
Peercy and Hesselink [41]	466, 541, 607	459, 515, 587, 663	
Kubota and Nishimura [37]	457, 532, 624		

recording a monochrome reflection hologram. The patterns are caused by interference between the two surfaces of the glass substrate. The multiple wavelengths present during the recording of a colour hologram causes individual moiré patterns corresponding to the different wavelengths to be superimposed, resulting in an almost moiré-free plate.

5.4.3 Illumination of Colour Holograms

The human observer will only correctly perceive colour and structural information in a colour hologram if the hologram is illuminated by the illuminant with which it was designed to be replayed. In addition, the intensity of diffracted light must be rather greater than the intensity of ambient light. The subject of illumination will be discussed at length in Chapter 13, but for now, we should underline the fact that a colour hologram can only be expected to replay properly if it is illuminated correctly. In Chapter 11, we shall derive expressions for the blurring in a general hologram and we shall see that the angular size of the light-source is critical to preserving structural information. Up until now, most colour reflection holograms have been illuminated with white-light halogen sources, but LED lighting is a strong candidate for the colour holograms of the future. However, narrow-band sources will bring their own problems—although more efficient, they will place higher demands on emulsion swelling and shrinkage at processing. High-power RGB laser sources are also being designed for projection televisions and are expected to provide a more expensive solution for super large depth digital colour holographic displays such as holographic windows—more about this in Chapter 14.

5.4.4 Demands on Lasers Required

To be able to record high-quality colour holograms, a minimum of three laser wavelengths (RGB) are needed. Each laser emission must be very stable.[*] From the experimenter's point of view, it is very difficult to obtain white, as variation in the output power in any of the lasers will result in a colour shift and thus a colour error in the recorded image. Generally, panchromatic emulsions are much less sensitive than their monochromatic cousins—as such, larger power CW lasers are required to record even quite small objects. As with any form of analogue holography, each laser must have good temporal coherence—at least several times the scene depth.

5.5 Setup for Recording Colour Holograms

5.5.1 Colour Transmission Holograms

Off-axis transmission colour holograms may soon be a reality. The main problems are a suitable laser illumination source (small red, green and blue solid-state sources ideally having a bandwidth of several nanometres) and the complex arrangement required to display such a hologram to avoid cross-talk. The laser safety aspect of the display system must be considered as well—that is, it is essential that the observer avoids looking directly into the reconstructing laser reference beams. However, the quality of a deep-scene hologram illuminated with the required amount of laser wavelengths would be very impressive. One should remember that a transmission hologram places far less demand on the recording material and, as a result, one needs less laser power for a given scene depth.

One variant of the colour transmission hologram is where the reference illuminating beams are brought in from three sides of a thick plastic block. The three-colour laser transmission hologram is optically mounted to the front of this block. The result is that each reference beam is internally reflected and dumped on the blackened black surface of the block. This type of hologram, which is really more suited to digital data due to implicit distortions, would be expected to have very good image contrast and, of course, no laser eye safety problems.

5.5.2 Colour Reflection Holograms

A typical recording setup for a three-colour Denisyuk reflection hologram utilising red, green and blue CW lasers is illustrated in Figure 5.23. The three laser beams are combined using two dichroic mirrors and pass through the same beam expander and spatial filter. The resulting white laser beam illuminates both the holographic plate and the object itself through the plate. Each of the three primary laser wavelengths forms its own individual interference pattern in the emulsion; all these patterns are recorded simultaneously during the exposure. In this way, three holographic images (red, green and blue) are effectively superimposed in the emulsion. For most of the colour holograms recorded by one of the authors (HB), the following three primary laser wavelengths were employed:

- 476 nm (provided by an argon ion laser)
- 532 nm (provided by a CW frequency-doubled Nd:YAG laser)
- 647 nm (provided by a krypton ion laser)

By using dichroic beam combiners, simultaneous exposure of the holographic plate can be performed. This makes it possible to independently control the RGB ratio and the overall exposure energy in the emulsion. The RGB ratio can be varied by individually changing the output power of the lasers, whereas the overall exposure energy is controlled solely by the exposure time.

---

[*] From an absolute point of view one could compensate for power variations by using different shutter exposure tunes for each colour, but from a purely practical point of view it is much easier to have stable power emission at each colour.

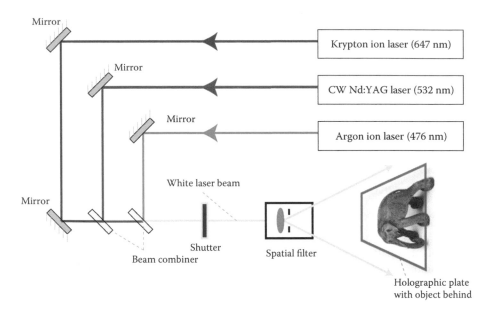

FIGURE 5.23   Denisyuk single-beam setup for recording colour reflection holograms.

In Figure 5.23, the object is shown positioned upside down. This is usual practice for holographers—working on an optical table with the beams running parallel with the table surface makes it easy to create the 45° overhead reference beam suitable for displaying the finished hologram. Of course, this can also be accomplished by introducing an overhead front-silvered mirror and installing the object in a normal position on the table surface. Every optical element (such as a mirror for example) introduced in the reference beam after the spatial filter is a source of noise. For example, dust particles on the mirror surface appear as disturbing dots on the hologram surface. Therefore, if possible, it is recommended to avoid using any optical elements in the reference beam after the spatial filter. In Figure 5.24, a *Russian Egg* is shown installed under the recording glass plate.

FIGURE 5.24   *Russian Egg* positioned upside-down on the recording table with the recording glass plate in front of it.

It is highly recommended to separate the room in which the lasers are located with their power supplies and the colour hologram recording room. The recording of analogue colour holograms is very sensitive to vibrations, air turbulence and noise sources (like from water-cooled ion lasers, for example). It is therefore best if the recording room is completely "silent" and during the recording, any existing air-conditioning in the room must be switched off. In Figures 5.25 through 5.29, two colour recording facilities are depicted. The Centre for Modern Optics' laboratory at Glyndŵr University in Wales (now closed) and the HOLOS facility in Fitzwilliam, New Hampshire (also now closed).

FIGURE 5.25   Laser room at the CMO laboratory with the author (HB) behind the laser table.

FIGURE 5.26   White laser beam passing through a hole in the wall between the CMO laser and recording rooms.

FIGURE 5.27   CMO recording room with the vibration-isolated table and the air-turbulence Styrofoam isolation sheets surrounding the recording setup.

FIGURE 5.28   HOLOS laser room with Qiang Huang behind the laser table.

5.5.3 Exposure of Colour Holograms

To determine the correct exposure time, it is necessary to know the sensitivity of the recording material. Because most of the panchromatic holographic recording materials are not isochromatic, one needs to establish the sensitivity at the three recording wavelengths. As an example, one particular batch number of the Slavich PFG-03C plates were determined to have the following sensitivity values:

- Blue (476 nm) 1480 μJ/cm^2
- Green (532 nm) 1410 μJ/cm^2
- Red (476 nm) 1084 μJ/cm^2

These values pertain to a particular processing technique, which will be described later. Very often, the PFG-03C emulsion has a higher red sensitivity as compared with the green and blue sensitivities.

To find the correct ratios between the recording wavelengths, the following technique can be used: Set the lasers to the same output power density as measured at the hologram recording position (this can be

FIGURE 5.29 Large recording table in the HOLOS recording room. The table is made of I-beams with 2-in. iron sheet metal slabs forming the table surface. Passive vibration isolation is provided between the I-beams with heavy-duty rubber inner tubes. This table was made for recording Denisyuk colour holograms of up to 1 m².

done using a laser power meter). Without recording a hologram, expose a third of the plate to be tested (cover the other two-thirds of the plate with black cardboard). Expose the test plate using one of the laser beams (red) for a given time (often a few seconds) slightly moving the sample during the recording. Make sure that this area of the plate has a mark (e.g., **R**). After this, the procedure is repeated with the second laser beam (green) exposing one of the previously covered areas (mask this with a **G** mark) of the test plate using *the same* exposure time as for the red. The red exposed area has to be covered not to be affected by the green exposure. Finally, the third area of the test plate is exposed with the third laser beam (blue) using the third, previously covered part (mask this with a **B** mark), again using the same exposure time. The red and green areas have to be covered during this third exposure. The test plate is then developed in a black-and-white developer and fixed. The exposure time should be chosen so that the optical density of this plate can easily be measured using a densitometer. If the recording material is isochromatic, of course, all three fields will have the same optical density after processing. However, in most cases, the **R**, **G** and **B** fields will have different optical densities. The test is repeated with the same exposure time, but now the laser power is increased for the colour regions (R, G or B) having lower optical density. After several such iterations, all three areas will have the same optical density after exposure

FIGURE 5.30 Emulsion colour sensitivity test plate marked with optical density values.

and processing (Figure 5.30) and the output power of the three lasers will be set for the material in question such that the correct white balance is obtained in a recorded colour hologram. Small adjustments of the laser output power may however be required to obtain the correct white colour balance. Nevertheless, this test procedure is the fastest way to set the lasers for recording colour holograms and to ensure that the exposure time is the same for all three laser wavelengths. What remains to be done is to find the overall correct exposure time to obtain the highest diffraction efficiency in the final colour hologram using simultaneous exposure. This requires a few test holograms with different exposure times.

5.5.4 Processing Recorded Colour Holograms

For colour holograms recorded on the Slavich PFG-03C emulsion, the processing set out in Table 5.4 is recommended. The final colour of the emulsion is completely clear (no stain caused by the developer or the bleach bath) and no emulsion shrinkage—these are both critical items for colour rendition. The processing consists of a prehardening step, a developing step and a rehalogenating bleach step.

The bleach can be used a few minutes after being mixed (enough oxidation of the developing agent metol must take place). Dilute 1 + 2 parts distilled water for use. More details on processing silver halide holograms and the chemicals used can be found in Bjelkhagen [46]. The prehardening bath is used to make sure the emulsion in the dried hologram after processing is very hard. One could suggest using a tanning developer such as a pyrogallol. The problem here is that the emulsion inherits a brown stain that

TABLE 5.4

Processing Colour Holograms

Formaldehyde hardener, 19°C ± 1°C	6 min
Rinse in a cold-water bath 10°C–15°C	10–20 s
CWC2 developer, 19°C ± 1°C	3 min
Wash in cold running water, 10°C–15°C	5 min
PBU-metol bleach, 19°C ± 1°C	5 min
Wash in cold running water, 10°C–15°C	10 min
Gently shower in deionised water	10–30 s
Kodak Photoflo (~1 + 200), 19°C ± 1°C	1 min

Allow to air-dry for 2 to 4 h with a good flow of cool (~20°C), clean air passing over the emulsion side of the plate.

The recipes for the hardener, the developer and bleach solutions are listed here:

Hardener:

Formaldehyde 37% (formalin)	10 mL (10.2 g)
Potassium bromide	2 g
Sodium carbonate (anhydrous)	5 g
Distilled water	1 L

Developer CWC2

Catechol	10 g
Ascorbic acid	5 g
Sodium sulphite (anhydrous)	5 g
Urea	50 g
Sodium carbonate (anhydrous)	30 g
Distilled water	1 L

Bleach solution PBU-metol

Cupric bromide	1 g
Potassium persulphate	10 g
Citric acid	50 g
Potassium bromide	20 g
Distilled water	1 L
Add 1 g metol [*p*-methylaminophenol sulphate]	

FIGURE 5.31   The upper part of this Sfera-S plate was not prehardened before development which caused the emulsion to shrink and deform.

is very difficult to get rid of without affecting the emulsion thickness. The reader may recall having seen early colour holograms looking rather brownish rather than being absolutely clear. One could also suggest making the emulsion very hard during manufacturing. The problem here is that if it is already hard before processing, it is not possible for the developer to penetrate the emulsion and only the upper part of the emulsion is developed, resulting in very low diffraction efficiency. Some readers may remember what happened when John Webster asked Agfa to increase the hardness of their holographic emulsion for his nuclear fuel element inspection transmission holograms; this made it impossible to record reflection holograms on these emulsions. For colour holograms, one needs an emulsion that is rather soft; then, by using the prehardening step before development, high diffraction efficiency can be achieved combined with a very hard emulsion in the finished hologram free of any emulsion shrinkage. Figure 5.31 demonstrates what happens when part of a Sfera-S emulsion is not hardened before development.

5.5.5 Sealing of Colour Holograms

It is important to protect colour holograms recorded in silver halide emulsions. To prevent any emulsion thickness variations occurring (mainly shrinkage), which may affect colour rendition of the holographic image, the emulsion needs to be protected. Similarly, like DCG holograms, humidity variations affect the silver halide emulsion as well, even when the final colour hologram emulsion is highly hardened during processing. A further problem is that holograms on display get heated by the spotlights illuminating them and this can cause the emulsion to shrink as well. This effect will, however, be less pronounced using the new LED lights as explained in Chapter 13.

A colour reflection hologram can be protected by a clean glass plate cemented to the emulsion of the hologram using an optical cement or epoxy. It is important to use an optical cement that does not affect the emulsion—for example, by causing shrinkage or swelling. The first stage of the process is, when the plate is still wet, to completely scrape off 5 to 6 mm of the emulsion around the edges. The purpose here is to prevent moisture from penetrating through the thin gelatin layer around the edges when it has been sealed to the cover glass. The hologram is then laminated to a clean glass cover with, for example, the VITRALIT 6127 optical cement from Eurobond Adhesives Ltd. [47]. VITRALIT 6127 needs UV curing to harden. To avoid strong UV exposure (which can cause printout of the bleached silver halide emulsion), it is sufficient to expose the hologram with the cover plate to sunlight or strong white light.

The hardening takes place in less than 1 min. With such a short exposure time, using a UV lamp is also possible. Colour holograms sealed with VITRALIT 6127 have been on display for very long periods without any deterioration of the holographic image.

Often, it is desirable to blacken the backside of a colour reflection hologram. The most common way of doing this is to cover the emulsion with black paint. There are various spray paints that can be used for this purpose, but caution must be exercised when choosing a particular paint, as some paints will react with the emulsion. Instead of using spray paint, one can employ the silk screen coating technique with acrylic screen printing black ink—this gives a very uniform and thick protective layer. However, because colour holograms on glass substrates should really be protected with a cover glass plate, it not advisable to attach any black coating to the emulsion. Instead, the black coating can then be applied to the cover plate and this means that the selection of the black coating is no longer critical.

Lamination of a colour reflection hologram recorded on film substrate can be done by using black Plexiglas with a clear adhesive from MACtac, for example, or the clear adhesive from 3M—the R948312P product. Sealing film holograms to black Plexiglas is a convenient way of combining blackening and mounting of film colour holograms. In particular, large-format film holograms need to be attached to a flat solid support for proper illumination. It needs to be pointed out that if colour film holograms are not perfectly sealed in this way, they may change colour due to large humidity variations or excessive heating. The film substrate (triacetate or polyester) does not provide a hermetic seal.

5.5.6 Recorded and Evaluated Holograms

Colour reflection holograms were recorded in the new panchromatic SilverCross emulsion [48]. The holograms were recorded in the same Denisyuk setup as previously described at the CMO laboratories with the same three RGB laser wavelengths. The processing method of Table 5.4 was used. It is important to point out that due to the ultrafine grains of the recording material, the holograms demonstrate very low light scattering in the blue region of the spectrum. An important aspect of the hologram is its ability to retain the spectral information of the image recorded. To retain this information, the processed holographic emulsion must not introduce changes to the interference pattern. If the emulsion shrinks or swells during the processing of the recorded hologram, it will distort the spectral information and it will introduce a colour shift. The spectral stability of the hologram is mainly dependent on the material that is used during the recording. The SilverCross emulsion is designed to provide high spectral stability and thus minimal colour distortion. Figure 5.32 shows a spectrogram taken by a spectrophotometer from the hologram of an elephant model. The model (left) and the hologram (right) are shown in the figure. It can be observed that the hologram, when illuminated by white halogen light, replays the spectral information of the recorded image, at the exact wavelengths used to record it (476, 532 and 647 nm). This indicates that no shrinkage of the emulsion has taken place during the recording.

In Figure 5.32, the measured white balance point coordinates at the elephant object are:

- $x = 0.42$
- $y = 0.40$

The corresponding measured white balance point coordinates at the hologram image are:

- $x = 0.38$
- $y = 0.42$

Many colour holograms have been recorded on the SilverCross emulsion to demonstrate its capabilities. Illustrated in Figure 5.33 is a photograph of a colour hologram of a blue object that can only be recorded in an ultrafine-grain emulsion. The recorded object is of the Franklin Mint decorative plate *Princess of the Iris* by M. Nolte.

One of the problems in colour analogue holography is the difficulty in obtaining an acceptable white colour in a hologram recorded of white objects. Very small variations in any of the red, green or blue

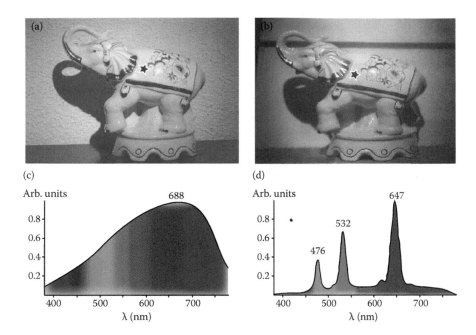

FIGURE 5.32 Evaluation demonstration of a colour hologram made using the SilverCross emulsion: the spectrum is measured at the white part of the elephant object (a) marked with a red star and the spectrum (c) is shown under the photo of the elephant. The spectrum (d) is measured at the corresponding point at the holographic image (b) of the elephant, also marked with a red star and is shown under the photo of the hologram image.

laser powers make the image turn slightly reddish or greenish. With the technique described previously (the RGB optical density test method), combined with extremely stable output powers from the RGB lasers, it is possible to get a white object to actually appear white in the recorded colour hologram (Figure 5.34).

Another problem that affects the possibility of obtaining a true white colour is connected to the processing of the hologram. If, for example, a developer containing pyrogallol is used, it is common to

FIGURE 5.33 SilverCross colour hologram of the decorative plate *Princess of the Iris*. The Franklin Mint Royal Doulton plate by M. Nolte.

FIGURE 5.34 Colour hologram showing the possibility of recording true white objects.

observe a brownish tint caused by the stained emulsion. Colloidal developing techniques also cause a reddish or brownish tint, as observed in many early Russian colour holograms. It is absolutely necessary that the developed and bleached emulsion is stain-free, which means that the hologram plate should be absolutely clear on processing.

5.6 Full-Colour Pulsed Portraiture

We have concentrated on the use of mostly CW lasers in this chapter to generate colour reflection holograms of real-world objects. One application of colour holography that requires pulsed lasers, however, is portraiture. The Geola organisation still manufactures monochrome holographic portraiture systems based on neodymium lasers. These machines can make large pulsed hologram masters as well as the final white light-viewable H_2 reflection copies. There is continuing mild interest in such systems from portrait holographers, artists, archaeologists and even medical scientists.

Very often, you will hear the comment that monochrome pulsed portraits are too "waxy" or give the impression of "ghosts". A large part of this is due to the monochrome nature of these reflection holograms. However, what if one could use RGB pulsed lasers to record true-colour reflection portraits? Surely, this would resolve many of the criticisms. Geola actually set out to do precisely this in the early 1990s [49] but dropped the project after several years in favour of digital colour holography. Nevertheless, they succeeded in developing a high-energy two-colour Raman laser and recorded some two-colour holograms—it had been hoped that a three-colour RGB laser would have been possible using the Raman concept, but Geola never managed to attain an adequate beam distribution in the blue.

One of the complicating factors of colour reflection portraiture is that colour cross-talk in the H_1 master hologram must be controlled either by using a reflection master or by using substantially different reference beam angles for the three colours in a transmission geometry. The H_2 reflection master requires an ultrafine grain emulsion and so necessitates much larger exposure energies—not only does this push the energies required for a suitable RGB pulsed laser to approximately 3×10 J, but this also means that the human subject is irradiated with approximately 30 times the laser energy as compared with a conventional monochrome portraiture scenario. It is possible to achieve such exposure within ocular safety standards, but the larger and more numerous diffusers severely complicate the system. The transmission master has the great advantage that a fine-grain emulsion can be used with an antihalation coating to control scattering. This means that one only irradiates the human subject with around three times the energy used in monochrome portraiture ($\sim 3 \times 1$ J). However, the H_2 copy is now rather more complicated and great care must be used to register the different colour images by producing exactly conjugate beams to the three original transmission reference beams. Additionally, an ultrafine grain emulsion is required for the final reflection H_2 if it is not to suffer from undue scattering. This means that several joules of energy in each colour are required to produce a 30 cm × 40 cm colour H_2. True-colour reflection holographic portraiture therefore looks extremely costly whichever way it is done! Geola dropped the project in the 1990s not because it could not make the lasers but because such systems would clearly not be commercial.

We shall see in Chapters 6 through 10 that the alternative technique of digital holography can in fact offer a rather better solution to full-colour portraiture. High-end structured-light quasi-real-time image acquisition systems combined with RGB pulsed-laser digital holographic printers can now print large holograms point by point; these holograms are at a stage where they are starting to become indistinguishable from full-colour analogue holograms. Of course, the great advantage here is that the image data is collected without laser illumination of the subject.

REFERENCES

1. E. N. Leith and J. Upatnieks, "Wavefront reconstruction with diffused illumination and three-dimensional objects," *J. Opt. Soc. Am.* **54**, 1295–1301 (1964).
2. L. Mandel, "Color imagery by wavefront reconstruction," *J. Opt. Soc. Am.* **55**, 1697–1698 (1965).
3. A. W. Lohmann, "Reconstruction of vectorial wavefronts," *Appl. Opt.* **4**, 1667–1668 (1965).
4. K. S. Pennington and L. H. Lin, "Multicolor wavefront reconstruction," *Appl. Phys. Lett.* **7**, 56–57 (1965).
5. R. J. Collier and K. S. Pennington, "Multicolor imaging from holograms formed on two-dimensional media," *Appl. Opt.* **6**, 1091–1095 (1967).
6. L. H. Lin, K. S. Pennington, G. W. Stroke, and A. E. Labeyrie, "Multicolor holographic image reconstruction with white-light illumination," *Bell Syst. Tech. J.* **45**, 659–661 (1966).
7. A. A. Friesem and R. J. Fedorowicz, "Recent advances in multicolor wavefront reconstruction," *Appl. Opt.* **5**, 1085–1086 (1966).
8. J. Upatnieks, J. Marks and R. Fedorowicz, "Color holograms for white light reconstruction," *Appl. Phys. Lett.* **8**, 286–287 (1966).
9. G. W. Stroke and R. G. Zech, "White-light reconstruction of color images from black-and-white volume holograms recorded on sheet film," *Appl. Phys. Lett.* **9**, 215–217 (1966).
10. E. Marom, "Color imagery by wavefront reconstruction," *J. Opt. Soc. Am.* **57**, 101–102 (1967).
11. A. A. Friesem and R. J. Fedorowicz, "Multicolor wavefront reconstruction," *Appl. Opt.* **6**, 529–536 (1967).
12. T. Kubota and T. Ose, "Lippmann color holograms recorded in methylene-blue-sensitized dichromated gelatin," *Opt. Lett.* **4**, 289–201 (1979).
13. T. Kubota, "Recording of high quality color holograms," *Appl. Opt.* **25**, 4141–4145 (1986).
14. G. A. Sobolev and O. B. Serov, "Recording color reflection holograms," *Sov. Tech. Phys. Lett.* **6**, 314–315 (1980).
15. V. P. Smaev, V. Z. Bryskin, E. M. Znamenskaya, A. M. Kursakova and I. B. Shakhova, "Features of the recording of holograms on a two-layer photographic material," *Sov. J. Opt. Technol.* **53**, 287–290 (1986).
16. V. Sainov, S. Sainov and H. Bjelkhagen, "Color reflection holography," in *Practical Holography*, T. H. Jeong and J. E. Ludman eds., Proc. SPIE **615**, 88–91 (1986).
17. P. Hariharan, "Colour holography," in *Progress in Optics* **20**, 263–324 North-Holland, Amsterdam (1983).
18. P. M. Hubel and L. Solymar, "Color reflection holography: theory and experiment," *Appl. Opt.* **30**, 4190–4203 (1991).
19. P. M. Hubel and A. A. Ward, "Color reflection holography," in *Practical Holography III*, S. A. Benton. Proc. SPIE **1051**, 18–24 (1989).
20. P. M. Hubel, "Recent advances in color reflection holography," in *Practical Holography V*, S. A. Benton ed., Proc. SPIE **1461**, 167–174 (1991).
21. K. S. Pennington, J. S. Harper and F. P. Laming, "New phototechnology suitable for recording phase holograms and similar information in hardened gelatin," *Appl. Phys. Lett.* **18**, 80–84 (1971).
22. Yu. E. Usanov and M. K. Shevtsov, "Principles of fabricating micropore silver-halide-gelatin holograms," *Opt. Spectrosc. (USSR)* **69**, 112–114 (1990).

23. Yu. E. Usanov, M. K. Shevtsov, N. L. Kosobokova and E. A. Kirienko, "Mechanism for forming a micro-void structure and methods for obtaining silver-halide gelatin holograms," *Opt. Spectrosc. (USSR)* **71**, 375–379 (1991).

24. Yu. E. Usanov and M. K. Shevtsov, "The volume reflection SHG holograms: principles and mechanism of microcavity structure formation," in *Holographic Imaging and Materials*, T. H. Jeong ed., Proc. SPIE **2043**, 52–56, (1994).

25. J. M. Kim, B. Choi, Y. Choi, S. I. Kim, J. M. Kim, H. I. Bjelkhagen and N. J. Phillips, "Transmission and reflection SHSG holograms," in *Practical Holography XV and Holographic Materials VII*, S. A. Benton, S. H. Stevenson and T. J. Trout eds., Proc. SPIE **4296**, 213–225 (2001).

26. J. M. Kim, B. S. Choi, S. I. Kim, H. I. Bjelkhagen and N. J. Phillips, "Holographic optical elements recorded in silver halide sensitized gelatin emulsions: Part 1. Transmission holographic optical elements," *Appl. Opt.* **40**, 622–632 (2001).

27. J. M. Kim, B. S. Choi, Y. S. Choi, J. M. Kim, H. I. Bjelkhagen and N. J. Phillips, "Holographic optical elements recorded in silver halide sensitized gelatin emulsions: Part 2. Reflection holographic optical elements," *App. Opt.* **41**, 1522–1533 (2002).

28. H. I. Bjelkhagen and D. Vukičević, "Lippmann color holography in a single-layer silver-halide emulsion," in *Fifth Int'l Symposium on Display Holography*, T. H. Jeong ed., Proc. SPIE **2333**, 34–48 (1995).

29. H. I. Bjelkhagen, T. H. Jeong, and D. Vukičević, "Color reflection holograms recorded in an ultrahigh-resolution single-layer silver halide emulsion," *J. Imaging Sci. Technol.* **40**, 134–146 (1996).

30. Y. Gentet and P. Gentet, ""Ultimate" emulsion and its applications: a laboratory-made silver halide emulsion of optimized quality for monochromatic pulsed and full color holography," in *HOLOGRAPHY 2000*, T. H. Jeong and W. K. Sobotka eds., Proc. SPIE **4149**, 56–62 (2000).

31. M. K. Shevtsov, "Diffraction efficiency of phase holograms for exposure superposition," *Sov. J. Opt. Technol.* **52**, 1–3 (1985).

32. D. Brotherton-Ratcliffe, "A treatment of the general volume holographic grating as an array of stacked mirrors," *J. Mod. Opt.* **59**, 1113–1132 (2012).

33. G. Ya. Buimistryuk and A. Ya. Dmitriev, "Selection of laser emission wavelengths to obtain color holographic images (in Russian)" *Izv. VUZ Priborostr. (USSR),* **25**, 79–82 (1982).

34. K. Bazargan, "Choice of laser wavelengths for recording true-colour holograms," in *Holographic Systems, Components and Applications*, J. S. Dainty ed., Proc. IEE, No. 311, 49–50 (1989).

35. K. Bazargan, "Factors affecting the choice of optimum recording wavelengths in true-color holography," in *Int'l Symposium on Display Holography*, T. H. Jeong ed., Proc, SPIE **1600**, 178–181 (1992).

36. T. Kubota and M. Nishimura, "Recording and demonstration of cultural assets by color holography (I) - Analysis for the optimum color reproduction," *J. Soc. Photogr. Sci. Tech. Jpn.* **53**, 291–296 (1990).

37. T. Kubota and M. Nishimura, "Recording and demonstration of cultural assets by color holography (II) - Recording method of hologram for optimizing the color reproduction," *J. Soc. Photogr. Sci. Tech. Jpn.* **53**, 297–302 (1990).

38. W. T. Wintringham, "Color television and colorimetry," in Proc. IRE **39**, 1135–1172 (1951).

39. M. R. Pointer, "The gamut of real surface colours," *Color Res. Appl.* **5** (No.3), 145–155 (1980).

40. W.A. Thornton, "Luminosity and color-rendering capability of white light," *J. Opt. Soc. Am.* **61**, 1155–1163 (1971).

41. M. S. Peercy and L. Hesselink, "Wavelength selection for true-color holography," *Appl. Opt.* **33**, 6811–6817 (1994).

42. Kubota, E. Takabayashi, T. Kashiwagi, M. Watanabe and K. Ueda, "Color reflection holography using four recording wavelengths," in *Practical Holography XV and Holographic Materials VII*, S. A. Benton, S. H. Stevenson and T. J. Trout eds., Proc. SPIE **4296**, 126–133 (2001).

43. E. Mirlis, H. I. Bjelkhagen and M. Turner, "Selection of optimum wavelength for holography recording," in *Practical Holography XIX: Materials and Applications*, T. H. Jeong, and H. I. Bjelkhagen eds., Proc. SPIE **5742**, 113–118 (2005).

44. H. I. Bjelkhagen and E. Mirlis, "Color holography to produce highly realistic three-dimensional images," *Appl. Opt.* **47**, A123–A133 (2008).

45. J. Hartong, J. Sadi, M. Torzynski and D. Vukičević, "Speckle phase averaging in high-resolution color holography," *J. Opt. Soc. Am. A* **14**, 405–410 (1997).

46. H. I. Bjelkhagen, *Silver Halide Recording Materials for Holography and Their Processing*, Springer Series in Optical Sciences, Vol. **66**, Springer-Verlag, Heidelberg, New York (1993).

47. Eurobond Adhesives Ltd., UK; www.eurobond-adhesives.co.uk. (Sept. 2012)

48. H. I. Bjelkhagen, P. G. Crosby, D. P. M. Green, E. Mirlis and N. J. Phillips, "Fabrication of ultra-fine-grain silver halide recording material for color holography," in *Practical Holography XXII: Materials and Applications*, H. I. Bjelkhagen and R. K. Kostuk eds., Proc. SPIE **6912**, 09-1–14 (2008).

49. D. Ratcliffe, "New pulsed laser systems," in *Holography Marketplace*, Eight Edition, A. Rhody, and F. Ross eds., Ross Books, Berkely, CA (1999) pp. 98–102.

6

Pulsed Lasers for Holography

6.1 Introduction

Pulsed lasers were employed in holography from very early on. Their ability to capture holograms of human subjects and to "freeze" time fascinated people. Nevertheless, as we have seen in Chapters 3 to 5, historically, the continuous wave (CW) laser proved extremely convenient to use: more wavelengths were available in CW, there were fewer ocular safety risks, it was easier to obtain an acceptable spatial beam distribution, photosensitive materials worked better and you could see what was happening in real time! But with time, this bias towards CW lasers has changed. There has been great progress in the field of pulsed lasers. Most importantly, there is now a far larger range of wavelengths available. Higher pulse energies, greater coherence, better pulse stability and higher repetition rates are all now possible. In the future, compact diode pumping can be expected to totally replace the older and bulky flash pumping that is still in use in most pulsed holography lasers today.

A typical pulsed Q-switched laser possesses a laser emission that lasts several nanoseconds. The use of such light to record a hologram therefore implies implicit interferometric stability. An object moving at velocity v will move a distance $s = v\tau$ in a time τ. If one assumes that during a holographic exposure, objects may only move less than one-tenth of a wavelength of light, then the maximum tolerable velocity of an object in a pulsed holographic recording is given by

$$v < \frac{\lambda}{10\tau} \tag{6.1}$$

If one assumes a typical pulse length of $\tau = 30$ ns and a wavelength of $\lambda \sim 532$ nm, this tells us that the object's velocity must be less than 1.8 m/s. Clearly, this means that all normal objects and scenes, including humans, can be captured as pulsed holograms without concern as to movement or vibration.

The essential freedom from vibration and movement conferred by the use of a pulsed laser is the principal reason that pulsed lasers are so interesting to holography. Picosecond pulsed lasers can even be used to record a bullet in flight: we can turn Equation 6.1 around to write an expression for the maximum pulse duration acceptable for a given velocity. Accordingly, a bullet travelling at 300 m/s requires a maximum pulse length of 180 ps.

Besides being critical in human portraiture, immunity from vibration and movement are also critical for applications such as recording holograms of delicate museum objects, paintings or biological samples. Full-colour analogue holography (see Chapter 5) has made significant improvements over the last few decades and, as such, applications requiring the use of pulsed lasers have become significantly more pertinent—all the more so because large improvements have also occurred in the illumination sources required to display full-colour holograms (see Chapter 13).

Perhaps the most important application of pulsed lasers, however, is in the new field of ultra-realistic digital holographic printing. In the direct-write digital holography method, first introduced by Yamaguchi et al. [1] in the early 1990s, a reflection hologram is recorded as a rectangular matrix of abutting, usually square, microholograms known as hogels [2] or holopixels [3]. These hogels are generated by a conventional object beam and reference beam. Digital data is encoded onto the object beam by passing it through a spatial light modulator connected to a computer. Typical hogel sizes are approximately 0.5 to

1 mm^2. Zebra Imaging Inc. [4] was the first to extend Yamaguchi's work to three-colour reflection holograms using CW lasers. Zebra Imaging produced direct-write digital holograms, which fundamentally changed the way people thought about holograms. However, writing more than one million microholograms was nevertheless problematic with CW lasers because of the fundamental sensitivity to motion. In 1999, the Geola organisation solved these problems with the introduction of a direct-write digital holography scheme using a compact RGB pulsed laser [5,6]. By using 40 ns pulses each of around only 1 mJ, much less average laser power was required, meaning that the Geola printer could be quite compact. The lack of implicit sensitivity to vibration made the pulsed laser printers clearly much more suitable for commercial exploitation and allowed much higher print speeds. However, Geola had to design and produce its own RGB pulsed lasers that would be suitable for digital holography. The main issue here was to achieve critical pulse-to-pulse stability in all three colour channels over many tens of millions of pulses—one wrong pulse in a million meant that the hologram had to be scrapped. This was a significantly complicated task.

There is an important difference between CW and pulsed holography lasers. Today, there are many commercial manufacturers of CW lasers suitable for holography. The holographer therefore seldom finds it necessary to make such lasers himself or to have a detailed knowledge of their construction—it is rather easier to simply select an appropriate commercial model. With pulsed lasers, however, by far the greatest commercial application is industrial processing. All too often, commercial companies will not offer the specifications (most notably temporal coherence, spatial distribution, pulse-to-pulse stability and beam pointing stability) required in holographic applications. When contracting a company either to make a customised pulsed laser or to actually build the laser yourself, it is extremely useful to possess knowledge of the fundamental design issues. In Chapter 3, we simply reviewed the available CW laser types and the basic physics behind their operation without going into deeper discussion. In this chapter, however, we will delve much deeper into the design and construction of the most important types of pulsed lasers. Most often, we will concentrate on the optical schemes required to assure parameters suitable for holography. However, where appropriate, we will also briefly introduce the basic atomic physics behind the different active materials used. A more in-depth discussion of the atomic and quantum physics of relevant materials can be found in various textbooks [7–9]. More details on laser engineering and pumping technology can also be found in the excellent book by Walter Koechner [10].

6.2 Ruby Laser

Ruby, $Cr^{3+}:Al_2O_3$, is a naturally occurring crystal. It is a variant of the mineral corundum, Al_2O_3 —or sapphire—in which some of the Al^{3+} ions have been replaced by Cr^{3+}. Ruby crystals suitable for use in lasers are usually produced artificially, however. This is done by adding small amounts of Cr_2O_3 to a highly purified melt of Al_2O_3. The crystal is then grown using the Czochralski method.

The ruby laser is a three-level system. Figure 6.1 shows an energy level diagram of the important features. The ground state of Cr^{3+} is $^4A_2$. A population inversion is created by pumping the Cr^{3+} ions from the ground state to the broad pump bands, $^4T_2$ and $^4T_1$. From here, the ions decay extremely rapidly to the metastable state, ^2E, which has a fluorescent lifetime of 3 ms. ^2E is in fact a doublet, and as such, there are actually two transitions to the ground state from the metastable state. These are the R_1 and R_2 emission lines of ruby and correspond to 592.9 and 694.3 nm. At room temperature, the R_1 attains lasing threshold before R_2; R_2 population transfer then occurs very rapidly from the upper to lower metastable level, which effectively leads to the entire population inversion decaying through R_1 at 694.3 nm.

The width of R_1 is 330 GHz at room temperature and is homogeneously broadened by interaction of the Cr^{3+} ions with lattice vibrations. Because of the fundamental degeneracy of the ground state, laser amplification only occurs if the R_1 level is at least half as densely populated as the ground state. This leads to a quite severe and fundamental constraint on the laser pumping power to achieve transparency; as a result, care must be taken to ensure that the laser crystal is uniformly illuminated by the pump, as any area in shadow will fall below the transparency condition and induce high optical absorption.

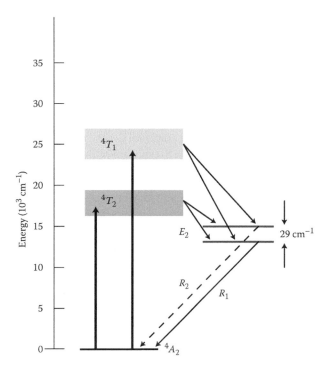

FIGURE 6.1 Energy level diagram of ruby.

Aside from being an innately three-level system and, as a consequence, requiring large optical pumping levels, the ruby laser benefits from its wide pumping bands; this makes pumping by xenon flashlamp relatively straightforward. Q-switching, which is usually arranged using a Pockels cell or (in older lasers) a Kerr cell, can produce large energies at some tens of nanoseconds. A master oscillator power amplifier (MOPA) architecture is often adopted. Figure 6.2 shows a photograph of a modern commercial ruby laser comprising a linear oscillator and an amplifier. Output energy at 694.3 nm is typically 3 J at 35 ns. The ruby laser also lends itself to a double pulsed output.

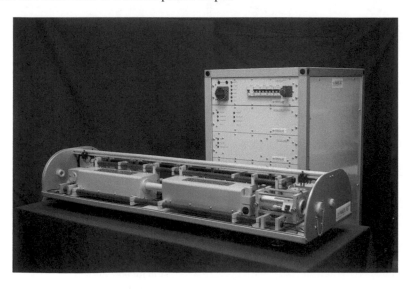

FIGURE 6.2 Modern commercial ruby laser (model HLSR-30 from InnoLas (UK)).

6.2.1 Practical Design of Ruby Lasers Suitable for Holography

Figure 6.3 illustrates Theodore "Ted" H. Maiman's first ruby laser [11]. This used a pink ruby rod (1 cm × 1.5 cm) and a helical xenon flashlamp from General Electric, Cleveland, OH. JK Lasers, Rugby, UK (founded in 1971) was the first company to offer a range of commercial ruby lasers suitable for holography. We review here the design of the System 2000 laser, which JK made in the 1980s. This was a laser producing up to 30 J of energy in a single pulse using an oscillator and three amplifiers. JK also made lasers with fewer or no amplifiers giving outputs of approximately 50 mJ, 1 J and 10 J. Today, lasers similar to these are produced by InnoLas (UK) Ltd., Rugby, UK (Table 6.1).

Figure 6.4 shows the basic optical scheme of the System 2000 ruby laser from JK. The heart of the laser is a linear cavity oscillator built around a 4 in. × 1/4 in. ruby crystal from Union Carbide, Piscataway, NJ. The ruby crystal is pumped by multiple linear xenon flashlamps within a ceramic reflector. This was a significant advance over the previous helical flashlamps. Large currents flowing in the helical lamps led to plasma instabilities, which stressed the glass tubes significantly, reducing their lifetime compared with the linear tubes.

The oscillator cavity is approximately 40 cm in length and has a planoconcave configuration with a rear mirror of 100% reflectivity and an output coupler with a transmission of 20%. The oscillator contains a 1.7 mm brass aperture to assure the generation of TEM_{00} only, a linear polariser, two etalons and a KDP Pockels Q-switch. The solid etalons, which assure good temporal coherence (>1 m), are 10 mm thick with 65% reflective coatings and 2.25 mm thick with 40% reflective coatings, respectively. The oscillator produces approximately 50 mJ of TEM_{00} single frequency light at 694.3 nm in a 20 to 30 ns pulse. Note that the ruby rod is cut at a slight angle to stop laser oscillation from its ends.

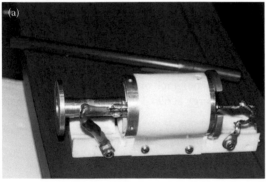

FIGURE 6.3   (a) Maiman's original 1960 ruby laser outside (b) and disassembled.

TABLE 6.1

Main Parameters of Modern Commercial Ruby Lasers for Holography Applications from InnoLas (UK) Ltd.

	HLS-R20	**HLS-R30**	**HLS-R40**
Max. energy of single pulse	1 J	3 J	10 J
Laser head (H × W × L)	26 × 44 × 97 cm	26 × 44 × 137 cm	26 × 44 × 137 cm
Power supply (W × D × H)	60 × 80 × 130 cm	60 × 80 × 130 cm	60 × 80 × 150 cm
Wavelength		694.3 nm	
Pulse repetition rate		4 pulses/min	
Pulse duration		25 ns	
Coherence length		>1 m for 90% of shots	
Electrical supply		90-255VAC 50–60 Hz	
Max. power consumption		5 kW	

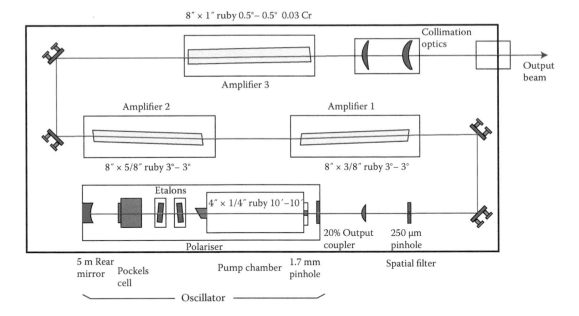

8″ × 1″ ruby 0.5°– 0.5° 0.03 Cr

FIGURE 6.4 Optical scheme of a JK System 2000 ruby laser producing 30 J of output energy at 694.3 nm.

The System 2000 laser has three amplifiers in addition to the laser oscillator. The oscillator output beam is first cleaned by spatial filtering. This comprises a small positive lens ($f = +150$ mm) and a 250 μm diamond pinhole. Three single-pass amplifiers are then used to further increase the energy of the pulse. The first amplifier is 8 in. × 3/8 in., and brings the pulse energy up to approximately 1 J. The second amplifier is 8 in. × 5/8 in. producing 10 J, and the third is 8 in. × 1 in. producing up to 30 J.

6.2.2 Pulse Lengthening in the Ruby Laser

Most Q-switched ruby lasers produce output pulses of approximately 20 to 30 ns duration. Some applications, however, require much longer pulses. As we pointed out at the start of this chapter, a 30 ns pulse length means that one can tolerate object speeds within a holographic setup of several metres per second. There are, however, many occasions in which the maximum object speeds are much slower than this. Reciprocity failure in silver halides is an important issue for finer grain emulsions and some materials, such as photopolymer and dichromated gelatin, are also much more sensitive to longer pulses. In addition, fibre optics can be used to carry coherent laser pulses at longer pulse durations. In the 1980s, one particular application focussed exceptional interest on generating longer pulses from a high-energy ruby laser. This was the application of bubble chamber holography, which was used for a time to record elementary particle tracks at Fermilab, Bataria, IL [12–14]. We shall discuss this application in Chapter 14, but for now, we shall briefly review the technique that was successfully implemented on a JK System 2000 laser whereby significantly longer pulses were generated [15].

Longer pulse lengths can be achieved in a variety of ways in pulsed lasers [16]. However, the most successful has been through fast feedback control of the Pockels Q-switch. The Fermilab team modified their JK System 2000 laser by replacing the rear oscillator mirror with one having 20% transmittance. They also replaced the standard Q-switch with a LaserMetrics model 1042 Pockels cell, which comprised two KDP crystals in parallel. This model was characterised by the relatively low quarter-wave voltage of 1200 V, which was important for designing the fast electronics.

The Fermilab pulse-stretching circuit is shown in Figure 6.5. It is composed of two parts

- fast feedback
- clamping

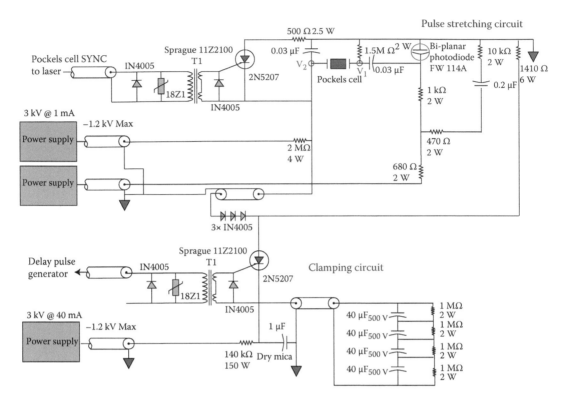

FIGURE 6.5 Fermilab pulse stretching circuit implemented on a JK System 2000 laser by the bubble chamber holography team [15].

Operation starts with the quarter-wave voltage being applied at V_2 and zero at V_1. At a predetermined time after the oscillator flashlamp ignition, the System 2000 generates a CMOS signal that triggers the feedback Silicon Controlled Rectifier (SCR) (2N5207) changing the potential of V_2 to zero and opening the Q-switch. Up until now, this is just the standard operation of the laser. However, a fast photodiode (ITT FW114A) now measures the light escaping from the 80% oscillator rear mirror and produces a voltage at V_1, which follows the instantaneous output power. In this way, the output power is controlled to a predefined value by the time-varying voltage at V_1.

The desired laser pulse length is programmed into a LeCroy 222 delay generator. When time is up, this generator produces a TTL pulse which activates the clamp SCR (2N5207), bringing the V_2 voltage back to the quarter-wave voltage. Lasing then ends within ~2 μs, with V_1 returning to zero. Within approximately 3 ms, the current in both SCRs decreases to lower than holding values, turning them off.

The Fermilab team observed that for 40 μs pulses, an amplitude modulation of less than 5% was produced. The pulses could be successfully amplified to approximately 8 J. Coherence lengths of as long as 11 m for 2.5 μs pulses were measured. Similar fast-feedback schemes have since been used in a variety of holography lasers to control Pockels cells. The Geola organisation, for example, routinely uses very fast prelasing control in neodymium lasers at a variety of wavelengths (1 and 1.3 μm) to ensure that the correct mode is present in the laser cavity at very low levels before opening the Q-switch properly. This technique guarantees stable pulses on demand with extremely small jitter.

6.3 Flashlamp-Pumped Lasers Based on Crystals Doped with Neodymium at 1 μm

Another common lasing material is made by doping either glass or one of several crystals with the rare earth metal neodymium. In each case, the active ions are Nd^{3+}. The most common crystals used for pulsed lasers suitable for holography are yttrium aluminium garnet (YAG or $Y_3Al_5O_{12}$), yttrium lithium fluoride (YLF or $YLiF_4$) and yttrium aluminium perovskite (YAP, YALO or $Nd:YAlO_3$). Both silicate and phosphate glasses are also routinely used. All these materials are available today in relatively large sizes and can be lamp-pumped using xenon discharge lamps. Yttrium vanadate (YVO_4) is often the preferred crystal for diode-pumped CW systems, but this material cannot be grown efficiently to sizes required for lamp pumping.

6.3.1 Nd:YAG

An energy level diagram of the Nd:YAG system is shown in Figure 6.6. The diagrams applicable to other crystal and glass hosts are extremely similar. However, the exact energy levels depend on the host's structure. The crystal field of the host splits each manifold into $(J + 1/2)$ levels, where J is the principal quantum number of the manifold. Therefore, the $^4I_{9/2}$ manifold is split into five Stark levels, which we label Z_1 to Z_5. Likewise, the $^4I_{11/2}$ manifold is split into six Stark levels, Y_1 to Y_6. Because each crystal host generates a slightly different field, the Stark splitting varies from host to host.

An important consequence of this multiplicity of states is that the fluorescence line shape is inhomogeneously broadened and exhibits significant asymmetry. This is even more apparent in glass, being an amorphous material; here, each site is different, and as a consequence, each of the energy levels is slightly different. The linewidth of the 1.06 μm YAG transition ($\delta\lambda_{YAG} \sim 7$ nm) is therefore much smaller than the corresponding linewidth in silicate ($\delta\lambda_{S.GLASS} \sim 30$ nm) or phosphate ($\delta\lambda_{P.GLASS} \sim 20$ nm) glass.

In the crystal hosts, laser transitions occur between the $^4F_{3/2}$ manifold and the $^4I_{13/2}$, $^4I_{11/2}$ or $^4I_{9/2}$ manifolds. In glass, only $^4F_{3/2} \rightarrow {}^4I_{11/2}$ occurs. The main laser transition of Nd:YAG occurs between the upper Stark level R_2 of $^4F_{3/2}$ and the third Stark level, Y_3 of $^4I_{11/2}$. This gives rise to an emission at 1064 nm. At

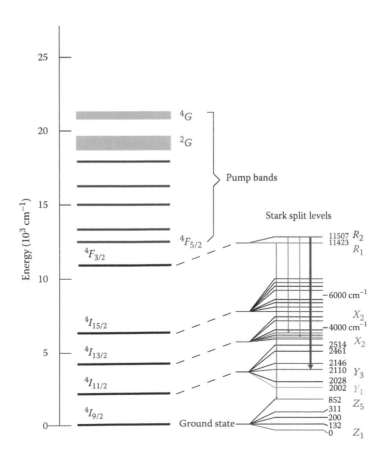

FIGURE 6.6 Energy level diagram of Nd:YAG.

room temperature, only 40% of the population inversion is in R_2, but as lasing occurs, this is constantly replenished by thermal transitions from the lower Stark level R_1.

The ground state of Nd:YAG is the $^4I_{9/2}$ level. Because the terminal level of the 1064 nm transition is $^4I_{11/2}$ (Y_3) and this is 2110 cm^{-1} above the ground state, the population of $^4I_{11/2}$ (Y_3) through thermal transitions from the ground state is effectively zero [as the Boltzmann factor $\exp(\Delta E/kT) \sim \exp(-10)$]. Nd:YAG at 1064 nm is therefore an intrinsically four-level laser and, unlike the ruby laser, lasing threshold can easily be achieved at modest pumping.

The pump bands for Nd:YAG start at $^4F_{5/2}$ and continue upwards. They are unfortunately not intrinsically broad as in the ruby laser, and as a result, one has to work hard with flashlamp pumping to match the pump to the laser absorption. Nevertheless, the fact that Nd:YAG is a four-level laser system means that much less power is required to create a population inversion and this greatly mitigates the pumping issue. The fluorescence lifetime of the $^4F_{3/2}$ manifold in YAG is approximately 230 µs, meaning that a variety of pump methods can be used, including flashlamps and semiconductor diodes. Once excited to any of the pump levels, ions decay efficiently to the upper lasing level.

6.3.2 Nd:YAP and Nd:YLF

Nd:YLF and Nd:YAP share principally the same features as Nd:YAG. As we have already mentioned, the main laser transitions are, however, a little shifted due to the effect of the different host. YLF emits at 1047 nm (π)/1053 nm (σ) and YAP at 1079.6 nm. YLF has a natural birefringence [which is why the main line is split into the ordinary (σ) and extraordinary (π) components] that dominates any thermally induced birefringence; its polarised output eliminates thermal depolarisation losses characteristic of YAG. Diode pumping is easier with YLF as its fluorescence lifetime of 485 µs is more than twice that of YAG. Although the gain is lower in YLF, energy storage is rather better. As a result, large-energy Q-switched output can be attained with YLF (up to 0.5 J from a single oscillator). Finally, YLF has a weaker thermal lens than YAG, making it sometimes preferable as a lasing material compared with YAG when the laser oscillator must work at different repetition rates. However, the material properties of YAG are clearly superior to YLF; YLF is a softer crystal than YAG and has a lower thermal conductivity.

YAP shows similar mechanical properties to YAG but does not suffer from the stress birefringence of YAG. It is a good material for high average powers, having a similar slope efficiency to YAG at 1 µm. Its fluorescence lifetime is only a little smaller than YAG at 170 µs. Like YLF, YAP has different refractive and thermal/mechanical properties in different directions. As with YAG and YLF, YAP has a transition, $^4F_{3/2} \rightarrow {}^4I_{13/2}$, producing emission near 1.3 µm. We shall discuss such emissions in relation to red and blue lasers in the next section.

6.3.3 Nd:Glass

Nd:glass has advantages and disadvantages when compared with the crystal hosts. The major advantage is that it can be doped at high concentrations with excellent uniformity and it can be produced in extremely large sizes to diffraction-limited quality. As we have seen, the linewidth of glass is much wider than in crystal hosts. Although this increases the lasing threshold, it permits larger energy storage for a given linear amplification coefficient and also allows the efficient amplification of shorter pulses. As a consequence, extremely high-energy pulsed emissions can be created with a glass laser and indeed such lasers are used for nuclear fusion applications [17]. The main disadvantage of glass is that its thermal conductivity is poor (around five to six times smaller than YLF for example), and as a consequence of this, only small laser repetition rates can be achieved at high energy.

Glass and crystal lasers complement each other in a very real way. As we have mentioned previously, the thermal conductivity of glass is much smaller than YAG, YLF or YAP. The crystal hosts are therefore far superior to glass for the generation of high repetition rate emissions. In addition, the crystal hosts, having a narrower linewidth, require lower pumping to threshold. The laser transition wavelength in glass and the crystals differs by up to 6 nm, but this is well within the typical 20 to 30 nm linewidth of glass. As we shall see below, a familiar configuration is the hybrid MOPA scheme, in which a crystal host is used as the laser oscillator and glass amplifiers are then employed to boost the energy. Both silicate

TABLE 6.2

Parameters of Neodymium-Doped Glasses

	Silicate Glass (Q-246, KIGRE)	Phosphate Glass (Q-88, KIGRE)
Peak wavelength (nm)	1062	1054
Cross-section ($\times 10^{-20}$ cm^{-2})	2.9	4.0
Fluorescence lifetime (μs)	340	330
Linewidth (nm)	27.7	21.9
Thermal conductivity (W/m°C)	1.30	0.84
Thermal expansion (10^{-7}/°C)	90	104
Young's modulus (kg/mm^2)	8570	7123

and phosphate glasses are routinely used. Table 6.2 summarises the important optical parameters of typical commercial glasses. As a general rule, silicate glass (1062 nm) seems better matched to Nd:YAG at 1064 nm. Conversely, phosphate glass at 1054 nm would seem better matched to Nd:YLF at 1053 nm. However, in practice, phosphate glass is often the most energy-efficient choice for both crystals.

6.3.4 Nd^{3+}-Doped Ceramic YAG

Another material, Nd^{3+}-doped ceramic YAG, is becoming increasingly popular. These laser rods have very similar properties to Nd:YAG crystals but, like glass, can be made in much larger sizes. Having good thermal properties, neodymium ceramic YAG may be used in place of glass amplifiers to allow high repetition rates. Nanocrystalline technology and the vacuum sintering method are used to fabricate such "synthetic" Nd:YAG crystals. Nd^{3+}-doped ceramic YAG is probably the most important innovation in the field of laser material fabrication technology in the last decade. Such ceramic laser crystals exhibit several key advantages over conventional single crystal growth technologies. Most importantly, the possibility of growing large samples quickly and at a low cost means that the mass production of this material is possible. On the other hand, the possibility of increasing the neodymium doping concentration (>4% doping is possible) compared with melt-grown technologies makes it possible to miniaturise the laser materials and points the way to new applications, such as single-mode microchip lasers.

6.3.5 Design of Commercial Neodymium Holography Lasers

Commercial neodymium pulsed lasers, oscillating at 1 μm and frequency-doubled to the green, were first used in holography to provide a more efficient* and more appropriate source of laser radiation for human and animal portraiture; previously, such applications had been catered for by the deep red emission of the ruby laser. Using glass amplifiers and a neodymium crystal oscillator, both pumped by liquid-cooled xenon flashlamps, a high pulse energy with excellent beam parameters can be efficiently obtained. The lasers are typically passively Q-switched for holographic applications. With electro-optic Q-switching and injection seeding [18–21], twin oscillator–amplifier configurations can produce the controllable mutually coherent double pulse emissions required by applications such as interferometry and holographic particle image velocimetry [22].

We describe here the construction of a family of Nd:YLF/Nd:phosphate glass lasers [23–25], which were designed especially for display holography by the Geola organisation. These lasers, which were first produced commercially in 1997, are currently in use by many holographers throughout the world and are still manufactured by Geola.

Each of the lasers is based on a high-stability single transverse and longitudinal mode ring cavity master oscillator and, depending on the model, can produce output energies from 1 to 5 J at the second harmonic wavelength of λ = 526.5 nm. The lasers incorporate a fast repetition mode, which can be used

---

* Frequency-doubled Nd:YLF/phosphate glass lasers are roughly four to five times more efficient than ruby in terms of electrical energy used per optical joule generated.

for the alignment of external optical elements. This mode only activates the oscillator and, at an adjustable frequency of up to 2 Hz, produces a bright green beam that has identical propagation characteristics to the main high-energy beam. This feature is a great improvement over the more common technique (often used with ruby lasers, for example) of using an additional CW alignment laser that inevitably is of a different wavelength and has different beam parameters compared with the main laser emission.

The lasers are built on a honeycomb stainless optical base that is mounted inside a separate laser case on a floating three-point suspension system. The master oscillator optical scheme is designed using a vertical cavity to minimise the influence of thermal bending. A breadboard thermosensor permits precise temperature equalisation of the liquid lamp coolant to the temperature of the optical breadboard. This

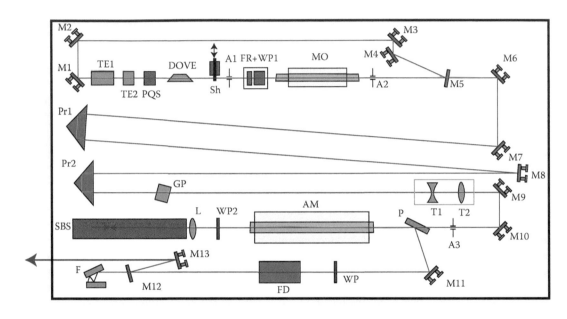

FIGURE 6.7   Optical schematic of the G2J Laser, a 2 J hybrid Nd:YLF/phosphate glass laser made by the Geola organisation.

FIGURE 6.8   Photograph of the G2J holography Laser. The laser produces 2 J at 527 nm. (Courtesy of Geola Digital UAB.)

reduces the influence of local heating of the breadboard by the pump chambers and preserves the laser scheme from thermal bending. The thermal walk-off of all adjustable optical mounts is designed to be minimal and each mount is tested in an interferometer before assembly.

Figure 6.7 shows the optical scheme of a G2J Laser. A photograph of its external appearance is shown in Figure 6.8. This laser produces up to 2 J of laser emission at 526.5 nm using a single Nd:YLF oscillator and a single double-pass Nd:phosphate glass amplifier with stimulated Brillouin scattering (SBS)* mirror.

6.3.5.1 Design and Operation of Master Oscillator

The ring master oscillator is defined by the output coupler (M5) and the four high-reflection mirrors (M1–M4). The resonator incorporates a Dove prism (mounted horizontally) and has an optical path length of 1.6 m. The Dove prism is used to reduce the sensitivity of the master oscillator to detuning of the cavity mirrors. A Nd:YLF laser rod (doped at 1.1%) measuring Ø5 × 80 mm is mounted in a stainless steel pump chamber incorporating a diffuse ceramic reflector. The rod is pumped by a linear flashlamp (5 × 76 XFP CFQ from Heraeus Noblelight, Great Britain). The close-coupled diffuse reflector design provides uniform rod pumping and ultraviolet (UV) filtration. The laser rod must be tuned rotationally in such a way that the polarisation of transmitted radiation remains invariant. This is the best way to suppress polarisation losses in the birefringent YLF crystal. An intracavity Faraday rotator (FR), together with a half-wave plate (WP1) totally suppress the generation of the parasitic reversed ring cavity mode.

Single longitudinal mode (SLM) generation is achieved with the use of a ring cavity, a LiF:F_2^- passive Q-switch and two tilted intracavity etalons (TE1 and TE2). A travelling-wave oscillator eliminates "spatial hole burning" caused by standing waves in a conventional linear oscillator. The passive Q-switch has frequency-selective properties due to its comparatively slow rate of change in transmission (i.e., its slow opening characteristics). The tilted etalons (TE1 and TE2) then provide final suppression of any satellite longitudinal modes.

Transverse mode selection is provided by two intracavity apertures (A1 and A2). A spherical mirror is used as the output mirror (M5). The master oscillator produces polarised pulses of single longitudinal and transverse mode with smooth temporal and spatial shapes. The output energy of these pulses exceeds 60 mJ at a wavelength of $\lambda = 1053$ nm; the pulse duration (full-width half-maximum, FWHM) is $\tau = 35$ ns. The intracavity beam shutter provides reliable blocking of the laser beam.

6.3.5.2 Amplification

A beam delivery scheme improves the oscillator output beam quality by natural "diffractive cleaning". The output beam from the master oscillator is reflected by the mirrors M6 and M7. It is then reflected by Pr1, M8 and Pr2 before striking the parallel glass plate (GP). The glass plate provides a smooth beam translation in the x and y axes, and so allows fine beam adjustment at the phosphate glass laser rod in the amplifier pump chamber (AM). The total beam pass distance is approximately 4.5 m.

After cleaning, the beam expansion telescope (lenses T1 and T2) expands the oscillator beam to optimally fill the Nd:phosphate glass amplifier (AM). This leads to efficient energy extraction and preserves good beam quality. The telescope T1 and T2 employs precision x–y translation lens mounts and a precision collimation adjustment.

The broadened laser beam is directed by the mirrors M9 and M10 to the amplifier scheme, which consists of a polariser (P), the amplifier pump chamber (AM), a quarter-wave plate (WP), a positive lens (L) and a liquid mirror (SBS). An apodising aperture (A3) cuts off any remaining wings present in the transverse beam distribution at the input to the phosphate glass amplifier.

High-quality spatial and temporal distributions (Figure 6.9) are assured by a two-pass phosphate-glass amplifier (AM) design incorporating phase-conjugation by stimulated Brillouin scattering (SBS).

The high-efficiency glazed Al-ceramic reflector in the amplifier pump chamber has a common channel for lamp and rod cooling, ensuring extremely uniform pumping from a single flashlamp. The G2J model uses a Ø12 × 300 mm GLS32 phosphate glass rod in the amplifier pump chamber. A xenon filled linear flashlamp INP13-250 (Zenit Company, Moscow) provides UV cut-off to increase the lifetime of the laser rod.

---

* SBS mirrors are liquid mirrors that reflect a phase-conjugate of the incident beam.

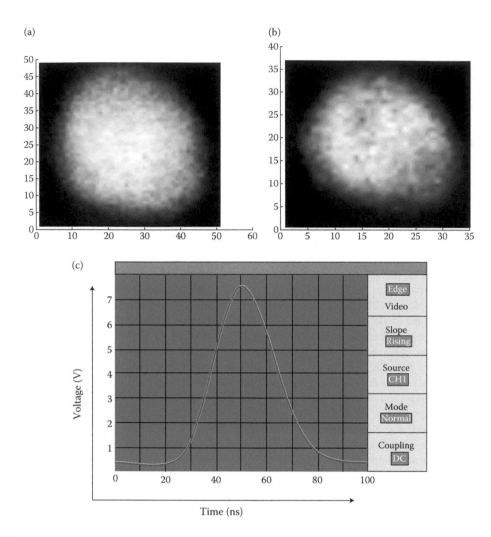

FIGURE 6.9 (a) Spatial beam distribution of the G2J Laser in near-field and (b) at 10 m from the laser head; (c) temporal distribution of the pulse.

The quarter-wave plate (WP2) and thin-film polariser (P) together form an optical separator which ensures that the second-pass (backwardly travelling) beam through the amplifier is diverted into the output channel (at P) and not back into the master oscillator. The quarter-wave plate ensures exact 90° polarisation rotation after the second pass. This suppresses any radiation going back into the master oscillator scheme. The use of a Brillouin cell (SBS) is very important for two reasons. First, SBS allows the formation of a diffraction-limited beam by compensation of the aberrations in the wave front, which are produced by the hot Nd:phosphate glass rod. This assures identical beam divergence and propagation direction in both the high-repetition low-energy alignment mode and in the usual high-energy, low-repetition mode. The second reason is the greater energy extraction possible with a double-pass scheme without self-excitation. Here, the Brillouin mirror serves as a selective reflector that reflects only a coherent signal and not the noise from any amplified spontaneous emission. As a result, around half of the stored energy in the Nd:phosphate glass rod is depleted.

The rotated second-pass radiation is reflected at the polariser (P) and directed to the mirror (M11). After the quarter-wave plate (WP), the radiation is converted to the second harmonic by a frequency doubler (FD). A deuterated potassium dihydrogen phosphate (DKDP) crystal is usually employed as the harmonic generator. This crystal must be sealed in a temperature-controlled dry cell because it is hydroscopic. A precision-engineered angle-tuning mechanism based on a simple mechanical design ensures excellent long-term stability of the harmonic output. This technique permits an energy conversion

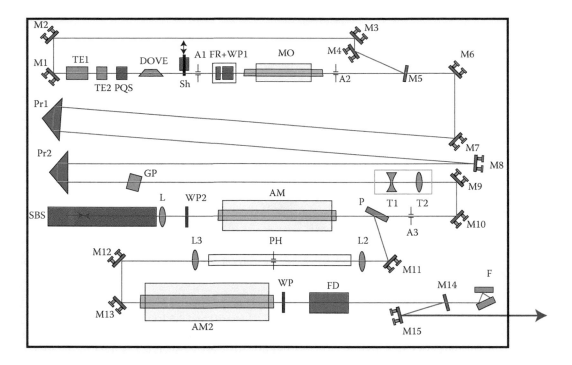

FIGURE 6.10 Optical schematic of the G5J Laser, a hybrid Nd:YLF/phosphate glass holography laser with an output of 5 J at 527 nm made by the Geola organisation.

efficiency to 526.5 nm of up to 60%. Harmonic separation is achieved by the pair of dichroic mirrors (M12 and M13), giving more than 99.7% separation. The residual part of the radiation at the basic harmonic wavelength is absorbed by two neutral density filters at the retroabsorption stage (F).

Figure 6.10 shows the optical scheme of a higher energy version of the laser capable of producing 5 J of output at 526.5 nm. This laser additionally uses a final stage large diameter Nd:phosphate glass amplifier (Ø20 mm × 300 mm) producing laser pulse amplification with almost complete stored-energy depletion. This amplifier is single pass and uses image plane translation lenses (L2 and L3) incorporating vacuum spatial filtering (PH). Vacuum filtering assures effective decoupling between amplifiers, thus preventing unwanted lasing, while also improving beam quality.

Two of the most important parameters for a holographic laser are beam distribution and coherence. The lasers described here have output pulse durations of approximately 35 ns with a coherence length of more than 3 m. The spatial profiles are quasi-Gaussian and have been optimised for holography.

6.3.5.3 Pump Electronics

The power supplies required for driving the flashlamps of the master oscillator and amplifiers each consist of a capacitor bank, a capacitor charging circuit, a trigger and pulse-forming module and a driving controller. The main capacitor bank is connected in series through an inductor directly to the flashlamp. The value of inductance essentially defines the characteristic pumping time. In the oscillator power supply, a high-frequency converter based on bipolar transistors with isolated input/output is used as the capacitor-charging unit. This unit has a power density of 0.3 W/cm^3 and presents a smooth noiseless load to the mains supply. The maximum output voltage is 1200 V with a stability of ±0.15%. The charge rate of the module is 520 J/s. Typical capacitor bank and inductor values for the oscillator lamp are 100 μF and 100 μH, respectively.

In the amplifier power supplies, a direct 50 Hz high-voltage network transformer incorporating recti-fier, and driven by a thyristor switch, is used as the capacitor charging unit. Such an electronic system ensures ultrahigh reliability. The voltage stability is ±0.2%. The amplifier in the G2J Laser uses a capacitor bank of 500 μF and an inductance of 100 μH, whereas the second amplifier in the G5J laser uses

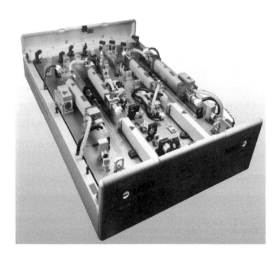

FIGURE 6.11   Photograph of the G5J Twin, a laser with two mutually coherent 5J outputs at 527 nm. (Courtesy of Geola Digital UAB.)

726 μF and 100 μH. Typical operating voltages of the two amplifiers are 1.7 kV (ampl) and 2.3 kV (amp2). Maximum stored energies are, respectively, 2.2 kJ (ampl) and 3.3 kJ (amp2).

The trigger modules of the various power supplies provide reliable operation of the commercially available flashlamps (INP5-75, INP13-250* and INP16-250). Series triggering is used for ignition of the flashlamps. The oscillator module provides an 18 kV triggering pulse for initial plasma creation—a "simmer" circuit then maintains ionisation; both amplifier modules provide a 25 kV triggering pulse. The pulse-forming modules are of a fixed pulsewidth type. The modules consist of a capacitor bank, an inductor and a series-triggering transformer with driving circuits. The discharge pulse duration is optimised for two constraints: (i) optimal efficient performance, which is produced by matching the intrinsic fluorescence lifetime of the Nd:YLF crystal or of the Nd:phosphate glass rod, and (ii) maintaining a sufficient flashlamp lifetime, which usually equates to not pumping too quickly.

6.3.5.4 Cooling

The xenon flashlamps used to excite laser operation produce considerable heat. As with the ruby laser, this heat must be dissipated by a liquid cooling circuit. Both the G2J and G5J lasers use either the PS1222CO unit from the Lithuanian Company, Ekspla UAB or equivalent units made by the German Company, Termotek, AG for this purpose. This 2 kW water/water type heat exchangers provide excellent heat removal from the operating laser. A magnetic drive centrifugal pump is used for the circulation of the cooling liquid, which is usually a mixture of deionised water and ethanol. This pump provides noiseless operation. The cooling unit stabilises the temperature of the cooling liquid in the internal loop with an accuracy of 0.2°C.

6.3.6 Towards Even Higher Energy

Some advanced applications in holography require higher energy than the lasers described previously can provide. Although it is perfectly possible to make lasers with larger glass amplifiers, the cost can rapidly get out of hand. Often, the output laser beam in holography is split into a reference beam and several object beams. As a result, it turns out to be cheaper to produce lasers with multiple mutually coherent outputs. Such lasers can be constructed by using a single laser oscillator feeding one or more separate amplifier chains. Figure 6.11 shows an example of a commercial holography laser based on this principle. The laser produces a total of 10 J at 526.5 nm in two 5 J beams. The master oscillator is in fact an exact copy of the G2J/G5J oscillator. The output beam energy is then increased by a single-pass

---

* Note that the INP13-250 lamp is used without simmer.

Nd:YLF preamplifier before being split and fed into identical amplifier and frequency conversion chains. The amplifiers and frequency convertors are identical to those in the G5J laser discussed previously. Care must be taken to use the correct highly purified SBS fluid for the two-pass amplifiers to assure small and equal frequency shifts from the SBS cells. If this is not done, the mutual coherence of the beams can be compromised. The Geola G10J-Twin has an auto and mutual coherence of greater than 10 m. The concept can be extended to virtually any number of channels.

6.3.7 Applications

Geola has produced (and still produces) semiautomatic holographic portraiture systems based on its neodymium glass lasers [26–31] (Figure 6.12). These systems are able to generate both the transmission

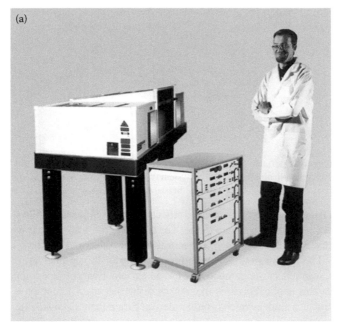

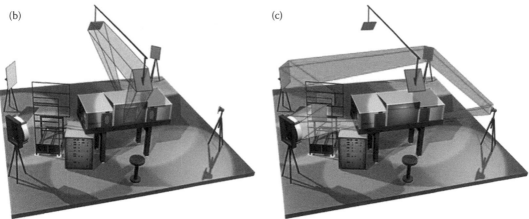

FIGURE 6.12 (a) Photograph of a commercial pulsed holographic portraiture system based on a 2 J Nd:YLF/glass laser; (b) mastering mode in which a pulsed H_1 transmission hologram is made; (c) copying mode, in which the H_1 hologram is transferred to a white light-viewable H_2 reflection hologram. The system automatically switches between modes (a) and (b), and sets all beam ratios and energies according to switch settings by the operator. Later models featured full computer control and automatic beam alignment. (Panel a, courtesy of Geola Digital UAB.)

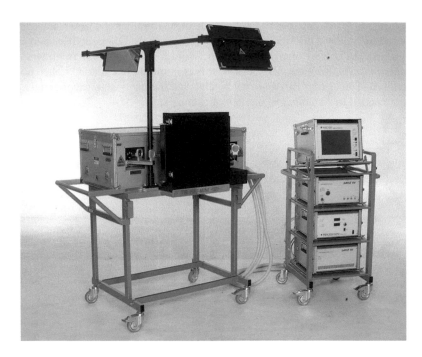

FIGURE 6.13 Photograph of a portable pulsed holographic recording system based on a 2 J Nd:YLF/glass laser with special shutter allowing operation in ambient light. The system is designed for use in medical topometry, producing highly accurate three-dimensional computer models of patients' heads. (Courtesy of Geola Digital UAB).

master hologram and the final white light-viewable reflection copy. Options also allow rainbow holograms to be produced up to sizes larger than a square metre. The company has also worked with the Ceasar Institute in Germany to produce a medical topometry system that is portable and can be used in normal daylight [32,33] (Figure 6.13). The system works by recording a laser transmission hologram of a patient's head—the hologram is then scanned into a computer where an extremely accurate computer model of the head is generated. In effect, this application uses the ultra-realistic nature of holographic imaging to produce a three-dimensional computer model, which is rather more accurate than that available with other techniques.

6.4 High-Energy Two-Colour Emission: Stimulated Raman Amplification

A high-energy red beam at 674 nm may be efficiently produced by using the second harmonic output of a standard neodymium-pulsed laser to pump a simple laser based on stimulated Raman scattering (SRS) in pressured hydrogen gas. In such a way, an intrinsically two-coloured red–green pulsed laser beam may be created far more economically than by mixing the beams of a ruby and a neodymium laser. It is worth pointing out that red–green holograms can sometimes give a very real impression of actually being full-colour holograms. Of course, this is not always the case, but given that the recording of a blue component is so much more problematic in pulsed holography due to the problem of Rayleigh scattering (which is present in all emulsions having a high enough sensitivity to avoid ocular damage to the human subject), there has been significant interest in such red–green concepts. Although present work has concentrated on SRS in hydrogen, replacement of the hydrogen with deuterium can be expected to produce a red-pulsed source of acceptable quality for holography at approximately 627 nm. Although in principle, anti-Stokes generation from a 527 nm pump can produce a blue signal, in practice, the quality of this beam has thus far been found insufficient for serious holographic applications.

SRS is well known as a simple and effective frequency conversion method. In its simplest form, a single cell known as an SRS-generator is filled with a compressed molecular gas such as hydrogen, and pump radiation from a Nd:YAG laser is then focussed to a point inside this cell. Under pumping by low-energy second harmonic (532 nm) pulses (pulse energy, $W_L < 50$ mJ), an energy conversion efficiency to Stokes radiation of up to 55% can be achieved in such simple schemes [34]. However, any further increase in the pumping energy leads to a decrease in the Stokes energy. Beyond a certain pumping threshold, any excess energy from the pump preferentially drives other competing non-linear processes in the SRS medium such as the generation of higher Stokes and anti-Stokes components, four-wave interactions and electro-optic breakdown. As the pump energy is increased, the probability of excitation of adjacent transverse SRS modes also increases.

The solution to producing a high-energy Stokes pulse suitable for holography is to use a more general scheme of SRS conversion incorporating the concept of a separate generator and amplifier [35]. In this case, a single mode Stokes seed with a smooth temporal shape is produced from a SRS generator cell by pumping just above the threshold. Further conversion then takes place in a separate Raman amplifier cell in which the collimated Stokes seed and the pump beam are collinearly mixed. To maintain a narrow spectral linewidth, compressed hydrogen is used as the SRS-active medium as this gas possesses the smallest known linewidth for spontaneous Raman scattering. Operation is in the collisional Dicke-narrowed region in which the spontaneous scattering linewidth can reach a value as low as 0.009 cm^{-1} (or 270 MHz) at a hydrogen pressure of only 3 to 4 atm. This linewidth can be additionally reduced during the SRS process by preferential amplification of the central spectral components.

If an optimised generation of Stokes radiation is to occur, the diameters of the interacting beams must be carefully chosen; in addition, the delay of the pump pulse relative to the Stokes seed must be optimised, and the hydrogen pressures in the generator and amplifier cells must be carefully equalised. Such optimisation allows conversion efficiencies close to the theoretical limit for the normal non-super-regenerative regime of Raman amplification. Practically, a maximum energy conversion efficiency to Stokes radiation of $\eta = 68\%$ has been attained to date. This is close to the theoretical limit of $\eta = 78\%$ given by the Manley–Rowe relation.

6.4.1 SRS Red–Green Pulsed Laser

We shall describe here a prototype SRS pulsed laser system, which was developed at Geola in the mid-1990s [35]. This laser was capable of producing 1 J of red radiation at 674 nm from a green pump pulse at 527 nm, at an efficiency of 68%.* Figure 6.14 shows the optical schematic of the system. A hybrid Nd:YLF/Nd:phosphate glass laser (a pre-commercial version of the Geola G5J model) is used as the pumping source for the forward Raman generator (RC1) and Raman amplifier (RC2). A pumping energy of $W_L \geq 5J$ at a wavelength of 526.5 nm is available from the Nd:YLF laser. The temporal shape of a typical pump pulse is shown in Figure 6.15. The pulses are of a duration of 25 ns and exhibit a rather sharper leading edge and a rather more gently sloping tail due to the influence of amplification and Brillouin compression processes within the pump laser. The pump beam has a super-Gaussian intensity profile.

The output beam from the pump is divided into two parts by the polariser (P1). The ratio of the energy in the two beams is controlled precisely by the wave plate (WP1). The smaller of the two beams, after polarisation cleaning by the polarisers (P2–P4), is focussed by the lens (L1) to a point at the centre of Raman generator cell (RC1) which is filled with compressed hydrogen. The pressure in the Raman generator is 4 atm. Here, a single-mode Stokes seed is excited by pumping just over the threshold. This seed is then collimated by the lens (L2) and injected into the two-pass Raman amplifier (RC2) by the mirrors M1 and M2. The larger part of the second harmonic radiation serves as the pump for this amplifier. This part passes through the polariser (P1) and after restoration of its initial polarisation by the half-wave plate (WP2) passes through to the delay line M4 to M8. Image plane translation by the lenses L3, L4 and the pinhole (PH) is used to preserve the spatial distribution of the pump beam after several metres of optical delay. The collimated collinear pump and the Stokes beams are combined by the mirror (M3) and are directed into the forward Raman amplifier RC2 (1 m length). The windows of the RC2 cell are anti-reflection coated for both the wavelength of the pump and for the Stokes radiation at 674 nm. The exact alignment of the pump and seed Stokes beams must be performed using a He–Ne laser.

---

* We should mention that as early as in 1986, workers at Imperial College, in collaboration with Ilford, had succeeded in producing a small full-colour test hologram through Raman conversion of 532 nm radiation in high-pressure H_2. However, the quality of the hologram was reported to be bad and the work was never published.

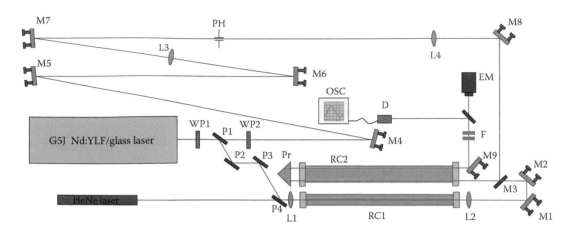

FIGURE 6.14   Optical schematic for forward Raman amplification of Stokes pulses in the Dicke-narrowed line of hydrogen.

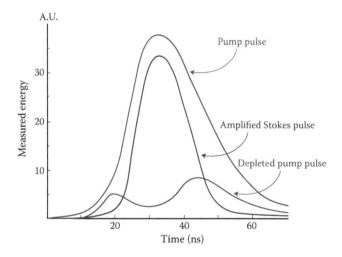

FIGURE 6.15   Typical temporal shapes of the pump pulse (red), the amplified Stokes pulse (blue) and the depleted pump pulse (green) for a conversion efficiency from 527 to 674 nm of greater than 55%.

After a first pass through the Raman amplifier, the transmitted beams are directed back into RC2 by the prism (Pr) for a second pass. M9 then directs the output beam, which consists of the depleted pump beam collinearly mixed with the Stokes beam, for analysis or use. The filters (F) weaken the output radiation and transmit either the Stokes or the depleted pump. The pulse energy is measured by an energy meter (EM). The temporal pulse shapes are measured by a fast oscilloscope (OSC) and a photodiode (D).

After optimisation of pump focussing in the generator cell (RC1) by selection of the lens (L1) and the optimal selection of input energy by rotation of the half-wave plate (WP1), a single-mode Stokes seed ($\lambda_{1S} = 674$ nm) with smooth temporal shape (Figure 6.16, blue curve) and Gaussian transverse beam profile is excited. The seed Stokes pulse typically has a rather sharper leading edge, a duration of approximately 10 to 15 ns (FWHM) and an energy of up to 1.5 mJ. The SRS generation threshold in the Raman generator should be exceeded by less than a factor of 1.4 to 1.7. Upon increasing the pumping to a factor of more than 2 over the threshold, deterioration of the temporal shape of the seed due to the onset of modulation typical of the transient SRS conversion regime is observed [34,36–38].

Due to the transient SRS conversion regime, the Raman amplification efficiency is strongly dependent on the delay between the pump pulses driving the amplifier RC2 and those driving the generator RC1. This delay is caused by the development of a phonon wave in the SRS-active medium provoked by the

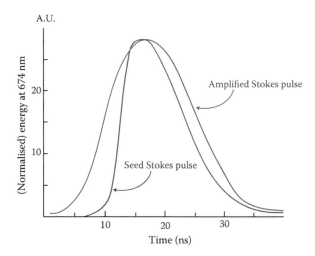

FIGURE 6.16 Temporal shapes of the seed Stokes pulse (blue) and the amplified Stokes pulse (red).

pump. Consequently, the Stokes pulse is emitted from the generator cell only at the end of the pumping pulse. In addition, the Raman amplification efficiency in the amplifier cell is maximal in the case that the seed Stokes pulse considerably precedes the instant corresponding to maximum pump intensity. Under a hydrogen pressure of 4 atm, the optimal delay amounts to approximately 20 ns. Detuning from this optimal delay of ±4 ns leads to a 15% decrease in the Raman amplification efficiency. A similar strong dependence of the Raman amplification efficiency can be observed for a difference in the hydrogen pressures of the generator and amplifier cells. The Stokes shift, as well as the spontaneous Raman scattering linewidth, is hydrogen pressure-dependent [39]. A pressure disparity in the cells of only 1 atm leads to a detuning of the seed spectrum from the centre of the SRS amplification gain profile of one-half of this profile width. Under such detuning, the efficiency of Raman amplification is reduced by 20%. Hence, to achieve maximal efficiency, the pressure in the cells must be equalised exactly. This can be assured by using a flexible hose that effectively connects the generator (RC1) and amplifier (RC2) cells.

Figure 6.17 shows the typical experimental dependence of the energy conversion efficiency in the Raman amplifier on pump energy W_L. It can be seen that even under pump energies of $W_L > 0.7$ J, the conversion efficiency goes into saturation. In optimal conditions, using a pump energy of $W_L > 1$ J, a

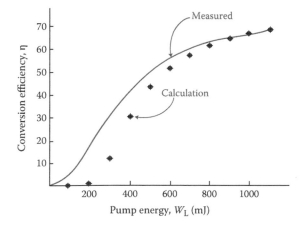

FIGURE 6.17 Typical experimental dependence of the energy conversion efficiency in the Raman amplifier with pump energy W_L for the case of a Raman seed energy of 1 mJ. The expected efficiencies as calculated from a three-dimensional computer simulation of forward transient SRS are also shown.

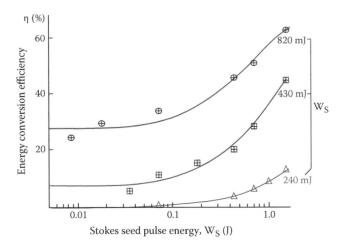

FIGURE 6.18 Dependence of Raman amplification efficiency versus injected seed energy measured for various values of the pump energy.

Stokes conversion with an efficiency of 68% can be expected. Self-excitation in the Raman amplifier at a pump energy of less than 1.1 J should not be observed in the absence of a Stokes seed pulse.

Figure 6.18 shows the dependence of the Raman amplification efficiency versus the injected seed energy measured under various values of the pump energy. As can be seen from these curves, the SRS conversion efficiency depends comparatively weakly on the energy of the injected seed over a wide range. The coherent properties of the radiation produced by the SRS converter are strongly controlled by the temporal shape of the pulses and their transverse beam profiles. Figure 6.15 shows typical shapes of the pump pulse (red curve), the amplified Stokes pulse (blue curve) and the depleted pump pulse (green curve) for a conversion efficiency of more than 55%. The amplified Stokes pulse has a smooth temporal shape and a duration of 18 ns, 30% lower than the pump pulse duration but 1.3 to 1.5 times higher than the seed Stokes pulse duration (Figure 6.16). The temporal shape of the Stokes pulses after amplification (Figure 6.16) becomes more symmetrical. The depleted pump pulses exhibit a large intensity hole close to the centre, which clearly demonstrates the high-energy conversion during Raman amplification. The transverse beam distribution of the amplified Stokes radiation generally mimics the pump beam distribution. When a pinhole is not used, diffraction rings on the pump beam are also seen on the amplified Stokes beam distribution.

Because the pump beam is severely depleted by the creation of the Stokes beam, it is usually necessary to dump this beam using a dichroic mirror and to then remix the Stokes beam with a clean collinear beam at 527 nm, again using dichroic mirrors. This assures a two-colour red–green beam of the correct spatial distribution for holography.

6.5 Pulsed RGB Lasers—Neodymium Lasers at 1.3 μm

Nd:YAG, Nd:YLF and Nd:YAP can all be made to produce laser action at 1.3 μm with the relatively strong $^4F_{3/2} \rightarrow {}^4I_{3/2}$ transition. In YAG, there are two lines that have very similar cross-sections at room temperature. These are the $R_2 \rightarrow X_1$ transition at 1319 nm and the $R_2 \rightarrow X_3$ transition at 1338 nm. The cross-section of both these transitions is approximately three times smaller than the main $R_2 \rightarrow Y_3$ transition at 1064 nm. Frequency doubling and tripling allow conversion of the 1.3 μm emission to red and blue outputs. Workers at Geola in Lithuania were the first to demonstrate that the 1.3 μm transitions in neodymium could be used to make a pulsed RGB laser suitable for digital holography [6]. Later work by the same group showed that much higher energies could be obtained by amplification of the 1.3 μm line but that such higher energy lasers, having energies up to 1 J in the red, green and blue, would be rather expensive due to the smaller gain available at 1.3 µm. Here, we shall only briefly mention amplification strategies. We shall concentrate instead on the relatively low energy pulsed lasers that can be realised most easily and efficiently using the 1.3 µm line. The largest application for these lasers is in digital holographic printing, where they have been used very successfully.

6.5.1 Dual-Ring Cavity Pulsed RGB Nd:YLF/Nd:YAG Laser

Historically, the first true RGB pulsed holography laser was a laser based on two separate ring resonators (Figure 6.19). The first resonator comprised a standard ring cavity oscillator using Nd:YLF and passive Q-switching very similar to that described in Section 6.3.5.1. Frequency doubling was used to convert the infrared emission at 1053 nm to green at 526.5 nm. The second resonator was based on a Nd:YAG crystal lasing at 1319 nm. Passive Q-switching was accomplished by V:YAG, which, at the time, was an extremely rare crystal to find with the correct parameters. Frequency doubling and tripling of the 1.3 µm emission were organised by lithium triborate (LBO) crystals. The laser produced highly coherent red (659.5 nm), green (526.5 nm) and blue (439.6 nm) TEM_{00}-like emissions of 1.8, 3.4 and 1.6 mJ, respectively, at pulsed durations of 45 ns (green) and 60 ns (red and blue). The laser maximum repetition rate was 15 Hz.

Figure 6.20 shows an optical schematic of the RGB laser. The first resonator, which is used for green light generation, is defined by the four rear high-reflection mirrors (M1g, M2g, M3g and M4g) and a meniscus output coupler (M5g). The active element (MOg) comprises a 4 mm diameter cylindrical Nd:YLF crystal that is 95 mm long with a 1% doping level, lasing at 1053 nm. The crystal is bevelled at each end at 3° to prevent reflection from these surfaces, creating an unwanted laser resonator. The crystal is pumped by a 75 mm xenon flashlamp, Samarium filters and ceramic reflectors housed in a custom pump chamber. Giant pulse operation is assured by a Cr:YAG passive Q-switch (Qg) with initial transmission of 47%. A 2.7 mm intracavity aperture (A1g) restricts lasing to TEM_{00}. A Dove prism (DPg) is likewise used to improve horizontal stabilisation of the vertical ring cavity. A Faraday rotator (FRg) and wave plate (WP1g) assure unidirectional lasing by introducing a direction-dependent loss. To achieve SLM operation, two air-spaced etalons are required: TE1g with a free spectral range of 3.26 GHz and

FIGURE 6.19 Original dual-ring cavity pulsed RGB holography laser from Geola (circa 2000).

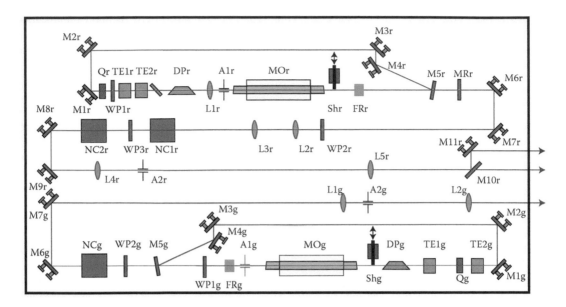

FIGURE 6.20 Optical schematic of the dual-ring cavity RGB pulsed holography laser.

TE2g with a free spectral range of 50 GHz. Frequency conversion to 526.5 nm is assured by an oven-mounted potassium dihydrogen phosphate (KTP) crystal (NCg) with wave plate (WP2g) as required for type II phase matching.

The second resonator, which is used for red and blue light generation, is defined by the four rear high-reflection mirrors (M1r, M2r, M3r and M4r) and a meniscus output coupler (M5r) having a small positive

TABLE 6.3

Parameters of First Commercial Dual-Ring Cavity RGB Pulsed Holography Laser

Parameter	Value
Master oscillator (1053 nm) flashlamp pump energy (threshold)	25 J
Measured master oscillator output energy (1053 nm)	27 mJ
Measured output energy (526.5 nm)	3.4 mJ
Energy stability of output pulses (526.5 nm) for 1200 pulses (σ)	±4%
Master oscillator (1319 nm) flashlamp pump energy (threshold)	28 J
Measured master oscillator output energy (1319 nm)	18 mJ
Measured output energy (659.5 nm)	1.8 mJ
Energy stability of output pulses (659.5 nm) for 1200 pulses (σ)	±4%
Measured output energy (439.7 nm)	1.6 mJ
Energy stability of output pulses (439.7 nm) for 1200 pulses (σ)	±8%
Coolant temperature setting	25°C
Coolant reservoir capacity:	4 L
External water temperature	<18°C
External water pressure	>150 kPa
Flow rate	<5 L/min
Cooling liquid type:	Ethanol 20% distilled water solution
Power consumption (220/240 V AC50/60 Hz, single phase):	<1250 W
Power supply PS-2241M:	<1000 W
Cooling unit PS1222CO:	2000 W
Grounding	<0.5 Ω

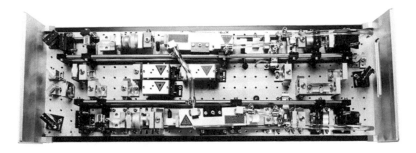

FIGURE 6.21   Later model dual-ring cavity pulsed holography RGB laser utilising super-invar resonators (circa 2002).

curvature. The active element (MOr) comprises a 4 mm diameter cylindrical Nd:YAG crystal that is 79 mm long with a 1% doping level, lasing at 1319 nm. As with the Nd:YLF crystal of the first resonator, the Nd:YAG crystal is bevelled at each end at 3° and is pumped by a 75 mm xenon flashlamp, Samarium filters and ceramic reflectors housed in a custom pump chamber. It is also antireflective (AR) coated for 1064 nm. Giant pulse operation is assured by a V:YAG passive Q-switch (Qr) with initial transmission of 53%. As above, TEM_{00} oscillation is arranged by a 2.7 mm intracavity aperture (A1r) and a Dove prism (DPr) is used to improve horizontal stabilisation. Unidirectional lasing is guaranteed by a return mirror (MRr) in addition to the Faraday rotator (FRr) and wave plate (WP1r). SLM operation requires, as with the green resonator, the use of two etalons: TE1r, a quartz etalon, and TE2g, an air-spaced etalon with a free spectral range of 96 GHz. The polariser (Pr) selects the correct cavity polarisation. Frequency conversion to 659.5 nm is assured by an oven-mounted KTP crystal (NC1r). The red and infrared signals are then combined in an LBO crystal (NC2r) where a third harmonic generation (THG) produces emission at 439.6 nm.

Table 6.3 lists the main parameters of the laser. To attain adequate stability for the intended application of digital holography, the resonators were mounted on a temperature-controlled aluminium breadboard. Later models (circa 2002) featured super-invar resonators (Figure 6.21).

6.5.2 Dual Linear-Cavity Pulsed RGB Nd:YAG Laser

To improve the stability and repetition rate of the pulsed RGB laser discussed in the previous section, the UK company, Geola Technologies Ltd., working in association with the Canadian company, XYZ Imaging Inc., produced a second-generation pulsed RGB laser in 2004 (Figure 6.22). This laser was based on twin linear Nd:YAG resonators, and replaced the passive Q-switching at 1.3 μm with the better-known crystal Co:MALO. Output was at 532 nm (6 mJ/35 ns), 659.5 nm (4 mJ/50 ns) and 438.6 nm (2.8 mJ/50 ns). The laser also incorporated electronic feedback of the cavity length to stabilise both

FIGURE 6.22   Photograph of the interior of a dual linear-cavity RGB pulsed holography laser manufactured by Geola Technologies (circa 2004).

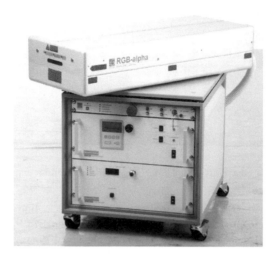

FIGURE 6.23   Modern commercial RGB pulsed holography laser from Geola.

frequency and pulse energy output; this was a technique that made it realistically possible to avoid essentially any bad pulses [40] when using the laser to write composite reflection holograms comprising well over a million separate hogels (see Appendix 3).

Usual operation of these lasers is at 30 Hz, but stable operation has been tested to 50 Hz. With appropriate pump chambers, operation should be possible with the same design at up to 120 Hz. The company Geola Digital UAB currently offers various commercial laser models based on this design using Nd:YAG, Nd:YLF and Nd:YAP crystals (Figure 6.23). Table 6.4 lists the principal characteristic parameters of these commercial laser systems.* In 2009, workers at Geola also succeeded in producing an electro-optically Q-switched version of the 1.3 μm resonator for applications in which extremely accurate pulse triggering is required [41].

Figure 6.24 shows the optical scheme for the dual-channel 2004 Nd:YAG RGB pulsed laser of Figure 6.22. The laser consists of two channels. One channel (the "G channel") generates 1064 nm laser light; green light (532 nm) is then achieved after frequency conversion. Another channel (the "R + B channel") generates 1319 nm laser light; red (660 nm) and blue (440 nm) are then achieved after second harmonic generation (SHG) and third harmonic generation (THG).

The optical schemes of the 1064 nm channel and the 1319 nm channel are nearly identical to each other, except that the position of the passive Q-switch is different. For economy of space and time, we will describe only the 1319 nm channel in detail.

The linear cavity master oscillator of the 1319 nm channel is built using a two-mirror scheme comprising an output coupler (M1r) and high-reflectivity mirror (M2r). A 4 mm diameter Nd:YAG laser rod 95 mm long is mounted in a pump chamber with a close-coupled diffuse reflector and is pumped by a xenon-filled linear flashlamp (NL7054 from Heraeus Noblelight). The close-coupled diffuse reflector provides uniform rod pumping and UV filtration. Transverse mode selection and thermal lens compensation is provided by an intracavity telescope (lens L1r and L2r). This telescope also helps to shorten the cavity length and to increase the TEM_{00} mode size several times inside the Nd:YAG active rod (MOr), so that it can produce a high-output energy at low pumping level. SLM generation is achieved with the use of two etalons (etalon TE1r and etalon TE2r), a passive Q-switch (Qr), which is placed between the intracavity telescope (while in the 1064 nm channel, the passive Q-switch is placed near the output coupler) and a rear cavity mirror operating with active feedback. To improve SLM stability, the etalon (TE2r) is placed in a temperature-controlled oven. In addition, two quarter-wave plates (WP1r and WP2r) are used to avoid spatial hole burning in the Nd:YAG laser rod and to improve the longitudinal mode selection.

---

* Very recent research at Geola has demonstrated that good lasing can be achieved at the Nd:YAG transition of 1356 nm using lamp pumping. Frequency doubling to 678 nm and tripling to 452 nm are possible.

TABLE 6.4

Specifications of Modern Commercial Pulsed RGB Holography Lasers from Geola Digital UAB

Nd:YLF	RGB-α-1353			RGB-α-1347		
Wavelength (nm)	657	527	438	657	524	438
Energy (mJ)	3.5	6.0	2.5	3.5	5.0	2.5
Pulse duration (ns)	35	35	30	35	35	30
Nd:YAP	**RGB-α-1379**			**RGB-α-1379F**		
Wavelength	671	540	447	1079	1341	
Energy	3.5	5.0	2.5	20	18	
Pulse duration	45	35	40	40	55	
Nd:YAG	**RGB-α-1964**			**RGB-α-3864**		
Wavelength	660	532	440	669	532	446
Energy	4.0	6.0	2.8	3.5	6.0	2.5
Pulse duration	50	35	50	50	35	50
Parameter				**Value**		
Typical energy stability (SD)				<3% over 10,000 pulses (without active stabilisation.)		
Beam divergence				Diffraction limited		
Linewidth				<0.003 cm^{-1}		
Beam diameter (1/e$_2$)				5...9 mm		
Beam profile				Near-Gaussian in near field, Gaussian in far field		
Repetition rate (Hz)				10/20/30/40/50		
Beam pointing stability (μrad)				<150		
Polarisation				Horizontal or vertical >90%		
Q-Switching				Passive or electro-optical		
Jitter				5 μs (passive Q-switching); 1 ns (EO Q-switching)		
Triggering				External/internal		
Laser head (L × W × H)				860 × 360 × 180 mm		
Power supply (L × W × H)				600 × 550 × 550 mm		

In the 1319 nm channel, the output energy and the pulse duration can be changed by moving the position of the Q-switch (Qr). The master oscillator generates horizontally polarised output pulses [determined by polariser (Pr) and the quarter-wave plates] with an energy of approximately 20 mJ and a pulse duration of approximately 70 ns, whereas in the G channel, fundamental output is approximately 13 mJ with pulse duration of approximately 45 ns. The beam shutter (SHTr) provides blocking of the laser beam on demand.

The output beam from the output coupler is reflected by a 45° mirror (M3r) before passing through an antireflection-coated window to reach another 45° mirror (M4r). A photodiode captures the reflected beam, providing information about pulse shape and output energy; these are used for active feedback of the cavity length through the piezo element PZTr and monitoring.

The half-wave plate after mirror (M4r) changes the horizontally polarised output beam by 45°, which is necessary for SHG conversion. The negative lens (L3r) and positive lens (L4r) together form a telescope for beam condensation, which allows high conversion efficiency in the SHG and THG crystals.

The SHG crystal is a type II [1319(e) + 1319(o)→660(e)], non-critical temperature phase-matching crystal, whose phase-matching temperature for 1319 nm is approximately 42°C. The crystal is cut along the z axis ($\theta = 0°$, $\varphi = 0°$). Because of the absence of walk-off effects in non-critical phase-matching, it is possible to use a longer crystal to achieve higher conversion efficiency. In this case, the crystal is purposely placed so that its SHG beam (660 nm) is horizontally polarised.

The THG crystal is a type I [1319(o) + 660(o)→440(e)], phase-matching crystal, whose phase-matching angle is $\theta = 90°$, $\varphi = 21.1°$. In this case, the THG beam (440 nm) is vertically polarised. The ratio of blue

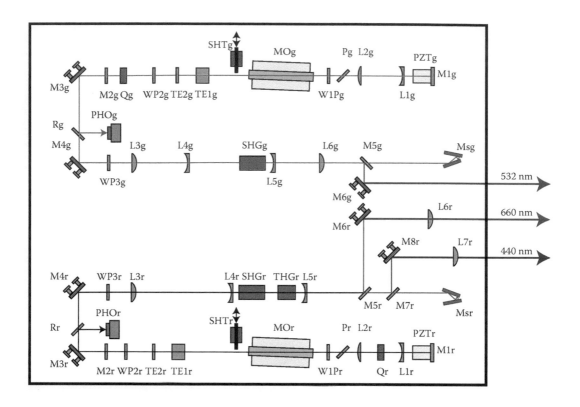

FIGURE 6.24 Optical schematic of the dual linear-cavity pulsed RGB holography laser in Figure 6.22.

(440 nm) and red (660 nm) beam output energy can be adjusted by slightly rotating the half-wave plate (WP3r).

Directly after the non-linear crystals, the fundamental beam, red beam and blue beam are collinear. The red beam is separated out and magnified by beam splitter M5r and M6r and telescope L5r/L6r, whereas the blue beam is treated by M7r and M8r, and L5r/L7r. The remaining fundamental beam is dumped to the beam diffuser (Msr).

The optical scheme of the G channel is almost the same as the R + B channel, except that, first, the position of passive Q-switch is different and, second, only one non-linear crystal is necessary for frequency conversion: a KTP type II phase-matching crystal is used to convert the 1064 nm infrared laser light into green light.

As with the dual-ring cavity laser, temperature stabilisation of the aluminium breadboard is vital for the type of stable output required for digital holography. This is arranged by a series of 27 small temperature controllers and heating pads capable of keeping every part of the breadboard at a constant temperature to within ±0.02°C (Figure 6.25). Both the linear resonators are also built using a construction employing three super-invar rods of ultra-low expansion coefficient. Final alignment of the laser must be done with its lid on such that the temperature stabilises around the resonators and is stable. This is accomplished via small holes in the lid and special tools.

The laser is able to operate at the same longitudinal mode at 1 μm and at 1.3 μm for a period of easily more than 30 min. However, after this period, mode-hops occur as the laser cavity length changes despite temperature control and the presence of super-invar resonators. This is usually caused by the efficiency of the flashlamp pumping changing in time as the lamp slowly degrades, producing different thermal loads on the active elements. Active stabilisation of the cavity lengths of both resonators proves extremely effective at stopping these mode-hops and allows continuous SLM operation at very stable energies for tens of millions of pulses. Cavity active stabilisation is covered in detail in Appendix 3. Very simply, it involves mounting the rear mirror on a piezo-crystal, or alternatively, inserting a thermally

FIGURE 6.25 Optical base of an RGB pulsed laser showing the mounting of multiple temperature controllers as well as liquid cooling and high-voltage connections.

controlled glass wedge within the cavity. The energy output from the cavity is then used to change the cavity length by either applying a voltage to the piezo-crystal or by heating the intracavity wedge. By using an algorithm that continuously optimises the output energy, the laser resonator is essentially always kept at its optimum position for oscillation at the incumbent longitudinal mode. The technique only fails when lamp degradation has continued to such an extent that the piezo or the wedge reaches their limit in producing a compensating cavity length change. Because one can tell when this is about to happen, it is possible to simply stop the laser and automatically recalibrate the active stabilisation system. Practically, when this happens, the optimal pumping voltage must be recalculated because the lasing threshold will have changed with lamp wear. Again, in modern commercial lasers, an automatic program takes care of these recalibrations.

6.5.3 Short-Cavity Pulsed RGB Lasers

All the lasers we have described previously have relatively large cavity lengths. As we have seen, one has to work hard to stabilise these lasers sufficiently for the testing application of digital holography. This is because long cavities are intrinsically unstable to thermal and mechanical stresses. One way forward is therefore to radically reduce the cavity length to produce what could be termed a short-cavity pulsed laser. At first sight, one might be tempted to raise two major objections to this idea. These are that a shorter cavity would be expected to naturally produce shorter pulse durations and also that a shorter cavity would have a smaller TEM_{00} mode volume. A shorter pulse length is a potential problem as the most reliable photosensitive panchromatic silver halide materials tend to work only well above 30 ns. The smaller mode volume is also a potential problem because it can be expected to produce a lower energy pulse.

As it turns out, neither of the above problems are serious. By choosing the resonator parameters appropriately, and by employing passive Q-switches of an appropriate initial transmission, pulse lengths greater than 35 ns can be generated without problem. In addition, the output energy at 1 and 1.3 μm is still sufficient for many applications in digital holography [42].

6.5.3.1 Green Lasers

Figure 6.26 shows a schematic of a typical 1064 nm Nd:YAG short-cavity resonator. Photographs of the resonator are shown in Figure 6.27a to c. A rear cavity mirror and resonant output coupler are held in L-shaped steel tilted mirror holders and mounted on super-invar rods. The length of the linear cavity is 128 mm. Suspension of the cavity mirror holders using flat rigid springs on the rod structure allows precise X–Y alignment while isolating the sensitive resonator from baseplate temperature variations. A small stainless steel pump chamber (visible in Figure 6.27b) with diffuse ceramic reflector is mounted on

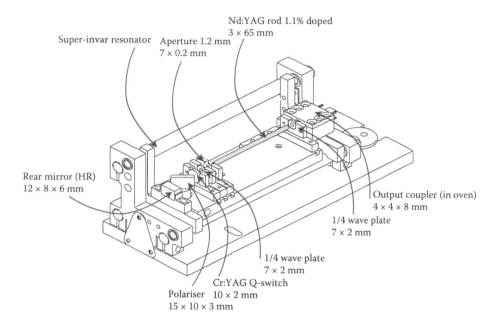

FIGURE 6.26   Short Nd:YAG resonant cavity oscillator (*l* = 128 mm).

FIGURE 6.27   Three photographs of a short-cavity Nd:YAG oscillator. (a) View of external thermally stabilised cover; (b) cover removed showing stainless steel (lamp) pump chamber mounted within the super-invar resonator; (c) pump chamber removed exposing resonator and optics.

a steel baseplate in line with an intracavity aperture, a polariser, two quarter-wave plates and a Cr:YAG passive Q-switching crystal. The Nd:YAG active laser rod (Ø3 mm × 65 mm) was excited by a Ø5 mm × 45 mm xenon-filled linear flashlamp with UV cut-off using Sm-doped glass. A laser power supply supplying pulses of up to 18 J at a repetition rate of up to 50 Hz is used to drive the pumping lamp. As usual, the function of the aperture is to restrict laser oscillation to TEM_{00}. The polariser likewise assures a single linear polarisation. The quarter-wave plates transform the cavity radiation to circular polarisation within the active element thus preventing spatial hole burning. The output coupler is usually made from an uncoated block of BK7 glass with parallel faces mounted in a temperature-controlled oven. This economical design allows frequency selection ensuring short-term SLM operation by acting as an etalon. It also allows cavity length control via computer adjustment of the oven temperature, which is required for long-term frequency and energy pulse stability.

By choosing a Cr:YAG Q-switch of an initial transmission of 76%, 43 ns pulses of SLM TEM_{00} linearly polarised radiation at 1064 nm can be generated at a lamp energy of 10 J per pulse. The energy from each pulse is approximately 1 mJ. Frequency conversion using a small reducing telescope and a type II oven-mounted KTP crystal (35.4°C) then produces stable pulses of 400 µJ at 532 nm. This is more than enough for modern panchromatic silver halide materials, which have a sensitivity of approximately 600 to 1500 µJ/cm^{-2} depending on the exact type. For a hogel of 1 mm^2, theoretically, one only then needs a pulse energy of 6 to 15 µJ. Of course, optical losses in the digital printer can increase this somewhat, but clearly, 400 µJ is far more than required.

For higher energy applications, more energy can be generated from the same laser resonator by using a Q-switch with lower initial transmission at the price of a reduced pulse duration (Figure 6.28). In addition, by placing the Q-switch between the two quarter-wave plates, greater pulse length can be attained for a given energy. In fact, pulse durations of up to 200 ns at output energies of 0.35 mJ can be attained in this way.

The short-cavity resonator is extremely stable when its temperature is properly stabilised. This is usually done by surrounding it in a metallic case that is kept at a uniform temperature by 10 or 20 individual thermocontrollers. The temperature of the output coupler is then fixed to an optimum level and this allows the resonator to produce tens of millions of pulses at the same longitudinal mode. Over 100 million pulses, it is usual to achieve a peak-to-peak energy stability of ±3.5%.

Higher energy emission at arbitrary pulse duration requires amplification. Figure 6.29 shows an example of a commercial laser (model G-MINI-B10) made by Geola Technologies Ltd., which combines a short-cavity resonator as described here with a simple lamp-pumped one-pass Nd:YAG amplifier (6 mm × 100 mm) giving more than 10 mJ at 532 nm with a pulsed length of approximately 12 ns. Typical output

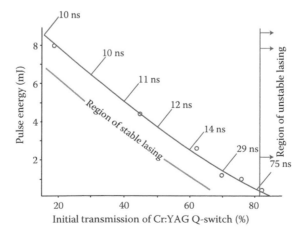

FIGURE 6.28 Pulse energy at 1064 nm versus initial transmission of the Cr:YAG Q-switch in a 128 mm short-cavity lamp-pumped Nd:YAG oscillator. Also shown are the corresponding pulse durations.

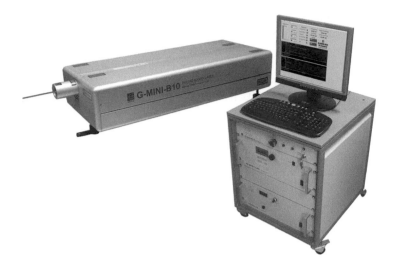

FIGURE 6.29 Commercial laser system based on the short-cavity oscillator concept designed for holography applications. The G-MINI-B10 shown here from Geola also incorporates a single-pass amplifier and frequency-doubling optics, producing an ultra-stable 532 nm output of >10 mJ per pulse at up to 10 Hz with a coherence length of >1 m.

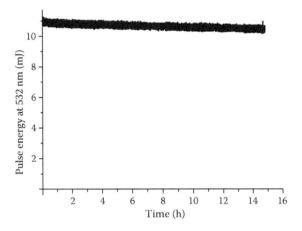

FIGURE 6.30 Typical plot of output pulse energy at 532 nm versus time of a G-MINI-B10 laser from Geola (all pulses plotted) showing ultra-stable operation. Note the gradual drop in energy due to wear of the amplifier flashlamp. Laser repetition rate is 10 Hz.

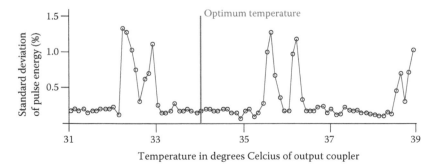

FIGURE 6.31 Computer scan of the output-coupler temperature of the short-cavity oscillator in a G-MINI-B10 showing stable and unstable regions. To guarantee long-term stability, the temperature must be set to the centre of the most stable region.

stability at 10 Hz is shown in Figure 6.30. To prevent mode-hops, the temperature of the output coupler must be set to an optimum level. This is determined by a computerised scan of oscillator output energy versus output coupler temperature (Figure 6.31).

6.5.3.2 Red and Blue Lasers

The short-cavity laser resonator described in Section 6.5.3.1 may be made to produce laser emission at 1.3 μm. Instead of the Cr:YAG Q-switch, a V:YAG or Co:MALO crystal is used. Because the gain at 1.3 μm is rather smaller than at 1 μm, a rather higher reflectivity must be used for the output coupler. Optics must be AR coated for 1064 nm as well as 1319 nm; otherwise, the stronger 1064 nm line will oscillate parasitically. The Nd:YAG rod must also be cut at 3° at both ends for the same reason. Finally, a rather longer excitation pulse must be used to excite the xenon lamp because too short a pulse will preferentially drive the 1338 nm line.

As with 1064 nm, the pulse duration of emission can be effectively tuned using the initial transmission of the passive Q-switch. For example, an energy of 1.2 mJ can be obtained at 45 ns. Typical energy stabilities for a resonator without active temperature stabilisation of the resonator cover or output coupler are 0.67% over 1000 pulses with a PTP stability of 3.7% [42]. With full temperature stabilisation, we again obtain single mode stable lasing for tens of millions of pulses.

To make a red laser emitting at 660 nm, a type II LBO crystal is used. Typical output characteristics are identical to the 532 nm laser. To make a blue laser emitting at 440 nm, however, the red 660 nm signal must be mixed with the 1319 nm signal in a type I LBO crystal using critical phase matching. The simplest way to organise this entails using two LBO crystals mounted very close to one another. Typical output characteristics for the blue laser are again very similar to the 660 nm laser except that a smaller conversion efficiency limits the energy to typically 300 μJ. Because a blue laser first entails the production of a red emission, and then mixing this signal with the infrared, a blue laser will always produce a residual red signal. By adjusting the incident polarisation to the non-linear conversion optics, a collinear red–blue laser may be made with a variable red–blue ratio.

6.5.4 Amplification at 1.3 μm: Higher Energy Emissions in the Red and Blue

Amplification at 1.3 μm is possible but quite inefficient using small seed energies. For example, starting with a 32 ns TEM_{00} SLM pulse from a short-cavity oscillator at 1319 nm of 1.2 mJ, one can expect to obtain only 8.5 mJ after two-pass amplification using a 4 mm × 130 mm Nd:YAG 1.1% doped rod with standard lamp pumping [42]. Unfortunately, one cannot practically use an SBS mirror in the two-pass scheme as the one-pass energy is well below the threshold of any useful SBS liquid.

Luckily, amplification at higher seed energies is rather easier. For example, a long-cavity 1319 nm resonator as described in Section 6.5.2 produces 17 mJ TEM_{00} SLM pulses of approximately 60 ns. These pulses may be amplified by a single 6 mm × 100 mm Nd:YAG 1.1% doped rod in a two-pass SBS scheme using $SiCl_4$ to produce 50 mJ. The SBS liquid mirror assures a good quality spatial distribution. This can be employed only because of the higher seed energy: typical one-pass gain is 2.4× and SBS reflectivity at 34 mJ is 60%. Using multiple amplifiers, energies up to 500 mJ and beyond should be easily attainable.

6.6 Pulsed Holography Lasers Based on Titanium Sapphire (Ti:Al₂O₃) and Cr:LiSAF

An alternative method to produce a blue-pulsed laser beam that is useful for holography is through the vibronic system of the titanium sapphire laser with amplification using Cr:LiSAF. This 860 nm laser can be frequency-doubled to produce a 430 nm blue beam.

Ti:Al₂O₃ laser crystals are grown using the Czochralski method and consist of sapphire doped with 0.1% Ti^{3+}. The Ti^{3+} ions replace Al^{3+} ions. An energy level diagram of Ti:Al₂O₃ is shown in Figure 6.32. Laser transition occurs between the ^2E excited state and the $^2T_2$ ground state. A large difference in the

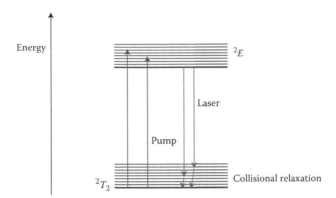

FIGURE 6.32 Energy level diagram for titanium sapphire (Ti:Al$_2$O$_3$).

electron energy distributions between these two levels and also the fact that there exists a strong interaction between the Ti atoms and the host crystal makes the transition fundamentally broad. Lasing is therefore possible over the wide band from 670 to 1070 nm, with fluorescence peaking at 780 nm. Although the material exhibits a large gain cross-section similar to Nd:YAG, it has a very small fluorescent lifetime (3.2 µs), making it rather inefficient for lamp pumping.

A similar vibronic laser to Ti:Al$_2$O$_3$ is the crystal Cr:LiSAF (Cr^{3+} LiSrAlF$_6$). This material has a much larger fluorescence lifetime of 67 µs, making it far more efficient for lamp pumping. Lasing is between the $^4T_2$ level and $^4A_2$, giving rise to a peak emission at 830 nm. Cr:LiSAF has a cross-section of around eight times less than Ti:Al$_2$O$_3$. Nevertheless, this material can be made into very large crystals, making it rather suitable for use in lamp-pumped amplifiers.

In 1996, Lutz, Albe and Tribillon [43] produced a blue-pulsed laser suitable for holography* that was based on a Ti:Al$_2$O$_3$ oscillator and a Cr:LiSAF amplifier. The laser was lamp-pumped and produced an energy of 70 mJ at 430 nm, with a pulse duration of 110 ns. The coherence length of the laser was shown to be more than 1 m.

Figure 6.33 shows a diagram of the Ti:Al$_2$O$_3$/Cr:LiSAF laser. The laser consists of three parts:

- A Ti:Al$_2$O$_3$ laser oscillator
- A Cr:LiSAF laser amplifier
- A frequency doubling system

The oscillator is of a linear type defined by mirrors M1 and M2, and has a cavity length of 1.45 m. The Ti:Al$_2$O$_3$ crystal is pumped by a flashlamp using single-pulse gain-switched operation. This simple mode does not require a Pockels cell; rather, the Ti:Al$_2$O$_3$ laser, when pumped by a single lamp pulse, simply produces a much shortened 110 ns laser pulse. Aperture A1 (2.5 mm) ensures that only the TEM$_{00}$ mode oscillates. A SF4 Brewster-incident prism (P1) is used to select horizontal polarisation and to obtain a lasing wavelength of 830 nm. SLM operation is then obtained using the three etalons (ET1–ET3). The oscillator produces an energy of 4 mJ per pulse at 110 ns, an M^2 of 2.6 and a measured bandwidth of 250 MHz.

The amplifier is a simple three-pass design using a Cr:LiSAF crystal (Ø6 mm × 101.6 mm) doped with 0.8% Cr and mounted in a double-lamp pump chamber. A pump energy of 156 J is able to produce an amplified output pulse energy of 160 mJ at 860 nm.

After amplification, the beam is frequency-doubled using a crystal of KNbO$_3$ (11.4 mm × 10 mm × 12 mm) with AR coatings for both 860 nm and 430 nm. Type I phase matching allows an energy of 70 mJ

---

* Transmission and reflection holograms were produced using this laser on silver halide emulsions.

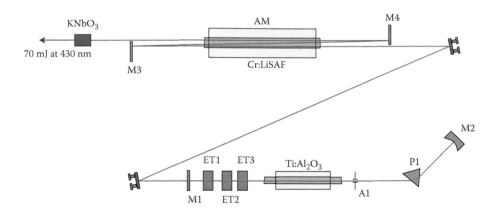

FIGURE 6.33 Optical scheme for a blue-pulsed laser based on a Ti:Al₂O₃ oscillator with amplification by Cr:LiSAF.

to be obtained at 430 nm. This type of laser should lend itself very easily to the production of the higher energies required for such applications as colour holographic portraiture.

6.7 Chromium Forsterite

Digital holographic printing is served extremely well by the neodymium lasers using the 1 and 1.3 μm transitions. However, the red radiation available from the best crystal hosts is still characterised by a rather large wavelength (657 nm). As we shall see in Chapter 13, modern red light-emitting diode illumination sources usually have wavelengths rather lower than this. As a result, for future digital printers, there is a desire to identify a laser solution having a wavelength of approximately 627 nm. One way to realise this is using the material chromium forsterite ($Cr^{4+}:Mg_2SiO_4$). This is a vibronic system, like titanium sapphire, and possesses a wide band of laser transitions centred at approximately 1235 nm. Frequency doubling by KTP can be used to produce an emission of approximately 630 nm. By suitable design of the laser cavity and insertion of frequency-selective elements, a coherence length of several centimetres can be attained, which is sufficient for many applications in digital holography. This laser is usually gain-switched by pumping with a single pulse nanosecond Nd:YAG laser at either 1064 or 532 nm.

Figure 6.34 shows a simplified energy level diagram of the Cr^{4+} ion in forsterite. In the tetrahedral field of this crystal, the free ion ^3F state is split into three states: $^3A_2$, $^3T_2$ and $^3T_1$ [44,45]. Lasing occurs through the $^3T_2$–$^3A_2$ vibronic transition. The fluorescence lifetime of the $^3T_2$ state is 2.7 μs. Excited state absorption of pump and laser radiation occurs from $^3T_2$ to $^3T_1$; this is followed by fast relaxation back to the $^3T_2$ state. The main pumping bands are at 850 to 1200 nm, 600 to 850 nm and 350 to 550 nm.

A commercial system is currently available from the Byelorussian company LOTIS TII. This is the LT-2212G (Figure 6.35), which is designed to be pumped by the LS-2132 Nd:YAG laser. The system will produce several millijoules of useful energy in a 10 ns pulse at 627 nm (Figure 6.36). A linewidth of less than 0.01 nm is achieved by the use of a grating within the cavity. Unfortunately, a pulse duration of 10 ns can be rather short for exposing panchromatic silver halide materials.* In addition, initial unpublished tests done with the laser at the Centre for Laser Photonics in Wales in 2009 showed marginal shot-to-shot stability characteristics. As such more work needs to be done on this type of laser before it can be considered as a viable solution for digital holography.

---

* Double pulse pumping can be used to achieve longer output pulses.

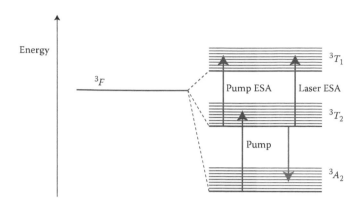

FIGURE 6.34   Simplified energy level diagram for $Cr^{4+}:Mg_2SiO_4$.

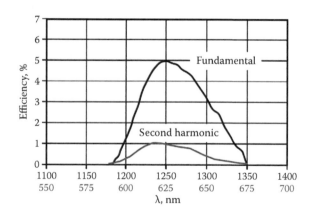

FIGURE 6.35   LT-2212G chromium forsterite laser from LOTIS TII.

FIGURE 6.36   Tuning curve for the LT-2212G laser (pumped at 1064 nm with 150 mJ using a LOTIS TII LS-2132 Nd:YAG laser).

6.7.1 Injection Seeding

One strategy for improving both the linewidth and the stability of the $Cr^{4+}:Mg_2SiO_4$ laser may very possibly be through the use of injection seeding. Although no publications have appeared to date concerning such seeding in $Cr^{4+}:Mg_2SiO_4$, there has been clear success with the $Ti:Al_2O_3$ system and this is very similar to $Cr^{4+}:Mg_2SiO_4$ [46]. In particular, a simple injection-seeded $Ti:Al_2O_3$ pulsed laser has been demonstrated which produces 4 mJ of TEM_{00} SLM pulses of 30 ns duration at 780 nm using a CW diode laser as the seeder [46]. This laser does not require other frequency-selective elements (Figure 6.37).

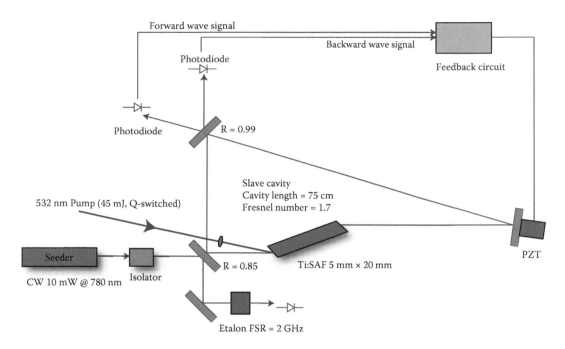

FIGURE 6.37 Optical scheme for a titanium sapphire diode injection-seeded SLM TEM00 pulsed laser pumped by a 532 nm Q-switched Nd:YAG laser. Two diodes measure the forward and backward cavity travelling waves and the piezo mirror (PZT), is used to optimise cavity length, guaranteeing SLM operation in the forward direction only. The authors suggest that this scheme might well be applied to $Cr^{4+}:Mg_2SiO_4$.

6.8 Pulsed Diode-Pumped Lasers for Holography and the Future

All the lasers we have reviewed in this chapter have, up until now, either been lamp-pumped or must be pumped using another laser. There is a very good reason for this. Lamp pumping continues to offer the most economic and sensible pumping solution for high-energy nanosecond pulsed lasers. Even for the lower energies required in digital holographic printing applications, lamp pumping is still almost always the most economical means. Nevertheless, progress in digital holographic printing has meant that the repetition rates required in this sector are increasing and, at the same time, the energy requirement is actually decreasing.* At approximately more than 200 Hz,† it becomes impractical to use lamp pumping and diode pumping becomes necessary.

Single-shot high-energy pulsed lasers still continue to be required in holography and in other interferential applications. Today, lamp pumping remains the only viable solution to these needs. Xenon lamps produce much larger instantaneous powers than diodes. For example, the typical pump time of a xenon lamp in a Q-switched Nd:YAG laser, such as the G5J, which we reviewed previously (producing 10 J at 1053 nm), is 350 μs. Within this time, approximately 1 kJ is dumped into the lamp, giving an instantaneous power of nearly 3 MW. This is many orders of magnitude greater than current diodes are capable of generating.‡ Diodes are of course much better matched to the absorption characteristics of many lasers. Their narrow band emissions can therefore be far more effective in pumping a population inversion than the brute force wide-spectrum lamp. However, this simply cannot compete with the huge advantage that large xenon lamps possess in peak power.

---

* We shall see in the next chapter that a higher resolution hologram with a smaller hogel leads to a smaller pulsed energy requirement.
† Often, the practical limit can be nearer to 100 Hz.
‡ Diodes are often monolithically packaged into "bars" for efficient pumping—a 1 cm bar typically produces around 50 W/ cm at an efficiency of 50% to 60%.

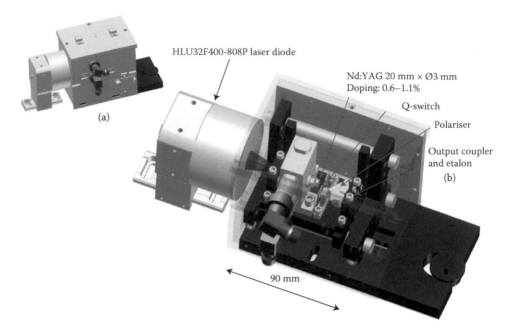

HLU32F400-808P laser diode

Nd:YAG 20 mm × Ø3 mm
Doping: 0.6–1.1%

Q-switch

Polariser

Output coupler
and etalon
(b)

(a)

90 mm

FIGURE 6.38 Prototype Nd:YAG pulsed DPSS laser pumped by a 40 W HLU32F400-808P laser diode driven by an LDD50 power supply (LIMO GmbH). Operation at both 1064 and 1319 nm was successfully achieved.

One sector that is attempting to change this equation is the inertial confinement nuclear fusion community. Current facilities, such as the National Ignition Facility (NIF), Laser Megajoule and Vulcan, use large neodymium glass lasers (essentially larger versions of the G5J) to achieve the enormous pulse energies (~2 MJ per pulse at the third harmonic of 1053 nm) required to heat a deuterium/tritium target to nuclear ignition. However, these lasers cannot be fired rapidly because of the thermal characteristics of the phosphate glass amplifiers; in addition, the laser wall-plug efficiency is extremely poor due to the intrinsic use of lamp pumping. The European project High Power Laser Energy Research (HiPER) Facility has identified key areas in the development of high-energy diode-pumped pulsed lasers. The eventual requirement is for a 200 kJ, 5 ns laser, but initially, the more moderate target of achieving a pulse energy of 10 kJ has been proposed. Currently, there are four such diode-pumped lasers worldwide that are capable of generating more than 100 J per pulse [17].

So in fact, it is possible today to use diode pumping in virtually all the holography lasers we have reviewed in this chapter. The reason this is generally not done is simply related to the cost of the diode stacks. However, as the price of diodes continues to fall, sooner or later we can expect high-energy diode-pumped solid-state (DPSS) lasers to replace most, if not all, of the current lamp-pumped solutions for holographic applications.

Some initial unpublished research into pulsed nanosecond DPSS RGB lasers for holography based on the 1 and 1.3 µm laser transitions in Nd:YAG was carried out in 2007 at the Centre for Laser Photonics in North Wales. Figure 6.38 shows a schematic diagram of the type of laser investigated. This was an end-pumped configuration using a 40 W, 808 nm diode source. TEM_{00} SLM Q-switched pulses* were attained at both 1064 and 1319 nm, having durations from 6 to 10 ns and pulse energies up to 300 µJ. This work was not continued but showed that a small DPSS RGB holography laser suitable for digital holographic printing should be relatively easy to produce.

Perhaps the greatest advantage of diode pumping is that you do not end up injecting large quantities of unuseful energy into the active medium. In lamp pumping, most of the energy ends up heating the active crystal and, in the case of Nd:YAG, introducing stress birefringence. One has to work very hard to get rid of this extra energy and to stabilise the laser in order to achieve the required stability. Diode pumping

---

* Cr:YAG was used for 1064 nm and V:YAG for 1319 nm.

circumvents this problem by simply not injecting so much waste heat. As a result, lasing stability is naturally much better. Better values of M^2 can also be attained and lifetime is superior.

One final clear advantage that diodes hold over lamps is that some weaker transitions just cannot be pumped by lamps. The much narrower band of the diode can then be instrumental in pumping such lines. For example, the $^4F_{3/2} - {}^4I_{9/2}$ ($R_1 - Z_5$) transition at 946 nm in Nd:YAG [47] is interesting as a source of blue radiation for holography (946 nm/2 = 473 nm), but its cross-section is around six times lower than the main 1064 nm line. The lower laser level of the 946 nm system is only just above the ground state and it has a thermal population of nearly 1%. This thermal population increases the laser threshold significantly and the laser essentially becomes a quasi-three-level system.

Other examples are the 942 nm transition of Nd:GSAG [48,49] and, of particular interest, the 935 nm transition of Nd:YGG ($Y_3Ga_5O_{12}$) [50]. All these transitions may be doubled to achieve useful blue emissions suitable for holography. They are especially interesting too as they achieve a better match with current light-emitting diodes which are starting to become the solution of choice for illumination of holograms (see Chapter 13).

A material of particular interest for diode pumping is Nd:YVO$_4$ (neodymium-doped yttrium vanadate). It has a stimulated emission cross-section five times that of YAG at 1064 nm, and a strong broadband absorption at 809 nm (7 times higher than YAG in the π direction). Like the other neodymium-doped crystal hosts, it can also be used to generate 1.3 µm radiation. As such, this material would be a good choice for a next-generation pulsed DPSS RGB holography laser.

REFERENCES

1. M. Yamaguchi, T. Koyama, H. Endoh, N. Ohyama, S. Takahashi and F. Iwatal, "Development of full-parallax HoloPrinter," in *Practical Holography IX*, S. A. Benton ed., Proc. SPIE **2406**, 50–56 (1995).

2. M. Lucente, *Diffraction-Specific Fringe Computation for Electro-Holography*, Ph. D. Thesis, Dept. of Electrical Engineering and Computer Science, Massachusetts Institute of Technology, (1994).

3. D. Brotherton-Ratcliffe, F. M. Vergnes, A. Rodin and M. Grichine, *Method and Apparatus to Print Holograms*, Lithuanian Patent LT4842, (1999).

4. M. Klug, M. Holzbach and A. Ferdman, *Method and Apparatus for Recording 1-Step Full-Color Full-Parallax Holographic Stereograms*, US Patent 6,330,088 (filed 1998, granted 2001).

5. D. Brotherton-Ratcliffe, F. M. Vergnes, A. Rodin and M. Grichine, *Holographic Printer*, US Patent 7,800,803 (filed 1999, granted 2010).

6. A. Rodin F. M. Vergnes, D. Brotherton-Ratcliffe, *Pulsed Multiple Colour Laser*, EU Patent EPO 1236073 (2000).

7. A. Yariv, *Quantum Electronics*, 4th Edn.,Wiley New York (1991).

8. A. E. Siegman, *Lasers*, University Science Books, Mill Valley CA (1986).

9. H. Haken, *Laser Theory*, Springer, Berlin Heidelberg (1984).

10. W. Koechner, *Solid State Engineering*, 6th Edn., Springer, Berlin Heidelberg (2006).

11. T. H. Maiman, "Stimulated optical radiation in ruby," *Nature* **187** (4736) 493–494 (1960).

12. H. Bingham, J. Lys, L. Verluyten, S. Willocq, J. Moreels, K. Geissler, G. Harigel, D. R. O. Morrison, F. Bellinger, H. I. Bjelkhagen, H. Carter, J. Ellermeier, J. Foglesong, J. Hawkins, J. Kilmer, T. Kovarik, W. Smart, J. Urbin, L. Voyvodic, E. Wesly, W. Williams, R. Cence, M. Peters, R. Burnstein, R. Naon, P. Nailor, M. Aderholz, G. Corrigan, R. Plano, R. L. Sekulin, S. Sewell, B. Brucker, H. Akbari, R. Milburn, D. Passmore, and J. Schneps, "Holography of particle tracks in the Fermilab 15-Foot Bubble Chamber," *Nucl. Instr. and Meth.* **A297**, 364–389 (1990).

13. H. Akbari and H. Bjelkhagen, "Pulsed holography for particle detection in bubble chambers," *Opt. Laser Technol.* **19**, 249–255 (1987).

14. R. Naon, H. Bjelkhagen, R. Burnstein and L. Voyvodic, "A system for viewing holograms," *Nucl. Instr. and Meth.* **A283**, 24–36 (1989).

15. G. Harigel, C. Baltay, M. Bregman, M. Hibbs, A. Schaffer, H. I. Bjelkhagen, J. Hawkins, W. Williams, P. Nailor, R. Michaels and H. Akbari, "Pulse Stretching in a Q-switched ruby laser for bubble chamber holography," *Appl. Opt.* **25**, 4102–4110 (1986).

16. V. A. Arsenev, I. N. Matveev and N. D. Ustinov, "Nanosecond and Microsecond pulse generation in solid-state lasers," *Soviet J. of Quantum Electron.* **7**, 1321 (1977).

17. N. Alexander, F. Amiranoff, P. Auger, S. Atzeni, H. Azechi, V. Bagnoud, P. Balcou, J. Badziak, D. Batani, C. Bellei, D. Besnard, R. Bingham, J. Breil, M. Borghesi, S. Borneis, A. Caruso, J. C. Chanteloup, R. J. Clarke, J. L. Collier, J. R. Davies, J.-P. Dufour, M. Dunne, P. Estrailler, R. L. Evans, M. Fajardo, R. Fedosejevs, G. Figueria, J. Fils, J. L. Feugeas, M. Galimberti, J.-C. Gauthier, A. Giulietti, L. A. Gizzi, D. Goodin, G. Gregori, S. Gus'kov, L. Hallo, C. Hermandez-Gomez, D. Hoffman, J. Honrubia, S. Jacquemot, M. Key, J. Kilkenny, R. Kingham, M. Koenig, F. Kovacs, A. McEvoy, P. McKenna, J. T. Mendonca, J. Meyer-ter-Vehn, K. Mima, G. Morou, S. Moustazis, Z. Najmudin, P. Nickles, D. Neely, P. Norreys, M. Olazabal, A. Offenberger, N. Papodogianis, J. M. Perlado, J. Ramirez, R. Ramis, Y. Rhee, X. Ribeyre, A. Robinson, K. Rohlena, S. Rose, M. Roth, C. Rouyer, C. Rulliere, B. Rus, W. Sandner, A. Schiavi, G. Schurtz, A. Sergeev, M. Sherlock, L. Silva, R Smith, G. Sorasio, C. Strangio, H. Takabe, M. Tatarakis, V. Tikhonchuk, M. Tolley, M. Vaselli, P. Velarde, T. Winstone, K. Witte, J. Wolowski, N. Woolsey, B. Wyborn, M. Zepf and J. Zhang, "HiPER: The European high power laser energy research facility: technical background and conceptual design," *DOI: RAL Technical Reports*, RAL-TR-2007-008 (2007).

18. J. E. Bjorkholm and H. G. Danielmeyer, "Frequency control of a pulsed optical parametric oscillator by radiation injection," *Appl. Phys. Lett.* **15**, 171–174 (1969).

19. U. Ganiel, A. Hardy and D. Treves, "Analysis of injection locking in pulsed dye laser systems," *IEEE J. Quantum Electron.* **12**, 704–716 (1976).

20. N. P. Barnes, J. A. Williams, J. C. Barnes and G. E. Lockard, "A self injection locked, Q-switched, line-narrowed Ti:Al_2O_3 laser," *IEEE J. Quantum Electron.* **24**, 1021–1028 (1988).

21. S. Basu and R. L. Byr, "Short pulse injection seeding of Q-switched Nd:glass laser oscillators – theory and experiment," *IEEE J. Quantum Electron.* **26**, 149–157 (1990).

22. D. H. Barnhart, R. J. Adrian and G. C. Papen, "Phase conjugate holographic system for high resolution particle image velocimetry," *Appl. Opt.* **33**, 7159–7170 (1994).

23. A. M. Rodin and D. Brotherton-Ratcliffe, "Compact 16 Joule phase-conjugated SLM Nd:YLF/ Nd:Phosphate glass laser," in *XIV Lithuanian-Byelorussian workshop on Lasers and Optical Nonlinearity*, Proc. Preila, Lithuania, (1999) pp. 51–52.

24. A. S. Dementjev, A. M. Rodin, M. V. Grichine and D. Brotherton-Ratcliffe, "High-energy phase-conjugated Nd:Glass laser for the pulsed holography," in *Spinduliuotes ir medziagos saveika, Konferencijos pranesimu medziaga, Kaunas, Technologija*, Proc. (1999) pp. 249–252.

25. M. V. Grichine, D. Brotherton-Ratcliffe and A. M. Rodin, "Design of a family of advanced Nd:YLF/ Phosphate glass lasers for pulsed holography," in *Sixth Int'l Symposium on Display Holography*, T. H. Jeong and H. I. Bjelkhagen eds., Proc. SPIE **3358**, 194–202 (1998).

26. M. V. Grichine, D. Brotherton-Ratcliffe and G. R. Skokov, "An integrated pulsed-holography system for mastering and transferring onto AGFA or VR-P emulsions," in *Sixth Int'l Symposium on Display Holography*, T. H. Jeong and H. I. Bjelkhagen eds., Proc. SPIE **3358**, 203–210 (1998).

27. A. M. Rodin, D. Brotherton-Ratcliffe and G. R. Skokov, "An automated system for the production of image-planed white-light-viewable holograms by pulsed laser," in *Practical Holography XIII*, S. A. Benton ed., Proc. SPIE **3637**, 141–147 (1999).

28. D. Brotherton-Ratcliffe, A. M. Rodin and S. J. Zacharovas, "Evolution of automatic turn-key systems for the production of rainbow and reflection hologram runs by pulsed laser," in *Holography 2000*, T. H. Jeong and W. K. Sobotka eds., Proc. SPIE **4149**, 359–366 (2000).

29. A. M. Rodin, D. Brotherton-Ratcliffe and R. Rus, "Large-format automated pulsed holography camera system," in *Optical Organic and Inorganic Materials*, S. P. Asmontas and J. Gradauskas eds., Proc. SPIE **4415**, 39–43 (2001).

30. A. M. Rodin, D. Brotherton-Ratcliffe and R. Rus, "Large-format automated pulsed holography camera system," in *Abstracts of the 2nd International Conference Advanced Optical Materials and Devices*, Vilnius, Lithuania, (2000) p. 60.

31. A. M. Rodin, S. J. Zacharovas, D. Brotherton-Ratcliffe and F. R. Vergnes, "A mobile system for 3D capture of ultrafast events," in *High-Speed Imaging and Sequence Analysis III*, SPIE **4308**, (2001) Presented, but not published, can be obtained from: www.geola.com.

32. G. Gudaitis, S. J. Zacharovas, R. Bakanas, D. Brotherton-Ratcliffe, S. Hirsch, S. Frey, A. Thelen, L. Ladriere and P. Hering, "Portable holographic camera system HSF-MINI application for 3D measurement in medicine," in *2nd Int'l Conference "HOLOEXPO-2005, Science and Practice, Holography in Russia and Abroad*, Moscow, Russia (2005) pp. 27–29.

33. R. Bakanas, G. A. Gudaitis, S. J. Zacharovas, D. Brotherton-Ratcliffe, S. Hirsch, S. Frey, A. Thelen, N. Ladrière and P. Hering, "Using a portable holographic camera in cosmetology," *J. Opt. Technol.* **73**, 457–461 (2006).

34. P. A. Apanasevich, D. E. Gahovich, A. S. Grabchikov, Ju. J. Djakov, I. N. Zmakin, V. P. Kozich, G. G. Kot, S. Ju. Nikitin and V. A. Orlovich, "Backward SRS in the Conditions of Short Pump Focusing," *Izvestiya Akademil Nauk USSR, Seriya Fizicheskay*, **53** (6), 1031–1037 (1989).

35. A. M. Rodin, A. S. Dement'ev, M. V. Grichine and D. Brotherton-Ratcliffe, "High efficiency Raman Amplification in the Dicke Narrowed Line of Hydrogen," in *Lasers and Electro-Optics (CLEO/Europe'96) Technical Digest*, Hamburg, Germany, (1996) p. 155.

36. G. L. Kachen and W. H. Lowdermilk, "Self induced gain and loss modulation in coherent, transient Raman pulse propagation," *Phys. Rev. A*, **14**, 1472–1474 (1976).

37. V. G. Bespalov and D. I. Staselko, "The influence of stimulated Raman scattering on pump radiation coherency in the saturation regime," *Opt. and Spectrosc.* **61**, 153–158 (1986).

38. I. A. Kulagin and T. Usmanov, "Analysis of self-induced amplitude-phase wave modulation in the processes of transient amplification of the stokes component," *Opt. and Spectrosc.* **80**, 944–947 (1996).

39. W. K. Bischel and M. J. Dyer, "Wavelength dependence of the absolute raman gain coefficient for the Q(1) transition in H_2," *J. Opt. Soc. Am. B*, **3**, 677–682 (1986).

40. D. Brotherton-Ratcliffe, *Laser*, US Patent 7,852,887 (filed 2002, granted 2008).

41. R. Bakanas and J. Pileckas, "Frequency doubled pulsed single longitudinal mode Nd:YAG laser at 1319 nm with pulse build-up negative feedback controls," in *Solid State Lasers XIX: Technology and Device*, W. A. Clarkson, N. Hodgson and R. K. Shori eds., Proc. SPIE **7578** 75780V-1-6 (2010).

42. D. Brotherton-Ratcliffe, "Large format digital colour holograms using RGB pulsed laser technology," in *7th Int'l Symposium on Display Holography – Advances in Display Holography*, H. I Bjelkhagen ed., Proc., River Valley Press, Iowa City, IA (2006) pp. 200–208.

43. Y. Lutz, F. Albe and J.-L. Tribillon, "Etude et réalisation d'un laser à solide pulsé émettant dans le bleu pour l'holographie couleur d'objets dynamiques", *C.R. Acad. Sci. Paris*, t. **323**, Série Iib, 465–471, (1996).

44. W. Jia, H. Liu, S. Jaffe and W. M. Yen, "Spectroscopy of Cr^{3+} and Cr^{4+} ions in forsterite," *Phys. Rev. B*, **43**, 5234–5242 (1991).

45. T. S. Rose, R. A. Fields, M. H. Whitmore and D. J. Singel, "Optical Zeeman spectroscopy of the near-infrared lasing center in chromium:for-sterite," *J. Opt. Soc. Am. B* **11**, 428–435 (1994).

46. T. D. Raymond and A. V. Smith, "Injection-seeded Titanium doped Sapphire laser," *Opt. Lett.* **16**, 33–35 (1991).

47. T. J. Axenson, N. P. Barnes, D. J. Reichle and E. E. Köhler, "High-energy Q-switched 0.946 μm solid-state diode pumped laser," *J. Opt. Soc. Am. B* **19**, 1535–1538 (2002).

48. F. Kallmeyer, M. Dziedzina, X. Wang, H. J. Eichler, C. Czeranowsky, B. Ileri, K. Petermann and G. Huber, "Nd:GSAG-pulsed laser operation at 943 nm and crystal growth," *Appl. Phys. B.* **89**, 305–310 (2007).

49. Z. Lin, X. Wang, F. Kallmeyer, H. J. Eichler, C. Gao, "Single-frequency operation of a tunable injection-seeded Nd:GSAG Q-switched laser around 942nm," *Opt. Expr.* **18**, 6131–6136 (2010).

50. J. Löhring, A. Meissner, D. Hoffmann, A. Fix, G. Ehret and M. Alpers, "Diode-pumped single-frequency-Nd:YGG-MOPA for water–vapor DIAL measurements: design, setup and performance," *Appl. Phys. B* **102**, 917–935 (2011).

7

Digital Colour Holography

7.1 Introduction

The earliest approach to "digital holography" sought to calculate numerically the complex wave front scattered by a virtual object [1,2]. By combining this wave front with a virtual reference beam, the resultant interference pattern could be calculated within a given medium. From the mid-1960s onwards, researchers used a combination of printing and photographic reduction methods to produce crude synthetic holograms from such calculated patterns. The technique of numerically synthesising and then physically encoding the interference pattern corresponding to a virtual object has today come to be known as computer-generated holography (CGH). The availability of comparatively cheap computational power has led to CGH now being used routinely to record high-quality transmission holographic gratings encoding either image or non-image data using electron beam lithography. Modern techniques allow the mass replication of such gratings for applications such as holographic security features and holographic optical elements (HOEs).

CGH is ideally suited to transmission holography because, in this case, the holographic information contained in the interference pattern is essentially two-dimensional (2D) in nature. In contrast, reflection holography intrinsically requires three-dimensional (3D) information; as such, no commercially realistic solution exists today for writing CGH data in the form of a reflection hologram. We have, however, seen in the preceding chapters that reflection holography is far more suited than transmission holography for the task of recording and playback of high-fidelity, distortion-free full-colour images. Historically, this led researchers to look for an alternative technique for digital holography: one that would be more naturally suited to the creation of reflection gratings.

7.2 Holographic Stereograms

As early as 1967, Pole [3] had been working on a technique that was rather different from CGH. In Pole's experiments, he was able to create a crude reflection hologram based on multiple photographs. It was intrinsically a two-step process: first, a 2D matrix of small lenslets was used to image photographs of an object taken from many different horizontal and vertical perspectives. The second step was an optical transfer of the lenslet matrix to a reflection hologram. Upon viewing this reflection hologram, the eye coincided with a virtual image of the lenslets and a 3D effect was perceived. Pole reported that the resulting "holographic stereograms" exhibited full three-dimensionality, exactly like ordinary holograms, but that the large inactive area between the lenslets caused image degradation akin to viewing an ordinary hologram through a coarse grid structure. He concluded that the optimum lenslet size would be equal to the diameter of the human eye so as to best accommodate the compromise of sampling and depth of field inherent to the new display.

In 1969, DeBitetto [4] reported an alternative system in which a masked holographic plate was sequentially exposed to different perspective view images.* This solved the resolution problem inherent in Pole's work, as with a contact aperture, the holographic exposures could be spaced with virtually no inactive area between them. Subsequent work by King et al. [5] reported the production of a white

---

* The images were projected with laser light onto a diffusive screen in front of the masked holographic plate.

light-viewable image plane hologram (an H_2) from a DeBitetto type (H_1) master. However, because the sequential exposure of the component holograms of a DeBitetto type H_1 required much more recording time than Pole's technique, the vertical parallax information was discarded, thus reducing the number of necessary exposures.

During the 1970s, Lloyd Cross [6] and others, inspired by the invention of the rainbow hologram [7], tackled the problem of generating holographic stereograms in yet another way. Here, transmission holography was used to produce bright rainbow holograms (without vertical parallax) using large cylindrical lenses for recording and cylindrical films for display. However, this type of system, although popular for a time, proved ultimately to be rather inferior to the DeBitetto/King approach.

By the early 1990s, most large stereograms had therefore started to be recorded as reflection holograms using the DeBitetto/King model. In 1991, Walter Spierings and his company, the Dutch Holographic Company B.V. introduced the first full-colour reflection stereograms [8] (see Chapter 1, Figure 1.45).* Although impressive, these holograms were still derived from analogue photographic data. The transition to digital data, however, was already starting. Stephen Benton and his group at the Massachusetts Institute of Technology were probably the earliest workers in this field—and certainly the most influential. In 1991, Halle et al. [9] described the ultragram. The invention allowed one to record a two-step holographic stereogram with an arbitrary transfer distance. This was the first use of digital image distortion techniques and provided a clear reason for going "fully digital". The advent of digital cameras and cheap spatial light modulators in recent years has only reinforced this doctrine. The original DeBitetto/King model is still used successfully today to produce full-colour horizontal parallax reflective holographic stereograms from digital camera or computer data.

7.3 One-Step Digital Holograms

With the advent of digital spatial light modulators (SLMs), a different avenue became available to create a high-resolution reflection hologram from computer or camera data. This was one-step or direct-write digital holography (DWDH). In analogue holography, an interference pattern is created by the superposition of the wave fronts of an object and reference wave within a photosensitive plate or film. CGH seeks to synthesise this pattern numerically. However, suppose that instead of calculating such a global interference pattern all at once, one breaks down the problem into writing only a small element of a hologram at a time. In other words, we consider the required hologram as being composed of a plurality of small microholograms arranged in the form of an (x, y) grid. The problem now reduces to writing sequentially each such microhologram. Of course, we could still calculate the interference pattern of each such microhologram via the methods of CGH and write them using electron beam lithography. However, with the advent of liquid crystal displays (LCDs), a far simpler solution became possible: a reference and object beam could be made to intersect at the surface of a photosensitive material to directly create the microhologram. The object beam is encoded with image data by being made to pass through a spatial light modulator such as a LCD and a lens system. A step-and-repeat mechanism then writes a plurality of juxtaposed microholograms (these came to be known as holographic pixels or hogels [10]).

There are many advantages to DWDH. The first and most evident is that, unlike CGH, it lends itself naturally to colour reflection holography; by using three or more laser wavelengths, full-colour reflective hogels can be written at the same physical location. The spatial frequencies within the hogels are indeed very large, but by using the natural interference process to generate each hogel, one is freed from having to use costly techniques such as electron beam lithography to attain the required high spatial resolution. Because the hogel is inevitably chosen to be small (usually in the range of 0.1 mm diameter to several millimetres) only small lasers are potentially required.

In fact, it turns out that depending on what image data is recorded, master holograms can also be generated in this fashion, hogel by hogel. These holograms can then be optically transferred to an H_2. When this is done, the technique is known as master-write digital holography (MWDH). MWDH is effectively

---

* These types of holograms were generated in a two-step H_1:H_2 process. The H_1 was made by laser projection of multiple analogue camera perspectives onto a diffusion screen—this process came to be known as MPGH.

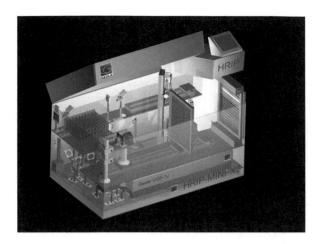

FIGURE 7.1   Early concept design (circa late 1990s) of a digital multiple photo-generated holography printer from Geola.

a DeBitetto/King full-parallax H_1 with a small square or hexagonal aperture. However, in modern systems, the aperture and the diffusion screen are usually replaced by a lens system.

In 1988, Yamaguchi et al. [11] became the first group to report experimental demonstration of DWDH. In their one-colour system, a 2D perspective sequence was generated. This was then image-processed to form an array of hogel mask frames, which were recorded on video tape and downloaded one-by-one to a twisted nematic LCD. A laser beam was used to illuminate the LCD and a lens system employed to record a volume reflection hologram of the Fourier transform of each mask. Each such hologram constituted a hogel and, by sequentially advancing the holographic plate between exposures, a matrix of abutting hogels was created. The resulting hologram reconstructed an accurate full-parallax view of the original scene. The process seemed promising, but the 320 × 240 hogel array required many hours to record.

In the late 1990s, Klug et al. [12], working at the US company Zebra Imaging Inc., extended the technique of Yamaguchi et al. to large-format, full-colour reflection holography. Zebra Imaging proved beyond a doubt that the DWDH technique was capable of generating large-format digital colour holograms of a quality never before imagined. In 1999, Brotherton-Ratcliffe et al. [13–16], working at the Lithuanian company Geola, subsequently demonstrated that the technique could be made to work much faster and more reliably using pulsed RGB lasers.

During the last decade, pulsed laser DWDH has been developed and used commercially by several companies, most notably Geola, XYZ Imaging Inc.* and Zebra Imaging. More recently, a dual-mode printer capable of writing holograms under both DWDH and CGH has been described by Kang et al. [17]. In 2009, a pulsed laser DWDH system was also developed, allowing the rapid generation of erasable digital holograms on photorefractive polymer [18].

7.4 A Simple DWDH Printer

7.4.1 Optical Scheme

Some early DWDH printers used a recording scheme very similar to the original DeBitetto scheme. A simplified diagram of such a printer is shown in Figure 7.2. A continuous wave (CW) laser is used to produce a reference and object beam through the use of a polarising beam splitter (PB). The reference to object ratio is controlled by the 1/2 wave plate (WP1) and the polarisations in the two beam paths are equalised by WP2. The object beam scheme basically consists of a projection system based on an

---

* More recently trading under the name Rabbitholes Media Inc.

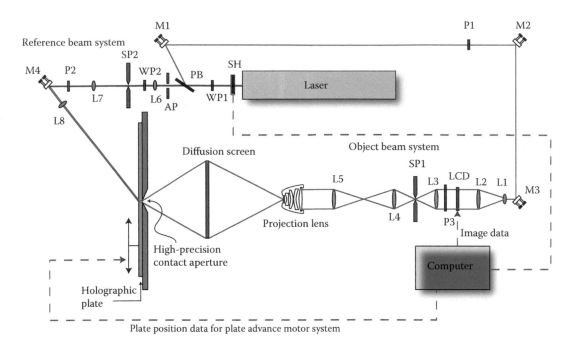

FIGURE 7.2 Simple DWDH monochromatic reflection hologram printer based on a CW laser. The object beam system comprises an LCD projector and a diffusing screen. A contact aperture is used to define the hogel. The device operates using a simple step-and-repeat sequence.

LCD panel. The beam polarisation is first cleaned up by the thin-film polariser (P1) before being collimated and expanded by the telescope (L1/L2). The LCD is a twisted nematic display, which modifies the polarisation of the radiation passing through it by its changeable birefringence. The polariser (P3) converts this polarisation change into an amplitude modulation. High-frequency noise is then removed from the transmitted object beam by the spatial filter (L3/SP1) and an image of the LCD is relayed to the projection lens (L4/L5). The projection lens then forms a high-quality distortion-free image of the LCD on the diffusion screen. Because the holographic plate is masked by a precision square contact aperture that defines the hogel, only the photosensitive emulsion directly within this hogel aperture can "see" the diffusing screen.

The reference beam system consists of three lenses: an aperture, a spatial filter and a polariser. The spatial filter (SP2) cleans out high-frequency structure in the beam and the polariser (P2) ensures a linear polarisation exactly matching the object beam. The lenses (L6–L8) and aperture (AP) define the shape of the collimated reference beam at the emulsion surface; this is usually chosen to be very close to the object beam hogel shape.

7.4.2 Speckle Blur

Because the object beam projection/diffusing system is a coherent system, it is subject to speckle. The speckle size will increase as the hogel size decreases, leading to a loss of angular resolution in the hologram. If unchecked, this will induce image blurring. Image blurring and speckle are treated in Chapter 11. The simplest solution to controlling speckle blurring is to incorporate an additional diffusing element upstream of the projection system.

7.4.3 Operation

The printer works by using a simple step-and-repeat procedure. The shutter (SH) is opened for a predetermined period of time and a hogel is written. The shutter is then closed and the holographic plate is

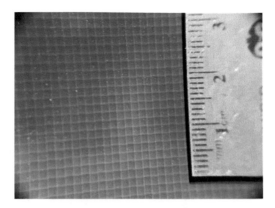

FIGURE 7.3 Magnified view of a small section of a DWDH hologram showing the matrix of hogels or elementary holograms (here, the hogels are 1.6 mm in diameter). Simple DWDH printers print one hogel at a time. At the end of each line, the printer drops to the next line and changes direction. Common hogel sizes range from just over 100 µm to several millimetres. It is also possible to print hexagonal hogels.

advanced. After a settling period, which is required for the system to reach interferometric stability, the shutter is again opened for the predetermined exposure time and the next hogel is written. The plate is moved one column at a time until a full line of hogels is finished. It then drops down a line and starts off in the other direction (Figure 7.3).

7.4.4 Image Data

The image data required by any DWDH printer can be derived from real-world images using devices such as holocams or structured light camera systems (Chapter 10). However, significant image processing needs to occur to get the data into a form ready for printing hogels. We shall discuss image-processing methods for both DWDH and MWDH in Chapters 8 and 9.

7.4.5 Deficiencies

There are several problems with this simple DWDH printer. First, the use of a contact aperture to define the hogel is difficult. Many emulsions are physically sensitive and the aperture must of course be in intimate contact with this sensitive surface. This usually means that an electromechanical system must lift the aperture away from the emulsion when the plate is moved and then gently push it back against the emulsion before exposure.* However, this takes a lot of time, and with the typical hogel size being less than 1 mm², even a 30 cm × 40 cm hologram can require 120,000 hogels. Another problem with contact apertures is that the final DWDH hologram can exhibit a clear grid-like structure. It can also be difficult to stop scattered light from actually recording a hologram of each aperture itself. Finally it can be difficult to guarantee exact alignment and proper contact at each exposure.

The use of a classic diffusing screen and an object beam projection system allows the recording of wide angle-of-view holograms with undistorted images. However, only an extremely small part of the object beam is actually used to expose the hogel! The result is that a large power laser is required and one ends up throwing away 99% of the power. Holographic diffusers can greatly improve this situation as they can diffuse the light into a small predetermined area that can be matched to the hogel. However, there is still a general problem with any type of diffusing system. This is the propensity of such systems to induce image blurring. The problem is that if the diffuser is too small and too close to the hogel, or if the hogel is too large, then the rays connecting a projected LCD pixel on the diffusing screen and any

---

* The commercial printer marketed by XYZ Imaging Inc. in 2005 actually used a system whereby the AgX film was sucked by a vacuum system onto an aperture in front of the writing optics. The function of the aperture was not to apodise the light beam in this printer but to ensure a flat film surface. There was indeed contact between the emulsion surface as it was dragged over the aperture and this caused many small scratches.

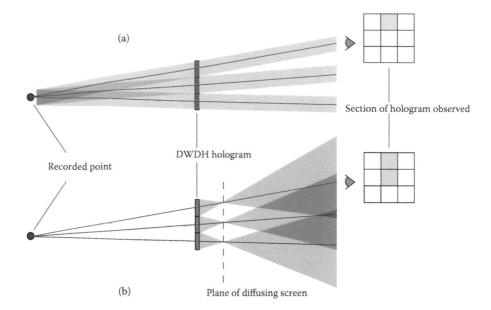

(a)

Section of hologram observed

Recorded point

DWDH hologram

(b)   Plane of diffusing screen

FIGURE 7.4   Diagram illustrating the induced blurring in a DWDH hologram, which occurs when too small a diffusion screen is used to write the hologram. Ideally, the ray bundles connecting a recording point with each hogel should be non-diverging as shown in (a). When too small a diffuser is used to record the hologram as in (b), the ray bundles become strongly diverging and, as a result, hogel bleeding occurs. The diagram only shows the central ray bundles for clarity—which is why there seem to be gaps in the viewing scenario in (a). In general, each hogel will be associated with just enough collimated ray bundles to fill these gaps, but then any divergence of the ray bundles in excess of the eye's angular resolution will lead to hogel bleeding as soon as the observation distance is greater than the diffuser distance.

hogel will, in general, form a cone with a relatively large angle—and the larger this angle, the greater the image blurring and the smaller the in-focus depth of the hologram (Figure 7.4). It can therefore be difficult to make compact printers capable of producing higher-resolution wide-angle holograms using diffusing screens.

A final problem with this simple printer is the time required to print a hologram. Interferometric stability must be present at each exposure. This translates into a long step-and-repeat time even when a high-power laser is used. With the added complication of a contact aperture, this can lead to print times of days for small holograms. Realistically, any small disturbance within this period of time is likely to create a badly exposed hogel and can effectively ruin the hologram.

7.5 Modern DWDH Printers

Modern DWDH printers use a number of major improvements over the simple system described in the previous section. First, and most importantly, small-pulsed lasers are often used to solve the stability problems inherent to the use of CW lasers. Second, the hogel is usually formed using an optical system rather than relying on an awkward contact aperture combined with a projection/diffusion scheme. This then allows the step-and-repeat plate movement system to be replaced by a constant velocity system.

7.5.1 Use of Pulsed Lasers

The use of nanosecond-pulsed lasers in DWDH printers can completely solve the problems of interferometric stability and low printing speed that plague CW laser printers. This is of fundamental importance because this means that small-power lasers may be used to print large holograms at reasonable times. With a pulsed laser, the holographic exposure is effectively done in such a small period of time that there is no need to let the system settle.

We shall see in the next section that the problematic contact apertures often used in early printers were quickly discarded and, as such, it became possible to simply move the holographic plate at a constant velocity while a constant repetition rate laser wrote sequential hogels. The maximum rate at which hogels can be written using this system is determined by the duration of the laser pulse, τ. If we demand that within this duration, the holographic plate may only move by one-tenth of a wavelength of light, then for a hogel of diameter, δ, the maximum hogel write rate is given by the simple formula

$$f = \frac{\lambda}{10\delta\tau} \tag{7.1}$$

For a pulse duration of 40 ns, which is typical of a Q-switched pulsed laser, and a hogel size of 0.5 mm, this equates to a rate of nearly 2.7 kHz at 532 nm! The pulse energy required to expose a single hogel is also extremely small. Taking a film sensitivity value of 2000 μJ/cm^2, for example, one can see that the ballpark figure for the energy per hogel is approximately 10 μJ. This assumes, of course, that one has an optical system that (unlike the diffuser system described previously) does not waste energy. If one assumes a realistic hogel write rate of 100 Hz, then the power requirement on the pulsed laser is only 1 mW!

This calculation can be compared with a CW laser system, which also functions with a constant velocity plate displacement system at a hogel write rate of 100 Hz. Here, a laser shutter must constrain the exposure time so that movement of less than one-tenth of a wavelength occurs during the exposure. The exposure must thus be limited to approximately 1 μs. To get sufficient energy for the exposure, this then requires a laser having a CW power of 10 W—or 10,000 times more than that required of the pulsed laser!

7.5.1.1 Microsecond Pulsed Lasers

There is therefore an overwhelming case for the use of pulsed lasers in DWDH printers. Notwithstanding this, there is one major problem here! Some of the best photosensitive materials for colour holography are not properly sensitive to nanosecond laser pulses. For example, dichromated gelatin is a superb photosensitive material that can be used for colour holography—but it produces poor results with nanosecond pulses. Photopolymers, which are wonderfully convenient due to their freedom from wet chemical development, can also fall into this category.

One positive indication is that some of these materials, particularly photopolymers, can show good sensitivity to multiple nanosecond pulses. Therefore, future DWDH printers may well use pulsed lasers producing emissions of a few microseconds' duration—or nanosecond pulse trains with envelopes stretching to several microseconds. Although write rates somewhat lower than 100 Hz may be required, we shall see later that there are methods for writing multiple hogels with every laser pulse. However, microsecond lasers are unfortunately more complex to produce than nanosecond lasers and usually require complex fast-switching high-voltage electronics. We have already reviewed a simple version of a pulse stretcher as applied to a ruby laser in Chapter 6. Recent work at the Geola organisation has also tentatively demonstrated the feasibility of active Q-switching systems for microsecond RGB-pulsed lasers. To date, however, no prototype printer using these longer pulse-length lasers has been tested.

7.5.2 Lens-Based Printers

Special lens systems are frequently used to replace the contact aperture and projection/diffusion scheme described above. Such lens systems create the hogel optically by focussing the light transmitted by the spatial light modulator into a narrow waist. The light distribution at the hogel then effectively becomes the Fourier transform of the distribution at the SLM. Such lens systems can also create a greatly enlarged image of the SLM downstream of the Fourier plane (Figure 7.5).

The contact aperture and projection/diffusion system can therefore be conveniently replaced by a compact non-contact optical system. The exact shape of the hogel can be precisely defined by an aperture placed at any optical plane, which is conjugate to the Fourier plane of the main objective lens.

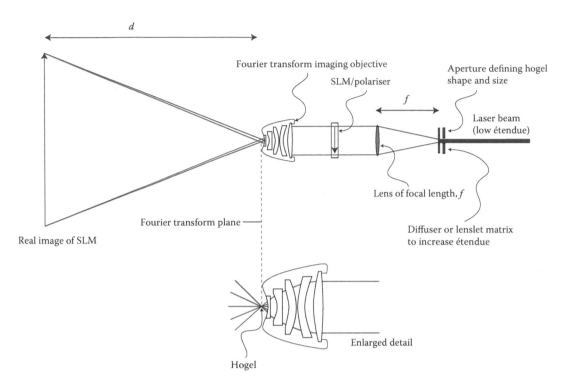

FIGURE 7.5 Lens-based hogel-forming system. Here, a Fourier transform imaging objective defines the hogel shape and also creates an image of the spatial light modulator downstream of the Fourier plane. The distance, d, should be large to avoid blurring of the hologram. The hogel shape may be conveniently defined by the shape of an aperture placed at a conjugate plane.

This type of system has great advantages. There is no need for any contact aperture and almost all the energy transmitted through the spatial light modulator is now used to record the hogel. The laser energy required is therefore extremely small. There are, however, two disadvantages we should mention. Both are related to the fact that it is difficult to design high-resolution Fourier transform imaging systems of high numerical aperture.

Unfortunately, the angular field of view of a DWDH hologram is defined by the conic angle of convergence of the focussed light forming the hogel. Usually, this angle must be large—in the order of 100° if the hologram is to be viewed from a variety of positions. This means that the numerical aperture of the Fourier transform optic must also be very large. However, as the numerical aperture increases, optical aberrations increase rapidly. These aberrations decrease the resolution of the optical system, which in turn induces image blurring in the final hologram.

It turns out that one can significantly increase the resolution of the optical system at high numerical apertures if one accepts a finite fifth coefficient (Barrel distortion) or, in other words, if one accepts that the image of the spatial light modulator produced by the optical system will become rather distorted (but not blurred). In Chapters 8 and 9, we shall discuss this problem, and in particular, we shall see how it can be dealt with in the context of image processing.

The second problem that arises through the need for a high numerical aperture system is related to the fact that higher resolution is always available from a monochromatic Fourier optical system as compared with that available from the corresponding apochromatic system. This usually results in three separate optical schemes being adopted in a DWDH printer, one for each primary colour. Such printers can be classified as triple-beam systems because they use three separate object/reference beam pairs to write three primary colour hogels in different physical locations of the holographic film. Of course, if large image depth or large fields of view are not required in a DWDH hologram, then it becomes possible to employ a single apochromatic optical system and to print single RGB hogels, one at a time.

7.5.3 Speckle in Lens-Based Printers

When a Fourier transform lens system is used in a DWDH printer, the hogel size must be controlled by the étendue of the object beam. An aperture in contact with a holographic diffuser and a Fourier transforming lens may therefore be conveniently used to define both the average étendue and the exact hogel shape. This system also provides ray averaging in that a single point at the image plane of the lens system is now connected to a single point on the spatial light modulator by multiple rays that travel different paths through the optical system. However, for small hogel sizes, too little averaging may be available due to the correspondingly small aperture size. In this case, speckle will appear at the real image of the spatial light modulator downstream of the lens system. Once again, this speckle can degrade the image by inducing image blur. Increasing the diffuser size, as can be done in a lensless printer (according to our previous discussions), is of course not an option here as the diffuser size is now directly coupled to the hogel size. One effective solution is to use a microlens array instead of a diffuser. By choosing the pitch of the lenslet matrix to be larger than a certain critical amount, lower spatial frequencies, which are predominantly responsible for the visible speckle and induced blurring, are eliminated. Of course, if too large a pitch is selected, then given that the area of the lenslet matrix is fixed by the hogel size, inefficient averaging will occur and again the image quality will be degraded. Nevertheless, for most hogel sizes, the lenslet matrix approach works well. An alternative solution that is sometimes adopted is the use of quasi-random phase plates, which are used to randomise the phase at the spatial light modulator.

7.5.4 Triple-Beam Printers

In its simplest form, an RGB triple-beam DWDH printer comprises three relatively identical optical channels—one for each of the three primary colours. Each optical channel comprises a laser emitting at a primary wavelength and an optical system for forming an object beam and a reference beam that are brought into physical coincidence at the surface of the photosensitive material where a hogel is formed.

7.5.4.1 Hogel-Writing Sequence

The hogel-writing sequence is illustrated in Figures 7.6 and 7.7. At first, the film or plate is moved at a constant speed and the lasers triggered at a constant interval such that a row of hogels is created for each of the three primary colours, each hogel being horizontally juxtaposed with respect to its neighbour. This is illustrated in Figure 7.6a.

Next, laser emission is blocked and the electromechanical system winds up the film or plate by an amount equal to the hogel diameter. The process described previously then restarts as the plate/film is moved to the right and a new line of hogels for each colour is formed under the previous line as shown in Figure 7.6b. The writing process continues in this way as illustrated in Figure 7.6c, d and e. Figure 7.6f shows the situation after the sixth line has been written.

The writing sequence continues in Figure 7.7. In Figure 7.7a, seven lines have been written. In Figure 7.7b, the eighth line of red hogels is seen to overprint the first line of green hogels. Similarly, the eighth line of green hogels overprints the first line of blue hogels. This overwriting process continues with further lines being overwritten until the blue hogels, which have already been overwritten by the green hogels, now start to be overwritten by the red hogels. As shown in Figure 7.7f, this process produces hogels that have all three primary colours.

There are several points to make about this writing procedure. Clearly, the distances between the centres of the writing locations for each colour channel have to be an integral multiple of the hogel diameter for the different colour hogels to coincide; this practically then leads to a constraint on the hogel diameter in a given printer. Typical hogel sizes are 1.6 and 0.8 mm. A typical distance between the red and green or green and blue writing locations in modern printers is 80 mm, corresponding to between 50 and 100 hogel diameters.*

---

* A distance between red and green writing locations of 7-hogel diameters was used for illustration purposes only in Figures 7.6 and 7.7. In modern printers, the figure is closer to 50 to 100.

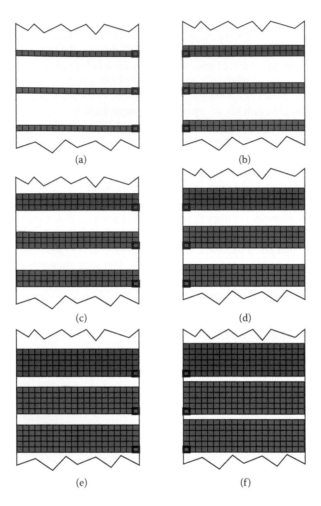

FIGURE 7.6 Hogel-writing sequence in a triple-beam DWDH printer. In (a) the photosensitive film has just been moved from right to left and a row of hogels written by each objective (the black square). In (b) the film has been moved up one step and then moved horizontally to the right while another line of hogels is written. The remaining figures show the progression of this process with three lines (c), four lines (d), five lines (e) and six lines (f) being written.

In practice, it is common for the writing process to start with the top of the photosensitive material directly under the bottom blue writing head. Writing starts only with the blue channel activated; only when the photosensitive material has moved up and the green head starts to overwrite the blue hogels is the green channel actually switched on. Likewise, the red channel is activated only when the red head actually starts to overwrite the blue/green hogels. At the end of the printing process, a similar inverse process is enacted whereby the blue and then the green heads are deactivated before the red head terminates the last line. In this way, all hogels printed contain the three primary colours.

7.5.4.2 Basic Systems in RGB-Pulsed Laser Triple-Beam Printers

Figure 7.8 shows a schematic optical and control diagram of a DWDH printer manufactured in 2001 by the Geola organisation for the production of large-format RGB horizontal parallax-only (HPO) reflection holograms (Figure 7.8). We will use this printer to illustrate how a modern pulsed laser triple-beam DWDH printer works before going on to discuss more complex variants. For clarity, Figure 7.8 shows only one colour channel. In the printer itself, there are three such colour channels (red, green and blue) which are schematically identical. The printer is designed to print reflection type hogels onto silver halide plates of a size up to 800 mm × 800 mm, using three pairs of object and reference beams powered

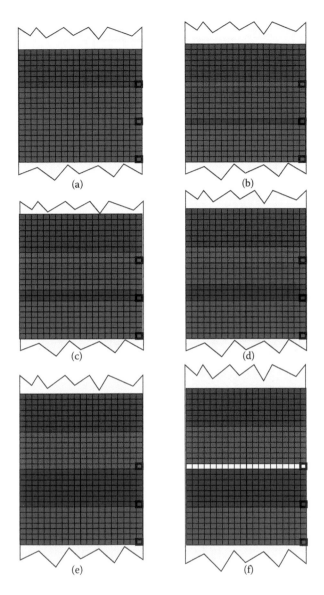

FIGURE 7.7 Hogel-writing sequence continued from Figure 7.6. In (a), seven lines of hogels have been written in each colour. In (b), the different colours start to overlap. Now a line of red hogels has been written over the green and a line of green hogels over the blue. By (e), additional lines have been written, and in (f), the first line of overlapping red, green and blue hogels (shown in white) is produced.

by an RGB-pulsed laser operating at 15 Hz. As with all triple-beam printers, each object and reference beam pair is made to intersect at a given location on the photosensitive film.

7.5.4.2.1 Control and Video Image Stream System

The printer is controlled by a DELL precision workstation 530 computer with twin Intel Xeon 1.4 GHz processors, Matrix Millennium G450 graphics card, an SCSI Raid HDD of 160 MB and an additional SCSI HDD of 73.4 GB and 1 GB of RDRAM running MS Windows 2000 Professional. An XVGA video signal connects the computer to a video splitter. This splitter drives a display monitor in addition to a CRI graphics controller card which feeds three Sony XGA1 1024 × 768 LCD panels for object beam data encoding. The printer includes many motorised microrotation and microtranslation stages for the

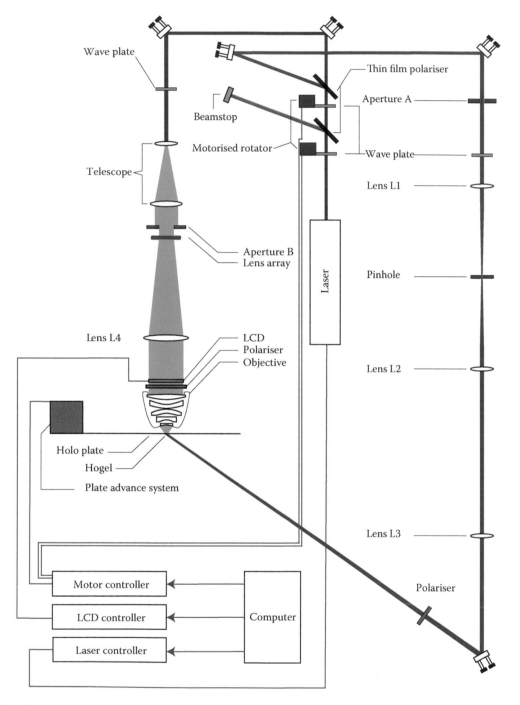

FIGURE 7.8 Optical and control schematic for the 2001 RGB-pulsed laser DWDH printer manufactured by Geola.

automatic adjustment of beam energies and ratios, in addition to electromechanical shutters. Controllers for these components are mounted with the main control computer in a large control rack (Figure 7.10).

7.5.4.2.2 *Mechanical Plate Displacement System*

The mechanical plate movement system comprises a vertical translator and a horizontal translator. The vertical translator consists of an LF6 200-mm width rail from the German Company Isel Germany AG

FIGURE 7.9 Photograph of the 2001 RGB-pulsed laser DWDH printer manufactured by Geola. Note the large plate-holder with 2D electromechanical displacement system in the foreground. To the right of the photograph can be seen the reference beam system. In the background are the main optical unit and laser (left) and the electronic control rack (right).

FIGURE 7.10 Photograph of the main control rack of the 2001 Geola printer showing the computer (top) and electronic controllers (bottom). To the left can be seen a separate smaller rack for the laser power supplies.

and a 16 × 5 ball screw spindle motor with gearbox (Shrittmotorantiesbstmodule 2430 Ncm, Amph Z-Achse, Dir. Antrieb m.Bef.Flan.ZF3 reicht). The horizontal translator comprises an Isel LF4 rail with 16 × 5 ball screw MS160 spindle motor. A master microprocessor-based controller (mounted in the main control rack) controls the triggering of both laser and stage motion.

7.5.4.2.3 System Architecture

The printer is limited to producing HPO holograms.* HPO holograms require much less memory storage and this allows a simplified computer system to be used, with image files being uploaded over a network.

---

* In later models, this was upgraded to also allow the creation of full-parallax holograms.

Holograms are created using perspective view information, which is generated either from a holocam (Chapter 10) or through commercially available computer modelling programs. The perspective information is uploaded to the control computer via a 1 GB intranet service where it is queued for processing. Computer software is based on a ".COM" architecture and comprises two main modules. The first module—the RFM—deals with the queuing of jobs for processing. It also undertakes the pixel swapping algorithms, optical distortion algorithms and gamma corrections required to convert the perspective view data to the actual data required by the LCDs. The output of the RFM is a folder containing the compressed LCD data files for each line in the hologram and a command sequence for every motor control required for the setup and printing of the hologram. The mathematical algorithms required for the distortion and pixel-swapping routines will be covered in Chapters 8 and 9. The second software module is the PMC, which controls printer operation, manages the print queue and prints each hologram using the data generated by the RFM.

7.5.4.2.4 *Laser and Optical System*

The optical system of the Geola printer is built around a dual ring-cavity neodymium RGB-pulsed laser emitting separate red, green and blue beams at wavelengths of 660, 526.5 and 439.5 nm. The pulse energies available are between 3 and 5 mJ at each colour and a repetition rate of 15 Hz is standard. We have already reviewed this laser in Section 6.5.1.

The laser beam is first attenuated to a desired level through a 1/2 wave plate and Brewster angle polariser pair. The wave plates for each colour channel are mounted in precision electromechanical rotation stages that are driven by stepper motors. These motors are controlled by standard motor controllers that are in turn connected to the printer's dedicated control computer. Calibration tables are defined which allow the control computer to quickly select a given red, green or blue laser energy. Excess unwanted energy is absorbed in a beamstop.

A similar 1/2 wave plate and Brewster angle polariser pair is used to divide the main laser beam into a reference and object beam. In Figure 7.8, the reference beam is coloured red and the object beam is coloured blue. The wave plates are mounted as before in precision electromechanical rotation stages; with appropriate calibration, the control computer is then capable of commanding not only an exact laser energy for a given colour but now also an exact reference energy and an exact object energy. Two further fixed wave plates, one in the reference beam and the other in the object beam, are used to tune the polarisation to the desired direction. Figure 7.11 shows a photograph of the main optics unit of the Geola printer with the energy and ratio control system visible in the foreground to the left.

7.5.4.2.5 *Object Beam Subsystem*

A simple telescope (Figure 7.8) expands and collimates each laser beam. The beam then illuminates a microlens array that is apodised by the aperture (B). The function of the lens array is twofold. First, it creates a light source that has a larger étendue, which is directly controllable by the size of aperture (B). Second, it produces a clean approximately top hat spatial distribution. The focal length of the lenslets in the lens array is chosen to create a gradual expansion of the object beam and lens (L4) is positioned at a distance of approximately one focal length from the lens array. This ensures that in the case that aperture (B) is very small, the beam after (L4) is collimated.

The object beam now illuminates a Sony XGA1* twisted-nematic LCD panel where the digital image data are encoded onto the object beam. The LCD panel has an active area of 38.8 mm by 27.6 mm and a resolution of 1024 × 768 pixels. A polariser is required to convert the data, which is written initially by the LCD as changes in the polarisation vector, to amplitude modulation.

The final element in the object beam system is a high numerical aperture Fourier transforming objective lens system. This acts to strongly focus the object beam down to form the hogel and is a key part of the system. It must create a high-fidelity image of the LCD at a distance equal to or greater than the expected viewing distance of the final hologram and must have a Fourier plane at approximately 5 mm downstream of the lens. The conic angle of focus from the objective to the hogel is 105° in the Geola

---

* In later versions of the printer, the three XGA1 LCDs were updated to higher resolution Sony LCX028ALT panels. These were then mounted in ovens to assure operation at a higher hogel write rate of 30 Hz.

FIGURE 7.11 Photograph of the main optics unit of the 2001 Geola printer showing how the three colour channels are stacked one on top of the other. The laser is visible towards the rear left and the energy control systems are situated in the left foreground. Also visible to the right behind the large glass plate are the three hogel-writing optical objectives. The optics and motorised rotation stages visible to the left form part of the triple reference beam system.

printer and this fixes the intrinsic field of view of the hologram at 105° (Figure 7.12). For hogel sizes from 0.8 to 1.6 mm, the objective has a resolution capable of resolving the LCD pixels at all angles.

The footprint of the object beam at the surface of the photosensitive material is determined by the shape and size of the aperture (B). This is because aperture (B) is at essentially a plane conjugate to the emulsion plane. To see this, remember that the objective lens system produces a Fourier transform of the LCD plane at the emulsion plane. However, the Fourier plane of (L4) is at the lens array and so L4 approximately induces an inverse Fourier transform of the lens array plane at the LCD plane. The two transforms therefore cancel leading to the emulsion plane and the lens–array plane being conjugate. This is a useful feature in that aperture (B) can be controlled automatically to change the hogel size.

7.5.4.2.6 Reference Beam Subsystem

The function of the reference beam optical system in the Geola printer is to produce a clean collimated beam of the correct polarisation that illuminates the hogel from a given fixed angle with a defined spatial distribution. A 1/2 wave plate is used to rotate the polarisation to the desired angle, and a polariser is used to remove any elliptical component. Generally, one wants to minimise reflection from the photosensitive

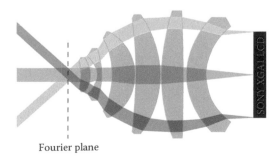

Fourier plane

FIGURE 7.12 The 105° hogel-writing objective lens system used in the 2001 Geola printer. Note that the hogel is formed at the Fourier plane. Note also that the ray bundles emanating from a specific LCD pixel pass through the hogel as approximately collimated beams.

material or from the glass or film onto which it has been coated. The polarisation of the electric field is therefore usually chosen to be in the plane parallel to the photosensitive material. If Brewster's angle is chosen as the incidence angle, this then eliminates any unwanted reflections completely. The axial object polarisation must of course be tuned to exactly the same angle as the reference polarisation to achieve maximum interference.

The positive lenses L1 and L2 constitute a Kepler telescope. This telescope changes the diameter of the laser beam by a factor

$$M = f_2/f_1 \tag{7.2}$$

where f_1 and f_2 are the focal lengths of L1 and L2, respectively. The separation of L1 and L2 must be chosen as

$$s = f_1 + f_2 \tag{7.3}$$

to ensure beam collimation. By placing a pinhole at the focus of L1, the beam may be cleaned very effectively and higher spatial frequencies removed. In practice, the focal length f_1 is chosen to be 20 to 50 times larger than the laser beam diameter to minimise aberrations and also to avoid breakdown of the air at the pinhole due to the presence of a high electric field. The focal length of L2 is then determined by the required beam magnification and the separation (s) is fixed by beam collimation.

The function of the two remaining elements in the reference beam system, aperture A and lens L3, is to shape the footprint of the reference beam at the hogel to a desired form. Without these elements, the reference beam would strike the photosensitive material at the incidence angle, forming an elliptical shape, the eccentricity of which is determined by this angle. This is not what is required because, ideally, the reference beam footprint should be matched to that of the object beam. In practice, one actually wants to make the reference beam just a little larger so that the object beam never falls outside the reference footprint. For a 0.8 mm diameter hogel, the reference beam footprint is therefore chosen to be a circle of 0.9 mm diameter, which leads to a small bleed from hogel to hogel.

The desired reference beam footprint is accomplished by arranging lens L3 so that it forms an approximate image of aperture A at the surface of the photosensitive material. The shape of aperture A is then designed such that the required (usually square) distribution of light is obtained at the hogel.

The Kepler telescope relays an image of aperture A by a distance

$$R = -d + f_1(M+1) + \frac{f_1 M \left(f_1(M+1) - Md \right)}{f_1 + 2M \left(f_1 - d \right)} \tag{7.4}$$

where d is distance from aperture A to L1. By choosing d carefully, the relayed image of aperture A can easily be positioned to the right of the Kepler telescope. Then, it is simply a question of choosing the focal length (f_3) and position of the positive lens (L3) such that the relayed image is in turn imaged to a location approximately coincident with the hogel. To do this, one simply uses the Gaussian form of the thin lens equation

$$\frac{1}{t'} + \frac{1}{t} = \frac{1}{f_3} \tag{7.5}$$

where t and t' are, respectively, the (positive valued) separation between the relayed image and L3 and that between L3 and the hogel. If t and t' are chosen to be too large then diffraction will wash out the image at the hogel completely. On the other hand, if they are chosen too small, then the reference beam will not be sufficiently collimated.

Because the hogel is created by the coherent interference of the object and reference beams, it is important that the optical path lengths of both object and reference beams are as similar as possible. Although the intrinsic coherence length of the laser is theoretically greater than 1 m, any discrepancy in the path lengths will in practice tend to reduce the overall diffractive efficiency of the hogel.

7.5.4.2.7 Energy Requirements

The 2001 Geola printer required approximately 1 mJ of energy per pulse per primary colour channel to print 1.6 mm diameter hogels using a panchromatic silver halide material having a sensitivity of approximately 1000 $\mu J/cm^2$. This is, in fact, much bigger than the theoretically required energy of approximately 30 μJ—largely because many of the optical systems were simply not optimised.

7.5.4.2.8 Alignment

Triple-beam printers are fundamentally more difficult to align than apochromatic single-beam printers. The problem is that in triple-beam printers, the virtual image of the red, green and blue spatial light modulators, downstream of the writing objectives, are separated by the distance between the objectives. Although it is relatively easy to align the actual objectives in the x, y and z directions, it is rather more difficult to ensure the correct orientation of the emerging light.

An effective alignment process for triple-beam printers is to focus in the objectives to form an image of each SLM, for example at 50 cm from the Fourier plane. A graticule is then loaded onto each of the SLMs and a high-energy object beam is used to project an image of this graticule onto an exactly perpendicular target. By taking digital photographs of the target with the three projected graticules and analysing these using a computer, the optical system may be aligned very accurately. It is important to point out that it is not sufficient that the objectives alone point in the same direction. One must also ensure that the SLM is exactly centred with respect to the objective and is not at an angle. In the case that the alignment is not done properly, holograms will show misaligned red, green and blue images. Almost always, an iteration process is required to ascertain whether alignment has been successful. This entails recording a test hologram of a grid structure and analysing whether a given colour channel needs readjustment. In film printers, it is imperative that the normal vector of the emulsion surface at the red, green and blue hogel write locations be the same, otherwise even a perfectly aligned optical system will lead to displaced colours. Although it is possible to numerically recalibrate a physically misaligned optical system by predistorting the SLM image data, this generally introduces some noise into the final 3D image and, as such, it is always strongly recommended to properly align the optical system.

In apochromatic and monochromatic printers, alignment is much easier because there is only one hogel write location and one writing objective. One therefore only needs to verify that a good image is present downstream of the Fourier plane. Colour slip is immediately obvious and can be corrected for by looking at the projected image while adjusting the optics.

In addition to the alignment of the object beam system, in all DWDH printers, the footprint of each object beam must coincide with the corresponding reference beam footprint at the holographic film surface. If this does not occur, then at best, a full hogel will not be formed, and at worst, there will be no hologram at all. For small hogel sizes, it can be quite tricky to ensure proper object/reference alignment. The usual way of doing this is to replace the square aperture used in the printer object beam system (which produces a square hogel) with a very small circular aperture. This essentially forms a very dim "point hogel" which scatters on the emulsion surface (one usually uses old film or an old plate). By reducing the energy per pulse, it is safe for an observer to then look into the writing objective where a luminous point will appear at a certain location at the emulsion surface. This is the centre of the hogel. It is then relatively easy to align the reference beam such that the point appears exactly in the centre of the larger reference square. When glass plates are used, it is vital that the system be recalibrated for the exact thickness of the glass used. It is useless aligning the object and reference writing beams at one thickness and then recording with a slightly different thickness plate—Snell's law will act to misalign the system and hogels will not be recorded properly.

7.5.4.2.9 Conjugate and Non-Conjugate Operation Geometries

Lens-based printers usually produce an image of each spatial light modulator downstream of the hogel. In contrast, diffusion screen systems always produce an image upstream. We have already discussed that a problem with diffusion screen systems is that the screen must be relatively large and at a good distance from the hogel if image blurring is not to result. In both lensless and lens-based hogel-forming systems, one has the choice of replaying the hologram with a conjugate or non-conjugate reference beam

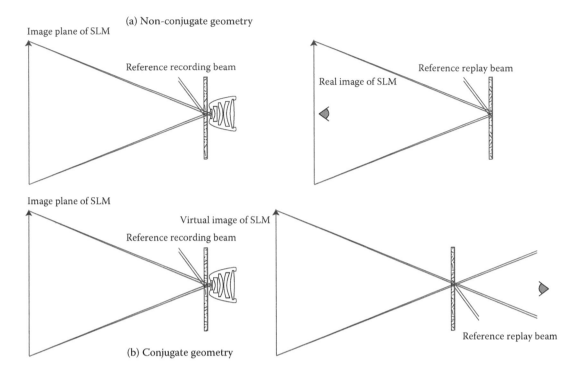

FIGURE 7.13 (a) Non-conjugate and (b) conjugate recording/replay geometries in a lens-based DWDH printer.

(Figure 7.13). If the image plane of the SLM is upstream* of the hogel at recording, then a non-conjugate reference beam at replay will produce a virtual image of the SLM behind the hologram. Likewise, a conjugate replay beam will produce a real image of the SLM, for each hogel, in front of the hologram.

The choice of which geometry to use is related to the printer construction. In the 2001 Geola printer, the glass holographic plates are always mounted with the emulsion facing towards the hogel-writing objectives. This is essentially a necessity, as the large plates are generally of a thickness that is greater than the distance between the physical end of the objectives and the Fourier plane. After processing, the sensitive emulsion surface must be at the rear of the hologram—blackening is then applied both to protect the emulsion and to improve the viewing characteristics of the hologram. This means that illumination of the hologram must be by a non-conjugate beam. When a diffusion screen system is employed, the same logic would dictate that a conjugate replay geometry be used.

If the image plane of the SLM is at a relatively small distance from the hologram and the hologram is to be viewed close-up, then it is generally better to ensure that a real image of the SLM is located on the viewing side of the hologram at replay. This will lead to the best image quality. However, it is often the case that the image plane distance can be made very large, and in this case, the choice of conjugate or non-conjugate replay geometries simply depends on the most convenient way to record and replay the hologram. In Chapters 8 and 9, we will see that the image transformations necessary to convert perspective view information into the SLM mask file information depend critically on whether the printer is designed for conjugate or non-conjugate operation.

7.5.4.2.10 Laser Stability Issues

By far the largest problem encountered with the 2001 Geola printer was related to the OEM RGB-pulsed laser. Although the laser's stability was relatively good from an absolute point of view, the occasional bad pulse often ruined a hologram after hours of writing—simply because a hologram could contain nearly one million hogels. For this reason, a new type of laser was developed by XYZ Imaging Inc. and Geola Technologies

---

* Note that Figure 7.13 shows the case of the image plane of the SLM being located downstream of the hogel.

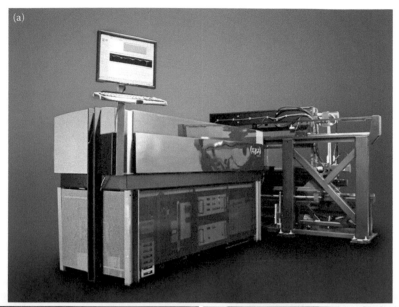

FIGURE 7.14 (a) Commercial DWDH triple-beam printer made by XYZ Imaging. (b) A shot of the interior showing the optics and the RGB-pulsed laser. (c) The three hogel-writing objectives, each surrounded by a vacuum unit, whose function is to pull the film flat at exactly the correct distance from the object. In practice, the optimum distance is a little downstream of the Fourier plane.

Ltd. This laser was built around twin linear telescopic cavities, rather than the initial ring cavity geometry. Another passive Q-switch, Cobalt MALO, was used in the 1319 nm channel and Nd:YLF was abandoned for Nd:YAG. This laser (see Section 6.5.2 in Chapter 6) was capable of operation at faster speeds—initially up to 30 Hz. The final piece of the puzzle was an active cavity length stabilisation scheme (see Appendix 3), which dramatically improved laser stability and allowed large DWDH holograms to be routinely produced.

7.5.4.2.11 Commercial DWDH Printers Based on 2001 Geola Printer

In 2004 to 2005, the company XYZ Imaging produced a commercial DWDH triple-beam printer based on the original Geola design (Figure 7.14). This was a film-based device* capable of writing DWDH holograms up to 1.1 m in width at hogel sizes of 0.8 or 1.6 mm. To keep the film at exactly the correct distance from the writing objectives, a vacuum system was used to suck the film onto a flat surface immediately in front of the three objectives (Figure 7.14c). XYZ Imaging also developed an automatic chemical processor. Photographs of holograms produced on the XYZ Imaging printer and on various other DWDH printers are shown in Chapters 10 and 14.

---

* The printer was designed to work with panchromatic silver halide film produced by the Russian Company Sfera-S.

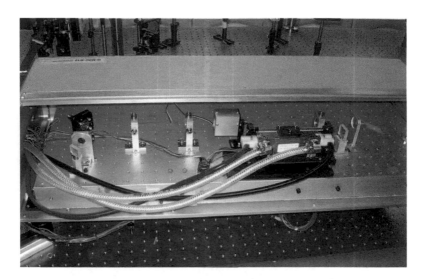

FIGURE 7.15   Short-cavity 532 nm Nd:YAG pulsed laser used in a 2006 prototype DWDH printer based on an LCOS display at Sussex University. The actual laser is visible in the centre of the picture. The optics in the larger laser case, are for frequency doubling and beam collimation. The main printer optics are visible in the background. The laser delivered stable pulses up to an energy of 1 mJ at a repetition rate of up to 50 Hz.

7.5.5 Printers Based on Liquid Crystal on Silicon Displays

John Tapsell [19], working at Sussex University in 2006, converted an old monochromatic DWDH printer supplied by Geola to work with a liquid crystal on silicon (LCOS) device. Marcin Lesniewski designed a telecentric afocal reversing system for the printer so that his 105° Fourier transform objective, which was used in the XYZ Imaging commercial printers, could be used with a BR768HC LCOS panel from Brillian. The printer used a short-cavity 532 nm, 30 ns pulsed Nd:YAG laser (Figure 7.15) of the type described in Section 6.5.3.1. The 768×1280 LCOS panel measured 17.91 mm diagonally and had a 12 µm pixel pitch. The fill factor was 92% with a reflectivity of 71% and a frame rate of 120 Hz. Small monochromatic DWDH reflection holograms were recorded with the system with a write rate of up to 40 Hz and a hogel size down to 300 µm. Very little energy was required and the 2000:1 contrast ratio available from the LCOS display produced a better quality image than available with comparative tests using an XGA1 Sony LCD panel. Figure 7.16 shows a diagram of the Sussex printer.

Geola Technologies Ltd used the experience gained from working on the LCOS printer at the engineering school of Sussex University in 2006 to come up with a concept design for a large-format RGB triple-beam DWDH LCOS printer. This design was subsequently used as the basis for the construction of a commercial printer built by the Centre for Laser Photonics in North Wales* for the production of metre-square full-colour reflection master holograms. Figure 7.17 shows a schematic of the design and Figure 7.18 shows a 3D visualisation.

The use of LCOS panels in DWDH printers is relatively simple. The slightly increased complexity of the object beam optical system is well merited due to the clear advantages offered by these panels in terms of higher switching speed (up to 200 Hz), better contrast (typically 2000:1) and superior efficiency (>70%). Table 7.1 lists the lenses used in the telecentric afocal reversing system and the Fourier transforming objective employed in the triple-beam printer.

---

* The Centre for Laser Photonics was a joint venture between Geola Technologies Ltd and Optropreneurs Ltd., which was operational between 2006 and 2010.

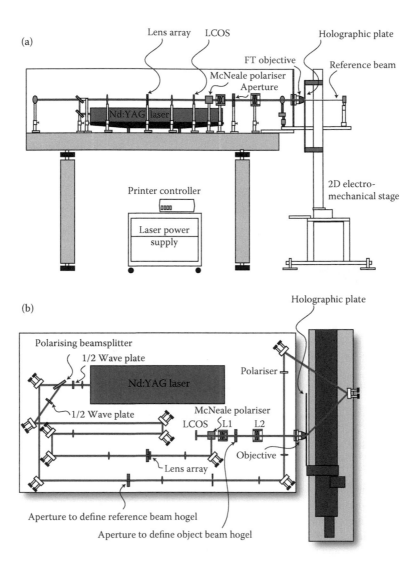

FIGURE 7.16 Sussex University LCOS printer. (a) Side view of the printer and (b) optical schematic. Note that the lens systems (L1 and L2), together with a meniscus field curvature correction lens next to the LCOS, form an afocal telecentric reversing system.

7.5.6 Printers Incorporating Variable Reference Beam Systems

All the printer schemes that we have reviewed above have used static reference beams. This is the simplest case to arrange optically. However, a static reference beam means that the written DWDH hologram must be replayed with a collimated light source if one is to avoid injecting any aberration into the hologram. However, this is rarely practical—in practice, the hologram must be illuminated by a point source relatively close to the display. One solution around this problem is to numerically predistort the image data to exactly counteract the induced aberration. We shall derive the equations needed for this purpose in Chapter 11 and discuss their solution in Appendix 4. For colour reflection holograms, both chromatic and geometric predistortion of the image data are required. Unfortunately, there is a rather strict limit on how much predistortion can be applied successfully and it gets more difficult the larger the hologram and the greater its field of view. Therefore, although numerical predistortion can certainly help, it is only very

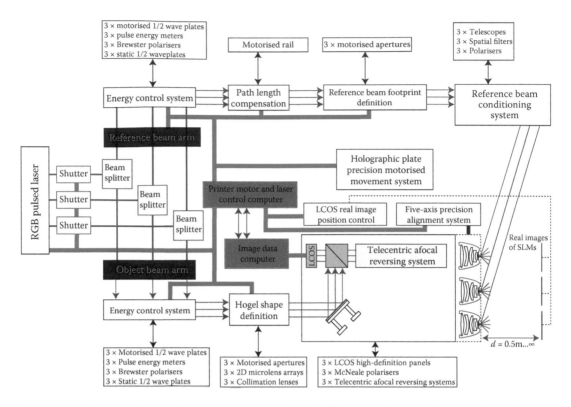

FIGURE 7.17 Systems schematic for a large, triple-beam DWDH LCOS printer that was built by the Centre for Laser Photonics in Wales. The design includes separate computers for image display and machine control for eventual operation at 120 Hz. The main trigger signal is originated by the LCOS controller and this is used to drive the main mechanical stage and the laser. A five-axis alignment system is used for easy alignment of the three objectives.

rarely capable of completely solving the problem of induced aberration due to a disparity in the reference recording and replay geometries.

The solution is to incorporate a variable-angle reference beam system into the printer. Each hogel can then be recorded using a software-selectable altitudinal and azimuthal angle. By choosing these angles carefully for each hogel, any type of macroscopic reference recording beam may be synthesised. In this way, a hologram may be produced so that it replays perfectly for a given location of the illuminating point source. In addition, under certain circumstances, small variations of the reference angle at each hogel may be combined favourably with numerical image predistortion to enhance the angle of view available from a given printer.

The downside to variable-angle reference beam systems is that they are rather more complex than static reference beam systems and can, if not designed properly, induce various problems including blurring and dimming into the hologram. The basic issue is that the object and reference beam footprints at the emulsion surface must overlap quite precisely. For small hogels (and they can go down to <250 μm), it can be a difficult enough task to arrange for proper footprint matching with a static reference beam, let alone for a 2D variable reference beam.

The simplest solution is to use a 2D-gimballed precision rotation stage to deflect the laser beam to a second 2D-gimballed rotation stage, itself mounted on a 2D translation stage. A computer then calculates the rotation angles and translation distances such that the reference beam strikes the hogel at given altitudinal and azimuthal angles. Of course, as the angles change, so the footprint at the emulsion surface also changes—this then needs to be compensated by the 2D rotation of a square aperture upstream of the hogel placed within a weak image-relaying system.* Figure 7.19 shows a schematic of this system.

---

* An LCD may also be used here as a programmable mask of variable shape.

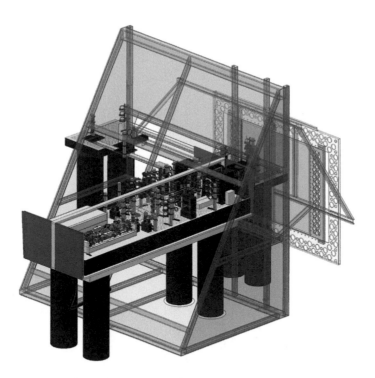

FIGURE 7.18 3D computer drawing of the large LCOS printer from Figure 7.17. Note the two lasers at the rear of the optical unit. The intention was to use a pulsed chromium forsterite laser for the red channel (627 nm) and to use an additional dual-channel 446 nm/532 nm pulsed Nd:YAG laser for the red and blue channels. In the actual printer, this design was modified to include only a single standard 440/532/660 nm RGB laser.

To work properly, this scheme must be very compact as the precision of the beam footprint alignment scales strongly with size.

Another type of variable reference beam system is based on a lens system and a single small-angle 2D-gimballed precision rotation stage as illustrated in Figure 7.20. Here, the centre of the reference beam is always aligned with the centre of the object beam and a small change in rotation angle produced by the rotation stage leads to a much larger change in angle at the hogel. The advantage of this type of system is that the footprint alignment is much more stable. The disadvantage is that if a large range of angles is required, then aberration in the lens system can induce blurring into the hologram. In addition, the footprint shape often changes in a non-linear way at large angles, requiring the use of an SLM as the apodising element. Figure 7.21 shows a photograph of a lens-based variable reference beam system in a recent triple-beam printer.

7.5.7 HPO Printers

The SLM mask file patterns used in triple-beam printers to print DWDH HPO holograms (ignoring numerical distortion correction for finite optical objective distortion and viewing window functions) are independent of the vertical coordinate. Horizontal information (as typified by the central row in the SLM) is essentially repeated in all rows within the viewing window. The vertical coordinate of the SLM is thus used in a very simplistic manner in these printers to induce a vertical divergence of rays at the hogel.* However, it is possible to use the vertical coordinate of the SLM to encode multiple hogels; to do

---

* Of course, when a full-parallax hologram (or indeed a rainbow hologram) is being written, this is not the case.

TABLE 7.1

Lens Parameters for the Telecentric Afocal Reversing System and Fourier Transform Objective Used in the Triple-Beam RGB DWDH Printer Manufactured by the Centre for Laser Photonics in Wales (2009)

No.	Green Channel EFL = −7.669				Red Channel EFL = −7.671				Blue Channel EFL = −7.716			
	Radius (mm)	Clear diameter (mm)	Separation (mm)	Material	Radius (mm)	Clear diameter (mm)	Separation (mm)	Material	Radius (mm)	Clear diameter (mm)	Separation (mm)	Material
1	Plane	2.301	4	Air	Plane	2.301	4	Air	Plane	2.318	4	Air
2	−20.34	9.562	3.07	S-SF6	−19.476	9.563	3.15	S-SF6	−21.69849	9.627	3.05	S-SF6
3	−9.616	11.63	1.93	Air	−9.3	11.678	1.8	Air	−10.03	11.678	2	Air
4	−7.6	12.36	1.45	S-SF6	−7.465	12.309	1.54	S-SF6	−7.852	12.474	1.45	S-SF6
5	−26.03	17.598	3.45	Air	−25.54	17.781	3.36	Air	−29.51541	17.734	3.45	Air
6	−25.027	24.134	7.45	S-SF6	−25.36	24.173	7.5	S-SF6	−27.31236	24.774	7.53	S-SF6
7	−16.144	28.39	0.3	Air	−16.144	28.396	0.3	Air	−16.707	28.995	0.3	Air
8	−201.01914	37.924	7.8	S-SF6	−131.52	37.312	7.64	S-SF6	−142.99318	38.172	7.87	S-SF6
9	−35.57	39.759	0.3	Air	−34.04	39.35	0.3	Air	−34.95308	40.243	0.3	Air
10	59.7	42.1	7.03	S-SF6	60.9	42.341	7.16	S-SF6	63.76992	42.852	6.83	S-SF6
11	1310.14201	41.384	1.27	Air	−568.9	41.829	1.28	Air	−1469.7098	42.281	1.3	Air
12	27.27	38.142	6.15	S-SF6	27.27	38.281	6.15	S-SF6	27.29649	38.636	6.15	S-SF6
13	20.51	32.078	14.72777	Air	20.51	32.205	14.82	Air	20.51	32.379	15.27	Air
14	Plane	46	246.59396	Air	Plane	64	246.59396	Air	Plane	64	246.59396	Air
15	693	64	4.33	S-SF5	693	64	4.33	S-SF5	693	64	4.33	S-SF5

Radius	Thickness	Semi-Diameter	Glass
224.9	64	7.33	S-BK7
-304.8	64	0.5	Air
693	64	4.33	S-SF5
224.9	64	7.33	S-BK7
-304.8	64	250.1	Air
Plane	50.895	73.58	Air
-29.51	35.98	15.2	J-BAF7
-37.1	45.5	0.4	Air
483.10001	47	6.4	J-SK4
-110.15	47	0.2	Air
129.42	47	9.8	J-SK12
-80.91	47	4	J-SF14
Plane	47	50	Air
Plane	30	30	S-BK7
Plane	30	36.68504	Air
-65.845	30	2	S-SF5
Plane	30	1.67108	Air
Thickness	808.279		

Radius	Thickness	Semi-Diameter	Glass
224.9	64	7.5	Air
693	64	4.33	S-SF5
224.9	64	7.33	S-BK7
-304.8	64	250.1	Air
Plane	50.895	73.58	Air
-29.51	35.98	15.2	J-BAF7
-37.1	45.5	0.4	Air
483.10001	47	6.4	J-SK4
-110.15	47	0.2	Air
129.42	47	9.8	J-SK12
-80.91	47	4	J-SF14
Plane	47	50	Air
Plane	30	30	S-BK7
Plane	30	36.68504	Air
-65.845	30	2	S-SF5
Plane	30	1.23033	Air
Thickness	807.779		

No.	Radius	Thickness	Semi-Diameter	Glass
16	224.9	64	7.33	S-BK7
17	-304.8	64	0.5	Air
18	693	64	4.33	S-SF5
19	224.9	64	7.33	S-BK7
20	-304.8	64	250.1	Air
21	Plane	50.895	73.58	Air
22	-29.51	35.98	15.2	J-BAF7
23	-37.1	45.5	0.4	Air
24	483.10001	47	6.4	J-SK4
25	-110.15	47	0.2	Air
26	129.42	47	9.8	J-SK12
27	-80.91	47	4	J-SF14
28	Plane	47	50	Air
29	Plane	30	30	S-BK7
30	Plane	30	36.68504	Air
31	-65.845	30	2	S-SF5
32	Plane	30	1.15834	Air
	Thickness	807.707		

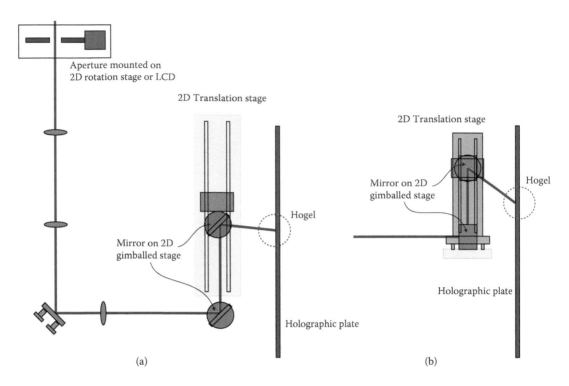

FIGURE 7.19 Simple optical scheme for automatically changing the reference beam angle (altitude and azimuth) at each hogel—view from the top (a) and from the side (b). Note that for fast printing, each motor controller must be preprogrammed with exact position data versus time. Before hogel-writing commences, the motors will need to backtrack a little and then start an acceleration sequence. Thereafter, the velocity of each stage will, in general, be a non-linear but smooth function of time. This ensures that the reference beam attains the correct angles and proper footprint alignment at each hogel at just the correct time without introducing mechanical transients into the system.

this, one must delegate the job of creating a vertical divergence of rays at the hogel to another system. The Fourier transforming objective then becomes a cylindrical lens system and the form of the hogel becomes an elongated column, the length of which, in the simplest variant, is equal to the SLM panel height, although a telescope may easily be used to modify this. A vertical diffusing element must be used in contact with the holographic emulsion and a modified "elongated column" reference beam must be employed. Klug and Kihara [20] described a variant of this system in 1995.

In 1998, Shirakura et al. [21], working at Sony Corporation, designed and built an integrated one-step CW laser HPO DWDH monochrome reflection hologram portraiture printer using this concept (Figure 7.22). The system consisted of a charge-coupled device (CCD) camera for image capture, a high-speed image processing device and a desktop DWDH HPO digital holographic printer. The portraits were delivered as an HPO 3D image (78 mm × 59 mm) and recorded on DuPont photopolymer film (HRF700XO71-20). The CCD camera unit moved along a straight track from right to left, driven by a stepping motor. There was another stepping motor in the camera unit to move a 2/3 in. CCD unit anti-parallel to the direction in which the camera moved; these two motors were synchronised so that the optical axis was always pointing directly at the object.

The hologram was recorded by projection of the digital images through a 510-K pixel thin-film transistor monochrome LCD using one-dimensional image compression with a cylindrical lens. The images displayed on the LCD were calculated from the perspective images available from the camera. The CCD camera recorded 295 2D images (640 pixels by 480 pixels) which were captured in 7.5 s of shooting (30 frames/s). A 400 mW frequency-doubled CW Nd:YAG laser of 532 nm was used as the recording light source in the tabletop printer, which measured only 1100 mm × 700 mm × 300 mm. Each column hogel (0.2 mm × 78 mm) was exposed onto the photopolymer in 0.25 s. A diffuser was attached to the LCD to

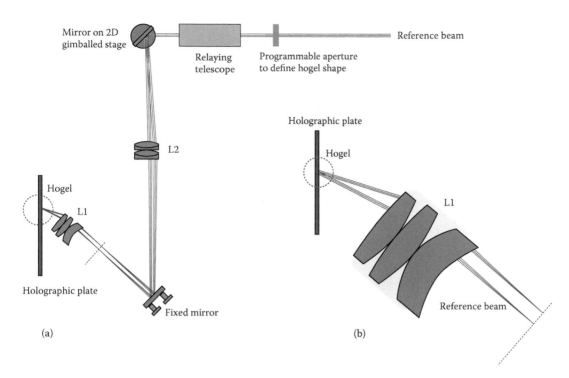

FIGURE 7.20 Lens-based optical scheme (a) for automatically changing the reference beam angle (in altitude and azimuth). Details of the main reference objective and the rays affecting the hogel under two different angles of incidence are shown (b). A typical system may be able to achieve a variation ±25° at the hogel (in both vertical and horizontal angles) for an angular variation at the gimballed rotation stage of ±1.5°. Due to intrinsic aberration of systems that can cope with large angle variations, a programmable aperture such as an LCD may be used in junction with a relaying telescope to ensure a proper hogel shape at all times.

make the beam intensity more uniform within the width of an elemental hologram and a slit, placed at an optical plane conjugate to the film plane, was used to form the hogel. The cylindrical focussing lens gave the holograms a horizontal field of view of 57°. A vertical diffuser in contact with the film likewise ensured a vertical viewing angle of 40°. The entire printing time of the 295 column hogels took only 147 s.

Clearly, a triple-beam pulsed laser DWDH printer designed using this concept could be expected to print HPO holograms much faster than the type of DWDH printers we have been discussing up until now. Nevertheless, in practice, it can be difficult to stop the hologram from looking "banded" and there are issues associated with the use of a contact diffuser.

7.5.8 Single-Beam RGB Printers

Single-beam DWDH printers can be constructed using apochromatic lens systems if resolution or angle of view can be sacrificed. Often, source size and chromatic blurring significantly limit the available depth in a display hologram. One may not then need the increased resolution available from a monochromatic system, and as such, it makes sense to design the printer using a single hogel write head. As we have discussed previously, this enormously simplifies the task of aligning the component colours.

7.5.8.1 Screen-Based Hogel Formation Systems

When limited depth in a hologram is acceptable, a lensless solution forming a single RGB hogel can also be used. In fact, we began our discussion of DWDH printers by presenting just such a system, based on a diffusion screen (Figure 7.2). We did, however, mention that there were several problems associated

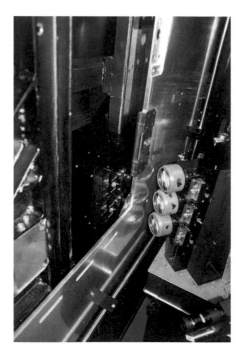

FIGURE 7.21 Photograph of a lens-based variable reference beam system in a modern large-format DWDH printer. To the left of the plate carrier (just visible) are two of the three object beam Fourier transform lenses. To the right of the plate carrier are the three lens systems for the reference beam angle control.

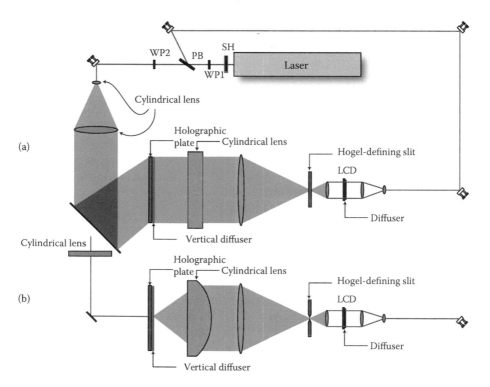

FIGURE 7.22 Simplified optical schematic of the 1998 Sony Corporation portrait printer, which printed small monochrome DWDH HPO reflection holograms as elongated "column" hogels using a 400 mW CW Nd:YAG laser in under three minutes. (a) Side view and (b) overhead view.

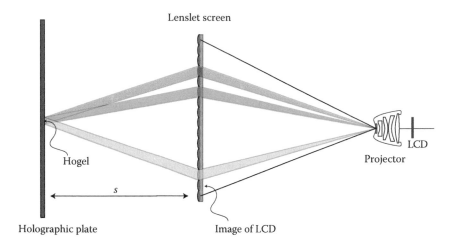

FIGURE 7.23 Diagram of an object beam hogel formation system based on a lenslet array. The size of the lenslets is greatly exaggerated.

with the simple diffusing screen. In particular, one needs a contact aperture to form the hogel; also, the energy efficiency is extremely poor. However, both these issues can be resolved by using a lenslet matrix to focus a 2D real image into a hogel (Figure 7.23). Nevertheless, there are two remaining problems. The first is illustrated in Figure 7.24. If the lenslet screen is too close and the lenslets are too large, then there will be viewing zones with no images. This produces the effect of image points flickering as an observer walks past the hologram. To avoid this, the lenslet size must be reduced, but this introduces a divergence into the ray bundles connecting each lenslet to each hogel. Any such divergence of a large enough magnitude will introduce blurring into points within the hologram beyond a certain depth. As we shall study in Chapter 11, a general paraxial formula for the critical depth, beyond which (interior) blurring occurs due to any form of ray bundle divergence is given by

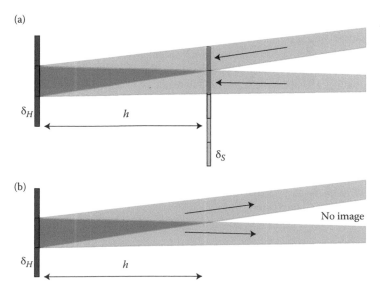

FIGURE 7.24 Diagram illustrating (a) the recording and (b) playback of a hogel using a lenslet screen. If the lenslet screen is too close, and the lenslets are too large, then there will be viewing zones with no images.

$$d \sim \frac{h\delta\theta_{Eye}}{\delta\varphi - \delta\theta_{Eye}} \qquad \forall \delta\varphi > \delta\theta_{Eye}$$

$$\sim \infty \qquad \forall \delta\varphi \le \delta\theta_{Eye} \tag{7.6}$$

where h is the viewing distance from the eye to the hogel, $\delta\varphi$ is the ray bundle divergence and $\delta\theta_{Eye}$ is the angular resolution of the human eye. In the limit that the lenslet size is much smaller than the hogel size, which will usually be correct for a close screen, the ray bundle divergence induced in the present case is simply $\delta\varphi = \delta_H/s$, where s is the screen to hogel distance.

As an example, let us take a hogel diameter of 0.5 mm. Then, if we place the screen at a distance of $s = 5$ cm from the hogel, at a viewing distance of $h = 0.5$ m, we will observe blurring at a distance of $d = 6$ cm into the hologram. This is clearly not very acceptable. At a distance of $s = 15$ cm, things are rather better. Resolvable image blurring then starts at approximately $d = 22$ cm for $h = 0.5$ m. Decreasing the hogel size makes things even better. If one goes to a hogel size of 250 μm, then a screen placed at 15 cm will only induce image burring at a depth of more than 75 cm for $h = 0.5$ m. However, we must be careful here because decreasing the hogel size will induce a second type of blurring—digital diffractive blurring. The critical distance at which digital diffractive blurring operates (for wavelengths, λ) is given by

$$d \sim \frac{h\delta\theta_{Eye}}{\lambda/\delta_H - \delta\theta_{Eye}} \qquad \forall \lambda/\delta_H > \delta\theta_{Eye}$$

$$\sim \infty \qquad \forall \lambda/\delta_H \le \delta\theta_{Eye} \tag{7.7}$$

We shall discuss this type of blurring in Chapter 11—but basically, it is caused by the innate diffractive property of a small source (the hogel). Plugging in the numbers for the case of interest, we obtain a value for d of 44 cm for $h = 0.5$ m.* Therefore, for a 250 μm hogel, digital diffractive blurring is more limiting at $h = 0.5$ m than the blurring induced by a close recording screen. In fact, digital diffractive blurring gets worse as you get closer to the hologram, and thus, one wants to avoid using too small a hogel. By ensuring that the hogel size is greater than or equal to 0.5 mm, this type of blurring is eliminated for the human observer with normal eyesight.

As long as the field of view of the hologram is not too great, it can therefore be feasible to use a lenslet screen to form the hogel. Good quality small-hogel (albeit relatively shallow) holograms can be made in this way within the design remit of a compact printer. However, we will now illustrate why this technique is not so appropriate for the case of ultra-realistic holograms of great depth and field of view. Let us again take a hogel diameter of 0.5 mm; anyhow, we cannot use a smaller diameter without incurring digital diffractive blurring. We now demand no induced blurring from any viewing distance. However, to guarantee this, we need to place the screen at a distance of at least 500 mm from the hogel. At a field of view of 130°, this leads to a screen that is more than 2 m wide!

Holographic diffusers and holographic optical elements may also be employed usefully as hogel-forming devices. These elements are usually used as a more convenient form of the lenslet matrix screen. Most screen-based hogel production techniques have two main potential advantages. The first is that they can usually be used in an apochromatic or single-beam printing system. The second is that, even for high fields of view, they have the potential of not inducing any image distortion into the hogel. In high-numerical aperture, lens-based systems, one must inevitably tolerate such induced distortion, which is caused by a finite fifth Seidel coefficient.

7.5.9 Ultra-Realistic Printers

Ultra-realistic printers are DWDH printers capable of producing full-colour high virtual volume (HVV) displays. HVV holograms are digital full-colour reflection holograms which, when illuminated correctly,

---

* We assume that the average human eye can resolve 1 mm separations at a distance of 1 m.

exhibit essentially no perceivable image blurring or distortion. For a printer to be capable of writing HVV displays, it must have the following characteristics:

- Rigid high-precision printing medium such as photosensitive glass plates
- High-precision 2D electromechanical plate translation stage
- Hogel-writing SLMs with a sufficiently high pixel count
- Hogel-forming optical system with a sufficiently high resolution and sufficiently high numerical aperture
- Reference beam system with a sufficiently low divergence

In addition to these constraints, the photosensitive material must be capable of supporting a high spatial frequency, of not changing its physical size upon processing* and of producing a good diffractive response—this is especially needed if the field of view at replay is required to be large as in the case of holographic window-type displays. The image data and image processing must also be able to produce a data set that either matches the optical resolution of the printer or betters that of the human eye. For an HVV hologram to actually generate a proper "HVV" image, the hologram must be illuminated properly. This means that the diameter of the illuminating source must be smaller than 1 mm for every 1 m that the source is diagonally distant from the hologram. The spectral width of each colour illuminating the hologram should also be less than 1 nm. Any larger than this and there will be induced chromatic aberration (unless Bragg selection is able to mitigate this—which is unlikely). However, much below 2 nm, speckle blur becomes a concern. We shall see in Chapter 13 that speckle may be essentially eliminated using devices that induce a fast temporal modulation in the phase of the illuminating light.

The 2001 Geola printer, which we described previously, is not an ultra-realistic printer. It uses an LCD having a horizontal pixel count of 1024 and a paraxial field of view of 86° (note that the non-paraxial field of view is nearly 105° due to a finite fifth coefficient at large angles). This endows each hologram written with an angular resolution of approximately $\delta\varphi = 1.8$ mrad—which is nearly two times the human eye resolution. Following Equation 7.6, the maximum clear depth that the holograms can display is given by the paraxial rule

$$d \sim \frac{h\delta\theta_{Eye}}{\delta\varphi - \delta\theta_{Eye}} \tag{7.8}$$

which, for a viewing distance of $h = 1$ m, comes out at approximately 1.25 m.

The 2001 Geola printer could be potentially modified by replacing the 1024×768 LCD display with a 1080p panel. This would solve the SLM insufficiency problem for paraxial viewing because the resolution of the Fourier lens system is easily sufficient to resolve the 1080p panel. However, at higher angles, the resolution falls below the pixel size and again blurring due to insufficient objective resolution is injected into the hologram, fundamentally limiting the virtual volume.

In thinking about the design of the ultra-realistic DWDH printer, we come up against two conflicting processes. On the one hand, we generally wish to increase the field of view of the hologram. This is certainly the case for "holographic window"-type displays in which the idea is to mimic a window. To do this, we are obliged to use a higher numerical aperture objective. However, on the other hand, we must now increase the resolution of the objective to be able to resolve more SLM pixels. Further work needs to be done in investigating how far both conditions may be satisfied in a single compound lens system. Current objectives made at Geola use SF6 glass. It is possible that by using higher index glasses, a higher numerical aperture might be attained at sufficiently high resolution.

However, even if this were the case, current HD SLMs do not have the pixel counts required for large fields of view. We shall see in Section 14.4.1 of Chapter 14, that at least four 1080p panels are required to write an HDD hologram with a field of view of $100° \times 120°$. Although it is possible to tile these displays together using prisms and a variant of the telecentric afocal reversing system described previously, the

---

* Although a physical change in the thickness of the emulsion and a change in the refractive index on processing can be compensated for using numerical image processing algorithms, this inevitably leads to the introduction of some noise.

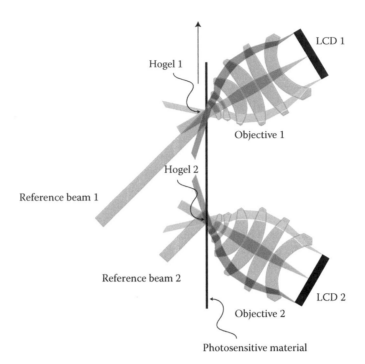

FIGURE 7.25 Writing a very-wide angle hologram using multiple objectives. Objective 1 writes hogel 1 and objective 2 writes hogel 2. Then, as the film advances, hogel 2 is overwritten by objective 1 again. In this way, hogels can be built up in several steps.

proposed UHDTV 4320p standard, which has a pixel count of 7680 × 4320, would certainly make the optics rather simpler.

7.5.9.1 3N-Objective Printers

Beyond a certain field of view, it becomes impractical to write HVV holograms using single compound lenses for each primary colour. With a sufficient index modulation in the photosensitive material (see Chapters 11 and 12 for the theory behind this), one can, however, envisage writing hogels in angular segments as illustrated in Figure 7.25. The idea is basically an extension of the RGB triple-beam printer concept—except that here, one would use, in the simplest variant, an array of 2 × 3 objectives, two for each colour. These two objectives would be angled such that, together, a greater horizontal field of view could be covered. Special care is needed with the numerical image processing in the overlapping regions, as the rays from the two SLM/objective systems do not of course align. In some cases, it may therefore be better to use a 3 × 3 system (or a $3^2 × 3$ system in the case of 2D angle extension) rather than a 2 × 3 system.

7.5.10 DWDH Transmission Hologram Printers

Geola has run a number of research projects using DWDH transmission printers since 1999. All these devices have been monochromatic pulsed laser printers operating at either 532 or 440 nm. The optical schematic is just the same as in Figure 7.8, with the single exception that now the reference beam impinges onto the hogel from the same side as the object beam. It can often be a little tricky getting the reference beam in, as there is not much space between the physical end of the objective and the photosensitive plate. For wide-angle objectives, one usually uses an angle of incidence that is a little larger to cope with this. Alternatively, the reference beam can be brought in through the main objective as shown in Figure 7.26.

The main interest in DWDH transmission holograms is that full-colour rainbow, achromatic and mixed rainbow–achromatic holograms may be generated from digital data in a single printing step using

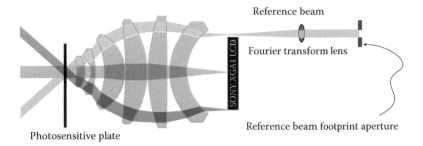

Reference beam

Fourier transform lens

SONY XGA1 LCD

Reference beam footprint aperture

Photosensitive plate

FIGURE 7.26 DWDH transmission hologram hogel formation. Here, the reference beam is actually brought in through the Fourier transform objective.

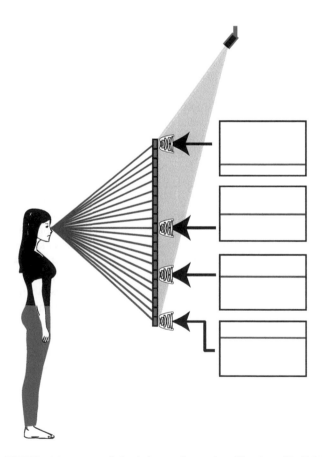

FIGURE 7.27 A 1-slit DWDH rainbow transmission hologram is seen here illuminated by light of only one colour. The observer, as positioned, sees a monochromatic holographic image. On illumination by white light, this image is available at different heights where it replays now with different colours. The four rectangles to the right of the diagram illustrate the data displayed on the printer LCD when the indicated hogels are written. The LCD data for each hogel is in the form of a line which is modulated by image data specific to that hogel. The vertical height of a given line on the LCD is determined by the vertical height of the hogel in the hologram in such a way that a rainbow viewing slit is synthesised as shown. The technique can be extended to any number of rainbow slits (using only a single colour laser) to produce full-colour DWDH rainbow transmission or achromatic transmission holograms.

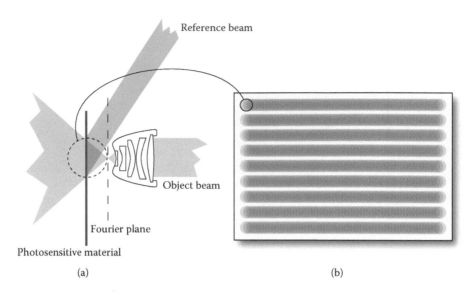

FIGURE 7.28 MWDH often uses overlapping hogels away from the Fourier plan (a). Typical hogel footprints, (b), are shown here for an MWDH transmission H_1 hologram designed for transfer to a HPO reflection H_2 to be illuminated by broadband illumination. Note the distance between hogel lines—dispersion in the vertical direction makes the gaps invisible in the final H_2. Note also the circular hogel shape—in MWDH one is not constrained to use a square or hexagonal footprint.

only a single-colour laser (see Figure 7.27). In principle, these printers are much simpler than the triple-beam reflection hologram printers. By introducing the reference beam into the writing objective, one essentially only has optics on one side of the photosensitive material—and by only needing one laser, the optical scheme can be made very small. Alignment of the reference and object beam is also much easier. Variable reference beam systems can also be incorporated with relative ease. Very compact printers, the size of normal photocopiers, should be achievable using this technique if processing-free materials, such as monochromatic photopolymers, are used. In Chapter 8, we shall study the image-processing algorithms required to write full-colour rainbow and achromatic (i.e., black-and-white) holograms.

Small transmission rainbow and achromatic DWDH holograms have applications in document security. The Geola organisation is currently able to produce such holograms using a 440 nm pulsed laser at a hogel size of 250 µm. The holograms are then transferred to photoresist to make the embossed shims.

Large transmission rainbow and achromatic DWDH holograms have potential applications in advertising and display. They are particularly useful as shop window displays as the images can project outside the shop and into the street. With reflection holograms, a light is needed on the same side as the viewer so this is not possible.

Finally, it is possible to write full-colour, full-parallax transmission holograms using either a single-colour laser or by using three lasers in a triple-beam configuration. However, such holograms must be illuminated by three different colours from substantially different angles to eliminate the cross-talk images. As we shall see in Chapter 11, the volume transmission hologram has greater angle selectivity than the corresponding reflection hologram, allowing angle discrimination to be used more easily.

7.6 MWDH Printers

MWDH is the technique of writing first an H_1 hologram using digital image data and then optically transferring the H_1 to a white light-viewable H_2. In many ways, MWDH is similar to the technique of multiple photo-generated holography pioneered by Spierings and van Nuland [8]. However, no diffusion screen is used, hogels are written as spots rather than long, thin rectangles and digital data replace the photographs. MWDH can also be used to create full-parallax holograms of great depth, which is difficult to arrange using multiple photo-generated holography.

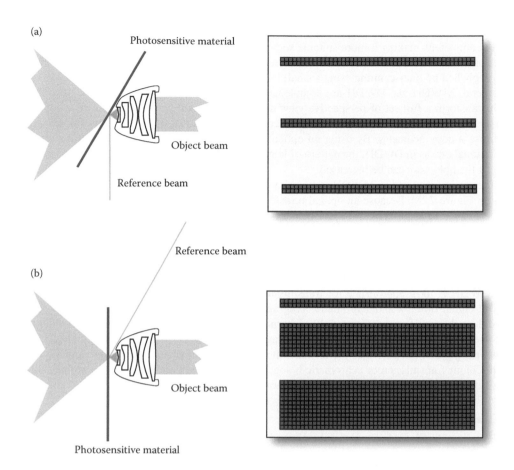

FIGURE 7.29 Writing an MWDH transmission H_1 rainbow master hologram with a monochromatic laser using a plate holder tilted at the achromatic angle (a) and a standard perpendicular plate-holder (b). In (a), the red, green and blue "slits" are the same width because the H_1 is tilted at the achromatic angle. In (b), two of the slits are much thicker as multiple rows of hogels are used to synthesise virtual slits at the correct distance behind the plate. When viewed under laser light, these wider bands would therefore seem to contain an image of a slit behind the hologram. Note of course that the size of the hogel has been greatly exaggerated in the diagram.

From an optical point of view, an MWDH printer is virtually the same as a DWDH printer. The major difference is therefore in the image data. Often, the data required by the SLM(s) in an MWDH printer are just the perspective view data available from a tracking camera. In a DWDH printer, the hogel data must be calculated from the perspective view data by a mathematical algorithm that changes the optical plane. The optical transfer from H_1 to H_2 fulfils this role in MWDH.

Basically, any type of hologram may be made using MWDH or DWDH. The decision as to which technique to use usually comes down to the speed of printing and the ease of copying. A 0.5 mm hogel DWDH hologram, 1 m × 1 m, takes 11 hours to write at a hogel write speed of 100 RGB hogels per second, but if the hologram is an HPO hologram, then the corresponding H_1 hologram can take only a very small fraction of this time to write. This is because an HPO hologram does not usually require such a high vertical hogel density.*

Copying of DWDH holograms is usually done through a contact–copy method. However, with this method, it is not possible to adjust the ratio of the object and reference beams at the copy—as this is defined by the diffractive response of the master hologram. However, certain materials may require a

---

* This is the case when a broadband illumination source is used.

higher modulation to record a proper copy hologram. The H_1/H_2 distance transfer process solves this problem completely, making it more suitable sometimes for the rapid production of copies. The disadvantage of course is that a full-aperture transfer requires a lot of energy, whereas the contact scheme can be accomplished by line-scanning using a small laser.

In general, MWDH and DWDH are complementary techniques. With full-parallax data, it is possible to transform a full set of perspective view data to any optical plane. In this way, computational and optical image-plane transformations can be combined as desired. This can be useful, for example, to optimise a copy geometry. By using an optical image plane transformation rather than an entirely computational one as in DWDH, the pattern of hogels on the physical plate becomes defocussed and the quality of the hologram can be increased.

Another technique used in MWDH is to write overlapping hogels downstream of the Fourier plane (shown in Figure 7.28). Because an optical transfer will be used to convert the H_1 to an H_2, the loss of diffractive efficiency caused by this overlap does not really matter. The quality of the final H_2 holograms can be somewhat increased using this technique. Another advantage is that for transmission systems, where it is not possible to bring the reference beam through the writing objective, there is now extra space to accomplish this.

Like DWDH printers, MWDH printers can produce either transmission or reflection holograms. Most often, MWDH is most appropriate for transmission holograms written using a monochromatic laser. Here, full-colour rainbow and achromatic holograms can easily be produced using a plate holder that is tilted at the achromatic angle (Figure 7.29). Alternatively, a standard perpendicular plate holder may be used if special image processing transformations are employed. Geola makes commercial H_1/H_2 transfer systems using green-pulsed lasers for formats up to 1 m × 1.5 m (Figure 6.12). One should also note that full-colour transmission or achromatic holograms can be produced using only a single "slit" with MWDH if an RGB laser is used in place of the monochromatic laser.

7.7 Copying Full-Colour DWDH Holograms

As we have already mentioned, printing full-parallax, ultra-realistic DWDH holograms is slow! For the technology to become commercially interesting, either the laser and print speed must be increased dramatically or one needs to develop a technique to copy the DWDH holograms produced. Certainly, print speed may realistically be increased by a certain amount. Current pulsed laser systems can be redesigned to work at up to 120 Hz with flash pumping. Beyond this, diode-pumped laser solutions may be expected to produce repetition rates that can be actually as high as required. However, no SLM technology exists at this moment to practically produce a printer with a repetition rate of greater than 200 hogels per second. To go faster than this requires multiple write heads that, although possible, will increase the price point of any printer rather dramatically.

Here, we present the results of recent experiments (2006–2010) carried out by the Geola organisation to produce high-quality holographic copies of digital master holograms written with a DWDH printer [22]. A standard RGB-pulsed laser (the same as that used to record the DWDH master holograms) was used in the line-scanning contact copying system. Figure 7.30 shows a simplified optical scheme of the experimental setup.

So that a good quality copy may be produced from a reflection master, it is vital that the DWDH master hologram replays at exactly its recording wavelengths. To ensure this, the emulsion must be processed in a special way and care must be taken with regard to ambient humidity and temperature during the entire process.

With reference to Figure 7.30, each of the laser beams (11) passes through computer-controlled wave plates (12) and polarisers (13). By rotating the wave plates (12), the colour balance of the hologram copy may be adjusted. The beams are now cleaned by spatial filters (14) and a proper polarisation is ensured by polarisation correctors (15). The beams are then directed by mirrors (16) to a three-colour combiner-deflector system (17), after which they are shaped by a shaping/deflection system (18) into a narrow elongated and slightly oval achromatic slit. This achromatic beam is then reflected by the flat mirror (22) to illuminate the non-exposed photosensitive material (1) as a reference beam. Part of the achromatic beam

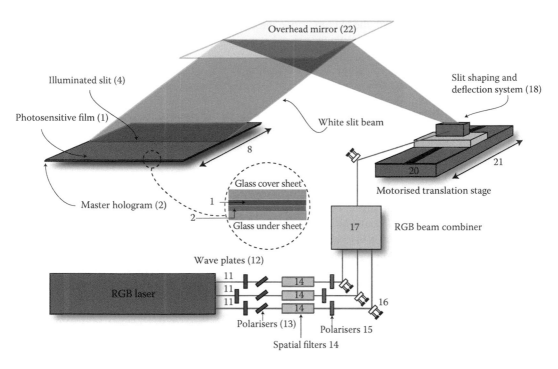

FIGURE 7.30 Simplified optical scheme of a line-scanning RGB film hologram contact copier. The system, which uses multiple pulses from an RGB-pulsed laser, produces RGB reflection film copies from a DWDH master (film) reflection hologram.

FIGURE 7.31 (a) Master DWDH hologram, 0.8 mm hogel, 20 × 30 cm; (b) three-colour copy on PFG03CN silver halide; (c) three-colour copy on Bayer Photopolymer. (Hologram image designed in 3D StudioMax. Courtesy of Razvan Maftei, 2005.)

is, however, transmitted through the photosensitive material and onto the master hologram (2) where it generates a diffractive reflection forming the object beam.

The zone illuminated by the laser slit beam (4) is transverse to the scanning movement of the slit. The laser radiation shaping/deflection system (18) is fixed onto the computer-controlled linear translation stage (20) to ensure an even movement in direction (21). At the same time, the linear translation stage (20) ensures movement of the light slit in direction (8), which is perpendicular to the longitudinal axis of the illumination slit. This ensures an even exposure of both the non-exposed photosensitive material and of the master hologram, giving, in turn, an even recording of the reconstructed master hologram.

Experimental results showed that existing silver halide photoemulsions (Geola tried both PFG-03CN from Sfera-S and the SilverCross emulsion) do not allow the production of full-colour contact copies having a diffraction efficiency greater than the diffraction efficiency of the master hologram. The best result that Geola was able to achieve was a colour copy with a relative diffraction efficiency (as a percentage of the master diffraction efficiency) of 100% in red and green and 50% in blue. Despite the less than perfect result in blue, the quality and brightness obtained were still judged adequate for commercialisation. An interesting observation is that if the copy is made with only two colours, its diffraction efficiency can reach 150% of the master hologram efficiency (using AgX).

A panchromatic photopolymer material from Bayer was also used to record high-quality copy holograms. Here, relative diffraction efficiencies in the red and green of well over 100% were obtained. Figure 7.31a, b and c show the experimental results for both AgX and photopolymer. All of Geola's work has thus far concentrated on AgX film masters. The use of glass-plate masters should, of course, substantially improve the image quality. Additionally, the use of SHSG processing could significantly improve copy efficiency.

REFERENCES

1. B. R. Brown and A. W. Lohmann, "Complex spatial filtering with binary masks," *Appl. Opt.* **5**, 967–969 (1966).
2. L. B. Lesem, P. M. Hirsch and J. A. Jordan, "Scientific applications: Computer synthesis of holograms for 3-D display," *Commun. ACM* **11**, 661–674 (1968).
3. R. V. Pole, "3-D imagery and holograms of objects illuminated in white light," *Appl. Phys. Lett.* **10** (1), 20–22 (1967).
4. D. J. DeBitetto, "Holographic panoramic stereograms synthesized from white light recordings," *Appl. Opt.* **8**, 1740–1741 (1969).
5. M. King, A. Noll and D. Berry, "A new approach to computer-generated holography," *Appl. Opt.* **9**, 471–475 (1970).
6. L. Cross, "The Multiplex technique for cylindrical holographic stereograms," in SPIE San Diego August Seminar (1977) [Presented but not published].
7. S. A. Benton, "Hologram reconstructions with extended incoherent sources," *J. Opt. Soc. Am.* **59**, 1545–1546A (1969).
8. W. Spierings and E. van Nuland, "Calculating the right perspectives for multiple photo-generated holograms," in *Int'l Symposium on Display Holography*, T .H. Jeong ed., Proc. SPIE **1600**, 96–108 (1992).
9. M. Halle, S. Benton, M. Klug and J. Underkoffler, "The Ultragram: A generalized holographic stereogram," in *Practical Holography V*, S. A. Benton ed., Proc. SPIE **1461**, 142-155 (1991).
10. M. Lucente, *"Diffraction-Specific Fringe Computation for Electro-Holography"*, Ph. D. Thesis, Dept. of Electrical Engineering and Computer Science, MIT, (1994).
11. M. Yamaguchi, N. Ohyama and T. Honda, "Holographic 3-D printer," in *Practical Holography IV*, S. A. Benton ed., Proc. SPIE **1212**, 84-92 (1990).
12. M. Klug, M. Holzbach and A. Ferdman, *Method and apparatus for recording 1-step full-color full-parallax holographic stereograms*, US Patent 6,330,088 (filed 1998, granted 2001).
13. D. Brotherton-Ratcliffe, S. J. Zacharovas, R. J. Bakanas, J. Pileckas, A. Nikolskij and J. Kuchin, "Digital holographic printing using pulsed RGB lasers," *Opt. Eng.* **50**, 091307-1-9 (2011).
14. D. Brotherton-Ratcliffe, F. M. Vergnes, A. Rodin and M. Grichine M, *Method and apparatus to print Holograms,* Lithuanian Patent LT4842 (1999).

15. D. Brotherton-Ratcliffe, F. M. Vergnes, A. Rodin and M. Grichine, *Holographic printer*, US Patent 7,800,803 (filed 1999, granted 2010).

16. A. Rodin, F. M. Vergnes and D. Brotherton-Ratcliffe, *Pulsed multiple colour laser*, EU Patent EPO 1236073 (2001).

17. D.-K. Kang, M. A. Rivera and M. L. Cruz-López, "New fully functioning digital hologram recording system and its applications" *Opt. Eng.* **49**, 105802-1-9 (2010).

18. P.-A. Blanche, A. Bablumian, R. Voorakaranam, C. Christenson, W. Lin, T. Gu, D. Flores, P. Wang, W.-Y. Hsieh, M. Kathaperumal, B. Rachwal, O. Siddiqui, J. Thomas, R. A. Norwood, M. Yamamoto and N. Peyghambarian, "Holographic three-dimensional telepresence using large-area photorefractive polymer," *Nature* **468**, 80–83, (4 Nov. 2010).

19. J. Tapsell, *"Direct-write digital holography"*, PhD Thesis, Univ. of Sussex, UK (2008).

20. M. Klug and N. Kihara, "Reseau full-colour one-step holographic stereograms," in *Fifth Int'l Symposium on Display Holography*, T. H. Jeong ed., Proc. SPIE **2333**, 411–417 (1995).

21. A. Shirakura, N. Kihara and S. Baba, "Instant Holographic Portrait Printing System," in *Practical Holography XII,* S. A. Benton ed., Proc. SPIE **3293**, 248-255 (1998).

22. I. Lancaster, "Geola builds holographic copier," *Holography News*, Vol. **23**, (3) 4, (2009).

8

Digital Holographic Printing: Data Preparation, Theory and Algorithms

8.1 Introduction

In the last chapter, we have seen how digital holographic printers are physically put together from the points of view of optical schemes, mechanical design and laser choice. In this chapter, we will concentrate on the subject of digital data processing. In general, we will start with digital image data that is derived from either a computer model or from a large array (or matrix) of digital cameras. We must then transform this data into a form suitable for actually printing a hologram [1,2].

Most of this chapter and the next describe a system for keeping track of and manipulating image data. More often than not, we actually start with the data we need! Unfortunately, this data is often in the wrong order and, as such, needs reordering and rescaling before a hologram can be written. From an absolute point of view, the algorithms and mathematics required to do this are extremely simple. However, we will need to keep track of many pointers in a fashion dictated by an accurate analysis of the ray geometries of the image model and the hologram, and this may well give the impression that things are more complicated than they actually are!

Because there is such a large difference between the cases of horizontal parallax-only (HPO) holograms and full-parallax holograms—not only in their optical properties and applications but also in the computational algorithms required to generate the printer write-data—we shall concentrate mostly on HPO holograms in this chapter, leaving a proper discussion of the full-parallax case to Chapter 9.

8.2 Basic Considerations

Every hologram seeks to display an image of a three-dimensional object, which is defined within a certain three-dimensional space. This is illustrated in Figure 8.1a. This figure may be interpreted in two different ways. In the first interpretation, object $\mathcal{A}$ is placed behind a glass window $\mathcal{H}$. In the second interpretation, $\mathcal{H}$ is a hologram and $\mathcal{A}$ is the holographic reproduction of the object of the first interpretation. In either case, assuming that the hologram is recorded and illuminated in a perfect fashion, an observer would find it hard to distinguish between the two interpretations, as the radiation field emanating from plane $\mathcal{H}$ must be ostensibly identical in both cases.

A well-known principle of optics is Huygens' principle [e.g., 3]. Formally, this principle, which can be directly derived from the Maxwell equations, can be used to analyse Figure 8.1. Huygens' principle tells us that to completely define the radiation field over $\mathcal{H}$, which emanates from region $\mathcal{A}$, knowledge of its distribution is sufficient at any plane ($\mathcal{P}1$, $\mathcal{P}2$, $\mathcal{P}3$, etc.). We therefore envisage the situation depicted in Figure 8.1b, in which a real object ($\mathcal{O}$) appears located behind plane $\mathcal{H}$. We would like to create a hologram that reproduces the radiation field passing through $\mathcal{H}$. Although we could just measure the radiation field at $\mathcal{H}$, it is more useful to consider measuring the field on a different plane. We therefore imagine an additional plane ($\mathcal{C}$) located in front of and perpendicular to $\mathcal{H}$. To measure the radiation field at $\mathcal{C}$, we imagine this plane to be covered with a matrix of perfect "pinhole" cameras all orthogonally directed with respect to plane $\mathcal{C}$; each camera records the luminous intensity it receives as a function of vertical

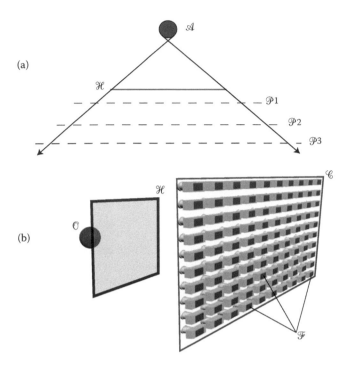

FIGURE 8.1   (a) In one interpretation of this diagram $\mathscr{A}$ may be regarded as an object and $\mathscr{H}$ as a glass window. In an alternative interpretation $\mathscr{A}$ is a holographic image and $\mathscr{H}$ is a hologram. The light-field at $\mathscr{H}$ due to $\mathscr{A}$ is the same in both cases. If we have a full knowledge of the light-field on any of the planes ($\mathscr{P}1$, $\mathscr{P}2$ or $\mathscr{P}3$) then the distribution of light at $\mathscr{H}$ is uniquely defined. (b) We can define the (incoherent) light-field produced by a hologram or object by arranging a matrix of idealised closely-spaced cameras on plane $\mathscr{C}$.

and horizontal angles. This information will then be encoded onto each of the planes ($\mathscr{F}_{\xi\zeta}$), which are none other than the planes of the film from each camera.

Following this logic, the optical information represented by the recorded light intensity distributions present on the totality of the planes ($\mathscr{F}_{\xi\zeta}$) formally represents all the information we need to reconstruct the (incoherent) radiation field produced by the original object passing through the hologram plane $\mathscr{H}$. Whether we use a real object and take a matrix of real photographs or whether we take a virtual object in a computer and take "virtual" photographs does not change the mathematics required.

In the following sections, we will formulate mathematically how the camera information from the totality of the planes ($\mathscr{F}_{\xi\zeta}$) may be used to calculate the optical information required by a digital holographic printer to print a hologram which reproduces a given radiation field.

8.3 Coordinate Systems

To record any type of digital hologram, we must consider an image acquisition scheme and a writing scheme. This is illustrated in Figure 8.2. For simplicity, we consider a pinhole model for both the recording camera(s) and for the printer's optical writing head. This is equivalent to using a geometric optics approximation with the assumption of a small-aperture camera and a "point" hogel. Figure 8.2a depicts a generic image acquisition system consisting of a large matrix of parallel forward-facing identical cameras distributed evenly over plane $\mathscr{C}$ situated at some distance from the object to be recorded. Associated with each camera is a film plane ($\mathscr{F}$), which is the plane where the camera produces a focussed image of

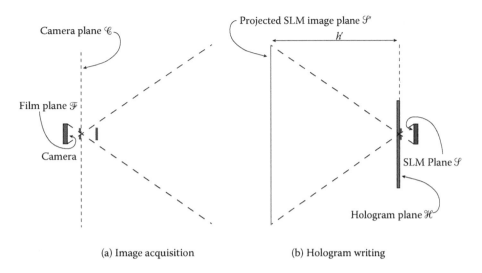

(a) Image acquisition   (b) Hologram writing

FIGURE 8.2 Generic image acquisition (a) and hologram-writing (b) geometries.

the object. We will also define the paraxially projected image plane of $\mathscr{F}$, which we shall label $\mathscr{F}'$. The film plane is assumed to be rectangular as in a usual camera and, accordingly, we associate a horizontal and vertical field of view with each of the cameras.

Figure 8.2b depicts a generic hologram-writing scenario in which, as mentioned previously, we have replaced the complex optical system of the printer with a simple pinhole located at the hologram plane. In addition to the hologram plane, we identify two additional planes; these are the spatial light modulator (SLM) plane, $\mathscr{S}$, where data are placed to encode the writing laser beam with the required information to form the hologram, and the projected image plane of the SLM, $\mathscr{S}'$, which is a plane at some distance downstream or upstream of the hogel. In general, the image transformation from $\mathscr{S}$ to $\mathscr{S}'$ will be non-paraxial. However, we shall limit all our discussions to the paraxial case until we introduce non-paraxial printer optics in the last part of this chapter. As with the cameras, we assume a rectangular SLM and, accordingly, we are able to associate with the printer writing system both a horizontal and vertical field of view.

Most of what follows in this chapter and in Chapter 9 is concerned with the computational techniques of converting camera information to the data required by the printer SLM to generate the various types of holograms. In the generic image acquisition and writing scenarios described previously, we have identified six principal planes: these are the hologram plane, the SLM plane, the projected SLM image plane, the camera plane, the film plane and the paraxially projected film plane. To discuss data processing, we must first characterise these planes.

We will start with the three principal planes involved in image acquisition (Figure 8.3). We let the camera plane $\mathscr{C}$ be parameterised by the Cartesian coordinate system (ξ,ζ). The origin of this and all other Cartesian planes used in this book is taken to be located at the lower left-hand corner of the plane as we view the plane from the right-hand side. The dimensions of plane $\mathscr{C}$ are defined as $D_\xi \times D_\zeta$. We let each of the film planes ($\mathscr{F}$) be defined by the Cartesian coordinate system (x, y); the dimensions of $\mathscr{F}$ are taken to be $D_x \times D_y$. Likewise, $\mathscr{F}'$ will be described by the Cartesian coordinate system (x', y') and the dimensions of $\mathscr{F}'$ are taken to be $D_x' \times D_y'$.

We now treat the three principal planes involved with the process of writing the hologram (Figure 8.4). We describe the hologram plane $\mathscr{H}$ by a Cartesian coordinate system (X, Y) having dimensions $D_X \times D_Y$. The SLM plane $\mathscr{S}$ is likewise described by the Cartesian system (U,V) with dimensions $\Pi \times \Sigma$. The projected SLM plane $\mathscr{S}'$ is defined by the Cartesian system (U', V') with dimensions $\Pi' \times \Sigma'$.

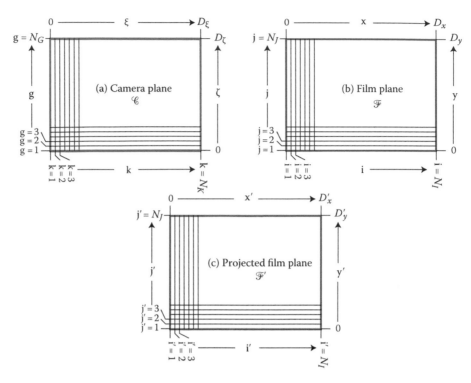

FIGURE 8.3 Characterisation of the principal image acquisition planes: (a) camera plane, (b) film plane and (c) projected film plane.

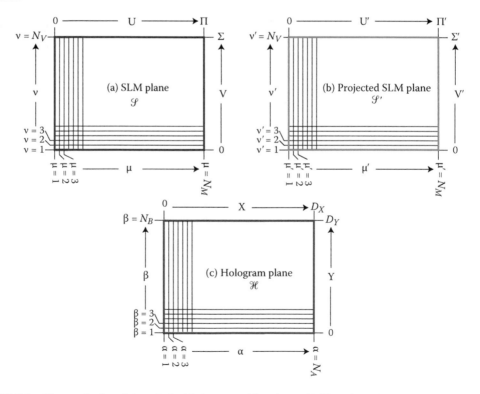

FIGURE 8.4 Characterisation of the principal hologram-writing planes: (a) SLM plane, (b) projected SLM plane and (c) hologram plane.

8.4 Coordinate Meshes

Having set up the coordinate systems, we must now discretise these systems because we wish to consider the case of writing a hologram, which is built up of digital hogels, and because plane $\mathscr{C}$ is populated with a finite number of digital cameras.

Starting with the camera plane ($\mathscr{C}$), we discretise the coordinates (ξ,ζ) in terms of positive integers k and g:

$$\xi = \frac{(k-1)D_\xi}{N_K - 1} \quad k = 1,2,3,...,N_K \tag{8.1}$$

$$\zeta = \frac{(g-1)D_\zeta}{N_G - 1} \quad g = 1,2,3,...,N_G \tag{8.2}$$

The parameters (k,g) then effectively describe a set of discrete positions in which a camera may be placed on the camera plane.

Likewise, we discretise the film plane coordinates (x,y) in terms of positive integers i and j:

$$x = \frac{(i-1)D_x}{N_I - 1} \quad i = 1,2,3,...,N_I \tag{8.3}$$

$$y = \frac{(j-1)D_y}{N_J - 1} \quad j = 1,2,3,...,N_J \tag{8.4}$$

In exactly the same manner as before, the parameters (i,j) divide up the film plane into discrete locations. Finally, we discretise the projected film plane coordinates (x', y') in terms of positive integers i' and j':

$$x' = \frac{(i'-1)D'_x}{N_I - 1} \quad i' = 1,2,3,...,N_I \tag{8.5}$$

$$y' = \frac{(j'-1)D'_y}{N_J - 1} \quad j' = 1,2,3,...,N_J \tag{8.6}$$

Turning now to the principal writing planes, we introduce discretised coordinates, α and β, to describe the location of each hogel on the hologram plane in terms of the continuous hologram coordinates (X,Y):

$$X = \frac{(\alpha-1)D_X}{N_A - 1} \quad \alpha = 1,2,3,...,N_A \tag{8.7}$$

$$Y = \frac{(\beta-1)D_Y}{N_B - 1} \quad \beta = 1,2,3,...,N_B \tag{8.8}$$

The SLM plane is parameterised by the integers (μ,ν) and by the continuous coordinates (U,V):

$$U = \frac{(\mu-1)\Pi}{N_M - 1} \quad \mu = 1,2,3,...,N_M \tag{8.9}$$

$$V = \frac{(\nu-1)\Sigma}{N_V - 1} \quad \nu = 1,2,3,...,N_V \tag{8.10}$$

The projected image plane, $\mathscr{S}'$, is likewise parameterised by

$$U' = \frac{(\mu' - 1)\Pi'}{N_M - 1} \quad \mu' = 1, 2, 3, \ldots, N_M \tag{8.11}$$

$$V' = \frac{(v' - 1)\Sigma'}{N_V - 1} \quad v' = 1, 2, 3, \ldots, N_V \tag{8.12}$$

The parameters Π and Σ effectively define, respectively, the horizontal and vertical fields of view of the printer writing head. The optical objective in the printer writing head is usually circularly symmetric, but when it is combined with a rectangular SLM, one obtains a different field of view in the horizontal and vertical directions. We may therefore write Π' and Σ' in terms of these angles:

$$\Pi' = 2h' \tan\left(\frac{\Psi_{PH}}{2}\right) \tag{8.13}$$

$$\Sigma' = 2h' \tan\left(\frac{\Psi_{PV}}{2}\right), \tag{8.14}$$

where h' is the distance from the hologram plane, $\mathscr{H}$, to the projected SLM image plane, $\mathscr{S}'$.

For each primary colour, we may define the totality of optical information measured by the cameras by the optical intensity ($\mathbf{I}$) of each of the pixels on each of the camera film planes. $\mathbf{I}$ is a four-dimensional object as it is a function of the integer parameters k, g, i and j. We will therefore write it as $^{kg}\mathbf{I}_{ij}$.

We may also define the totality of optical information for a given primary colour that we wish to write to the hologram as the optical intensity ($\mathbf{S}$) of each of the SLM pixels required to write each of the hogels. Like $\mathbf{I}$, $\mathbf{S}$ is a four-dimensional object as it is a function of the integer parameters μ, v, α and β. We will therefore write it as $^{\mu v}\mathbf{S}_{\alpha\beta}$. The question is now simple: how can we best calculate $\mathbf{S}$ from $\mathbf{I}$?

8.5 Independent Primary Colours

As we have seen in Chapter 7, most current holographic printers use three primary colours to generate full-colour reflection holograms. Some types of holograms, such as full-colour rainbow holograms, require only a single-colour writing laser but nevertheless require image data in three or more primary colours. We have also mentioned in Chapter 5 that the use of more than three primary colours in any type of full-colour hologram can be beneficial in producing a better spectral replay. In the following sections, we shall treat each individual primary colour separately. In many cases, we will then simply be able to apply the same algorithms to each primary colour. Of course, separate γ corrections will need to be applied for proper colour balance, but this is trivial. In some cases, however, cross-coupling occurs between the colour data, for example, when the primary hologram recording and replay wavelengths differ, when the recording reference beam geometry does not match the replay illumination geometry, or when the holographic photosensitive material changes its properties after writing. We will return to such cross-coupling a little later.

8.6 Viewing Plane

To write a digital hologram, a data set for each of the primary colours needs to be calculated for every hogel. This data set populates the writing SLM at the moment that the hogel is written. Clearly, we wish to calculate this data set from camera data ($^{kg}\mathbf{I}_{ij}$). However, before we can do this, we must consider

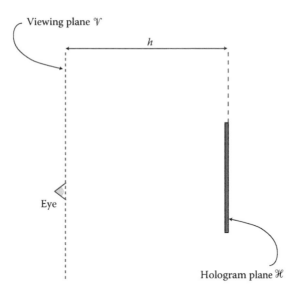

FIGURE 8.5　Hologram viewing geometry.

how we are going to view the final hologram. Figure 8.5 shows a simplified diagram of the viewing process. In particular, we consider a plane ($\mathcal{V}$), parallel to $\mathcal{S}$ and $\mathcal{S}'$, situated a distance h away from $\mathcal{H}$ at which we want to observe the hologram. It is important to realise that for HPO holograms, there is a unique viewing distance. Only at this distance does the viewer observe an undistorted image. For full-parallax holograms, a viewer may, in general, observe the undistorted hologram from any point in space. However, even here, it is extremely useful to define a specific viewing plane as we can then seek to match the radiation field characteristic of the acquired image data and the field produced by the hologram at replay at this plane. The mathematical operation of matching the fields in this way essentially defines the data required to write the hologram. In the case of a full-parallax hologram, this process then assures the matching of the light-field at all other planes automatically as required by Huygens' principle.

8.7 Simple Cases

There are two simple cases in which **S** may be trivially derived from **I**. The first is the case of a direct-write digital holography (DWDH) analogue of a Denisyuk reflection hologram and the second is the case of master-write digital holography (MWDH), where a master H_1 hologram is written. In both cases, we will assume that the printer optical objective is paraxial. We shall discuss non-paraxial objectives in Section 8.10.

8.7.1 Full-Parallax DWDH "Denisyuk" Reflection Hologram

First, we arrange for the camera or cameras, which acquire the image data, to have identical horizontal and vertical fields of view as those possessed by the printer writing head and for the SLM and charge-coupled device (CCD) dimensions to correspond, that is

$$\Psi_{CH} = \Psi_{PH} \equiv \Psi_H$$

$$\Psi_{CV} = \Psi_{PV} \equiv \Psi_V$$

$$N_M = N_I$$

$$N_V = N_J$$

(8.15)

Generally, we must collect image data by either arranging a large matrix of cameras situated on the camera plane or by using a single camera, which we move in such a fashion that it takes sequential pictures from all locations on the defined integer grid at the camera plane. Of course, if the three-dimensional object that we want to digitise is not static, then it is difficult to use the latter choice. We shall discuss image acquisition in greater detail in Chapter 10, but for now, we should keep in mind that the single camera may be either a real camera or a virtual camera within a computer-aided design (CAD) program.

We now seek to superimpose the various planes involved in image acquisition, hologram writing and hologram viewing with the intention of matching, at the viewing plane, the light-field emanating from the original object, the light-field used to write the hologram and the light-field observed upon viewing the hologram. This amounts to superimposing the diagrams of Figure 8.2a and b and Figure 8.5. This is shown in Figure 8.6. Note that we have chosen to collocate the camera plane ($\mathcal{C}$) with the hologram plane ($\mathcal{H}$). In addition, the projected SLM plane $\mathcal{S}'$, the projected film plane $\mathcal{F}'$ and the viewing plane ($\mathcal{V}$) have been collocated and we have taken $h' = h$. This choice of collocation means that the camera pinhole and the hogel are also collocated.

Having chosen to configure the various principal planes in this fashion, we see that a test ray emanating from the object, which passes through a given point on $\mathcal{H}$, will intersect exactly the same pixel address on the projected film plane as on the projected SLM plane or, in other words, for the viewer to see a light-field exactly reproducing the acquired light-field that emanated from the original object in plane $\mathcal{V}$, the data in $\mathcal{S}'$ must correspond exactly with the camera data recorded on $\mathcal{F}'$. More specifically

$$^{\mu'\nu'}\mathbf{S}'_{\alpha\beta} = {}^{\alpha\beta}\mathbf{I}'_{\mu'\nu'} \tag{8.16}$$

where, as previously,

$$
\begin{aligned}
\mu' &= 1,2,3,...,N_M & \nu' &= 1,2,3,...,N_V \\
\alpha &= 1,2,3,...,N_A & \beta &= 1,2,3,...,N_B
\end{aligned}
\tag{8.17}
$$

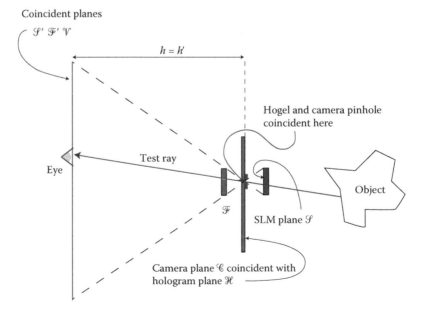

FIGURE 8.6 Writing a DWDH Denisyuk hologram.

Then, because $\mathscr{F}'$, $\mathscr{S}'$ and $\mathscr{F}$ are downstream of the hogel and camera pinhole, whereas $\mathscr{S}$ is located upstream, this leads to the following transformation rule between $\mathbf{I}$ and $\mathbf{S}$:

$$^{\mu\nu}\mathbf{S}_{\alpha\beta} = {}^{\alpha\beta}\mathbf{I}_{N_M-1-\mu,N_V-1-\nu} \tag{8.18}$$

This transformation describes the SLM data for each primary colour required to write a full-parallax reflection hologram, which is a direct analogue to the Denisyuk type of analogue reflection hologram. The image appears entirely behind the hologram plane as with the Denisyuk hologram and is visible at any angle within the field of view of the recording camera. Note that if a real camera is used with a finite aperture, then blurring will occur because any real camera will only focus a certain range of depths properly onto its film plane. In CAD programs, one usually uses a camera of effectively zero aperture, which gives infinite depth. Nevertheless, just like a real Denisyuk hologram, the final digital hologram will be subject to inherent image blurring if it is illuminated by a source of spatially or temporally incoherent radiation (i.e., a normal white-light source). By reducing the hogel size to submillimetre dimensions, it can be difficult to tell this type of digital hologram from a true Denisyuk hologram.

8.7.2 MWDH Master Hologram

The second simple case we will examine is that of a master (MWDH) hologram. This is essentially the analogue of the H_1 hologram of classical analogue holography [4]. This hologram can be HPO or full-parallax, and it can be of transmission or reflection type. If the hologram is HPO, it will generally be in the form of a slit, which is recorded at a distance from the object that corresponds to the viewing distance of the hologram. In the full-parallax case, the recording distance is chosen so that subsequent optical transfer to an H_2 hologram can provide a final image that appears partly in front of the holographic plate. By writing a reflection master, several primary colours may be encoded onto the same plate and a full-colour H_2 hologram produced using well-known techniques. For the purposes of this section, we shall treat the case of a full-parallax master H_1 hologram and derive the trivial transformations between the acquired image data $\mathbf{I}$ and the hologram write-data $\mathbf{S}$.

As before we must arrange for the acquisition camera to have an identical horizontal and vertical field of view as those possessed by the printer writing head (i.e., Equation 8.15 applies). We then associate the various principal planes of acquisition, writing and viewing. This is a little more complex this time, as we must consider the H_1:H_2 transfer process as well. Figure 8.7a illustrates the writing of the H_1 reflection hologram. Figure 8.7b then illustrates the H_1:H_2 transfer. Finally, Figure 8.7c illustrates the viewing of the H_2 reflection hologram. From these figures, it can be seen that both the camera plane ($\mathscr{C}$) and the viewing plane ($\mathscr{V}$) are collocated with the hologram plane which we can now label $\mathscr{H}_1$ for clarity. This means that we must take $h = 0$ of course. The transfer process is completely standard and upon viewing, a faithful reproduction of the original object is viewable at $\mathscr{V}$ in the case of an HPO hologram or at any location within the camera field of view in the full-parallax case.

Clearly, the geometry of Figure 8.7 implies that the light-fields emitted by object and hologram at, respectively, $\mathscr{F}'$ and $\mathscr{S}'$ are the same and accordingly the transformation between $\mathbf{S}'$ and $\mathbf{I}'$ is:

$$^{\mu'\nu'}\mathbf{S}'_{\alpha\beta} = {}^{\alpha\beta}\mathbf{I}'_{\mu'\nu'} \tag{8.19}$$

Because $\mathscr{F}$ and $\mathscr{S}$ are collocated, this then gives simply

$$^{\mu\nu}\mathbf{S}_{\alpha\beta} = {}^{\alpha\beta}\mathbf{I}_{\mu\nu}. \tag{8.20}$$

8.8 Image-Planed DWDH HPO Holograms

We will now turn our attention to the less trivial case of the $\mathbf{I}$-to-$\mathbf{S}$ transformations describing DWDH HPO reflection holograms. These types of holograms are characterised by the assumption that the camera and viewing planes must coincide such that $h = h'$. This allows all the optical write data for

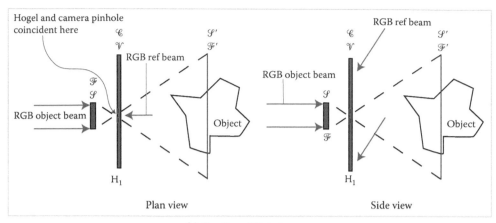

(a) Creation of H$_1$ hologram

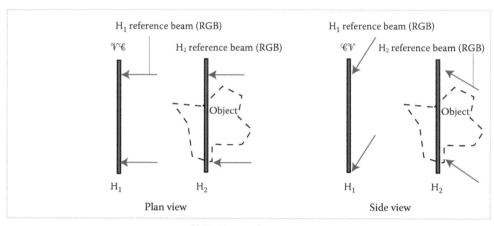

(b) H$_1$:H$_2$ transfer process

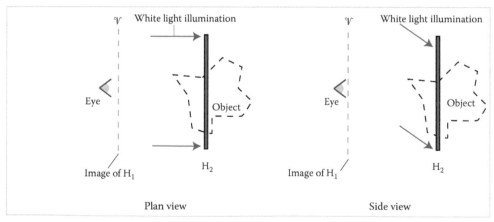

(c) Viewing the H$_2$

FIGURE 8.7   Writing an MWDH H$_1$ reflection master hologram: (a) writing the H$_1$ hologram, (b) H$_1$:H$_2$ transfer process and (c) viewing the H$_2$ reflection hologram.

the hologram to be calculated using only a linear array of cameras instead of a two-dimensional matrix, which is required in the full-parallax case. The camera data required for an HPO hologram is therefore associated with a three-dimensional object ($^{k}\mathbf{I}_{ij}$) rather than the four-dimensional object ($^{kg}\mathbf{I}_{ij}$), which we have introduced above and which is required for the general full-parallax case. In addition, we shall see that the form of $^{\mu\nu}\mathbf{S}_{\alpha\beta}$ also essentially becomes three-dimensional in the HPO case.

In terms of computer memory requirements, HPO represents a considerable saving. Take for example the case of a 1 m × 1 m DWDH full-parallax colour reflection hologram, which might typically require an array of 1000 × 1000 cameras, each of a resolution of 1024 × 768. The total uncompressed information in $\mathbf{I}$ is approximately 2.4 TB. For the HPO case, this information decreases by a factor of 1000 to 2.4 GB!

We must remember, however, that there is a price to pay for such a spectacular reduction in information. An HPO hologram only replays a non-distorted image when the viewer's eyes coincide geometrically with the linear array of cameras. As the viewer approaches and withdraws from the hologram, the image becomes distorted, and as the viewer moves his or her head up and down, instead of seeing over and underneath an object, the object just appears to rotate such that it always presents the same orientation to the viewer. If the camera distance (h) is large enough, there nevertheless exists a significant viewing region around this distance where distortion is difficult to notice. In general, because the distortion becomes increasingly difficult to perceive as the observation distance is increased, it is normally possible to achieve practical distortion-free viewing from the camera plane to infinity if h is rather larger than the characteristic hologram dimension.

In addition, there are two significant further advantages inherent to reflection HPO holograms when illuminated by usual broadband sources. The first is that very large image depths may be recorded and replayed using such holograms, making them an obvious choice for large-format display applications. This is not usually the case for modern full-parallax reflection holograms, which must be recorded onto an emulsion of finite thickness and are usually illuminated by broadband sources; here, as we shall discuss in Chapters 9 and 11, chromatic blurring can severely limit the clear image depth. The second advantage is that multiple broadband illumination sources may be used to illuminate an HPO reflection hologram without introducing image blurring. We shall discuss this aspect in detail in Chapter 13.

We should also mention that one can indeed discuss curved camera trajectories instead of linear ones. After all, initial work on stereograms [5] used a fixed camera and a rotating object—although most often the final stereogram was itself actually curved. Nevertheless, we find little benefit in an HPO hologram that has a curved optimum viewing plane in place of a linear one. In the case of full-parallax holograms, however, the camera plane is essentially independent of the viewing plane and the hologram can (ideally) be viewed without distortion at any location; in Chapter 10, we shall see how the results of this chapter can then be usefully extended to the cases of cylindrical and spherical camera surfaces.

8.8.1 Printer, Camera and Viewing Window Options

There are several choices as to how a single parallax DWDH hologram may be physically written. First, we may decide to keep the SLM static or we may use an objective with a larger entrance pupil and opt to move the SLM within this pupil. Second, various formats of computer data, which effectively correspond to different camera geometries, may be used. A simple translating camera (for example, a real physical camera) will produce, in general, a different $^{k}\mathbf{I}_{ij}$ data set than a specially programmed virtual camera. Either a real or a virtual camera may rotate as it translates, and this too produces a different $^{k}\mathbf{I}_{ij}$. Finally, different viewing window geometries may be selected. If we elect to use the full field of view of the printer optical objective when writing each hogel, then we will have a different result than if we constrain our viewing window to a well-defined rectangle at the viewing plane.

All the above choices must be made according to the specific application at hand. In the following sections, we will treat the most important major cases and derive for each geometry the transformations necessary to convert the single parallax camera data, $^{k}\mathbf{I}_{ij}$, into the SLM data, $^{\mu\nu}\mathbf{S}_{\alpha\beta}$. We will only treat the case of a static SLM in this book because the historical motivation for moving the SLM has largely been overtaken by modern advances in high-definition SLM technology.

8.8.2 General Rectangular Viewing Window

8.8.2.1 Simple Translating Camera

In this section, a specific image data model, $^k\mathbf{I}_{ij}$, will be assumed. This model relates to a simple translating camera that follows a horizontal trajectory through the midpoint of the hologram viewing plane. Figure 8.8 shows a side (a) and plan (b) view of the geometry. The viewing zone of the hologram is defined (at $\mathcal{V}$) to be a rectangle having horizontal dimension W and vertical dimension H. This means that $D_\xi = W$.

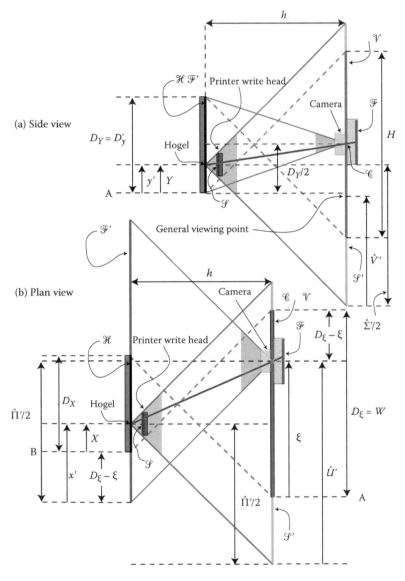

FIGURE 8.8 Writing a DWDH HPO reflection hologram. Ray geometry for (a) side view and (b) plan view. Note the conjugate SLM position.

To simplify the geometry of Figure 8.8, we chose to illustrate a (symmetrically) restricted field of view of the printer's optical objective. We define the horizontal and vertical components of this restricted field of view as

$$\tan\left(\frac{\hat{\Psi}_{PH}}{2}\right) = \frac{\hat{\Pi}'}{2h} = \frac{W + D_X}{2h}$$

$$\tan\left(\frac{\hat{\Psi}_{PV}}{2}\right) = \frac{\hat{\Sigma}'}{2h} = \frac{H + D_Y}{2h} \qquad (8.21)$$

Of course, we are implicitly assuming here that $\Psi_{PH} \geq \hat{\Psi}_{PH}$ and $\Psi_{PV} \geq \hat{\Psi}_{PV}$ because, clearly, one cannot restrict an angle to a larger angle! For clarity, we shall label the SLM data, when we calculate it using restricted fields of view, as $\hat{\mathbf{S}}$ to distinguish it from $\mathbf{S}$, which refers to data corresponding to the full field of view of the printer. Likewise, we will introduce $\hat{N}_M$ and $\hat{N}_V$ as the reduced SLM integer dimensions and $\hat{U}'$ and $\hat{V}'$ as the restricted real dimensions of the projected SLM. Finally, we introduce the restricted integer coordinates of the SLM, $\hat{\mu}'$ and $\hat{v}'$:

$$\hat{U}' = \frac{(\hat{\mu}' - 1)\hat{\Pi}'}{\hat{N}_M - 1} \quad \hat{\mu}' = 1,2,3,...,\hat{N}_M \qquad (8.22)$$

$$\hat{V}' = \frac{(\hat{v}' - 1)\hat{\Sigma}'}{\hat{N}_V - 1} \quad \hat{v}' = 1,2,3,...,\hat{N}_V \qquad (8.23)$$

We choose a camera whose horizontal field of view exactly matches the restricted horizontal field of view of the printer so that $\Psi_{CH} = \hat{\Psi}_{PH}$. If we were to use a larger camera angle, we would simply end up collecting image data that we did not need! Using the same philosophy, we use a vertical camera field of view that is defined as

$$\tan\left(\frac{\Psi_{CV}}{2}\right) = \frac{D_Y}{2h}. \qquad (8.24)$$

The difference in the choice between horizontal and vertical angles is because we are considering an HPO hologram and so the camera always stays vertically at the midpoint height of the hologram. This is in contrast with the horizontal behaviour of the camera, which consists of linear tracking and therefore requires a larger field of view.

We are now ready to examine Figure 8.8 in detail. Let us look at the side view in (a) to start with. Here, we have superimposed the hologram, the printer SLM and the camera. The camera is fixed at a height corresponding to the centre of the hologram and points straight forward. Its field of view is illustrated in blue and can be seen to just cover the hologram plane—that is, vertically, the projected film plane ($\mathcal{F}'$) is perfectly coincident with $\mathcal{H}$. In the plan view of (b), $\mathcal{F}'$ can be seen to be much larger than $\mathcal{H}$, as the camera, when on one extreme end of its track, must still cover all of $\mathcal{H}$. This is a characteristic of a simple translating camera and one can immediately see that this type of camera is inefficient, as it produces data that falls outside the field of $\mathcal{H}$—such data is, of course, useless.

We consider an arbitrary hogel on the hologram plane. The diagrams illustrate (in green) the field of view of the SLM and the printer objective during the printing process of this hogel. By considering the projected SLM plane $\mathcal{S}'$ located at the camera distance (h), one can see that each hogel clearly has the capacity to deliver radiation to every point within the viewing zone. The magenta line connecting the hogel with the camera is a test ray. We will use this test ray to match the radiation fields at the viewing line ($\mathcal{V}$) during image acquisition and during hologram writing to relate camera data to SLM data. Specifically, we will demand that the angle and position of this arbitrary test ray at the viewing line ($\mathcal{V}$) be exactly the same for the case of image acquisition and for the case of writing the hologram. This then makes sure that the hologram replays an identical copy of the light-field emitted by the original object.

In fact, we will cheat a little and not match angles and positions at $\mathcal{V}$. Instead, we will opt for the equivalent process of matching positional coordinates (only) at two planes, $\mathcal{H}$ and $\mathcal{V}$. Because we are considering an HPO hologram, we must seek to match acquisition plane coordinates and hologram-writing coordinates at the hogel location, both horizontally and vertically. In addition, we must match acquisition plane horizontal coordinates and hologram-writing horizontal coordinates at the camera location. These three matching operations will define the three freedoms implicit in the transformation between the two three-dimensional objects ($\mathbf{I}$ and $\hat{\mathbf{S}}$).

The simplest matching operation is that of the vertical ($\mathcal{F}'$) coordinate at the hogel with that of $\mathcal{H}$. This gives

$$y' = Y, \tag{8.25}$$

which leads to the index law through Equations 8.4 and 8.8

$$j' = \left\| \frac{(\beta - 1)(N_J - 1)}{N_B - 1} \right\| + 1 \tag{8.26}$$

Note that we use the notation of two sets of double lines enclosing a real expression to indicate the nearest integer operation—that is, in terms of the floor function $\|x\| \equiv \lfloor x + 1/2 \rfloor$. Now, because we have chosen to place $\mathcal{F}'$ downstream of the camera pinhole, the $\mathcal{F}$ coordinates are related to their primed counterparts by the simple relation

$$\begin{pmatrix} i \\ j \end{pmatrix} = \begin{pmatrix} N_I - i' + 1 \\ N_J - j' + 1 \end{pmatrix} \tag{8.27}$$

We can therefore use this to convert Equation 8.26 to the more useful form:

$$j = N_J - \left\| \frac{(\beta - 1)(N_J - 1)}{N_B - 1} \right\| \tag{8.28}$$

Of course, Equation 8.27 assumes that we start with data from a normal inverting camera. This is not always the case as both virtual cameras in CAD programs and modern digital cameras are fundamentally non-inverting. In such a case, Equation 8.27 must be replaced by

$$\begin{pmatrix} i \\ j \end{pmatrix} = \begin{pmatrix} i' \\ j' \end{pmatrix} \tag{8.29}$$

and Equation 8.28 becomes

$$j = \left\| \frac{(\beta - 1)(N_J - 1)}{N_B - 1} \right\| + 1 \tag{8.30}$$

Nevertheless, we shall continue with the case of an inverting camera and use Equations 8.27 and 8.28 in the following sections. Two further matching laws are then clear from Figure 8.8b.

$$X = x' - (W - \xi). \tag{8.31}$$

$$\hat{U}' - \xi = D_X - X \tag{8.32}$$

These are best combined to give

$$\xi = \hat{U}' - D_X + X \tag{8.33}$$

and

$$x' = W + D_X - \hat{U}' \tag{8.34}$$

We can then derive the corresponding index transformations using Equations 8.1, 8.5, 8.7 and 8.22:

$$k = 1 + \left\| \frac{(N_K - 1)}{W} D_X \left[\frac{(\hat{\mu}' - 1)}{(\hat{N}_M - 1)} \frac{\hat{\Pi}'}{D_X} - 1 + \frac{(\alpha - 1)}{(N_A - 1)} \right] \right\|$$

$$= 1 + \left\| \frac{(N_K - 1)}{W} D_X \left[\frac{(\hat{\mu} - 1)}{(\hat{N}_M - 1)} \left(\frac{W}{D_X} + 1 \right) + \frac{(\alpha - N_A)}{(N_A - 1)} \right] \right\| \tag{8.35}$$

$$i' = 1 + \left\| \frac{(N_I - 1)}{D_X} \left(W + D_X - \frac{(\hat{\mu}' - 1)\hat{\Pi}'}{(\hat{N}_M - 1)} \right) \right\|$$

$$= 1 + \left\| \frac{(N_I - 1)(\hat{N}_M - \hat{\mu})}{(\hat{N}_M - 1)} \right\| \tag{8.36}$$

We have used the fact that $\hat{\mu} = \hat{\mu}'$ in the second line of both Equations 8.35 and 8.36. This comes from the choice of locating both $\mathscr{S}$ and $\mathscr{S}'$ upstream of the hogel (we refer to this as the "conjugate SLM position"). Again, we should comment on this as positioning the SLM to the right of the hologram is simply a choice. The alternative ("the non-conjugate SLM position") is to position the SLM to the left (with both the recording and replay reference beams from the right), in which case the coordinates of $\mathscr{S}$ and $\mathscr{S}'$ are related by

$$\begin{pmatrix} \hat{\mu} \\ \hat{v} \end{pmatrix} = \begin{pmatrix} \hat{N}_M - \hat{\mu}' + 1 \\ \hat{N}_V - \hat{v}' + 1 \end{pmatrix} \tag{8.37}$$

This is illustrated in Figure 8.9. As discussed in Chapter 7, the choice of a conjugate or non-conjugate SLM position depends largely on the optical design of the printer. When using a non-collimated reference beam (i.e., one which is characterised by different azimuthal and altitudinal reference angles for each hogel), one must be careful in the conjugate SLM case that the reconstruction reference geometry is geometrically conjugate to the recording geometry. In this case, the recording beam is from the left with

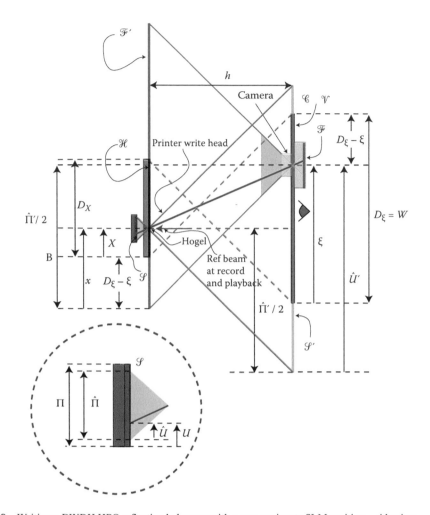

FIGURE 8.9   Writing a DWDH HPO reflection hologram with a non-conjugate SLM position—side view.

replay from the right. In the non-conjugate case, the recording reference beam and the replay reference beam are identical and both illuminate the hogel from the right.

We now use Equation 8.27, which describes an inverting camera to arrive at the unprimed version of Equation 8.36:

$$i = N_I - \left\| \frac{(N_I - 1)(\hat{N}_M - \hat{\mu})}{(\hat{N}_M - 1)} \right\| \tag{8.38}$$

Equations 8.28, 8.35 and 8.38 provide us with the necessary rules to calculate $\hat{\mathbf{S}}$ at the viewing line. This corresponds to knowing $^{\hat{\mu}\hat{v}}\hat{\mathbf{S}}_{\alpha\beta}$ for only a single value of the index $\hat{v}$. Because we are seeking to construct an HPO hologram, we can simply copy this information for all other values of $\hat{v}$. This will mean that each hogel emits a light field that uniformly sweeps the vertical direction—which is precisely what we mean by an HPO hologram. If we wish to limit the vertical field of view of the hologram as a function of the hogel coordinate Y such as to create a viewing window of height H then we can impose the condition that

$$^{\hat{\mu}\hat{v}}\hat{\mathbf{S}}_{\alpha\beta} = {}^{k}\mathbf{I}_{ij} \quad \forall \left\{ \hat{V}' \in \mathbb{R} \,\middle|\, D_Y - Y \le \hat{V}' \le D_Y - Y + H \right\}$$

$$= 0 \quad \text{otherwise} \tag{8.39}$$

This condition on $\hat{V}'$ translates into the following condition on $\hat{v}$:

$$\left\|\frac{D_Y}{D_Y+H}\frac{(\hat{N}_V-1)(N_B-\beta)}{(N_B-1)}\right\|+1\leq\hat{v}\leq1+\left\|\frac{D_Y}{D_Y+H}(\hat{N}_V-1)\left\{\frac{N_B-\beta}{N_B-1}+\frac{H}{D_Y}\right\}\right\| \tag{8.40}$$

If we do not impose this condition, then we will generate a rolling vertical window. This has the effect that at the optimum viewing distance, there will be positions where we see only half the hologram and that as we change the position of our head up and down, so we will see more or less of the hologram. Of course, even with Condition 8.40, when the observer is significantly further away from the hologram than $\mathscr{V}$, he will observe such a rolling window. However, the important point to understand is that HPO holograms are designed to be viewed from (approximately at least) a given distance and as such it makes good sense to create a proper rectangular window at this distance. The perceived brightness of a hogel is after all a direct function of its solid angle of emittance and it makes little sense to therefore put energy and brightness into angular emission zones where only a proportion of the hologram is visible—often, it is therefore much better to maximise the brightness in a clearly defined viewing zone.

We must also be attentive to the horizontal coordinates of the viewing window boundary. In general, Equation 8.35 can produce values of k outside the lower limit of 1 and the upper limit of N_K. To constrain k properly and to assure the proper horizontal definition of the viewing window, we must use a condition exactly similar to Equation 8.40. This condition can of course simply be derived by setting $k = 1$ and $k = N_K$ in Equation 8.35:

$$\left\|\frac{D_X}{D_X+W}\frac{(\hat{N}_M-1)(N_A-\alpha)}{(N_A-1)}\right\|+1\leq\hat{\mu}\leq1+\left\|\frac{D_X}{D_X+W}(\hat{N}_M-1)\left\{\frac{N_A-\alpha}{N_A-1}+\frac{W}{D_X}\right\}\right\| \tag{8.41}$$

We are now in a position to write down the full I-to-$\hat{S}$ transformation for a monochromatic DWDH HPO reflection hologram for the case of a simple translating camera and a paraxial printer objective:

$$^{\hat{\mu}\hat{v}}\hat{S}_{\alpha\beta}={}^k I_{ij} \quad \forall\hat{v}\left\{\hat{v}\in\mathbb{N}\left|\left\|\frac{D_Y}{D_Y+H}\frac{(\hat{N}_V-1)(N_B-\beta)}{(N_B-1)}\right\|+1\leq\hat{v}\leq1+\left\|\frac{D_Y}{D_Y+H}(\hat{N}_V-1)\left\{\frac{N_B-\beta}{N_B-1}+\frac{H}{D_Y}\right\}\right\|\right.\right\}$$

$$\forall\hat{\mu}\left\{\hat{\mu}\in\mathbb{N}\left|\left\|\frac{D_X}{D_X+W}\frac{(\hat{N}_M-1)(N_A-\alpha)}{(N_A-1)}\right\|+1\leq\hat{\mu}\leq1+\left\|\frac{D_X}{D_X+W}(\hat{N}_M-1)\left\{\frac{N_A-\alpha}{N_A-1}+\frac{W}{D_X}\right\}\right\|\right.\right\}$$

$$\forall\alpha\left\{\alpha\in\mathbb{N}\left|\alpha\leq N_A\right.\right\} \quad \forall\beta\left\{\beta\in\mathbb{N}\left|\beta\leq N_B\right.\right\}$$

$$=0 \qquad \text{otherwise} \tag{8.42}$$

where

$$i=N_I-\left\|\frac{(N_I-1)(\hat{N}_M-\hat{\mu})}{(\hat{N}_M-1)}\right\| \qquad j=N_J-\left\|\frac{(\beta-1)(N_J-1)}{N_B-1}\right\|$$

$$k=1+\left\|\frac{(N_K-1)}{W}D_X\left[\frac{(\hat{\mu}-1)}{(\hat{N}_M-1)}\left(\frac{W}{D_X}+1\right)+\frac{(\alpha-N_A)}{(N_A-1)}\right]\right\| \tag{8.43}$$

Remember that this transformation is valid for an inverting camera and a conjugate SLM position; it is also valid for the restricted field of view of the printer, not the actual field of view. To formulate the transformation for **S** rather than for **Ŝ**, we write the projected SLM coordinates in terms of their restricted counterparts (Figure 8.10):

$$U' = \hat{U}' + \frac{1}{2}\left(\Pi' - \hat{\Pi}'\right)$$

$$V' = \hat{V}' + \frac{1}{2}\left(\Sigma' - \hat{\Sigma}'\right) \tag{8.44}$$

We then use Equations 8.11, 8.12, 8.22 and 8.23 to arrive at the transformation rules between μ and $\hat{\mu}$ and between ν and $\hat{\nu}$:

$$\hat{\mu} = \hat{\mu}' = \left\| \frac{\left(\hat{N}_M + 1\right)}{2} + \frac{h\left(\hat{N}_M - 1\right)\tan\left(\dfrac{\Psi_{PH}}{2}\right)}{\left(D_X + W\right)\left(N_M - 1\right)}\left(2\mu - N_M - 1\right) \right\|$$

$$\hat{\nu} = \hat{\nu}' = \left\| \frac{\left(\hat{N}_V + 1\right)}{2} + \frac{h\left(\hat{N}_V - 1\right)\tan\left(\dfrac{\Psi_{PV}}{2}\right)}{\left(D_Y + H\right)\left(N_V - 1\right)}\left(2\nu - N_V - 1\right) \right\| \tag{8.45}$$

The values of $\hat{\mu}$ and $\hat{\nu}$ must, however, be between 1 and $\hat{N}_M$ and $\hat{N}_V$, respectively. This means that there are minimum and maximum values for μ and ν outside which **S** = 0 (this was after all why we introduced these restricted values!). If we denote the minimum index value by a subscript 1 and the maximum by 2, we can write expressions for these minimum and maximum values:

$$\mu_1 = \left\| \frac{N_M + 1}{2} - \frac{\left(D_X + W\right)\left(N_M - 1\right)}{4h\tan\left(\dfrac{\Psi_{PH}}{2}\right)} \right\| \qquad \mu_2 = \left\| \frac{N_M + 1}{2} + \frac{\left(D_X + W\right)\left(N_M - 1\right)}{4h\tan\left(\dfrac{\Psi_{PH}}{2}\right)} \right\| \tag{8.46}$$

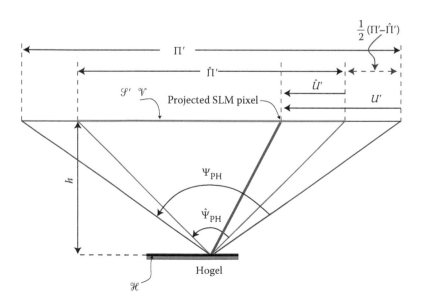

FIGURE 8.10   Plan view of the hologram showing the relationship between restricted and unrestricted SLM coordinate systems.

and

$$v_1 = \left\| \frac{N_V + 1}{2} - \frac{(D_Y + H)(N_V - 1)}{4h \tan\left(\dfrac{\Psi_{PV}}{2}\right)} \right\| \qquad v_2 = \left\| \frac{N_V + 1}{2} + \frac{(D_Y + H)(N_V - 1)}{4h \tan\left(\dfrac{\Psi_{PV}}{2}\right)} \right\| \tag{8.47}$$

Of course, the viewing window constraints (Equations 8.40 and 8.41) must always constitute more constraining conditions than the conditions $\mu_1 \le \mu \le \mu_2$ and $v_1 \le v \le v_2$; the viewing window conditions can be written in terms of (μ, v) by the use of Equation 8.45 in Equations 8.40 and 8.41, giving

$$N_{U1} \le \mu \le N_{U2}$$
$$N_{V1} \le v \le N_{V2} \tag{8.48}$$

where

$$N_{U1}(\alpha) = \left\| \frac{N_M + 1}{2} + \frac{(N_M - 1)}{2h \tan\left(\dfrac{\Psi_{PH}}{2}\right)} \left\{ D_X \frac{(N_A - \alpha)}{N_A - 1} - \frac{(D_X + W)}{2} \right\} \right\|$$

$$N_{U2}(\alpha) = \left\| \frac{N_M + 1}{2} + \frac{(N_M - 1)}{2h \tan\left(\dfrac{\Psi_{PH}}{2}\right)} \left\{ D_X \left\{ \frac{(N_A - \alpha)}{N_A - 1} + \frac{W}{D_X} \right\} - \frac{(D_X + W)}{2} \right\} \right\| \tag{8.49}$$

and

$$N_{V1}(\beta) = \left\| \frac{N_V + 1}{2} + \frac{(N_V - 1)}{2h \tan\left(\dfrac{\Psi_{PV}}{2}\right)} \left\{ D_Y \frac{(N_B - \beta)}{N_B - 1} - \frac{(D_Y + H)}{2} \right\} \right\|$$

$$N_{V2}(\beta) = \left\| \frac{N_V + 1}{2} + \frac{(N_V - 1)}{2h \tan\left(\dfrac{\Psi_{PV}}{2}\right)} \left\{ D_Y \left\{ \frac{(N_B - \beta)}{N_B - 1} + \frac{H}{D_Y} \right\} - \frac{(D_Y + H)}{2} \right\} \right\| \tag{8.50}$$

From a practical point of view, these (or equivalently, Equations 8.40 and 8.41) are the constraints we will need. Nevertheless, Equations 8.46 and 8.47 can be useful when using restricted variables as offset indices to address **S** directly.*

---

* In this case, it is good practice to fine-tune W and H such that the offset indices are integers.

We can now define **S** in terms of $\hat{\mathbf{S}}$:

$$^{\mu\nu}\mathbf{S}_{\alpha\beta} = {}^{\hat{\mu}\hat{\nu}}\hat{\mathbf{S}}_{\alpha\beta} \qquad \forall\alpha\{\alpha \in \mathbb{N} | \alpha \leq N_A\} \qquad\qquad \forall\beta\{\beta \in \mathbb{N} | \beta \leq N_B\}$$

$$\forall\mu\{\mu \in \mathbb{N} | N_{U1}(\alpha) \leq \mu \leq N_{U2}(\alpha)\} \quad \forall\nu\{\nu \in \mathbb{N} | N_{V1}(\beta) \leq \nu \leq N_{V2}(\beta)\}$$

$$= 0 \quad \text{otherwise} \tag{8.51}$$

where $\hat{\mu}$ and $\hat{\nu}$ are given by Equation 8.45. Alternatively, we can substitute Equation 8.45 into Equations 8.42 and 8.43 to give the direct **I**-to-**S** transformation:

$$^{\mu\nu}\mathbf{S}_{\alpha\beta} = {}^{k}\mathbf{I}_{ij} \qquad \forall\mu\{\mu \in \mathbb{N} | N_{U1}(\alpha) \leq \mu \leq N_{U2}(\alpha)\} \; \forall\nu\{\nu \in \mathbb{N} | N_{V1}(\beta) \leq \nu \leq N_{V2}(\beta)\}$$

$$\forall\alpha\{\alpha \in \mathbb{N} | \alpha \leq N_A\} \qquad \forall\beta\{\beta \in \mathbb{N} | \beta \leq N_B\}$$

$$= 0 \qquad \text{otherwise} \tag{8.52}$$

where

$$i = N_I - \left\| (N_I - 1)\left\{ \frac{1}{2} - \frac{h(2\mu - N_M - 1)}{(D_X + W)(N_M - 1)} \tan\left(\frac{\Psi_{\text{PH}}}{2}\right) \right\} \right\| \tag{8.53}$$

$$j = N_J - \left\| \frac{(\beta - 1)(N_J - 1)}{N_B - 1} \right\| \tag{8.54}$$

$$k = 1 + \left\| \frac{(N_K - 1)(W + D_X)}{W}\left\{ \frac{1}{2} + \frac{h(2\mu - N_M - 1)}{(D_X + W)(N_M - 1)} \tan\left(\frac{\Psi_{\text{PH}}}{2}\right) + \frac{D_X(\alpha - N_A)}{(W + D_X)(N_A - 1)} \right\} \right\| \tag{8.55}$$

You will notice that all mention of restricted values has now disappeared from this transformation. These values might therefore be regarded as simply a useful tool to arrive at the general **I**-to-**S** transformation. However, very often it is more efficient to calculate $\hat{\mathbf{S}}$, as **S** usually contains a high percentage of constantly zero elements (often >70%). Computationally, it can therefore make better sense to calculate only $\hat{\mathbf{S}}$ for each hogel and to then only update these elements on the SLM. This is particularly true when HPO SLM data is calculated from camera data in real time between successive hogel-write operations; one then has very limited time available to calculate the data.*

8.8.3 Centred Camera Configuration

It will be clear from Figure 8.8b that most of the camera film plane is not actually used in computing the SLM data, as much of the projected film plane, $\mathscr{F}'$, falls outside the hologram plane $\mathscr{H}$. This is the major disadvantage of the simple translating camera configuration. We shall see in Chapter 10 how a combination of translation and rotation of the camera can be used in real-world image acquisition systems to

---

* We shall introduce shortly an even more efficient system known as the centred SLM coordinate system. However, this system has variable rather than fixed offset indices and this can induce digital noise—more on this later.

improve this situation dramatically—but at a cost. Here, we shall describe a much simpler technique which is available for use with most virtual cameras within CAD software and which has essentially no disadvantages. This is the centred camera configuration. In Chapter 10, we shall also look at how this configuration can be programmed under the 3D Studio Max software platform.

The centred camera configuration is essentially an idealised camera with a translating film plane. In the case of an HPO reflection hologram with a fixed rectangular viewing box located at distance h from the hologram, this consists of a camera whose field of view is chosen to be

$$\Psi_{CH} = 2\tan^{-1}\left(\frac{D_X + W}{2h}\right)$$

$$\Psi_{CV} = 2\tan^{-1}\left(\frac{D_Y}{2h}\right) \tag{8.56}$$

This is of course exactly the same as the restricted field of view that we introduced in the last section for a simple translating camera. The camera is, however, now chosen to have a (horizontally) smaller film plane, which we shall label, $\hat{\mathscr{F}}$, such that not all the light gathered by the camera falls on this plane. The horizontal dimensions of $\hat{\mathscr{F}}$ are chosen such that when positioned in front of the hologram at the centre of the camera track, an image of the hologram just fills $\hat{\mathscr{F}}$. As the camera is moved either to the right or to the left of this position, $\hat{\mathscr{F}}$ is automatically shifted horizontally such that the image of the hologram always exactly fills $\hat{\mathscr{F}}$. In this way, $\hat{\mathscr{F}}'$ is always exactly collocated with $\mathscr{H}$.

Mathematically, we may describe the centred camera configuration for HPO holograms by a requirement on the size of $\hat{\mathscr{F}}'$ within $\mathscr{F}'$ and by the required horizontal translation of $\hat{\mathscr{F}}'$ in terms of the usual x coordinate of $\mathscr{F}'$. Mathematically, this may be expressed as

$$x_1 = W - \xi$$

$$x_2 = W - \xi + D_X \tag{8.57}$$

where the subscript 1 indicates where $\hat{\mathscr{F}}'$ starts and the subscript 2 indicates where $\hat{\mathscr{F}}'$ stops. We shall come back to this in Chapter 10 when we discuss how to program a centred camera configuration. For our present purposes, however, we will simply assume that we are given $\mathbf{I}$ in terms of $\hat{\mathscr{F}}$ rather than in terms of $\mathscr{F}$. We shall then use the coordinates x and y (and their corresponding integer partners i and j) simply to refer to $\hat{\mathscr{F}}$ rather than to $\mathscr{F}$. Accordingly, we shall put $D_x = D_X$ in addition to the condition $D_y = D_Y$, which we have used throughout our discussion of HPO holograms. It is important to understand that in the HPO simple translating camera configuration, D_x is uniquely defined by the horizontal field of view of the camera, Ψ_{CH}, whereas in the HPO centred camera configuration, the relation $D_x = D_X$ defines a subset of the total horizontal optical field defined by Ψ_{CH}.

The centred camera configuration simplifies the horizontal geometry (the plan view) significantly; the vertical geometry (the side view) remains identical to the case of the HPO simple translating camera (Figure 8.8a). Figure 8.11 shows a plan view of the geometry for the centred camera configuration. As previously, we use a test ray (shown in magenta) to identify rays of $\mathbf{I}$ and $\mathbf{S}$ and to match the light-fields at the viewing plane. We do this by matching $\mathbf{I}$ and $\mathbf{S}$ at the indicated hogel on $\mathscr{H}/\hat{\mathscr{F}}$ and at the indicated camera position on $\mathscr{C}/\mathscr{P}$, which is equivalent to matching direction and angle of all rays on $\mathscr{V}$.

The matching conditions at the hogel give the trivial relations

$$X = x' \; ; \; Y = y' \tag{8.58}$$

The conditions at the camera pinhole (Equation 8.32) remain unchanged:

$$\hat{U}' - \xi = D_X - X \tag{8.59}$$

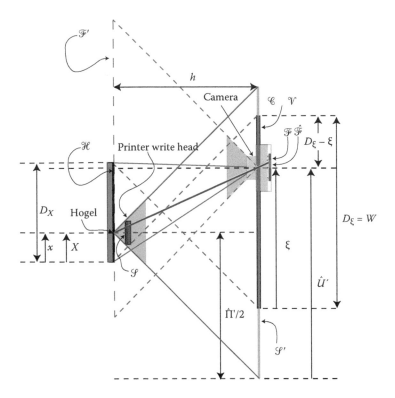

FIGURE 8.11 Plan view of ray geometry for a centred camera.

The corresponding index transformations are then derived using Equations 8.1, 8.5, 8.7, 8.8, 8.22 and 8.27:

$$i = N_I - \left\| \frac{(\alpha - 1)(N_I - 1)}{(N_A - 1)} \right\|, \quad j = N_J - \left\| \frac{(\beta - 1)(N_J - 1)}{(N_B - 1)} \right\|$$

$$k = 1 + \left\| \frac{(N_K - 1)}{W} D_X \left[\frac{(\hat{\mu} - 1)}{(\hat{N}_M - 1)} \left(\frac{W}{D_X} + 1 \right) + \frac{(\alpha - N_A)}{(N_A - 1)} \right] \right\| \tag{8.60}$$

The paraxial **I**-to-**Ŝ** transformation for a monochromatic DWDH HPO reflection hologram for the case of the centred camera configuration and a general rectangular window can therefore be written:

$$^{\hat{\mu}\hat{v}}\hat{\mathbf{S}}_{\alpha\beta} = {}^{k}\mathbf{I}_{ij} \quad \forall \hat{v} \left\{ \hat{v} \in \mathbb{N} \left\| \frac{D_Y}{D_Y + H} \frac{(\hat{N}_v - 1)(N_B - \beta)}{(N_B - 1)} \right\| + 1 \le \hat{v} \le 1 + \left\| \frac{D_Y}{D_Y + H} (\hat{N}_v - 1) \left\{ \frac{N_B - \beta}{N_B - 1} + \frac{H}{D_Y} \right\} \right\| \right\}$$

$$\forall \hat{\mu} \left\{ \hat{\mu} \in \mathbb{N} \left\| \frac{D_X}{D_X + W} \frac{(\hat{N}_M - 1)(N_A - \alpha)}{(N_A - 1)} \right\| + 1 \le \hat{\mu} \le 1 + \left\| \frac{D_X}{D_X + W} (\hat{N}_M - 1) \left\{ \frac{N_A - \alpha}{N_A - 1} + \frac{W}{D_X} \right\} \right\| \right\}$$

$$\forall \alpha \left\{ \alpha \in \mathbb{N} | \alpha \le N_A \right\} \quad \forall \beta \left\{ \beta \in \mathbb{N} | \beta \le N_B \right\}$$

$$= 0 \quad \text{otherwise} \tag{8.61}$$

where i, j and k are given by Equation 8.60. We can also use Equation 8.45 in Equation 8.60 to arrive at the corresponding direct **I**-to-**S** transformation:

$$^{\mu\nu}\mathbf{S}_{\alpha\beta} = {}^{k}\mathbf{I}_{ij} \qquad \forall\alpha\{\alpha \in \mathbb{N}|\alpha \le N_A\} \qquad \forall\beta\{\beta \in \mathbb{N}|\beta \le N_B\}$$

$$\forall\mu\{\mu \in \mathbb{N}|N_{U1}(\alpha) \le \mu \le N_{U2}(\alpha)\} \quad \forall\nu\{\nu \in \mathbb{N}|N_{V1}(\beta) \le \nu \le N_{V2}(\beta)\}$$

$$= 0 \qquad \text{otherwise} \tag{8.62}$$

where

$$i = N_I - \left\|\frac{(\alpha-1)(N_I-1)}{N_A-1}\right\|, \qquad j = N_J - \left\|\frac{(\beta-1)(N_J-1)}{N_B-1}\right\|$$

$$k = 1 + \left\|\frac{(N_K-1)}{W}\left\{\frac{D_X+W}{2} + \frac{h(2\mu-N_M-1)}{(N_M-1)}\tan\left(\frac{\Psi_{PH}}{2}\right) + \frac{(\alpha-N_A)D_X}{N_A-1}\right\}\right\| \tag{8.63}$$

and where N_{U1}, N_{U2}, N_{V1} and N_{V2} are given by Equations 8.49 and 8.50. This transformation is the main result of the present section. Again, as with all the transformations described here, it pertains to the case of an inverting camera and to a conjugate SLM position.

Before terminating our discussion on centred cameras, we will, for completeness, show how camera data from a simple translating camera can be converted to the data from a centred camera. To do this, we will introduce new x' and i' coordinates for the centred camera plane, $\hat{\mathscr{F}}'$

$$\hat{x}' = \frac{(\hat{i}'-1)}{(\hat{N}_I-1)}D_X \tag{8.64}$$

We will keep the original coordinates (x', i') for the simple translating camera plane $\mathscr{F}'$. Comparing Figure 8.8b and Figure 8.11, one can easily see that

$$\hat{x}' = x' - (W - \xi) \tag{8.65}$$

Equations 8.5, 8.64 and 8.65 then tell us that

$$i' = 1 + \left\|\frac{(N_I-1)W}{W+D_X}\left\{\frac{(\hat{i}'-1)}{(\hat{N}_I-1)}\frac{D_X}{W} + \frac{(N_K-k)}{(N_K-1)}\right\}\right\| \tag{8.66}$$

or equivalently,

$$i = N_I - \left\|\frac{(N_I-1)W}{W+D_X}\left\{\frac{(\hat{N}_I-\hat{i})}{(\hat{N}_I-1)}\frac{D_X}{W} + \frac{(N_K-k)}{(N_K-1)}\right\}\right\| \tag{8.67}$$

Usually, one would also try and choose

$$\frac{\hat{N}_I-1}{N_I-1} = \frac{D_X}{W+D_X} \tag{8.68}$$

for obvious reasons. If we now denote $\mathbf{I}$ as the camera data for the simple translating camera, and $\hat{\mathbf{I}}$ for corresponding data pertaining to a centred camera, then we can define $\hat{\mathbf{I}}$ in terms of $\mathbf{I}$ as follows:

$$^{k}\hat{\mathbf{I}}_{\hat{i}j} = {}^{k}\mathbf{I}_{ij} \quad \forall \hat{i}\left\{\hat{i} \in \mathbb{N} \middle| \hat{i} \leq \hat{N}_{I}\right\}$$

$$\forall j\left\{j \in \mathbb{N} \middle| j \leq N_{J}\right\}$$

$$\forall k\left\{k \in \mathbb{N} \middle| k \leq N_{K}\right\} \tag{8.69}$$

where i is given by Equation 8.67.

8.8.4 Centred SLM Configuration

We mentioned in previous sections that the system of restricted SLM coordinates allowed us to economise memory and processing time by not bothering to calculate and store image data that were always going to be zero-valued. The SLM was then updated using the restricted data set $\hat{\mathbf{S}}$ together with constant offset relations that defined which subsection of the SLM address space $\hat{\mathbf{S}}$ pertained. We can actually extend this logic one step further and define projected centred SLM coordinates $(\bar{U}', \bar{V}')$ as illustrated in Figure 8.12. The centred SLM geometry corresponds exactly to the centred camera geometry; the coordinates now only span the camera plane. In this way, absolutely no non-zero values are stored. The only disadvantage is that the required offset relations are now not constant but change with the hogel being printed. It is therefore necessary to calculate these offset relations in real time or to keep a record of them before printing. The choice of whether to use restricted SLM coordinates or centred SLM coordinates usually comes down to the design of the SLM electronics. Restricted coordinates allow the exterior

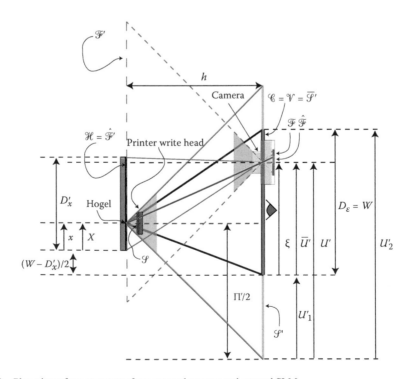

FIGURE 8.12 Plan view of ray geometry for a centred camera and centred SLM.

region of the SLM to be cleared and then only the fixed interior region updated from hogel to hogel. Centred coordinates require the SLM to be cleared before writing the data set. We shall see in Chapter 9 that the centred SLM geometry is particularly useful in the context of full-parallax holograms.

The centred SLM coordinates are defined by the simple relation

$$\bar{U}' = \xi \tag{8.70}$$

Extending our standard notation, this leads to the integer rule

$$k = 1 + \left\| \frac{\Pi'(N_K - 1)}{W(N_M - 1)} (\bar{\mu}' - 1) \right\| \tag{8.71}$$

which, in turn, leads to the following paraxial **I**-to-$\bar{\mathbf{S}}$ transformation valid for a centred inverting camera and a conjugate SLM position

$$^{\bar{\mu}\bar{\nu}}\bar{\mathbf{S}}_{\alpha\beta} = {}^k\mathbf{I}_{ij} \qquad \forall \bar{\nu} \left\{ \bar{\nu} \in \mathbb{N} \middle| 1 \le \bar{\nu} \le 1 + \left\| \frac{H(N_V - 1)}{\Sigma'} \right\| \right\}$$

$$\forall \bar{\mu} \left\{ \bar{\mu} \in \mathbb{N} \middle| 1 \le \bar{\mu} \le 1 + \left\| \frac{W(N_M - 1)}{\Pi'} \right\| \right\}$$

$$\forall \alpha \left\{ \alpha \in \mathbb{N} \middle| \alpha \le N_A \right\} \tag{8.72}$$

$$\forall \beta \left\{ \beta \in \mathbb{N} \middle| \beta \le N_B \right\}$$

$$= 0 \qquad \text{otherwise}$$

where

$$i = N_I - \left\| \frac{(\alpha - 1)(N_I - 1)}{(N_A - 1)} \right\|$$

$$j = N_J - \left\| \frac{(\beta - 1)(N_J - 1)}{(N_B - 1)} \right\|$$

$$k = 1 + \left\| \frac{\Pi'(N_K - 1)}{W(N_M - 1)} (\bar{\mu}' - 1) \right\| \tag{8.73}$$

Under special circumstances, we can sometimes arrange that

$$\frac{\Pi'}{W} = \frac{(N_M - 1)}{(N_K - 1)} \tag{8.74}$$

in which case for a non-inverting camera, the transformation simplifies dramatically to

$$\overline{^{\mu\nu}}\mathbf{S}_{\alpha\beta} = {}^{k}\mathbf{I}_{ij} \qquad \forall \overline{\nu}\left\{\overline{\nu} \in \mathbb{N} \left| 1 \le \overline{\nu} \le 1 + \left\| \frac{H(N_V - 1)}{\Sigma'} \right\| \right. \right\}$$

$$\forall \overline{\mu}\left\{\overline{\mu} \in \mathbb{N} \left| 1 \le \overline{\mu} = k \le N_K \right. \right\}$$

$$\forall \alpha\left\{\alpha \in \mathbb{N} \left| \alpha = i \le N_A \right. \right\}$$

$$\forall \beta\left\{\beta \in \mathbb{N} \left| \beta = j \le N_B \right. \right\}$$

$$= 0 \qquad \text{otherwise} \tag{8.75}$$

In other words, $\overline{^{\mu\nu}}\mathbf{S}_{\alpha\beta}$ is identical to ${}^{k}\mathbf{I}_{ij}$! The conditions required for this are that the integer coordinate meshes at recording and hologram-writing line up. We shall discuss this problem of coordinate line-up at length in the next section, but for now, we should just note that such a line-up occurs more easily in the context of full-parallax holograms as we shall see in Chapter 9.

The centred SLM transformation needs to be used with offset relations so that we can relate the actual SLM memory map $\mathbf{S}$ to the centred data set $\overline{\mathbf{S}}$. We can work out these offset relations by relating U' to $\overline{U}'$ with the help of Figure 8.12:

$$U' = \overline{U}' + U_1' = \overline{U}' + \Pi'/2 - X - W/2 + D_X/2 \tag{8.76}$$

Or, in terms of integer coordinates

$$\mu = \overline{\mu}' + \left\| \frac{1}{2}(N_M - 1)\left(1 + \frac{D_X - W}{\Pi'}\right) - \frac{(\alpha - 1)D_X(N_M - 1)}{\Pi'(N_A - 1)} \right\| \tag{8.77}$$

Then, the offset coordinates, which tell us which indices μ_1 and μ_2 in $\mathbf{S}$ correspond, respectively, to the indices $\overline{\mu}' = 1$ and $\overline{\mu}' = \left\| W(N_M - 1)/\Pi' \right\| + 1$ in $\overline{\mathbf{S}}$, are defined as

$$\mu_1 = 1 + \left\| \frac{1}{2}(N_M - 1)\left(1 + \frac{D_X - W}{\Pi'}\right) - \frac{(\alpha - 1)D_X(N_M - 1)}{\Pi'(N_A - 1)} \right\|$$

$$\mu_2 = 1 + \left\| \frac{W}{\Pi'}(N_M - 1) + \frac{1}{2}(N_M - 1)\left(1 + \frac{D_X - W}{\Pi'}\right) - \frac{(\alpha - 1)D_X(N_M - 1)}{\Pi'(N_A - 1)} \right\| \tag{8.78}$$

In an exactly similar fashion, the vertical offset coordinates are given by

$$\nu_1 = 1 + \left\| \frac{1}{2}(N_V - 1)\left(1 + \frac{D_Y - H}{\Sigma'}\right) - \frac{(\beta - 1)D_Y(N_V - 1)}{\Sigma'(N_B - 1)} \right\|$$

$$\nu_2 = 1 + \left\| \frac{H}{\Sigma'}(N_V - 1) + \frac{1}{2}(N_V - 1)\left(1 + \frac{D_Y - H}{\Sigma'}\right) - \frac{(\beta - 1)D_Y(N_V - 1)}{\Sigma'(N_B - 1)} \right\| \tag{8.79}$$

We can now appreciate a weakness of the centred SLM coordinate system. The offset index relations contain a truncation error, which will change from hogel to hogel. Because, in general, the $\mathbf{I}$-to-$\mathbf{S}$ transformation itself also contains an interpolation error, we see that with the centred SLM coordinate system, we can end up with two sequential errors. The index relations (Equations 8.46 and 8.47), which are relevant to restricted SLM coordinates, are invariant under a change of hogel coordinates; more importantly, they can also easily be made free of truncation error by choosing W and H appropriately.*

8.8.5 Fundamental Integer Constraints

Let us review the simple procedure that we have used to derive the **I**-to-**S** transformations in the previous sections. We started by defining various planes—the camera plane, the viewing plane, the hologram plane, etc.—we then described the general positions on these planes using two sets of coordinates. The first set is real and the second set is composed of integers. Of course, we want to define integer coordinates because our data is in digital form. The SLM is a digital device, as is the CCD detector of a digital camera; we divide the hologram into digital "hogels" and we move the camera from one discrete location to another. However, nothing says that the digital data of the acquisition system must line up and coincide exactly with the digital data that must be written to the hologram. Such lining up of the digital data is actually a powerful constraint that we may or may not be able to impose on a real printer. The fact that, in general, the digitisation of camera data ($\mathscr{C}$ and $\mathscr{F}$) does not line up with the digitisation of hologram data ($\mathscr{H}$ and $\mathscr{S}$) creates optical noise in the hologram. In the case of HPO holograms, this optical noise is most evident as dark vertical lines traversing the hologram. We will have more to say about such interpolation noise a little later.

Before embarking on a discussion of how we might force our digital data to line up, it is worth pointing out that most digital holographic printers do not have paraxial optical objectives. This means that the projection of the SLM, more often than not, takes the form of an intrinsically curvilinear coordinate system. From a practical point of view, this means that one simply cannot make the Cartesian integer coordinates of the SLM line up with the camera coordinates. In Section 8.10 and in the next chapter, we will discuss the non-paraxial projections of $\mathscr{S}$ to $\mathscr{S}'$ caused by objective distortion and holographic aberrations, and we will derive practical **I**-to-**S** transformations that reduce interpolation noise. We shall see that the main idea is to incorporate the inherent optical aberrations directly into the **I**-to-**S** transformations and to perform a single nearest integer operation rather than performing successive nearest integer operations.

In the case that there are no aberrations in the printer and in the hologram, it makes sense to impose conditions on the **I**-to-**S** transformation such that the various integer coordinate meshes line up. There are basically three conditions that lead to a quantisation of the distance, h, in terms of the (integer) quantum number, n. These rules are best understood from Figures 8.13 and 8.14, where a simple case is examined for, respectively, $n = 1$ and $n = 2$. The diagrams are plan views of the hologram and viewing/camera planes for a centred camera configuration using restricted angles of view.

The first condition is that the camera spacing must be equal to the hologram pitch. This can be written as

$$\frac{W}{N_K - 1} = \frac{D_X}{N_A - 1} \tag{8.80}$$

The second condition is that the (horizontal) pitch of the projected camera film plane at the hologram surface must equal an integer multiple of the hogel pitch (note that the centred camera condition has been applied here) or

$$\frac{D_X - m_X \delta}{N_I - 1} = n\delta$$

$$\frac{D_Y}{N_J - 1} = \delta \tag{8.81}$$

---

* If N_M and N_V are large, which is of course always the case, then Equations 8.46 and 8.47 act to introduce a fine-scale quantisation into W and H. Thus, one does not have to change the window by much to ensure that the index laws for conversion of $\hat{\mathbf{S}}$ to $\mathbf{S}$ are exact.

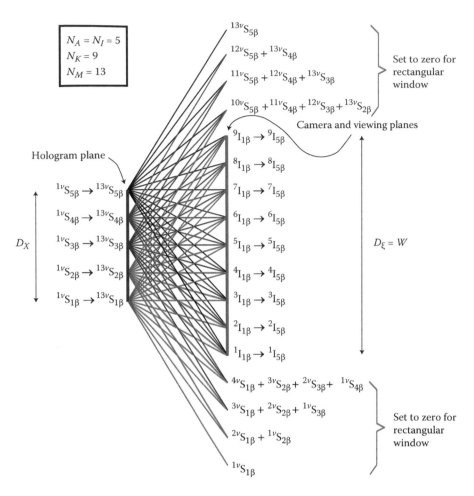

FIGURE 8.13 Illustration of exact ray correspondence between **I** and **S** for the simplified case of five hogels with a quantum number of $n = 1$.

where n is an integer greater or equal to 1 and δ is the hogel diameter defined as

$$\delta = \frac{D_X}{N_A - 1} = \frac{D_Y}{N_B - 1} \tag{8.82}$$

The m parameter reflects the fact that camera data is not required in all cases at those film plane pixels corresponding to the last $n - 1$ hogels closest to the hologram boundary. This is just modular arithmetic; imagine that you want to divide a series of seven points/hogels into a line of points having twice the spacing. If you want to make the second series line up with the first, then there are just two principal ways of doing this which together use up all the points. We could choose points 1, 3, 5 and 7 for our second series or we could choose the series 2, 4 and 6. Both series have double the spacing of the original series, but the dimension of the new series is 4 in the first case and 3 in the second case. This is what happens with the projected film mesh—this mesh often finishes at $n - 1$ or fewer pixels from the edge of the hologram mesh. If you spend a little time with a pen and paper, you will be able to see that*

$$m_X = N_A - 1 - \left\lfloor \frac{N_A - k}{n} \right\rfloor n - \left\lfloor \frac{k-1}{n} \right\rfloor n$$

$$\tag{8.83}$$

---

* We have assumed that both N_A and N_M are odd here.

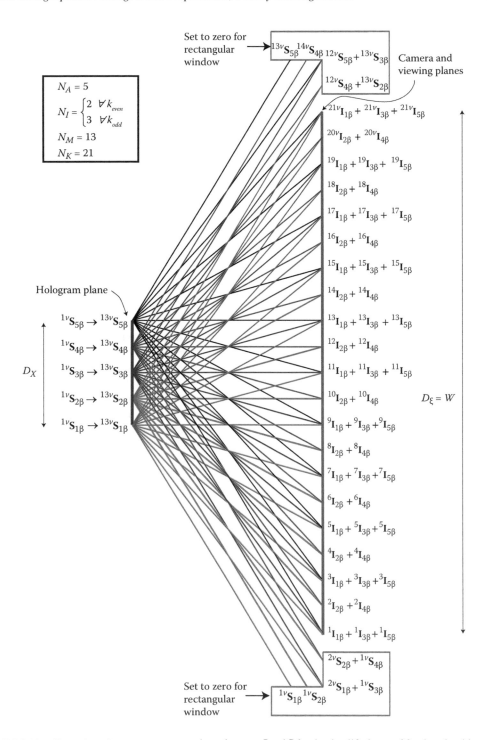

FIGURE 8.14 Illustration of exact ray correspondence between **I** and **S** for the simplified case of five hogels with a quantum number of $n = 2$.

This has the interesting consequence that N_I becomes a function of k. Equation 8.81 tells us that

$$N_I = 1 + \left\lfloor \frac{N_A - k}{n} \right\rfloor + \left\lfloor \frac{k - 1}{n} \right\rfloor$$

$$N_J = N_B$$

(8.84)

By comparing Figures 8.13 and 8.14, you can see how this works. In the $n = 1$ case of Figure 8.13, each camera position is connected to all five hogels. However, in the $n = 2$ case of Figure 8.14 each camera position is connected by either two (even k) or three (odd k) rays to every other hogel. The odd values of k correspond to a film plane size equal to the entire hologram dimension, but the even k values correspond to only 1/2 the hologram dimension.

The third and final condition is that the pitch of the projected SLM plane at the camera plane must be equal to the same integral multiple, n of the camera spacing or

$$\frac{\Pi'}{N_M - 1} = \frac{\hat{\Pi}'}{\hat{N}_M - 1} = n \frac{W}{N_K - 1}$$

(8.85)

To see how these conditions work, we must think which parameters are fixed by their absolute nature and which may be changed to accommodate the conditions. In general, we start with a knowledge of N_M and N_V as these numbers are determined by the printer SLM. We will also know the desired hologram size, defined by D_X and D_Y, and the hogel diameter, δ. This then defines the number of hogels horizontally, N_A, and the number of hogels vertically, N_B. We regard all these numbers as being fixed in an absolute way. Condition 8.80 then tells us that we must choose the viewing window width, W, such that the quantity W/δ is an exact integer. This is usually a very acceptable constraint because adjusting the viewing width by a very minimal amount is, for almost all applications, simply without consequence.

Having now chosen W, D_X and N_A, we define N_K (using Equation 8.80) as

$$N_K = 1 + \frac{W}{\delta} = \frac{W}{D_X}(N_A - 1) + 1$$

(8.86)

Condition 8.81 then tells us the values of N_I (for each value of k) and N_J. With a numerical camera within a CAD program, a camera can usually be configured in a very flexible fashion, and so again, this step is not a problem. We shall discuss real cameras in Chapter 10.

Finally, we come to Condition 8.85, which represents by far the most stringent condition and effectively defines a unique value of h for each n:

$$h = \frac{(N_M - 1)}{(N_A - 1)} \frac{n D_X}{2 \tan\left(\frac{\Psi_{PH}}{2}\right)}$$

(8.87)

Table 8.1 shows how this works for a typical printer in two cases. The first case is a 600 mm × 400 mm landscape format hologram having a hogel diameter of 1 mm. The second case pertains to the same format but with a hogel size of 0.5 mm. Note that the values of D_X and D_Y are always a hogel less than the full-hologram size as the D values measure distance between hogel centres rather than hogel edges. The table shows clearly how the value of the integer parameter n quantises the viewing distance h. The bottom line is that forcing the integer coordinates of our camera acquisition and hologram recording planes to align has a significant price. This is the quantisation of the camera/viewing distance. For large hogel

TABLE 8.1

Table Showing the Quantised Distance, h, for Two HPO Holograms with Different Quantum Numbers, n

N_M	N_V	D_X	D_Y	Ψ_{PH}	Ψ_{PV}	W	N_A	N_B	n	N_K	$N_I (k=1)$	N_J	h	$N_I N_J N_K \cdot 10^{-9}$
1280	1024	599	399	105.2	92.6	800	600	400	1	801	600	400	488.9355	192
1280	1024	599	399	105.2	92.6	2199	600	400	2	2200	300	400	977.8709	264
1280	1024	599	399	105.2	92.6	3598	600	400	3	3599	200	400	1466.806	288
1280	1024	599	399	105.2	92.6	4997	600	400	4	4998	150	400	1955.742	300
1280	1024	599.5	399.5	105.2	92.6	800	1200	800	1	1601	1200	800	244.4677	1537
1280	1024	599.5	399.5	105.2	92.6	2200	1200	800	2	4400	600	800	488.9355	2112
1280	1024	599.5	399.5	105.2	92.6	4999	1200	800	4	9998	300	800	977.8709	2400
1280	1024	599.5	399.5	105.2	92.6	6398	1200	800	5	12797	240	800	1222.339	2457
1280	1024	599.5	399.5	105.2	92.6	7798	1200	800	6	15596	200	800	1466.806	2495
1280	1024	599.5	399.5	105.2	92.6	9197	1200	800	7	18395	172	800	1711.274	2531
1280	1024	599.5	399.5	105.2	92.6	10597	1200	800	8	21194	150	800	1955.742	2544

Note: The holograms are both 600×400 mm. The first has a hogel size of 1 mm (first 4 lines of table). The second has a hogel size of 0.5 mm (remaining eight lines of table). Note that angles are in degrees and dimensional distances in mm.

sizes, the spacing between the various "n" quantum levels can be large, making it difficult to use this procedure. However, as can be seen from Table 8.1, for smaller hogel sizes, the situation does improve. Of course, the great advantage of using lined-up coordinate meshes is that one escapes completely from injecting interpolation noise into the hologram. We shall see later that the most common aberrations vanish when a hogel is viewed perpendicularly to the hologram surface; by using the procedure outlined here, we can then often, even for a real printer and a real hologram, effectively abolish interpolation noise in the central (and most visually obvious) zone of the hologram's light emission cone.

Finally, we should note that when $n > 1$, the camera data is effectively squashed by a factor n in the horizontal direction. Indeed the information content of a single frame, $\mathbf{I}_{ij}$, is reduced by a factor of n. However the centred camera frame essentially dithers as k changes and this means that both the angular resolution and spatial resolution of the hologram are independent of n. Note that for full parallax holograms, the aspect ratio of the camera data is unchanged for any n.

8.9 Rainbow and Achromatic Transmission Holograms

Rainbow [6] and achromatic [7] holograms can both be generated using the two-step MWDH method or directly using one-step DWDH. Because these holograms are single-parallax holograms, there is a large potential advantage in using MWDH to generate the H_1 hologram (which is simply a one-dimensional "slit" hologram in the rainbow case) and then producing the H_2 by traditional optical transfer. The total information printed using this technique can be many orders of magnitude less than DWDH.

8.9.1 MWDH

The most obvious way to produce full-colour analogue rainbow holograms is by using a single on-axis RGB slit reflection H_1 using three (R, G and B) lasers. An RGB transfer process of this single slit H_1 then produces a full-colour transmission rainbow H_2 hologram. In digital holography, however, we are free to use another method that requires a single laser only. In this variant, three (or more) slit masters are recorded (usually transmission holograms) using specially processed RGB image data from a simple translating (or translating/rotating) camera which moves on a rail in front of the real or virtual object. A single-wavelength transfer is then made in which the three H_1 masters are geometrically staggered according to their desired replay colour and aligned on a plane that is tilted vertically at the achromatic angle with respect to the H_2 normal. This well-known technique was first practised using analogue film cameras and by projecting laser light through the diapositive sequence obtained onto a diffusing screen which was then holographed [8]. In Chapter 7, we described how the more modern MWDH technique naturally lends itself to the production of H_1 holograms required for the generation of full-colour rainbow H_2 holograms. In particular, we saw how this could be arranged, from an optics point of view, through the generation of a single RGB H_1 or via multiple component-colour H_1 holograms (see Figure 7.29). In Chapter 7, however, we did not discuss the image data transformations required; we shall study this aspect now. Before starting this discussion, we should note in passing that the technique of using a single laser can indeed be applied in analogue holography as well, but here, it can be difficult to optically introduce the required image transformations into each of the off-axis H_1 masters—and this usually leads to image deregistration between colours.

8.9.1.1 Vertically Aligned RGB H_1 Master

We will consider first the case of using three or more lasers each operating at a different primary wavelength and writing a single RGB reflection H_1 hologram suitable for RGB transfer with the same wavelengths to a final H_2 transmission rainbow hologram. This is a simple task, but it will serve to set the stage for the more complicated case of component-colour H_1 holograms and will also serve to set up the required notation. Plan and side views of the three critical stages—writing the H_1, producing the H_2 and viewing the H_2—are shown in Figure 8.15. Because the rainbow hologram is intrinsically an HPO hologram, we must generate the $\mathbf{I}$ data along a camera track, which is placed at a distance from the (virtual or real) object that is

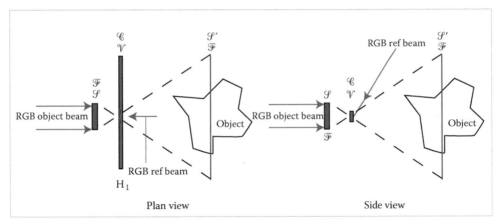

(a) Creation of H$_1$ hologram

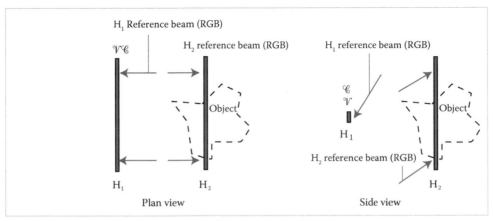

(b) H$_1$:H$_2$ transfer process

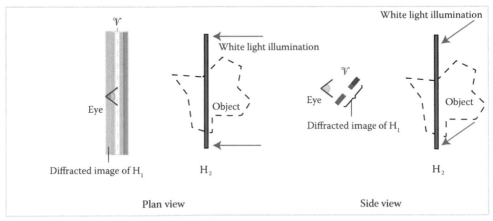

(c) Viewing the H$_2$

FIGURE 8.15 The creation of an MWDH rainbow master hologram using red, green and blue lasers: (a) the creation of the RGB H$_1$ reflection hologram, (b) the H$_1$:H$_2$ transfer using an RGB laser, and (c) viewing the final H$_2$ rainbow transmission hologram.

equal to the distance of the intended viewing plane from the holographic image of the object in the final H_2. This is similar to the situation we discussed in Section 8.7.2, except that there, we were considering a two-dimensional grid of camera positions. We will use the integer coordinates (i, j, k) to describe such a simple one-dimensional (forward facing) translating camera with inverted image and we will denote $^k\mathbf{I}_{ij}^R$, $^k\mathbf{I}_{ij}^G$ and $^k\mathbf{I}_{ij}^B$ as the red, green and blue primary-colour component data produced by it. As in Section 8.7.2, we shall assume that the horizontal and vertical fields of view of the camera are identical to those of the printer and that the number of pixels on the printer SLM and on the camera CCD is identical or

$$\Psi_{CH} = \Psi_{PH} \equiv \Psi_H$$

$$\Psi_{CV} = \Psi_{PV} \equiv \Psi_V$$

$$N_M = N_I \tag{8.88}$$

$$N_V = N_J$$

We shall describe the red, green and blue SLM data as, respectively, $^{\mu\nu}\mathbf{S}_{\alpha\beta}^R$, $^{\mu\nu}\mathbf{S}_{\alpha\beta}^G$ and $^{\mu\nu}\mathbf{S}_{\alpha\beta}^B$ and we shall assume a non-conjugate SLM configuration. Then we may define the SLM data using the following trivial transformation:

$$^{\mu\nu}\mathbf{S}_{\alpha\beta}^C = {}^{\alpha}\mathbf{I}_{\mu j}^C \qquad \forall\beta\left\{\beta \in \mathbb{N} \middle| 1 \le \beta \le N_B\right\}$$

$$\forall\mu\left\{\mu \in \mathbb{N} \middle| \mu \le N_M\right\} \qquad \forall\alpha\left\{\alpha \in \mathbb{N} \middle| \alpha \le N_A\right\}$$

$$\forall\nu\left\{\nu \in \mathbb{N} \middle| \nu \le N_V\right\} \qquad \forall C\left\{C \in \{R,G,B\}\right\} \tag{8.89}$$

$$= 0 \qquad \text{otherwise}$$

where

$$j = \nu + \left\Vert\left(\beta - \left\{\frac{N_B+1}{2}\right\}\right)(N_J - 1)\frac{\delta}{2h\tan\left(\dfrac{\Psi_V}{2}\right)}\right\Vert \tag{8.90}$$

and the integer parameter $N_B \ge 1$ controls the vertical size of the H_1 slit. The last term in Equation 8.90 essentially shifts the camera image up or down for slits below or above the camera rail, thus ensuring that each such shifted slit hologram projects a real image of exactly the same camera data onto the H_2 plane in perfect alignment (see Figure 8.16, from which it is evident that $V' + Y - Y_C = y'$, which implies the integer rule if we assume a symmetric hogel). As discussed in Chapter 7, the plane of recording of the H_1 hologram may be somewhat displaced from the Fourier plane of the printer objective in MWDH and, in this case, small values of N_B close or equal to 1 may be employed. Alternatively, the printer may be operated with the Fourier plane of the objective/SLM being coincident with the hologram plane, in which case, larger values of N_B can be used. The larger the value of N_B, the greater will be the colour desaturation observed in the final H_2 and the lower will be the intrinsic image noise. However, the price to pay for this is that chromatic blurring increases with N_B. As with usual analogue rainbow holograms, a compromise must therefore be adopted and a practical value for N_B chosen for each situation. This is because we have considered the case of the H_1 hologram being perpendicular to the optical axis of the printer writing head; this is, after all, the most usual case for an RGB printer. You may remember that we discussed in Chapter 7 that an alternative existed and that was to tilt the H_1 plate at the achromatic angle as defined by Equation 8.92 below. This allows us to potentially decouple chromatic blurring from saturation and can lead to a better hologram when large image depths are required. To be in a position to write down the transformation for this case, we will first study the case of using only a single wavelength to record three H_1 slit masters.

Figure 8.17 shows the three critical steps involved: (a) writing the three H_1 holograms on a single tilted plate, (b) transferring the H_1 holograms to an H_2 rainbow hologram and (c) viewing. We shall assume

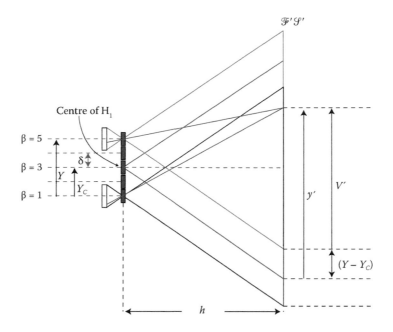

FIGURE 8.16 Side view of the vertically aligned H_1 showing its individual hogels and the light cones generated by the writing process. The central light cone must coincide with the camera light cone. All light cones must project an image at $\mathscr{F}'$ that exactly corresponds to the image produced by the central light cone.

that the RGB image data is collected, as in our previous discussions, by a simply translating rail-mounted camera. This data is then colour-separated and each of the component colours is processed and used to generate a slit master. The image processing is chosen such that a given slit master, if illuminated under the transfer geometry of Figure 8.17b and by its designed replay wavelength, will produce a real image of the aperture of the slit-master at $\mathscr{V}$ coincident with the camera track and further, each hogel of this master will project a focussed image of the original camera data onto the H_2 plane. The image processing must above all ensure that each of the processed R, G and B component-colour images, corresponding to a given camera position and as projected by the corresponding $3N_B$ hogels associated with this camera position, align precisely at the H_2 plane.

8.9.1.2 Component-Colour H_1 Masters

To understand how to write down the **I**-to-**S** transformations for three-component colour H_1 holograms, we must use results from the paraxial hologram theory that is described in Chapter 11. This theory describes how an object point, holographically recorded at a first wavelength, maps to an image point when the hologram is replayed at a second wavelength. Figure 8.17a shows the geometry we need to consider. The three H_1 slits are located at coordinates (y_B, z_B), (y_G, z_G) and (y_R, z_R) with the H_2 plane (which is coincident with $\mathscr{S}'$ and $\mathscr{F}'$ in all cases) being at $z = 0$. The centre of the H_2 is taken as $(0,0)$ meaning that all z values of interest are negative; positive y means that the point is above the H_2 mid-plane.

Each of the slit holograms is recorded using light at a wavelength of λ_W. We wish to arrange that, when we illuminate the recorded red H_1 with light at λ_R, the image of the red slit will move onto the axis at $z = -h$. Likewise, when we illuminate the recorded green or blue H_1 with light at, respectively, λ_G or λ_B, the image of the green or blue slit must also move onto the axis at $z = -h$. Note also that any ray intersecting a given point on a given H_1 must always intersect with the image of the H_1 at the same relative position along the horizontal or x dimension of the H_1 when this H_1 is replayed at its design wavelength.

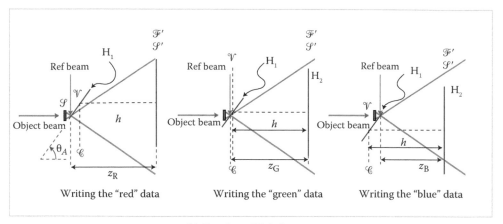

(a) Creation of H_1 hologram

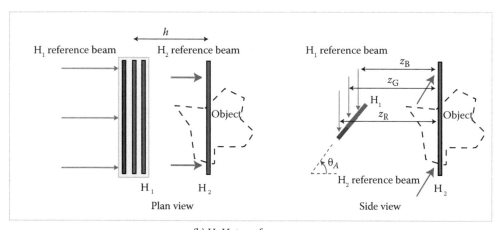

(b) H_1:H_2 transfer process

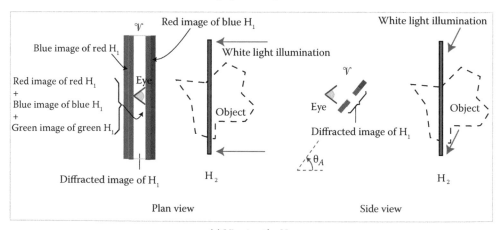

(c) Viewing the H_2

FIGURE 8.17 The creation of an MWDH rainbow master hologram using a single laser: (a) creation of the red, green and blue H_1 holograms, (b) the H_1:H_2 transfer using a single colour laser, and (c) viewing the final H_2 rainbow transmission hologram. Note the inverting camera and the non-conjugate SLM configuration.

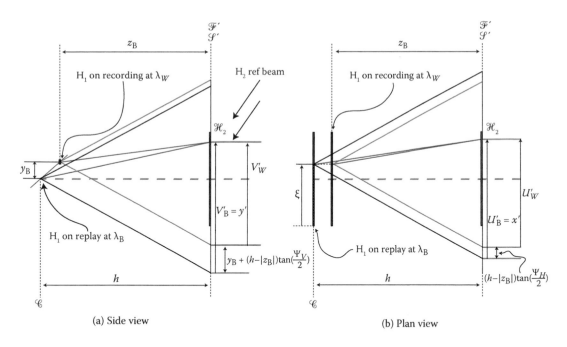

(a) Side view   (b) Plan view

FIGURE 8.18 Coordinate systems (U'_W, V'_W) and (U'_B, V'_B) characterising the projected SLM plane showing rays at off-axis recording (λ_W) and on-axis replay (λ_B) for (a) side view and (b) plan view.

Using the results from Chapter 11 (Equation 11.20)* and taking h to be a positive value, we can see that the first condition described above is satisfied (in terms of the following 3×2 matrices) when

$$\begin{pmatrix} y_R & z_R \\ y_G & z_G \\ y_B & z_B \end{pmatrix} = -\frac{h}{\lambda_W} \begin{pmatrix} (\lambda_R - \lambda_W)\sin\theta & \lambda_R \\ (\lambda_G - \lambda_W)\sin\theta & \lambda_G \\ (\lambda_B - \lambda_W)\sin\theta & \lambda_B \end{pmatrix} \tag{8.91}$$

This implies that the R, G and B H_1 holograms are aligned at an angle equal to the achromatic angle defined as

$$\theta_A = \tan^{-1}(\sin\theta) \tag{8.92}$$

where θ represents the angle of reference (and illumination) incidence. To understand the image processing required, we consider two cases: (i) writing the blue H_1 at λ_W and (ii) how this H_1 replays at λ_B. With this in view, we set up separate real Cartesian coordinate systems (U'_W, V'_W) and (U'_B, V'_B), respectively, for the projected SLM plane $\mathscr{S}'$ in both cases as illustrated in Figure 8.18. For a test ray written at λ_W off-axis to then replay at λ_B the correct camera data on-axis, we require that

$$U'_W = x' - \left(h - |z_B|\right)\tan\left(\frac{\Psi_H}{2}\right)$$

$$V'_W = y' - y_B - \left(h - |z_B|\right)\tan\left(\frac{\Psi_V}{2}\right) \tag{8.93}$$

---

* Note that we have transformed to the coordinates we are using here which are different from those used in Equation 11.20.

We must then be careful to discretise the $\left(U'_W, V'_W\right)$ system in the following way:

$$U'_W = \frac{2|z_B|(\mu'-1)}{N_M-1}\tan\left(\frac{\Psi_H}{2}\right)$$

$$V'_W = \frac{2|z_B|(v'-1)}{N_V-1}\tan\left(\frac{\Psi_V}{2}\right) \tag{8.94}$$

This takes into account the fact that the effective field of view of the printer objective is decreased on replay due to the change in wavelength. This then leads us, with the help of Equations 8.88 and 8.91, to the index rules between (i', j') and (μ', v'):

$$i' = 1 + \left\|\frac{\lambda_B}{\lambda_W}(\mu'-1) + \frac{(N_M-1)(\lambda_W-\lambda_B)}{2\lambda_W}\right\|$$

$$j' = 1 + \left\|\frac{\lambda_B}{\lambda_W}(v'-1) + \frac{(N_V-1)(\lambda_W-\lambda_B)}{2\lambda_W}\left(1 + \sin\theta\cot\left(\frac{\Psi_V}{2}\right)\right)\right\| \tag{8.95}$$

We may now convert these equations to relations for the un-primed integers (i,j,μ,v) by application of the inverting camera Equation 8.27 and the corresponding relations for a non-conjugate SLM—$\mu' = N_M - \mu + 1$ and $v' = N_V - v + 1$. Finally we can use these index laws and Equation 8.89 with $NB = 1$ to arrive at the following transformation:

$$^{\mu v}\mathbf{S}^C_\alpha = {}^\alpha\mathbf{I}^C_{ij} \qquad \forall\mu\left\{\mu \in \mathbb{N}\middle|\mu \leq N_M\right\} \qquad \forall\alpha\left\{\alpha \in \mathbb{N}\middle|\alpha \leq N_A\right\}$$

$$\forall v\left\{v \in \mathbb{N}\middle|v \leq N_V\right\} \qquad \forall C\left\{C \in \{R,G,B\}\right\} \tag{8.96}$$

$$= 0 \qquad \text{otherwise}$$

where

$$i = N_M - \left\|\frac{\lambda_C}{\lambda_W}(N_M-\mu) + \frac{(N_M-1)(\lambda_W-\lambda_C)}{2\lambda_W}\right\|$$

$$j = N_V - \left\|\frac{\lambda_C}{\lambda_W}(N_V-v) + \frac{(N_V-1)(\lambda_W-\lambda_C)}{2\lambda_W}\left(1 + \sin\theta\cot\left(\frac{\Psi_V}{2}\right)\right)\right\| \tag{8.97}$$

We have omitted the subscript β from $\mathbf{S}$ on purpose, as this transformation describes R, G and B H_1 slit holograms only one hogel high. This is the case of maximum possible saturation. To reduce saturation and to increase image fidelity without introducing chromatic blurring, we must write each H_1 with N_B rows of hogels. In general, we may choose the row-to-row spacing to be different from the hogel diameter as in the final H_2 the pattern of hogels will be located on the viewing plane. The smaller the row-to-row separation, the greater the overlap of the hogels will be. This will decrease diffractive efficiency of the H_1 somewhat but will also tend to smooth out the "hogelisation". The final choice of the number of rows N_B to print and the row-to-row spacing must be made according to the requirements of image fidelity, required final image saturation and image clarity.

We should note also that Equation 8.97 may lead to values of i and j greater than the maximum, or less than the minimum permitted values if $\lambda_C > \lambda_R$. This is because the effective field of view changes with wavelength.

8.9.1.3 *Achromatically Tilted Component-Colour H₁ Masters*

Because we have assumed that the photosensitive plate is tilted at the achromatic angle to the optical axis of the printer writing head, we characterise the row-to-row separation by a quantity δ_V. At $\beta = 1$ with $N_B = 1$ we have seen above that the blue H_1 slit must be located at coordinates (y_B, z_B). At a general value of β when $N_B \neq 1$, these coordinates will change to

$$
\begin{pmatrix} y_B(\beta) \\ z_B(\beta) \end{pmatrix} = \begin{pmatrix} y_B(\cent) + \left(\beta - \dfrac{N_B+1}{2}\right)\delta_V \sin\theta_A \\ z_B(\cent) + \left(\beta - \dfrac{N_B+1}{2}\right)\delta_V \cos\theta_A \end{pmatrix} = \begin{pmatrix} -\dfrac{h}{\lambda_W}(\lambda_B - \lambda_W)\sin\theta + \left(\beta - \dfrac{N_B+1}{2}\right)\delta_V \sin\theta_A \\ -\dfrac{h\lambda_B}{\lambda_W} + \left(\beta - \dfrac{N_B+1}{2}\right)\delta_V \cos\theta_A \end{pmatrix}
$$

(8.98)

where $(y_B(\cent), z_B(\cent))$ locates the centre of the blue slit and where we have used Equation 8.91 to deduce the second step. We can now proceed in the same way and use these expressions in Equation 8.93 to obtain

$$
i' = \left\| \left\{ \frac{\lambda_B}{\lambda_W} - \left(\beta - \frac{N_B+1}{2}\right)\frac{\delta_V}{h}\cos\theta_A \right\} \left\{ \mu' - \frac{N_M+1}{2} \right\} + \frac{N_M+1}{2} \right\|
$$

$$
j' = 1 + \left\| \begin{array}{l} \left\{ \dfrac{\lambda_B}{\lambda_W} - \left(\beta - \dfrac{N_B+1}{2}\right)\dfrac{\delta_V}{h}\cos\theta_A \right\} \left\{ \nu' - \dfrac{N_V+1}{2} \right\} \\ + \dfrac{(N_V-1)}{2} \left\{ \left[\dfrac{\lambda_W - \lambda_B}{\lambda_W}\sin\theta + \left(\beta - \dfrac{N_B+1}{2}\right)\dfrac{\delta_V}{h}\sin\theta_A \right]\cot\left(\dfrac{\Psi_V}{2}\right) + 1 \right\} \end{array} \right\|
$$

(8.99)

The **I**-to-**S** transformation for the MWDH achromatically tilted H_1 rainbow hologram with a thick slit of N_B hogels high and a row-to-row spacing of δ_V is then

$$
{}^{\mu\nu}\mathbf{S}^C_{\alpha\beta} = {}^{\alpha}\mathbf{I}^C_{ij} \qquad \forall\mu\left\{\mu \in \mathbb{N} \middle| \mu \leq N_M\right\} \qquad \forall\alpha\left\{\alpha \in \mathbb{N} \middle| \alpha \leq N_A\right\}
$$

$$
\forall\nu\left\{\nu \in \mathbb{N} \middle| \nu \leq N_V\right\} \qquad \forall C\left\{C \in \{R,G,B\}\right\}
$$

$$
\forall\beta\left\{\beta \in \mathbb{N} \middle| \beta \leq N_B\right\}
$$

(8.100)

$$
= 0 \qquad \text{otherwise}
$$

where

$$
i = \left\| \left\{ \frac{\lambda_C}{\lambda_W} - \left(\beta - \frac{N_B+1}{2}\right)\frac{\delta_V}{h}\cos\theta_A \right\} \left\{ \mu - \frac{N_M+1}{2} \right\} + \frac{N_M+1}{2} \right\|
$$

$$
j = \left\| \begin{array}{l} \left\{ \dfrac{\lambda_C}{\lambda_W} - \left(\beta - \dfrac{N_B+1}{2}\right)\dfrac{\delta_V}{h}\cos\theta_A \right\} \left\{ \nu - \dfrac{N_V+1}{2} \right\} + \dfrac{N_V+1}{2} \\ - \dfrac{(N_V-1)}{2} \left\{ \left[\dfrac{\lambda_W - \lambda_C}{\lambda_W}\sin\theta + \left(\beta - \dfrac{N_B+1}{2}\right)\dfrac{\delta_V}{h}\sin\theta_A \right]\cot\left(\dfrac{\Psi_V}{2}\right) \right\} \end{array} \right\|
$$

(8.101)

The positional coordinates of the three slits during the H_2 transfer are defined by Equation 8.91.

8.9.1.4 Achromatically Tilted RGB H₁ Master

We can now return to the MWDH case of writing a single H_1 with red, green and blue lasers. We discussed this case previously in the context of the photosensitive plate being aligned perpendicular to the optical axis of the printer writing head. We noted that in this case, when we wrote more than one row of hogels, then inevitably we would increase chromatic blurring. By tilting the photosensitive plate at the achromatic angle, we can largely avoid this. We can see immediately how to write the **I**-to-**S** transformation for this case from Equation 8.101 by simply putting $\lambda_W = \lambda_C$:

$$^{\mu\nu}\mathbf{S}_{\alpha\beta}^C = {}^\alpha \mathbf{I}_{ij}^C \qquad \forall \mu\{\mu \in \mathbb{N}|\mu \leq N_M\} \qquad \forall \alpha\{\alpha \in \mathbb{N}|\alpha \leq N_A\}$$

$$\forall \nu\{\nu \in \mathbb{N}|\nu \leq N_V\} \qquad \forall C\{C \in \{R,G,B\}\}$$

$$\forall \beta\{\beta \in \mathbb{N}|\beta \leq N_B\} \tag{8.102}$$

$$= 0 \qquad \text{otherwise}$$

where

$$i = \left\| \left\{1 - \left(\beta - \frac{N_B+1}{2}\right)\frac{\delta_V}{h}\cos\theta_A\right\}\left\{\mu - \frac{N_M+1}{2}\right\} + \frac{N_M+1}{2} \right\| \tag{8.103}$$

$$j = \left\| \begin{array}{l} \left\{1 - \left(\beta - \frac{N_B+1}{2}\right)\frac{\delta_V}{h}\cos\theta_A\right\}\left\{\nu - \frac{N_V+1}{2}\right\} \\ -\frac{(N_V-1)}{2}\left\{\left(\beta - \frac{N_B+1}{2}\right)\frac{\delta_V}{h}\sin\theta_A\cot\left(\frac{\Psi_V}{2}\right)-1\right\} \end{array} \right\| \tag{8.104}$$

Note that by assuming a symmetric hogel—that is, if $\delta_V = \delta$—and by taking $\theta_A = \pi/2$, we retrieve, as expected, the transformation of Equation 8.89 and Equation 8.90, which is pertinent to the case of RGB vertically tilted slits.

8.9.2 MWDH Achromats

By using achromatic camera data and a single achromatically tilted H_1 with a very large value of δ_V, extreme desaturation may be achieved. This then changes the nature of the final transmission H_2 to that of an achromatic hologram. The hologram no longer changes colour as the observer moves his head up and down; rather, it appears uniformly achromatic with only the extreme ends of the vertical viewing window being coloured. This type of hologram can be produced using the transformation of Equations 8.102 to 8.104, with one writing laser wavelength and one achromatic camera data set.

8.9.3 DWDH

Writing digital rainbow and achromat holograms using MWDH is a good idea when large quantities of such holograms are required and the substantial extra cost of the transfer apparatus is merited. For small quantities, however, DWDH offers a cheaper alternative that can generate the final rainbow or achromatic hologram in a single step and without any transfer apparatus at all! In Section 7.5.10, we discussed briefly how a digital printer must be set up to create this type of hologram—we saw that, essentially, the only substantial detail was that the reference beam had to be made to be incident to the hogel on the same side of the photosensitive substrate; this is because we wish to produce a transmission hologram. As with full-colour reflection holograms, the photoplate is mounted in a normal fashion, orthogonal to the optical axis. The technique then consists of writing hogel by hogel and generating the entire rainbow hologram as a two-dimensional grid.

As with MWDH, there is a basic choice of using either one laser wavelength for the recording or using three laser wavelengths. The physics is exactly the same as with the MWDH case, but with DWDH, we simply seek to computationally synthesise the three viewing slits of a rainbow hologram by numerical image processing of the camera data instead of doing this by physically creating three H_1 holograms and then optically changing the image plane through a transfer process. Essentially, the process then splits into two geometries: the vertical geometry, which defines the slits, and the horizontal geometry, which defines the image transformations of the camera data. The horizontal geometry for both the one wavelength and the three wavelength cases is identical to MWDH. This is obvious because the physical slit H_1 holograms of MWDH are identical to the "slits" of DWDH and the same information must therefore be recorded. However, the vertical geometry of the DWDH technique needs to be derived specifically.

8.9.3.1 Synthesis of the Vertical Slit

We will start by calculating how to define a general virtual viewing slit located on an achromatically tilted plane centred on the camera rail. We will need this information for both the one wavelength and the three wavelength cases. Figure 8.19 shows a diagram of the vertical geometry. We consider two hogels and the ray intersections connecting their SLM planes with the desired slit location. Clearly, a viewing slit is created by only writing lines of information to the SLM at each hogel and by changing the height of the line according to a special rule. Thicker slits can then be built up by superposition in a trivial way. We can find this special rule by considering the projection of the rays emanating from the slit onto the SLM plane associated with each hogel and writing down the corresponding V' coordinate on $\mathscr{S}'$:

$$V'(y_s, z_s) = \frac{\Sigma'}{2} + \frac{h}{z_s}\left(y_s + \frac{D_Y}{2} - Y\right)$$

$$= h\tan\left(\frac{\Psi_{PV}}{2}\right) + \frac{h}{z_s}\left(y_s + \frac{D_Y}{2} - Y\right) \qquad (8.105)$$

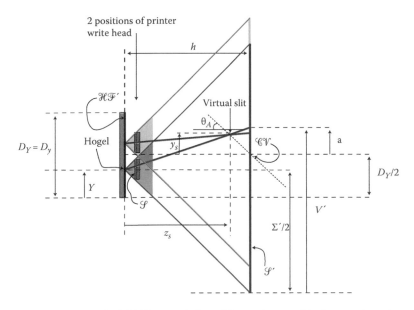

FIGURE 8.19 Vertical geometry of a DWDH transmission rainbow hologram: this diagram shows how a virtual slit is synthesised.

This then leads to the general integer coordinate rule

$$v_s(\beta) = v' = \left\| \frac{N_V + 1}{2} + \frac{(N_V - 1)}{2z_s \tan\left(\dfrac{\Psi_{PV}}{2}\right)} \left\{ y_s + \frac{D_Y}{2(N_B - 1)}(N_B - 2\beta + 1) \right\} \right\| \tag{8.106}$$

This tells us that to create a viewing slit of unit size at the location (y_s, z_s), we need to write image information to the SLM corresponding to each hogel in the form of a single line of unit height and at the pixel height location given by Equation 8.106. Each such line will, of course, be modulated by the camera information so as to create the actual image.

We must now extend this rule to the case of a broad slit of width δ_s centred on $(\langle y_s \rangle),(\langle z_s \rangle)$ and composed of points (y_s, z_s). Instead of drawing a one pixel-high line on each SLM exposure, we will now have to draw a rectangular region composed of a number of abutting lines, each of which will be modulated with differently processed camera data in the horizontal dimension. The first task is therefore to determine the maximum (N_2) and minimum (N_1) values of v that determine the top and bottom boundaries of the rectangular region. These are given by simply putting

$$\begin{pmatrix} y_s \\ z_s \end{pmatrix} \rightarrow \begin{pmatrix} y_s \pm \dfrac{\delta_s}{2}\sin(\theta_A) \\ z_s \mp \dfrac{\delta_s}{2}\cos(\theta_A) \end{pmatrix} \tag{8.107}$$

in Equation 8.106 giving,

$$N_1(\beta) = \left\| \frac{N_V + 1}{2} + \frac{(N_V - 1)}{2\left\{\langle z_s \rangle + \dfrac{\delta_s}{2}\cos(\theta_A)\right\}\tan\left(\dfrac{\Psi_{PV}}{2}\right)} \left\{ \langle y_s \rangle - \frac{\delta_s}{2}\sin(\theta_A) + \frac{D_Y}{2(N_B - 1)}(N_B - 2\beta + 1) \right\} \right\|$$

$$N_2(\beta) = \left\| \frac{N_V + 1}{2} + \frac{(N_V - 1)}{2\left\{\langle z_s \rangle - \dfrac{\delta_s}{2}\cos(\theta_A)\right\}\tan\left(\dfrac{\Psi_{PV}}{2}\right)} \left\{ \langle y_s \rangle + \frac{\delta_s}{2}\sin(\theta_A) + \frac{D_Y}{2(N_B - 1)}(N_B - 2\beta + 1) \right\} \right\|$$

$$(8.108)$$

This tells us that the SLM region of $N_1 \leq v \leq N_2$ for each hogel must be filled with the processed image data that we wish to view within the finite slit centred at $(\langle y_s \rangle),(\langle z_s \rangle)$. This now brings us to the image information required. We shall deal first with the case of writing a rainbow hologram with three laser wavelengths. As with the corresponding case in MWDH, we shall assume that all three slits are designed to play back on-axis and, as such, we consider three superimposed slits (R, G and B) of width δ_s centred on $(\langle y_s \rangle),(\langle z_s \rangle) = (0, h)$. Each slit is then recorded and viewed at its own wavelength. The fact that the slits are wide simply decreases saturation.

8.9.3.2 Transformations Required When Using an RGB Laser

The **I**-to-**S** transformation required here is similar to the HPO reflection hologram case. If we take a centred camera configuration, we must simply take into consideration the fact that any and all slits are characterised by the locus of points (y_s, z_s) in the y–z plane such that

$$\frac{dy_s}{dz_s} = -\sin\theta = -\tan\theta_A \tag{8.109}$$

Note that this equation differs from Equation 8.92 by a minus sign as we are considering here the geometry of Figure 8.20, in which the viewing slit is located to the right of the coordinate origin at a positive z value. The rays emanating from different (y_s, z_s) points (but having equal horizontal coordinates along the slit) must then all be associated with the same camera position and be characterised by the equations $x = X$ and $y = Y$. This can be understood in several ways. The first is that diffraction occurs at the hologram surface (i.e., at each hogel) on replay and acts paraxially to induce a rotation of rays about the x axis of the hogel. This rotation brings rays from one (y_s, z_s) point to another in the achromatic plane. It is therefore clear that image registration in a direction perpendicular to the parallax direction can only be assured if rays connecting a given hogel with both (y_s, z_s) points at recording correspond to exactly the same camera position and exactly the same camera film plane coordinates. An alternative way of looking at this is to think of the analogy with writing an achromatically tilted H_1 using the old method of projecting camera data (usually diapositives) onto a diffusion screen using a laser object beam (Figure 8.21). Here, the tilted H_1 is placed in front of the diffusion screen. It is immediately clear that any two vertically separated points on the tilted H_1 will then see the same image displayed on the diffusion screen.

The one thing we need to be careful about is that when calculating the SLM data for a hogel corresponding to a certain slit location (y_s, z_s), we must replace h in Equation 8.63 by z_s. This is because the slit is physically at this distance on writing even if at replay diffraction refocusses it to a distance of h. Accordingly, we can write down the **I**-to-**S** transformation for the generation of a general achromatically tilted slit of width δ_s centred at $(\langle y_s \rangle),(\langle z_s \rangle)$ on recording. This is

$$^{\mu\nu}\mathbf{S}_{\alpha\beta} = {}^k\mathbf{I}_{ij} \quad \forall\mu\left\{\mu \in \mathbb{N} \big| N_{U1}(\alpha) \leq \mu \leq N_{U2}(\alpha)\right\} \quad \forall\nu\left\{\nu \in \mathbb{N} \big| N_1(\beta) \leq \nu \leq N_2(\beta)\right\}$$

$$\forall\alpha\left\{\alpha \in \mathbb{N} \big| \alpha \leq N_A\right\} \qquad \forall\beta\left\{\beta \in \mathbb{N} \big| \beta \leq N_B\right\} \tag{8.110}$$

$$= 0 \qquad \text{otherwise}$$

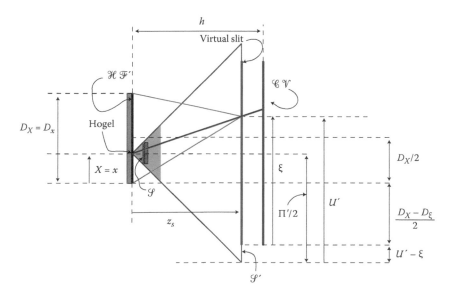

FIGURE 8.20 Horizontal geometry of a DWDH rainbow transmission hologram made with a single-colour laser. The rays intersecting a given point (y_s, z_s) on the virtual slit (on writing and replay) must be characterised by the equations $x = X$ and $y = Y$. In addition, the k equation is given by replacing h by z_s in Equation 8.63.

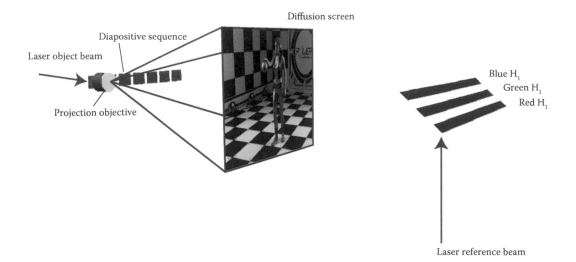

FIGURE 8.21 Original technique of writing a transmission rainbow hologram based on data from diapositives. The single-colour laser beam is split into reference and object beams. The object beam is used to project the image of a diapositive onto the diffusion screen and the three slit masters in turn record the red, green and blue information as different black and white slides are selected. All three slits see colour-separated but registered images on the diffusion screen.

where

$$i = 1 + \left\| \frac{(\alpha - 1)(N_I - 1)}{N_A - 1} \right\| \qquad j = 1 + \left\| \frac{(\beta - 1)(N_J - 1)}{N_B - 1} \right\| \tag{8.111}$$

$$k = 1 + \left\| \frac{(N_K - 1)}{W} \left\{ \frac{D_X + W}{2} + \frac{z_s(2\mu - N_M - 1)}{(N_M - 1)} \tan\left(\frac{\Psi_{PH}}{2} \right) + \frac{(\alpha - N_A)D_X}{N_A - 1} \right\} \right\| \tag{8.112}$$

$$z_s = \left\{ \frac{\dfrac{D_Y(N_B - 2\beta + 1)}{2(N_B - 1)} + \langle y_s \rangle + \tan\theta_A \langle z_s \rangle}{\dfrac{\tan\left(\dfrac{\Psi_{PV}}{2} \right)(2\nu - 1 - N_V)}{(N_V - 1)} + \tan\theta_A} \right\} \tag{8.113}$$

and where

$$N_1(\beta) = \left\| \frac{N_V + 1}{2} + \frac{(N_V - 1)}{\{2\langle z_s \rangle + \delta_s \cos(\theta_A)\} \tan\left(\dfrac{\Psi_{PV}}{2} \right)} \left\{ \langle y_s \rangle - \frac{\delta_s}{2} \sin(\theta_A) + \frac{D_Y}{2(N_B - 1)}(N_B - 2\beta + 1) \right\} \right\| \tag{8.114}$$

$$N_2(\beta) = \left\| \frac{N_V + 1}{2} + \frac{\left(N_V - 1\right)}{\left\{2\langle z_s \rangle - \delta_s \cos\left(\theta_A\right)\right\} \tan\left(\dfrac{\Psi_{PV}}{2}\right)} \left\{ \langle y_s \rangle + \frac{\delta_s}{2} \sin\left(\theta_A\right) + \frac{D_Y}{2\left(N_B - 1\right)}\left(N_B - 2\beta + 1\right)\right\} \right\|$$

(8.115)

$$N_{U1}(\alpha) = \left\| \frac{N_M + 1}{2} + \frac{\left(N_M - 1\right)}{2z_s \tan\left(\dfrac{\Psi_{PH}}{2}\right)} \left\{ D_X \frac{\left(N_A - \alpha\right)}{N_A - 1} - \frac{\left(D_X + W\right)}{2}\right\} \right\|$$

(8.116)

$$N_{U2}(\alpha) = \left\| \frac{N_M + 1}{2} + \frac{\left(N_M - 1\right)}{2z_s \tan\left(\dfrac{\Psi_{PH}}{2}\right)} \left\{ D_X \left\{ \frac{\left(N_A - \alpha\right)}{N_A - 1} + \frac{W}{D_X} \right\} - \frac{\left(D_X + W\right)}{2}\right\} \right\|$$

(8.117)

Here, we have obtained Equation 8.113 by using Equation 8.106 and the equation of the achromatic plane $y_s = -\tan\theta_A z_s + \langle y_s \rangle + \tan\theta_A \langle z_s \rangle$ to solve for z_s. The parameters N_{U1} and N_{U2} are derived from Equation 8.50 by replacing h with z_s.

This transformation is the basis of DWDH rainbow and achromatic holography. By setting $\langle y_s \rangle = 0$ and $\langle z_s \rangle = h$, we arrive directly at the case of a DWDH rainbow hologram written by three R, G and B lasers. Each of the three primary colour channels has an identical transformation. By using only a single wavelength and a single achromatic data set, and also using $\langle y_s \rangle = 0$ and $\langle z_s \rangle = h$, we obtain, if δ_s is large, the transformation describing an achromatic hologram.

8.9.3.3 Transformations for One-Colour Laser

We can also use the general transformation above to discuss the case of the generation of the three slits required to create a full-colour rainbow hologram with only one laser wavelength. Here, we write the three slits, each of width, δ_s, on the achromatic plane intersecting $(y_s, z_s) = (0, h)$ at wavelength λ_W. The central coordinates of each of these slits are $(\langle y_R \rangle, \langle z_R \rangle)$, $(\langle y_G \rangle), (\langle z_G \rangle)$ and $(\langle y_B \rangle), (\langle z_B \rangle)$. On replay, each of the slits is designed to diffract, respectively, onto axis at the wavelengths $(\lambda_R, \lambda_G, \lambda_B)$. We should therefore be able to use Equation 8.91 to calculate the required central slit positions which, upon substitution in Equation 8.113, should lead us to the required transformation for the full-colour three-slit rainbow hologram recorded at a single wavelength. We must, however, be careful that in the present geometry z_s is positive and so the appropriate form of Equation 8.91 is actually

$$\begin{pmatrix} y_R & z_R \\ y_G & z_G \\ y_B & z_B \end{pmatrix} = \frac{h}{\lambda_W} \begin{pmatrix} \left(\lambda_W - \lambda_R\right)\sin\theta & \lambda_R \\ \left(\lambda_W - \lambda_G\right)\sin\theta & \lambda_G \\ \left(\lambda_W - \lambda_B\right)\sin\theta & \lambda_B \end{pmatrix}$$

(8.118)

Using this in Equation 8.113 to define $\langle y_s \rangle$ and $\langle z_s \rangle$, we arrive at the required transformation:

$$^{\mu\nu}\mathbf{S}_{\alpha\beta} = \sum_C {}^{\mu\nu_C}\mathbf{S}_{\alpha\beta}^C$$

(8.119)

where

$$^{\mu\nu_C}\mathbf{S}_{\alpha\beta}^C = {}^k\mathbf{I}_{ij}^C \qquad \forall\mu\left\{\mu\in\mathbb{N}\,\middle|\,N_{U1}(\alpha)\le\mu\le N_{U2}(\alpha)\right\} \quad \forall\nu_C\left\{\nu_C\in\mathbb{N}\,\middle|\,N_1^C(\beta)\le\nu_C\le N_2^C(\beta)\right\}$$

$$\forall\alpha\left\{\alpha\in\mathbb{N}\,\middle|\,\alpha\le N_A\right\} \quad \forall\beta\left\{\beta\in\mathbb{N}\,\middle|\,\beta\le N_B\right\} \quad \forall C\left\{C\in\{R,G,B\}\right\} \tag{8.120}$$

$$= 0 \qquad \text{otherwise}$$

and where

$$i = 1 + \left\|\frac{(\alpha-1)(N_I-1)}{N_A-1}\right\|, \quad j = 1 + \left\|\frac{(\beta-1)(N_J-1)}{N_B-1}\right\| \tag{8.121}$$

$$k = 1 + \left\|\frac{(N_K-1)}{W}\left\{\frac{D_X+W}{2}+z_C\frac{(2\mu-N_M-1)\tan\left(\dfrac{\Psi_{PH}}{2}\right)}{(N_M-1)}+\frac{(\alpha-N_A)D_X}{N_A-1}\right\}\right\| \tag{8.122}$$

$$z_C = \left\{\frac{\dfrac{D_Y(N_B-2\beta+1)}{2(N_B-1)}+\langle y_C\rangle+\tan\theta_A\langle z_C\rangle}{\dfrac{\tan\left(\dfrac{\Psi_{PV}}{2}\right)(2\nu_C-1-N_V)}{(N_V-1)}+\tan\theta_A}\right\} \tag{8.123}$$

$$N_1^C(\beta) = \left\|\frac{N_V+1}{2}+\frac{(N_V-1)}{\left\{2\langle z_C\rangle+\delta_s\cos(\theta_A)\right\}\tan\left(\dfrac{\Psi_{PV}}{2}\right)}\left\{\langle y_C\rangle-\frac{\delta_s}{2}\sin(\theta_A)+\frac{D_Y}{2(N_B-1)}(N_B-2\beta+1)\right\}\right\| \tag{8.124}$$

$$N_2^C(\beta) = \left\|\frac{N_V+1}{2}+\frac{(N_V-1)}{\left\{2\langle z_C\rangle-\delta_s\cos(\theta_A)\right\}\tan\left(\dfrac{\Psi_{PV}}{2}\right)}\left\{\langle y_C\rangle+\frac{\delta_s}{2}\sin(\theta_A)+\frac{D_Y}{2(N_B-1)}(N_B-2\beta+1)\right\}\right\| \tag{8.125}$$

$$N_{U1}(\alpha) = \left\|\frac{N_M+1}{2}+\frac{(N_M-1)}{2z_C\tan\left(\dfrac{\Psi_{PH}}{2}\right)}\left\{D_X\frac{(N_A-\alpha)}{N_A-1}-\frac{(D_X+W)}{2}\right\}\right\| \tag{8.126}$$

$$N_{U2}(\alpha) = \left\|\frac{N_M+1}{2}+\frac{(N_M-1)}{2z_C\tan\left(\dfrac{\Psi_{PH}}{2}\right)}\left\{D_X\left\{\frac{(N_A-\alpha)}{N_A-1}+\frac{W}{D_X}\right\}-\frac{(D_X+W)}{2}\right\}\right\| \tag{8.127}$$

with

$$\langle y_C \rangle = \frac{h}{\lambda_W}\left(\lambda_W - \lambda_C\right)\tan\theta_A$$

$$\langle z_C \rangle = \frac{h\lambda_C}{\lambda_W}$$

(8.128)

Note that this transformation is only valid when δ_s is small enough such that none of the slits overlap.

8.10 Correcting for Inherent Distortion in Printer Optical Objectives

Up until now, we have discussed **I**-to-**S** transformations pertinent to the case of a printer with a paraxial optical system and a perfect aberration-free system of recording, processing and playback of the hologram. In some cases, this approximation is genuinely all one needs—particularly when the angles of view are small. However, in many cases, fundamental aberrations are present and these must be taken into account. The most basic of these is the distortion induced by a printer objective having a high numerical aperture and a finite fifth Siedel coefficient. Such objectives allow extremely high resolution at the same time as affording a large angle of view. As such, they are fairly fundamental to printer design. As we have already mentioned, interpolation error is a significant worry in formulating the **I**-to-**S** transformation as generally the various integer coordinate meshes associated with image acquisition and hologram writing do not line up. The naïve scheme of using a paraxial **I**-to-**S** transformation and then predistorting each SLM frame to correct for a finite fifth coefficient often ends up by introducing interpolation noise two times over. It therefore turns out to be much more efficient to combine the pre-distortion of data with the paraxial **I**-to-**S** transformation to create a single non-paraxial transformation. In Appendix 4, we discuss how this idea can be extended to all holographic aberrations including those induced by the processing chemistry.

8.10.1 Setting up the Formalism

We will start by setting up some coordinate systems. Let (ρ,θ) represent right-handed cylindrical polar coordinates of a pixel on the SLM with origin at the centre of the SLM. Laser light traversing the SLM is focussed by the optical objective to form an image downstream of the hogel position on a plane whose normal vector is parallel to the axial propagation vector. Let (r,ϑ) be right-handed cylindrical polar coordinates describing the position of the pixel on this image plane with $r = 0$ representing the optical axis. The objective distortion may then be characterised by the following relation

$$\rho = r\left\{1 - \mathcal{D}(r)\right\}$$

$$\theta = \vartheta + \pi$$

(8.129)

Alternatively, an inverse formalism can be used

$$r = \rho\left\{1 - \mathcal{G}(\rho)\right\}$$

$$\vartheta = \theta + \pi$$

(8.130)

Now, we define x and y to be the right-handed Cartesian coordinates of a pixel on the SLM. As before, we take the origin to be at the centre of the SLM. Likewise, we introduce x' and y' as the Cartesian coordinates of the projected image plane with origin as the optical axis.

We define the aspect ratio of the SLM as its useful height divided by its useful width. In general, the rectangular SLM will either underfill or overfill the circular aperture of the objective. Here we will assume that there is an overfill in the horizontal direction leading to "dead" areas of the SLM (Figure 8.22). We label the absolute width and height of the SLM by w_s and h_s where $w_s > h_s$. We will assume of course that the SLM is mounted symmetrically with respect to the objective. We then label the length of the dead zone to one side of the SLM as x_d. Using these labels, we may write the following simple expression for the SLM aspect ratio:

$$a_s = \frac{h_s}{w_s - 2x_d}. \tag{8.131}$$

It is useful to now introduce non-dimensional coordinates for both the SLM plane and the projected SLM image plane. On the SLM plane, we introduce the coordinates $(\hat{x}, \hat{y})$, which are defined as

$$\hat{x} \equiv \frac{2a_s x}{h_s}$$

$$\hat{y} \equiv \frac{2a_s y}{h_s} \tag{8.132}$$

On the projected SLM image plane, we introduce the coordinates $(\hat{x}', \hat{y}')$, which are likewise defined as

$$\hat{x}' \equiv \frac{2a_s x'}{h_s\left[1 - \mathcal{G}\left(w_s/2 - x_d\right)\right]}$$

$$\hat{y}' \equiv \frac{2a_s y'}{h_s\left[1 - \mathcal{G}\left(w_s/2 - x_d\right)\right]} \tag{8.133}$$

Figure 8.23 shows a diagram of the projected image plane of the SLM (a) and the SLM plane (b). To use the maximum field of view of the printer, we must specify the image data over a grid within the red rectangle shown in (a). This image data must then be distorted using an inverse transform such that we write data to the SLM that will then be transformed to form effectively undistorted data. The red contour in (b) shows the effect of such predistortion on the contour of the outer data rectangle of (a). Note that the aspect ratio of the image data in (a) is equal to that of the SLM as the distortion transformation does not affect angle.

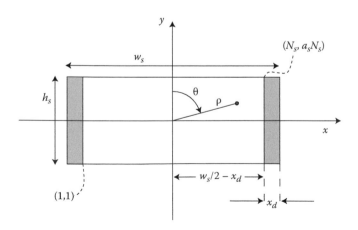

FIGURE 8.22 The printer SLM and the parameters that describe it. The physical SLM has height, h_s and width, w_s. However we envisage the SLM overfilling the aperture of the objective giving rise to two dead zones of length x_d.

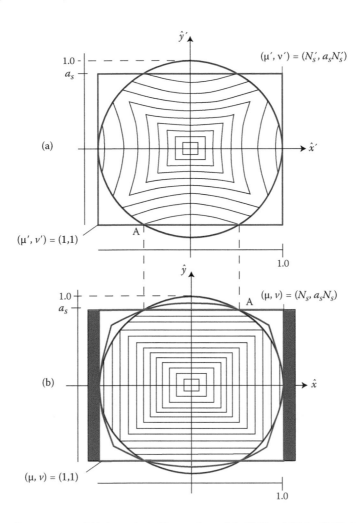

FIGURE 8.23 Coordinate systems of (a) the projected image plane of the SLM and (b) the SLM plane.

8.10.2 Data Predistortion

Data predistortion seeks to introduce an inverse distortion into the input image data such that the optical distortion introduced by the objective exactly cancels this out. The result is therefore that the objective appears completely paraxial. To formalise the predistortion process, we must introduce two coordinate systems. These describe image data on the SLM and on the projected plane of the SLM downstream of the hogel.

Image data must be known over the red rectangular region of Figure 8.23a. We therefore divide up this region of the projected image plane using a uniform (x,y) Cartesian grid of integer dimensions $(N'_s, a_s N'_s)$ and denote the irradiance distribution over this grid by $\mathbf{S}'_{\mu'v'}$, where

$$
\begin{aligned}
\hat{x}' &= \frac{2(\mu'-1)}{N'_s - 1} - 1 \quad \forall \mu' = 1,2,3,...,N'_s \\
\hat{y}' &= \frac{2a_s(v'-1)}{a_s N'_s - 1} - a_s \quad \forall v' = 1,2,3,...,a_s N'_s
\end{aligned}
\tag{8.134}
$$

The point $(\mu'v') = (1,1)$ is taken to be the bottom left-hand corner of the rectangle in Figure 8.23a and the point $(\mu'v') = (N'_s, a_s N'_s)$ corresponds to the top right-hand-corner.

In an exactly similar manner to above, we discretise the useful aperture of the SLM plane as defined by the green rectangle in Figure 8.23b. We denote the irradiance distribution over this grid by $\mathbf{S}_{\mu\nu}$ where

$$\hat{x} = \frac{2(\mu-1)}{N_s-1} - 1 \quad \forall \mu = 1,2,3,...,N_s$$

$$\hat{y} = \frac{2a_s(\nu-1)}{a_s N_s - 1} - a_s \quad \forall \nu = 1,2,3,...,a_s N_s \tag{8.135}$$

Following our convention, we take the point $(\mu,\nu) = (1,1)$ to be the bottom left-hand corner of the green rectangle in Figure 8.23b and the point $(\mu,\nu) = (N_s, a_s N_s)$ to correspond to the top right-hand-corner.

Now, in the case that the objective was perfectly paraxial, we would have for $N_s' = N_s$

$$\mathbf{S}_{\mu\nu} = \mathbf{S}_{\mu'\nu'}' \quad \forall \left\{ \mu',\nu' \in \mathbb{N} \middle| \mu' \leq N_s, \nu' \leq a_s N_s \right\} \tag{8.136}$$

where

$$\mu' = N_s + 1 - \mu$$

$$\nu' = aN_s + 1 - \nu \tag{8.137}$$

Equation 8.137 just describes an image inversion from the primed to the unprimed plane as the primed and unprimed planes have been assumed to be on opposite sides of the hogel—this is just because we have chosen to study the case in which the SLM is in the non-conjugate position. Of course, in the case that the primed and unprimed planes are on the same side as the hogel, as in the case of Figure 8.19, for example (the conjugate SLM position), then Equation 8.137 must be replaced by the simpler relation

$$\mu' = \mu$$

$$\nu' = \nu \tag{8.138}$$

Anyhow, in the non-paraxial case (and for the non-conjugate SLM position), a point $(\hat{x}, \hat{y})$ on the SLM is related to a point $(\hat{x}', \hat{y}')$ on the projected image plane by

$$\hat{x}' = -\hat{x} \frac{1 - \mathcal{G}(\rho)}{1 - \mathcal{G}(w_s/2 - x_d)}$$

$$\hat{y}' = -\hat{y} \frac{1 - \mathcal{G}(\rho)}{1 - \mathcal{G}(w_s/2 - x_d)} \tag{8.139}$$

or in terms of the discrete coordinates

$$\mu' = \frac{N_s'+1}{2} + \left(\frac{N_s+1}{2} - \mu \right) \left\{ \frac{N_s'-1}{N_s-1} \right\} \frac{1 - \mathcal{G}(\rho)}{1 - \mathcal{G}(w_s/2 - x_d)}$$

$$\nu' = \frac{a_s N_s'+1}{2} + \left(\frac{a_s N_s+1}{2} - \nu \right) \left\{ \frac{N_s'-1}{N_s-1} \right\} \frac{1 - \mathcal{G}(\rho)}{1 - \mathcal{G}(w_s/2 - x_d)} \tag{8.140}$$

It is usually more convenient at this point to introduce a non-dimensional radius

$$\hat{\rho} \equiv \frac{2a_s}{h_s} \rho \tag{8.141}$$

and an alternative normalised distortion function

$$\mathcal{K}(\hat{\rho}) \equiv \frac{1 - \mathcal{G}(\rho)}{1 - \mathcal{G}\left(w_s/2 - x_d\right)}. \tag{8.142}$$

This function can be easily generated using ray tracing of the printer optical system and may then be conveniently expanded using a least-squares fit as a truncated one-dimensional power series

$$\mathcal{K}(\hat{\rho}) \equiv \sum_{\sigma=0}^{N_E} k_\sigma \hat{\rho}^\sigma. \tag{8.143}$$

A typical example of this distortion function is shown in Figure 8.24. Note that by a sensible choice of N_s' and N_s the coordinate systems (μ, v) and (μ', v') may be made to line up in the central field where $\mathcal{K}(\rho) \sim 0$. To achieve an undistorted data set at the projected image plane, we must then load our SLM with information $\mathbf{S}_{\mu v}$, which is calculated by the transformation

$$\mathbf{S}_{\mu v} = \mathbf{S}'_{\mu' v'} \qquad \forall \left\{\mu, v, \mu', v' \in \mathbb{N} \middle| \mu, \mu' \le N_s; v, v' \le a_s N_s\right\}$$

$$= 0 \qquad \forall \left\{\mu, v, \mu', v' \in \mathbb{N} \middle| \mu \le N_s; v \le a_s N_s; \mu' > N_s\right\} \tag{8.144}$$

$$= 0 \qquad \forall \left\{\mu, v, \mu', v' \in \mathbb{N} \middle| \mu \le N_s; v \le a_s N_s; v' > a_s N_s\right\}$$

where

$$\mu' = \left\| \frac{N_s'+1}{2} + \left(\frac{N_s+1}{2} - \mu\right)\left\{\frac{N_s'-1}{N_s-1}\right\}\mathcal{K}(\hat{\rho})\right\|$$

$$v' = \left\| \frac{a_s N_s'+1}{2} + \left(\frac{a_s N_s+1}{2} - v\right)\left\{\frac{N_s'-1}{N_s-1}\right\}\mathcal{K}(\hat{\rho})\right\| \tag{8.145}$$

$$\hat{\rho} = \left\{\left[2\frac{(\mu-1)}{N_s-1} - 1\right]^2 + a_s^2\left[2\frac{(v-1)}{a_s N_s-1} - 1\right]^2\right\}^{1/2}$$

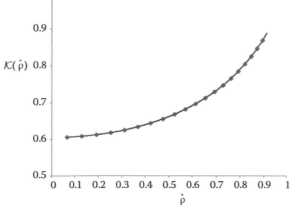

FIGURE 8.24 Example of the measured distortion function of a typical printer objective along with a 6th order power law fit. The data are from a GEOLA blue (437.7 nm) objective apodised by a sony XGA1 LCD with horizontal paraxial field of view of 85 degrees. The function $\mathcal{K}(\hat{\rho})/\mathcal{K}(0)$ essentially measures the change in effective pitch of paraxial and non-paraxial integer coordinate systems at $\mathcal{S}'$.

Setting $\mathcal{K}(\hat{\rho}) = 1$ and $N'_s = N_s$, we retrieve the paraxial case of Equations 8.136 and 8.137. For a conjugate SLM position, the first two relations of Equation 8.145 must be replaced by

$$\mu' = \left\| \frac{N'_s+1}{2} + \left(\mu - \frac{N_s+1}{2} \right) \left\{ \frac{N'_s-1}{N_s-1} \right\} \mathcal{K}(\hat{\rho}) \right\|$$

$$\nu' = \left\| \frac{a_s N'_s+1}{2} + \left(\nu - \frac{a_s N_s+1}{2} \right) \left\{ \frac{N'_s-1}{N_s-1} \right\} \mathcal{K}(\hat{\rho}) \right\|$$

(8.146)

when $\mathcal{K}(\hat{\rho}) = 1$ and $N'_s = N_s$, this then leads to Equation 8.138 as expected.

The transformation of Equations 8.144 and 8.145 allow us to use a paraxial transformation on **I** to arrive at **S'** and then from **S'** to calculate the required SLM data **S**. As we have already discussed, however, this procedure often introduces interpolation errors two times over. The exception is if we specifically align our various integer coordinate meshes and choose a specific quantised value of h such that **S'** is free of interpolation error as described in Section 8.8.5. In this case, Equation 8.144 can be implemented in a real-time fashion within the printing process itself. As a result, from the point of view of data preparation, the task then becomes simply the generation of paraxial SLM data. One should note that Equation 8.144 is intrinsically two-dimensional and as written, uses nearest-neighbour interpolation. This transformation may be rewritten using bilinear and bicubic interpolation.* These more advanced types of interpolation can often still be used in real-time within the printer writing software and act to lessen the introduced interpolation noise, particularly at high spatial frequencies.

In many cases, one simply cannot fulfil the requirements necessary to abolish interpolation error in the calculation of the paraxial SLM data and, in this case, it makes good sense to formulate a single non-paraxial **I**-to-**S** transformation. We cannot avoid the introduction of interpolation noise on the application of this transformation, but we will always introduce less noise in a single nearest-neighbour transform than if we use two sequential transforms; that is, if g and α are integers and f_1 and f_2 are arbitrary but well-behaved real functions, then generally, $g = \left\| f_1 \left(\left\| f_2(\alpha) \right\| \right) \right\|$ can be a poor approximation to $g = \left\| f_1 \left(f_2(\alpha) \right) \right\|$. One might have thought that replacing the nearest-neighbour interpolation in an **I**-to-**S** transformation with a higher order interpolation could offer a good alternative approach. However, HPO holograms practically require three-dimensional interpolation schemes and full-parallax holograms, four-dimensional schemes. As such, they are difficult to implement, extremely time-consuming with current computing technology and of questionable advantage for the moment.

8.10.3 Non-Paraxial I-to-S Transformations

Any of the previous paraxial transformations that we have introduced giving **S'** in terms of **I** may be trivially combined with Equations 8.144 to 8.146, to give a single non-paraxial transformation by simple substitution. For example, the transformation of Equation 8.62 and Equation 8.63 can be written in terms of (μ', ν') in the following trivial way:

$$^{\mu'\nu'}\mathbf{S}'_{\alpha\beta} = {}^k\mathbf{I}_{ij} \qquad \forall \mu' \left\{ \mu' \in \mathbb{N} \middle| N_{U1}(\alpha) \le \mu' \le N_{U2}(\alpha) \right\}$$

$$\forall \nu' \left\{ \nu' \in \mathbb{N} \middle| N_{V1}(\beta) \le \nu' \le N_{V2}(\beta) \right\}$$

$$\forall \alpha \left\{ \alpha \in \mathbb{N} \middle| \alpha \le N_A \right\}$$

(8.147)

$$\forall \beta \left\{ \beta \in \mathbb{N} \middle| \beta \le N_B \right\}$$

$$= 0 \qquad \text{otherwise}$$

---

* We discuss higher order interpolation in Chapter 9 and in Appendix 7.

where

$$i = N_I - \left\| \frac{(\alpha - 1)(N_I - 1)}{N_A - 1} \right\|, \quad j = N_J - \left\| \frac{(\beta - 1)(N_J - 1)}{N_B - 1} \right\| \tag{8.148}$$

$$k = 1 + \left\| \frac{(N_K - 1)}{W} \left\{ \frac{D_X + W}{2} + \frac{h(2\mu' - N_M - 1)}{(N_M - 1)} \tan\left(\frac{\Psi'_{PH}}{2} \right) + \frac{(\alpha - N_A)D_X}{N_A - 1} \right\} \right\| \tag{8.149}$$

and where N_{U1}, N_{U2}, N_{V1} and N_{V2} are given by Equations 8.49 and 8.50. The parameter Ψ'_{PH} is the non-paraxial field of view and is defined as

$$\tan\left(\frac{\Psi'_{PH}}{2} \right) = \frac{1}{\mathcal{K}(0)} \tan\left(\frac{\Psi_{PH}}{2} \right) \tag{8.150}$$

Using a conjugate SLM position, we then simply substitute Equation 8.146 into Equation 8.149 with $N_s = N'_s = N_M$ to give

$$k = 1 + \left\| \frac{(N_K - 1)}{W} \left\{ \frac{D_X + W}{2} + \frac{(2\mu - N_M - 1)\mathcal{K}(\hat{\rho})}{(N_M - 1)} h \tan\left(\frac{\Psi'_{PH}}{2} \right) + \frac{(\alpha - N_A)D_X}{N_A - 1} \right\} \right\| \tag{8.151}$$

where

$$\hat{\rho} = \left\{ \left[2\frac{(\mu - 1)}{N_M - 1} - 1 \right]^2 + a_s^2 \left[2\frac{(v - 1)}{a_s N_M - 1} - 1 \right]^2 \right\}^{1/2} \tag{8.152}$$

and

$$a_s N_M = N_V \tag{8.153}$$

Now, it is usual that the non-paraxial distortion is not greater than about 20%. If this is the case, we can usually ignore its effect on the viewing window boundary and write the full non-paraxial transformation as

$$^{\mu v}\mathbf{S}_{\alpha\beta} = {}^k\mathbf{I}_{ij} \qquad \forall \mu \left\{ \mu \in \mathbb{N} \middle| N_{U1}(\alpha) \le \mu \le N_{U2}(\alpha) \right\}$$

$$\forall v \left\{ v \in \mathbb{N} \middle| N_{V1}(\beta) \le v \le N_{V2}(\beta) \right\}$$

$$\forall \alpha \left\{ \alpha \in \mathbb{N} \middle| \alpha \le N_A \right\} \tag{8.154}$$

$$\forall \beta \left\{ \beta \in \mathbb{N} \middle| \beta \le N_B \right\}$$

$$= 0 \qquad \text{otherwise}$$

where

$$i = N_I - \left\| \frac{(\alpha - 1)(N_I - 1)}{N_A - 1} \right\|, \quad j = N_J - \left\| \frac{(\beta - 1)(N_J - 1)}{N_B - 1} \right\| \tag{8.155}$$

$$k = 1 + \left\| \frac{(N_K - 1)}{W} \left\{ \frac{D_X + W}{2} + \frac{(2\mu - N_M - 1)\mathcal{K}(\hat{\rho})}{(N_M - 1)} h \tan\left(\frac{\Psi'_{PH}}{2} \right) + \frac{(\alpha - N_A)D_X}{N_A - 1} \right\} \right\| \tag{8.156}$$

In the central paraxial region where $\hat{\rho}$ is small this equation is identical to the paraxial equation but clearly as $\hat{\rho}$ increases the rule between k and μ is now modified. This transformation will, however, yield a viewing window that is a distorted rectangle. To arrive at a transform with a perfectly rectangular window, we must transform the (μ', ν') inequality of Equation 8.147. To transform this inequality, one must solve a number of non-linear equations to determine the viewing boundary in terms of (μ,ν) for each hogel. Because of the form of Equation 8.130, for a given ν, there will always be either 0, 1 or 2 values of μ. We therefore adopt the strategy of determining the topmost and bottommost values of ν on the window boundary; we then calculate pairs of μ values for all rows between these two extremes. This allows us to perform the calculation of **S** on a row-by-row basis.

The first step is then to calculate the topmost and bottommost values of ν on the boundary. To do this, we start with coordinates of the top left-hand corner:

$$\mu = N_{U1} ; \quad \nu = N_{V2} \tag{8.157}$$

and the top right-hand corner

$$\mu = N_{U2} ; \quad \nu = N_{V2} \tag{8.158}$$

where N_{U1}, N_{U2} and N_{V2} are defined by Equations 8.49 and 8.50 but with non-paraxial angles (e.g., Equation 8.150) replacing paraxial angles. Then, we solve the following $N_{U2} - N_{U1} + 1$ non-linear equations to give the ν coordinates of the top window boundary:

$$N'_{VT}(\mu) = \|\nu\| \; \forall \mu \left\{ \mu \in \mathbb{N} \middle| N_{U1} \leq \mu \leq N_{U2} \right\}$$

where

$$N_{V2} = \frac{N_V + 1}{2} + \left(\nu - \frac{N_V + 1}{2} \right) \mathcal{K} \left(\left\{ \left[2\frac{(\mu - 1)}{N_M - 1} - 1 \right]^2 + a_s^2 \left[2\frac{(\nu - 1)}{N_V - 1} - 1 \right]^2 \right\}^{1/2} \right) \tag{8.159}$$

We then calculate the maximum value of $N'_{VT}(\mu)$, which is the topmost ν coordinate of the viewing window:

$$N'_{V2}(\alpha,\beta) \equiv \max \left\{ N'_{VT}\left(\mu = N_{U1} \right), N'_{VT}\left(\mu = N_{U1} + 1 \right), ..., N'_{VT}\left(\mu = N_{U2} \right) \right\} \tag{8.160}$$

A similar procedure then defines the minimum value of $N'_{VB}(\mu)$, which is the bottommost ν coordinate of the viewing window:

$$N'_{VI}(\alpha,\beta) \equiv \min \left\{ N'_{VB}\left(\mu = N_{U1} \right), N'_{VB}\left(\mu = N_{U1} + 1 \right), ..., N'_{VB}\left(\mu = N_{U2} \right) \right\} \tag{8.161}$$

Equation 8.159 and its partner, which defines $N'_{VB}(\mu)$, are typically solved computationally using a Newton–Raphson iteration. Note that we have deleted the nearest integer operators so that the equations are well behaved—this just means that we treat ν as real for the purpose of solving the equation and then we simply use $\|\nu\|$.

We now solve a further $(N_{V2} - N_{VI} + 1)$ equations for the left-hand μ coordinate of each row of the viewing window and a similar number for the right-hand coordinate. These are, respectively

(i) $N'_{U1}(\nu) = \|\mu\| \forall \nu \left\{ \nu \in \mathbb{N} \middle| N'_{VI} \leq \nu \leq N'_{V2} \right\}$ where

$$N_{U1} = \frac{N_M + 1}{2} + \left(\mu - \frac{N_M + 1}{2} \right) \mathcal{K} \left(\left\{ \left[2\frac{(\mu - 1)}{N_M - 1} - 1 \right]^2 + a_s^2 \left[2\frac{(\nu - 1)}{N_V - 1} - 1 \right]^2 \right\}^{1/2} \right) \tag{8.162}$$

(ii) $N'_{U2}(v) = \|\mu\| \forall v \left\{ v \in \mathbb{N} \big| N'_{VI} \leq v \leq N'_{V2} \right\}$ where

$$N_{U2} = \frac{N_M+1}{2} + \left(\mu - \frac{N_M+1}{2}\right)\mathcal{K}\left(\left\{\left[2\frac{(\mu-1)}{N_M-1}-1\right]^2 + a_s^2\left[2\frac{(v-1)}{N_V-1}-1\right]^2\right\}^{1/2}\right) \qquad (8.163)$$

The exact **I**-to-**S** transformation can then be formulated in the following way:

$$^{\mu v}\mathbf{S}_{\alpha\beta} = {}^{k}\mathbf{I}_{ij} \qquad \forall \mu \left\{\mu \in \mathbb{N} \big| N'_{U1}(\alpha,v) \leq \mu \leq N'_{U2}(\alpha,v)\right\}$$

$$\forall v \left\{v \in \mathbb{N} \big| N'_{V1}(\alpha,\beta) \leq v \leq N'_{V2}(\alpha,\beta)\right\}$$

$$\forall \alpha \left\{\alpha \in \mathbb{N} \big| \alpha \leq N_A\right\}$$

$$\forall \beta \left\{\beta \in \mathbb{N} \big| \beta \leq N_B\right\} \qquad (8.164)$$

$$\forall k \left\{k \in \mathbb{N} \big| k \leq N_K\right\}$$

$$= 0 \qquad \text{otherwise}$$

where

$$i = N_I - \left\|\frac{(\alpha-1)(N_I-1)}{N_A-1}\right\|, \quad j = N_J - \left\|\frac{(\beta-1)(N_J-1)}{N_B-1}\right\| \qquad (8.165)$$

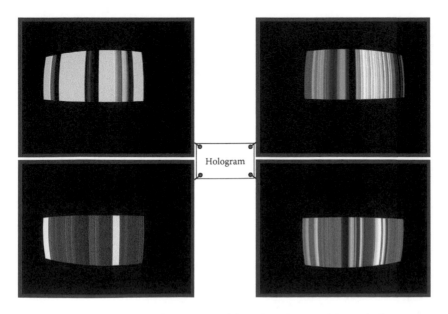

FIGURE 8.25 Typical SLM data, prepared using a non-paraxial transformation, pertaining to the four extreme corner hogels of an HPO reflection hologram with a rectangular viewing window. Note that the SLM is in the non-conjugate position here.

$$k = 1 + \left\| \frac{\left(N_K - 1\right)}{W} \left\{ \frac{D_X + W}{2} + \frac{\left(2\mu - N_M - 1\right)\mathcal{K}(\hat{\rho})}{\left(N_M - 1\right)} h \tan\left(\frac{\Psi'_{PH}}{2}\right) + \frac{\left(\alpha - N_A\right)D_X}{N_A - 1} \right\} \right\| \qquad (8.166)$$

A different technique which can be used to define an exactly rectangular viewing window is, for each hogel, to calculate (μ',ν') for every (μ,ν) on the SLM using Equation 8.146. By testing (μ',ν') against the required rectangle, the parameters $N'_{V1}(\alpha,\beta), N'_{V2}(\alpha,\beta), N'_{U1}(\alpha,\nu), N'_{U2}(\alpha,\nu)$ can once again be defined. Precalculation of these boundary functions is often advantageous can be performed with either technique. Figure 8.25 shows typical SLM data pertaining to the four extreme corner hogels of an HPO reflection hologram with a rectangular viewing window. The distortion function used is that of Figure 8.24.

REFERENCES

1. D. Brotherton-Ratcliffe and A. Rodin, *Holographic printer*, US patent 7,161,722 (filed 2002, granted 2007).
2. D. Brotherton-Ratcliffe, A. Rodin and L. Hrynkiw, *Method of writing a composite 1-step hologram*, US patent 7,333,252 (filed 2002, granted 2008).
3. R. Guenther, *Modern Optics*, John Wiley & Sons, New York (1990).
4. G. Saxby, *Practical Holography*, 3rd Edn, IOP Publishing Ltd., London, UK (2004).
5. L. Cross, "The multiplex technique for cylindrical holographic stereograms," in SPIE San Diego August Seminar (1977). [Presented, but not published]
6. S. A. Benton, "Hologram reconstructions with extended incoherent sources," *J. Opt. Soc. Am.* **59**, 1545–1546A (1969).
7. S. A. Benton, "Achromatic images from white light transmission holograms," *J. Opt. Soc. Am.* **68**, 1441A (1978).
8. K. Ohnuma and F. Iwata, "Color rainbow hologram and color reproduction," *Appl. Opt.* **27**, 3859–3863 (1988).

9

Digital Holographic Printing: Computational Methods for Full-Parallax Holograms

9.1 Introduction

In Chapter 8, we presented a detailed discussion of the typical image-processing algorithms which are required for the conversion of raw camera data into a form suitable for the generation of both direct-write and master-write (DWDH and MWDH) digital holograms. We paid particular attention to horizontal parallax-only (HPO) holograms. The algorithms for this type of hologram can all be implemented on a single PC with several gigabytes of memory. In all but the largest DWDH HPO holograms, all data can be read directly, and in its entirety, into the computer's memory and the transformations applied. One therefore only needs to read the raw camera data one time, process it, and then write it out one time. With the advent of 64-bit PCs, high-resolution DWDH HPO holograms of even many tens of square metres in size present no particular difficulty in such an implementation.

In contrast to HPO holograms, even small full-parallax holograms can require just too much image data for all the data to be loaded at one time into main memory. For example, a 40 cm × 30 cm RGB DWDH full-parallax reflection hologram with a maximal viewing window, good angular resolution and a 0.5 mm-diameter hogel can be expected to require approximately 1 to 2 TB of memory storage when used with a printer spatial light modulator (SLM) of 1024 × 1280. This is beyond the practical main memory limitations of modern PCs. As a result, it is not possible to read all image data into main memory, apply the appropriate image processing transformation and then write the data out. Rather if one follows the techniques used in HPO holography directly, one ends up reading and writing data to and from the disk many times. This then leads to enormous inefficiency and prohibitively long processing times.

The intrinsically large image data set characteristic of full-parallax display holograms leads to various problems. In Chapter 10, we shall discuss how one copes with generating such large data sets in typical computer modelling programs and how one collects such data from real-world objects. The sheer size of the data can present transport and storage problems for companies which print digital holograms. The image data for even large HPO holograms can be compressed and sent over the Internet without any problem. In contrast, at the time of writing, even small full-parallax data sets can rarely be sent electronically. Archiving customer data effectively often requires petabyte storage facilities.

There are various strategies one can apply to circumvent the issue of image processing of intrinsically large data sets. Perhaps the most obvious is the use of MWDH in place of DWDH. Here, an H_1 master is written using the direct camera data. In Chapter 8, we derived the **I**-to-**S** transformation required (Equation 8.20). Another strategy is to write a DWDH Denisyuk hologram according to Section 8.7.1. Again, this only requires raw camera data. Both these techniques have their applications and merits. RGB DWDH reflection Denisyuk holograms are an excellent choice when no image projection in front of the hologram plane is required. MWDH is particularly suited for medium-sized holograms—but if the final H_2 hologram is too small, then one runs into resolution and aberration problems with the printer write-head.* At the other end of the scale, one ends up requiring an extremely large RGB laser to achieve the optical H_1:H_2 transfer.

---

* Essentially, one can end up requiring a very high resolution SLM and then only using a small fraction of it.

At the end of the day, many full-parallax holograms require fundamental image data reorganisation. This is because the raw image data is inevitably available at one plane and we would simply like to write a hologram at another plane. The most usual reason for this situation is apparent from analogue holography. Here, an H_1 hologram is made and then optically transferred to an H_2 hologram, thereby effectively changing the image plane. As a result of this procedure, the image of the final H_2 hologram can be made to bisect the physical hologram surface. With digital holograms, we have just the same problem. To end up with an image that bisects the physical hologram plane, we must apply an image-planing transformation to our raw camera data, the only difference being that, in this case, the transformation is mathematical and not optical.

One rather different way around the data reorganisation problem is available when special computer modelling software is used to generate the three-dimensional (3D) model. Most commercially available modelling software programs generate conventional camera image data only. They do this by calculating the intensity values of many light rays emanating from the 3D model given various assumptions about the optical properties of the model. Groups of rays are then calculated to define a camera image. From an absolute point of view, one can imagine using this same model to calculate the intensity of all those rays required to directly define a hogel. This comes down to simply asking the software to work in a different order in calculating rays. There is, however, a sizeable problem here as most current commercial software packages are configured and optimised to calculate rays in the groups associated with normal cameras rather than with hogels. The result is that either the software works many hundreds of times slower when forced to calculate hogels directly or it simply cannot be configured in this fashion. Nevertheless, software packages that are able to calculate hogel data directly from a 3D computer model have been written and, although not commercially available, are in use today by some groups working in digital holography. The technology behind this type of software is outside the scope of this book, and the interested reader is referred to the published literature [1–5]. The advantage of this technique is that it enables one to directly calculate the required rendered images for each hogel from a compact 3D computer model. In addition, such software can, in principle, correct for many aberrations including optical aberration in the writing objective. In this light, and from a truly absolute point of view, it must be regarded theoretically as the best solution for generating virtual data sets for full-parallax holograms. However, from a practical point of view, the number of people needing this type of special modelling program is relatively small. As such, the more standard commercially available programs for 3D computer modelling, although they lack the facility to generate direct hogel data, inevitably provide a better solution for the generation of state-of-the-art render data today. This is the approach we take in this book; it is also the approach that companies such as Geola Digital UAB have taken since the early days of digital display holography.

9.2 Practical Strategies for Changing the Image Plane

In Chapter 8, we discussed the alignment of the integer coordinate meshes concerned with data acquisition and hologram writing. We saw how, in the context of HPO holograms, the viewing distance became quantised when one demanded that these coordinate meshes lined up. From a practical point of view, given the specifications of a printer and the specifications of a desired HPO hologram, the most common situation we encounter is that such quantisation is unacceptable.* In other words, we find ourselves wanting to define a viewing plane distance that inevitably leads to a non-integer mapping between indices representing acquisition and writing coordinate systems. This is perhaps inevitable in the context of HPO holography because the optimal viewing distance at observation time is equal to the camera distance at acquisition time. With full-parallax digital holography, however, we are in a much better position due to Huygens' principle; here, we can view the hologram where we like and we can put the camera plane where we like. Of course, we may find ourselves in a situation in which we have image data from a physical camera (as opposed to a virtual camera) at a fixed plane and in this case we have to either adapt

---

* We should make the point that in the case that the write-head of a digital printer is designed to function for only a selected set of specific hologram formats, it may be possible to design it such that the rules of quantisation are acceptable for this restricted set. This solution trades flexibility for a reduction in interpolation noise.

the hologram parameters so that the camera distance obeys the quantisation rules or use more advanced interpolation mathematics - or accept a certain amount of interpolation noise.

9.2.1 Camera Definition

In the case of a virtual camera, by far the best and most practical solution for generating a full-parallax hologram is to prepare the data set at the first quantised value of the camera distance (i.e., in the language of Section 8.8.5, at a value of h corresponding to $n = 1$). At this distance, the size of a projected SLM pixel is the same as the hogel diameter and assuming a centered camera configuration

$$N_I = N_A$$
$$N_J = N_B \tag{9.1}$$

We must further arrange that the camera spacing is equal to the hogel diameter according to Equation 8.80. This amounts to the minor restriction of the window width W being an integer multiple of the hogel diameter. Similarly, we need to add a vertical constraint

$$\frac{H}{N_G - 1} = \frac{D_Y}{N_B - 1} \tag{9.2}$$

which then leads to a corresponding constraint on the window height, H. With these constraints, one has all one needs to define the camera and its animation under a centred camera configuration for a hologram of size $(D_X \times D_Y) = (N_A - 1)\delta \times (N_B - 1)\delta$. The relevant camera parameters are

$$h = \frac{\delta}{2}\left(N_M - 1\right)\cot\left(\Psi_{PH}/2\right) = \frac{\delta}{2}\left(N_V - 1\right)\cot\left(\Psi_{PV}/2\right) \tag{9.3}$$

$$N_I = N_A; \quad N_J = N_B \tag{9.4}$$

$$N_K = 1 + \frac{W}{\delta}; \quad N_G = 1 + \frac{H}{\delta} \tag{9.5}$$

where δ represents the hogel diameter and the camera/viewing window is defined by

$$W \equiv m_1\delta \leq \delta(N_M - N_A); \quad H = m_2\delta \leq \delta(N_V - N_B) \tag{9.6}$$

with m_1 and m_2 non-zero positive integers. The parameters Ψ_{PH} and Ψ_{PV} are defined by the printer writing objective and N_M and N_V by the printer SLM. The first-quantised h value given in Equation 9.3 is then seen to be an intrinsic property of the printer for a given hogel diameter.

The above camera definition produces a single-colour image data set ${}^{kg}\mathbf{I}_{ij}$ for $1 \leq k \leq N_K$, $1 \leq g \leq N_G$, $1 \leq i \leq N_I$ and $1 \leq j \leq N_J$. To illustrate this, let us use the example of the 40 cm × 30 cm, 0.5 mm-hogel hologram we discussed in Section 9.1. Here, $(D_X, D_Y) = (399.5, 299.5$ mm$)$ as we always calculate distances from hogel centre to hogel centre. This then gives $N_A = 800$ and $N_B = 600$. With a horizontal printer field of view (FOV) of 85° and a printer SLM with pixel dimensions of 1280 × 1024, the $n = 1$ h value is 348.95 mm. If we choose a window size at this distance of 50 cm × 40 cm, then we must use $N_K = 1001$ and $N_G = 801$. Using Equation 9.4, we can calculate the memory requirement for the combined red, green and blue component image data (${}^{kg}\mathbf{I}_{ij}$). This is simply

$$M = 3N_A N_B N_K N_G \tag{9.7}$$

which works out at just over 1 TB. In Chapter 10, we shall look at how to generate this data using computer modelling software. We shall also see how it may be best derived from physical camera data. For now, however, we shall simply assume that we have this data on disk and we will direct our attention to the problem of changing the image plane—that is, we will examine how this data can be used to calculate a new data set corresponding to a camera position defined by $h = 0$.

9.2.2 Changing the Image Plane—Two-Step I-to-S Transformations

As we have discussed in the previous section, the size of $^{kg}\mathbf{I}_{ij}$ makes it practically impossible to read into memory at one time. The data is organised in $N_K \times N_G$ files each of size $3N_A N_B$. Each hogel requires just one RGB irradiance value from each of these 800,000 odd files. Opening 800,000 files and closing them each time we want to define a hogel is of course not an option. So converting the data to the form wanted in one step seems impossible. However, not all is lost, as we can do the next best thing: convert the data in two steps. This turns out to be actually very efficient, as we shall see.

The idea is as follows: we start with our data ($^{kg}\mathbf{I}_{ij}$) arranged in $N_K \times N_G$ RGB disk files. Let us denote this data as $^{kg}\mathbf{I}_{[ij]}$ to emphasise that the object is divided into files and that each file has an i and a j index. The first step is for each index $k = \bar{k}$ to read into main memory the N_G files described by $^{\bar{k}g}\mathbf{I}_{[ij]}$. The memory load for each such operation is just a little under 1.2 GB. Then, we simply reorganise the data and write it out into files described not by i and j but now by i and g. In other words, we make the transformation:

$$^{\bar{k}g}\mathbf{I}_{[ij]} \rightarrow {}^{\bar{k}}\left[{}^{g}\mathbf{I}_{i}\right]_{j} \quad \forall \bar{k}\left\{\bar{k} \in \mathbb{N} \middle| \bar{k} \leq N_K\right\} \tag{9.8}$$

This then gives us $N_K \times N_J$ files, each of which we denote by $^{g}\mathbf{I}_{i}$. Taking a typical SATA 3 drive speed of 300 MB/s, we can see that at the time of writing, a fairly standard PC is able to perform this transformation in somewhat less than 3 h.

Having arranged our data in this new form, we will see that the required **I**-to-**S** transformation becomes much easier to implement. At this stage, we need to go back to our derivation of the paraxial **I**-to-**S** transformation for the monochromatic HPO DWDH reflection hologram with a centred camera configuration and written in restricted coordinates (see Equations 8.60 and 8.61). We can generalise this transformation to the full-parallax case and our specific (non-inverting) camera configuration fairly simply*:

$$^{\hat{\mu}\hat{v}}\hat{\mathbf{S}}_{\alpha\beta} = {}^{k}\left[{}^{g}\mathbf{I}_{i}\right]_{j} \quad \forall \hat{v}\left\{\hat{v} \in \mathbb{N} \middle| N_B - \beta + 1 \leq \hat{v} \leq N_B - \beta + N_G\right\}$$

$$\forall \hat{\mu}\left\{\hat{\mu} \in \mathbb{N} \middle| N_A - \alpha + 1 \leq \hat{\mu} \leq N_A - \alpha + N_K\right\} \tag{9.9}$$

$$\forall \alpha\left\{\alpha \in \mathbb{N} \middle| \alpha \leq N_A\right\} \quad \forall \beta\left\{\beta \in \mathbb{N} \middle| \beta \leq N_B\right\}$$

$$= 0 \qquad \text{otherwise}$$

where

$$i = \alpha \quad j = \beta \quad k = \hat{\mu} + \alpha - N_A \quad g = \hat{v} + \beta - N_B \tag{9.10}$$

Given ${}^{k}\left[{}^{g}\mathbf{I}_{i}\right]_{j}$ instead of $^{kg}\mathbf{I}_{[ij]}$, it should be immediately obvious that we can now perform this transformation one β value at a time. In other words, we can work out the required hogel write data one hologram line at a time. Thus, for $\beta = 1$ we read into main memory the N_K RGB files ${}^{k}\left[{}^{g}\mathbf{I}_{i}\right]_{1}$. In our example, these files will take up a memory space of approximately 1.9 GB. We can then perform the transformation (Equations 9.9 and 9.10) in main memory and write out a whole line of hogel data, $^{\hat{\mu}\hat{v}}\hat{\mathbf{S}}_{\alpha 1}$. We then proceed to $j = \beta = 2$ to write out the second line and so forth. Each line of hogels can be expected to take approximately 6 s to calculate with modern SATA disk drives. This means that for RGB hogel write speeds of less than 100 Hz, a single PC is able to cope with performing the required **I**-to-**S** transformation on the image data and in writing the processed data to the printer SLM.

You will note that we have used restricted coordinates to characterise the SLM plane $\mathcal{S}$ as we wish to minimise memory load as much as possible. Anyway, all values of μ and ν outside their restricted limits are zero all the time, so it makes little sense to allow space for them in our calculation. Instead, we use the fixed offset expressions (Equations 8.46 and 8.47) as pointers to our SLM memory map and then employ directly $^{\hat{\mu}\hat{v}}\hat{\mathbf{S}}_{\alpha\beta}$ to update the SLM. Figure 9.1 provides a summary of the two-step **I**-to-**S** transformation and hogel-writing sequence using restricted coordinates.

---

* This transformation is valid for a conjugate SLM position. See Section 8.8.2.1.

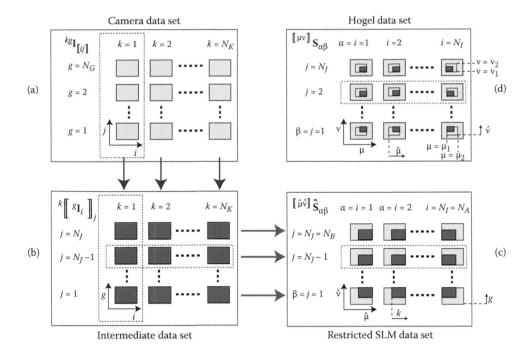

FIGURE 9.1 Illustration of the two-step paraxial objective **I**-to-**S** transformation described by Equations 9.8 and 9.9 using restricted SLM coordinates. This transformation converts camera image data corresponding to a given primary colour to the hogel data required to write a DWDH full-parallax, full-colour reflection hologram. The initial camera data (a) is composed of an $N_K \times N_G$ matrix of standard camera image files, each of dimensions $N_I \times N_J$ collected at the special $n = 1$ plane. The first step in the processing sequence is to load each column of this matrix into main memory and then to write it out to disk in the form of N_J files, each of dimensions $N_I \times N_G$. This process occurs for each of the N_K matrix columns leading to the intermediate $N_K \times N_J$ matrix of files shown in (b). Each file is of dimensions $N_I \times N_G$. The second step is to read each row of the intermediate matrix into main memory. Equations 9.9 and 9.10, with index rules, can then be used simply to populate $^{[\hat{\mu}\hat{v}]}\hat{\mathbf{S}}_{\alpha\beta}$ row by row (c). The SLM defining each hogel can finally be updated (d) using the corresponding file in the restricted SLM data set with the offset rules from Equations 8.46 and 8.47.

In Chapter 8, we introduced a system of coordinates that was even more efficient than the restricted SLM coordinates—this was the centred SLM configuration with its associated data set $\overline{\mathbf{S}}$. To use the data set $\overline{\mathbf{S}}$ to update the real SLM data, $\mathbf{S}$, we need to use variable index offset relations that change from hogel to hogel. As we saw in Chapter 8, it is not usually possible in HPO holography to achieve an alignment of the coordinate meshes, and as such, the centred SLM configuration can reduce image fidelity through the introduction of extra interpolation noise.

In full-parallax holography, the viewing plane is not collocated with the camera plane and as discussed previously, we can usually achieve a coordinate alignment. This makes the centred SLM configuration especially useful as it provides the greatest economy in memory. In its simplest form, we can simply replace Equation 9.9 by a step exactly analogous to Equation 9.8:

$$^{k}\Big[\, ^{g}\mathbf{I}_i \,\Big]_j \rightarrow \, ^{[kg]}\mathbf{I}_{ij} \tag{9.11}$$

In other words, we simply read in for $j = 1$ the N_K files $^{k}\Big[\, ^{g}\mathbf{I}_i \,\Big]_1$ and then write out the N_I files $^{[kg]}\mathbf{I}_{i1}$. We then repeat this exercise for all j. The final data set $^{[kg]}\mathbf{I}_{ij}$ is then actually identical to $\overline{\mathbf{S}}$—that is,

$$^{\overline{\mu v}}\mathbf{S}_{\alpha\beta} = \, ^{kg}\mathbf{I}_{ij} \,\, \forall \overline{\mu} = k, \overline{v} = g, \alpha = i, \beta = j \tag{9.12}$$

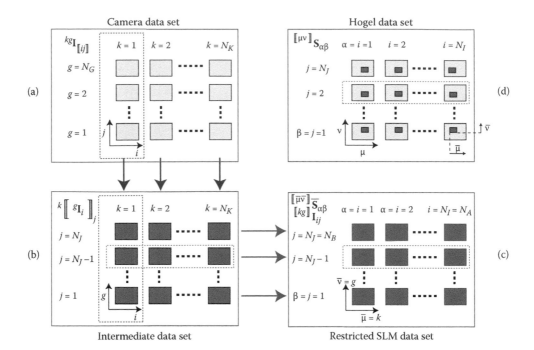

FIGURE 9.2 Illustration of the two-step paraxial objective **I**-to-**S** transformation described by Equations 9.8, 9.11, 9.12 and 9.13 using centred SLM coordinates. This transformation converts camera image data corresponding to a given primary colour to the hogel data required to write a DWDH full-parallax, full-colour reflection hologram. The initial camera data (a) is composed of an $N_K \times N_G$ matrix of standard camera image files, each of dimensions $N_I \times N_J$ collected at the special $n = 1$ plane. The first step in the processing sequence is to load each column of this matrix into main memory and then to write it out to disk in the form of N_J files, each of dimensions $N_I \times N_G$. This process occurs for each of the N_K matrix columns leading to the intermediate $N_K \times N_J$ matrix of files shown in (b). Each file is of dimensions $N_I \times N_G$. The second step is to read each row of the intermediate matrix into main memory and to write it out to disk in the form of $N_I = N_A$ files of dimensions $N_K \times N_G$. This process occurs for each of the $N_J = N_B$ matrix rows leading to the new $N_A \times N_B$ matrix of files $\left({}^{[\overline{\mu v}]}\overline{\mathbf{S}}_{\alpha\beta} = {}^{[kg]}\mathbf{I}_{ij} \right)$ shown in (c). Each file is of dimensions $N_K \times N_G$. The SLM defining each hogel can finally be updated (d) using the corresponding file in the centred SLM data set with the variables offset rules from Equation 9.13.

Therefore, to calculate $\overline{\mathbf{S}}$, all we need to do is to essentially swap two indices of the object **I**—this is done by the two independent transforms (Equations 9.8 and 9.11).

The index relations required to compute **S** from $\overline{\mathbf{S}}$ are those of Equations 8.78 and 8.79. Under the $n = 1$ quantum conditions, these reduce to

$$\mu_1 = 2 - \alpha + \left\| \frac{1}{2}\left(N_A - N_K + N_M - 1\right) \right\| \quad \mu_2 = N_K + 1 - \alpha + \left\| \frac{1}{2}\left(N_A - N_K + N_M - 1\right) \right\|$$
$$\nu_1 = 2 - \beta + \left\| \frac{1}{2}\left(N_B - N_G + N_V - 1\right) \right\| \quad \nu_2 = N_G + 1 - \beta + \left\| \frac{1}{2}\left(N_B - N_G + N_V - 1\right) \right\| \tag{9.13}$$

To avoid truncation errors, one should further ensure that $N_A - N_K + N_M$ and $N_B - N_G + N_V$ are odd numbers. Figure 9.2 illustrates the processing sequence when using the centred SLM configuration.

9.2.3 General Full-Parallax Paraxial Objective Transformations

In cases in which the available camera data does not coincide with any of the quantised values of h, we must use a direct generalisation of the HPO transformations reviewed in Chapter 8. The two-stage process described above can of course be used in this case, only that the second stage transformation is modified from that of Equations 9.9 and 9.10. In terms of restricted SLM coordinates, the appropriate monochromatic **I**-to-**S** DWDH transformation for a centred camera can be written as follows:

$$
{}^{\hat{\mu}\hat{v}}\hat{S}_{\alpha\beta} = {}^{kg}I_{ij} \quad \forall\hat{v}\left\{ \hat{v} \in \mathbb{N} \left\| \frac{D_Y(\hat{N}_V - 1)(N_B - \beta)}{(D_Y + H)(N_B - 1)} \right\| + 1 \le \hat{v} \le 1 + \left\| \frac{D_Y(\hat{N}_V - 1)}{D_Y + H} \left\{ \frac{N_B - \beta}{N_B - 1} + \frac{H}{D_Y} \right\} \right\| \right\}
$$

$$
\forall\hat{\mu}\left\{ \hat{\mu} \in \mathbb{N} \left\| \frac{D_X(\hat{N}_M - 1)(N_A - \alpha)}{(D_X + W)(N_A - 1)} \right\| + 1 \le \hat{\mu} \le 1 + \left\| \frac{D_X(\hat{N}_M - 1)}{D_X + W} \left\{ \frac{N_A - \alpha}{N_A - 1} + \frac{W}{D_X} \right\} \right\| \right\}
$$

$$
\forall\alpha\left\{ \alpha \in \mathbb{N} \middle| \alpha \le N_A \right\} \quad \forall\beta\left\{ \beta \in \mathbb{N} \middle| \beta \le N_B \right\}
$$

$$
= 0 \qquad \text{otherwise}
$$

(9.14)

where

$$
i = N_I - \left\| \frac{(\alpha - 1)(N_I - 1)}{(N_A - 1)} \right\|
$$

(9.15)

$$
j = N_B - \left\| \frac{(\beta - 1)(N_J - 1)}{(N_B - 1)} \right\|
$$

(9.16)

$$
k = 1 + \left\| \frac{(N_K - 1)}{W} D_X \left[\frac{(\hat{\mu} - 1)}{(\hat{N}_M - 1)} \left(\frac{W}{D_X} + 1 \right) + \frac{(\alpha - N_A)}{(N_A - 1)} \right] \right\|
$$

(9.17)

$$
g = 1 + \left\| \frac{(N_G - 1)}{H} D_Y \left[\frac{(\hat{v} - 1)}{(\hat{N}_V - 1)} \left(\frac{H}{D_Y} + 1 \right) + \frac{(\beta - N_B)}{(N_B - 1)} \right] \right\|
$$

(9.18)

This transformation is valid for an inverted camera and a conjugate SLM position and should be used in conjunction with the previous offset Equations 8.46 and 8.47. Using centred SLM coordinates, the corresponding transformation (also pertaining to an inverting camera and a conjugate SLM position) can be written*

$$
{}^{\bar{\mu}\bar{v}}\bar{S}_{\alpha\beta} = {}^{kg}I_{ij} \quad \forall\bar{v}\left\{ \bar{v} \in \mathbb{N} \middle| 1 \le \bar{v} \le 1 + \left\| \frac{H(N_V - 1)}{\Sigma'} \right\| \right\}
$$

$$
\forall\bar{\mu}\left\{ \bar{\mu} \in \mathbb{N} \middle| 1 \le \bar{\mu} \le 1 + \left\| \frac{W(N_M - 1)}{\Pi'} \right\| \right\}
$$

$$
\forall\alpha\left\{ \alpha \in \mathbb{N} \middle| \alpha \le N_A \right\}
$$

(9.19)

$$
\forall\beta\left\{ \beta \in \mathbb{N} \middle| \beta \le N_B \right\}
$$

$$
= 0 \qquad \text{otherwise}
$$

---

* Here we have assumed that $\Pi'/(N_M - 1) = \bar{\Pi}'/(\bar{N}_M - 1)$ and that $\Sigma'/(N_V - 1) = \bar{\Sigma}'/(\bar{N}_V - 1)$.

where

$$
i = N_I - \left\| \frac{(\alpha - 1)(N_I - 1)}{(N_A - 1)} \right\| \qquad j = N_J - \left\| \frac{(\beta - 1)(N_J - 1)}{(N_B - 1)} \right\|
$$

$$
k = 1 + \left\| \frac{\Pi'(N_K - 1)}{W(N_M - 1)}(\bar{\mu} - 1) \right\| \qquad g = 1 + \left\| \frac{\Sigma'(N_G - 1)}{H(N_V - 1)}(\bar{\nu} - 1) \right\|
$$

(9.20)

Once again, the offset index relations required to compute $\mathbf{S}$ from $\overline{\mathbf{S}}$ are those of Equations 8.78 and 8.79.

9.2.4 Non-Paraxial Printer Objectives

By using an intrinsically two-step process, we have seen how a single PC can calculate exact $h = 0$ camera data from a given data set characteristic of a first quantised camera distance. This data can then be naturally adjusted to compensate for the fifth Seidel aberration of a non-paraxial objective using bilinear or bicubic interpolation before writing to the printer SLM. This process allows low-noise digital holograms to be produced efficiently, having an image projection up to a value equal to the first quantised value of h. If larger image projections are required, then one can either decide to go to $n = 2$ or increase the hogel size. Alternatively, one may simply decide to use a non-paraxial general transformation as mentioned in Section 8.10. For example, Expressions 8.154 to 8.156 may be generalised to the full-parallax case by rewriting Equation 8.154 as

$$
{}^{\mu\nu}\mathbf{S}_{\alpha\beta} = {}^{kg}\mathbf{I}_{ij} \quad \forall\mu\left\{\mu \in \mathbb{N} \middle| N_{U1}(\alpha) \le \mu \le N_{U2}(\alpha)\right\} \qquad \forall\alpha\left\{\alpha \in \mathbb{N} \middle| \alpha \le N_A\right\}
$$

$$
\forall\nu\left\{\nu \in \mathbb{N} \middle| N_{V1}(\beta) \le \nu \le N_{V2}(\beta)\right\} \qquad \forall\beta\left\{\beta \in \mathbb{N} \middle| \beta \le N_B\right\}
$$

(9.21)

$$
= 0 \qquad \text{otherwise}
$$

and introducing another "g" equation:

$$
g = 1 + \left\| \frac{(N_G - 1)}{H}\left\{ \frac{D_Y + H}{2} + \frac{(2\nu - N_V - 1)\mathcal{K}(\hat{\rho})}{(N_V - 1)} h \tan\left(\frac{\Psi_{PV}}{2}\right) + \frac{(\beta - N_B)D_Y}{N_B - 1} \right\} \right\|
$$

(9.22)

As before, the two-step method then applies. We shall see in Chapter 10 that even physical camera data can nearly always be processed to produce the required data set defined in Section 9.2.1. As we have already mentioned, a great advantage of exact transformations such as Equations 9.9 and 9.10, or their higher-order equivalents is that each file ${}^{[\bar{\mu}\bar{\nu}]}\hat{\mathbf{S}}$ can be corrected efficiently for printer objective distortion using bilinear or bicubic interpolation.

9.2.5 Larger Holograms

The two-step process that we have described above for generating the image data required for DWDH full-parallax holograms can be applied to holograms of reasonable size. As we have seen, a high-resolution (0.5 mm-hogel) 40 cm × 30 cm DWDH hologram with a typical window requires around a 3 h preprocessing cycle and a 1.5 h write cycle. A 64-bit PC with approximately 6 GB of main memory would be recommended for this type of task. Larger DWDH holograms can increase the memory requirement quite fast however. For example, a 1 m × 1 m, 0.5 mm-hogel DWDH three-colour reflection hologram with a maximum (unapodised) window assuming a horizontal printer FOV of 85° and a printer SLM with pixel dimensions of 1280 × 1024 would entail loading *step 1* files into the main memory amounting to a total size of around 15 GB. The *step 2* files would also be approximately of this size. Although it is just about possible for 64-bit PCs to cope with these numbers, the disk read times do become rather long.

For larger holograms, it therefore makes sense to divide the hologram into smaller zones and to calculate the data required for each zone separately. The simplest way to do this is to divide the hologram into several vertical slices. Because the two-stage calculation process that we have described works by hologram lines, the size of the *step 1* files (assuming an $n = 1$ camera distance) is $3N_A N_B N_G$ and the size of the *step 2* files is likewise $3N_A N_G N_K$.* By reducing N_A, we therefore reduce both the *step 1* and *step 2* file sizes. Of course, we are at liberty to define a two-step process which works by columns rather than lines; in this case, we would then seek to reduce N_B rather than N_A. In either case, each slice of the hologram can now be calculated by a different PC. Not only is the memory requirement of that PC reduced linearly with the number of slices employed, but also the total time taken to calculate the data is also reduced linearly. There are two relatively minor complications to this method. The first is that the original data set must be made available to all PCs, and the second is that once each PC has calculated its slice of the hologram, these slices usually have to be integrated so that the printer is able to print a full line of hogels.

For extremely large and wide-angle full-parallax holograms with small hogel sizes, the sliced two-step technique that we have just described may nevertheless become inadequate. In this case, we must adopt the technique of dividing both *the hologram and its viewing window* into individual equal areas. As before, the original data must now be distributed to all PCs in the network and each PC is then responsible for calculating its own partial hogel set using the standard line-based two-step technique described in Section 9.2.2. After calculation, a given PC will contain partial hogel data for a section of the hologram and corresponding only to a certain viewing zone. The data from the different PCs must therefore be integrated by the printer before hogel lines may be printed.

9.2.6 Rectangular Viewing Windows

We saw in Chapter 8 that writing an HPO digital hologram with a well-defined rectangular viewing window made good sense for a variety of reasons. The most important of these reasons is that an HPO hologram has, by definition, a preferred viewing distance as the camera plane and viewing plane are collocated. Only at this plane will the hologram look truly "right". It therefore makes sense to define a rectangular window at this unique distance such that an observer here will see the whole image or nothing—depending on whether he or she is laterally in or out of the defined viewing zone. This is just good sense as any hologram that emits over a wide solid angle will be dimmer than one of similar diffractive properties but which focusses the light into a well-defined rectangular window.

With full-parallax holograms, this logic remains only partly valid. Certainly, if we spread the diffracted light too widely over a solid angle, we are still going to end up with a dim hologram. However, defining a window at a certain fixed distance from the hologram can lead to problems with viewing. This is illustrated in Figure 9.3, in which it is clear that there are zones where only a partial image of the hologram is available. In HPO holography, one is always near the special camera/viewing plane and so these partial image zones are not so important: you just cannot go up to a 1 m × 1 m HPO hologram and look at it at a 30 cm distance when it has been written for viewing at 1 m. Because of the lack of a vertical parallax, the image simply appears to disintegrate. However, this is just the advantage of the full-parallax hologram! The image is designed for viewing over a wide range of distances.

The upshot of all this is that one has to be more careful with window definitions in full-parallax display holography than with HPO holography. It can therefore be desirable to use a rectangular window at a distance that is relatively large compared with the hologram size. If we elect not to use a window at all and simply use the full SLM field for every hogel, then this is equivalent to a rectangular window at infinity. However, this will often produce too dim a hologram[†] and we need to at least restrict the vertical

---

* This is true even for the case of an unrestricted viewing window where we use all the printer SLM—in this case, however, the files $^{kg}I_{[ij]}$ are not all the same size and in general N_I and N_J become functions of k and g.

[†] This depends on the type of photosensitive material used and the type of illumination.

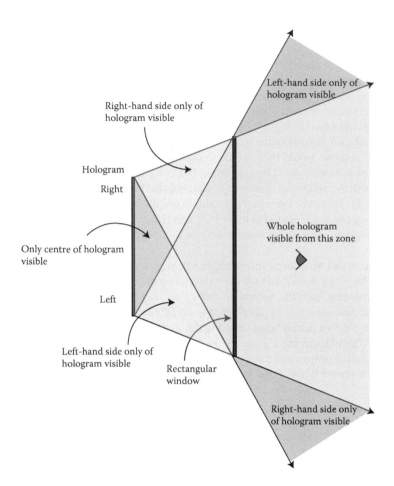

FIGURE 9.3 Plan view of a hologram made with a defined rectangular window. Note that an observer who is not at the same distance as the window location can find himself or herself in a zone where only part of the holographic image is visible.

field. We may therefore sometimes wish to use an astigmatic window in which the horizontal focus is at a greater distance than the vertical focus.

Another very important point to remember is that we do not always require the window distance in full-parallax holography to coincide with the camera plane. The most usual reason for this is that if one is forced to use a small camera distance by the quantisation rules, then the window will be just too close to the hologram. In fact, this is frequently the case with full-parallax digital holograms.

An alternative to using the entire SLM for each hogel is to use only a constant restricted area in its centre. Figure 9.4 illustrates the viewing geometry for this case. Mathematically, this case is equivalent to the largest possible infinitely distant rectangular window we could generate using the restricted field of view corresponding to the restricted SLM area.

9.2.6.1 General Astigmatic Rectangular Viewing Window

We shall now discuss how to define a general astigmatic rectangular viewing window situated at a distance that is greater than the camera plane. As we mentioned previously, this is frequently the most desired method to compensate for a small camera distance imposed by quantisation. Figure 9.5 shows a plan view of the ray geometry for a DWDH hologram with a centred camera and centred SLM

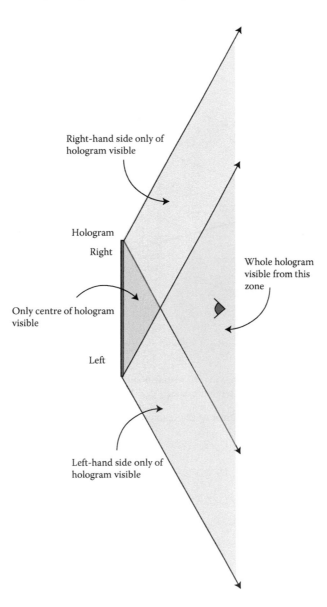

FIGURE 9.4 Plan view of a hologram with an unrestricted viewing angle. Note that an observer can still find himself or herself in a zone where only part of the holographic image is available.

configuration incorporating a rectangular viewing window at a distance $h_{w1} \neq h$. As in Chapter 8, the coordinates ξ and $\bar{U}'$ can be defined to be the same as long as we choose

$$D_\xi = W \frac{h}{h_{w1}} + D_X \left(1 - \frac{h}{h_{w1}} \right) \tag{9.23}$$

Likewise, if we define the vertical window distance as h_{w2}, then the corresponding vertical coordinates ζ and $\bar{V}'$ can be defined to be the same as long as

$$D_\zeta = H \frac{h}{h_{w2}} + D_Y \left(1 - \frac{h}{h_{w2}} \right) \tag{9.24}$$

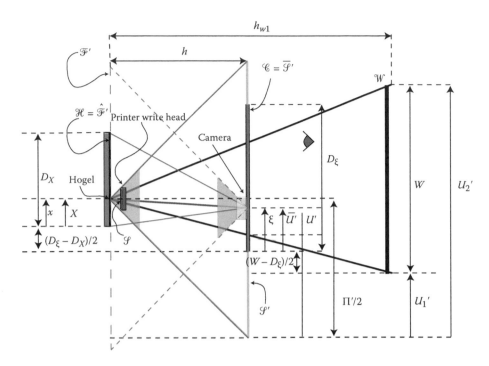

FIGURE 9.5 Plan view of ray geometry showing the writing of a DWDH hologram with a rectangular window at a distance $h_{w1} > h$. Coordinates corresponding to a centred camera $(x = X)$ and to a centred SLM $(\xi = \bar{U}')$ are used.

There is, however, a sizeable difference to the transformation of Equations 9.19 and 9.20. Defining

$$
\begin{aligned}
{}^1D_\xi &\equiv D_\xi/2 - (W + D_X)h/(2h_{w1}) + D_X/2 \\
{}^2D_\xi &\equiv D_\xi/2 + (W + D_X)h/(2h_{w1}) - D_X/2 \\
{}^1D_\zeta &\equiv D_\zeta/2 - (H + D_Y)h/(2h_{w2}) + D_Y/2 \\
{}^2D_\zeta &\equiv D_\zeta/2 + (H + D_Y)h/(2h_{w2}) - D_Y/2
\end{aligned}
\tag{9.25}
$$

then only within the limits

$$
\begin{aligned}
{}^1D_\xi &\le \bar{U}' \le {}^2D_\xi \\
{}^1D_\zeta &\le \bar{V}' \le {}^2D_\zeta
\end{aligned}
\tag{9.26}
$$

will the projected centred film plane, $\hat{\bar{\mathscr{F}}}$, be congruent to the hologram plane, $\mathscr{H}$. In other words, only in this region is $D_x = D_X$ and $D_y = D_Y$. Outside this region, either $D_x < D_X$, $D_y < D_Y$ or both. In fact, the camera plane is, in general, split up into nine different zones as illustrated in Figure 9.6.

There are two ways to proceed with calculating ${}^{\overline{\mu v}}\mathbf{S}_{\alpha\beta}$ from ${}^{kg}\mathbf{I}_{ij}$. The first is a simple brute force method where we use a data set ${}^{kg}\mathbf{I}_{[ij]}$ in which the files $\mathbf{I}_{[ij]}$ all have dimensions of $N_I \times N_J$ and all follow the usual definition of a centred camera, which states that the film plane $\hat{\bar{\mathscr{F}}}$ is congruent to $\mathscr{H}$. We then impose the following window limits on $\bar{\mu}$ and $\bar{v}$:[*]

$$
\bar{\mu}_1 = 1 + \left\| \frac{(N_M - 1)}{\Pi'} \left\{ \left(\frac{h_{w1} - h}{h_{w1}} \right) \left[\frac{\alpha - 1}{N_A - 1} D_X + \frac{W - D_X}{2} \right] - \frac{(W - D_\xi)}{2} \right\} \right\|
\tag{9.27}
$$

---

[*] Here we have assumed that $\Pi'/(N_M - 1) = \bar{\Pi}'/(\bar{N}_M - 1)$ and that $\Sigma'/(N_V - 1) = \bar{\Sigma}'/(\bar{N}_V - 1)$.

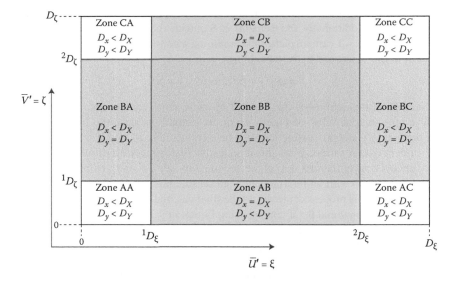

FIGURE 9.6 The camera plane divides into nine zones when a rectangular viewing window is not in the same plane. A camera in zone BB is required to define (projected) image data over the entire hologram surface. However, a camera in the other eight zones is only required to define image data over a partial region.

$$\bar{\mu}_2 = 1 + \left\| \frac{(N_M - 1)}{\Pi'} \left\{ \left(\frac{h_{w1} - h}{h_{w1}} \right) \left[\frac{\alpha - 1}{N_A - 1} D_X - \frac{W + D_X}{2} \right] + \frac{(W + D_\xi)}{2} \right\} \right\| \tag{9.28}$$

$$\bar{v}_1 = 1 + \left\| \frac{(N_V - 1)}{\Sigma'} \left\{ \left(\frac{h_{w2} - h}{h_{w2}} \right) \left[\frac{\beta - 1}{N_B - 1} D_Y + \frac{H - D_Y}{2} \right] - \frac{(H - D_\zeta)}{2} \right\} \right\| \tag{9.29}$$

$$\bar{v}_2 = 1 + \left\| \frac{(N_V - 1)}{\Sigma'} \left\{ \left(\frac{h_{w2} - h}{h_{w2}} \right) \left[\frac{\beta - 1}{N_B - 1} D_Y - \frac{H + D_Y}{2} \right] + \frac{(H + D_\zeta)}{2} \right\} \right\| \tag{9.30}$$

We can therefore write the **I**-to-**S** transformation between $\bar{\mathbf{S}}$ and **I** as

$$^{\bar{\mu}\bar{v}}\bar{\mathbf{S}}_{\alpha\beta} = {}^{kg}\mathbf{I}_{ij} \qquad \forall \bar{\mu}\left\{ \bar{\mu} \in \mathbb{N} \middle| \bar{\mu}_1 \leq \bar{\mu} \leq \bar{\mu}_2 \right\} \qquad \forall \alpha \left\{ \alpha \in \mathbb{N} \middle| \alpha \leq N_A \right\}$$

$$\forall \bar{v}\left\{ \bar{v} \in \mathbb{N} \middle| \bar{v}_1 \leq \bar{v} \leq \bar{v}_2 \right\} \qquad \forall \beta \left\{ \beta \in \mathbb{N} \middle| \beta \leq N_B \right\} \tag{9.31}$$

$$= 0 \qquad \text{otherwise}$$

where $\bar{\mu}_1$, $\bar{\mu}_2$, $\bar{v}_1$ and $\bar{v}_2$ are given by Equations 9.27 to 9.30 and where (for the case of an inverting camera and conjugate SLM position)

$$i = N_I - \left\| \frac{(\alpha - 1)(N_I - 1)}{(N_A - 1)} \right\| \qquad j = N_J - \left\| \frac{(\beta - 1)(N_J - 1)}{(N_B - 1)} \right\|$$

$$k = 1 + \left\| \frac{\Pi'(N_K - 1)}{W(N_M - 1)} (\bar{\mu} - 1) \right\| \qquad g = 1 + \left\| \frac{\Sigma'(N_G - 1)}{H(N_V - 1)} (\bar{v} - 1) \right\| \tag{9.32}$$

Before updating the SLM with $\overline{\mathbf{S}}$, and prior to a hogel-write operation, the index rules (Equations 8.78 and 8.79) must be used. Additionally, the elements of $\mathbf{S}$ outside the area specified by these indices must be blanked. If we use the $n = 1$ camera plane (and assuming a non-inverting camera and $N_I = N_A$ and $N_J = N_B$), then Equation 9.32 is reduced to the trivial form

$$i = \alpha \quad j = \beta \quad k = \overline{\mu}' \quad g = \overline{\nu}' \tag{9.33}$$

The disadvantage of this scheme is that we end up using more camera data than we actually need. This is because only in zone "BB" of Figure 9.6 do we actually require the full film-plane data. Not surprisingly, the memory and processing load can be significantly reduced if we only process the strict minimum of camera data required. What this amounts to is starting with different-sized files ${}^{kg}\mathbf{I}_{[ij]}$ for each k and g. When we apply the first step of the general two-step $\mathbf{I}$-to-$\mathbf{S}$ transformation as illustrated in Figure 9.2, we then end up with an intermediate data set that now also contains files of differing sizes as its matrix elements. This process continues to the final calculated data set, ${}^{[kg]}\mathbf{I}_{ij} = {}^{[\overline{\mu}\overline{\nu}]}\overline{\mathbf{S}}_{\alpha\beta}$, where the size of each file depends on α and β. Of course, using Equation 9.31 directly gets one to the same result by calculating and then just ignoring the redundant data.

We shall now describe briefly the steps required in an optimised algorithm. As before, we use a centred camera and centred SLM. We use a camera distance, h, given by the $n = 1$ constraints. The camera plane is defined by Equation 9.23 and 9.24 in terms of the desired astigmatic window $W \times H$. Assuming that we are going to generate the required 3D image data using a computer modelling program, our very first step must be to define the camera geometry. In zone BB of Figure 9.6, the camera constitutes a completely standard centred configuration for $h_{n=1}$ as summarised in Section 9.2.1. In particular, $\hat{\mathcal{F}}$ is congruent to the hologram plane, $\mathcal{H}$, and each rendered film frame must be of integer dimensions $N_I = N_A$ and $N_J = N_B$. In this way, each film frame pixel corresponds to a partner hogel.

In zones other than zone BB, only a subset of the film plane needs be populated with image data. The reason for this is apparent from Figure 9.5, in which it can be seen that from certain camera positions, there are simply no rays connecting the camera to some portions of the hologram plane. Figure 9.7 shows a labelling scheme for the image subset characteristic of the nine camera plane zones. For a general zone ZZ, we require image data only in the region

$$
\begin{aligned}
{}^{1ZZ}N_I \geq i' \geq {}^{2ZZ}N_I \\
{}^{1ZZ}N_J \geq j' \geq {}^{2ZZ}N_J
\end{aligned}
\tag{9.34}
$$

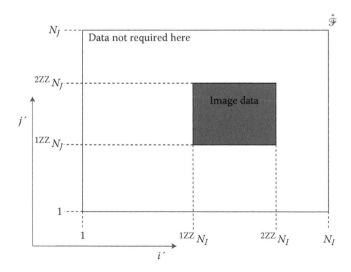

FIGURE 9.7   This diagram shows the centred film plane, $\hat{\mathcal{F}}$, corresponding to a camera located within a general zone, ZZ, of Figure 9.6. The yellow rectangle indicates the required subsection of $\hat{\mathcal{F}}$ that is required to be populated with image data. The white region of $\mathcal{F}$ is not required and so does not need to be stored on disk.

where i' and j' are the standard projected centred film coordinates. These integer values can be calculated by examining the limiting rays as illustrated by the example in Figure 9.8 for zones AA, AB and AC. From this diagram, we see that

$$Y_{max} = \frac{h_{2w}}{h_{2w} - h}\zeta \tag{9.35}$$

This limiting value corresponds to the integer coordinate

$$j' = 1 + \left(\frac{g-1}{N_G - 1}\right)\frac{D_\zeta h_{2w}}{D_Y\left(h_{2w} - h\right)}\left(N_B - 1\right) \tag{9.36}$$

Using this and similar arguments for the other zones and the horizontal geometry, we are led to the following results:

$$^{1AZ}N_J = {}^{1BZ}N_J = 1$$

$$^{1CZ}N_J = 1 + \left(N_B - 1\right)\left\{1 - \frac{D_\zeta}{D_Y}\left(\frac{N_G - g}{N_G - 1}\right)\frac{h_{2w}}{\left(h_{2w} - h\right)}\right\}$$

$$^{2AZ}N_J = 1 + \left(\frac{g-1}{N_G - 1}\right)\frac{D_\zeta h_{2w}}{D_Y\left(h_{2w} - h\right)}\left(N_B - 1\right) \tag{9.37}$$

$$^{2BZ}N_J = {}^{2CZ}N_J = N_B$$

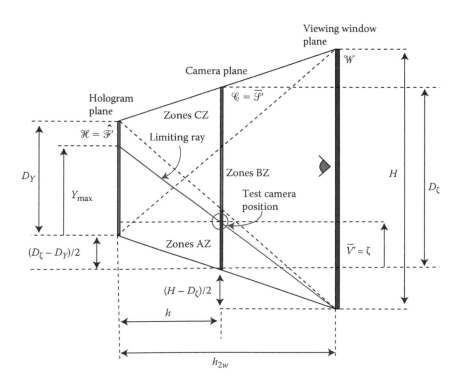

FIGURE 9.8 Side view of the ray geometry showing the limiting ray for a given value of g.

$$^{1ZA}N_I = {}^{1ZB}N_I = 1$$

$$^{1ZC}N_I = 1 + \left(N_A - 1\right)\left\{1 - \frac{D_\xi}{D_X}\left(\frac{N_K - k}{N_K - 1}\right)\frac{h_{1w}}{\left(h_{1w} - h\right)}\right\}$$

$$^{2ZA}N_I = 1 + \left(\frac{k-1}{N_K - 1}\right)\frac{D_\xi h_{1w}}{D_X\left(h_{1w} - h\right)}\left(N_A - 1\right)$$

$$^{2ZB}N_I = {}^{2ZC}N_I = N_A$$

(9.38)

With the inequalities (Equation 9.34), these labels now define the subset of the centred film plane (i', j') required for each camera shot (k,g). This information may be programmed into a virtual camera in computer modelling software so as to produce image files of only the correct subset of the full centred camera frame.

Having rendered the data, we must then apply the general two-step processing scheme. The easiest way to do this is to define a cubic array in the main memory, which we can label $\mathcal{A}(g,i',j')$.* We start with $k = 1$ and read into memory the files $^{1g}\mathbf{I}_{[i'j']}$ for all g. Now for every g outside of zone BB, the size of the file will not be $N_A \times N_B$ and so we will need to use Equation 9.37 to calculate the correct location within the $\mathcal{A}(g,i',j')$ to place each file. In general, we will load the files according to the rule

$$\mathcal{A}\left(i' + {}^{1ZZ}N_I(k) - 1, j' + {}^{1ZZ}N_J(g) - 1, g\right) = {}^{kg}\mathbf{I}_{i'j'}$$

(9.39)

with

$$i' \le {}^{2ZZ}N_I - {}^{1ZZ}N_I + 1$$
$$j' \le {}^{2ZZ}N_J - {}^{1ZZ}N_J + 1$$

(9.40)

Note that we have purposely labelled the camera zones using a two-letter label "ZZ" as the first label is defined by the index g and the second by k. Given g and k, one therefore knows the label "ZZ" and then Equations 9.37 and 9.38 may be used to calculate the various ^{ZZ}N values in Equations 9.39 and 9.40. This method does leave some blank spaces in the memory but because the array is overwritten by the next value of k, this is not a major concern. Of course, one may define a complex non-cubic array structure using Equations 9.37 and 9.38, but this is usually unnecessary.

Having now entered data into memory, we can simply write out the required intermediate files ${}^{k}\left[{}^{g}\mathbf{I}_{i'}\right]_j$. Like the original film files, these files are of variable size depending on the indices k and j. If we define a virtual plane $\mathcal{J}$ as representing the plane spanned by the coordinates g and i or, equivalently, ζ and x, then we can draw a similar diagram to Figure 9.7 for these coordinates, but this time showing the subspace (i,g) instead of (i,j). This is shown in Figure 9.9, where we introduce the new quantities $^{1ZZ}N_G$ and $^{2ZZ}N_G$, which are simply the inverse functions of $^{1ZZ}N_J$ and $^{2ZZ}N_J$:

$$^{1ZZ}N_G = 1 + \left(\frac{j'-1}{N_B - 1}\right)\frac{D_Y\left(h_{2w} - h\right)}{D_\zeta h_{2w}}\left(N_G - 1\right)$$

$$^{2ZZ}N_G = N_G + \left(N_G - 1\right)\frac{D_Y}{D_\zeta}\left(\frac{j' - N_B}{N_B - 1}\right)\frac{\left(h_{2w} - h\right)}{h_{2w}}$$

(9.41)

We then use the rule

$${}^{k}\left[{}^{g}\mathbf{I}_{i'}\right]_{j'} = \mathcal{A}\left(g + {}^{1ZZ}N_G(j') - 1, i' + {}^{1ZZ}N_I(k) - 1, j'\right)$$

(9.42)

---

* This is simply a more convenient notation for $\mathcal{A}_{gi'j'}$.

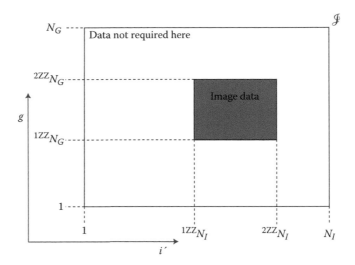

FIGURE 9.9 This diagram shows the virtual (x,ζ) plane, $\mathcal{J}$, corresponding to a camera located within a general zone, ZZ, of Figure 9.6. The yellow rectangle indicates the required subsection of $\mathcal{J}$ that is required to be populated with image data. The white region of $\mathcal{J}$ is not required and so does not need to be stored on disk.

with

$$i' \le {}^{2ZZ}N_I - {}^{1ZZ}N_I + 1$$
$$g \le {}^{2ZZ}N_G - {}^{1ZZ}N_G + 1 \tag{9.43}$$

to define all the intermediate files for $k = 1$. This process is then repeated for all remaining values of k, at which point we will have a full intermediate data set on disk.

The next stage is to define a new array, which we can label $\mathcal{B}(k,g,i')$. We load all $j' = 1$ files into this array using the rule

$$\mathcal{B}\left(k + {}^{1ZZ}N_K(i') - 1, g + {}^{1ZZ}N_G(j') - 1, i'\right) = {}^{k}\left[{}^{g}\mathbf{I}_{i'}\right]_{j'} \tag{9.44}$$

with

$$k \le {}^{2ZZ}N_K - {}^{1ZZ}N_K + 1$$
$$g \le {}^{2ZZ}N_G - {}^{1ZZ}N_G + 1 \tag{9.45}$$

Here, we have defined the inverse functions

$${}^{1ZZ}N_K = 1 + \left(\frac{i'-1}{N_A-1}\right)\frac{D_X(h_{1w}-h)}{D_\xi h_{1w}}(N_K - 1)$$

$${}^{2ZZ}N_K = N_K + (N_K - 1)\frac{D_X}{D_\xi}\left(\frac{i'-N_A}{N_A-1}\right)\frac{(h_{1w}-h)}{h_{1w}} \tag{9.46}$$

Finally, we can write out the final data set $^{[\overline{\mu\nu}]}\overline{S}_{\alpha\beta} = {}^{[kg]}I_{i'j'}$ using the rule

$$^{[\overline{\mu\nu}]}\overline{S}_{\alpha\beta} = {}^{[kg]}I_{i'j'} = \mathcal{B}(k + {}^{1ZZ}N_K(i') - 1, g + {}^{1ZZ}N_G(g) - 1, i') \qquad (9.47)$$

with

$$\begin{aligned} k &\leq {}^{2ZZ}N_K - {}^{1ZZ}N_K + 1 \\ g &\leq {}^{2ZZ}N_G - {}^{1ZZ}N_G + 1 \end{aligned} \qquad (9.48)$$

Upon repeating this last cycle for each value of the index j', we will have completed the two-step optimised **I**-to-**S** transformation.

9.3 Resolution Requirements of DWDH Reflection Holograms

In Chapter 11, we shall review various effects that limit the practical resolution of display holograms. In particular, we shall see that the four main effects that limit the resolution of DWDH reflection holograms are source-size blurring (see Section 11.2.2 and 11.11), chromatic blurring (see Sections 11.2.3 and 11.11), speckle (see Section 11.3) and digital diffractive blurring (see Section 11.11.1).* Source-size blurring is due to the intrinsic angular size of the illuminating source and, from a practical point of view, is not really affected by realistic emulsion characteristics. Chromatic blurring is due to the finite bandwidth of the illuminating source but is strongly controlled by emulsion thickness. Speckle is caused by too small a bandwidth in the illuminating source. Finally, digital diffractive blurring is caused by hogels that are generally small and thus display inherent diffraction.

We shall see in Chapter 11 that the human eye ultimately controls the best observable resolution of any display hologram. Although this varies from person to person, from a practical point of view, we can say that the human eye can just resolve two lines spaced around 1 mm from the other at a distance of 1 m. This corresponds to an angle of $\Delta\theta \sim 0.06°$. When we use an illuminating light source having too large a physical diameter, too broad a spectrum or too narrow a spectrum, the resolution of the hologram will fall below the limit discernible by the human eye. However, when we choose the illumination source correctly, the observable resolution of the hologram will only be determined by the resolution of the human eye. We shall now examine the values of various critical digital parameters that are required to meet this most stringent limit.

We shall start by making the observation that if the hologram is to be viewed from a distance of more than 1 m, then a hogel size of less than 1 mm will not be resolvable by the average human eye. With HPO holography, in which the camera plane and viewing plane are collocated, we can make use of this fact because, effectively, there is no point in using a smaller hogel diameter than is resolvable by the average human eye at the viewing plane. After all, when one observes an HPO hologram from significantly closer than the viewing plane, the image effectively falls to pieces. Seeing that one seldom wants to make the viewing distance less than approximately 40 cm in HPO holography, a hogel size of approximately 0.5 mm is usually adequate in any situation.

The situation with full-parallax holograms is, however, completely different because these types of holograms can be viewed without distortion at any distance. One therefore often wants to reduce, as much as possible, the hogel size in full-parallax holography. Unfortunately, as we shall see in Chapter 11, digital diffractive blurring increases as the hogel size is diminished. For instance, a square hogel 0.1 mm in diameter will exhibit a diffractive blurring angle of approximately four times greater than the resolution angle of the human eye. The hogel size starts to limit the optimal resolution at diameters

---

* This ignores insufficiencies in the perspective image data, printer SLM insufficiency, aberrations induced by chemical processing and printer optical aberrations.

of approximately 0.5 mm; below this value, one simply has to choose between good resolution close-up to the hologram at the expense of induced blurring at an ever-diminishing depth. This is a fundamental constraint of DWDH holography because the hologram is composed of a matrix of hogels that have all been recorded incoherently.* As such, a full-colour, full-parallax DWDH hologram will never quite have the resolution of a high-quality mirror or a physical glass window. In these cases, almost perfect images of zero to infinite depth may be seen at all distances. In the case of the DWDH hologram, which displays essentially perfect images over a range of viewing distances from infinity down to around half a metre, observation at a closer distance will inevitably reveal a substandard resolution at the hologram surface. Nevertheless, such holograms are capable of producing extremely impressive displays when illuminated with the proper illumination source. One should also point out that too large a hogel size produces a loss of resolution, predominantly at the hologram surface. When one focuses on a distant object with one's head close to the hologram, then the finite size of the hogels simply acts as an out-of-focus grill through which an effectively perfect image is seen. By paying close attention to how the hogels are written, this "grill" may be greatly minimised.

In addition to controlling the hogel size, the resolution of the human eye also puts a constraint on the number of camera views, N_K and N_G, and on the dimensions of the printer SLM, N_M and N_V. If the camera plane is taken at an h value of 1 m, then clearly we will require a spacing between camera positions of a little less than 1 mm to accommodate the average human eye. The maximum values of N_K and N_G are therefore given by dividing the window size by this distance. For a metre square hologram with a rectangular camera plane at a distance of 1 m and of a size of say 1 m × 2 m, this then equates to a camera array of $N_K \times N_G = 2000 \times 1000$ positions.

The number of pixels required of the printer SLM is similarly constrained. At a distance of 1 m, the SLM should project an image of itself with a characteristic spacing of a little less than 1 mm. In other words, two adjacent pixels should be separated by a little less than 0.06°. Given that we usually want a total horizontal field of view of around 100°, we can estimate that N_M should ideally be around 1700.

Of course, when we design a hologram to be illuminated by a light that has a greater than ideal source size or a greater than ideal bandwidth, we can relax the amount of digital data required. Values for N_K and N_G and also for the required pixel dimensions of the SLM must then be calculated using the formulae for induced blurring, which we cover in Chapter 11.

9.4 DWDH Transmission Holograms

The most successful type of ultra-realistic full-colour digital display hologram in commercial production today is the reflection hologram. Various companies including the US company Zebra Imaging Inc., the Canadian company XYZ Imaging Inc., and the Lithuanian firm Geola UAB, have all produced high-resolution large-format full-colour holograms showing many metres of image depth. On the other hand, the older large-format laser transmission holograms created by the likes of Nick Phillips and Paula Dawson produced perhaps even more stunning and deeper images using monochromatic analogue holography.

With the advent of modern laser diodes and light-emitting diodes, digital laser transmission holography now constitutes a promising and yet largely unexplored field potentially capable of producing spectacular full-colour displays. The digital image processing that we have reviewed in this chapter and in Chapter 8 is exactly the same for the monochromatic DWDH reflection hologram as it is for the full-aperture monochromatic DWDH transmission hologram. The only difference is that the reference beam must be on the same size as the object beam in the transmission scenario and the appropriate equations relevant to transmission holograms derived in Chapter 11 should be used for image predistortion.

---

* It is worth noting here that, in principle, one can record the hogels coherently by ensuring that the printer is interferometrically stable. This usually requires a slow step-and-repeat system incorporating fringe-locking technology. For most practical display applications, the recording of such mutually coherent hogels is just too slow.

The great advantage of the transmission hologram over the reflection hologram is that the illumination light is behind the hologram rather than in front of it. By using a compact recording geometry in which the illumination light is injected into the hologram at an angle greater than the critical angle,* it is possible to make sure that no zeroth order light is transmitted by the hologram on replay. This means that a hologram may be illuminated by laser and yet no harmful reference beam makes it through the hologram. The fact that the geometry is compact further means that a compact rear-illuminated display can be practically constructed displaying near-perfect image quality. This is likely to be a deciding factor in future applications such as integrated holographic windows, walls and floors.

Recording full-colour digital images using transmission holography of course requires a slightly different technique from that used in reflection holography. Because transmission holograms of a "normal" thickness are much less discriminating in wavelength than reflection holograms, the technique of using collinear RGB reference recording and reconstruction beams leads simply to multiple parasitic images with transmission holography. In Chapter 11, however, we shall show that a typical transmission hologram recorded on a typical silver halide emulsion of 7 µm thickness can be expected to exhibit a reference angle discrimination of approximately 15°. This is sufficiently small to allow the different coloured reference beams at recording and illumination to be arranged at differing altitudinal angles. Alternatively, the reference beams may be arranged at substantially differing azimuthal angles.

Transmission holography is likely to represent the technique of choice for future real-time holographic displays. Reflection holography has two disadvantages: first, the reference beam is in front of the hologram and so one cannot construct a fully integrated unit, and second, the reflection grating is intrinsically thick. Future submicron real-time SLM displays might realistically be expected to simply replace the fixed emulsion in transmission holographic technology, effectively converting the ultra-realistic static holographic display into a true 3D holographic television display. Colour control could then be performed using time multiplexing.

9.5 MWDH Reflection Holograms

In this chapter, we have concentrated on the DWDH technique and in particular on computational methods for preparing the required image data for this type of hologram. Arguably, DWDH is the most appropriate technique to write large-format full-parallax holograms. The reason for this is that one only needs a small laser to write each small hogel and it is easy to build such lasers to operate at relatively high repetition rates. We have seen that copying full-colour DWDH reflection holograms can be done effectively in some cases using a quasi-contact method—but at the time of writing, there are still some problems in commercialising such copying technology. As a result, DWDH holograms remain expensive.

Both full-parallax and HPO digital holograms may be generated via the MWDH technique. Before finishing this chapter, we will therefore say a few words about MWDH holograms for completeness. There are advantages and disadvantages to the MWDH technique. First, if a large hologram is to be produced, then a large RGB copy laser is required to convert the MWDH hologram to the final white light-viewable reflection hologram. Unfortunately, such lasers, although possible to build, are still very expensive at the time of writing. The speed at which copies can be produced is, however, usually far superior to DWDH combined with a contact-copy scenario. Furthermore, with MWDH, as we saw in Chapter 8, no computational change to the image plane is required—thus camera data can be used immediately in the printer without lengthy processing.

In small to medium-sized holograms, MWDH can indeed sometimes offer a superior solution over DWDH because a large costly RGB laser is not required and the intrinsic distance-copy process of

---

* This is usually arranged by writing the DWDH hologram using a large reference angle and then inducing an emulsion shrinkage upon processing. Replay of the hologram then requires side-illumination with the reference replay angle greater than that required for total internal reflection.

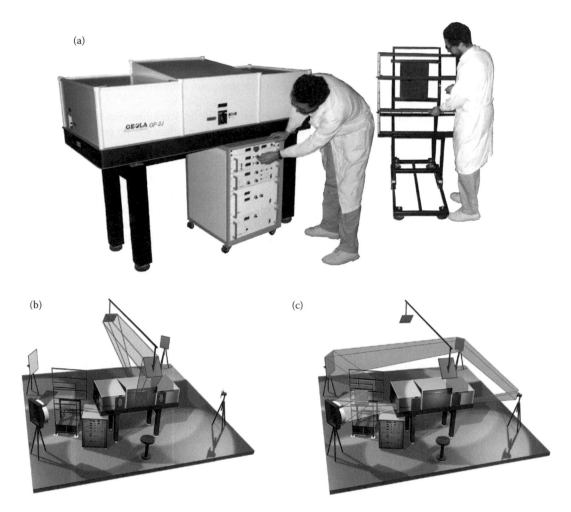

FIGURE 9.10 Photograph of a commercial large-format analogue holographic camera (made by Geola Digital UAB). The system (a), which is based on a 2J Nd:glass laser emitting at 526.5 nm, is capable of recording analogue master holograms as well as producing monochromatic reflection, achromatic or rainbow transfers of analogue or MWDH masters to an H_2 copy size of up to 0.8 m × 1.0 m. Diagrams (b) and (c) illustrate the operation. The creation of an analogue transmission H_1 master hologram is shown in (b). A full-aperture copy of an H_1 to a monochromatic reflection H_2 is shown in (c). Geola makes larger models based on 8J lasers capable of producing H_2 copies to 1.0 m × 1.5 m. (Photographs courtesy of Geola Digital UAB.)

MWDH offers more control over the quality and brightness of the final H_2 hologram.* Large-format MWDH HPO holograms of the achromatic and rainbow variety are also of significant interest as they can be copied to an H_2 hologram using commercially available large-format copying systems based on neodymium glass lasers (Figure 9.10).

9.6 Full-Parallax and HPO DWDH

We now come to the end of our discussion of the data preparation issues concerning full-parallax and HPO digital display holography. We have seen that the full-parallax and HPO cases are first and foremost distinguished simply by the amount of data implicit in the definition of the hologram. High-definition metre-square HPO holograms are usually characterised by image data of several gigabytes, whereas

---

* Ultimately, the choice of photosensitive material plays a vital role in defining the optimal technique for printing a digital hologram.

the corresponding full-parallax case is characterised by image data of several terabytes. However, we have also seen that there is a fundamental optical difference between HPO and full-parallax holograms. HPO holograms only present a faithful reproduction of the recorded image at a unique line in space that coincides with the camera line. Full-parallax holograms do not suffer from this restriction and, as such, at least theoretically, an aberration-free image is available at any point within the emission region of the hologram.

We have also seen in Chapter 8 how, in general, the various discrete coordinate meshes that define the image data, the digital hologram itself and the writing of the image data by the printer do not line up. This leads to the introduction of noise into the DWDH hologram due to the truncation operations in converting look-up indices required to connect image data to the data written on the hologram.

The reason for such a lack of alignment is twofold. The first is that all printer objectives usually have a finite fifth coefficient, which gives rise to Barrel distortion. The second is that even for the case of a paraxial printer objective, the various coordinate meshes only line up under certain conditions and then only at quantised values of the camera distance parameter, h. For HPO holograms, which must be viewed at the camera distance, all too often, it is therefore just impossible to use such quantised values. The result is that here one must simply accept that the coordinate meshes do not line up.

In Chapter 8, we showed how to correct the printer write data for finite objective distortion. Two methods were discussed in the context of DWDH. The first was to formulate a single "non-paraxial" image transformation capable of converting camera data to actual printer data and incorporating an inverse distortion to correct for the implicit distortion present in the printer. This approach ensures the injection of the minimum amount of interpolation or truncation noise into the hologram. The alternative is to use two separate sequential transformations, the first to convert camera data to paraxial printer data and the second to compensate for the finite distortion of the objective. The first transformation is intrinsically three-dimensional in the case of HPO holograms and four-dimensional in the case of full-parallax holograms, so one is pretty much forced to use nearest-neighbour interpolation. However, the second transformation is two-dimensional and is therefore amenable to bilinear or bicubic interpolation.

In the case of DWDH HPO holograms, by far the best practical way to deal with the non-alignment of coordinate meshes and with interpolation noise is to use oversampling. One therefore maintains complete freedom in defining the camera track and all hologram parameters at the cost of using more camera positions than normally required and, in some cases, in generating more camera data than required for each camera frame. Typical oversampling ratios can be anything from two to ten times. When the 3D image is created by computer modelling software, one usually only needs to oversample the camera positions, as a centred camera frame can be programmed to exactly line up with the hologram hogel mesh. If this cannot be arranged, as will be the case with a real physical camera (we shall study this case in more detail in Chapter 10), then typically bilinear or bicubic interpolation is used to define the proper centred camera frames from the oversampled camera frame data. In either case, a combined **I**-to-**S** transformation is then used to calculate the final printer data. Because the intrinsic data set of HPO holograms is much smaller than full-parallax holograms, the oversampling technique is extremely effective.

Clearly, oversampling is rarely an option with full-parallax holograms! However, here the fact that the viewing plane is not collocated with the camera plane changes everything. Unlike HPO holograms, it is frequently possible to choose a camera distance that obeys the quantisation rules and as such to achieve perfect alignment of the coordinate meshes under the paraxial objective approximation. Then simple bilinear or bicubic interpolation can be used to calculate the final non-paraxial printer data.

Of course, even with a minimal data set, full-parallax data processing requires a lot of memory. As such, we have seen that today it is simply not practical to use a one-step **I**-to-**S** transformation as can be applied, even with oversampling, in HPO holography. However, by splitting the problem into two steps, we have seen that in fact the processing required can usually be performed within a very reasonable time on a single PC. Of course, the problem lends itself very simply to parallelisation if increased speed is required.

MWDH inevitably presents a far simpler data-processing problem because the camera data is usually the same as or quite close to the paraxial printer data. For both full-parallax and HPO, the final non-paraxial printer data can be calculated very effectively using oversampling of frame data and bilinear/bicubic interpolation.

REFERENCES

1. S. A. Benton and V. M. Bove Jr., *Holographic Imaging*, Wiley, Hoboken, NJ (2008).
2. M. A. Teitel, *Anamorphic raytracing for synthetic alcove holographic stereograms*, Master's thesis, Massachusetts Institute of Technology, Cambridge, MA (1986).
3. M. Holzbach, *Three-dimensional image processing for synthetic holographic stereograms*, Master's thesis, Massachusetts Institute of Technology, Cambridge, MA (1986).
4. M. W. Halle and A. B. Kropp, "Fast computer graphics rendering for full parallax spatial displays," in *Practical Holography XI and Holographic Materials III*, S. A. Benton and T. J. Trout eds., Proc. of SPIE **3011**, 105–482 (1997).
5. M. E. Holzbach and D. T. Chen, *Rendering methods for full parallax autostereoscopic displays,* US Patent 6,366,370 (filed 1999, granted 2002).

10

Image Data Creation and Acquisition for Digital Display Holograms

10.1 Introduction

The image data required by digital display holograms must be derived either from a physical scene or from a virtual computer model. In this chapter, we shall consider both scenarios. In the case of image data corresponding to a physical scene, we shall see that there are different preferred techniques of data acquisition depending on whether a horizontal parallax-only (HPO) hologram or a full-parallax hologram is to be written. The most common device used is, however, some type of holocam. These devices are based on a mechanically animated high-resolution digital camera. We shall discuss in detail both one-dimensional and two-dimensional holocam systems. Although these systems are, in some respects, intrinsically simple, they alone are capable of recording images of sufficient accuracy for such applications as the holographic archival of cultural heritage or biological specimens.

In the context of computer modelling, we shall discuss typical solutions to setting up and animating a virtual camera using the well-known program 3D StudioMax. Two examples will be discussed in detail. The first is a simple animation procedure that can be graphically programmed to generate the images necessary for HPO holograms. The second discusses the use of the MAXScript scripting language to program a virtual two-dimensional holocam system within Max. We chose Max because we like it; however, there are many three-dimensional (3D) design programs available today. One example of an open-source program of particular interest at the time of writing is Blender. Hopefully, you will be able to apply the general ideas presented in Max to other platforms without much difficulty.

Although nearly all of this chapter is about either real or virtual holocam systems, we will spend a little time talking about structured-light scanning. We would classify this type of data acquisition system as a *secondary* system because, with such systems, one must first generate a 3D computer model of a real-life scene and then one uses a virtual holocam to generate the perspective view data required by the hologram. This type of system is particularly interesting when one wants to record intrinsically dynamic data with full-parallax information or when real and virtual objects must be integrated into the same hologram.

10.2 Image Acquisition from a Physical Scene: HPO Holograms

To write an HPO hologram, we require camera image data taken along a centred horizontal line in front of the object or scene we wish to record. Figure 10.1 shows an illustration of the basic process. A digital camera is mounted on a precision linear translation rail and a stepper or servo motor is used to move it along the rail at a constant velocity. A computer generally controls both the camera motion and also acquires the frame data from the camera.

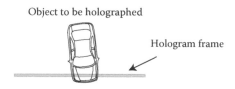

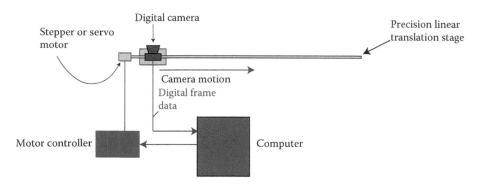

FIGURE 10.1   Basic scenario of image acquisition of a real scene: a digital camera is mounted on a precision linear translation rail and a stepper or servo motor is used to move it along the rail at a constant velocity. A computer generally controls both the camera motion and also acquires the frame data from the camera.

10.2.1 Simple Translating Camera

The simplest way to generate the required camera data is to arrange for a front-facing digital camera to be mounted on an automated linear translation stage. Such a configuration, which we referred to in Chapter 8 as the *simple translating camera*, is illustrated in Figure 10.2. This is the simplest form of a holocam device. A motor drives the camera at a constant velocity from one end of the rail to the other. The simple translating camera, however, suffers from an important defect. The field of view (FOV) of

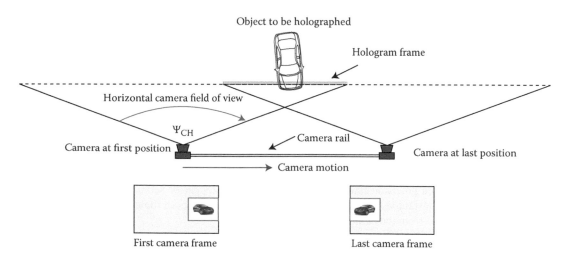

FIGURE 10.2   Simple translating camera configuration. The camera simply moves from one side of the rail to the other at a constant speed taking photographs at a fixed rate. The camera needs to have a large FOV to collect all the data required. The horizontal FOV, Ψ_{CH}, is shown for the first and last pictures. The vertical FOV (not shown) is Ψ_{CV}.

the camera objective must usually be very large to include all the data required by the hologram. As a consequence, only a small part of each camera frame actually falls within the optical field of the hologram and so most of the camera data is not used. A simple example will make this very clear. Suppose we want to record a 1 m × 1 m HPO digital hologram with a rectangular window of a horizontal size of 2 m situated at a distance of $h = 1$ m. If the hologram image to object size is 1:1, we will then need to place our camera rail at a distance of 1 m from the centre of the object (assuming here that the centre of the object coincides with the centre of the hologram). The camera rail will of course need to be 2 m in length. Simple geometry then tells us that the minimum horizontal FOV of the camera must be 112.6°. If we take the aspect ratio of the camera charge-coupled device (CCD) to be 1:0.75, then each frame contains only 14.8% of useful data; the other 85.2% falls outside the hologram optical field! Such a large FOV is of course possible with modern camera lenses, but invariably, there will be significant distortion. Such distortion may usually be compensated for computationally, but the price to pay is a greater requirement on the CCD resolution. Given that one is only using approximately 15% of the image field, one ends up requiring an extremely large CCD. Even worse, if one extends the viewing zone to 3 m wide, the useful CCD area now falls to 1/12th!

This is not to say that the simple translating camera is never useful. It can be used for small holograms of limited FOV very successfully, but as soon as the hologram size or the required holographic FOV becomes large, a simple translating camera provides a rather inefficient means for the recording image data.

10.2.2 Hybrid Translating/Rotating Camera with Fixed Target

A better solution to efficient image acquisition is to use a camera that translates and rotates at the same time [1]. Figure 10.3 illustrates the general idea. By arranging the camera in such a way that it always points to a position in space corresponding to the centre of the physical hologram surface, the FOV required of the camera becomes much smaller. In addition, much more of each camera frame carries useful image data. Taking the example used above of a 1 m × 1 m hologram with a 2 m camera track at $h = 1$ m, the required horizontal camera FOV of 112.6° drops to only 56.14°.*

We shall now look at the geometry of the hybrid camera and see how the data produced by such a configuration can be converted to the data corresponding to the ideal centred camera configuration of Section 8.8.3. Figures 10.4a and b show different sectional views through the relevant geometry, and Figure 10.4c shows a projection onto the camera plane of the hologram frame. As in Chapters 8 and 9, we use the label, ξ, to represent distance along the camera track starting from an origin at the extreme left-hand side. We shall now define two right-handed orthogonal Cartesian coordinates systems. The first is the system (x,y), which describes the raw camera frame. The second is the system $(\hat{x}, \hat{y})$, which describes the "centred" camera frame as defined in Section 8.8.3. Figure 10.5 shows the plane (x,y) and the projection of $(\hat{x}, \hat{y})$ onto this plane. We shall also need the 3D orthogonal Cartesian system $(\mathcal{X}, \mathcal{Y}, \mathcal{Z})$. These are the world coordinates whose origin is the extreme left-hand side of the camera rail. The coordinate $\mathcal{X}$ points along the rail to the right, the coordinate $\mathcal{Y}$ points upwards, and the coordinate $\mathcal{Z}$ points from the object to the rail.

With reference to Figures 10.4 and 10.5, we now define, in world coordinates, the direction vector of a ray that intersects both the camera origin, O, and a general point on $(\hat{x}, \hat{y})$:

$$\mathbf{k} = \begin{pmatrix} W/2 - D_X/2 - \xi + \hat{x} \\ \hat{y} - D_Y/2 \\ -h \end{pmatrix} \tag{10.1}$$

---

* In fact, in this case, the FOV of the camera is more strongly controlled by the vertical geometry—here, we require a vertical FOV of 58.6°, which given an aspect ratio of 1:0.75, corresponds to an effective horizontal FOV requirement of 73.6°.

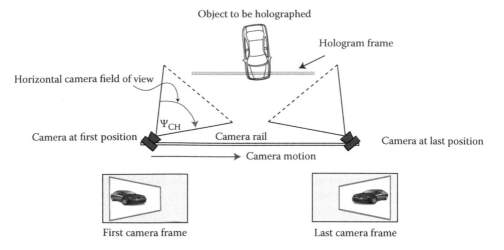

FIGURE 10.3 Hybrid translating/rotating camera configuration with fixed central target. By arranging that the camera always points to a position in space corresponding to the centre of the physical hologram surface, the FOV required of the camera becomes much smaller. This type of system can usually produce higher-quality data sets.

We also define a general point on the raw camera frame:

$$
\begin{pmatrix} x \\ y \\ z \end{pmatrix} = \begin{pmatrix} h\left\{ \dfrac{x}{\sqrt{h^2 + (W/2 - \xi)^2}} - \tan\left(\dfrac{\Psi_{CH}}{2}\right) \right\} + W/2 \\[2mm] y - \sqrt{h^2 + (W/2 - \xi)^2}\ \tan\left(\dfrac{\Psi_{CV}}{2}\right) \\[2mm] -h + (W/2 - \xi)\left\{ \dfrac{x}{\sqrt{h^2 + (W/2 - \xi)^2}} - \tan\left(\dfrac{\Psi_{CH}}{2}\right) \right\} \end{pmatrix} \tag{10.2}
$$

It is then easy to see that we can define the intersection of our test ray with the raw camera plane by extending its direction vector by a factor λ

$$
\begin{pmatrix} \xi \\ 0 \\ 0 \end{pmatrix} + \lambda \begin{pmatrix} W/2 - D_X/2 - \xi + \hat{x} \\ \hat{y} - D_Y/2 \\ -h \end{pmatrix} = \begin{pmatrix} h\left\{ \dfrac{x}{\sqrt{h^2 + (W/2 - \xi)^2}} - \tan\left(\dfrac{\Psi_{CH}}{2}\right) \right\} + W/2 \\[2mm] y - \sqrt{h^2 + (W/2 - \xi)^2}\ \tan\left(\dfrac{\Psi_{CV}}{2}\right) \\[2mm] -h + (W/2 - \xi)\left\{ \dfrac{x}{\sqrt{h^2 + (W/2 - \xi)^2}} - \tan\left(\dfrac{\Psi_{CH}}{2}\right) \right\} \end{pmatrix}
$$

$$(10.3)$$

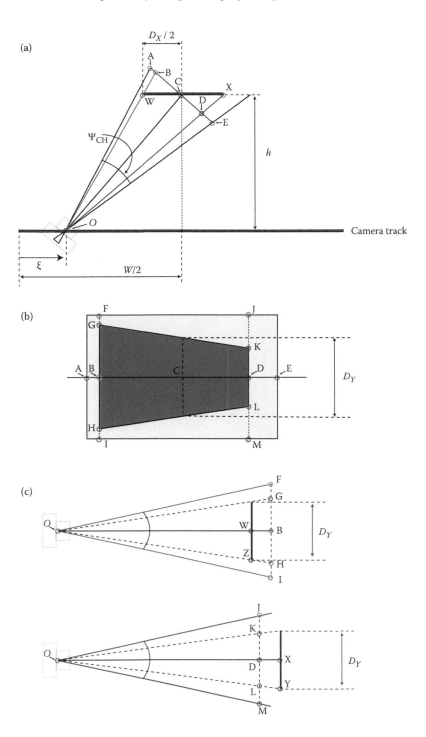

FIGURE 10.4 Geometry of hybrid translating/rotating camera configuration with fixed central target. (a) Plan view showing a midsection through the geometry. Camera track of length W is shown in green, hologram frame in red. The centre of the object is labelled C. Plane labelled ABCDE is the plane of the projected camera frame. This plane is also shown two-dimensionally in (b). The hologram frame and hologram surface are projected onto the projected camera plane in yellow. Two vertical slices are shown in (c) of the geometry from camera point O along the blue and red lines of (a). Letters indicate the correspondence of the points marked on the various diagrams.

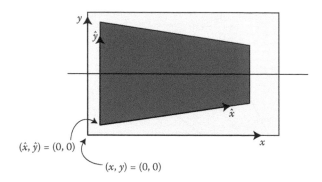

FIGURE 10.5 Projected raw camera frame plane described by the coordinates (x,y) and the projection onto this plane of the "hologram" plane (shown in yellow) described by the coordinates $(\hat{x}, \hat{y})$. The hologram plane is defined by the area in the real object scene that corresponds exactly, in position and dimension, to the hologram surface in the final holographic reproduction.

This yields the correspondence between the (x,y) and $(\hat{x}, \hat{y})$ systems:

$$\hat{x} = D_X/2 + \xi - W/2 + \frac{h\left\{\dfrac{x}{\sqrt{h^2 + (W/2-\xi)^2}} - \tan\left(\dfrac{\Psi_{CH}}{2}\right)\right\} + (W/2-\xi)}{1 - \dfrac{(W/2-\xi)}{h}\left\{\dfrac{x}{\sqrt{h^2 + (W/2-\xi)^2}} - \tan\left(\dfrac{\Psi_{CH}}{2}\right)\right\}} \tag{10.4}$$

$$\hat{y} = \frac{D_Y}{2} + \frac{\left(y - \sqrt{h^2 + (W/2-\xi)^2}\,\tan\left(\dfrac{\Psi_{CV}}{2}\right)\right)}{1 - \dfrac{(W/2-\xi)}{h}\left\{\dfrac{x}{\sqrt{h^2 + (W/2-\xi)^2}} - \tan\left(\dfrac{\Psi_{CH}}{2}\right)\right\}} \tag{10.5}$$

We can also express such correspondence equally well using the inverse relations

$$x = \sqrt{h^2 + (W/2-\xi)^2}\left\{\tan\frac{\Psi_{CH}}{2} + \frac{\hat{x} - \dfrac{D_X}{2}}{h + \dfrac{(W/2-\xi)}{h}\left[\hat{x} - \dfrac{D_X}{2} + W/2 - \xi\right]}\right\} \tag{10.6}$$

$$y = \left(\hat{y} - \frac{D_Y}{2}\right)\left[1 - \frac{(W/2-\xi)}{h}\left\{\frac{\hat{x} - \dfrac{D_X}{2}}{h + \dfrac{(W/2-\xi)}{h}\left[\hat{x} - \dfrac{D_X}{2} + W/2 - \xi\right]}\right\}\right]$$
$$+ \sqrt{h^2 + (W/2-\xi)^2}\,\tan\left(\frac{\Psi_{CV}}{2}\right) \tag{10.7}$$

To calculate the centred camera data set, $^{k}\hat{\mathbf{I}}_{\hat{i}\hat{j}}$, from the raw camera data set, $^{k}\mathbf{I}_{ij}$, we must now introduce integer coordinates (i,j) and $(\hat{i},\hat{j})$ corresponding to the real systems (x,y) and $(\hat{x},\hat{y})$ according to Section 8.8.3:

$$x = 2\sqrt{h^2 + (W/2 - \xi)^2}\ \tan\left(\frac{\Psi_{\mathrm{CH}}}{2}\right)\left(\frac{i-1}{N_I - 1}\right) \quad i = 1,2,3,...,N_I$$

$$y = 2\sqrt{h^2 + (W/2 - \xi)^2}\ \tan\left(\frac{\Psi_{\mathrm{CV}}}{2}\right)\left(\frac{j-1}{N_J - 1}\right) \quad j = 1,2,3,...,N_J$$

(10.8)

$$\hat{x} = D_X\left(\frac{\hat{i}-1}{\hat{N}_I - 1}\right) \quad \hat{i} = 1,2,3,...,\hat{N}_I$$

$$\hat{y} = D_Y\left(\frac{\hat{j}-1}{\hat{N}_J - 1}\right) \quad \hat{j} = 1,2,3,...,\hat{N}_J$$

(10.9)

We must also digitise the camera position

$$\xi = \frac{(k-1)}{(N_K - 1)} D_\xi \quad k = 1,2,3,...,N_K$$

(10.10)

We can then use Equations 10.6 through 10.9 to define the two real functions, $\Omega(\hat{i},\xi)$ and $\Upsilon(\hat{j},\xi)$, which constitute real estimates of the integers i and j:

$$\Omega(\hat{i},\xi) = \frac{1}{2}(N_I + 1)$$

$$+ \frac{D_X h(2\hat{i} - \hat{N}_I - 1)(N_I - 1)\cot\left(\dfrac{\Psi_{\mathrm{CH}}}{2}\right)}{4(\hat{N}_I - 1)h^2 + 2(W/2 - \xi)\left[D_X(2\hat{i} - \hat{N}_I - 1) + (\hat{N}_I - 1)(W - 2\xi)\right]}$$

(10.11)

$$\Upsilon(\hat{i}\hat{j},\xi) = \frac{1}{2}(N_J + 1)$$

$$+ \frac{D_Y(\hat{N}_I - 1)(N_J - 1)(2\hat{j} - \hat{N}_J - 1)\sqrt{h^2 + (W/2 - \xi)^2}\ \cot\left(\dfrac{\Psi_{\mathrm{CV}}}{2}\right)}{(\hat{N}_J - 1)\left\{4h^2(\hat{N}_I - 1) + (W - 2\xi)\left[D_X(2\hat{i} - \hat{N}_I - 1) + (\hat{N}_I - 1)(W - 2\xi)\right]\right\}}$$

(10.12)

These functions may then be used to calculate the centred camera data set through the transformation

$$^{k}\hat{\mathbf{I}}_{\hat{i}\hat{j}} = {}^{k}\mathbf{I}_{ij} \quad \forall \hat{i}\left\{\hat{i} \in \mathbb{N} \middle| \hat{i} \leq \hat{N}_I\right\} \quad \forall \hat{j}\left\{\hat{j} \in \mathbb{N} \middle| \hat{j} \leq \hat{N}_J\right\} \quad \forall k\left\{k \in \mathbb{N} \middle| k \leq N_K\right\}$$

(10.13)

where, if we use nearest integer interpolation,

$$i \equiv \left\| \Omega\left(\hat{i}, \xi\right) \right\|$$
$$j \equiv \left\| \Upsilon\left(\hat{j}, \xi\right) \right\|$$
(10.14)

Alternatively, we may use a bilinear interpolation,[*] in which case, Equations 10.13 and 10.14 must be replaced by

$$
\begin{aligned}
{}^{k}\mathbf{I}_{\hat{i}\hat{j}} ={}& \frac{{}^{k}\mathbf{I}_{\lfloor\Omega\rfloor\lfloor\Upsilon\rfloor}}{\left(\lceil\Omega\rceil-\lfloor\Omega\rfloor\right)\left(\lceil\Upsilon\rceil-\lfloor\Upsilon\rfloor\right)}\left(\lceil\Omega\rceil-\Omega\right)\left(\lceil\Upsilon\rceil-\Upsilon\right) \\[2mm]
&+ \frac{{}^{k}\mathbf{I}_{\lceil\Omega\rceil\lfloor\Upsilon\rfloor}}{\left(\lceil\Omega\rceil-\lfloor\Omega\rfloor\right)\left(\lceil\Upsilon\rceil-\lfloor\Upsilon\rfloor\right)}\left(\Omega-\lfloor\Omega\rfloor\right)\left(\lceil\Upsilon\rceil-\Upsilon\right) \\[2mm]
&+ \frac{{}^{k}\mathbf{I}_{\lfloor\Omega\rfloor\lceil\Upsilon\rceil}}{\left(\lceil\Omega\rceil-\lfloor\Omega\rfloor\right)\left(\lceil\Upsilon\rceil-\lfloor\Upsilon\rfloor\right)}\left(\lceil\Omega\rceil-\Omega\right)\left(\Upsilon-\lfloor\Upsilon\rfloor\right) \\[2mm]
&+ \frac{{}^{k}\mathbf{I}_{\lceil\Omega\rceil\lceil\Upsilon\rceil}}{\left(\lceil\Omega\rceil-\lfloor\Omega\rfloor\right)\left(\lceil\Upsilon\rceil-\lfloor\Upsilon\rfloor\right)}\left(\Omega-\lfloor\Omega\rfloor\right)\left(\Upsilon-\lfloor\Upsilon\rfloor\right)
\end{aligned}
$$
(10.15)

The above transformations let us calculate a centred camera data set from raw camera data when we are explicitly given the camera parameters—that is, when we know the horizontal and vertical FOVs of the camera (Ψ_{CH}, Ψ_{CV}) and the numerical dimensions of the camera CCD ($N_I \times N_J$). But, how does one calculate the most desirable horizontal and vertical FOVs?

To answer this question, we consider the behaviour of the lower left-hand corner of the projection of the centred data set onto the raw camera frame for the left-hand half of the camera trajectory. The minimum horizontal FOV that our camera may possess without clipping any required data is then given by the constraint

$$x = 0 \quad \text{if} \quad \left(\hat{x}, \hat{y}\right) = (0,0) \quad \forall \xi \in \left\{ \Re \middle| 0 \le \xi \le W/2 \right\}$$
(10.16)

This leads to a constraint for each camera position

$$\Psi_{CH} \ge 2\tan^{-1}\left\{ \frac{D_X}{2h + \dfrac{2}{h}(W/2 - \xi)(W/2 - \xi - D_X/2)} \right\}$$
(10.17)

The value of ($W/2 - \xi$) corresponding to the most stringent limit is given by differentiating this expression with respect to ($W/2 - \xi$). This gives ($W/2 - \xi$) = $D_X/4$, which then defines the general limit

$$\Psi_{CH} \ge 2\tan^{-1}\left\{ \frac{8hD_X}{16h^2 - D_X^2} \right\}$$
(10.18)

---

[*] See Appendix 7 for a more detailed description of bilinear and bicubic interpolation methods.

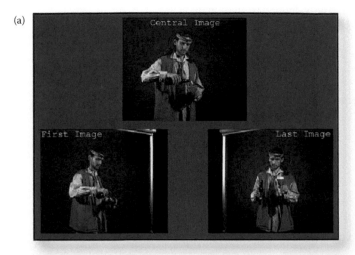

FIGURE 10.6 Example of the data processing required by a hybrid translator/rotator system. (a) Some hundreds to some thousands of raw camera frames are processed using Equation 10.15 to produce a centred data set (b), which is then used to print the final DWDH hologram (c).

The vertical FOV is similarly defined by the condition

$$y = 0 \quad \text{if} \quad (\hat{x}, \hat{y}) = (0,0) \quad \forall \xi \in \left\{ \Re \middle| 0 \leq \xi \leq W/2 \right\} \tag{10.19}$$

which leads to the general limit

$$\Psi_{CV} \geq 2 \tan^{-1} \left\{ \frac{D_Y \sqrt{h^2 + \beta^2}}{2(h^2 + \beta^2 - \beta D_X/2)} \right\} \tag{10.20}$$

where

$$\beta = \frac{2(6^{1/3} h^2) - \left\{ \sqrt{81 D_X{}^2 h^4 + 48 h^6} - 9 D_X h^2 \right\}^{2/3}}{6^{2/3} \left\{ \sqrt{81 D_X{}^2 h^4 + 48 h^6} - 9 D_X h^2 \right\}^{1/3}} \tag{10.21}$$

Equations 10.18 and 10.20 then define the most desirable FOV parameters for a centre-targeted fixed-zoom camera. If one uses these values, then the raw camera data will contain the maximum amount of useful image information. Of course, any camera is characterised by a certain aspect ratio determined by its CCD sensor and as such only one of these constraints will, in practice, be attainable with a spherical optical zoom lens.

10.2.3 Hybrid Translating/Rotating Camera with Optimised Target

The hybrid translating/rotating camera can be somewhat improved by optimising the target point. In the previous section, we considered the case in which the camera targeted the centre of the hologram. We will now consider moving the target position as a function of camera position. This is shown in Figure 10.7, where we have introduced the parameter $\rho(\xi)$ to describe a changing target. As before, the direction

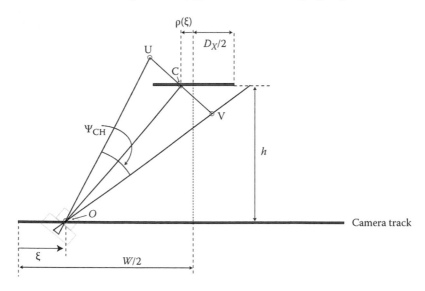

FIGURE 10.7 Geometry of hybrid translating/rotating camera when target position is now moved as a function of camera position. The geometry shown corresponds to a plan view of a meridional section. The target position is characterised by the function $\rho(\xi)$.

of a test ray, expressed in world coordinates that intersects both the camera origin, O, and a general point on $(\hat{x}, \hat{y})$ is given by

$$\mathbf{k} = \begin{pmatrix} W/2 - D_X/2 - \xi + \hat{x} \\ \hat{y} - D_Y/2 \\ -h \end{pmatrix} \tag{10.22}$$

However, now a general point on the (x,y) plane is

$$\begin{pmatrix} x_w \\ y_w \\ z_w \end{pmatrix} = \begin{pmatrix} h\left\{ \dfrac{x}{\sqrt{h^2 + (W/2 - \xi - \rho(\xi))^2}} - \tan\left(\dfrac{\Psi_{CH}}{2}\right) \right\} + W/2 - \rho(\xi) \\ y - \sqrt{h^2 + (W/2 - \xi - \rho(\xi))^2}\, \tan\left(\dfrac{\Psi_{CH}}{2}\right) \\ -h + (W/2 - \xi - \rho(\xi))\left\{ \dfrac{x}{\sqrt{h^2 + (W/2 - \xi - \rho(\xi))^2}} - \tan\left(\dfrac{\Psi_{CH}}{2}\right) \right\} \end{pmatrix} \tag{10.23}$$

Following our previous methodology we extend the test ray by a factor λ to arrive at the following three equations relating the (x,y) and $(\hat{x}, \hat{y})$ systems

$$\begin{pmatrix} \xi \\ 0 \\ 0 \end{pmatrix} + \lambda \begin{pmatrix} W/2 - D_X/2 - \xi + \hat{x} \\ \hat{y} - D_Y/2 \\ -h \end{pmatrix} = \begin{pmatrix} h\left\{ \dfrac{x}{\sqrt{h^2 + (W/2 - \xi - \rho(\xi))^2}} - \tan\left(\dfrac{\Psi_{CH}}{2}\right) \right\} + W/2 - \rho(\xi) \\ y - \sqrt{h^2 + (W/2 - \xi - \rho(\xi))^2}\, \tan\left(\dfrac{\Psi_{CV}}{2}\right) \\ -h + (W/2 - \xi - \rho(\xi))\left\{ \dfrac{x}{\sqrt{h^2 + (W/2 - \xi - \rho(\xi))^2}} - \tan\left(\dfrac{\Psi_{CH}}{2}\right) \right\} \end{pmatrix} \tag{10.24}$$

These equations then give the following rules:

$$\hat{x} = D_X/2 - \beta - \frac{h^2\left\{ \dfrac{x}{\sqrt{h^2 + (\beta - \rho)^2}} - \tan\left(\dfrac{\Psi_{CH}}{2}\right) \right\} + h(\beta - \rho)}{(\beta - \rho)\left\{ \dfrac{x}{\sqrt{h^2 + (\beta - \rho)^2}} - \tan\left(\dfrac{\Psi_{CH}}{2}\right) \right\} - h} \tag{10.25}$$

$$\hat{y} = \frac{D_Y}{2} + \frac{y - \sqrt{h^2 + (\beta - \rho)^2}\, \tan\left(\dfrac{\Psi_{CV}}{2}\right)}{1 - \dfrac{(\beta - \rho)}{h}\left\{ \dfrac{x}{\sqrt{h^2 + (\beta - \rho)^2}} - \tan\left(\dfrac{\Psi_{CH}}{2}\right) \right\}} \tag{10.26}$$

where

$$\beta = W/2 - \xi \tag{10.27}$$

Alternatively, we can write these relations in an inverse fashion:

$$x = \sqrt{h^2 + (\beta - \rho)^2} \left\{ \frac{h(\hat{x} - D_X/2 + \rho)}{\left(h^2 + (\hat{x} - D_X/2 + \beta)(\beta - \rho)\right)} + \tan\left(\frac{\Psi_{CH}}{2}\right) \right\} \qquad (10.28)$$

$$y = \left(\hat{y} - \frac{D_Y}{2}\right)\left[1 - \frac{(\beta - \rho)}{h}\left\{\frac{x}{\sqrt{h^2 + (\beta - \rho)^2}} - \tan\left(\frac{\Psi_{CH}}{2}\right)\right\}\right] + \sqrt{h^2 + (\beta - \rho)^2}\,\tan\left(\frac{\Psi_{CV}}{2}\right) \quad (10.29)$$

Equations 10.28 and 10.29 may then be used to define the centred camera data set in terms of the raw camera data using exactly the same method as described in the previous section. However, this assumes that we know the camera pointing function $\rho(\xi)$.

To calculate $\rho(\xi)$, we follow a similar optimisation procedure as introduced for the case of the centrally targeted camera. In the previous section, we required (in the horizontal direction)

$$x = 0 \quad \text{if} \quad (\hat{x}, \hat{y}) = (0,0) \quad \forall \xi \in \left\{\Re \middle| 0 \le \xi \le W/2\right\} \qquad (10.30)$$

This gives an expression similar to before

$$\tan\left(\frac{\Psi_{CH}}{2}\right) \ge \frac{h(\rho - D_X/2)}{(D_X/2 - \beta)(\beta - \rho) - h^2} \qquad (10.31)$$

But now we have the supplementary condition

$$\hat{x} = D_x \quad \text{if} \quad (x,y) = \left(2\sqrt{h^2 + (\beta - \rho)^2}\,\tan\left(\frac{\Psi_{CH}}{2}\right), 0\right) \quad \forall \beta \in \left\{\Re \middle| -W/2 \le \beta \le 0\right\} \quad (10.32)$$

This condition, which makes sure that the right-hand side of the raw and centred camera frames line up, yields a further equation

$$\tan\left(\frac{\Psi_{CH}}{2}\right) = \frac{h(D_X + 2\rho)}{(D_X + 2\beta)(\beta - \rho) + 2h^2} \qquad (10.33)$$

Together, Equations 10.31 and 10.33 define ρ and Ψ_{CH} as functions of β

$$\rho = \frac{4h^2 + D_X^2 + 4\beta^2 - \sqrt{16h^2 D_X^2 + (-4h^2 + D_X^2 - 4\beta^2)^2}}{8\beta},$$

$$\tan\left(\frac{\Psi_{CH}}{2}\right) = \frac{-4h^2 + D_X^2 - 4\beta^2 + \sqrt{16h^2 D_X^2 + (-4h^2 + D_X^2 - 4\beta^2)^2}}{4hD_X} \qquad (10.34)$$

By differentiating the second expression with respect to β, we can find which value of β corresponds to the greatest required value of $\tan\left(\dfrac{\Psi_{CH}}{2}\right)$. This is at $\beta = 0$ where $\rho = 0$ and $\tan(\Psi_{CH}/2) = D_X/2h$. Therefore, by moving the target point of the camera according to the rule

$$\rho = \frac{4h^2 + D_X^2 + 4\beta^2 - \sqrt{16h^2 D_X^2 + (-4h^2 + D_X^2 - 4\beta^2)^2}}{8\beta} \tag{10.35}$$

we are able to use a horizontal FOV given by

$$\Psi_{CH} = 2\tan^{-1}\{D_X/2h\} \tag{10.36}$$

This is better than when the camera is targeted to the centre of the hologram as in this case ($\rho = 0$) we must use an FOV of

$$\Psi_{CH} \ge 2\tan^{-1}\left\{\frac{8hD_X}{16h^2 - D_X^2}\right\} \tag{10.37}$$

The horizontal linear resolution of the acquired data can therefore be improved by a factor, f, by following the modified rule for the target point where

$$f = \frac{\dfrac{8hD_X}{16h^2 - D_X^2}}{\dfrac{D_X}{2h}} = \frac{16h^2}{16h^2 - D_X^2} \tag{10.38}$$

For a 1 m-wide hologram with a viewing distance of 1 m, this amounts to a potential increase in resolution of approximately 7%. The effect is most useful for wide holograms. For example, a 2 m-wide hologram with a viewing distance of 1 m can benefit from a 33% increase in resolution.

To determine the most desirable vertical FOV, we use the constraint we mentioned previously, namely,

$$y = 0 \quad \text{if} \quad (\hat{x},\hat{y}) = (0,0) \quad \forall \xi \in \left\{\Re \mid 0 \le \xi \le W/2\right\} \tag{10.39}$$

This leads to the relation

$$\tan\left(\frac{\Psi_{CV}}{2}\right) = \frac{D_Y\sqrt{h^2 + (\beta - \rho)^2}}{2\{h^2 + (\beta - \rho)(\beta - D_X/2)\}} \tag{10.40}$$

This must be maximised numerically as no analytic solution exists; that is, one must solve

$$\frac{\partial}{\partial\beta}\left\{\frac{D_Y\sqrt{h^2 + (\beta - \rho(\beta))^2}}{2\{h^2 + (\beta - \rho(\beta))(\beta - D_X/2)\}}\right\} = 0 \tag{10.41}$$

over the range $-W/2 \le \beta \le 0$.

Figure 10.8 shows a comparison of what the raw camera frames look like using the three methods we have discussed—the simply translating camera (a), the hybrid rotating/translating camera with fixed central target (b) and finally the rotating/translating case with an optimised target (c). Clearly, the most inefficient of the three variants is the simple translating camera, followed by the fixed-target hybrid, with the most efficient variant being the optimised target hybrid camera. Figure 10.9 shows the camera target position of Figure 10.8c versus the camera position. Also shown is the theoretical minimum vertical FOV required versus camera position (derived from Equation 10.40). The value used in Figure 10.8c corresponds to the greatest value of this FOV, which occurs just below $\xi = 1$.*

---

* You will note that although we have plotted in Figure 10.9 $\Psi(\xi)$ for $0 \le \xi \le 3$ m, Equation 10.40 only gives meaningful values for $0 \le \xi \le 1.5$ m here.

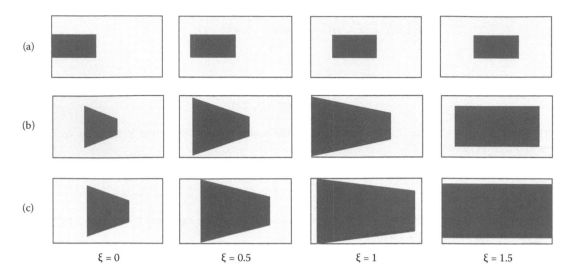

FIGURE 10.8 Comparison of raw camera frames using the three methods discussed for the case of a hologram 1 m high by 2 m wide using a camera distance of 1 m and a rail length of 3 m—(a) the simply translating camera, (b) the hybrid translating/rotating camera with fixed central target, and finally, (c) the translating/rotating case with an optimised target. All cases use optimally calculated horizontal camera FOVs. Case (a) uses a vertical FOV calculated to maintain the CCD aspect ratio equal to that of cases (b) and (c).

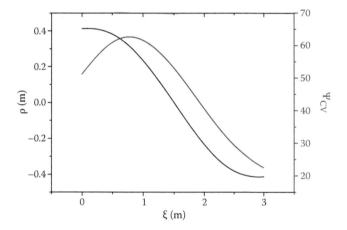

FIGURE 10.9 Optimal camera target position, ρ, of Equation 10.35 used in Figure 10.8c versus camera position, ξ. Also shown is the theoretical minimum vertical FOV required versus camera position (derived from Equation 10.40). The value used in Figure 10.8c corresponds to the greatest value of this FOV, which occurs just below $\xi = 1$.

10.2.4 Resolution Requirements

In Chapter 9, we saw that to meet the resolution requirements of the human eye, the distance between camera positions along the camera line had to be less than 1 mm per unit hologram-camera distance. In other words, if one wants to view a hologram at a distance of 0.5 m, then the camera line will need to be located at $h = 0.5$ m and one will need 2000 camera positions per metre length of camera track. We also saw in Chapter 9 that oversampling of camera positions was desirable in HPO holography due to the intrinsic misalignment of the various coordinate meshes associated with the printer image data and the raw camera data. When large maximum-resolution HPO holograms are to be written with horizontal windows of several metres in length, it is not uncommon to require more than 10,000 camera frames.

Attention must also be given to the resolution of the CCD used in the camera. We have seen in the previous sections how a hybrid rotating/translating camera with an optimised tracking target can allow one to use the CCD much more efficiently than in the case of a simply translating camera—and this generally translates to a much better hologram for a given camera CCD. Whatever the configuration used, however, one must try to ensure that the camera resolution is several times better than the minimum angular spacing between hogels. In practice, this can be more difficult to achieve at large angles of view as will be clear from the $\xi = 0$ diagrams of Figure 10.8a, b or c.

10.2.5 Image Reduction and Magnification

The holographic image may be reduced or magnified by proportionately reducing or expanding all physical sizes at recording with respect to the corresponding values at hologram replay. Let us denote the distance between the camera track centre and the hologram centre as h_1, and the corresponding viewing distance on replay of the hologram by h. Similarly, let us denote the physical camera rail length as W_1 and the width of the viewing box on replay by W. Finally, we denote the hologram frame size on recording by $D_{X1} \times D_{Y1}$ and by $D_X \times D_Y$ at replay. Then we can define the magnification of the holographic image as

$$M = h/h_1 = W/W_1 = D_X/D_{X1} = D_Y/D_{Y1} \qquad (10.42)$$

If $M = 2$, then the holographic image will appear twice as big as the actual physical object. Of course, to make large reductions—for example, reducing a building to a 30×40 cm hologram—takes a very large rail system. Large magnifications can be achieved using microstages and this is a definite area of future interest in the field of biological specimen archival.

10.2.6 Commercial Holocam Systems

At the time of writing, several companies offered commercial holocam systems for sale. Most notable are the UK company Spatial Imaging Ltd. and the Lithuanian company Geola Digital UAB. Both companies use the efficient hybrid rotating/translating geometry. Figure 10.10 shows a standard studio system manufactured by Geola, which was built around a 4 m electromechanical rail system. The rail was built by the German company Isel. Early models of this device used high-resolution, high-speed CMOS camera systems requiring bright studio lights. However, in the past few years, camera technology has improved radically, with the result that all of Geola's current models now only require normal room lighting.

Figure 10.11 shows a portable holocam system that has proven to be the most popular type of system. The system is based on a short electromechanical rail just over 1 m in length, mounted on two photographic tripods. The holocam is supplied with its own laptop running highly automated proprietary software that takes care of alignment, shooting and data processing. A typical camera now being shipped with the device (Figure 10.12) is the Xacti HD electronic camera with a pixel resolution of 1920×1080, capable of operation at 60 frames per second. The system can be transported easily in the back of a car and assembled in a few minutes. Operation is likewise extremely easy and the resulting files can be sent via the Internet straight to Geola for printing. The electromechanical rail is made by Isel in Germany and camera rotation is assured by a precision electromechanical rotation stage produced by the Lithuanian company Standa UAB.

One of the intrinsic problems with holocam systems is their cost. For top-quality results, the animation of the camera must be very precise, and as a result, the electromechanics required is often rather costly. A good camera can also be a costly element. To tackle this problem, Geola has just released a cheap holocam system (Figure 10.13), which currently markets for less than £1000. This is a small device that uses cheap components to ensure the required hybrid camera animation. Instead of the ultraprecise rail from Isel, the system uses a rather simpler rail in which the camera is mounted on a platform which has rubber wheels and which rides a type of "railway" track. The camera rotation is assured by a mechanical

FIGURE 10.10 Professional holocam system manufactured and available commercially from the Lithuanian company Geola Digital UAB. Designed for installation in photographic studios, this fully computer-controlled holocam, which is a hybrid translator/rotator system, comprises a 4 m-long precision electromechanical rail with high-resolution fast digital camera. It is suitable for the production of high-resolution image data sets required by high-quality large-format digital colour reflection holograms.

FIGURE 10.11 Portable commercial holocam system using hybrid translator/rotator configuration. This self-contained system is capable of easy transport and quick setup—ideal for capturing data sets when out of the studio. A Wi-Fi Internet connection allows the user to beam the data back for printing. Although not as large as the studio models offered by Geola, the company has sold many more of these smaller devices.

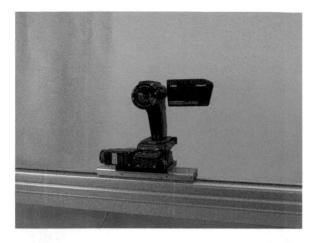

FIGURE 10.12 The Xacti HD electronic camera with a pixel resolution of 1920 × 1080 capable of operation at 60 frames per second was recommended by Geola Digital UAB for use in its portable holocam systems at the time of writing.

FIGURE 10.13 A new concept from Geola is the affordable holocam. Made from cheaper components, the unit allows smaller good-quality holographic portraits to be taken effectively. The system is a hybrid translator/rotator in which the rotation is assured mechanically by a metal arm instead of an expensive precision rotation stage. The unit, which comes with its own netbook, is completely self-contained and yet markets for less than £1000. The system can be supplied with a variety of digital cameras. (a) Side view showing rotator arm, (b) supplied netbook with installed proprietary software, (c) camera mount and (d) bottom view showing motor and wheel-based system for camera translation.

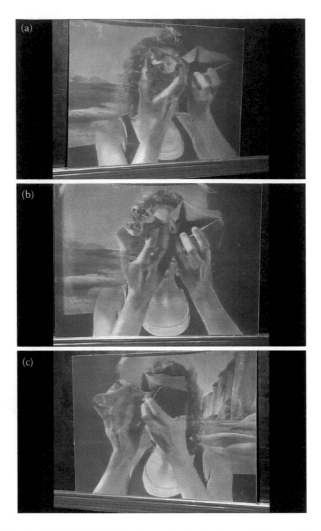

FIGURE 10.14 Example of an HPO DWDH full-colour reflection hologram (40 cm × 60 cm) created using a portable Geola holocam system. (a) View from the left, (b) view from the centre and (c) view from the right. (Artist: Martin Richardson 2010.)

arm that physically turns the camera as it rotates. This cheap holocam is principally useful for portraiture at a fixed format. As such, it is a little restrictive, but the results are not so bad.

10.2.7 Alternative Strategies—Rotating the Object

An alternative strategy to the use of rail-mounted camera systems is the use of a static camera and a rotating platform. This geometry can most successfully be used with small objects or human subjects destined for the creation of small- to medium-sized display holograms. The system has a number of advantages, the most obvious of which is the potential simplicity offered by a simple rotating platform— if the mass of the platform is sufficiently high, then manual rotation will often produce acceptable results. Motorisation is, however, preferable, but again, it is much simpler to achieve than any of the linear-rail options. The main disadvantage is that people look at a hologram from a constant perpendicular distance; when the hologram is mounted on a wall, for example, people tend to walk past it, describing a linear trajectory parallel to the hologram. For small angles of view, there is not a great difference between a linear viewing line and the curved viewing line characteristic of the rotating platform geometry. However, at larger angles (Figure 10.16), the difference becomes important and the normal observer finds that the HPO hologram distorts when viewed obliquely.

The image processing for the static camera and rotating platform configuration is a little different from the case of a linear rail as the viewing line becomes the segment of a circle. There are basically two parts to the image processing problem. The first is the transformation of the raw camera data into a centred camera data set. This is very similar to the problem we encountered in the context of a hybrid camera on a linear rail. In Section 10.3.1, we will solve this problem for the case of a horizontal rotating platform combined with a simple vertical tracking of the camera; this more general geometry is useful for full-parallax image acquisition. The present problem represents a specific case of this configuration, and as such, we shall simply refer the reader to Equations 10.71 through 10.75 where, for the problem at hand, we must put $D_\zeta = 0$.

The second part of the problem entails, for a given hogel, a derivation of the relationship between the horizontal spatial light modulator (SLM) coordinate and the camera shot coordinate. We will need this relationship to write down a modified **I**-to-**S** transformation appropriate for the rotating platform configuration. This is somewhat different from what we discussed in Chapter 8 in so much as the camera line and the viewing line are now segments of a circle rather than simple lines.

FIGURE 10.15 Example of an HPO DWDH full-colour reflection portrait (30 cm × 40 cm) created using a portable Geola holocam system. (a) View from the left, (b) view from the centre and (c) view from the right. (Copyright Julio Ruis, ITMA, Spain 2010.)

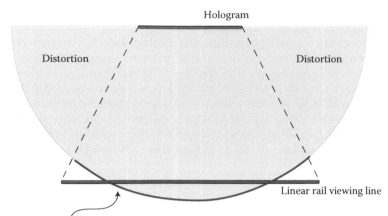

Curved viewing line characteristic of rotating platform configuration

FIGURE 10.16 Comparison of the geometry of a linear camera track and that of the curved track characteristic of the technique of a static camera and rotating platform. At oblique angles, the curved track differs substantially from the linear track, leading to unacceptable distortion in the HPO hologram.

Figure 10.17 shows a plan view of the geometry. We start by digitising the effective camera position* in terms of the rotation angle of the platform

$$\theta_\xi = \frac{(k-1)}{(N_K-1)}\Phi_C \tag{10.43}$$

This means that we take pictures at even angle increments; if the rotation platform rotates with a uniform angular velocity, then this is equivalent to the camera taking pictures at a constant rate. Next, we digitise the horizontal coordinate of a paraxial projection of the SLM onto a plane at an arbitrary perpendicular distance h^*:

$$U' = 2h^*\tan\left(\frac{\Psi_{PH}}{2}\right)\frac{(\mu'-1)}{(N_M-1)} \tag{10.44}$$

As before, we digitise the hologram and the film frame using

$$X = \frac{(\alpha-1)}{(N_A-1)}D_X \tag{10.45}$$

and

$$x' = \frac{(i'-1)}{(N_I-1)}D_x \tag{10.46}$$

Because we are using a centred camera configuration, we put

$$D_x = D_X$$
$$N_I = N_A \tag{10.47}$$

We now need one additional coordinate system. These are the world coordinates $(\mathcal{X},\mathcal{Z})$ which we shall choose to be right-handed, two-dimensional orthonormal Cartesian coordinates with origin located at the centre of the hologram. A test ray emanating from a hogel described by the horizontal coordinate, X, and associated with the horizontal SLM coordinate, U', will then intersect the camera circle at a

---

* For ease of analysis, we consider the equivalent case of a static platform and a targeted camera that rotates around the centre of the platform.

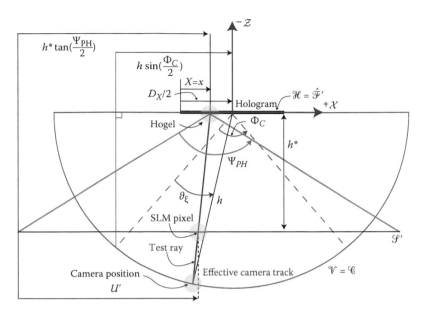

FIGURE 10.17 Plan view of the geometry characterising the static camera and rotating platform configuration. We use the equivalent circular camera track with centre-targeted camera. In particular, we consider the intersection of the purple test ray emanating from a given hogel with both the camera track and with a projection of the SLM plane at a perpendicular distance, h^*, from the hologram. This procedure leads us to the **I**-to-**S** transformation characteristic of the configuration.

horizontal world coordinate given by the following vector expression, which we obtain by simply extending the direction vector of the test ray by a distance, λ

$$\begin{pmatrix} \mathcal{X} \\ \sqrt{h^2 - \mathcal{X}^2} \end{pmatrix} = \begin{pmatrix} X - D_X/2 \\ 0 \end{pmatrix} + \lambda \begin{pmatrix} U' - h^* \tan(\Psi_{PH}/2) \\ h^* \end{pmatrix} \tag{10.48}$$

The z component immediately defines λ

$$\lambda = \frac{\sqrt{h^2 - \mathcal{X}^2}}{h^*} \tag{10.49}$$

whereupon we can use the x component to solve for the ray intersection

$$\mathcal{X} = \frac{X - \dfrac{D_X}{2} \pm \sqrt{\left(X - \dfrac{D_X}{2}\right)^2 + \left[1 + \dfrac{1}{h^{*2}}\left\{U' - h^*\tan\left(\dfrac{\Psi_{PH}}{2}\right)\right\}^2\right]\left[\dfrac{h^2}{h^{*2}}\left\{U' - h^*\tan\left(\dfrac{\Psi_{PH}}{2}\right)\right\}^2 - \left(X - \dfrac{D_X}{2}\right)^2\right]}}{1 + \dfrac{1}{h^{*2}}\left\{U' - h^*\tan\left(\dfrac{\Psi_{PH}}{2}\right)\right\}^2} \tag{10.50}$$

This may now be written in terms of the integer coordinates

$$\mathcal{X} = \frac{\left[\dfrac{D_X}{2}\right]\left(\dfrac{2\alpha - 1 - N_A}{(N_A - 1)}\right) \pm \sqrt{\varpi}}{\left(1 + \tan^2(\Psi_{PH}/2)\left\{\dfrac{2\mu' - 1 - N_M}{(N_M - 1)}\right\}^2\right)} \tag{10.51}$$

where

$$\varpi = \left[\frac{D_X}{2} \right]^2 \left(\frac{2\alpha - 1 - N_A}{(N_A - 1)} \right)^2$$

$$+ \left[h^2 \tan^2\left(\frac{\Psi_{PH}}{2} \right) \left\{ \frac{2\mu' - 1 - N_M}{(N_M - 1)} \right\}^2 - \left[\frac{D_X}{2} \right]^2 \left(\frac{2\alpha - 1 - N_A}{N_A - 1} \right)^2 \right] \times$$

$$\left[1 + \tan^2\left(\frac{\Psi_{PH}}{2} \right) \left\{ \frac{2\mu' - 1 - N_M}{(N_M - 1)} \right\}^2 \right] \tag{10.52}$$

The camera position, k, is then related to $\mathcal{X}$ by the simple expression

$$k = \left\| \frac{(N_K + 1)}{2} + \frac{(N_K - 1)}{\Phi_C} \sin^{-1}\left[\frac{\mathcal{X}}{h} \right] \right\| \tag{10.53}$$

This may be used in place of Equation 8.63 in the calculation of the relevant printer data set, $^{\mu\nu}\mathbf{S}_{\alpha\beta}$. The full **I**-to-**S** transformation for an inverting camera and a conjugate SLM position may be written as*

$$^{\mu\nu}\mathbf{S}_{\alpha\beta} = {}^k\hat{\mathbf{I}}_{\hat{i}\hat{j}} \quad \forall\alpha\left\{ \alpha \in \mathbb{N} \middle| \alpha \leq N_A \right\} \qquad \forall\beta\left\{ \beta \in \mathbb{N} \middle| \beta \leq N_B \right\}$$

$$\forall\mu\left\{ \mu \in \mathbb{N} \middle| N_{U1}(\alpha) \leq \mu \leq N_{U2}(\alpha) \right\} \quad \forall\nu\left\{ \nu \in \mathbb{N} \middle| N_{V1}(\beta) \leq \nu \leq N_{V2}(\beta) \right\} \tag{10.54}$$

$$= 0 \qquad \text{otherwise}$$

where

$$\hat{i} = N_I - \alpha + 1$$

$$\hat{j} = N_J - \beta + 1 \tag{10.55}$$

$$k = \left\| \frac{(N_K + 1)}{2} + \frac{(N_K - 1)}{\Phi_C} \sin^{-1}\left[\frac{\mathcal{X}}{h} \right] \right\|$$

and

$$N_{V1}(\beta) = \left\| \frac{N_V + 1}{2} + \frac{(N_V - 1)}{2h\tan\left(\frac{\Psi_{PV}}{2} \right)} \left\{ D_Y \frac{(N_B - \beta)}{N_B - 1} - \frac{(D_Y + H)}{2} \right\} \right\|$$

$$\tag{10.56}$$

$$N_{V2}(\beta) = \left\| \frac{N_V + 1}{2} + \frac{(N_V - 1)}{2h\tan\left(\frac{\Psi_{PV}}{2} \right)} \left\{ D_Y \left\{ \frac{(N_B - \beta)}{N_B - 1} + \frac{H}{D_Y} \right\} - \frac{(D_Y + H)}{2} \right\} \right\|$$

---

* Note that in Equation 10.56 we have chosen to define the vertical window focus at a distance of h from the hologram surface. Another popular choice is to set this focus to follow the camera track. One does this by simply replacing the two h variables in Equation 10.56 by $\sqrt{h^2 - \mathcal{X}^2}$. Note also that the variable $\mathcal{X}$ in Equation 10.55 must be calculated using Equations 10.51 and 10.52.

The limit expressions for $N_{U1}(\alpha)$ and $N_{U2}(\alpha)$ may be obtained by setting, respectively, $k = 1$ and $k = N_K$ in Equations 10.51 through 10.53:

$$N_{U\gamma} = \frac{N_M + 1}{2} + \frac{1}{2}\left\{ \mathcal{X}_\gamma + \frac{D_X}{2}\frac{(1 + N_A - 2\alpha)}{N_A - 1}\right\}\frac{\cot\left(\dfrac{\Psi_{PH}}{2}\right)(N_M - 1)}{\sqrt{h^2 - \mathcal{X}_\gamma^2}} \tag{10.57}$$

where

$$\mathcal{X}_1 = -h\sin\left(\frac{\Phi_C}{2}\right)$$

$$\mathcal{X}_2 = h\sin\left(\frac{\Phi_C}{2}\right) \tag{10.58}$$

10.3 Image Acquisition from a Physical Scene: Full-Parallax Digital Holograms

Perhaps the greatest difference between the problems of image data acquisition for HPO holograms and full-parallax holograms is that a scanning camera cannot usually be used in the latter case when the object is able to move. In the HPO case, a small amount of movement is often desired as this adds realism to the final hologram. For example, a human portrait may be made of a subject breaking into a smile as the camera traverses the rail from left to right. Then, on replay, the observer will observe, on traversing the viewing zone of the hologram, an exact replay of this animation. As long as the animation is "slow", then the horizontal parallax is not affected.* In full-parallax holography, the only way to possibly include animation is to use a plurality of vertically arranged cameras that track simultaneously along a rail. However, given the discussions of Section 10.2.4, it will be immediately obvious that this solution is going to be rather difficult for any but the simplest display holograms. Nevertheless, we should not discount this possibility because, for small depth holograms, the vertical parallax does not require a lot of sampling; as we shall see in the next chapter, chromatic blurring in reflection holograms operates principally in the vertical direction, blurring the vertical parallax. For wideband illumination, a large stacked array of compact cameras can therefore sometimes provide a solution when it is necessary to include moving objects. Clearly, however, in the context of ultra-realistic holographic images of large depth and wide angle of view, not only would one require typically thousands of cameras in a stack, but also each camera would either need to be extremely small or the stack extremely large.

Rail-mounted camera systems for full-parallax image data acquisition are therefore nearly always used with static objects. Applications particularly suited to rail-mounted camera acquisition are museum artefact archival and biological archival. Often, large magnifications are desirable (either up or down). The techniques presented in the last section transfer over directly to the full-parallax case with the camera now being mounted on a two-dimensional translating stage in place of a linear rail. In principle, each technique we discussed in the context of HPO holograms can be combined with another for the second dimension. Nine principal ways therefore exist to control the camera using a two-dimensional stage. In addition, a rotation platform may be used for either dimension leading to a

---

* Fast animation leads to coupling of time and parallax leading to unexpected effects in the 3D properties of the image.

total of 16 ways in which we might possibly contemplate recording a full-parallax data set. It is worth pointing out that the previous restrictions of the rotating platform technique that operate in the HPO case no longer apply to full-parallax applications. Here, due to Huygens' principle, we see no distortion at oblique angles. Accordingly, a rotating platform becomes an extremely useful tool for many full-parallax applications. We shall study only one case here for reasons of economy of space, but the method we describe can be mechanically applied to any of the 16 possible configurations. The case we will study is that of the horizontally positioned rotating platform combined with a simple vertically translating camera.

10.3.1 Horizontal Rotating Platform with Vertically Linear-Translating Camera

As in Section 10.2.7, a discussion of this geometry is naturally divided into two parts. The first part, sometimes referred to as "keystoning", is about transforming the raw camera data into a centred camera data set. The second part is about deriving the **I**-to-**S** transformation characteristic of the configuration. We shall start with the first problem and develop the mathematics required to define a centred camera data set.

10.3.1.1 Calculation of a Centred Camera Data Set

Figure 10.18 shows vertical and horizontal sections through the relevant geometry. We define 3D world coordinates as usual but with the difference now that their origin is at the centre of the hologram frame. A general point on the plane $(\hat{x}, \hat{y})$ may therefore be written in world coordinates as

$$\mathbf{Q}(\hat{x}, \hat{y}) = \hat{\mathbf{e}}_x \left(\hat{x} - \frac{D_X}{2} \right) + \hat{\mathbf{e}}_y \left(\hat{y} - \frac{D_Y}{2} \right) \tag{10.59}$$

where $\hat{\mathbf{e}}_x$, $\hat{\mathbf{e}}_y$ and $\hat{\mathbf{e}}_z$ are the world coordinate unit vectors. Likewise, a general camera position may be expressed as

$$\mathbf{C} = \hat{\mathbf{e}}_x h \sin(\vartheta_\xi - \Phi_C/2) + \hat{\mathbf{e}}_y (\zeta - D_\zeta/2) + \hat{\mathbf{e}}_z h \cos(\vartheta_\xi - \Phi_C/2) \tag{10.60}$$

Accordingly, the direction vector of a ray passing through both the camera point and an arbitrary point on the $(\hat{x}, \hat{y})$ plane can be written as

$$\mathbf{k} = \left\{ h \sin\left(\vartheta_\xi - \frac{\Phi_C}{2} \right) - \hat{x} + \frac{D_X}{2} \right\} \hat{\mathbf{e}}_x + \left\{ \zeta - \frac{D_\zeta}{2} - \hat{y} + \frac{D_Y}{2} \right\} \hat{\mathbf{e}}_y + \left\{ h \cos\left(\vartheta_\xi - \frac{\Phi_C}{2} \right) \right\} \hat{\mathbf{e}}_z \tag{10.61}$$

The unit vector $\hat{\mathbf{k}}_o$, representing the direction between the centre of the raw camera frame and the vertically central camera point, is then given by

$$\hat{\mathbf{k}}_o = \sin\left(\vartheta_\xi - \frac{\Phi_C}{2} \right) \hat{\mathbf{e}}_x + \cos\left(\vartheta_\xi - \frac{\Phi_C}{2} \right) \hat{\mathbf{e}}_z \tag{10.62}$$

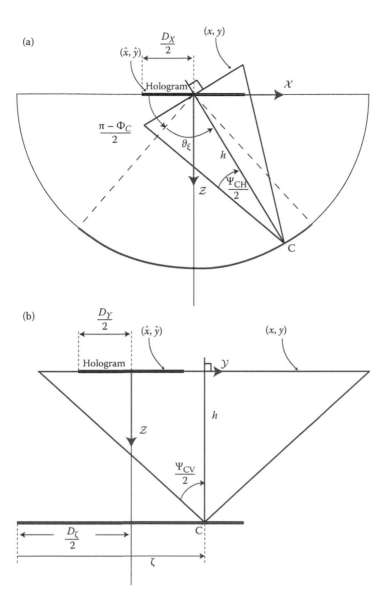

FIGURE 10.18 Full-parallax image data acquisition: geometry for a horizontal rotating platform with vertically forward-facing linear-translating camera. (a) Plan view and (b) side view.

We can use this vector to define the unit vectors of the (x,y) system as

$$
\hat{\mathbf{e}}_x = -\frac{\hat{\mathbf{k}}_o \times \hat{\mathbf{e}}_y}{\left|\hat{\mathbf{k}}_o \times \hat{\mathbf{e}}_y\right|} = -\frac{1}{\left|\hat{\mathbf{k}}_o \times \hat{\mathbf{e}}_y\right|}
\begin{vmatrix}
\hat{\mathbf{e}}_x & \hat{\mathbf{e}}_y & \hat{\mathbf{e}}_z \\
\sin\left(\vartheta_\xi - \dfrac{\Phi_C}{2}\right) & 0 & \cos\left(\vartheta_\xi - \dfrac{\Phi_C}{2}\right) \\
0 & 1 & 0
\end{vmatrix}
$$

$$
= \hat{\mathbf{e}}_x \cos\left(\vartheta_\xi - \frac{\Phi_C}{2}\right) - \sin\left(\vartheta_\xi - \frac{\Phi_C}{2}\right)\hat{\mathbf{e}}_z
\tag{10.63}
$$

and

$$\hat{\mathbf{e}}_y = \frac{\mathbf{k}_o \times \hat{\mathbf{e}}_x}{\left|\mathbf{k}_o \times \hat{\mathbf{e}}_x\right|} = \frac{1}{\left|\mathbf{k}_o \times \hat{\mathbf{e}}_x\right|} \begin{vmatrix} \hat{\mathbf{e}}_x & \hat{\mathbf{e}}_y & \hat{\mathbf{e}}_z \\ \sin\left(\vartheta_\xi - \dfrac{\Phi_C}{2}\right) & 0 & \cos\left(\vartheta_\xi - \dfrac{\Phi_C}{2}\right) \\ 1 & 0 & 0 \end{vmatrix} = \hat{\mathbf{e}}_y \tag{10.64}$$

A general point on the plane (x,y) may then be written as

$$\begin{aligned}
\mathbf{P}(x,y) = \mathbf{\Theta} + \hat{\mathbf{e}}_x x + \hat{\mathbf{e}}_y y &= \hat{\mathbf{e}}_x\left\{ x\cos\left(\vartheta_\xi - \frac{\Phi_C}{2}\right) - h\tan\left(\frac{\Psi_{CH}}{2}\right)\cos\left(\vartheta_\xi - \frac{\Phi_C}{2}\right) \right\} \\
&+ \hat{\mathbf{e}}_y\left\{ \zeta - D_\zeta/2 - h\tan\left(\frac{\Psi_{CV}}{2}\right) + y \right\} \\
&+ \hat{\mathbf{e}}_z\left\{ h\tan\left(\frac{\Psi_{CH}}{2}\right)\sin\left(\vartheta_\xi - \frac{\Phi_C}{2}\right) - x\sin\left(\vartheta_\xi - \frac{\Phi_C}{2}\right) \right\}
\end{aligned} \tag{10.65}$$

where $\mathbf{\Theta}$ is a world vector describing the origin of the raw camera frame

$$\begin{aligned}
\mathbf{\Theta} &= -\hat{\mathbf{e}}_x h\tan\left(\frac{\Psi_{CH}}{2}\right) - \hat{\mathbf{e}}_y h\tan\left(\frac{\Psi_{CH}}{2}\right) + (\zeta - D_\zeta/2)\hat{\mathbf{e}}_y \\
&= -\hat{\mathbf{e}}_x\left\{ h\tan\left(\frac{\Psi_{CH}}{2}\right)\cos\left(\vartheta_\xi - \frac{\Phi_C}{2}\right) \right\} + \hat{\mathbf{e}}_y\left\{ \zeta - D_\zeta/2 - h\tan\left(\frac{\Psi_{CV}}{2}\right) \right\} \\
&+ \hat{\mathbf{e}}_z\left\{ h\tan\left(\frac{\Psi_{CH}}{2}\right)\sin\left(\vartheta_\xi - \frac{\Phi_C}{2}\right) \right\}
\end{aligned} \tag{10.66}$$

We can now define the relationship between the (x,y) and $(\hat{x}, \hat{y})$ coordinates by considering the intersection of a test ray passing through the camera point and general points on both coordinate planes:

$$\mathbf{P}(x,y) + \lambda \mathbf{k} = \mathbf{Q}(\hat{x}, \hat{y}) \tag{10.67}$$

These three equations yield λ, the ray extension parameter, and the required coordinate transformation equations:

$$\begin{aligned}
x &= h\sec\left(\frac{\Psi_{CH}}{2}\right)\frac{\left(2h\sin\left(\dfrac{\Psi_{CH}}{2}\right) - \left(D_X - 2\hat{x}\right)\cos\left\{\dfrac{1}{2}(\Phi_C - \Psi_{CH}) - \vartheta_\xi\right\}\right)}{2h - \left(D_X - 2\hat{x}\right)\sin\left\{\dfrac{\Phi_C}{2} - \vartheta_\xi\right\}} \\
y &= h\tan\left(\frac{\Psi_{CV}}{2}\right) + \frac{h(2\hat{y} - D_Y + D_\zeta - 2\zeta)}{2h - (D_X - 2\hat{x})\sin\left\{\dfrac{\Phi_C}{2} - \vartheta_\xi\right\}}
\end{aligned} \tag{10.68}$$

To calculate the centred camera data set $({}^{k}\hat{\mathbf{I}}_{\hat{i}\hat{j}})$ from the raw camera data set $({}^{k}\mathbf{I}_{ij})$, we must now use the appropriate integer coordinates:

$$x = 2h\tan\left(\frac{\Psi_{CH}}{2}\right)\left(\frac{i-1}{N_I-1}\right) \quad i = 1,2,3,...,N_I$$

$$y = 2h\tan\left(\frac{\Psi_{CV}}{2}\right)\left(\frac{j-1}{N_J-1}\right) \quad j = 1,2,3,...,N_J$$

(10.69)

and

$$\theta_\xi = \frac{(k-1)}{(N_K-1)}\Phi_C \quad k = 1,2,3,...,N_K$$

$$\zeta = \frac{(g-1)}{(N_G-1)}D_\varsigma \quad g = 1,2,3,...,N_G$$

(10.70)

As in Section 10.2.2, we can then use Equations 10.68 through 10.70 with 10.9 to define the two real functions, $\Omega(\hat{i},k)$ and $\Upsilon(\hat{i},\hat{j},k,g)$, which constitute real estimates of the integers i and j:

$$\Omega(\hat{i},k) = \frac{1}{2} + (N_I+1)$$

$$+ \frac{D_X(N_I-1)(2\hat{i}-\hat{N}_I-1)\cos\left[\frac{\Phi_C}{2}\left\{\frac{(1-2k+N_K)}{1-N_K}\right\}\right]\cot\left(\frac{\Psi_{CH}}{2}\right)}{4h(\hat{N}_I-1)+2D_X(1-2\hat{i}+\hat{N}_I)\sin\left[\frac{\Phi_C}{2}\left\{\frac{(1-2k+N_K)}{1-N_K}\right\}\right]}$$

(10.71)

$$\Upsilon(\hat{i},\hat{j},k,g) = \frac{1}{2}(N_J+1)$$

$$+ \frac{(\hat{N}_I-1)(N_J-1)\left\{\frac{D_Y(N_G-1)(2\hat{j}-1-\hat{N}_J)}{(\hat{N}_J-1)} - D_\zeta(2g-1-N_G)\right\}\cot\left(\frac{\Psi_{CV}}{2}\right)}{2(N_G-1)\left\{2h(\hat{N}_I-1)+D_X(1-2\hat{i}+\hat{N}_I)\sin\left[\frac{\Phi_C}{2}\left\{\frac{1-2k+N_K}{1-N_K}\right\}\right]\right\}}$$

(10.72)

It is then a simple matter to define the centred data set via the transformation

$$^{kg}\hat{\mathbf{I}}_{\hat{i}\hat{j}} = {}^{kg}\mathbf{I}_{ij} \quad \forall\hat{i}\left\{\hat{i}\in\mathbb{N}\middle|\hat{i}\leq\hat{N}_I\right\} \quad \forall\hat{j}\left\{\hat{j}\in\mathbb{N}\middle|\hat{j}\leq\hat{N}_J\right\}$$

$$\forall k\left\{k\in\mathbb{N}\middle|k\leq N_K\right\} \quad \forall g\left\{g\in\mathbb{N}\middle|g\leq N_G\right\}$$

(10.73)

where, if we use nearest integer interpolation,

$$i \equiv \left\|\Omega(\hat{i},k)\right\|$$

$$j \equiv \left\|\Upsilon(\hat{i},\hat{j},k,g)\right\|$$

(10.74)

Or using bilinear interpolation,*

$$
{}^{kg}\mathbf{I}_{ij} = \frac{{}^{kg}\mathbf{I}_{\lfloor\Omega\rfloor\lfloor\Upsilon\rfloor}}{\big(\lceil\Omega\rceil-\lfloor\Omega\rfloor\big)\big(\lceil\Upsilon\rceil-\lfloor\Upsilon\rfloor\big)}\big(\lceil\Omega\rceil-\Omega\big)\big(\lceil\Upsilon\rceil-\Upsilon\big)
$$

$$
+\frac{{}^{kg}\mathbf{I}_{\lceil\Omega\rceil\lfloor\Upsilon\rfloor}}{\big(\lceil\Omega\rceil-\lfloor\Omega\rfloor\big)\big(\lceil\Upsilon\rceil-\lfloor\Upsilon\rfloor\big)}\big(\Omega-\lfloor\Omega\rfloor\big)\big(\lceil\Upsilon\rceil-\Upsilon\big)
$$

$$
+\frac{{}^{kg}\mathbf{I}_{\lfloor\Omega\rfloor\lceil\Upsilon\rceil}}{\big(\lceil\Omega\rceil-\lfloor\Omega\rfloor\big)\big(\lceil\Upsilon\rceil-\lfloor\Upsilon\rfloor\big)}\big(\lceil\Omega\rceil-\Omega\big)\big(\Upsilon-\lfloor\Upsilon\rfloor\big)
$$

$$
+\frac{{}^{kg}\mathbf{I}_{\lceil\Omega\rceil\lceil\Upsilon\rceil}}{\big(\lceil\Omega\rceil-\lfloor\Omega\rfloor\big)\big(\lceil\Upsilon\rceil-\lfloor\Upsilon\rfloor\big)}\big(\Omega-\lfloor\Omega\rfloor\big)\big(\Upsilon-\lfloor\Upsilon\rfloor\big)
\tag{10.75}
$$

Following the optimisation procedure introduced in Sections 10.2.2 and 10.2.3, we can now define the minimal fixed FOV a camera must have to just fit all of the image data required:

$$
\Psi_{\text{CH}} = 2\tan^{-1}\left\{\frac{D_X\sin\left(\dfrac{\Phi_C}{2}+\sin^{-1}\left[\dfrac{\sqrt{2}\sqrt{h^2\big(4h^2-D_X^2\big)\{1+\cos(\Phi_C)\}}-2D_X h\sin\left(\dfrac{\Phi_C}{2}\right)}{4h^2}\right]\right)}{2h-D_X\cos\left(\dfrac{\Phi_C}{2}+\sin^{-1}\left[\dfrac{\sqrt{2}\sqrt{h^2\big(4h^2-D_X^2\big)\{1+\cos(\Phi_C)\}}-2D_X h\sin\left(\dfrac{\Phi_C}{2}\right)}{4h^2}\right]\right)}\right\}
$$

$$
\Psi_{\text{CV}} = 2\tan^{-1}\left\{\frac{D_Y+D_\zeta}{2h-D_X\sin\left(\dfrac{\Phi_C}{2}\right)}\right\}
\tag{10.76}
$$

10.3.1.2 Derivation of the I-to-S Transformation

The derivation of the **I**-to-**S** transformation very closely follows the procedure we developed in Section 10.2.7. There are two obvious differences: the first is that we need to extend the world coordinates to the full three dimensional system $(\mathcal{X},\mathcal{Y},\mathcal{Z})$ as used in the last section. The second is the introduction of the vertical analogue of Equation 10.44:

$$
V' = 2h^*\tan\left(\frac{\Psi_{\text{PV}}}{2}\right)\frac{(v'-1)}{(N_V-1)}
\tag{10.77}
$$

With these changes in hand, we may rewrite Equation 10.48 as follows

$$
\begin{pmatrix} \mathcal{X} \\ \mathcal{Y} \\ \sqrt{h^2-\mathcal{X}^2} \end{pmatrix} = \begin{pmatrix} X-D_X/2 \\ Y-D_Y/2 \\ 0 \end{pmatrix} + \lambda\begin{pmatrix} U'-h^*\tan(\Psi_{\text{PH}}/2) \\ V'-h^*\tan(\Psi_{\text{PV}}/2) \\ h^* \end{pmatrix}
\tag{10.78}
$$

---

* See Appendix 7 for a more detailed description of bilinear and bicubic interpolation methods.

from which we derive

$$
\mathcal{X} = \frac{\left[\dfrac{D_X}{2}\right]\left(\dfrac{2\alpha - 1 - N_A}{(N_A - 1)}\right) \pm \sqrt{\varpi}}{\left\{1 + \tan^2\left(\dfrac{\Psi_{PH}}{2}\right)\left(2\dfrac{(\mu' - 1)}{(N_M - 1)} - 1\right)^2\right\}}
\tag{10.79}
$$

where

$$
\varpi = \left[\frac{D_X}{2}\right]^2 \left[\frac{2\alpha - 1 - N_A}{(N_A - 1)}\right]^2 + \left\{1 + \tan^2\left(\frac{\Psi_{PH}}{2}\right)\left(2\frac{(\mu' - 1)}{(N_M - 1)} - 1\right)^2\right\}
$$

$$
\times \left\{h^2 \tan^2\left(\frac{\Psi_{PH}}{2}\right)\left(2\frac{(\mu' - 1)}{(N_M - 1)} - 1\right)^2 - \left[\frac{D_X}{2}\right]^2 \left[\frac{2\alpha - 1 - N_A}{(N_A - 1)}\right]^2\right\}
\tag{10.80}
$$

and

$$
\mathcal{Y} = D_Y \frac{(2\beta - 1 - N_B)}{2(N_B - 1)} + \frac{\tan\left(\dfrac{\Psi_{PV}}{2}\right)(N_M - 1)(2\nu' - 1 - N_V)}{\tan\left(\dfrac{\Psi_{PH}}{2}\right)(N_V - 1)(2\mu' - 1 - N_M)}\left\{\mathcal{X} - D_X \frac{(2\alpha - 1 - N_A)}{2(N_A - 1)}\right\}
\tag{10.81}
$$

The camera position (k, g), is then related to $(\mathcal{X}, \mathcal{Y})$ by the expressions

$$
k = \left\|\frac{(N_K + 1)}{2} + \frac{(N_K - 1)}{\Phi_C}\sin^{-1}\left[\frac{\mathcal{X}}{h}\right]\right\|
$$

$$
g = \left\|1 + \mathcal{Y}\frac{(N_G - 1)}{D_\varsigma}\right\|
\tag{10.82}
$$

The full **I**-to-**S** transformation for an inverting camera and a conjugate SLM position may finally be written

$$
^{\mu\nu}\mathbf{S}_{\alpha\beta} = {}^{kg}\hat{\mathbf{I}}_{\hat{i}\hat{j}} \quad \forall\alpha\left\{\alpha \in \mathbb{N} \middle| \alpha \le N_A\right\} \qquad \forall\beta\left\{\beta \in \mathbb{N} \middle| \beta \le N_B\right\}
$$

$$
\forall\mu\left\{\mu \in \mathbb{N} \middle| N_{U1}(\alpha) \le \mu \le N_{U2}(\alpha)\right\} \quad \forall\nu\left\{\nu \in \mathbb{N} \middle| N_{V1}(\beta) \le \nu \le N_{V2}(\beta)\right\} \tag{10.83}
$$

$$
= 0 \qquad \text{otherwise}
$$

where

$$
\hat{i} = N_I - \alpha + 1
$$

$$
\hat{j} = N_J - \beta + 1
$$

$$
k = \left\|\frac{(N_K + 1)}{2} + \frac{(N_K - 1)}{\Phi_C}\sin^{-1}\left[\frac{\mathcal{X}}{h}\right]\right\|
$$

$$
g = \left\|1 + \mathcal{Y}\frac{(N_G - 1)}{D_\varsigma}\right\|
\tag{10.84}
$$

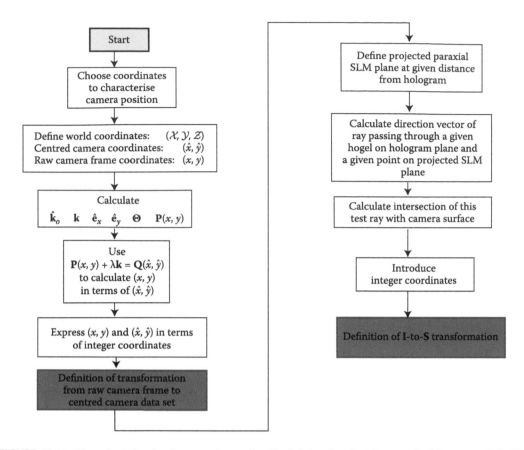

FIGURE 10.19 Flow chart showing the general procedure for defining the algorithms required for a general double-parallax data acquisition (holocam) system. In total, there are 16 basic variants. In each case, one must define an algorithm for the calculation of the centred camera data set and an **I**-to-**S** transformation.

The limit expressions for $N_{U1}(\alpha)$, $N_{U2}(\alpha)$, $N_{V1}(\beta)$ and $N_{V2}(\beta)$ may be obtained by setting, respectively, $k = 1$ and $k = N_K$ in Equations 10.79, 10.80 and 10.82 and $g = 1$ and $g = N_G$ in Equations 10.81 and 10.82.

The procedure that we have followed to analyse the present camera configuration is summarised in Figure 10.19. This method may be mechanically applied to any of the 16 possible camera configurations to define the algorithm for both the generation of the centred camera frame in addition to the required **I**-to-**S** transformation.

10.3.2 3D Structured-Light Scanners

For HPO applications, the holocam is really the instrument of choice. It provides a high-resolution data set ideal for the creation of this type of hologram and can be fabricated for a reasonable price. With double-parallax scanning, however, the situation becomes a little less clear. If the object is static, then a two-dimensional holocam may be used and indeed this technique will always provide the highest quality solution. This is because a holocam generates exactly the information required to produce the hologram. Applications, such as cultural heritage or specimen archival, will as such nearly always seek to use a two-dimensional holocam. For certain applications, however, there is sometimes an advantage in using an alternative method that seeks to create an intrinsically 3D model of the scene that is to be reproduced as a hologram. Although a rendering engine must then be used to generate the multiple photographs needed to create the hologram, this approach does allow real-world data to be seamlessly mixed with virtual computer models, something that can be tricky with holocam data. In fact, if the integration of

FIGURE 10.20 Photograph of the low-cost Pico Scan structured-light scanner from 4DDynamics. (Photograph supplied courtesy of 4DDynamics BVBA.)

virtual and real objects is a vital requirement, then such alternative methods may even be considered for HPO holograms.

A data acquisition system that is able to produce such 3D model files directly is called the "structured-light scanner". At the time of writing, the price-point for such scanners is dropping quickly and useful commercial devices are starting to appear. However, the time of data acquisition is still rather high for all but the highest-range models—usually around the 20 s mark, making such systems more suitable for static objects rather than for human subjects. An example of a low-cost system is the PicoScan product from the Belgian company 4DDynamics, which was priced at the time of writing at just under €3000 (Figure 10.20).

Structured-light scanners use a projector to project a pattern of lines onto the object to be scanned; a camera then photographs the distorted lines and uses this distortion data to infer the geometry through a simple mathematical transform. Problems occur when the structured light disappears down a hole and the camera or cameras cannot see what is happening in this region; reflective and translucent materials* provide additional problems. Equally, lines can cross-over and it can be difficult to tell which line is which. In such cases, the mathematical transformation becomes singular and it can be difficult or impossible to reconstruct the exact geometry. However, by using more complex patterns of the structured light, time modulation, many cameras and more complex image-processing algorithms, it is possible to address the most frequently encountered of these problems.

An application that has driven fast high-end, structured-light scanners is the film industry, where full-body scanning of human subjects is being increasingly called for. Systems here often integrate many cameras and projectors and a computer then seamlessly integrates all the data. Frequently, algorithms are used to correct for slight movements during the data acquisition period such as heartbeats or breathing. Scanning times can be reduced in these systems to well under a second. This, however, comes at a cost. Such systems are extremely expensive. For example, a minimum full-body scanner setup from 4DDynamics consists of four CX-PRO scanners (Figures 10.21 and 10.22) each costing approximately €15,000 with a typical integrated system including installation and software, coming in at just under the €100,000 mark. Although the output can be in a point-cloud format, most 4DDynamics users choose to output a mesh including textures. Postprocessing software, which is included by the company, permits the user to opt for vertex colour output or (UV) texture map to a watertight mesh. Depending on the settings, a merging of four raw body scans on a standard laptop takes approximately 6 min.

---

* Human skin can, in particular, present problems due to the phenomenon of subsurface scattering.

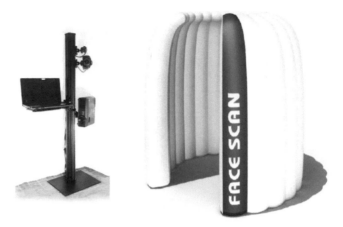

FIGURE 10.21   Photograph of the high-end CX-PRO 4DDynamics structured-light scanner suitable for body-scanning applications. (Photograph supplied courtesy of 4DDynamics BVBA.)

FIGURE 10.22   Rendered images produced from the 3D watertight mesh and texture maps generated by a CX-PRO 4DDynamics structured-light scanner. (Photograph supplied courtesy of 4DDynamics BVBA.)

10.4 Images Derived from a Virtual Computer Model: HPO Holograms

In this section, we will cover how the popular 3D design program 3D StudioMax version 8 can be programmed to produce a rendered image sequence suitable for the production of an HPO reflection hologram. Many other commercial programs are capable of generating the type of image data required for digital display holograms, but space prevents us from discussing each programming environment.

Before proceeding, however, we shall mention a few programs that are particularly suited to generating the image data for direct-write and master-write digital holograms (DWDH and MWDH). First and foremost is the open-source program Blender, which uses a scripting language based on Python. This is an excellent place to start with 3D modelling for digital holography. Other commercial programs to note at the time of writing are SoftImage, Cinema 4D, Maya and LightWave.

10.4.1 Data Preparation

Each digital holographic printer will have its own unique recommended parameter set. This will be available from the manufacturer or from the printer operator in the case of a service bureau. For the purposes of this section, we shall use information published by Geola Digital UAB, which operates

TABLE 10.1

Recommended Printing Parameters for HPO Reflection Holograms

Hologram Height (mm)	Hologram Width (mm)	Horizontal Camera FOV (°)	Camera Track (mm)	Camera Distance (mm)	Frames Rendered (High Angular Resolution)	Frames Rendered (Low Angular Resolution)	Image Size (0.8 mm Hogel)		Image Size (1.6 mm Hogel)	
							Pixels	Pixels	Pixels	Pixels
200	×300	82.79	793.6	619.0	497	—	1366	×250	683	×125
300	×400	82.72	1060.8	828.7	664	640	1826	×374	913	×187
490	×640	82.83	1704.0	1327.9	1066	640	2930	×612	1465	×306
640	×900	82.81	2395.2	1867.0	1498	640	4118	×800	2059	×400
999	×1250	82.83	3331.2	2595.8	2083	640	5726	×1248	2863	×624
999	×1500	82.85	4000.0	3115.0	2501	640	6874	×1248	3437	×624
300	×200	77.41	793.6	619.0	497	—	1242	×374	621	×187
490	×330	77.48	1300.8	1015.0	814	640	2038	×612	1019	×306
640	×490	79.07	1704.0	1327.9	1066	640	2742	×800	1371	×400
999	×750	78.83	2660.8	2073.3	1664	640	4262	×1248	2131	×624

Source: Courtesy of Geola Digital UAB.

a DWDH reflection holography printing service from its offices in Vilnius, Lithuania. Table 10.1 lists Geola's recommended parameter set. Here, the supported formats are listed as well as a series of parameters for each such format. These include the camera distance, h, the required camera track length, $D_\xi = W$, the horizontal camera FOV, Ψ_{CH}, the number of rendered frames, N_K and the pixel dimensions of the individual photographs, N_I and N_J. Note that Geola offers two hogel diameters, 0.8 and 1.6 mm. It also offers a high and low angular resolution for scenes with large or small depths.

To illustrate how a file is prepared in 3D StudioMax, we will treat in detail the case of a 20 cm × 30 cm landscape format high-resolution hologram having a hogel diameter of 0.8 mm. Table 10.1 may then be used to create any of the other formats by simply replacing the relevant parameters in the description given below.

10.4.2 Creating a 20 cm × 30 cm Landscape Hologram

We will review here the simplest technique in 3D StudioMax to create a data set for the *simply translating camera*. In Section 10.5, we will cover centred camera configurations in the context of double parallax. These techniques, which use the scripting capabilities of Max, can easily be applied to the HPO case. For now, however, we will use the animation facility in Max to generate the required HPO data set.

We start by launching 3D StudioMax with the scene we wish to use. Note that the model we use here (a Ducati motorbike) has been scaled to a larger dimension than required for the final holographic image. We shall therefore use a scale of 1:10 to reduce the image to an appropriate size for our present application. Note that the world coordinates (0,0,0) represent the exact centre of the hologram.

First, we must go to the time configuration panel to set the number of animation frames, N_K (right click on any animation control at the bottom right—i.e., the "►" button). The animation end time will need to be changed from 100 to 496. Press "OK" when you have done this. This will give a total of 497 frames in the animation sequence, which is what we require from Table 10.1 (Figure 10.23).

Now we create a "Free" camera that faces (0,0,0) and is positioned at a distance of 619 mm in front of the hologram.* You can do this by selecting the camera panel from the top right-hand graphical user interface (GUI) menu, selecting **Free** and then clicking anywhere in the "Front" window. Then, ensuring that the free movement icon is selected (the crossed arrows in the top row of large GUI icons) enter the (X,Y,Z) coordinates at the bottom centre of the screen as −396.8, −619.0 and 0.0.

---

* Using the 10:1 scale, we therefore set a distance of 619 cm in Max.

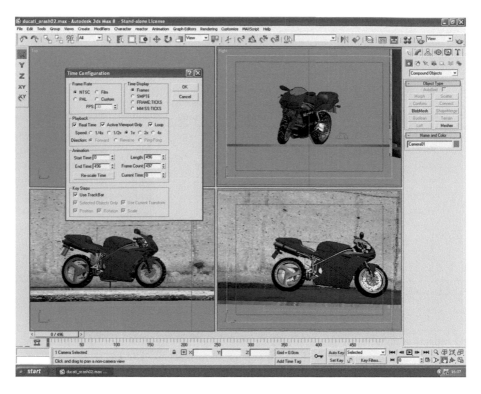

FIGURE 10.23   To set the number of frames in the animation sequence, open the time configuration dialogue in Max by right-clicking the "▶" button at the bottom right-hand corner of the GUI.

We can enter the starting *X* value for the camera animation: To do this, activate the **Set Key** toggle button* at the bottom left of the screen, type −396.8[†] in the central *X* box and click on the key icon to the immediate left of the **Set Key** toggle control. The default frame is 0, and so this sets frame 0 to an *X* value of −396.8. Move the slider now to go to frame 496; this time, enter a value of +396.8 in the *X* box and click on the key icon. Press the **Set Key** button again to exit the key framing mode. At this point, you should have a camera travelling from left to right from frame 0 to 496.[‡]

The next thing we need to do is to make the path of the camera linear. Right click on the camera and select the curve editor. In the curve editor, select World > Objects > Camera01 > Transform > Position > *X* position. The screen should then look like Figure 10.24. Make the curve linear by selecting the two end points in turn and clicking the **Set Tangents to Linear** icon[§] in the top menu. The graph should then look like that of Figure 10.25.

We now need to set up the FOV of the camera. To do this, select the camera and click on the modify panel.[¶] Change the **FOV** value to 82.79° and make sure the little icon to the left looks like two horizontal arrows (Figure 10.26). You can also tick the **Show Cone** option and select a **Near Range** of 0 cm and a **Far Range** of 619 cm. This will draw in each of the design windows the camera's cone of vision just up to the hologram surface.

---

* The animation slider bar will now turn red.
† Again, this is actually −396.8 cm given the 10:1 scale we are using.
‡ You can verify this by moving the slider and watching the camera animate.
§ This is the icon with a 45° red line.
¶ The modify panel is the icon that looks like a quarter tyre at the top right of the screen (second to the left, second row down).

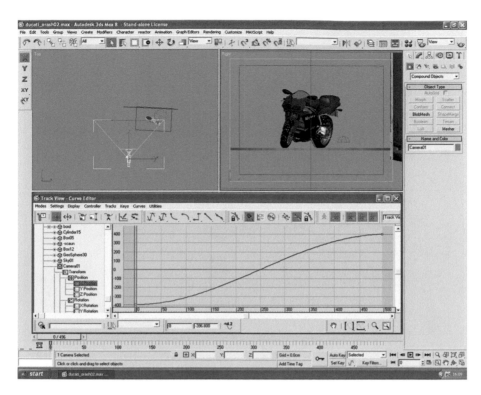

FIGURE 10.24   Setting the camera animation through the Curve Editor in Max, which can be accessed by right-clicking on the camera and selecting **Curve Editor** from the drop-down menu.

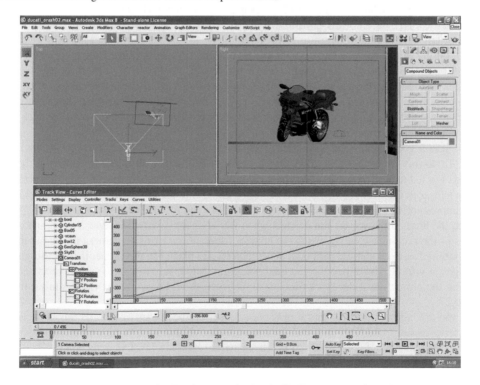

FIGURE 10.25   By selecting the two end-points in turn and using the **Set Tangents to Linear** button, the position animation of the camera becomes linear.

FIGURE 10.26 Setting the render parameters in Max from the render dialogue.

To set the camera resolution (Figure 10.26), open the render menu (F10 in windows or the teapot icon at the top of the GUI) and enter the **Output Size** panel **Width: 1366** and **Height: 250**. Choose the **Active Time Segment** option in the **Common Parameters** panel to tell Max to render all your files. Tick the **Save File** option in the render Output panel and then input a rendering directory by pressing the **Files** button. Make sure that the renderer is set to output bitmaps in 24 bits (16 M colours, no alpha channels). Also, check that the **ViewPort** at the bottom of the render panel is set to the active camera. You will need to define one of the viewports as the camera port (e.g., Camera01) before the drop-down list will allow you to select the camera. You do this by right-clicking a viewport name (at the top-left of each viewport and selecting the camera view from the Views entry).

Finally, it is a good idea to create a rectangular hologram frame to make sure everything is aligned properly. To do this, create a 300 × 200 rectangle centred at (0,0,0) with normal pointing in the Y direction.* This is used as a reference to check how the hologram will look. Everything that is in front of this rectangle (closer to the camera) will appear to "come out" of the hologram and everything behind it will give the impression of being "inside" the hologram. Obviously, you do not want to render this "hologram frame" on the final output so you can delete it when you are happy with the scene. You should check that the edges of the rectangle exactly match the camera frame at the extreme camera positions; this means that everything has been entered properly. When objects are placed in the scene, if something is clipped at the edges, it will also be clipped in the final hologram. Make sure to centre your objects accordingly. By moving the animation slider from frame 0 to 496, you should now obtain views similar to those of Figure 10.27.

When you are happy with the setup, simply press render on the render menu, and this will generate the required series of 496 24-bit BMP files each having dimensions of 1366 × 250. Figure 10.28 shows a selection of the 497 files generated. Figure 10.29 shows the actual 20 × 30 cm DWDH reflection hologram together with a 1 m × 1 m format made using the same Max file and the corresponding parameters listed in Table 10.1.

---

* You might want to extrude the rectangle and then use a Boolean operation under the compound object menu to create a more visible picture frame.

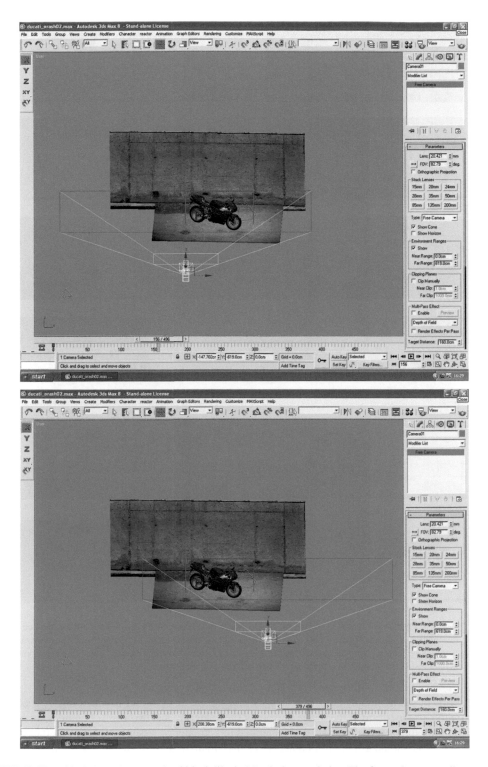

FIGURE 10.27 This is how the scene should look like in Max before rendering. The figure shows two diagrams, one with the animation slider at frame number 156 and the other at frame number 379. The first frame is frame 0 and the last frame is 496.

FIGURE 10.28 Six of the 497 rendered images of the Ducati motorbike produced by the animation sequence described in the text: (a) image 1, (b) image 100, (c) image 200, (d) image 300, (e) image 400 and (f) image 497. The camera geometry is a *simply translating camera*. Note the green frame that has been included to emphasise the hologram location within the data set. The data inside the green frame corresponds to a centred camera data set.

A good rule of thumb for any hologram is the following: for an optimum volume effect, generally one should aim to have one-third of the image in front of the hologram and two-thirds behind. Of course, this depends on the type of scene that is being generated to a greater or lesser extent. For a 30 cm × 20 cm hologram, the user should usually limit objects to a maximum of approximately 6 to 7 cm in front of the film; this limit not only ensures the sharpness of the object (particularly when using silver halide emulsions and non-diode illumination) but also prevents too much clipping.

For animation in an HPO hologram, it must be slow and subtle. It must not be treated like video (fast panning, transitions, and zooms are a definite no). Good examples are a cross-dissolve of an object to reveal the interior (given that the dissolve envelope is very long and linear), a character slowly waving hands, etc. Bad

FIGURE 10.29 Photograph of (a) final 20 cm × 30 cm HPO DWDH refection hologram and (b) photograph of a larger format (1 m × 1 m) HPO DWDH reflection hologram produced from the same Max file. Both holograms have a hogel diameter of 0.8 mm. (Image created in 3D StudioMax by Razvan Maftei, 2005. Photographs courtesy of Geola Technologies Ltd.)

FIGURE 10.30 Three photographs showing a large-format (1 m × 1.5 m) HPO DWDH hologram of a computer-created and completely life-like great white shark. The hologram was installed behind a metal cage in the SeaLife centre in Brighton, UK, to simulate viewing from an underwater cage. The hologram featured a slow animation as is clear from the pictures. (a) Installation of the hologram in 2005, (b) view from the right and (c) view from the left. (Image created in 3D StudioMax by Razvan Maftei, 2005. Photographs courtesy of Geola Technologies Ltd.)

examples are systems of movement that act to mimic the parallax; these will change the perceived depth of the hologram—sometimes radically! Also, every animation key frame should be linear; never use splines!

Holograms are replayed using light; they will reflect or transmit light information according to the data files used to print them; making a scene in darkness like a scene from *Batman* at night is unlikely to be as good as the cliché "bowl of fruits" given the performance of current silver halide emulsions. Scenes should therefore be bright and contrasted. Avoid using ambient lighting to compensate for the lack of overall brightness. This can kill the contrast of the scene. A good hologram always has a good lighting rig in the 3D creation software.

One further point to make about HPO holograms is that "streaking" across the physical surface of the hologram can appear. This phenomenon is caused by bright spots close to the hologram surface. Although such bright spots stand still as the observer moves from right to left, if a bright spot is situated slightly in front of the hologram surface, a corresponding high-intensity spot will appear to track from right to left on the hologram surface. This causes a line of high-intensity light at the hologram surface that exceeds the normal

FIGURE 10.31   Two HPO DWDH holograms created from computer models. The top image shows a hologram installed by Geola in the NIKE store in Oxford St., London in 2005. The bottom image is of an early HPO DWDH hologram created by XYZ Imaging Inc., circa 2003/2004, seen here with the company's founding president Eric Bosco. Note that the bottom image is made up from nine "tiles" which have been printed separately.

linear response of the emulsion, causing a "burn". As well as being unsightly, the hologram replay characteristics within the burnt area are modified and the colour balance and brightness of objects behind the burn can change when viewed through the overexposed area. This effect is similar to the effect seen in analogue holograms, in which light is focussed too much at a given point on the surface of the hologram leading to a burn pattern. The difference with HPO holograms is that the burn pattern is nearly always elongated into a line due to the HPO nature of the recording. As such, one needs to be extremely careful when using highly reflective surfaces in HPO holograms if employing recording materials with a limited dynamic recording range.

10.5 Images Derived from a Virtual Computer Model: Full-Parallax Holograms

We have seen in previous sections how a linear tracking camera can be programmed in 3D StudioMax to generate all the virtual camera frames required for an HPO hologram. In this section, we will discuss how to program a virtual camera that tracks in two dimensions to generate the camera frames required for a full-parallax digital hologram. To do this, we will use the scripting language of Max known as MAXScript. This approach is in fact often used with HPO holograms as well. In particular, MAXScript allows one to directly produce a *centred camera* output rather than a *simply translating camera* output. Most 3D design programs offer scripting languages and this is often the most convenient technique.

10.5.1 Creating a MAXScript GUI

We will now discuss how to program using MAXScript, a basic script that automatically generates the rendered camera images (using a centred camera configuration) required for the printing of a double parallax three-colour DWDH reflection hologram. To do this, we will use the results from Chapter 9 and, in particular, we will use a first quantised camera distance to achieve a DWDH data set free from interpolation errors.

First of all, Max must be started and the chosen 3D model file opened. Once this has been done, click on the **MAXScript** menu in the top bar of the Max GUI and select **New Script** from the drop-down menu. A blank MAXScript editing window will now appear. Select **Edit** from the top bar menu of the MAXScript edit window and choose **Edit Rollout**. This will bring up the Visual MAXScript editing window (Figure 10.32). This is where you will be able to design the GUI for the MAXScript program.

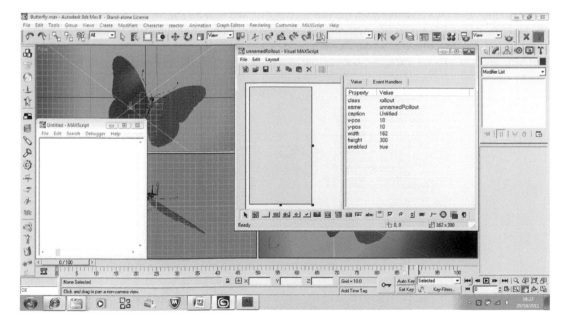

FIGURE 10.32 Visual MAXScript editing window.

The visual MAXScript window, by default, contains a blank panel and, at the bottom, a list of possible buttons and functions that can be added to this panel. The panel to the right lets you set the names and values of these controls. It also lets you tell Max what code to generate when an event is generated.

The first thing we shall do is to add a button. We will use this button to start the animation sequence of the rendering process, so we shall call it "Start Render". To add the button to the panel, simply click once on the blank square icon on the bottom menu row of the Visual MAXScript editing window. Then click, hold and drag out the shape of the button desired on the panel. Now go to the right-hand windowpane and select the **Value** tab. Here, you can fill out a name for the button—we shall use BRen—and the caption "Start Render" (Figure 10.33).

By clicking on the panel itself, you will bring up in the right-hand pane the details of the panel itself. We will change the name and caption to "DoubleParallaxRender".

Because it is good practice to save your work from time to time, click on the disk symbol in the top icon menu bar of the visual MAXScript editing window now or, alternatively, select **Save** from the drop-down **File** menu. You will see that this immediately generates the following code in the MAXScript text editing window:

```
Rollout DoubleParallaxRender "DoubleParallaxRender" width:162 height:300
(
Button BRen "Start Render" pos:[11,24] width:132 height:44
)
```

This code simply defines the button you have created. Braces indicate sections of self-contained code. To prompt Max to include code for an event handler, simply select the button and then choose the **Event Handler** tab in the right-hand windowpane. Tick the **pressed** option and then save again using the disk symbol. The text edit window should now look like this:

```
rollout DoubleParallaxRender "DoubleParallaxRender" width:162 height:300
(
        button BRen "Start Render" pos:[11,24] width:132 height:44
        on BRen pressed do
        (
        )
)
```

By placing your own code between the two new braces, you will be able to tell Max what to do when the button is pressed. For now, however, we will continue with populating the panel by adding three more buttons: "Make Directories", "Make Camera" and "Calculate", We will name these buttons "BMD", "BCam" and "BCal". For each, we must create event handlers (tick the **pressed** option). We will then

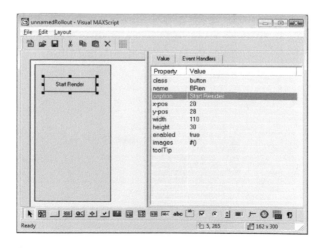

FIGURE 10.33 Programming the "Start Render" button in the Visual MAXScript editing window.

use the "Group Box" control to lay things out sensibly. This is the MAXScript control that looks like a little square with an "ab" at its centre top. Click on this control and then click, hold and drag until a box appears on the panel. The group box is useful for keeping bits of information together—you can change the caption in the **value** tab to the right of the visual MAXScript editor. We will want to make six group boxes which we will entitle "SLM Definition", "Printer Optics Definition", "Window Parameters", "Hologram Definition", "Calculated Parameters" and "Job Definition". We will then populate these group boxes with three types of control. These are the Spinner, the Edit Box and the Label. The spinner, which looks like an up-facing and down-facing arrowhead stacked one on top of the other, allows the controlled input of numeric values. You can specify in the **value** tab the name, caption, range, default value and the type of input. We shall always use the integer input type. The range format is (low value, high value, default value at time of execution).

The Edit Box is very similar to the spinner control except that we will be using it for text input. We will just need the name and caption property here. Finally, the Label control is useful as an indicator so that Max can display information on chosen variables while it is working. With two of the spinner controls, we will choose to restrict the valid input to even numbers. This is the case for the hologram size expressed in hogels (N_A and N_B). Although we could choose to accept odd numbers here, certain of the integer formulae change and therefore it makes sense in the context of this simple example to restrict ourselves to the even cases. To do this, we select the spinner control and then click on the event handler tab in the right-hand pane and tick **changed**. Upon saving, this then generates the following code:

```
on NA changed val do
(
)
```

We can then insert a command within the brackets to select only even input:

```
on NA changed val do
(
NA.value=2.0*floor(NA.value/2.0)
)
```

With these instructions in mind, you should be able to create the MAXScript panel shown in Figure 10.34. Table 10.2 lists the properties set for each control and Figure 10.35 lists the code Max generates to define the panel.

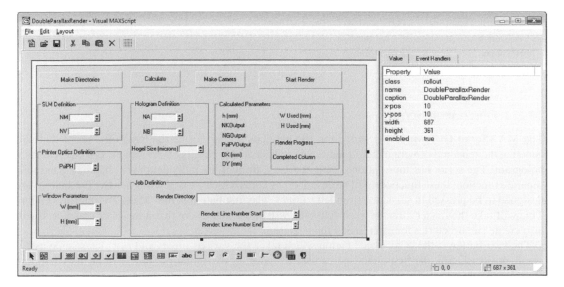

FIGURE 10.34 Finished MAXScript panel.

TABLE 10.2

MAXScript Parameters and Settings for the MAXScript Panel of Figure 10.34

Buttons

Name	Caption	Event Handler
BMD	Make Directories	Pressed ticked
BCal	Calculate	Pressed ticked
BCam	Make Camera	Pressed ticked
BRen	Start Render	Pressed ticked

Spinners

Name	Caption	Range	Type	Event Handler
NM	NM	[100, 5000, 1280]	Integer	Change ticked
NV	NV	[100, 5000, 1024]	Integer	Change ticked
NA	NA	[100, 3000, 600]	Integer	Change ticked
NB	NB	[100, 3000, 400]	Integer	Change ticked
PsiPH	PsiPH	[10, 120, 89]	Integer	
Win	W (mm)	[10, 5000, 600]	Integer	
Hin	H (mm)	[10, 5000, 400]	Integer	
Delta	Hogel size (μm)	[10, 3000, 800]	Integer	
NStartLine	Render: line number start	[1, 5000, 1]	Integer	
NstopLine	Render: line number stop	[1, 5000, 1]	Integer	

Edit Boxes

Name	Caption
RenderDir	Render directory

Labels

Name	Caption
hOutput	h (mm)
NKOutput	NKOutput
NGOutput	NGOutput
PsiPVOutput	PsiPVOutput
DXOutput	DX (mm)
DYOutput	DY (mm)
W_Output	W Used (mm)
H_Output	H Used (mm)

10.5.2 Function of Script

The MAXScript GUI window described in the previous section forms the basis of a script that will generate the rendered camera images required to make a double parallax three-colour DWDH reflection hologram. The script has four buttons, each of which, through its individual event handler, controls a separate function; these functions will be described by code, which we will write shortly. The buttons are designed to be pressed in sequence from left to right. The first button is the "Make Directories" button. This will take the render director we specify and will either make sure that it exists or create it. It will also create subdirectories for each line of the hologram.

The next button is the "Calculate" button which takes the various inputs and calculates some useful parameters including the first quantised camera distance, h, the actual physical size of the hologram, $D_X \times D_Y$, the actual window size which is subject to an adjustment, $W \times H$, the number of camera views required, $N_K \times N_G$, and the vertical printer FOV, Ψ_{PV}. The reason for displaying these values is to let the

FIGURE 10.35 MAXScript code corresponding to the panel in Figure 10.34.

operator decide if he or she wants to re-input certain values before continuing. The "Make Camera" button then creates a camera and assigns the active view to this camera.

The last button is the "Start Render" button, which starts the rendering process. The job definition window allows a region of lines to be assigned to the script. This facility is useful because in most cases a farm of many computers will be calculating the whole data set. In this way, one can easily set the lines that a given computer is responsible for rendering.

10.5.3 Global Variables

For different parts of the MAXScript program to communicate with each other, we must define variables with a global scope. This is done using the "global" command. We will need the following variables:

```
global W,h1,H,DX,DY,NK,NG,TopDirName
```

This line may be inserted anywhere in the main program, but it is good practice to put it at the top of the code.

10.5.4 Writing the "Make Directories" Event Handler

The code for the "Make Directories" event handler should be inserted between the two braces immediately following the statement "on BMDpressed do" in the MAXScript text editor window. To insert the code, you must first save and close the visual MAXScript editor window. The first code we will insert is the following:

```
TopDirName= "C;/"+RenderDir.Text
makedir TopDirName
```

The first line creates a text variable TopDirName and sets it to the string "C:/". It then takes the text data entered into the RenderDir Edit Box control on the panel at execution time and adds this to the string. The second line then uses the MAXScript control "makedir" to create a new directory with the name contained in TopDirName. If the directory already exists, nothing happens. The MAXScript commands are well documented—just click **Help** on the menu bar of the MAXScript text editor and then choose **Help** in the drop-down menu. This will take you directly to the MAXScript reference. Before being able to run the code, you will need to check at the very bottom of the file if the following code line is present:

```
"CreateDialog DoubleParallaxRender".
```

Sometimes, the visual editor forgets to add this or misses a letter or two at the end.

To run the code, click on the **File** menu at the top of the text editor and choose **Evaluate All** from the drop-down menu. The script will now run and the window we designed will appear. Type in the word "TestDir1" as the Render Directory and press **Make Directories**. Now go to C:/ and check that a new empty directory called TestDir1 has been created.

We will now create a subdirectory for each camera line rendered. In the panel we specified two variables—**NStartLine** and **NStopLine**—are the values of the index *g* for which we want to calculate render data. In each line directory, we will calculate N_K files. The following code uses a simple MAXScript "for" loop to accomplish the task of creating the subdirectories:

```
for i=NStartLine.value to NStopLine.value do
    (
    LineText=i as string
    DirName=TopDirName+"/Line"+LineText
    makedir DirName
    )
```

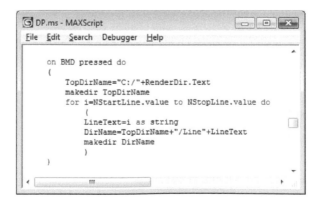

FIGURE 10.36 MAXScript code of the "Make Directories" event handler.

This code then creates subdirectories of the form "Line1", "Line2", "Line3", etc. in the parent directory specified. The full code of the "Make Directories" event handler is shown in Figure 10.36.

10.5.5 Writing the "Calculate" Event Handler

We now come to the next event handler. This code simply calculates a number of useful parameters that the user should check before initiating the render process. The formulae on which the calculations are based are all listed in Chapter 9, Section 2.

```
h1=Delta.value/2.0/1000.0*(NM.value-1)/tan(PsiPH.value/2.0)
hOutput.caption="h(mm)="+h1
```

This calculates the first quantised camera distance, h, through Equation 9.3. The values of N_K, N_G, W and H are likewise defined by Equations 9.5 and 9.6:

```
NK=2*floor(Win.value*1000.0/2.0/Delta.value) as integer
NG=2*floor(Hin.value*1000.0/2.0/Delta.value) as integer
NKText=NK as string
NGText=NG as string
NKOutput.caption="NK = "+NKText
NGOutput.caption="NG = "+NGText
W=(NK-1)*Delta.value/1000.0
H=(NG-1)*Delta.value/1000.0
WText=W as string
HText=H as string
W_Output.caption="W(mm)  = "+WText
H_Output.caption="H(mm)  = "+HText
```

Note that we have restricted N_K and N_G to be even as N_A and N_B have been restricted to even numbers. The vertical FOV is determined by the SLM aspect ratio and the given horizontal FOV:

$$\Psi_{CV} = 2\tan^{-1}\left\{ \frac{(N_V-1)}{(N_M-1)}\tan\left(\frac{\Psi_{CH}}{2}\right)\right\} \tag{10.85}$$

```
PsiPV=2.0*atan(float(NV.value-1)/float(NM.value-1)*tan(PsiPH.value/2.0))
PsiPVText=PsiPV as string
PsiPVOutput.caption="PsiPV = "+PsiPVText
```

Finally, D_X and D_Y are defined by

$$D_X = (N_A - 1)\delta$$
$$D_Y = (N_B - 1)\delta \tag{10.86}$$

where δ is the hogel diameter, giving:

```
DX=(NA.value-1)*Delta.value/1000.0
DY=(NB.value-1)*Delta.value/1000.0
DXText=DX as string
DYText=DY as string
DXOutput.caption="DX(mm)  = "+DXText
DYOutput.caption="DY(mm)  = "+DYtext
```

The full code of the "Calculate" event handler is shown in Figure 10.37.

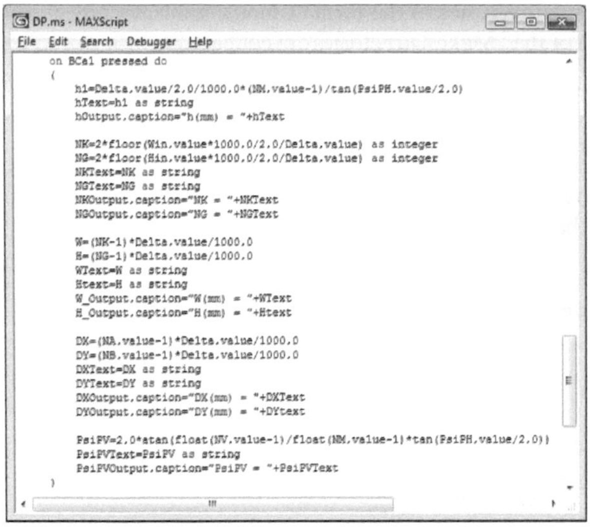

FIGURE 10.37   MAXScript code of the "Calculate" event handler.

10.5.6 Writing the "Make Camera" Event Handler

This event handler creates the camera. Its horizontal FOV is defined by the formula

$$\Psi_{CH} = 2\tan^{-1}\left\{\frac{W + D_X}{2h}\right\} \tag{10.87}$$

The camera can be created in MAXScript using the following commands:

```
PsiCH=2.0*atan((W+DX)/2.0/h1)
TargetCamera    NAME:"Cam1" \
                SHOWCONE: true \
                FOV:PsiCH \
                POS:[-W/2,-h1,-H/2.0] \
                TARGET: (targetObject NAME:"Tar1" POS: [-W/2,0.0,-H/2.0])
```

This creates a camera, Cam1, positioned at the starting location. A target point is also defined by the variable Tar1 and is set such that the camera faces straight ahead. Note the use of the backslash characters to spread the command out over several lines.

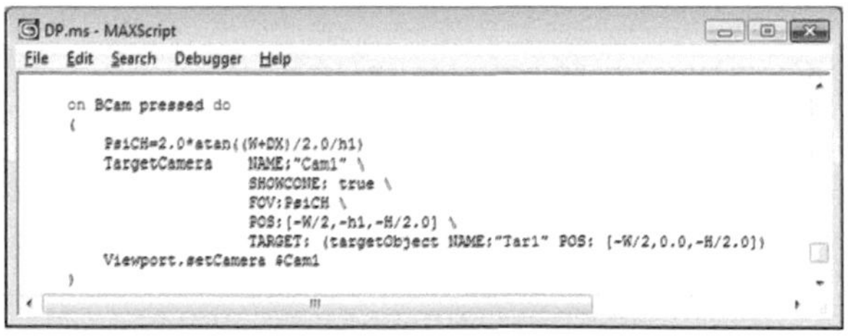

FIGURE 10.38   MAXScript code of the "Make Camera" event handler.

Because we will be using render commands that only work in the active viewport, we will also need to set the active viewport to the camera that we have just created:

```
Viewport.SetCamera $Cam1
```

The full code of the "Make Camera" event handler is shown in Figure 10.38.

10.5.7 Writing the "Start Render" Event Handler

This is the last and most complicated event handler. It must perform a two-dimensional loop over the lines, g, and the columns, k, of the window using the created camera to render each frame. The line loop is limited by the variables NStartLine and NStopLine. The column loop will, however, be the full range from 1 to N_G. At each value of k and g, only a region of the camera frame must be rendered. This is accomplished using the MAXScript command "render":

```
Render\
            OUTPUTFILE: FrameFileName\
            OUTPUTWIDTH:NI\
            OUTPUTHEIGHT:NJ \
            RENDERTYPE: #RegionCrop \
            REGION: #(N1x,N1y,N2x,N2y) \
            VFB:OFF
```

The size of the full camera frame is given by the parameters N_I and N_J, which are defined by

$$N_I = 1 + \frac{D_X + W}{\delta}$$

$$N_J = 1 + \frac{D_Y + H}{\delta}$$

(10.88)

Together with Equation 10.87, this means that the full camera frame will always contain the hologram frame as a complete subset, irrespective of the position of the camera in the window. However, at each camera position, we will only need to render a small part of the full camera frame to create a "centred camera" data set. The subsection of the full camera frame that we must render is therefore a function of the camera indices k and g, and is given by the simple formula:

$$N_{1X} = N_K - k + 1$$

$$N_{2X} = N_{1X} + N_A - 1$$

$$N_{1Y} = g$$

$$N_{2Y} = N_{1Y} + N_B - 1$$

(10.89)

The camera must be moved before each render operation by the commands:

```
$Cam1.pos=[-W/2+(k-1)*Delta.value/1000.0,-h1,-H/2+(g-1)*Delta.value/1000.0]
$Tar1.pos=[-W/2+(k-1)*Delta.value/1000.0,0,-H/2+(g-1)*Delta.value/1000.0]
```

Figure 10.39 shows a full listing of the "Start Render" event handler. There are several things in particular that we should mention. The first is that the render command can fill up memory quickly and as a result you need to periodically use the command gc(). Second, the render command, when it uses RenderCrop, contains a bug in that the last column and last row of pixels are incorrectly set to zero,

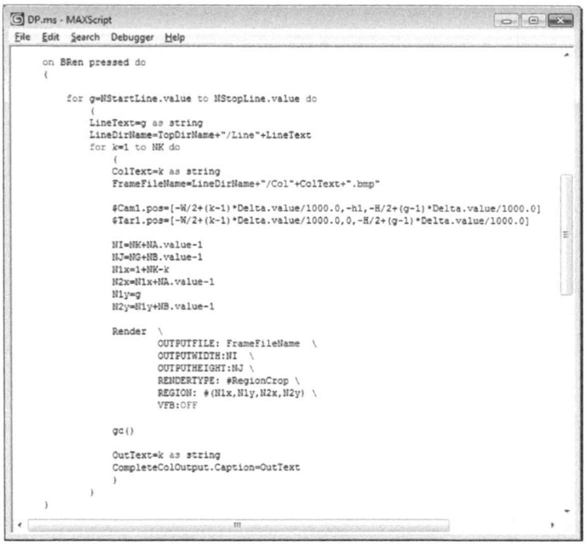

FIGURE 10.39 MAXScript code of the "Start Render" event handler.

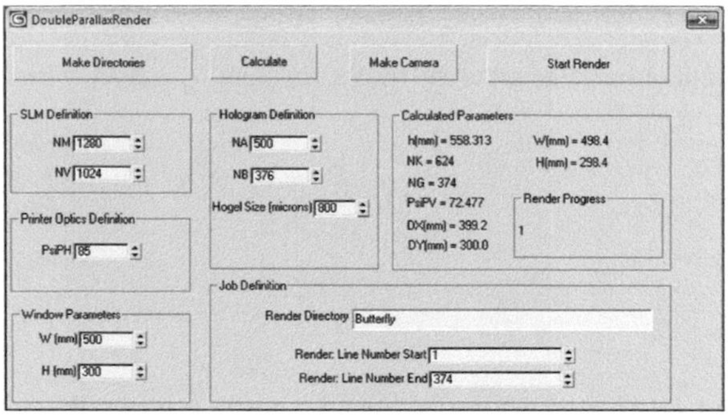

FIGURE 10.40 Final MAXScript panel filled out to generate the complete image set required for a typical 30 × 40 cm DWDH reflection hologram.

leading to one-pixel-thick black lines at the extreme edges of the camera frame.* For most applications, this has a minimal effect and practically one just specifies N_A and N_B to be two hogels larger than necessary.

In the code presented here, we have opted to write out the render data in the form of bitmap files. It is worth mentioning that Max supports the jpeg format. To write jpegs instead of bitmaps, you just need to change the code line

```
FrameFileName=LineDirName+"/Col"+ColText+".bmp"
```

to

```
FrameFileName=LineDirName+"/Col"+ColText+".jpg"
```

By selecting a reasonable compression setting, you can save a lot of space without much effect on the final hologram. You can change various options including compression in the Max render dialogue.

The script has been designed for system units of millimetres. Be sure to check that you have chosen this as your default preference in Max! Figure 10.40 shows a screenshot of the panel for the case of a

---

* This error is also present in all later versions of Max up until the time of writing.

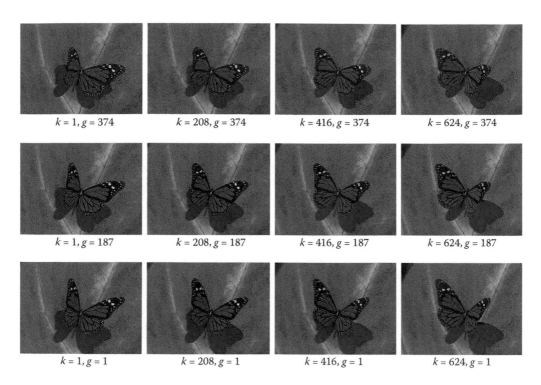

$k = 1, g = 374$	$k = 208, g = 374$	$k = 416, g = 374$	$k = 624, g = 374$
$k = 1, g = 187$	$k = 208, g = 187$	$k = 416, g = 187$	$k = 624, g = 187$
$k = 1, g = 1$	$k = 208, g = 1$	$k = 416, g = 1$	$k = 624, g = 1$

FIGURE 10.41 Selected camera frames generated by the MAXScript program with the settings of the panel shown in Figure 10.40. The total data set for the 30 cm × 40 cm hologram comprises 624 × 374 frames requiring 125 GB of disk space using the .bmp format.

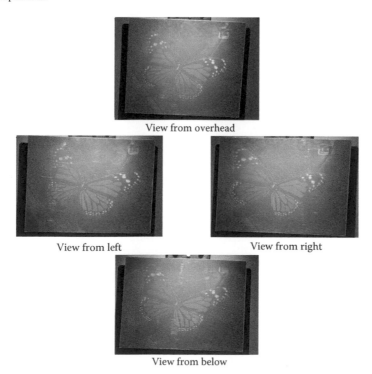

View from overhead

View from left   View from right

View from below

FIGURE 10.42 Several pictures of the final DWDH reflection hologram of Figures 10.40 and 10.41 printed on PFG3CN silver halide emulsion and illuminated by a 50 W MR-16 halogen lamp. (Photograph courtesy of Geola Digital UAB.)

typical 30 × 40 cm hologram. Figure 10.41 shows selected camera frames generated by the script, and Figure 10.42 shows several pictures of an actual DWDH reflection hologram printed using this data. A full listing of the complete MAXScript program is given in Appendix 5. The data set for the hologram comprises 624 × 374 individual frames requiring a total of 125 GB of disk space. By using jpegs with a compression setting of 75% instead of bitmaps, this can be reduced to just under 6 GB. Rendering with either format on a standard 2011 laptop takes approximately 38 h.

REFERENCE

1. D. Brotherton-Ratcliffe, A. Nikolskij, S. Zacharovas, J. Pileckas and R. Bakanas, *Image capture system for a digital holographic printer*, US Patent 8,154,584 (filed 2006, granted 2012).

11

Theoretical Basis for High-Fidelity Display Holograms

11.1 Introduction

This book is primarily about the creation of ultra-realistic images through holography. In recent years, the quality of images has progressed rapidly in the field, and we are optimistic that this process will continue. But why should we be confident of this? This chapter is partly a concise answer to this question. However, it also fulfils another vital role. It introduces, in some detail, the basic physical models that mathematically describe how a high-fidelity hologram actually functions. A clear understanding of these models is invaluable for any worker in the field. Besides providing formulae for vital processes, such as diffractive efficiency, image blurring, and aberration correction, a comprehensive understanding of the underlying physics allows the worker to develop a clear intuition of what is likely to work best.

We shall start our discussion of display holograms with the thin transmission hologram. This is the simplest place to start and the most productive. A simple first-order mathematical analysis shows why any hologram, thick or thin—reflective or transmissive—is subject to source size and chromatic blurring. Formulae for these effects and for dispersion in the transmission hologram are derived. This analysis forms the foundation for rainbow and achromatic holography.

In Section 11.4, we extend the discussion to volume phase holograms. Here, we derive the Bragg condition from a perturbation analysis of the phase function using spherical waves. We also derive both paraxial and fully non-paraxial equations describing the general reconstruction of images in both transmission and reflection geometries when different wavelengths and angles are used at recording and playback. The non-paraxial equations are of fundamental use in predistortion and in distortion correction algorithms employed in digital display holography. In Section 11.6, we describe the thick volume phase hologram in terms of a set of parallel-stacked mirrors (PSM). We develop the same non-paraxial equations as in Section 11.4 using this method, demonstrating the equivalence of the two pictures. We use a very basic discussion of the PSM model* to introduce the important differences between transmission and reflection holograms, and we estimate diffractive efficiencies as well as angular and wavelength sensitivities for both types.

In Section 11.10, we describe a more rigorous treatment of the problem of diffraction from volume phase holograms. This is Kogelnik's one-dimensional coupled wave theory [1]. We present the main results of this theory using the simpler σ-polarisation and a two-dimensional grating, although we later extend this somewhat briefly to the π-polarisation. This leads us to a discussion of the physical characteristics of suitable materials for ultra-realistic imaging. We also present a discussion of the N-coupled wave theory [2–4] because this analytical extension of Kogelnik's theory nicely demonstrates how holograms of diffuse images can be expected to attain almost perfect diffractive efficiency in the limit of a large amplitude reference recording beam. It also provides some intuition that multicolour volume phase holograms may be designed, at least in principle, to have high diffractive efficiencies—and indeed this will be confirmed in Chapter 12 when we study polychromatic gratings and holograms more rigorously using the PSM model.

In Section 11.11, we return to the intrinsic image blurring in analogue and digital holograms in the light of what we have learnt. We end the chapter with a brief discussion in Section 11.12 of computational methods for calculating the diffraction efficiency from planar gratings.

---

* We return to the parallel-stacked mirror model in Chapter 12, in which we present a detailed discussion of this theory.

11.2 Three-Dimensional Paraxial Theory of the Thin Transmission Hologram

We consider two infinitely thin holograms* each produced by the interference of a pair of spherical waves, one originating from an object point and the other originating from a reference point (Figure 11.1). In the first case, we take the object point to have coordinates (x_o, y_o, z_o) and the reference source to be located at (x_r, y_r, z_r). In the second case, the object point is located at (x_i, y_i, z_i) and the reference source is at (x_c, y_c, z_c). The hologram is, in both cases, located on the $x = 0$ plane.

Now, our two holograms will be essentially identical if the surface distribution of the time-averaged square of the electric field amplitude is the same for each. With this in mind, we will examine a zone near the coordinate origin without loss of generality. The time-averaged squared electric field distribution at a point on the first hologram $(0, y_h, z_h)$ is then given by

$$\left\langle \mathbf{E}^2(y_h, z_h) \right\rangle = \frac{1}{T} \int \left\{ \Re \left(\mathbf{E_r} \frac{e^{i\left[\omega t - \frac{2\pi}{\lambda_r}\rho_r(y_h, z_h)\right]}}{\rho_r(y_h, z_h)} + \mathbf{E_o} \frac{e^{i\left[\omega t - \frac{2\pi}{\lambda_r}\rho_o(y_h, z_h)\right]}}{\rho_o(y_h, z_h)} \right) \right\}^2 dt \qquad (11.1)$$

Likewise, the distribution on the second hologram may be written as

$$\left\langle \mathbf{E}^2(y_h, z_h) \right\rangle = \frac{1}{T} \int \left\{ \Re \left(\mathbf{E_c} \frac{e^{i\left[\omega t - \frac{2\pi}{\lambda_c}\rho_c(y_h, z_h)\right]}}{\rho_c(y_h, z_h)} + \mathbf{E_i} \frac{e^{i\left[\omega t - \frac{2\pi}{\lambda_c}\rho_i(y_h, z_h)\right]}}{\rho_i(y_h, z_h)} \right) \right\}^2 dt \qquad (11.2)$$

In both cases

$$\rho_\gamma(y_h, z_h) = \sqrt{x_\gamma^2 + (y_\gamma - y_h)^2 + (z_\gamma - z_h)^2} \qquad \forall \gamma \in \{r, o, c, i\} \qquad (11.3)$$

The E variables control the relative amplitude of each wave. Now, Equation (11.1) can be rewritten assuming $T \to \infty$ as

$$\left\langle \mathbf{E}^2(y_h, z_h) \right\rangle = \frac{1}{2} \left\{ \frac{|E_r|^2}{\rho_r^2(y_h, z_h)} + \frac{|E_o|^2}{\rho_o^2(y_h, z_h)} \right\}$$

$$+ \frac{\kappa_{ro}(y_h, z_h)\sqrt{E_r E_o E_r^* E_o^*}}{\rho_o(y_h, z_h)\rho_r(y_h, z_h)} \cos\left[\frac{2\pi}{\lambda_r}\rho_o(y_h, z_h) - \frac{2\pi}{\lambda_r}\rho_r(y_h, z_h) + \varsigma_{ro}(E_r, E_o) \right] \qquad (11.4)$$

where $\kappa_{ro}(x_h, z_h)$ is a function related to the (approximately linear) polarisation of the sources. Both the first term and the multiplier of the second term are slowly varying functions of y_h and z_h. The trigonometric term, on the other hand, varies extremely rapidly with y_h and z_h as its argument is very large. As a consequence, our two holograms can be considered identical if they share an identical trigonometric argument or if

$$\frac{2\pi}{\lambda_r}\{\rho_r - \rho_o\} = \pm\frac{2\pi}{\lambda_c}\{\rho_c - \rho_i\} \pm \{\varsigma_{ro} - \varsigma_{ci}\} \pm 2n\pi \qquad (11.5)$$

Here, the term $\varsigma_{ro} - \varsigma_{ci}$ is a constant phase term that depends only on the magnitudes of the reference and object waves. The term $2n\pi$, where n is any integer, simply reflects the periodic nature of the trigonometric function. Figure 11.2 shows several illustrations of typical fringe patterns calculated from Equation 11.4.

---

* In this chapter we will use the words "grating" and "hologram" interchangeably. In Chapter 12 we will make a distinction between these two terms when we specifically treat the hologram as the large N limit of a spatially multiplexed grating.

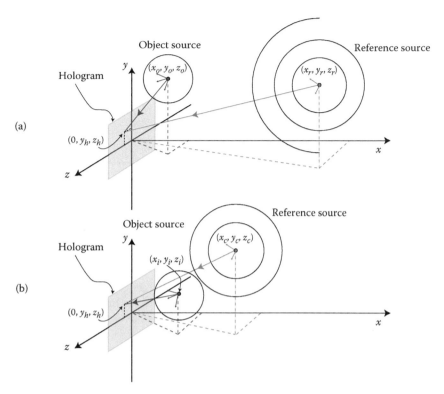

FIGURE 11.1 This diagram may be interpreted in two ways. In the first way, (a) and (b) show two different recording geometries for a thin transmission hologram. In each case, there is a single reference spherical wave and a single object spherical wave. The recording wavelength and position of the object and reference sources are different in each case. If we now assume that the fringe pattern within each transmission hologram produced by recording methods (a) and (b) are identical (i.e., if the fringe function $\langle \mathbf{E}^2(y_h, z_h) \rangle$ is the same for both scenarios), then we can interpret the two diagrams in a different fashion: (a) now represents the recording of a hologram; (b) now represents how it can be expected to replay at another wavelength and using another position for the reference source. Under this interpretation and as drawn, (b) represents the reconstruction of a virtual image situated at (x_i, y_i, z_i) when the hologram recorded in (a) is illuminated by a spherical source whose origin is (x_c, y_c, z_c). This dual interpretation requires the use of the fundamental holography theorem: that an ideal hologram reconstructed with exactly the same wavelength and geometry as used in recording will faithfully reproduce the original image wave.

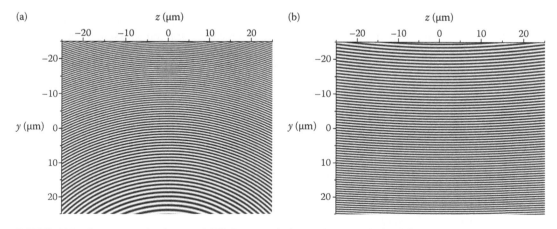

FIGURE 11.2 Some example pictures of $\langle \mathbf{E}^2 \rangle$ for transmission gratings as calculated from Equation 11.4. Recording wavelength is 532 nm. (a) Object sink located at $x_o = -100$ μm, $y_o = 0$ μm, $z_o = 0$ μm; reference source located at $x_r = 100$ μm, $y_r = 100$ μm, $z_o = 0$ μm. Object source located at $x_o = 100$ μm, $y_o = 0$ μm, $z_o = 0$ μm; (b) reference source located at $x_r = 100$ μm, $y_r = 100$ μm, $z_o = 0$ μm. Note the really quite small curvature of the fringes. For larger object and reference distances, the fringes will appear essentially straight on the scale of the recording wavelength.

We can use Equation 11.5 to develop a useful theory of the thin transmission hologram by expanding the general function ρ_γ in terms of a two-dimensional power series in the small quantities y_h and z_h:

$$\rho_\gamma = \sum_{m=0}^{M}\sum_{n=0}^{N}\rho_{\gamma mn}z_h^m y_h^n \qquad \forall \gamma \in \{r,o,c,i\} \tag{11.6}$$

Using Equation 11.3, we see that the lowest-order terms can be written as follows:

$$\rho_{\gamma 00} = \sqrt{x_\gamma^2 + y_\gamma^2 + z_\gamma^2}; \qquad \rho_{\gamma 10} = -\frac{z_\gamma}{\sqrt{x_\gamma^2 + y_\gamma^2 + z_\gamma^2}}$$

$$\rho_{\gamma 01} = -\frac{y_\gamma}{\sqrt{x_\gamma^2 + y_\gamma^2 + z_\gamma^2}}; \qquad \rho_{\gamma 11} = -\frac{y_\gamma z_\gamma}{\left(x_\gamma^2 + y_\gamma^2 + z_\gamma^2\right)^{3/2}} \tag{11.7}$$

Substitution of these expressions in Equation 11.5 leads to a set of non-linear equations for (x_i, y_i, z_i) in terms of (x_o, y_o, z_o). By imposing a paraxial condition on both the "i" and "o" source points, however, we can linearise these equations. In practice, we do this by expanding the coordinate pairs (y_i, z_i) and (y_o, z_o) as Taylor series and then truncating Equation 11.5 at quadratic orders in these four variables. The expansion orders $(m = 1, n = 0)$, $(m = 0, n = 1)$ and $(m = 1, n = 1)$ then lead to the following paraxial equations:

$$\frac{1}{x_i} = \frac{y_c z_c}{\left(x_c^2 + y_c^2 + z_c^2\right)^{3/2}} \mp \frac{\lambda_c}{\lambda_r}\left\{\frac{y_r z_r}{\left(x_r^2 + y_r^2 + z_r^2\right)^{3/2}} - \frac{1}{x_o}\right\} \tag{11.8}$$

$$\frac{y_i}{x_i} = \frac{y_c}{\sqrt{x_c^2 + y_c^2 + z_c^2}} \mp \frac{\lambda_c}{\lambda_r}\left\{\frac{y_r}{\sqrt{x_r^2 + y_r^2 + z_r^2}} - \frac{y_o}{x_o}\right\} \tag{11.9}$$

$$\frac{z_i}{x_i} = \frac{z_c}{\sqrt{x_c^2 + y_c^2 + z_c^2}} \mp \frac{\lambda_c}{\lambda_r}\left\{\frac{z_r}{\sqrt{x_r^2 + y_r^2 + z_r^2}} - \frac{z_o}{x_o}\right\} \tag{11.10}$$

where x_o, x_r and x_c are assumed to have positive values.* Note that the $(m = 0, n = 0)$ order defines the term $\{\varsigma_{ro} - \varsigma_{ci}\} + 2n\pi$ in Equation 11.5, and all higher orders simply restate the information contained in the first four orders. To recap, these expressions relate two identical thin transmission holograms. The first is recorded at a wavelength of λ_r using an object wave originating at (x_o, y_o, z_o) and a reference wave at (x_r, y_r, z_r). The second is recorded at a wavelength of λ_c using an object wave originating at (x_i, y_i, z_i) and a reference wave at (x_c, y_c, z_c). The holograms are identical because Equation 11.5 means that they both share identical interference patterns. Now, this is useful because a thin hologram recorded at a given wavelength with a given source and reference is known to replay an image, constituting a faithful reproduction, of the object source if illuminated by the same reference source at the same wavelength. We can demonstrate this easily enough for the case of an amplitude hologram (the case of the phase hologram is only a little more tedious [e.g., 5]).

Suppose we consider an arbitrary object wave of time dependence $e^{i\omega t}$

$$\mathbf{E}_o = \mathfrak{R}[\mathbf{A}(x, y, z)e^{-i\phi(x,y,z)}] \tag{11.11}$$

and an arbitrary reference wave

$$\mathbf{E}_r = \mathfrak{R}[\mathbf{B}(x, y, z)e^{-i\psi(x,y,z)}] \tag{11.12}$$

---

* In fact, if we take x_o to be negative, then this represents either an object sink or alternatively an object source but recorded in "reflection" mode. The thin transmission hologram makes no distinction between these two cases.

Then, on the surface of the photographic plate, we can expect the following average squared electric field distribution

$$\left\langle \mathbf{E}^2 \right\rangle \sim \frac{1}{2}\left(\left|\mathbf{A}\right|^2 + \left|\mathbf{B}\right|^2\right) + \frac{1}{2}\mathbf{A}^* \cdot \mathbf{B} e^{i(\phi-\psi)} + \frac{1}{2}\mathbf{A} \cdot \mathbf{B}^* e^{-i(\phi-\psi)} \quad (11.13)$$

If we assume that after development, the transmission of the plate is proportional to $\left\langle \mathbf{E}^2 \right\rangle$, then the complex light amplitude produced at the plate surface on illumination by the same reference wave will be

$$\mathbf{E}_i \propto \mathbf{B} e^{-i\psi}\left\langle \mathbf{E}^2 \right\rangle \sim \left\{\frac{\left|\mathbf{A}\right|^2 + \left|\mathbf{B}\right|^2}{2}\right\}\mathbf{B} e^{-i\psi} + \left\{\frac{1}{2}\mathbf{A}^* \cdot \mathbf{B}\right\}\mathbf{B} e^{i(\phi-2\psi)} + \left\{\frac{1}{2}\mathbf{A} \cdot \mathbf{B}^*\right\}\mathbf{B} e^{-i\phi} \quad (11.14)$$

If $|\mathbf{B}|$ is roughly constant over the plate and the polarisations of the waves do not vary, then the third term faithfully reproduces the original object wave. By choosing an off-axis geometry, the first and second terms may be angularly displaced from this third wave.

Hence, we are naturally led to an alternative interpretation of Equations 11.8 through 11.10. They can be regarded as specifying how a first hologram recorded at a wavelength λ_r using an object wave originating at (x_o, y_o, z_o) and a reference wave at (x_r, y_r, z_r) reproduce an image at (x_i, y_i, z_i) when illuminated by a wave originating at the reference point (x_c, y_c, z_c) at wavelength λ_c.

The presence of the second term in Equation 11.14 is related in the $\mp$ sign appearing in Equations 11.8 through 11.10. This sign is present in these equations because the interference pattern of Equation 11.4 is invariant under the transformation

$$\rho_o(y_h, z_h) - \rho_r(y_h, z_h) + \frac{\lambda_r}{2\pi}\varsigma_{ro}(E_r, E_o) \rightarrow -\rho_o(y_h, z_h) + \rho_r(y_h, z_h) - \frac{\lambda_r}{2\pi}\varsigma_{ro}(E_r, E_o) \quad (11.15)$$

This leads, in general, to two reconstruction images being present in a thin transmission hologram—the *virtual image* and the *real image*. Using the notation we have developed and with x_o, x_r and x_c positive and relating to sources rather than sinks, the negative sign in Equations 11.8 through 11.10 represents the virtual image (x_i positive) whereas a positive sign indicates the real image (x_i negative). In all cases, the negative sign represents the primary non-conjugate image and the positive sign the conjugate image.

The simultaneous reconstruction of both a virtual image and a real image is a property of the thin transmission hologram. When the reference beam is on-axis or nearly on-axis, both images will be present in the same line of view—however, as the reference beam angle is increased, the conjugate or parasitic image is quickly shifted to high angles where it either fails to reconstruct or suffers large aberration. The volume nature of most common transmission holograms, as we shall see shortly, also conspires to eliminate the conjugate image.

11.2.1 Collimated Reference and Object Beams

In many cases of interest, display holograms are written with essentially collimated reference and replay beams. The paraxial theory is enormously simplified under this approximation and it is therefore instructive to start any discussion of the thin transmission hologram here.

We will start by writing the two reference waves in Equations 11.8 through 11.10 in spherical polar coordinates (Figure 11.3):

$$\frac{1}{x_i} = \frac{\sin^2\theta_c \cos\phi_c \sin\phi_c}{r_c} \mp \frac{\lambda_c}{\lambda_r}\left\{\frac{\sin^2\theta_r \cos\phi_r \sin\phi_r}{r_r} - \frac{1}{x_o}\right\} \quad (11.16)$$

$$\frac{y_i}{x_i} = \sin\theta_c \cos\phi_c \mp \frac{\lambda_c}{\lambda_r}\left\{\sin\theta_r \cos\phi_r - \frac{y_o}{x_o}\right\} \quad (11.17)$$

$$\frac{z_i}{x_i} = \sin\theta_c \sin\phi_c \mp \frac{\lambda_c}{\lambda_r}\left\{\sin\theta_r \sin\phi_r - \frac{z_o}{x_o}\right\} \quad (11.18)$$

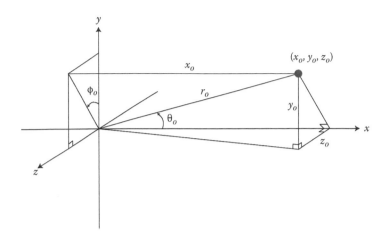

FIGURE 11.3 Recording geometry in spherical polar coordinates. The diagram shows how an object point (x_o, y_o, z_o) is defined by the equivalent right-handed spherical system (r_o, θ_o, ϕ_o). The hologram is located on the (y, z) plane at $x = 0$.

Here, we have used the relations

$$x_r = r_r \cos\theta_r; \qquad y_r = r_r \sin\theta_r \cos\phi_r; \qquad z_r = r_r \sin\theta_r \sin\phi_r$$
$$x_c = r_c \cos\theta_c; \qquad y_c = r_c \sin\theta_c \cos\phi_c; \qquad z_c = r_c \sin\theta_c \sin\phi_c \tag{11.19}$$

with $0 \le \theta \le \pi$, $0 \le \phi \le 2\pi$ and r positive.

In the case that $r_c = r_r \to \infty$ and $\phi_c = \phi_r = 0$ we can characterise the reference and replay waves by plane waves whose wave vectors have purely altitudinal angles of, respectively, θ_r and θ_c. Equations 11.16 through 11.18 then take the simple form*

$$x_i = x_o \frac{\lambda_r}{\lambda_c}; \qquad y_i = x_o \left(\frac{\lambda_r}{\lambda_c} \sin\theta_c - \sin\theta_r \right) + y_o; \qquad z_i = z_o \tag{11.20}$$

This tells us that if we record a single point (x_o, y_o, z_o) and then replay the hologram in white light (Figure 11.4), the virtual image of this point will be dispersed into a line defined by the equation

$$y_i = x_i \sin\theta_c - x_o \sin\theta_r + y_o \tag{11.21}$$

This line is orthogonal to the z direction; its gradient in the x,y plane is usually expressed as the tangent of the achromatic angle

$$\tan\theta_A \equiv \sin\theta_c \tag{11.22}$$

Equations 11.20 through 11.22 form the basis of rainbow and achromatic holography as discovered by Steven Benton in 1969 [6]. They show that dispersion only occurs along the achromatic plane and that the horizontal (z) plane is dispersion-free. This of course means that one can construct transmission holograms whose parallax is in the horizontal dimension (horizontal parallax-only [HPO] holograms) and where the vertical dimension is dispersed. These results were used in Chapter 8, in which they were essential in letting us define the digital image data transformations required for the printing of direct-write digital holography (DWDH) rainbow and achromatic holograms.

We should note that if the hologram replay beam is not collimated—that is, it is a spherical wave with finite r_c—then Equations 11.16 through 11.18 show us that there is now dispersion in all dimensions; as such, rainbow and achromatic holograms can be expected to suffer from blurring due to the coupling of the horizontal parallax with horizontal dispersion if the illumination source is placed too close to the

---

* Here, we have used the minus sign in Equations 11.16 to 11.18 to treat the primary image.

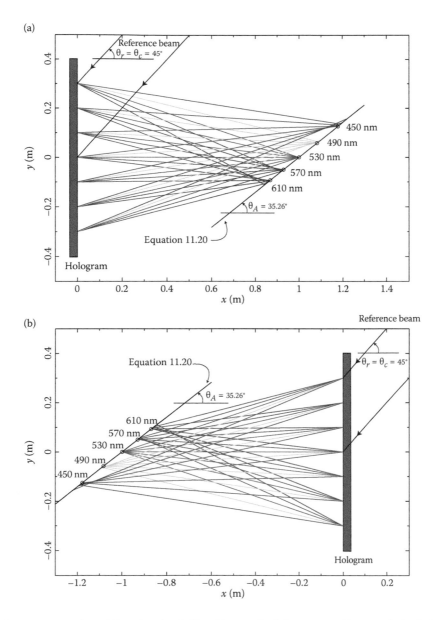

FIGURE 11.4 Reconstruction of (a) a virtual and (b) real image of a finite-aperture thin transmission hologram using identical geometry as at recording but at 5 different wavelengths. The reference beam is assumed to be collimated ($r_r = r_c \rightarrow \infty$) and axial ($\phi_r = \phi_c = 0$) and has an angle of incidence of $\theta_r = \theta_c = 45°$. In (a) a single object source at $(x_o, y_o, z_o) = (1.0, 0.0, 1.1 \text{ m})$ is recorded at 530nm and then replayed at (450 nm, 490 nm, 530 nm, 570 nm, 610 nm). In (b) a corresponding object sink at $(x_o, y_o, z_o) = (1.0, 0.0, 1.1 \text{ m})$ is similarly recorded and replayed. The rays represent fully non-paraxial calculations. Note that away from the centre of the hologram the rays do not intersect at a defined point as they are expected to do under the paraxial theory. The solid line represents the paraxial calculation of equation 11.21 and the circles are the paraxial points as calculated from 11.20. Note that the reconstructed conjugate images (not shown) in the replay scenarios of both (a) and (b) which correspond to the choice of the plus sign in Equations 11.16 through 11.18 (rather than the minus sign which has been assumed in Equations 11.20 through 11.22) lead to highly shifted images with $y < 1$ m.

hologram. Note, however, that if $\phi_c = 0$ and $r_c \to \infty$ but r_r and ϕ_r are finite, then the transverse z direction can once again be made dispersion-free. In this case, the image data for a rainbow or achromatic hologram will certainly need to be predistorted, but the important point is that the "mechanics" of these holograms depends more on the illumination rather than the recording geometry.

Equations 11.20 through 11.22 are equally valid for the case of positive x_o—in which case the image replays as a virtual image—and for negative x_o—in which case the image replays as a real image (see Figure 11.4).

11.2.2 Source-Size Blurring

We can use the paraxial theory to examine how a thin transmission hologram will blur due to an illumination source of finite size. Let us suppose that the illumination source has a finite extent which we define by the angle $\delta\theta_c$. We will make the approximation of an axial distant source on replay and recording:

$$\phi_c = \phi_r = 0; \quad r_c, r_r \to \infty \tag{11.23}$$

If the illumination source is circularly symmetric we may define the corresponding Cartesian uncertainties which describe the finite nature of the source as

$$\delta x \equiv r\delta\theta_c \sin \langle \theta_c \rangle$$
$$\delta y \equiv r\delta\theta_c \sin \langle \theta_c \rangle \tag{11.24}$$
$$\delta z \equiv r\delta\theta_c$$

where the operator $\langle \ \rangle$ indicates an average value for all rays. We may now use the Jacobian $\partial(x_i, y_i, z_i)/\partial(x_c, y_c, z_c)$, obtained by differentiating Equations (11.8) through (11.10), to map these uncertainties onto the image coordinates.

$$\delta x_i = \frac{\partial x_i}{\partial x_c}\delta x_c + \frac{\partial x_i}{\partial y_c}\delta y_c + \frac{\partial x_i}{\partial z_c}\delta z_c = 0$$

$$\delta y_i = \frac{\partial y_i}{\partial x_c}\delta x_c + \frac{\partial y_i}{\partial y_c}\delta y_c + \frac{\partial y_i}{\partial z_c}\delta z_c = \left\{\frac{\lambda_r}{\lambda_c}x_o\cos\langle\theta_c\rangle\right\}\delta\theta_c \tag{11.25}$$

$$\delta z_i = \frac{\partial z_i}{\partial x_c}\delta x_c + \frac{\partial z_i}{\partial y_c}\delta y_c + \frac{\partial z_i}{\partial z_c}\delta z_c = \left\{\frac{\lambda_r}{\lambda_c}x_o\right\}\delta\theta_c$$

When the recording and replay wavelengths are the same this then leads to an image point uncertainty due to the finite source of

$$\delta x_i = 0$$
$$\delta y_i = x_o\cos\langle\theta_c\rangle\delta\theta_c \tag{11.26}$$
$$\delta z_i = x_o\delta\theta_c$$

These Equations provide us with a simple estimation of the intrinsic blur in the transverse y and z directions caused by a finite illumination source size. Note that there is no longitudinal blur as x_i is independent of x_c, y_c, and z_c for distant axial sources (Equation 11.23).

As somewhat of an aside, it can be instructive to verify these equations using a computer. For simplicity, we assume the case of an amplitude transmission hologram in which the transmissivity, $\tau(y,z)$ is proportional to $\langle \mathbf{E}^2 \rangle$. We can then easily calculate $\tau(y,z)$ for a given object point and a given reference point according to Equation 11.4. Finally, we reconstruct the hologram with a spherical wave using a Kirchoff diffraction integral over the hologram surface:

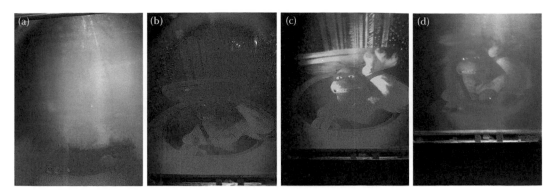

FIGURE 11.5 A single-slit analogue rainbow transmission hologram (60 × 80 cm) viewed from the front at various heights from (a) to (d) showing how the hologram changes colour due to the dispersion of Figure 11.4b. Note that each photograph shows more than a single colour for two reasons—the reference light is not precisely collimated and the photographs have not been taken precisely at the slit focus.

FIGURE 11.6 Same hologram as in Figure 11.5 but now illuminated from the other side, showing a dispersed virtual image of the master slit in accordance with Figure 11.4a.

$$E_i(x,y,z) = \frac{\lambda_c}{i} \int \tau(y,z) \frac{e^{i\frac{2\pi}{\lambda_c}(\rho_c+\rho_i)}}{\rho_c \rho_i} \mathcal{O}(\theta_c,\theta_i) dS \tag{11.27}$$

For small angles, we may approximate the obliquity factor to 1. Figure 11.7 shows one case of interest. The transmissivity[*] function of a small square hologram (50 μm × 50 μm) whose centre is located at (0,0,0) is calculated using a point reference source at $(x, y, z) = (10^{-2}, 10^{-2}, 0 \text{ m})$ and a point object sink at $(-10^{-4}, 0, 0 \text{ m})$. The hologram is then replayed using Equation 11.27 first for a point illumination source at $(10^{-2}, 10^{-2}, 0 \text{ m})$ and subsequently for a shifted source at $(10^{-2}, 1.1 \times 10^{-2}, 0 \text{ m})$ giving $\delta\theta \sim 0.0476$. The wavelength both at recording and at replay is assumed to be 532 nm. Figure 11.7a and b show plots of the calculated values of $\langle \mathbf{E}^2 \rangle$ versus the vertical coordinate y at the plane in front of the hologram defined by $x = -10^{-4} \text{ m}$ and $z = 0$. Graph (a) shows the case of the unshifted reference beam and, as expected, shows a strong reconstruction peak at $y = 0$. Graph (b) shows the case of the shifted reference beam, and this time, we see that the peak reconstruction is shifted to $y = -3.4$ μm; this is very close to the result predicted by Equation 11.26. Note that the width of the reconstruction peaks and the presence of the small secondary peaks are determined by the fact that the size of the hologram is only 50 μm.

---

[*] A 3% modulation of the transmissivity is assumed for the purposes of the calculation.

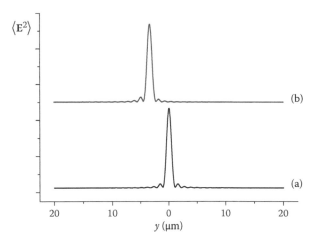

FIGURE 11.7 Diagram showing how a thin amplitude-modulated transmission hologram replays with a slightly shifted reference beam. The square hologram (50×50 μm) is recorded using a point reference source at $(x, y, z) = (10^4, 10^4, 0$ μm) and a point object sink at $(-10, 0, 0$ μm). The average squared electric field is calculated using Equation 11.4. The amplitude transmissivity of the processed hologram is assumed to be proportional to this calculated distribution. A Kirchhoff integral is then computationally evaluated for two cases: (a) spherical replay wave originating at $(x, y, z) = (10^4, 10^4, 0$ μm); (b) spherical replay wave originating at $(x, y, z) = (10^4, 1.1 \times 10^4, 0$ μm). The average squared electric field of the diffracted wave is plotted against the y coordinate at $x = -10$ μm and $z = 0$. The wavelength at recording and replay is 532 nm. As expected, when the replay and reference geometries are the same as in case (a), the reconstructed image point lies exactly at $(-10, 0, 0$ μm). With a shift in replay source as in case (b), the image point is also displaced. The amount of displacement is given by Equation 11.26.

To understand the significance of the relations in Equation 11.26 in terms of what a human observer actually "sees", we need to define the practical resolution of the human eye. The average human eye can resolve two points separated by 1 mm at a distance of approximately 1 m.[*] This corresponds to an angle of approximately $\delta\theta_{Eye} = 0.06°$.

Let us now consider the case of observing a transmission hologram from a distance h with the image being virtual and located behind the hologram surface. The source-size blurring of the image point in the vertical direction now corresponds to an angle of

$$\delta\theta_s \sim \frac{\delta y_i}{h + x_i}$$

$$\sim \delta\theta_c \frac{x_i \cos\langle\theta_c\rangle}{h + x_i} \tag{11.28}$$

When $h = 0$ (in which case we are viewing the hologram with our eyes pressed to the plate) or $x_i \gg h$ (in which case we are considering very large image depths), this reduces simply to

$$\delta\theta_s \sim \delta\theta_c \cos\langle\theta_c\rangle \tag{11.29}$$

This tells us that the virtual image of a transmission hologram will appear blurred to a close observer or at large depths unless the illumination source used to reconstruct it is characterised by

$$\delta\theta_s \leq \delta\theta_{Eye} \tag{11.30}$$

For all values of $h \neq 0$, Equations 11.28 and 11.30 show that at $x_i = 0$, the virtual image is always perfectly sharp. In addition, close behind the hologram, the image will always appear sharp because for any $\delta\theta_s$, there will always be a zone in which $\delta\theta_s < \delta\theta_{Eye}$. However, at a critical image depth, blurring will overtake the eye's resolution and deeper parts of the image will then always appear out of focus. This critical depth is given by

---

[*] This varies from person to person and also with the lighting conditions as the pupil size changes.

$$x_{\text{CRIT}} = \frac{h\delta\theta_{\text{Eye}}}{\delta\theta_c \cos\langle\theta_c\rangle - \delta\theta_{\text{Eye}}} \tag{11.31}$$

If we consider source-size blurring in the lateral (z) direction then the factor $\cos\langle\theta_c\rangle$ drops out in Equations 11.28, 11.29 and 11.31. For most light sources, lateral (z) blurring is therefore usually larger than the blurring in the vertical (y) dimension. Equation 11.28 shows that a real image in front of the hologram becomes blurred much more quickly than the virtual image behind the plate. Again, there is always a zone close to the hologram surface where the real image is sharp, but this zone will be shorter in the x dimension by a factor of

$$f = \frac{\delta\theta_c \cos\langle\theta_c\rangle - \delta\theta_{\text{Eye}}}{\delta\theta_c \cos\langle\theta_c\rangle + \delta\theta_{\text{Eye}}} \tag{11.32}$$

As before, if we consider lateral (z) blurring then the $\cos\langle\theta_c\rangle$ terms drop out. One can also note that at $h = x_i$, the blur becomes effectively infinite.

The above discussion is useful as it lets us begin to understand the "mechanics" of the thin transmission hologram. In particular, we see that any thin transmission hologram is highly sensitive to the physical size of the illumination source; as such, these holograms always require an illumination source of high spatial coherence. The discussion is also useful from a very practical point of view—as for digital rainbow and achromatic holograms, Equation 11.20 is precisely what is needed to formulate the required image data transformations.

Many transmission holograms are, however, not intrinsically thin and Bragg selection, which occurs due to the volume nature of any thick hologram, may in some cases somewhat mitigate the effects described by the source-size blurring formulae derived in this previous section. We shall see a little later that if we use a similar treatment to that discussed thus far in relation to the thin transmission hologram but this time for the case of an *infinitely* thick hologram (of almost vanishing permittivity modulation), then source-size blurring in this extreme case essentially disappears for monochromatic illumination. For most practical applications of display holography, however, current silver halide and photopolymer materials are just not thick enough to significantly alter source-size blurring. As a consequence, one of the fundamental constraints of ultra-realistic display holography is the spatial coherence of the illumination source.

11.2.3 Chromatic Blurring

When a thin transmission hologram is illuminated by a broadband source, each wavelength acts to form a virtual image of a given object point at a different location in space. This effect is known as chromatic blurring and is present in both transmission and reflection holograms.

We start by assuming that our illumination source is characterised by a spread of wavelengths $\delta\lambda$, that is,

$$\langle\lambda_c\rangle - \delta\lambda/2 \le \lambda_c \le \langle\lambda_c\rangle + \delta\lambda/2 \tag{11.33}$$

Equation 11.20 then defines the corresponding range inherited by x_i and y_i:*

$$\delta x_i = x_o \frac{\lambda_r}{\left(\langle\lambda_c\rangle - \delta\lambda/2\right)} - x_o \frac{\lambda_r}{\left(\langle\lambda_c\rangle + \delta\lambda/2\right)} \sim x_o \lambda_r \frac{\delta\lambda}{\langle\lambda_c\rangle^2}$$

$$\delta y_i = x_o \left(\frac{\lambda_r}{\left(\langle\lambda_c\rangle - \delta\lambda/2\right)}\sin\theta_c\right) - x_o \left(\frac{\lambda_r}{\left(\langle\lambda_c\rangle + \delta\lambda/2\right)}\sin\theta_c\right) \sim x_o \lambda_r \sin\theta_c \frac{\delta\lambda}{\langle\lambda_c\rangle^2} \tag{11.34}$$

$$\delta z_i = 0$$

---

* Note that we could of course have proceeded as in Equation 11.25 by simply differentiating Equation 11.20.

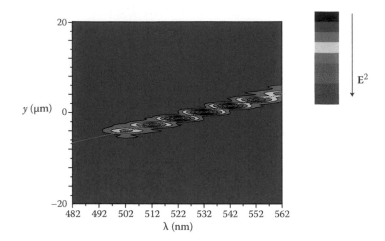

FIGURE 11.8 Contour plot of $\langle \mathbf{E}^2 \rangle$ for the hologram of Figure 11.7 illuminated by a discrete set of wavelengths. The dotted line is the theoretical prediction of the paraxial theory.

This is an important result because it shows that chromatic blurring only affects the vertical and depth dimensions of a thin transmission hologram under the paraxial plane wave approximation. Note, however, that as soon as spherical waves are used for the replay, this result no longer applies and chromatic blurring occurs in all three orthogonal directions.

Using the same methodology we employed with source-size blurring, we can analyse the consequences of chromatic blurring for the thin hologram. In fact, the behaviour is qualitatively identical. Blurring will always be greater for a real image in front of the hologram, and there will, in general, always exist a critical depth behind which all features in the hologram will appear blurred. This critical depth is

$$x_{\mathrm{CRIT}} = \frac{h\delta\theta_{\mathrm{Eye}}}{\dfrac{\delta\lambda}{\lambda}\sin\theta_c - \delta\theta_{\mathrm{Eye}}} \tag{11.35}$$

If $\delta\lambda/\lambda < \delta\theta_{\mathrm{Eye}}$ then the entire hologram will appear unblurred. The corresponding formula for the front-projected real image is

$$x_{\mathit{CRIT}} = \frac{h\delta\theta_{\mathrm{Eye}}}{\dfrac{\delta\lambda}{\lambda}\sin\theta_c + \delta\theta_{\mathrm{Eye}}}. \tag{11.36}$$

We can verify the relations in Equation 11.34 using a numerical evaluation of the Kirchhoff diffraction integral exactly as we did in Section 11.2.2. As an example, Figure 11.8 shows the numerically reconstructed signal of the hologram of Figure 11.7 for different reconstruction wavelengths at the object plane; the replay geometry is the same as that at recording. The dotted red line shows the theoretical fit produced by the paraxial theory, which is clearly in good agreement with the full diffractive simulation.

11.3 Laser Transmission Holograms and the Problem of Speckle

If a thin transmission hologram is illuminated by an effective point source of high temporal coherence at precisely its recording wavelength and recording geometry, the replayed image does not suffer from chromatic or source-size blurring. Large super-realistic monochromatic holograms have been made using this principle, often of very large size [7–9]. However, these types of holograms often suffer from speckle noise if the illumination source is too coherent. The major cause of speckle noise in laser transmission holograms is subjective speckle—this type of speckle depends on the observer and, in particular, the diameter of the effective entrance pupil of the human eye for normal distance observation of the hologram. Practically, this leads to an angular speckle size of

$$\delta\theta \sim 2.4\frac{\lambda}{D} \tag{11.37}$$

where D is the human eye diameter. The eye can resolve this and the result is an annoying dynamic random pattern superimposed on an otherwise clear hologram. From a theoretical point of view, the transmission hologram can record structures in objects down to the diffraction limit. However, in practice, speckle can severely limit this capability.

One effective way to reduce speckle is to use a lower temporal coherence of the replaying laser. Copper vapour lasers, for example, have been used very effectively in this regard. Of course, this introduces some chromatic blurring in the transmission hologram, which reduces image resolution, but a careful choice of the temporal coherence can produce stunning speckle-free images of great depth. The illumination source is of vital importance to ultra-realistic holographic imaging; we shall have more to say about this subject in Chapter 13.

11.4 Three-Dimensional Theory of the Thick Transmission Hologram

We can extend, in a simple fashion, our treatment of the thin transmission hologram to a thick hologram. To do this, we first assume that the hologram itself is surrounded by a thick dielectric having the same refractive index as the holographic emulsion and that all the source and reference points lie within this dielectric. Then, all we have to do is to use the following three-dimensional power series expansion of Equation 11.3 instead of our previous two-dimensional expansion:

$$\rho_\gamma = \sum_{m=0}^{M}\sum_{n=0}^{N}\sum_{p=0}^{P}\rho_{\gamma mnp}z_h^m y_h^n x_h^p \qquad \forall\gamma \in \{r,o,c,i\} \tag{11.38}$$

where

$$\rho_\gamma(x_h,y_h,z_h) = \sqrt{(x_\gamma - x_h)^2 + (y_\gamma - y_h)^2 + (z_\gamma - z_h)^2} \qquad \forall\gamma \in \{r,o,c,i\} \tag{11.39}$$

Mathematically, this is indeed simple—the only detail worth noting is that we must now regard all the Cartesian coordinates as representing an optical path rather than a pure distance—that is, $x \to nx$, $y \to ny$ and $z \to nz$ with n being the refractive index of the holographic emulsion; this is because we are now discussing the propagation of waves within a dielectric medium.

We should, however, be careful of what we are doing here physically. Indeed, we may still assert confidently that two holograms recorded each by a pair of different spherical waves are identical if and only if their time-averaged squared electric field distributions are identical. However, by including an expansion in x_h—or depth—it becomes more difficult to be confident of how a hologram will replay even under identical conditions in which it was recorded! This is because now that the hologram is thick, the replaying electromagnetic wave must propagate through a complex dielectric, and as it does so, its field distribution will inevitably change. Hence, inside the holographic emulsion, the replay wave will not be the same as the recording wave because the recording wave propagated through a uniform dielectric, not a hologram. To properly treat a thick hologram, we will therefore need to solve the wave equation within the holographic emulsion. We will do this later using the coupled wave theory. However, for now, we will continue with the simpler theory because not only will we see that it helps us understand the "mechanics" of the thick hologram, but we will also see that it produces formulae of significant practical use for digital holography.

Substitution of Equation 11.38 into Equation 11.5 defines the various ρ objects:

$$\rho_{\gamma001} = -\frac{x_\gamma}{\sqrt{x_\gamma^2 + y_\gamma^2 + z_\gamma^2}}; \qquad \rho_{\gamma010} = -\frac{y_\gamma}{\sqrt{x_\gamma^2 + y_\gamma^2 + z_\gamma^2}}$$

$$\rho_{\gamma100} = -\frac{z_\gamma}{\sqrt{x_\gamma^2 + y_\gamma^2 + z_\gamma^2}}; \qquad \rho_{\gamma110} = -\frac{y_\gamma z_\gamma}{\left(x_\gamma^2 + y_\gamma^2 + z_\gamma^2\right)^{3/2}} \qquad (11.40)$$

As before, we impose a paraxial condition on both the "i" and "o" source points by expanding the coordinate pairs (y_i, z_i) and (y_o, z_o) as Taylor series and then truncating Equation 11.5 at quadratic orders in these four variables. The expansion orders $(m = 1, n = 0, p = 0)$, $(m = 0, n = 1, p = 0)$ and $(m = 1, n = 1, p = 0)$ then lead directly to our previous equations (Equations 11.8 through 11.10). However, the order $(m = 0, n = 0, p = 1)$ leads to an additional equation:

$$\frac{\lambda_c}{\lambda_r} = \pm \left(\frac{x_c}{\sqrt{x_c^2 + y_c^2 + z_c^2}} - 1 \right) \left(\frac{x_r}{\sqrt{x_r^2 + y_r^2 + z_r^2}} - 1 \right)^{-1} \qquad (11.41)$$

This equation is the Bragg condition of a thick transmission hologram. It asserts that there is a unique replay wavelength for both the primary reconstructed image and the conjugate image of the hologram, which is determined by the recording geometry, replay geometry and recording wavelength. This is very different from the case of the thin transmission hologram in which any wavelength can be used to reconstruct the hologram. The Bragg effect is, however, disproportionately (from a practical point of view) important for reflection holograms which we shall study in Section 11.5. We shall see shortly that for all current types of transmission holograms used for display applications today, the Bragg effect is relatively weak and as such, for most common transmission holograms, Equation 11.41 constitutes at best an estimation of the wavelength of maximum diffractive replay.

11.4.1 Snell's Law at the Air–Hologram Boundary

Before we continue with a discussion of the consequences of the Bragg condition, we should remember that the thick hologram equations (Equations 11.8 through 11.10 and 11.41) that we have just derived pertain to the case of an infinitely thick dielectric of refractive index n. As such, the points (x_o, y_o, z_o), (x_r, y_r, z_r), (x_i, y_i, z_i) and (x_c, y_c, z_c) are all assumed to lie within the dielectric. If we wish to examine the case of these points being located in the air, and the hologram alone being located within the dielectric, then we must use Snell's law at the air–hologram boundary. Snell's law may be written in the coordinate system of Figure 11.3 as

$$n_1 \sin \theta_1 = n_2 \sin \theta_2 \qquad (11.42)$$

where n_1 and θ_1 are the refractive index and angle of incidence in medium 1, and n_2 and θ_2 are the refractive index and angle of incidence in medium 2.

At this point, it is instructive to write down the solution of Equation 11.5 using the relations in Equation 11.40 and the coordinates of Figure 11.3 without making a paraxial approximation for the object and image points. In this more general case, we obtain the following three relations for the primary nonconjugate image:

$$\frac{\sin^2 \theta_i \cos\phi_i \sin\phi_i}{r_i} = \frac{\sin^2 \theta_c \cos\phi_c \sin\phi_c}{r_c} - \frac{\lambda_c}{\lambda_r} \left\{ \frac{\sin^2 \theta_r \cos\phi_r \sin\phi_r}{r_r} - \frac{\sin^2 \theta_o \cos\phi_o \sin\phi_o}{r_o} \right\} \qquad (11.43)$$

$$\sin\theta_i \cos\phi_i = \sin\theta_c \cos\phi_c - \frac{\lambda_c}{\lambda_r} \{ \sin\theta_r \cos\phi_r - \sin\theta_o \cos\phi_o \} \qquad (11.44)$$

$$\sin\theta_i \sin\phi_i = \sin\theta_c \sin\phi_c - \frac{\lambda_c}{\lambda_r}\{\sin\theta_r \sin\phi_r - \sin\theta_o \sin\phi_o\} \tag{11.45}$$

In these equations, all angles are within the dielectric and the distances, r, are real distances rather than optical paths. Now, if we regard the hologram as being surrounded by air instead of a dielectric, then we must replace the angles and distances in Equations 11.43 through 11.45 using the transformation

$$\sin\theta_\gamma \rightarrow \frac{1}{n}\sin\theta_\gamma$$

$$\phi_\gamma \rightarrow \phi_\gamma \tag{11.46}$$

$$r_\gamma \rightarrow r_\gamma/n$$

If we apply these transformations to Equations 11.43 through 11.45, then we will arrive at a set of equations that describe a thick transmission hologram of finite refractive index surrounded by air with all angles and distances being those measured in the air. Of course, we implicitly assume a certain ordering here. The hologram is located at $x = 0$ and all source and reference points are regarded as being located in the air to the right of $x = 0$. To contemplate the replay of the hologram, we are obliged to regard its thickness as being small compared with all source and reference distances because an observer would need to be located to the left of the hologram. However, this is not a problem as long as the hologram thickness is assumed to be much greater than the recording and replay wavelengths.

It is immediately evident that Equations 11.43 through 11.45 are invariant under the action of Equation 11.46 and as such the paraxial forms Equations 11.8 through 11.10, or equivalently, Equations 11.16 through 11.18 are also invariant. As such, these equations may be read as describing the exterior ray angles of either a thick or a thin hologram surrounded by air. However, if we write the non-paraxial form of Equation 11.41 in spherical coordinates

$$\frac{\lambda_c}{\lambda_r} = \frac{(\cos\theta_c - \cos\theta_i)}{(\cos\theta_r - \cos\theta_o)} \tag{11.47}$$

it is clear that this equation is not invariant under the action of Equation 11.46. Indeed, it adopts the new paraxial form in terms of exterior angles:

$$\frac{\lambda_c}{\lambda_r} = \left(\sqrt{1 - \frac{1}{n^2}\sin^2\theta_c} - 1\right)\left(\sqrt{1 - \frac{1}{n^2}\sin^2\theta_r} - 1\right)^{-1} \tag{11.48}$$

or equivalently

$$\frac{\lambda_c}{\lambda_r} = \left(\sqrt{1 - \frac{y_c^2 + z_c^2}{n^2\left(x_c^2 + y_c^2 + z_c^2\right)}} - 1\right)\left(\sqrt{1 - \frac{y_r^2 + z_r^2}{n^2\left(x_r^2 + y_r^2 + z_r^2\right)}} - 1\right)^{-1} \tag{11.49}$$

11.4.2 Blurring in the Ultra-Thick Transmission Hologram

We mentioned in the previous section that the Bragg condition disproportionately affected the reflection hologram rather than the transmission hologram. Most display transmission holograms today are made on emulsions of less than 12 μm thickness, and as such, the Bragg condition does little to alter the spectral behaviour of these holograms over and above that of the standard thin transmission hologram model. The situation is, however, a little different if the emulsion is *much* thicker. In the ultra-thick transmission hologram (of almost vanishing permittivity modulation), Equation 11.48 defines a unique replay wavelength for the primary image at a given recording wavelength and at a given reference angle. As such, an ultra-thick transmission hologram illuminated by a broadband source of finite angular extent will still show dispersion along the achromatic plane—as for each θ_c within the range of the source $\langle\theta_c\rangle + \delta\theta_c/2 \geq \theta_c \geq \langle\theta_c\rangle - \delta\theta_c/2$, there will always exist a reconstruction wavelength that satisfies Equation 11.48. This means that the dispersion in the vertical direction will generally be limited by the angular size of the source—and the angular extent of this dispersion will be governed by Equation 11.29.

11.5 Reflection Holograms

The preceding analysis may be applied in exactly the same form to the case of a reflection hologram. However, now the object wave propagates in the reverse direction, so

$$\mathbf{E}_o(x_h, y_h, z_h, t) = \Re\left(\mathbf{E}_o \frac{e^{i\left[\omega t + \frac{2\pi}{\lambda_r}\rho_o(x_h, y_h, z_h)\right]}}{\rho_o(x_h, y_h, z_h)}\right) \tag{11.50}$$

This then changes the three-dimensional hologram fringe pattern to

$$\left\langle \mathbf{E}^2(x_h, y_h, z_h)\right\rangle = \frac{1}{2}\left\{\frac{|E_r|^2}{\rho_r^2(x_h, y_h, z_h)} + \frac{|E_o|^2}{\rho_o^2(x_h, y_h, z_h)}\right\}$$

$$+ \frac{\kappa_{ro}(y_h, z_h)\sqrt{E_r E_o E_r^* E_o^*}}{\rho_o(x_h, y_h, z_h)\rho_r(x_h, y_h, z_h)}\cos\left[\frac{2\pi}{\lambda_r}\rho_o(x_h, y_h, z_h) + \frac{2\pi}{\lambda_r}\rho_r(x_h, y_h, z_h) + \varsigma_{ro}(E_r, E_o)\right] \tag{11.51}$$

which, on application of the expansion in Equation 11.38, gives identical paraxial formulae to Equations 11.8 through 11.10 together with the paraxial Bragg condition, which differs from the transmission formula (Equation 11.49)

$$\frac{\lambda_c}{\lambda_r} = \left(\sqrt{1 - \frac{y_c^2 + z_c^2}{n^2\left(x_c^2 + y_c^2 + z_c^2\right)}} + 1\right)\left(\sqrt{1 - \frac{y_r^2 + z_r^2}{n^2\left(x_r^2 + y_r^2 + z_r^2\right)}} + 1\right)^{-1} \tag{11.52}$$

Note that the image and object points, (x_i, y_i, z_i) and (x_o, y_o, z_o), are now characterised by having negative values of x. Using the spherical coordinates of Figure 11.3, the associated non-paraxial equations in the plane wave limit $(r \to \infty)$ can be rewritten as

$$\sin\theta_i\cos\phi_i - \sin\theta_c\cos\phi_c = -\frac{\lambda_c}{\lambda_r}\{\sin\theta_r\cos\phi_r - \sin\theta_o\cos\phi_o\} \tag{11.53}$$

$$\sin\theta_i\sin\phi_i - \sin\theta_c\sin\phi_c = -\frac{\lambda_c}{\lambda_r}\{\sin\theta_r\sin\phi_r - \sin\theta_o\sin\phi_o\} \tag{11.54}$$

$$\frac{\lambda_c}{\lambda_r} = \left(\sqrt{1 - \frac{1}{n^2}\sin^2\theta_c} + \sqrt{1 - \frac{1}{n^2}\sin^2\theta_i}\right)\left(\sqrt{1 - \frac{1}{n^2}\sin^2\theta_r} + \sqrt{1 - \frac{1}{n^2}\sin^2\theta_o}\right)^{-1} \tag{11.55}$$

Equations 11.53 and 11.54 are identical to their transmission hologram counterparts, Equations 11.44 and 11.45, as we use the following convention:

- Transmission grating: $0 \leq \theta_o, \theta_i, \theta_r, \theta_c \leq \pi/2$; $0 \leq \phi_o, \phi_i, \phi_r, \phi_c \leq 2\pi$; $r \geq 0$
- Reflection grating: $0 \leq \theta_r, \theta_c \leq \pi/2$; $\pi/2 < \theta_o, \theta_i \leq \pi$; $0 \leq \phi_o, \phi_i, \phi_r, \phi_c \leq 2\pi$; $r \geq 0$

Equations 11.53 through 11.55 are particularly useful in digital display holography as they describe how a given recorded ray will replay in terms of angle and wavelength.

If we are considering only the two-dimensional case ($\phi = 0$), then we should use the following alternative definitions:

- Transmission grating: $-\pi/2 \leq \theta_o, \theta_i, \theta_r, \theta_c \leq \pi/2$; $r \geq 0$
- Reflection grating: $\pi/2 < \theta_o, \theta_i \leq 3\pi/2$; $r \geq 0$

The two-dimensional plane wave formulae for any hologram can now be written as

$$\sin\theta_i - \sin\theta_c = -\frac{\lambda_c}{\lambda_r}\{\sin\theta_r - \sin\theta_o\} \tag{11.56}$$

$$\frac{\lambda_c}{\lambda_r} = \left(\sqrt{1 - \frac{1}{n^2}\sin^2\theta_c} \pm \sqrt{1 - \frac{1}{n^2}\sin^2\theta_i}\right)\left(\sqrt{1 - \frac{1}{n^2}\sin^2\theta_r} \pm \sqrt{1 - \frac{1}{n^2}\sin^2\theta_o}\right)^{-1} \tag{11.57}$$

where the plus sign (+) in Equation 11.57 indicates a reflection hologram and the minus sign (−) indicates a transmission hologram.

Despite the apparent similarities, there is an important difference between thick transmission and thick reflection holograms. This is the orientation of the fringe pattern within the grating. In a thick hologram, the fringes are intrinsically three-dimensional and locally form nested planes. We shall see in the next section that a useful model of the thick hologram simply treats these nested planes as mirrors. Constructive and destructive interference then occurs between the various reflections from each such mirror or "Bragg plane". It is this interference process that determines the wavelength of replay, and it is the (two-dimensional) orientation of the planes that decides principally how a given ray will be "redirected" by the hologram on replay.

The orientation of the fringe planes in a transmission and reflection hologram is quite different (Figure 11.9). This effect is exacerbated by Snell's law, which acts to steepen the angle of incidence of a light ray within the emulsion layer. The end result is that transmission holograms are usually characterised by steep fringe planes whose normal vector is almost within the hologram plane; reflection

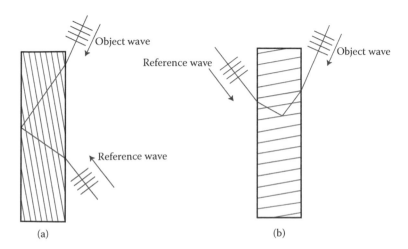

FIGURE 11.9 Typical fringe patterns for (a) a reflection grating and (b) a transmission grating.

holograms, on the other hand, are characterised by very shallow-angle fringe planes whose normal vector is more nearly aligned with the hologram normal. This has a profound effect on the importance of the Bragg selection process. The practical outcome is that common types of transmission volume holograms do not possess nearly as much (spectral) Bragg discrimination as their reflection counterparts. The result is that intermediate thickness transmission holograms lend themselves well to HPO rainbow and achromatic imaging, whereas their reflection counterparts lend themselves better to full-colour HPO imaging. In addition, full-aperture analogue (or full-parallax digital) reflection holograms can be expected to exhibit much deeper unblurred images when illuminated by white-light sources than their transmission counterparts. The higher (spectral) Bragg discrimination of the reflection hologram makes this hologram particularly interesting for large-format full-colour, full-parallax display imaging. Nonetheless, here, as we shall discuss in Chapter 13, very large depths can, from a practical point of view, only be attained by the use of special RGB illumination sources with narrow bands and relatively high spatial coherence.

11.6 A Simple Model of the Thick Hologram: Parallel Stacked Mirrors

The PSM model of a thick-phase hologram is a simple model that leads us to exactly the same results as those presented in the discussion of the previous section. We provide here a very basic and semi-heuristic discussion of this model. Such a discussion is primarily useful in terms of the simple but powerful picture it offers of what is actually going on in any volume transmission or reflection grating. In fact, the PSM model can be formulated in a much more rigorous way [10], but we shall wait until Chapter 12 before going into the extra detail required.

Figure 11.10 illustrates the interaction of two mutually coherent plane waves within a holographic emulsion. We can imagine these two waves as constituting an object and a reference wave. As drawn, the hologram is of the reflection type. The red lines indicate lines where the average square electric field will always be a minimum due to the interference of the two propagating waves. These are then the grating planes. On processing of the hologram, the square of the refractive index can be expected to exhibit a small sinusoidal modulation from plane to plane (this assumes that the permittivity of the medium varies as the time-averaged square of the electric field). On replay, one can expect an incident wave to be reflected from these grating planes. Of course, Maxwell's equations within a dielectric medium simply tell us that a wave will be partially reflected when it encounters any discontinuity in the refractive index [e.g., 11]. However, for the purpose of the present model, we shall make the plausible simplifying assumption that such reflection happens precisely and only at the indicated grating planes.

Analysing Figure 11.10, we see that the angle that the grating planes make to the object wave vector is given by

$$\psi = \frac{1}{2}\cos^{-1}\left(\hat{\mathbf{k}}_o \cdot \hat{\mathbf{k}}_r\right) \tag{11.58}$$

Here, $\hat{\mathbf{k}}_o$ and $\hat{\mathbf{k}}_r$ are the unit wave vectors of the object and reference waves. We can write them explicitly as Cartesian column vectors:

$$\hat{\mathbf{k}}_r = -\begin{pmatrix} \cos\theta_r \\ \sin\theta_r\cos\phi_r \\ \sin\theta_r\sin\phi_r \end{pmatrix}; \quad \hat{\mathbf{k}}_o = -\begin{pmatrix} \cos\theta_o \\ \sin\theta_o\cos\phi_o \\ \sin\theta_o\sin\phi_o \end{pmatrix} \tag{11.59}$$

As previously, we use the following convention (see Figure 11.3)

- Transmission grating: $0 \le \theta_o$, θ_i, θ_r, $\theta_c \le \pi/2$; $0 \le \phi_o$, ϕ_i, ϕ_r, $\phi_c \le 2\pi$; $r \ge 0$
- Reflection grating: $0 \le \theta_r$, $\theta_c \le \pi/2$; $\pi/2 < \theta_o$, $\theta_i \le \pi$; $0 \le \phi_o$, ϕ_i, ϕ_r, $\phi_c \le 2\pi$; $r \ge 0$

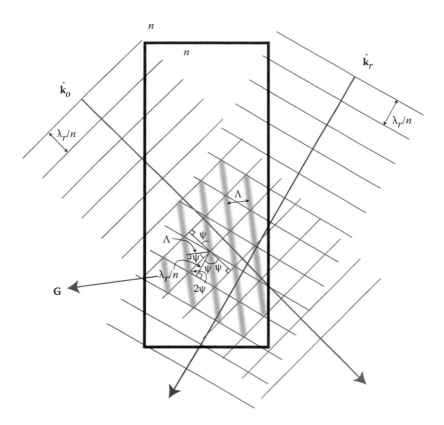

FIGURE 11.10 The interaction of two mutually coherent plane waves within a holographic emulsion with unit wave vectors $\hat{\mathbf{k}}_r$ and $\hat{\mathbf{k}}_o$. We can imagine these two waves as constituting, respectively, a reference and an object wave. As drawn, the hologram is of the reflection type. The red lines indicate lines of the grating planes in which the average square electric field will always be a minimum due to the interference of the two propagating waves.

From Figure 11.10 the grating planes can be seen to exactly bisect the angle between the two wave vectors. The distance between the grating planes is given by

$$\Lambda = \frac{\lambda_r}{n}\left|\frac{\cos\psi}{\sin(2\psi)}\right| = \frac{\lambda_r}{2n|\sin(\psi)|} = \frac{\lambda_r}{\sqrt{2}n\sqrt{1-\hat{\mathbf{k}}_o\cdot\hat{\mathbf{k}}_r}} \qquad (11.60)$$

The unit vector normal to the grating planes is given by

$$\hat{\mathbf{G}} = \frac{\left(\hat{\mathbf{k}}_r - \hat{\mathbf{k}}_o\right)}{\left|\hat{\mathbf{k}}_r - \hat{\mathbf{k}}_o\right|} \qquad (11.61)$$

It is also useful to define the grating vector as

$$\mathbf{G} \equiv \frac{2\pi}{\Lambda}\hat{\mathbf{G}} = \frac{2\pi n}{\lambda_r}\left(\hat{\mathbf{k}}_r - \hat{\mathbf{k}}_o\right) \qquad (11.62)$$

Figure 11.11 shows how we can imagine the hologram replay process. A plane wave with unit wave vector $\hat{\mathbf{k}}_c$ illuminates the hologram and is reflected from the grating planes producing another plane

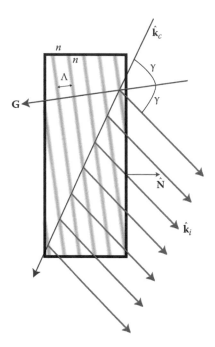

FIGURE 11.11 The hologram replay process. A plane wave with unit wave vector $\hat{\mathbf{k}}_c$ illuminates the grating of Figure 11.10 and is reflected from the grating planes producing another plane wave, this time with unit wave vector $\hat{\mathbf{k}}_i$.

wave, this time with unit wave vector $\hat{\mathbf{k}}_i$. The reflection process follows the normal law of classical reflection with angle of incidence equal to angle of reflection. This is most simply written as

$$\hat{\mathbf{k}}_i \times \hat{\mathbf{G}} = \hat{\mathbf{k}}_c \times \hat{\mathbf{G}} \tag{11.63}$$

This is a natural assumption to make because Equation 11.63 just comes from a plane wave solution of Maxwell's equations across a planar discontinuity in refractive index. Such an analysis [e.g., 11] shows that the amplitude reflection coefficient at near normal incidence for either π or σ-polarisations of the wave is given by

$$R \equiv \delta n / n \tag{11.64}$$

where $\delta n \ll n$ is the change in refractive index across the boundary.

Equation 11.63* can be solved for $\hat{\mathbf{k}}_i$

$$
\begin{aligned}
\hat{\mathbf{k}}_i &= \hat{\mathbf{k}}_c - 2\hat{\mathbf{G}}\left(\hat{\mathbf{k}}_c \cdot \hat{\mathbf{G}}\right) \\
&= \hat{\mathbf{k}}_c + \frac{\hat{\mathbf{k}}_c \cdot \left(\hat{\mathbf{k}}_r - \hat{\mathbf{k}}_o\right)}{\hat{\mathbf{k}}_o \cdot \hat{\mathbf{k}}_r - 1}\left(\hat{\mathbf{k}}_r - \hat{\mathbf{k}}_o\right)
\end{aligned}
\tag{11.65}
$$

One can immediately see that the angle (in azimuth and altitude) a reconstruction ray reflects off a single grating plane is solely determined, under the PSM model, by the classical law of reflection and is in no

---

* Note that we discard the trivial solution $\hat{\mathbf{k}}_i = \hat{\mathbf{k}}_c$.

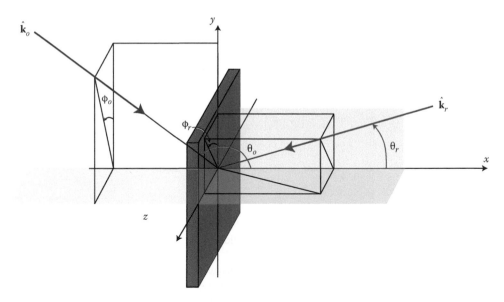

FIGURE 11.12 Altitudinal and azimuthal angles of incidence of the wave vectors of Figures 11.10 and 11.11.

way determined by the wavelength of the reconstructing wave. Of course, if reflections from successive grating planes produce waves that are in phase, then clearly the total reflected energy will be high; conversely, if the waves do not add up, they will tend to cancel. In fact, we can expect that if the parameter δn tends to zero with $(\delta n)d/\left(\Lambda\hat{\mathbf{G}}\cdot\hat{\mathbf{N}}\right)\gg 1$ (the thick hologram limit) then there will be an infinite sum of infinitesimal reflections from the grating, giving perfect discrimination and perfect diffraction.* The Bragg condition for constructive reflection can then be written simply as

$$\lambda_c/n = 2\Lambda\cos\gamma$$
$$= 2\Lambda\hat{\mathbf{k}}_c\cdot\hat{\mathbf{G}} \tag{11.66}$$
$$= \frac{\lambda_r}{n\left(1-\hat{\mathbf{k}}_o\cdot\hat{\mathbf{k}}_r\right)}\hat{\mathbf{k}}_c\cdot\left(\hat{\mathbf{k}}_r-\hat{\mathbf{k}}_o\right)$$

If we now combine Equations 11.65 and 11.66, we arrive at the equation

$$\frac{1}{\lambda_r}\left(\hat{\mathbf{k}}_r-\hat{\mathbf{k}}_o\right) = \frac{1}{\lambda_c}\left(\hat{\mathbf{k}}_c-\hat{\mathbf{k}}_i\right) \tag{11.67}$$

or defining $\mathbf{k} \equiv \dfrac{2\pi n}{\lambda}\hat{\mathbf{k}}$

$$\mathbf{k}_c - \mathbf{k}_i = \mathbf{k}_r - \mathbf{k}_o = \mathbf{G} \tag{11.68}$$

Equations 11.66 through 11.68 tell us how any thick hologram (either reflection or transmission) recorded at one wavelength with a given object and reference ray will replay. Equation 11.68 is particularly interesting because it shows us that if we illuminate a thick grating with a plane wave of wave vector $\mathbf{k}_c$, the diffracted response will be a plane wave of wave vector $\mathbf{k}_c - \mathbf{G}$. Using the expressions in

---

* At least for the σ-polarisation. The π-polarisation is slightly more complicated as here the electric field vectors of both the illumination wave and the diffracted signal wave are in the plane of incidence and so under certain circumstances may be mutually orthogonal. This effect is discussed in Section 11.10.6.1.

Equation 11.59 for the unit wave vectors, we can rewrite these equations in terms of the altitudinal and azimuthal angles of incidence of the rays* as illustrated in Figure 11.12.

$$\frac{1}{\lambda_r}(\sin\phi_o \sin\theta_o - \sin\phi_r \sin\theta_r) = \frac{1}{\lambda_c}(\sin\phi_i \sin\theta_i - \sin\phi_c \sin\theta_c) \qquad (11.69)$$

$$\frac{1}{\lambda_r}(\cos\phi_o \sin\theta_o - \cos\phi_r \sin\theta_r) = \frac{1}{\lambda_c}(\cos\phi_i \sin\theta_i - \cos\phi_c \sin\theta_c) \qquad (11.70)$$

$$\frac{1}{\lambda_r}(\cos\theta_o - \cos\theta_r) = \frac{1}{\lambda_c}(\cos\theta_i - \cos\theta_c) \qquad (11.71)$$

These equations should be familiar! They are just our previous non-paraxial Equations 11.47, 11.53 and 11.54. The PSM model can therefore be seen to lead to exactly the same results as our previous model, which sought to equate the average squared electric field distributions of two holograms recorded at different wavelengths and ray geometries.

Of course, we have not discussed the problem of Snell's law at the hologram–air boundary yet; as a result, all the wave vectors and angles discussed in the previous sections pertain to the holographic dielectric itself. However, converting the angles is simple, as we have seen in Section 11.4.1. As before, we see that Equations 11.69 and 11.70 are invariant under Snell's law. Equation 11.71 must be rewritten for exterior angles as

$$\frac{1}{\lambda_r}\left(\sqrt{1-\frac{1}{n^2}\sin^2\theta_o} \pm \sqrt{1-\frac{1}{n^2}\sin^2\theta_r}\right) = \frac{1}{\lambda_c}\left(\sqrt{1-\frac{1}{n^2}\sin^2\theta_i} \pm \sqrt{1-\frac{1}{n^2}\sin^2\theta_c}\right) \qquad (11.72)$$

where a "+" sign describes the reflection grating and a "−" sign the transmission grating. It is sometimes quite useful to write Snell's law in terms of vectors

$$\frac{1}{n}\hat{\mathbf{k}}_{ext} \times \hat{\mathbf{N}} = \hat{\mathbf{k}}_{int} \times \hat{\mathbf{N}} \qquad (11.73)$$

Here, $\hat{\mathbf{N}}$ is the unit normal vector of the hologram. This equation has solutions:

$$\hat{\mathbf{k}}_{ext} = n\hat{\mathbf{k}}_{int} - \hat{\mathbf{N}}\left\{n(\hat{\mathbf{k}}_{int} \cdot \hat{\mathbf{N}}) - \sqrt{(1-n^2) + n^2(\hat{\mathbf{k}}_{int} \cdot \hat{\mathbf{N}})^2}\right\}$$

$$\hat{\mathbf{k}}_{int} = \frac{1}{n}\hat{\mathbf{k}}_{ext} - \hat{\mathbf{N}}\left\{\frac{1}{n}(\hat{\mathbf{k}}_{ext} \cdot \hat{\mathbf{N}}) - \sqrt{\left(1-\frac{1}{n^2}\right) + \frac{1}{n^2}(\hat{\mathbf{k}}_{ext} \cdot \hat{\mathbf{N}})^2}\right\} \qquad (11.74)$$

Equations 11.69 through 11.72 constitute a useful mathematical model of the thick hologram. They are valid for all angles, not just paraxial angles. In particular, they provide a simple and effective method when digital image data must be recalculated to compensate for changes in replay wavelength and reference beam geometry in the digital hologram.

11.7 Holograms of Finite Thickness

In the previous section, we described a basic variant of the PSM model of the thick hologram. In the limit that the number of grating planes sampled by an illuminating wave is large the model predicted

---

* Note that these angles are internal angles within the grating dielectric.

an essentially perfect wavelength discrimination as defined by Equation 11.66. This led to the relation (Equation 11.68) showing that an incident reference wave $\mathbf{E} = \mathbf{E}_c \exp(i\mathbf{k}_c \cdot \mathbf{r})$ provoked the hologram to produce a response wave $\mathbf{E} = \mathbf{E}_i \exp(i\mathbf{k}_c \cdot \mathbf{r} - i\mathbf{G} \cdot \mathbf{r})$.

The basic PSM model is however not fully consistent for real gratings. For example, in the case of the thick reflection hologram, an illuminating plane wave will not sample all the grating planes because, at each plane, a proportion of its amplitude will be reflected. At a certain depth, the illuminating wave will be completely depleted and effectively the hologram will "look" to this wave like a grating of finite thickness. The problem then is that we cannot assume that the Bragg condition must be exactly satisfied and so in general there will be a band of wavelengths and angles under which the hologram will replay. To analyse this situation properly, we must formulate the problem in terms of two waves—an illuminating wave, which gradually becomes depleted during its passage through the hologram, and a response wave, which is gradually created by the depletion of the first wave. By solving Maxwell's equations within the grating and imposing proper boundary conditions at the hologram boundary, we can consistently* work out how this conversion process happens when the Bragg condition is satisfied only approximately. The first properly successful approach to this problem was published by Kogelnik [1] in 1969, and is known as the coupled wave theory. We shall describe this theory in detail in Section 11.10. A proper treatment of the PSM model may, however, also be used to solve this problem and has the important advantage of being able to treat full-colour gratings with great simplicity. We shall return to this more rigorous PSM theory in Chapter 12. For now, however, we shall restrict ourselves to making several naive (but nevertheless quite accurate) calculations of what a more basic version of the PSM model can say about a volume hologram of finite thickness.

Assuming a lossless phase hologram of finite thickness, we can use Equation 11.64 to estimate the strength of the response wave when we illuminate the hologram with a given illumination wave. Typical values of $\delta n/n$ for a modern panchromatic silver halide emulsion are in the region of $\delta n/n \sim 0.03$. The thickness of such emulsions is approximately 10 μm, giving approximately 20 grating planes for an unslanted reflection geometry. Coherent summation of the reflections then theoretically leads to a response of approximately 60% of the amplitude of the illuminating beam. Of course, this is a rather crude calculation, but it does serve to give some understanding that with reasonable modulation, volume phase holograms of the reflection variety do not have to be very thick to produce strong image signals on replay.

Another interesting calculation [4] is worthwhile. Suppose we do not exactly satisfy the Bragg condition within a hologram of N (unslanted) grating planes. For example, suppose our replay wavelength misses the Bragg wavelength by an amount $\delta\lambda$. In this case, the reflected signal from each grating plane will be dephased by $2\pi\,\delta\lambda/\lambda$. Coherently adding up the various reflections, we see that an estimate of the resultant image signal amplitude is given by

$$S \sim \frac{\delta n}{n} \left| \sum_{\beta=1}^{N} e^{i\beta(2\pi\delta\lambda/\lambda)} \right| = \frac{\delta n}{n} \left| \frac{\sin(N\pi\delta\lambda/\lambda)}{\sin(\pi\delta\lambda/\lambda)} \right| \tag{11.75}$$

For zero dephasing, we retrieve the relation we used above—$S = N\delta n/n$. At $N\delta\lambda/\lambda = 1$, however, S reaches zero. Now for an unslanted transmission grating, the number of grating planes crossed by a ray incident at angle θ_c (Figure 11.13) will be given by

$$N = \frac{d\tan\theta_c}{\Lambda} = \frac{|\mathbf{G}|d\tan\theta_c}{2\pi} \tag{11.76}$$

And so we can expect our hologram to replay over a range of wavelengths given by

$$\frac{\delta\lambda}{\lambda} \sim \frac{2\pi}{|\mathbf{G}|d\tan\theta_c} \sim \frac{\lambda}{2dn\sin\theta_c\tan\theta_c} \tag{11.77}$$

---

* One should note that for full consistency rigorous coupled wave theory must be used where higher diffraction orders are treated – see Appendix 8.

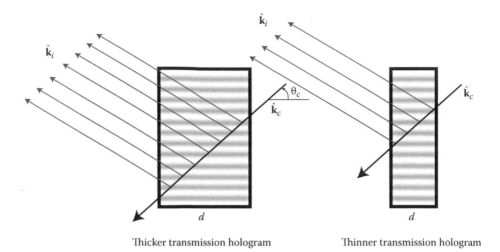

FIGURE 11.13 Replay from an unslanted transmission grating. The number of grating planes crossed by a ray incident at angle θ_c depends on the thickness of the grating.

Here, we have assumed that the hologram has been written at wavelength λ, which is also approximately the replay wavelength, and that the writing and replay reference angles are both equal to θ_c. For the corresponding case of the unslanted reflection grating, a ray transverses all the grating planes and so

$$N = \frac{d \tan \theta_c}{\Lambda} = \frac{|\mathbf{G}| d}{2\pi} \tag{11.78}$$

giving

$$\frac{\delta\lambda}{\lambda} \sim \frac{\lambda}{2dn \cos \theta_c} \tag{11.79}$$

Let us take the following as an example. We will use a wavelength of 532 nm and an emulsion of 12 μm thickness, recorded and replayed with an interior dielectric angle of $\theta_c = 15°$. We will assume an average index of $n = 1.5$. Then, for the transmission hologram, we estimate a fractional bandwidth of 21%. For the corresponding reflection hologram, the bandwidth comes out as 1.5%. We shall see later that these crude estimations of the PSM model are actually rather accurate, differing only from those produced by the coupled wave theory by 10% to 15%.

These calculations serve to give us a very good understanding of what is going on in the hologram. Specifically, we can see that a ray in a reflection hologram undergoes much more wavelength discrimination than a ray in the corresponding transmission case. This is because of the angle of the grating planes but also because of the distance between the grating planes (this is much smaller in the reflection case). These two effects multiply and lead to a much stronger wavelength selectivity in the reflection hologram.

One might have naively thought that because a ray in a transmission grating generally samples far fewer Bragg planes than in the corresponding reflection case, then the transmission hologram might well be fundamentally less diffractively efficient than the reflection hologram of a similar thickness. However, the Fresnel formulae tell us that the reflection coefficient at a discontinuous interface in permittivity strongly depends on incidence angle, with smaller angles giving substantially higher reflections. This exact effect operates in the transmission hologram. So even though a ray in this type of hologram samples fewer Bragg planes than a ray in the corresponding reflection case, reflection from any one Bragg plane, being at a much higher angle of incidence, means that ultimately there is little difference in the diffractive powers of similar thickness transmission and reflection holograms.

We should note here that amplitude transmission holograms are fundamentally less diffractive than phase holograms. This was demonstrated very effectively by Kogelnik's work [1]. One can show, however, that phase holograms must in many cases be fundamentally more noisy than amplitude holograms [5], and originally, this created the opinion (circa 1970) that phase holograms would always produce inferior quality images. However, great advances in processing and materials have occurred since—for example the discovery of reversal bleaching—and as a consequence, modern day phase holograms are largely regarded as being superior to amplitude holograms in virtually all applications of display holography. As a consequence, we will have little to say about amplitude holograms in this book.

11.8 Emulsion Swelling and Change in Refractive Index on Processing

In general, any thick hologram, whether of the transmission or reflection type, will undergo a change in both the refractive index and the emulsion thickness on processing. Equation 11.67 must then be changed to the following form:

$$\frac{n_r}{\lambda_r}\underline{\alpha}\cdot\left(\hat{\mathbf{k}}_r-\hat{\mathbf{k}}_o\right)=\frac{n_c}{\lambda_c}\left(\hat{\mathbf{k}}_c-\hat{\mathbf{k}}_i\right)$$

(11.80)

Here, n_r represents the refractive index on recording, n_c is the refractive index after processing and $\underline{\alpha} = \mathrm{diag}(\tau,1,1)$ is the emulsion swelling matrix with the scalar parameter τ representing the linear emulsion contraction in the x direction (normal to the hologram surface). The wave vectors in this formula are of course the wave vectors within the emulsion dielectric and Equations 11.73 and 11.74 must be used to convert them to a form valid outside the hologram–air boundary.

Emulsion swelling acts to change both the grating angle and the grating separation. Denoting unprimed quantities as pertaining to the situation before swelling and primed quantities for the situation after swelling, we can write

$$\hat{\mathbf{G}}'=\frac{\underline{\alpha}\cdot\hat{\mathbf{G}}}{\left|\underline{\alpha}\cdot\hat{\mathbf{G}}\right|}$$

$$\Lambda'=\Lambda\left|\underline{\alpha}\cdot\hat{\mathbf{G}}\right|^{-1}$$

(11.81)

By substitution of Equation 11.59 into Equation 11.80, we see once again that the y and z components of Equation 11.80 are Snell invariant. The x component must, however, be modified. Accordingly, in terms of exterior (air) angles, we can write down the general non-paraxial equations relating the recorded object and the reconstructed image rays:

$$\frac{n_r}{\lambda_r}(\sin\phi_o\sin\theta_o-\sin\phi_r\sin\theta_r)=\frac{n_c}{\lambda_c}(\sin\phi_i\sin\theta_i-\sin\phi_c\sin\theta_c)$$

(11.82)

$$\frac{n_r}{\lambda_r}(\cos\phi_o\sin\theta_o-\cos\phi_r\sin\theta_r)=\frac{n_c}{\lambda_c}(\cos\phi_i\sin\theta_i-\cos\phi_c\sin\theta_c)$$

(11.83)

$$\frac{n_r\tau}{\lambda_r}\left(\sqrt{1-\frac{1}{n_r^2}\sin^2\theta_o}\pm\sqrt{1-\frac{1}{n_r^2}\sin^2\theta_r}\right)=\frac{n_c}{\lambda_c}\left(\sqrt{1-\frac{1}{n_c^2}\sin^2\theta_i}\pm\sqrt{1-\frac{1}{n_c^2}\sin^2\theta_c}\right)$$

(11.84)

From the form of these equations, we can expect an emulsion thickness change and a change of refractive index on processing to lead to slightly different behaviours. However, both effects will change the replay characteristics of any transmission or reflection hologram. This will manifest itself as a different optimal replay wavelength being associated with a given replay geometry and a modification of the object point to image point mapping. It is perhaps worth pointing out that Equations 11.82 through 11.84 only have solutions for a given range of replay angles. Outside of these angles, there is no wavelength to satisfy the Bragg condition. This is particularly important for the reflection hologram.

11.9 Non-Paraxial Behaviour and Digital Image Predistortion

We have discussed at length the paraxial behaviour of both the thin and thick holograms. We have seen that as long as we replay our hologram with a point source, then a given object point will map into a well-defined image point. This is the case for both transmission and reflection holograms of both the thin and thick variety. For many holograms, however, we cannot make the paraxial approximation as the image and object points are characterised by large altitudinal angles. In this more general regime, Equations 11.82 through 11.84 show us that an object point will only map onto a well-defined image point if there is no swelling and no refractive index change on processing and, furthermore, if both the recording and replay geometries are the same and the replay wavelength is the same as the recording wavelength. If these rather stringent conditions are not satisfied, then we can expect different physical parts of the hologram to possess different object-image point mappings. The result is a chromatic and geometric aberration of the holographic image.

Non-paraxial image aberration may be very effectively corrected in digital holograms using the mathematical model we have developed. Often, digital reflection holograms are recorded at a slightly different wavelength than that desired to replay them. This may be due to intrinsic emulsion swelling with optimal chemistries or simply because available lasers have too high or too low a wavelength for optimal replay. Either way, without correction, the digital images suffer from noticeable chromatic and geometric aberration. Digital holograms are also very often written using a collimated reference beam—only to be illuminated using a spot lamp. We have seen in Chapter 7 that it is easier to use a collimated reference beam in a digital printer. In Appendix 4, we discuss in detail how the mathematical model of this chapter may be applied to the digital image data of a DWDH hologram to correct the induced geometric and chromatic aberrations, greatly enhancing the image realism.

11.10 Solving the Helmholtz Equation in Volume Gratings: Coupled Wave Theory

A simple but rather more rigorous approach to the volume hologram is to assume the existence of just two plane waves propagating in and outside a grating of finite thickness and to use the Helmholtz equation to calculate how a specific modulation in the dielectric permittivity intrinsically couples these waves. This approach originates from the field of acousto-optics but was first applied to holography by Kogelnik [1] in 1969. The first wave is assumed to be the illuminating reference wave and the second wave is the hologram's response or "signal" wave. The adoption of just two waves is made on the assumption that coupling to higher-order modes will be negligible. There is no rigorous mathematical proof for this per se. However numerical results from a fully accurate solution which is available from Moharam and Gaylord's rigorous coupled wave theory [12] (see Appendix 8) show good agreement with the two-wave theory for most practical gratings. The results of Kogelnik's coupled wave theory are simple and extremely useful for the purposes of display holography and holographic imaging.

11.10.1 One-Dimensional Coupled Wave Theory

Assuming a time dependence of $\sim\exp(i\omega t)$, we use Maxwell's equations and Ohm's law to write down the general wave equation in a dielectric in SI units:

$$\nabla \times (\nabla \times \mathbf{E}) - \gamma^2 \mathbf{E} = 0 \qquad (11.85)$$

where

$$\gamma^2 = i\omega\mu\sigma - \omega^2\mu\varepsilon \tag{11.86}$$

Here, μ is the permeability of the medium, ε its permittivity and σ represents its electrical conductivity. We shall now make two important assumptions. The first is that our grating is lossless, so $\sigma = 0$. The second is that the polarisation of our two waves is perpendicular to the grating vector or $\mathbf{E} \cdot \nabla\varepsilon = 0$. This allows the simplification of Equation 11.85 to the Helmholtz equation:

$$\nabla^2\mathbf{E} - \gamma^2\mathbf{E} = 0 \tag{11.87}$$

The assumption of small conductivity means that we restrict our analysis to ideal phase holograms with no absorption. The assumption that $\mathbf{E} \cdot \nabla\varepsilon = 0$ leads us to study gratings that have their grating vector in the propagation plane of two σ-polarised waves. These restrictions are not terribly constraining and the picture we get from the much-simplified analysis makes adopting them worthwhile. The interested reader is referred to Kogelnik [1] and Solymar and Cooke [4] for a discussion of more general models.

We will assume a one-dimensional grating extending from $x = 0$ to $x = d$. The relative permittivity is assumed to vary within the grating as

$$\varepsilon_r = \varepsilon_{r0} + \varepsilon_{r1} \cos \mathbf{G} \cdot \mathbf{r} \tag{11.88}$$

We may therefore write the γ parameter as

$$\gamma^2 \sim -\beta^2 - 4\kappa\beta\cos \mathbf{G} \cdot \mathbf{r} \tag{11.89}$$

with

$$\beta = \omega(\mu\varepsilon_0\varepsilon_{r0})^{1/2} \tag{11.90}$$

and where we have introduced Kogelnik's coupling constant

$$\kappa = \frac{1}{4}\frac{\varepsilon_{r1}}{\varepsilon_{r0}}\beta \sim \frac{1}{2}\left(\frac{\delta n}{n}\right)\beta n. \tag{11.91}$$

11.10.2 Solution with Perfect Bragg Compliance

In Section 11.6, we saw how the Bragg condition related the image and response wave vectors to the grating vector. Specifically, Equation 11.68 told us that

$$\mathbf{k}_i = \mathbf{k}_c - \mathbf{G} \tag{11.92}$$

The magnitude of both $\mathbf{k}_c$ and $\mathbf{k}_i$ was also shown to be exactly $\beta = 2\pi n/\lambda_c$. We will first investigate the case of perfect Bragg compliance; due to symmetry arguments, we would strongly expect the highest diffractive response to be produced by such compliance.* Accordingly, we will assume that

$$\gamma^2 = -\beta^2 - 4\kappa\beta \cos(\mathbf{k}_c - \mathbf{k}_i) \cdot \mathbf{r} \tag{11.93}$$

and that

$$\left|k_c\right| = \left|k_i\right| = \beta = \frac{2\pi n}{\lambda_c} \tag{11.94}$$

We now choose a very particular trial solution of the form

$$E_z = R(x)e^{-i\mathbf{k}_c \cdot \mathbf{r}} + S(x)e^{-i\mathbf{k}_i \cdot \mathbf{r}} \tag{11.95}$$

---

* This is almost always the case.

The first term represents the input wave and the second term represents the response wave (Figure 11.14). Both are plane waves. Note that the complex functions R and S are functions of x only—even though the wave vectors $\mathbf{k}_c$ and $\mathbf{k}_i$ have both x and y components. The grating is assumed to be surrounded by a dielectric having the same permittivity and permeability as the average values within the grating so as not to unduly complicate the boundary conditions. Within the external dielectric, both R and S are constants. The choice of just using two waves in the calculation—the absolute minimum—with only a one-dimensional behaviour was inspired by the work of Bhatia and Noble [13] and Phariseau [14] in the field of acousto-optics. However, one really proceeds here with the philosophy that if the trial solution of Equation 11.95 cannot sensibly satisfy the Helmholtz equation given in Equation 11.87 with sensible

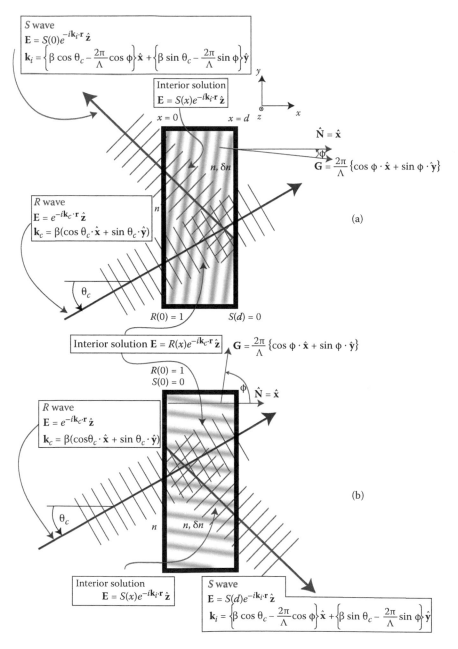

FIGURE 11.14 Diagram showing how the R and S waves of Kogelnik's coupled wave theory propagate within (a) a reflection and (b) a transmission grating.

boundary conditions, then the process will not provide a sensible answer. As we progress, we shall make more assumptions that appear at first sight to be at best plausible—but in the end, we shall see that the whole procedure provides a consistent method producing sensible results. With this in mind, we substitute Equations 11.93 through 11.95 into Equation 11.87 to obtain

$$e^{-i\mathbf{k}_c \cdot \mathbf{r}} \left\{ \frac{d^2 R}{dx^2} - 2ik_{cx} \frac{dR}{dx} + 2\beta\kappa S \right\} + e^{-i\mathbf{k}_i \cdot \mathbf{r}} \left\{ \frac{d^2 S}{dx^2} - 2ik_{ix} \frac{dS}{dx} + 2\beta\kappa R \right\}$$

$$+ 2\beta\kappa S e^{-i(2\mathbf{k}_i - \mathbf{k}_c) \cdot \mathbf{r}} + 2\beta\kappa R e^{-i(2\mathbf{k}_c - \mathbf{k}_i) \cdot \mathbf{r}} = 0 \tag{11.96}$$

Because we are assuming only two waves in the solution, we must now disregard the third and fourth terms of this expression assuming that they inherit only negligible energy from the primary modes. The next assumption is to neglect second-order derivatives on the premise that R and S are slowly varying functions. Given that the modulation of permittivity is small in all usual cases, this is quite plausible. With these assumptions in hand, Equation 11.96 implies the following two coupled first-order ordinary differential equations:

$$\frac{k_{cx}}{\beta} \frac{dR}{dx} + i\kappa S = 0 \tag{11.97}$$

$$\frac{k_{ix}}{\beta} \frac{dS}{dx} + i\kappa R = 0 \tag{11.98}$$

We can then use Equations 11.97 and 11.98 to write down identical uncoupled second-order differential equations for R and S:

$$\frac{d^2 R}{dx^2} + (\kappa^2 \sec\theta_c \sec\theta_i) R = 0$$

$$\frac{d^2 S}{dx^2} + (\kappa^2 \sec\theta_c \sec\theta_i) S = 0 \tag{11.99}$$

Here, the x component of the Bragg condition tells us that

$$\sec\theta_i = \left\{ \cos\theta_c - \frac{\lambda_c G_x}{2\pi n} \right\}^{-1} \tag{11.100}$$

Of course, if the grating has been written by a reference and object wave having angles of incidence of, respectively, θ_r and θ_o and at a wavelength of λ_r, then

$$G_x = \frac{2\pi n}{\lambda_r} (\cos\theta_r - \cos\theta_o) \tag{11.101}$$

11.10.3 Boundary Conditions

We have assumed that the R wave is the input wave and the S wave is the response. Normalising the input amplitude to unity, we can therefore write down different boundary conditions for transmission and reflection holograms. For transmission holograms, we must demand that $R(0) = 1$ (i.e., all the power is in the input wave at the point where the wave enters the grating) and $S(0) = 0$ (i.e., the power of the transmitted response wave must be zero at the input boundary as evidently no conversion has yet taken place). For reflection holograms, we demand that $R(0) = 1$ and $S(d) = 0$. Here, on the entrance boundary, we demand unit power in the input wave. Because the reflected response wave is travelling in the direction $x = d$ to $x = 0$, its amplitude must clearly be zero at the far boundary.

Armed with these boundary conditions, we can now solve Equation 11.99 (or equivalently Equations 11.97 and 11.98) for the transmission and reflection cases. For transmission holograms, we have

$$R = \cos\left\{\kappa x (\sec\theta_c \sec\theta_i)^{1/2}\right\}$$

$$S = -i\sqrt{\frac{\cos\theta_c}{\cos\theta_i}}\, \sin\left\{\kappa x (\sec\theta_c \sec\theta_i)^{1/2}\right\}$$

(11.102)

And for reflection holograms

$$R = \operatorname{sech}\left\{\kappa d \left(\sec\theta_c \left|\sec\theta_i\right|\right)^{1/2}\right\} \cosh\left\{\kappa(d-x)\left(\sec\theta_c \left|\sec\theta_i\right|\right)^{1/2}\right\}$$

$$S = -i\sqrt{\frac{\cos\theta_c}{|\cos\theta_i|}}\, \operatorname{sech}\left\{\kappa d \left(\sec\theta_c \left|\sec\theta_i\right|\right)^{1/2}\right\} \sinh\left\{\kappa(d-x)\left(\sec\theta_c \left|\sec\theta_i\right|\right)^{1/2}\right\}$$

(11.103)

These are remarkably simple solutions that paint a very logical picture. For the transmission case, we see that as the input wave enters the grating, it slowly donates power to the response wave which grows with increasing x. When the argument of the cosine function in Equation 11.102 reaches $\pi/2$, all power has been transferred to the S wave, which is now at a maximum. As x increases further, the waves change roles; the S wave now slowly donates power to a newly growing R wave. This process goes on until the waves exit the grating.

In the reflection case, the behaviour is different. Here, as one might well expect, there is simply a slow transfer of energy from the input wave to the reflected response wave. If the emulsion is thin, then the response wave is weak and most of the energy escapes as a transmitted R wave. If the emulsion is thick, then the amplitudes of both waves become exponentially small as x increases and all the energy is transferred from the R wave to the reflected S wave.

11.10.4 Power Conservation

Using Poynting's theorem, one can show that the power flowing along the x direction is given by

$$P = \cos\theta_c RR^* + \cos\theta_i SS^*$$

(11.104)

Multiplying Equations 11.97 and 11.98 by R^* and S^*, respectively, and then adding these equations and taking the real part, we see that $dP/dx = 0$. This is indeed indicative that the assumptions made were sensible.

11.10.5 Diffraction Efficiency

It is of particular interest to us to understand how efficient we can expect a holographic grating to be. With this in mind, we define the diffraction efficiency of a grating with reference wave of unit amplitude as

$$\eta = \frac{|k_{ix}|}{k_{cx}}SS^*$$

(11.105)

where S is evaluated on the exit boundary. It is now simple to use the forms for R and S given in Equations 11.102 and 11.103 to calculate the expected diffractive efficiencies for transmission and reflection holograms:

$$\eta_T = \sin^2\left\{\kappa d (\sec\theta_c \sec\theta_i)^{1/2}\right\}$$

(11.106)

$$\eta_R = \tanh^2\left\{\kappa d \left(\sec\theta_c \left|\sec\theta_i\right|\right)^{1/2}\right\}$$

(11.107)

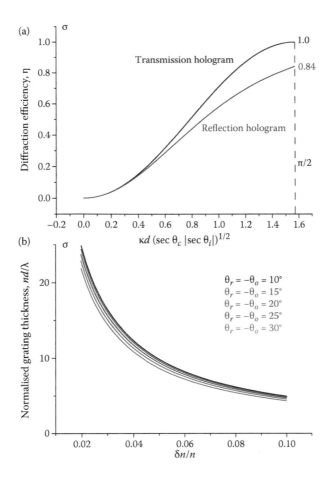

FIGURE 11.15 Perfect Bragg compliance: (a) diffractive replay efficiencies (σ-polarisation) of the transmission hologram (η_T) and the reflection hologram (η_R) versus the normalised grating thickness, $\kappa d(\sec \theta_c \sec \theta_i)^{1/2}$ according to Kogelnik's coupled wave theory. (b) Optimal value of the normalised grating thickness (providing $\eta_T = 1$) versus the modulation, $\delta n/n$ for a transmission hologram.

Figure 11.15a shows this graphically for $0 \le \kappa d(\sec \theta_c |\sec \theta_i|)^{1/2} \le \pi/2$. Clearly, for a small emulsion thickness or for a small permittivity modulation, the diffractive efficiencies of the reflection and transmission types of holograms are identical. As the parameter $\kappa d(\sec \theta_c |\sec \theta_i|)^{1/2}$ increases towards $\pi/2$, the transmission hologram becomes slightly more diffractive than its corresponding reflection counterpart. However, as we have remarked in the previous section, when $0 \le \kappa d(\sec \theta_c |\sec \theta_i|)^{1/2} > \pi/2$, the transmission hologram decreases in diffractive response whereas the corresponding reflection hologram continues to produce an increasing response. Figure 11.15b shows the relationship between the optimum grating thickness at which the diffractive response of the transmission hologram peaks and the grating modulation.

A recent photopolymer emulsion from the commercial manufacturer Bayer is characterised by a value of $\delta n \sim 0.037$ and an emulsion thickness of approximately 13.7 µm. This corresponds to a value of $\kappa d = 2.52$ at 633 nm, giving an ideal diffractive efficiency of approximately 97% in reflection mode. Actual measured values indicated 94% [15].

11.10.6 Small Departure from the Bragg Condition

To study the case of a small departure from the Bragg condition, we continue to use Equation 11.92, but now, we relax the condition that $|\mathbf{k}_i| = \beta$. This has the effect that the phases of the contributions of the signal wave from each Bragg plane do not add up coherently. We shall see that this choice leads naturally to the definition of an "off-Bragg" parameter, which allows us to quantify how much the Bragg condition is violated either in terms of wavelength or in terms of angle.

Proceeding in this fashion, Equation 11.97 remains the same, but Equation 11.98 generalises to

$$\frac{k_{ix}}{\beta} \frac{dS}{dx} + i \left(\frac{\beta^2 - |k_i|^2}{2\beta} \right) S + i\kappa R = 0 \tag{11.108}$$

We now define the "off-Bragg" or "dephasing" parameter

$$\vartheta = \frac{\beta^2 - |k_i|^2}{2\beta} = |\mathbf{G}| \cos(\phi - \theta_c) - \frac{|\mathbf{G}|^2}{2\beta} \tag{11.109}$$

where ϕ represents the slant angle between the grating normal and the grating vector (Figure 11.14). The value of ϑ is determined by the angle of incidence on reconstruction (θ_c) and by the wavelength of the illuminating light ($\lambda_c = 2\pi n/\beta$). Clearly, when $\vartheta = 0$, the Bragg condition is satisfied and $|\mathbf{k}_i| = \beta$. We define the obliquity factors:*

$$k_{ix}/\beta = (|k_{ix}|/\beta)\cos\theta_i \equiv c_S$$
$$k_{cx}/\beta = \cos\theta_c \equiv c_R \tag{11.110}$$

Then, as previously, we can solve Equations 11.97 and 11.108 to find expressions for the diffractive efficiency. For the transmission hologram, the result is

$$\eta_T = \frac{\sin^2 \left(\dfrac{\kappa^2 d^2}{c_R c_S} + \dfrac{d^2 \vartheta^2}{4c_S^2} \right)^{1/2}}{1 + \dfrac{\vartheta^2 c_R}{4c_S \kappa^2}} \tag{11.111}$$

And for the reflection hologram,

$$\eta_R = \left\{ 1 + \frac{1 - \dfrac{\vartheta^2 c_R}{4|c_S|\kappa^2}}{\sinh^2 \left(\dfrac{\kappa^2 d^2}{c_R |c_S|} - \dfrac{d^2 \vartheta^2}{4c_S^2} \right)^{1/2}} \right\}^{-1} \tag{11.112}$$

Clearly, for $\vartheta = 0$, these equations revert, respectively, to Equations 11.106 and 11.107. We can better understand the parameter ϑ if we imagine having recorded the grating we are now seeking to play back with an object beam at angle of incidence θ_o and with a reference beam at angle θ_r. The recording wavelength is λ_r and we take azimuthal angles $\phi_r = \phi_o = 0$. We assume no emulsion shrinkage and no change in average emulsion index. Then, we can write our various wave vectors as the following two vectors:†

---

* See Equations 11.113 and 11.114.
† Note the simple change of sign from our previous discussions (see Equation 11.59) in which the reference beam came from the right instead of the left.

$$\mathbf{k}_r = \frac{2\pi n}{\lambda_r} \begin{pmatrix} \cos\theta_r \\ \sin\theta_r \end{pmatrix} \; ; \; \mathbf{k}_o = \frac{2\pi n}{\lambda_r} \begin{pmatrix} \cos\theta_o \\ \sin\theta_o \end{pmatrix} \tag{11.113}$$

$$\mathbf{k}_c = \frac{2\pi n}{\lambda_c} \begin{pmatrix} \cos\theta_c \\ \sin\theta_c \end{pmatrix} \; ; \; \mathbf{k}_i = |k_i| \begin{pmatrix} \cos\theta_i \\ \sin\theta_i \end{pmatrix} \tag{11.114}$$

Equation 11.109 can now be written as

$$\vartheta = \frac{2\pi n}{\lambda_r}\{\cos(\theta_r - \theta_c) - \cos(\theta_o - \theta_c)\} - \frac{2\pi n \lambda_c}{\lambda_r^2}\{1 - \cos(\theta_r - \theta_o)\} \tag{11.115}$$

This tells us how the parameter ϑ behaves when $\lambda_c \neq \lambda_r$ and when $\theta_c \neq \theta_r$. Direct substitution of Equation 11.115 into Equations 11.111 and 11.112 leads trivially to general expressions for the diffractive response of a lossless holographic grating recorded with parameters $(\theta_r, \theta_o, \lambda_r)$ and replayed with (θ_c, λ_c).* These expressions are of prime importance for estimating the diffractive response of a grating.

11.10.6.1 Behaviour of the Lossless Transmission Hologram

It is instructive to study separately the effect of hologram replay at a differing wavelength and at a differing angle. We shall start with differing angles. To simplify things, we shall take the case of an unslanted grating and set $\theta_r = -\theta_o$. To eliminate the contribution in ϑ due to wavelength, we set $\lambda_c = \lambda_r$ whereupon Equation 11.115 becomes

$$\vartheta = \frac{4\pi n}{\lambda_r} \sin\theta_r \{\sin\theta_c - \sin\theta_r\} \tag{11.116}$$

The obliquity factors for the unslanted transmission grating are

$$c_R = \cos\theta_c \tag{11.117}$$

and

$$c_S = \frac{|k_{ix}|}{\beta}\cos\theta_i = \cos\theta_c - \frac{\lambda_c}{\lambda_r}(\cos\theta_r - \cos\theta_o) = \cos\theta_c \tag{11.118}$$

Substitution of Equations 11.116 through 11.118 into 11.111 then yields

$$\eta_T = \frac{\sin^2\left[\left(1 + \frac{4\sin^2\theta_r\{\sin\theta_c - \sin\theta_r\}^2}{(\delta n/n)^2}\right)^{1/2}\left(\frac{\delta n}{n}\right)\frac{\pi}{\cos\theta_c}\frac{nd}{\lambda_r}\right]}{1 + \frac{4\sin^2\theta_r\{\sin\theta_c - \sin\theta_r\}^2}{(\delta n/n)^2}} \tag{11.119}$$

This describes the diffractive response of the hologram in terms of the modulation $\delta n/n$, the record and illumination angles, θ_r and θ_c, and the normalised grating thickness, nd/λ_r. We can make one further useful simplification to this formula. The diffractive efficiency of a transmission hologram illuminated under perfect Bragg compliance is a maximum when

---

* One also needs Equation 11.110.

$$\left(\frac{\delta n}{n}\right)\frac{\pi}{\cos\theta_r}\frac{nd}{\lambda_r} = \frac{\pi}{2} \tag{11.120}$$

This was described in Figure 11.15. We can use this formula to define the thickness of the hologram. If we do this, we arrive at an expression for the diffractive response of the optimally thick transmission hologram:

$$\hat{\eta}_T = \frac{\sin^2\left[\left(1 + \frac{4\sin^2\theta_r\{\sin\theta_c - \sin\theta_r\}^2}{(\delta n/n)^2}\right)^{1/2}\frac{\pi\cos\theta_r}{2\cos\theta_c}\right]}{1 + \frac{4\sin^2\theta_r\{\sin\theta_c - \sin\theta_r\}^2}{(\delta n/n)^2}} \tag{11.121}$$

The function $\hat{\eta}_T(\theta_c)$ is plotted out in Figure 11.16a through 11.16c for various values of θ_r and for three cases of typical modulations. For a modulation characteristic of contemporary silver halide materials employed in display holography and at an internal recording angle of $\theta_r = 30°$ (equivalent to an external angle of approximately 45° at typical index values), we see that the full-width half maximum (FWHM) of $\hat{\eta}_T(\theta_c)$ is approximately 3°. We should not, however, expect our results here to be accurate as θ_c becomes too small. In the unslanted transmission grating the Bragg condition will in general be satisfied for $\theta_c = \pm\theta_r$. However we have implicitly assumed that only the solution $\theta_c = \theta_r$ corresponds to perfect Bragg compliance. We are of course at liberty to choose the other root. In this case we would define $\boldsymbol{k}_c - \boldsymbol{k}_i = -\boldsymbol{G}$ instead of $\boldsymbol{k}_c - \boldsymbol{k}_i = +\boldsymbol{G}$. When θ_c is relatively large there is a large angular spacing between the two perfect Bragg angles. As such, for all but the thinnest gratings, only one root produces significant diffraction at a given angle. If we were to replot Figure 11.16b for the case of $\theta_r = 10°$ assuming $\boldsymbol{k}_c - \boldsymbol{k}_i = -\boldsymbol{G}$ then we would see a peak value of η of approximately 0.1 which is about 10% of the value plotted. The corresponding graph at $\theta_r = 30°$ however shows negligible diffraction for the alternate solution. As $\theta_c \rightarrow 0$, the assumption of a single diffracted mode breaks down. Indeed at $\theta_c = 0$ we encounter a fundamentally new type of diffraction known as Raman-Nath diffraction.

To investigate the effect of fringe slant on the diffractive replay, in Figure 11.17a, we plot $\eta_T(\theta_c)$ of Equation 11.111 for various values of the object beam recording angle θ_o for a very specific case ($d = 7$ μm, $n = 1.5$, $\delta n/n = 0.02$, $\lambda_r = \lambda_c = 532$ nm, $\theta_r = 30°$). Clearly, there is very little change in the diffractive response of the hologram as the slant of the grating changes over the normal range used in display holography. However, one does see a broadening of the curve, indicating a decrease in angle selectivity of the grating, as the recording and reference beams come closer together.

At this point, it is interesting to expand our discussion of the coupled wave theory to the case of the π-polarisation. You will remember that we have thus far restricted ourselves to the σ-polarisation. In fact, Kogelnik and others have shown [1,4] that the only difference to the coupled wave equations for the σ-polarisation is that, for the π-polarisation, the constant κ must be replaced by

$$\hat{\kappa} = -\kappa\cos 2(\theta_c - \phi) \tag{11.122}$$

where ϕ is the grating slant angle defined by

$$\phi = \tan^{-1}\frac{G_y}{G_x} = \tan^{-1}\left\{\frac{\sin\theta_r - \sin\theta_o}{\cos\theta_r - \cos\theta_o}\right\} \tag{11.123}$$

Proving this result using the coupled wave theory is a little involved, and as such, the interested reader is referred to the references given.* However, we can easily use this to re-plot Figure 11.17a for the case of the π-polarisation. This is shown in Figure 11.17b. The graph is different from the σ-polarisation case because the diffractive efficiency drops somewhat as the object and reference beams become more

---

* We shall, however, use the PSM theory in Chapter 12 to prove this result. Whereas coupled wave theory is substantially easier and clearer to apply to the σ-polarisation, the PSM theory can be used to treat either polarisation with equal ease.

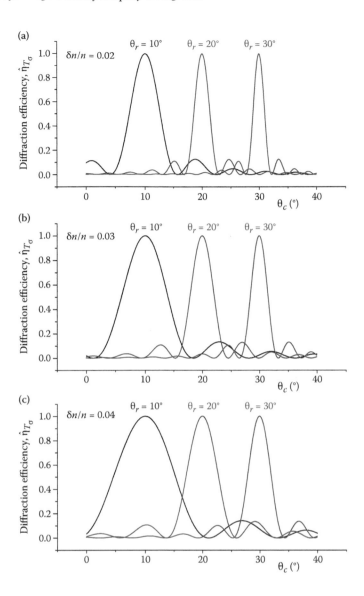

FIGURE 11.16 Optimal diffractive efficiency of the lossless unslanted transmission grating (σ-polarisation) as predicted by the coupled wave theory. The function $\hat{\eta}_r(\theta_c)$ is plotted for three values of $\theta_r = 10°$, $20°$, $30°$ and for three cases of typical modulations: (a) $\delta n/n = 0.02$; (b) $\delta n/n = 0.03$; (c) $\delta n/n = 0.04$. Note that all angles are within the dielectric.

angularly spaced during recording. Clearly, from Equations 11.111 and 11.122, we can expect the diffractive efficiency to drop to zero as the difference in angle at recording tends to 90°. Intuitively, we can understand this by the fact that when $\theta_c = \theta_r$ and $|\theta_o - \theta_r| = \pi/2$, we are essentially asking the hologram to reconstruct a signal beam having an electric field vector in a direction in which the illuminating beam has zero electric field.

Returning to the σ-polarisation, we now investigate the behaviour of the lossless transmission hologram to variation of the replay wavelength by setting $\theta_c = \theta_r$. As before, we will first assume that the grating is unslanted so that $\theta_r = -\theta_o$. Equation 11.115 then becomes

$$\vartheta = \frac{2\pi n}{\lambda_r}\{1 - \cos(2\theta_r)\}\left(1 - \frac{\lambda_c}{\lambda_r}\right) \tag{11.124}$$

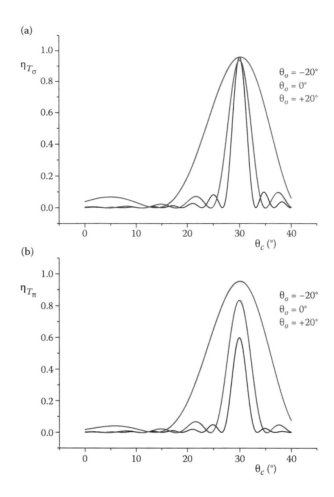

(a)

(b)

FIGURE 11.17 Diffractive efficiency versus replay angle, $\eta_T(\theta_c)$ as predicted by the coupled wave theory of the lossless transmission hologram for various values of the object beam recording angle $\theta_o = -20°$, $0°$, $+20°$ and for the grating parameters $d = 7$ μm, $n = 1.5$, $\delta n/n = 0.02$; $\lambda_r = \lambda_c = 532$ nm, $\theta_r = 30°$: (a) σ-polarisation and (b) π-polarisation. Angles quoted are within the dielectric.

Substituting this, Equation 11.117 and Equation 11.118 into Equation 11.111 yields

$$\eta_T = \frac{\sin^2\left\{\left(1 + \frac{\frac{\lambda_c^2}{\lambda_r^2}\{1-\cos(2\theta_r)\}^2\left(1-\frac{\lambda_c}{\lambda_r}\right)^2}{(\delta n/n)^2}\right)^{1/2} \frac{\pi}{\cos\theta_r}\left(\frac{\delta n}{n}\right)\frac{nd}{\lambda_c}\right\}}{\left(1 + \frac{\frac{\lambda_c^2}{\lambda_r^2}\{1-\cos(2\theta_r)\}^2\left(1-\frac{\lambda_c}{\lambda_r}\right)^2}{(\delta n/n)^2}\right)} \tag{11.125}$$

Using Equation 11.120, we can then write down an expression for the diffractive response versus replay wavelength for the optimally thick transmission hologram:

$$\hat{\eta}_T = \frac{\sin^2\left\{ \left[1 + \frac{\frac{\lambda_c^2}{\lambda_r^2}\{1 - \cos(2\theta_r)\}^2 \left(1 - \frac{\lambda_c}{\lambda_r}\right)^2}{(\delta n/n)^2} \right]^{1/2} \frac{\pi \lambda_r}{2\lambda_c} \right\}}{\left(1 + \frac{\frac{\lambda_c^2}{\lambda_r^2}\{1 - \cos(2\theta_r)\}^2 \left(1 - \frac{\lambda_c}{\lambda_r}\right)^2}{(\delta n/n)^2} \right)} \tag{11.126}$$

The function $\hat{\eta}_T(\lambda_c)$ is plotted out in Figure 11.18a through 11.18c for the same values of θ_r and $\delta n/n$ as used in Figure 11.16 and for three popular recording wavelengths. Clearly, the replay characteristics of the unslanted transmission grating are rather insensitive to replay wavelength. At modulations characteristic of contemporary silver halide materials used in display holography and at an internal recording angle of $\theta_r = -\theta_o = 30°$ (equivalent to an external angle of approximately 45° at typical index values), we see that the FWHM of $\hat{\eta}_T(\lambda_c)$ is well over 60 nm, and at values of $\theta_r = -\theta_o = 10°$, the FWHM is greater than the entire visible spectrum.

To get an understanding of the effect of fringe slant on the diffractive replay, we will again choose a specific case. Figure 11.19a shows a plot of $\eta_T(\lambda_c)$ calculated directly from Equation 11.111 for three values of the object beam recording angle, $\theta_o = (-20°, 0°, +20°)$ and for the following parameters: $d = 7$ μm, $n = 1.5$, $\delta n/n = 0.02$, $= \theta_r = \theta_c = 30°$. This is an interesting graph because it shows an effective blue shift on replay of the grating recorded at $\theta_r = 30°$, $\theta_o = 20°$, $\lambda_r = 532$ nm. The Bragg condition is however exactly satisfied at $\theta_c = 30°$, $\lambda_c = 532$ nm where $\vartheta = 0$, but this is not the peak of diffractive efficiency! This is because the reference and object beam are very close together during recording, leading to a large fringe spacing and a very broadband replay characteristic. As such, the diffractive efficiency depends more on $\kappa(\lambda_c)$ than on $\vartheta(\lambda_c)$; the dominant physical effect is therefore that a blue replay wavelength actually "feels" the grating to be thicker than the recording green wavelength and therefore exhibits a higher diffractive response. This is an effect of the thin grating only. As grating thickness is increased (with a proportionate decrease in modulation), the effect disappears. Figure 11.19b shows the corresponding case for the π-polarisation, which shows a lessening diffractive response when the angle between the electric field vectors of the diffracted and illuminating ray approaches 90°.

Such apparent violation, and it is of course only apparent, of the Bragg condition has a direct bearing on our prior discussion of the geometrical ray equations describing how a holographic image distorts when replayed. There, we implicitly assumed that the maximum diffractive ray was indeed to be associated with the ray that exactly satisfied the Bragg condition. However, we see here that for certain cases in moderately thin gratings, notably when the replay behaviour is extremely broadband, this assumption starts to break down.

11.10.6.2 Behaviour of the Lossless Reflection Hologram

We can usefully characterise the lossless reflection hologram in a very similar manner to the way we have proceeded in the previous section. We will start by restricting the discussion to the unslanted reflection grating for which $\theta_r = \pi - \theta_o$. Then, as previously, we will study the angle dependence by setting $\lambda_c = \lambda_r$. The obliquity factor c_R is still given by (11.117), but the form of c_S changes to

$$c_S = \cos \theta_c - 2\cos \theta_r \tag{11.127}$$

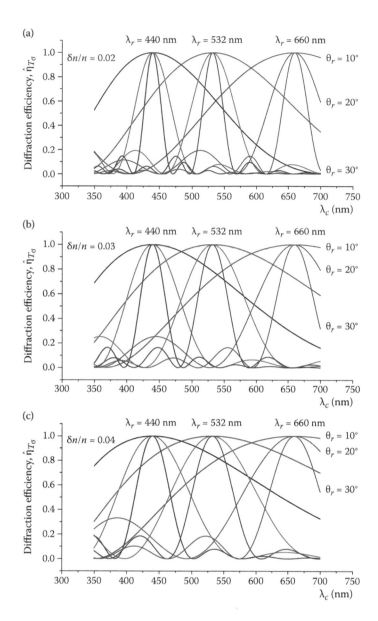

FIGURE 11.18 Optimal diffractive efficiency versus free-space replay wavelength, $\hat{\eta}_T(\lambda_c)$ of the lossless unslanted transmission grating (σ-polarisation) as predicted by the coupled wave theory for three values of $\theta_r = 10°, 20°, 30°$; three recording wavelengths, $\lambda_r = 440$ nm, $\lambda_r = 532$ nm and $\lambda_r = 660$ nm; and for three cases of typical modulations, (a) $\delta n/n = 0.02$, (b) $\delta n/n = 0.03$, (c) $\delta n/n = 0.04$. Angles quoted are within the dielectric.

The parameter ϑ also changes:

$$\vartheta = \frac{4\pi n}{\lambda_r}\cos\theta_r(\cos\theta_c - \cos\theta_r) \tag{11.128}$$

With these formulae, we can now rewrite Equation 11.112 to give a formula for the diffraction efficiency of the lossless unslanted reflection grating in terms of replay angle:

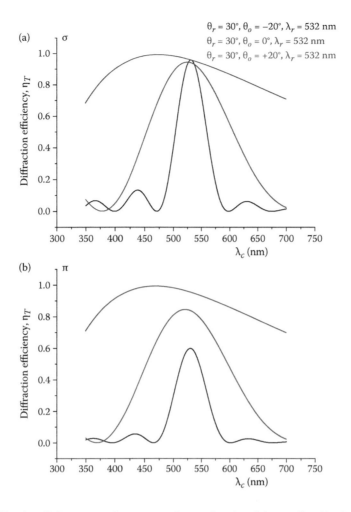

FIGURE 11.19 Diffractive efficiency versus free-space replay wavelength, $\eta_T(\lambda_c)$ as predicted by the coupled wave theory of the lossless transmission hologram for three values of the object beam recording angle, $\theta_o = (-20°, 0°, +20°)$ and for the following parameters: $d = 7\ \mu m$, $n = 1.5$, $\delta n/n = 0.02$, $\lambda_r = 532\ nm$, $\theta_r = \theta_c = 30°$): (a) σ-polarisation and (b) π-polarisation. Angles quoted are within the dielectric.

$$\eta_R = \frac{\sinh^2\left\{ \Upsilon \cdot \dfrac{\kappa d}{\sqrt{c_R |c_S|}} \right\}}{\sinh^2\left\{ \Upsilon \cdot \dfrac{\kappa d}{\sqrt{c_R |c_S|}} \right\} + \Upsilon^2} \tag{11.129}$$

where

$$\Upsilon = \left(1 - \frac{4(\cos\theta_c - \cos\theta_r)^2 \cos\theta_c \cos^2\theta_r}{\left|(\cos\theta_c - 2\cos\theta_r)\right|(\delta n/n)^2} \right)^{1/2} \tag{11.130}$$

and

$$\frac{\kappa d}{\sqrt{c_R |c_S|}} = \frac{\pi}{\sqrt{|(\cos\theta_c - 2\cos\theta_r)|\cos\theta_c}} \left(\frac{nd}{\lambda_r}\right)\left(\frac{\delta n}{n}\right)$$ (11.131)

Let us now define the thickness of the hologram to give a 90% diffractive efficiency at $\theta_c = \theta_r$:

$$\pi\left(\frac{nd}{\lambda_r}\right)\left(\frac{\delta n}{n}\right) = \cos\theta_r \tanh^{-1}\left\{\sqrt{0.9}\right\}$$ (11.132)

We then arrive at the formula for the diffractive efficiency versus angle $\hat{\eta}_{R90}(\theta_c)$ for a 90% optimised lossless unslanted reflection grating:

$$\hat{\eta}_{R90} = \frac{\sinh^2\left\{\Upsilon \dfrac{\cos\theta_r \tanh^{-1}\left\{\sqrt{0.9}\right\}}{\sqrt{|(\cos\theta_c - 2\cos\theta_r)|\cos\theta_c}}\right\}}{\sinh^2\left\{\Upsilon \dfrac{\cos\theta_r \tanh^{-1}\left\{\sqrt{0.9}\right\}}{\sqrt{|(\cos\theta_c - 2\cos\theta_r)|\cos\theta_c}}\right\} + \Upsilon^2}$$ (11.133)

The function $\hat{\eta}_{R90}(\theta_c)$ is plotted out in Figure 11.20a through 11.20c for various values of θ_r and for three cases of typical modulations. It is interesting to note that, generally, the reflection hologram is characterised by slightly poorer replay angle discrimination than the transmission hologram.

To investigate the importance of fringe slant, in Figure 11.21a, we plot $\eta_R(\theta_c)$ of Equation 11.112 for three object beam recording angles ($\theta_o = 160°$, $180°$, $200°$) and for the parameters ($d = 7$ μm, $n = 1.5$, $\delta n/n = 0.03$, $\lambda_r = \lambda_c = 532$ nm, $\theta_r = 30°$). For completeness, we plot the corresponding graph for the π-polarisation in Figure 11.21b. You will notice that the curves exhibit two peaks corresponding to a reflection of $\mathbf{k}_c$ about the grating vector $\mathbf{G}$. For instance, when $\theta_o = 160°$, the fringes are slanted at $\phi = 5°$ and the angle between $\mathbf{k}_r$ and $\mathbf{G}$ is $25°$. One would therefore expect a second peak at a replay angle of $-20°$ because at this angle, the angle between $\mathbf{k}_c$ and $\mathbf{G}$ is again $25°$. The two peaks for $\theta_o = 200°$ are close enough together that they merge into a single wide peak.

Finally, we come to the wavelength dependence of the lossless reflection grating. To study this, we first assume an unslanted grating and put $\theta_o = \pi - \theta_r$ and $\theta_c = \theta_r \geq 0$ with $\lambda_c \neq \lambda_r$. Following our previous work, we then arrive at a formula for the diffractive efficiency versus wavelength, $\hat{\eta}_{R90}(\lambda_c)$ for a 90% optimised lossless unslanted reflection grating:

$$\hat{\eta}_{R90} = \sinh^2\left\{\Upsilon \dfrac{\dfrac{\lambda_r}{\lambda_c}\tanh^{-1}\left\{\sqrt{0.9}\right\}}{\sqrt{\left|1 - 2\dfrac{\lambda_c}{\lambda_r}\right|}}\right\}\left[\sinh^2\left\{\Upsilon \dfrac{\dfrac{\lambda_r}{\lambda_c}\tanh^{-1}\left\{\sqrt{0.9}\right\}}{\sqrt{\left|1 - 2\dfrac{\lambda_c}{\lambda_r}\right|}}\right\} + \Upsilon^2\right]^{-1}$$ (11.134)

where

$$\Upsilon = \left\{1 - \frac{4\left(\dfrac{\lambda_c}{\lambda_r}\right)^2 \cos^4\theta_r \left[1 - \dfrac{\lambda_c}{\lambda_r}\right]^2}{\left|\left(1 - 2\dfrac{\lambda_c}{\lambda_r}\right)\right|\left(\dfrac{\delta n}{n}\right)^2}\right\}^{1/2}$$ (11.135)

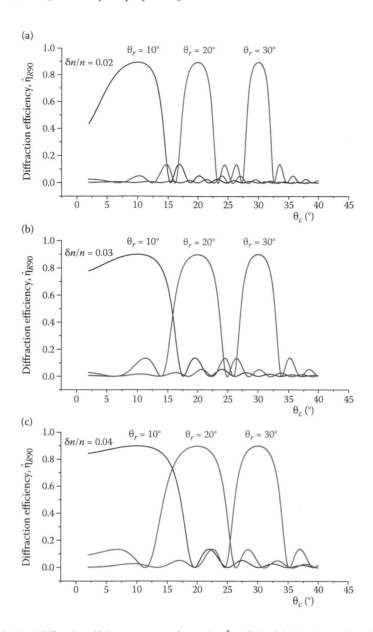

FIGURE 11.20 Optimal diffractive efficiency versus replay angle, $\hat{\eta}_{R90}(\theta_c)$ of the lossless unslanted reflection grating (σ-polarisation) as predicted by the coupled wave theory for three values of recording angle $\theta_r = 10°$, $20°$, $30°$ and for three cases of typical modulations: (a) $\delta n/n = 0.02$ (corresponding to a grating thickness of 10.1 μm for $\theta_r = 10°$, 9.65 μm for $\theta_r = 20°$ and 8.89 μm for $\theta_r = 30°$); (b) $\delta n/n = 0.03$ (corresponding to a grating thickness of 6.74 μm for $\theta_r = 10°$, 6.43 μm for $\theta_r = 20°$ and 5.93 μm for $\theta_r = 30°$); (c) $\delta n/n = 0.04$ (corresponding to a grating thickness of 5.05 μm for $\theta_r = 10°$, 4.82 μm for $\theta_r = 20°$ and 4.44 μm for $\theta_r = 30°$). The function $\hat{\eta}_{R90}(\theta_c)$ is chosen by defining the product of normalised grating thickness and index modulation such that $\hat{\eta}_{R90}(\theta_r) = 0.9$ for each value of θ_r plotted. All angles quoted are within the dielectric.

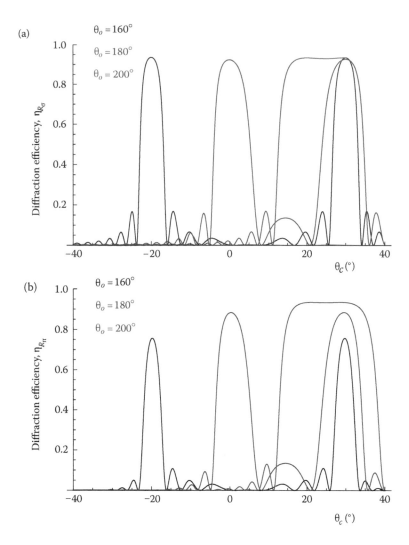

FIGURE 11.21 Diffractive efficiency versus replay angle, $\eta_R(\theta_c)$, as predicted by the coupled wave theory of the lossless reflection hologram for various values of the object beam recording angle $\theta_o = 160°$, $180°$, $200°$ and for the grating parameters $d = 7$ μm, $n = 1.5$, $\delta n/n = 0.03$; $\lambda_r = \lambda_c = 532$ nm, $\theta_r = 30°$: (a) σ-polarisation and (b) π-polarisation. All angles quoted are within the dielectric.

The function $\hat{\eta}_{R90}(\lambda_c)$ is plotted out in Figure 11.22a through 11.22c for the same values of θ_r and $\delta n/n$ as used in Figure 11.18 for the transmission case. Clearly, and as expected from our previous discussions, the unslanted reflection grating is far more sensitive to replay wavelength than its transmission counterpart. In addition, the replay bandwidth is much less sensitive to recording angle.

Figure 11.23a and b show the effect of fringe slant on the replay wavelength dependence of the reflection hologram for the σ- and π-polarisations, respectively. The graphs are plotted for a typical silver halide emulsion grating thickness of 7 μm and an index modulation of $\delta n/n = 0.03$, giving a maximum diffraction efficiency of more than 90% and a typical FWHM replay bandwidth of 25 nm.

It is worthwhile spending a little time trying to understand the basic reasons, in terms of the coupled wave theory, behind the angle and wavelength behaviour of the transmission and reflection holograms. To do this, we will assume that the illumination wave on playback is of magnitude $|k_i| = 2\pi n/\lambda_r + \Delta\beta$ and that its angle of incidence is $\theta_c = \theta_r + \Delta\theta$. Then, Equations 11.92, 11.109, 11.113 and 11.114 permit us to write down the following simple expression, which relates ϑ to $\Delta\theta_c$ and $\Delta\beta$:

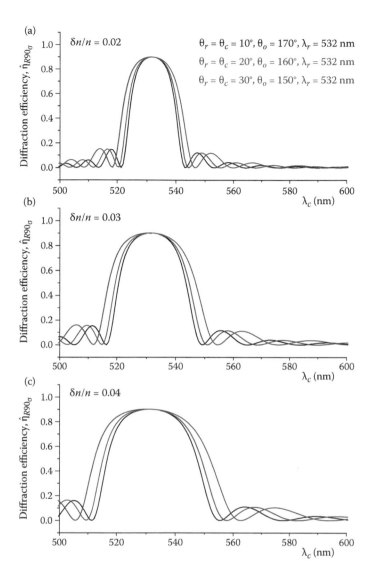

FIGURE 11.22 Optimal diffractive efficiency versus (freespace) replay wavelength, $\hat{\eta}_{R90}(\lambda_c)$, of the lossless unslanted reflection grating (σ-polarisation) as predicted by the coupled wave theory for three values of $\theta_r = 10°, 20°, 30°$, a recording wavelength of $\lambda_r = 532$ nm and for three cases of typical modulations: (a) $\delta n/n = 0.02$, (b) $\delta n/n = 0.03$ and (c) $\delta n/n = 0.04$. The function $\hat{\eta}_{R90}(\lambda_c)$ is defined by choosing the product of the normalised grating thickness and index modulation, such that $\hat{\eta}_{R90}(\lambda_r, \theta_r) = 0.9$ for each case of (θ_r, λ_r) plotted. Assuming an average index of $n = 1.5$, case (a) corresponds to a grating thickness of $d = 10.1$ μm for $\theta_r = 10°$, $d = 9.65$ μm for $\theta_r = 20°$ and $d = 8.89$ μm for $\theta_r = 30°$; likewise, case (b) corresponds to a grating thickness of $d = 6.74$ μm for $\theta_r = 10°$, $d = 6.43$ μm for $\theta_r = 20°$ and $d = 5.93$ μm for $\theta_r = 30°$; and case (c) corresponds to a grating thickness of $d = 5.05$ μm for $\theta_r = 10°$, $d = 4.82$ μm for $\theta_r = 20°$ and $d = 4.44$ μm for $\theta_r = 30°$. All angles internal.

$$\vartheta = \frac{2\pi n}{\lambda_r} \Delta\theta \sin(\theta_r - \theta_o) + \Delta\beta\{1 - \cos(\theta_r - \theta_o)\} \tag{11.136}$$

Now let us adopt a value of $\kappa d/\sqrt{c_R |c_S|} = \pi/2$. You will recall that this gives us perfect conversion from the R wave to the S wave in the transmission hologram when $\vartheta = 0$. It also corresponds to a diffractive efficiency for the reflection hologram of 0.84. We use Equations 11.111 and 11.112 to calculate the value of the dephasing parameter ϑ that is required to bring the diffraction to its first zero. This is given by

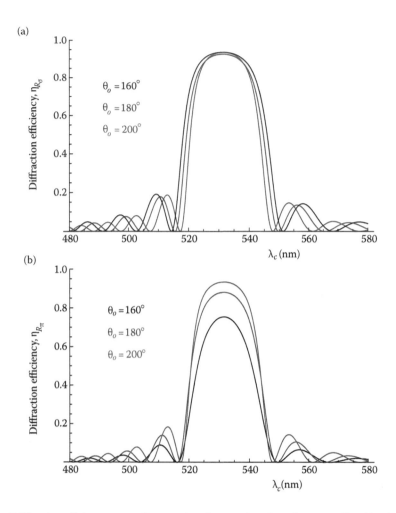

FIGURE 11.23 Diffractive efficiency versus (freespace) replay wavelength, $\eta_R(\lambda_c)$, as predicted by the coupled wave theory of the lossless reflection hologram for three values of the object beam recording angle, $\theta_o = (160°, 180°, 200°)$ and for the following parameters: $d = 7 \, \mu m$, $n = 1.5$, $\delta n/n = 0.03$, $\lambda_r = 532$, $\theta_r = \theta_c = 30°)$: (a) σ-polarisation and (b) π-polarisation. Angles quoted are internal.

$$\vartheta_T = \sqrt{3}\pi \frac{c_S}{d}$$

$$\vartheta_R = \sqrt{5}\pi \frac{|c_S|}{d}$$

(11.137)

We may then use Equation 11.136 to show that for the unslanted transmission hologram,[*]

$$\Delta\theta_T \sim \frac{\sqrt{3}}{2}\frac{\Lambda}{d} = \frac{\sqrt{3}}{4}\frac{\lambda}{dn}\csc\theta_r$$

(11.138)

$$\left(\frac{\Delta\lambda}{\lambda}\right)_T \sim \frac{\sqrt{3}}{2}\frac{\Lambda}{d}\cot\theta_r = \frac{\sqrt{3}}{4}\frac{\lambda}{dn}\cos\theta_r\csc^2\theta_r$$

(11.139)

---

[*] Note that Kogelnik [1] gives the following formulae for the FWHM: $\Delta\theta_{\mathrm{FWHM}} = \Lambda/d$; $\Delta\lambda_{\mathrm{FWHM}} = \cot\theta_c \cdot \Lambda/d$.

and for the corresponding reflection hologram,

$$\Delta\theta_R \sim \frac{\sqrt{5}}{2}\frac{\Lambda\cot\theta_r}{d} = \frac{\sqrt{5}}{4}\frac{\lambda}{dn}\csc\theta_r \tag{11.140}$$

$$\left(\frac{\Delta\lambda}{\lambda}\right)_R = \frac{\sqrt{5}}{2}\frac{\Lambda}{d} = \frac{\sqrt{5}}{4}\frac{\lambda\sec\theta_r}{dn} \tag{11.141}$$

We can now see why a transmission hologram is generally more selective in angle than a reflection hologram: $\Delta\theta_R/\Delta\theta_T = \sqrt{5/3}$, independent of wavelength and angle! Similarly, $\Delta\lambda_R/\Delta\lambda_T \sim \sqrt{5/3}\tan^2\theta_r$, which for small θ_r makes the reflection hologram much more selective than the corresponding transmission case. Note that the expressions in Equations 11.139 and 11.141 are remarkably close to the estimations of the PSM model that led us to Equations 11.77 and 11.79.

11.10.7 Effect of Loss in the Dielectric

Up until now, we have assumed that the electrical conductivity of the dielectric grating was zero. This led us to a form of the Helmholtz equation with real coefficients. It is, however, very straightforward to generalise the coupled wave theory to the case of a complex coefficient. To do this, we introduce the following companion equation to Equation 11.88:

$$\sigma = \sigma_o + \sigma_1\cos\mathbf{G}\cdot\mathbf{r} \tag{11.142}$$

Here, σ is the conductivity of the grating, which we divide into a constant and modulated part just as we did with the permittivity. If we then introduce the parameter

$$\alpha = \frac{\sigma_0}{2}\sqrt{\frac{\mu}{\epsilon_0\epsilon_1}} \tag{11.143}$$

we can generalise Equation 11.89 to

$$\gamma^2 = -\beta^2 + 2i\alpha\beta - 4\kappa\beta\cos\mathbf{G}\cdot\mathbf{r} \tag{11.144}$$

where Kogelnik's coupling constant now becomes

$$\kappa = \frac{1}{4}\frac{\epsilon_{r1}}{\epsilon_{r0}}\beta + i\frac{\alpha}{2}\frac{\sigma_1}{\sigma_0} \tag{11.145}$$

It is then a simple matter to solve the Helmholtz equation exactly as before by setting up two coupled waves, R and S, which are then described by the differential equations

$$c_R\frac{dR}{dx} + \alpha R + i\kappa S = 0 \tag{11.146}$$

$$c_S\frac{dS}{dx} + (\alpha + i\vartheta)S + i\kappa R = 0 \tag{11.147}$$

Using these equations, we can see that the energy balance of the lossy coupled wave model is now described by

$$\frac{d}{dx}(c_R RR^* + c_S SS^*) + 2\alpha(RR^* + SS^*) + i(\kappa - \kappa^*)(RS^* + R^*S) = 0 \tag{11.148}$$

The presence of the obliquity factors in the first term indicates that power flows along the x axis. The second and third terms correspond to the expected ohmic heating, σEE^*.

Equations 11.146 and 11.147 can be solved to find (rather complex) analytic expressions for S and R for both the transmission and reflection boundary conditions. We shall simply summarise the most important results here that pertain to unslanted mixed-phase amplitude gratings. In the case of the transmission hologram, the angular selectivity is essentially independent of loss. Kogelnik suggests that this is because absorption does not really alter the phase relationship between R and S. The only practical effect of including absorption then is to reduce the diffractive efficiency of the hologram as one would naively expect.

The effect of absorption on the unslanted reflection hologram is perhaps a little more noticeable. Here, loss tends to broaden the $\eta(\vartheta d)$ curve in the wings and steepen it towards the centre. Again, the principal effect is, however, the reduction of diffractive efficiency. Even for losses that reduce the efficiency by 10 times, the broadening in $\Delta\lambda$ or $\Delta\theta$ for typical display holography type gratings is rarely more than 10%. One further effect is worth mentioning however. Loss does reduce somewhat the significant side lobes in the $\eta(\vartheta d)$ curve present at higher values of the parameter κd.

The formulae we have derived in the last few sections very much form part of the essential toolkit for emulsion design in display holography. By choosing values of $\delta n/n$ and the emulsion thickness, we can now calculate the expected diffractive efficiencies and the expected angular and wavelength selectivities for either transmission or reflection geometries. Experience has shown that Kogelnik's coupled wave theory works extremely well in most cases of interest. More complex theoretical work, numerical simulation and experimental comparison all show that the theory is remarkably good for the low modulations normally encountered.

11.10.8 Recording of Complex Wave Fronts and Multiple Gratings

Throughout this chapter, we have examined how a single grating behaves. We have seen how such gratings can be expected to faithfully replay "image points" in three-dimensional space when we employ a spherical-wave formulism. Using a simpler plane wave formulism, we have then answered questions about the diffractive response and the angular and wavelength selectivities of the hologram. Here, we have seen that, with the right design for the emulsion in terms of permittivity modulation and thickness, we can expect to produce high-brightness reproductions of simple recorded waves.

Two critical questions remain unanswered, however. How does the hologram cope with complex wave fronts of the kind required to synthesise the three-dimensional image of a real object? And how does the hologram cope with the polychromatic gratings, which will inevitably be required for full-colour reflection holography?

11.10.9 Multiple Gratings Generated by Many Object Points

The question of spatially multiplexed gratings may be addressed quite simply by an extension of the coupled wave theory. This is the two-dimensional N-coupled wave theory [2]. We will concentrate first on the problem of recording a complex object composed of many object points. In the limit, we can regard this as describing the holographic recording of a diffuse object. Here, we will follow Solymar and Cooke [4] and assume a reference wave (labelled by subscript 0) and $N-1$ object waves of the form

$$E_j = A_{j0}a_j(x, y)\exp(-i\beta\rho_j) \tag{11.149}$$

Here, A_{j0} is a complex amplitude, a_j is the normalised amplitude distribution and ρ_j is the phase of the wave. We will assume an infinite two-dimensional slab of thickness d with z being an ignorable coordinate (Figure 11.24). As before, we take the σ-polarisation for simplicity.

We will now assume that the reference wave is large compared with the other waves. This is a vital assumption because, without it, radiation from individual points will interact together to form their own gratings. If we make this assumption, then the grating created will be free from cross-terms and we can write

$$\Delta\varepsilon = g\left|\sum_{j=0}^{N-1}E_j\right|^2 = \frac{1}{2}\sum_{j=1}^{N-1}a_o a_j\left\{\varepsilon_j\exp(-i\beta[\rho_0 - \rho_j]) + \varepsilon_j^*\exp(i\beta[\rho_0 - \rho_j])\right\} \tag{11.150}$$

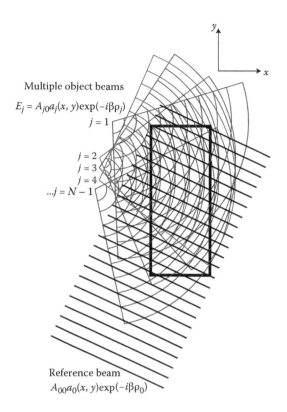

FIGURE 11.24 Two-dimensional N-coupled wave theory. An infinite two-dimensional slab grating is recorded with reference wave of complex amplitude A_{00} and j object waves of complex amplitude A_{j0}. All waves are assumed to be of σ-polarisation.

where

$$\varepsilon_j = 2gA_{00}A_{j0}^*$$ (11.151)

and where g is a constant. Of course, this expression would be exact if the individual object points were recorded sequentially rather than all at the same time. However, in the limit of a large-amplitude reference beam, it also provides a good approximation of the actual recorded grating. To see how this grating replays when illuminated by the original reference beam, we must solve the wave equation

$$\nabla^2 E + \beta^2(1 + \Delta\varepsilon/\varepsilon_{r0})E = 0$$ (11.152)

Clearly, the Bragg condition will be satisfied for each of the $N-1$ waves and so all will be reconstructed. Assuming no other waves are produced, we can then adopt a trial solution of the form

$$E = \sum_{j=0}^{N-1} A_j(x,y)a_j(x,y)\exp(-i\beta\rho_j)$$ (11.153)

One then proceeds exactly as with the one-dimensional coupled wave theory, equating coefficients of each exponential term to zero. Discarding second-order derivatives, assuming exact Bragg compliance for each signal wave, keeping only the signal waves corresponding to our trial function and using the geometric optics laws,

$$(\nabla\rho_j)^2 = 1$$

$$\nabla \cdot (a_j^2 \nabla\rho_j) = 0$$ (11.154)

we are led to the following set of coupled differential equations:

$$\nabla A_0 \cdot \nabla \rho_0 + i \sum_{j=1}^{N-1} \kappa_j a_j^2 A_j = 0 \tag{11.155}$$

$$\nabla A_j \cdot \nabla \rho_j + i \kappa_j^* a_0^2 A_0 = 0 \qquad \forall j \in \{1,2,3,...,(N-1)\} \tag{11.156}$$

where

$$\kappa_j = \frac{\varepsilon_j \beta}{4 \varepsilon_{r0}} \tag{11.157}$$

These equations can now be solved analytically for the case that all the waves are plane waves. In fact, this turns out to be a simple extension of the one-dimensional coupled wave theory. Taking the angle of incidence of the jth wave as θ_j, we may rewrite Equations 11.155 and 11.156 as

$$\cos \theta_0 \cdot \frac{dA_0}{dx} + i \sum_{j=1}^{N-1} \kappa_j A_j = 0 \tag{11.158}$$

and

$$\frac{dA_j}{dx} \cos \theta_j + i \kappa_j^* A_0 = 0 \quad \forall j \in \left\{1,2,3,...,(N-1)\right\} \tag{11.159}$$

Note that $a_j = 1 \; \forall j$. These equations may then be recast into a simple second-order ordinary differential equation:

$$\frac{d^2 A_0}{dx^2} + A_0 \left\{ \sec \theta_0 \sum_{j=1}^{N-1} \sec \theta_j |\kappa_j|^2 \right\} = 0 \tag{11.160}$$

Now the boundary conditions for a transmission hologram are $A_0 = 1$ and $A_j = 0 \; \forall j \neq 0$ at $x = 0$. This then yields the solution

$$A_0 = \cos(Lx)$$
$$A_j = -i \frac{\kappa_j^*}{L} \sin(Lx) \sec \theta_j \qquad \forall j \in \{1,2,3,...,(N-1)\} \tag{11.161}$$

where

$$L^2 = \sec \theta_0 \sum_{j=1}^{N-1} \left| \sec \theta_j \right| \left\| \kappa_j \right|^2 \tag{11.162}$$

If we take $j = 1$, then Equation 11.161 reduces to Equation 11.102—the standard result from the one-dimensional coupled wave theory. As we discussed earlier, in this case, when $Lx = \pi/2$, complete power transfer occurs between the provoking wave and the signal wave—that is, when $Lx = \pi/2$, $A_0 = 0$ and A_1 is a maximum. If we now, however, consider the case of $N > 1$, then $Lx = \pi/2$ and A_0 is still zero, but this time, the total power in this wave has been donated to the $N-1$ signal waves. We see very clearly then that the N-coupled wave theory predicts that complete power transfer should be possible from a single reference wave to many signal waves under the assumption of a strong reference beam at recording. From this, we can deduce that a hologram made up of many image points may theoretically at least be expected to play back with perfect diffractive efficiency under the correct conditions. From the above solution, we see that each of the plane waves reconstructed faithfully reproduces the directional information recorded. The relative amplitudes are, however, subject to some modification for using Equation 11.161, we see that

$$\frac{A_i}{A_j} = \frac{A_{i0}\cos\theta_j}{A_{j0}\cos\theta_i} \tag{11.163}$$

This means that the image is only reconstructed perfectly in amplitude in the central field (i.e., paraxially)—at higher angles, the amplitudes are modified according to an angle law. This is not, however, a great problem, as the angular field is usually restricted, and even at large angles of view, Snell's law acts to strongly confine the angular field within the emulsion layer of most display holograms. Finally, any residual amplitude modification can of course be compensated for in the case of digital holograms.

Exactly the same argument applies to reflection holograms. Here, the N-wave analytical solution for plane waves is

$$A_0 = \frac{\cosh(L\{d - x\})}{\cosh(Ld)}$$

$$A_j = -\frac{i\kappa_j^*}{L|\cos\theta_j|}\frac{\sinh(L\{d - x\})}{\cosh(Ld)} \tag{11.164}$$

If we define the diffractive efficiency of the jth wave as

$$\eta_j = \frac{|\cos\theta_j|}{|\cos\theta_0|}A_j(0)A_j^*(0) \tag{11.165}$$

then each individual wave will have a diffractive power given by

$$\eta_j = \frac{|k_j|^2}{L^2}|\sec\theta_j|\sec\theta_0\tanh^2(Ld) \tag{11.166}$$

or

$$\eta_j = \frac{|k_j|^2}{|\cos\theta_j|\sum_{j=1}^{N-1}|\sec\theta_j||k_j|^2}\tanh^2\left[d\sqrt{\sec\theta_0\sum_{j=1}^{N-1}|\sec\theta_j||k_j|^2}\right] \tag{11.167}$$

The total diffractive efficiency is likewise given by

$$\eta = \sum_{j=1}^{N-1}\eta_j = \tanh^2\left[d\sqrt{\sec\theta_0\sum_{j=1}^{N-1}|\sec\theta_j||k_j|^2}\right] \tag{11.168}$$

showing clearly that the $N - 1$ signal waves can inherit 100% of the power from the driving wave.

We should perhaps put in a slight word of caution here. N-wave theory as applied to plane waves does seem to offer an extremely simple model of holographic replay. Nevertheless, the coupling between the waves which we discard in N-wave theory is in some ways equivalent to a deviation from the Bragg condition in the remaining gratings—and this leads, in the real world, to imperfect power coupling between the reference reading wave and the signal waves.

A treatment of *N*-wave theory that deals with the non-compliance of the Bragg condition has been given by Peri and Friesem [16]. Cooke and Solymar [3] have further shown that analytic solutions exist for a version of the theory in which no couplings are ignored—in fact, in this work, too much coupling is included. The result is therefore to be regarded as a pessimistic estimate of any final diffractive efficiency. For small angles of incidence, Cooke and Solymar find that for a reference to object intensity ratio of 5, a maximum diffractive efficiency of 95.4% is found for a transmission grating. For equal object and reference intensities, this decreases to 80%.

11.10.10 Recording Multiple Colour Gratings

The problem of co-recording two or more gratings using different wavelengths of light is very similar to the problem of multiple gratings, which we discussed in the last section. The only difference is that now there will exist gratings within the volume hologram which do not satisfy Bragg compliance at all. A hologram made with a red reference and red object beam and a green reference and green object beam will lead to a permittivity distribution formally described by an equation similar to Equation 11.150. The main difference is that half of the signal waves will not reconstruct under stimulation by the red reference reading beam as the "green" gratings will not be in Bragg compliance for the red reference beam. If we assume naively that this non-compliance does not matter, then again, we are led to the conclusion that perfect diffractive conversion should be theoretically possible in a multiple-colour hologram. We shall see in the next chapter that the PSM model brings us to this same conclusion but rather more clearly.

The incoherent storage of many holograms is a subject of primary interest to the field of holographic memory. Here, a given material is characterised by its *M* number [17]. This is a number that characterises the dynamic range of the material—that is, how many optimally diffractive holograms may be incoherently multiplexed for a material of given thickness. The *M* number is defined as

$$M\,/\,\# = \sum_{i=1}^{M} \eta_i^{1/2} \tag{11.169}$$

where η_i is the maximum diffractive efficiency of the *i*th co-written grating.

Ulibarrena [18,19] has recently investigated the panchromatic silver halide emulsion, BB640, produced by the UK company Colour Holographics Ltd., and compared experimental measurements of the diffractive efficiency of three-colour multiplexed reflection gratings (442, 532 and 632.8 nm) made on this material with the predictions of the coupled wave theory. A best-fitting procedure of the measured and theoretically expected diffractive efficiency versus wavelength produced an estimated emulsion thickness of 7.3 μm and an index modulation of each component grating of $\delta n/n = 0.027/1.579$. The measured diffractive efficiency of unslanted gratings at each of their three peak replay wavelengths was more than 52% (for $\theta_r = \theta_c = 0°$). When only a one-colour grating was recorded in the emulsion, typical diffraction efficiencies exceeded 72%. Ulibarrena also recorded diffuse images and measured approximate diffractive efficiencies for three-colour BB640 reflection holograms of 40% at 632.8 nm, and 20% at 442 nm/532 nm.

11.10.11 Dispersion Equation Theory

In the previous sections, we have given a simple exposition of the coupled wave theory. As we have stated, the coupled wave theory is a simple but highly effective method of analysing a volume grating. In its two-wave form, it is not mathematically rigorous, but for all practical applications of display holography, it is extremely useful. The interested reader is referred to the excellent book by Solymar and Cooke [4] for a discussion of how this theory may be generalised to two and three dimensions. There are, however, some alternative approaches to the coupled wave theory.* In particular, the analysis techniques used in the field of x-ray and electron diffraction yielded an alternative method for analysing the volume

---

* One such variant is the PSM model, which we shall discuss in detail in Chapter 12.

hologram known as dispersion equation theory. This theory was first suggested in 1967 by Saccocio [20], two years before Kogelnik's theory was published. It was not until 1971, however, that Aristov and Shekhtman [21] outlined the first proper treatment of a two-wave and *N*-wave dispersion treatment of the volume hologram assuming perfect Bragg compliance. In 1976, Sheppard [22] produced a version in which Bragg compliance was not assumed, essentially reproducing an equivalent but alternative theory to Kogelnik's. Dispersion equation theory is, in many ways, very similar to the coupled wave theory. In the simplest variant, a trial solution of the wave equation assumes just two waves, *R* and *S*, just as in Kogelnik's theory. However, these two waves are assumed constant throughout the grating, leading to a set of algebraic equations rather than differential equations. Each of the two waves is then split into a further two component waves and these waves beat to produce the one-dimensional variation implicit to Kogelnik's theory. The results of dispersion equation theory are usually very close to the coupled wave theory, although there are differences. The theory is, however, fundamentally based on plane waves and as such cannot be generalised in the way the coupled wave theory can be. However, as a model of what is going on in the grating, dispersion equation theory offers an alternative way of looking at things, which is often useful.

11.11 Blurring Revisited

In Section 11.2.3, we worked out that the presence of a broadband of wavelengths at reconstruction would act to blur the image of a hologram in the vertical and depth directions. We saw that in the vertical direction, an image point in the thin transmission or reflection hologram would be blurred by an amount

$$\delta y_i \sim x_o \lambda_r \sin \theta_c \frac{\delta \lambda}{\left\langle \lambda_c \right\rangle^2} \tag{11.170}$$

We can now use Equation 11.141 to calculate, for the case of an unslanted reflection grating with $\kappa d / \sqrt{|c_R||c_S|} = \pi/2$ *, that at the point of diffraction, the reconstructed image ray will produce a cone of light having an angle of approximately[†]

$$\delta \theta \sim \frac{\sqrt{5}}{4} \frac{\lambda}{dn} \tan \theta_c \tag{11.171}$$

Note that this formula is different from Equation 11.140, which gave an estimate for the expected tolerance in reconstruction angle at a fixed wavelength. Equation 11.171 predicts that for an average fine-grain silver halide emulsion, such as the Slavich VRP-M, which has a thickness of approximately 7 μm, one can expect an angular blurring angle of around 1° at 532 nm or approximately 17 times that of the human eye's resolution.[‡] For the latest Bayer photopolymer, which is approximately twice as thick, the blurring halves, but it is still more than eight times worse than the practical human eye resolution. The effect of this chromatic blurring is to produce a finite depth and a finite front projection in which a clear image can be displayed by a reflection hologram illuminated by a broadband source. We saw in Section 11.2.3 that the critical defocussing distance behind the hologram was given by

$$x_{\text{CRIT}} = \frac{h \delta \theta_{\text{Eye}}}{\delta \theta - \delta \theta_{\text{Eye}}} \tag{11.172}$$

---

* This represents a pretty good design choice, giving a diffractive efficiency of 0.84 under the lossless coupled wave theory.
† Note that we could have derived a similar formula by considering $\theta_i(\lambda_c) = \tan^{-1}[k_{iy}/k_{ix}]$ under a variation of $\lambda_c \to \lambda_c + \delta \lambda_c$.

  This leads to $\delta \theta \sim \dfrac{\sqrt{5}}{2} \dfrac{\lambda}{dn} \sin \theta_r$.

‡ This is assuming an angle θ_c within the emulsion of 30°.

For the VRP-M material, at an observation distance of 1 m and at 532 nm, this gives a value of approximately 7 cm. The front projection is also around this figure leading to a total clear image depth of field of approximately 14 cm. For the Bayer photopolymer and for the ultrafine grain silver halide emulsion PFG-03CN from Sfera-S, this roughly doubles to a total clear field of 28 cm.

Of course, these calculations pertain to analogue reflection holograms and full-parallax digital reflection holograms. For HPO digital holograms illuminated by a distant point source, we have already seen that there is no chromatic blurring in the paraxial approximation. We come to an important conclusion then. For full-parallax reflection holographic images to be unaffected by the chromatic blurring inherent in broadband illumination, we must increase the emulsion thickness to more than 100 μm. The permittivity modulation must then decrease significantly so that $\kappa d / \sqrt{|c_R||c_S|}$ continues to be of the order of $\pi/2$. There is, however, a sizeable problem here. In materials such as silver halide, chemical processing needs to effect a change throughout the emulsion thickness, and for emulsions much thicker than approximately 15 μm, one starts to see problems of penetration of the processing chemicals [23]. As a result, non-uniform development occurs and it becomes difficult to guarantee the uniformity of the grating. Silver halides can be made sensitive to pulsed radiation, and as such, these materials have a great advantage for the generation of digital holograms. Realistically, they work best at lower thicknesses. Photopolymers, on the other hand, are usually not easily sensitised to pulsed laser radiation, but they are easier to produce in large thicknesses.* Bayer, for example, has recently tested a 50 μm material sensitive in the red and green spectral regions with a value of $\delta n \sim 0.012$ [24]. Such material may be expected to show a clear depth behind an analogue or full-parallax reflection hologram of more than 1 m when viewed from a distance of 1 m.

There is, however, another problem with thick emulsions. The very fact that they have such great wavelength selectivity means that they only use a fraction of any broadband light illuminating them! As we shall discuss in Chapter 13, modern light-emitting diode sources can currently produce bandwidths down to approximately 15 nm at the time of writing. A 50 μm emulsion has a value of $\delta\lambda$ at 532 nm of approximately 2.5 nm. So one could expect that maybe only 20% of the light will actually be used depending on the form of the distribution. As such, until smaller bandwidth light-emitting diodes or other cheap illumination sources become available, most full-parallax holograms are always going to be either dimmer or shallower than HPO holograms. The obvious exception is of course when lasers are used to illuminate the hologram.

We have used the coupled wave theory to derive expressions for the angular replay sensitivity of both the transmission and reflection hologram (Equations 11.138 and 11.140) in the absence of fringe slant. For the reflection case, we can use Equation 11.140 to see that for a typical silver halide emulsion of 7 μm thickness, we can expect an angular selectivity of the order of 5° depending on replay angle and wavelength. In fact, Equation 11.140 gives us an estimate of the half-width at zero diffractive efficiency. However, this figure is also very close to the FWHM as can be verified from the plots in Figures 11.20 and 11.21. This value is very much larger than the human eye value of $\delta\theta_{Eye} \sim 0.06°$ meaning that source-size blurring will be essentially unaffected by the Bragg effect of the volume grating. Unfortunately, the only way with these types of emulsion to avoid source-size blurring is to use a very compact light source!

11.11.1 Blurring in the Digital Hologram

We have discussed at length the method of DWDH printing in Chapters 7, 8 and 9. Here, the hologram is split into a matrix of abutting square microholograms or hogels. This technique has been extremely successful in generating both small and large full-colour digital holograms of excellent quality. In a certain obvious sense, just like standard two-dimensional dot matrix printing, the smaller the hogel, the better the image quality. Digital holograms have been produced with hogels down to almost 250 μm [25]. However, there are two principal disadvantages to reducing the hogel size. The first is the time required to write a hologram—clearly, if one reduces the linear dimension of the hogel by two times, then it is going to take four times as long to print the hologram. For large holograms, which may measure several metres in linear dimension, this is an important constraint. Usually, such holograms are viewed at large distances; as such, the observer will simply not notice the difference between a hogel size of 1 mm and 0.5 mm.

---

* One should note, however, that thicker photopolymers can suffer from absorption of the recording light by the photosensitiser.

The second disadvantage of reducing the hogel size is diffractive blurring. For a hogel diameter of d, each hogel will radiate a cone of light, due to diffraction, having an angle of $\delta\theta \sim \lambda/(\gamma d)$, in which the form factor γ is 1.0 for a square hogel and 1.22 for a circular hogel. This will lead to a blurring of the holographic image if $\delta\theta$ is greater than the smallest angle resolvable by the human eye. If $\delta\theta > \delta\theta_{Eye}$, then the hologram will exhibit a maximum "in focus" depth after which it will appear blurred in accordance with our discussion in Section 11.2.2. Taking the most constraining wavelength as 440 nm and estimating $\delta\theta_{Eye} \sim 0.06°$, we see that the critical square hogel diameter for the onset of diffractive blurring is approximately 0.5 mm. Of course, this figure depends strongly on $\delta\theta_{Eye}$ which varies somewhat depending on the person. As a result, it is not surprising that some observers report that large-depth and large-projection holograms made with 0.8 mm-diameter hogels appear more blurred than when written with 1.6 mm hogels. Of course this is under good illumination conditions which ensure that source-size and chromatic blurring are both less than $\delta\theta_{Eye}$.

Hogel sizes less than 0.5 mm can demonstrate sizeable blurring. For example, a square hogel of 0.1 mm diameter will show a value of $\delta\theta$ equal to around 0.26°. Using Equation 11.172, this then leads to a blur-free depth of 30 cm when the hologram is observed from a distance of 1 m.

11.12 Computational Methods of Calculating Diffractive Efficiency of Planar Gratings

11.12.1 Rigorous Coupled Wave Theory and Rigorous Modal Theory

In 1981, Moharam and Gaylord [12] described a rigorous coupled wave theory, which provided the first fully accurate and convenient computational solution to the problem of diffraction from slanted planar gratings. It was shown that Kogelnik's analytic theory provided a rather good description of the unslanted reflection grating, but at high index modulations ($\delta n/n \sim 0.16$), it somewhat overestimated the diffraction efficiency of the slanted grating. Differences in the transmission grating were shown to appear at a rather lower modulation ($\delta n/n \sim 0.06$). The main reason for the difference between Kogelnik's theory and the rigorous coupled wave theory in the slanted reflection grating is in the treatment of boundary diffraction. This is related to the neglect of the second-order derivatives in Kogelnik's theory. Differences in the transmission hologram arise because other diffractive orders become important.

An earlier approach to the computational solution of diffraction from planar gratings is the modal approach [26–36]. Both the rigorous coupled wave approach and the (rigorous) modal approach analyse the planar grating diffraction problem by solving the wave equation in three regions—the grating region and two exterior regions—and then matching the tangential electric and the magnetic fields at the two boundaries to determine the unknowns. The main difference between the two approaches is in the technique used to find solutions of the wave equation in the grating region. In the rigorous coupled wave approach, the system of coupled wave equations is usually formulated as a simple matrix relation for which the solution is obtained by calculating the eigenvalues and the eigenvectors. The modal approach requires a more complicated transcendental relationship in the form of a continued fraction expansion to be solved to find the wave numbers and their related coefficients, which are needed to satisfy the wave equation in the grating region. The primary difficulty when applying the modal approach to the analysis of slanted gratings is the difficulty in formulating a systematic technique capable of solving the general transcendental continued fraction relation. Rigorous coupled wave theory and its application to simple polychromatic and more complex spatially multiplexed gratings are described in more detail in Appendix 8.

11.12.2 Rigorous Chain Matrix Method

The rigorous coupled wave theory provides a useful way to calculate accurate diffraction efficiencies for any planar grating. Another method, the rigorous chain matrix method [37–41] can, however, be programmed extremely quickly on any PC for the case of the unslanted planar grating. The method is based on dividing up the grating into thin layers each of a given index. A general index profile can then be constructed by using many layers. For the case of the reflection grating, a reference wave, R, and a counterpropagating signal wave, S, are assumed. The tangential components of the electric and magnetic fields are then matched at each layer boundary.

At the mth layer, this gives

$$R_m = t_m^{-1}\left\{R_{m-1}e^{i\delta_{m-1}} + r_m S_{m-1}e^{-i\delta_{m-1}}\right\}$$
$$S_m = t_m^{-1}\left\{r_m R_{m-1}e^{i\delta_{m-1}} + S_{m-1}e^{-i\delta_{m-1}}\right\}$$

(11.173)

where for the π-polarisation, the reflection and transmission coefficients are

$$r_m = \frac{n_m \cos\phi_{m-1} - n_{m-1}\cos\phi_m}{n_m \cos\phi_{m-1} + n_{m-1}\cos\phi_m}$$
$$t_m = \frac{2n_m \cos\phi_m}{n_m \cos\phi_{m-1} + n_{m-1}\cos\phi_m}$$

(11.174)

and for the σ-polarisation

$$r_m = \frac{n_m \cos\phi_m - n_{m-1}\cos\phi_{m-1}}{n_m \cos\phi_m + n_{m-1}\cos\phi_{m-1}}$$
$$t_m = \frac{2n_m \cos\phi_m}{n_m \cos\phi_m + n_{m-1}\cos\phi_{m-1}}$$

(11.175)

Here, n_m and ϕ_m are, respectively, the index and ray angle of incidence at the mth layer. The phase parameters are likewise given by

$$\delta_m = \frac{2\pi n_m}{\lambda}\delta\cos\phi_m$$

(11.176)

with δ being the distance between successive layers. Finally, the angles of incidence at each layer are given by Snell's law:

$$n_m \sin\phi_m = n_{m-1}\sin\phi_{m-1}$$

(11.177)

Equation 11.173 may be written in the matrix form

$$\begin{pmatrix} R_m \\ S_m \end{pmatrix} = \frac{1}{t_m}M_{m-1}\begin{pmatrix} R_{m-1} \\ S_{m-1} \end{pmatrix}$$

(11.178)

The solution for a stack of k layers on a substrate may then be written using the chain matrix form

$$\begin{pmatrix} R_{k+1} \\ S_{k+1} \end{pmatrix} = \frac{M_k M_{k-1}....M_2 M_1}{t_k t_{k-1}....t_2 t_1}\begin{pmatrix} R_0 \\ S_0 \end{pmatrix} = \begin{pmatrix} a & b \\ c & d \end{pmatrix}\begin{pmatrix} R_0 \\ S_0 \end{pmatrix}$$

(11.179)

Using the boundary conditions, $R_0 = 1$ and $S_{k+1} = 0$, the diffraction efficiency of the reflection grating may then be calculated as

$$\eta = S_0 S_0^* = \left|\frac{c}{d}\right|^2$$

(11.180)

The chain matrix method provides an extremely practical method for calculating the diffraction efficiency of any unslanted planar grating—and this includes polychromatic gratings! The method can also be extended to a complex index to model lossy polychromatic gratings. In the next chapter, we shall see that a differential formulation of this method provides a powerful analytic alternative to the coupled wave theory.

REFERENCES

1. H. Kogelnik, "Coupled wave theory for thick hologram gratings," *Bell Syst. Tech. J.* **48**, 2909–2947 (1969).
2. L. Solymar, "Two-dimensional N-coupled-wave theory for volume holograms," *Opt. Commun.* **23**, 199–202 (1977).
3. D. J. Cooke and L. Solymar, "Comparison of two-wave geometrical optics and N-wave theories for volume phase holograms," *J. Opt. Soc. Am. B* **70**, 1631A (1980).
4. L. Solymar and D. J. Cooke, *Volume Holography and Volume Gratings*, Academic Press Inc., New York (1981).
5. H. M. Smith, *Principles of Holography*, 2nd Edn, John Wiley and Sons Inc. (1975).
6. S. A. Benton, "Hologram reconstructions with extended incoherent sources," *J. Opt. Soc. Am.* **59**, 1545–1546A (1969).
7. J. Wolff, N. Phillips and A. Furst, *Light Fantastic*, Bergström + Boyle Books Ltd, London (1977) and J. Wolff, N. Phillips and A. Furst. *Light Fantastic 2*, Bergström + Boyle Books Ltd, London (1978).
8. P. Dawson and P. A. Wilksch, "Laser transmission holograms maximum permissible exposure," in *Laser Florence 2009*, AIP Conf. Proc. **1226**, 147–154 (2009).
9. P. Dawson, *Virtual Encounters; Paula Dawson Holograms*, Macquarie Univ. and Newcastle Region Art Gallery, Australia (2010).
10. D. Brotherton-Ratcliffe, "A treatment of the general volume holographic grating as an array of parallel stacked mirrors," *J. Mod. Optic* **59**, 1113–1132 (2011).
11. J. D. Jackson, *Classical Electrodynamics*, 3rd Edn, John Wiley and Sons Inc. (1974).
12. M. G. Moharam and T. K. Gaylord, "Rigorous coupled wave analysis of planar grating diffraction," *J. Opt. Soc. Am.* **71**, 811–818.
13. A. B. Bhatia and W. J. Noble, "Diffraction of light by ultrasonic waves," *Proc. Roy. Soc.* **A220**, 356–385 (1953).
14. P. Phariseau, "On the diffraction of light by supersonic waves," *Proc. Ind. Acad. Sci.* **44A**, 165–170 (1956).
15. F.-K. Bruder, F. Deuber, T. Fäcke, R. Hagen, D. Hönel, D. Jurbergs, M. Kogure, T. Rölle and M.-S. Weiser, "Full-color self-processing holographic photopolymers with high sensitivity in red – the first class of instant holographic photopolymers," *J. Photopolym. Sci. Technol.* **22**(2) 257–260 (2009).
16. D. Peri and A. A. Friesem, "Volume hologram for image restoration," *J. Opt. Soc. Am.* **70**, 515–522 (1980).
17. H. J. Coufal, D. Psaltis and G. T. Sincerbox, *Holographic Data Storage*, Springer, New York (2000).
18. M. Ulibarrena, *Estudio y caracterización de la emulsión fotográfica de grano ultrafino BB640 como material de registro holográfico*, [Transl: Study and characterization of the ultra-fine-grain photographic emulsion BB640 as a holographic recording material] PhD Thesis, Univ. Miguel Hernandez, Elche, Spain (2003).
19. M. Ulibarrena, L. Carretero, R. F. Madrigal, S. Blaya and A. Fimia, "Multiple band holographic reflection gratings recorded in new ultra-fine-grain emulsion BBVPan," *Opt. Expr.* **11**, 3385–3392 (2003).
20. E. J. Saccocio, "Application of the dynamical theory of X-ray diffraction to holography," *J. Appl. Phys.* **38**, 3994–3998 (1967).
21. V. V. Aristov and V. Shekhtman, "Properties of three-dimensional holograms," *Sov. Phys. Uspekhi* **14**, 263–277 (1971).
22. C. J. R. Sheppard, "The application of the dynamical theory of X-ray diffraction to thick hologram gratings," *Int. J. Electron.* **41**, 365–373 (1976).
23. D. Dainton, A. R. Gattiker and W. O. Lock, "The processing of thick photographic emulsions," *Phil. Mag.*, (Series 7) **42**, Issue 327 (1951).
24. T. Fäcke, F.-K. Bruder, M.-S. Weiser, T. Rölle and D. Hönel, *Novel holographic media and photopolymers*, US patent application US 2011/0065827 A1 (2011).
25. S. Zacharovas, Geola Digital UAB, Private Communication (2012).
26. T. Tamir, H. C. Wang and A. A. Oliner, "Wave propagation in sinusoidally stratified dielectric media," *IEEE Trans. Microwave Theory Tech.* **MTT-12**, 323–335 (1964).
27. T. Tamir and H. C. Wang, "Scattering of electromagnetic waves by a sinusoidally stratified half-space: I. Formal solution and analysis approximations," *Can. J. Phys.* **44**, 2073–2094 (1966).

28. T. Tamir, "Scattering of electromagnetic waves by a sinusoidally stratified half-space: II. Diffraction aspects of the Rayleigh and Bragg wavelengths," *Can. J. Phys.* **44**, 2461–2494 (1966).

29. C. B. Burckhardt, "Diffraction of a plane wave at a sinusoidally stratified dielectric grating," *J. Opt. Soc. Am.* **56**, 1502–1509 (1966).

30. L. Bergstein and D. Kermisch, "Image storage and reconstruction in volume holography," *Proc. Symp. Modern Opt.* **17**, 655–680 (1967).

31. R. S. Chu and T. Tamir, "Guided-wave theory of light diffraction by acoustic microwaves," *IEEE Trans. Microwave Theory Tech.* **MTT-18**, 486–504 (1970).

32. R. S. Chu and T. Tamir, "Wave propagation and dispersion in space-time periodic media," *Proc. IEE* **119**, 797–806 (1972).

33. F. G. Kaspar, "Diffraction by thick periodically stratified gratings with complex dielectric constant," *J. Opt. Soc. Am.* **63**, 37–45 (1973).

34. S. T. Peng, T. Tamir, and H. L. Bertoni, "Theory of periodic dielectric waveguides," *IEEE Trans. Microwave Theory Tech.* **MTT-23**, 123–133 (1975).

35. R. S. Chu and J. A. Kong, "Modal theory of spatially periodic media," *IEEE Trans. Microwave Theory Tech.* **MTT-25**, 18–24 (1977).

36. R. Magnusson and T. K. Gaylord, "Equivalence of multiwave coupled-wave theory and modal theory of periodic-media diffraction," *J. Opt. Soc. Am.* **68**, 1777–1779 (1978).

37. O. S. Heavens, "Optical Properties of Thin Films," Reports on Progress in Physics, Vol **XXIII**, p. 1, (1960).

38. F. Abeles, "Recherches sur la propagation des ondes électromagnétiques sinusoïdales dans les milieux stratifiés. Application aux couches minces," *Ann. Phys. (Paris)* **5**, 596–640 (1950).

39. M. G. Moharam and T. K. Gaylord, "Chain-matrix analysis of arbitrary-thickness dielectric reflection gratings," *J. Opt. Soc. Am.* **72**, 187–190 (1982).

40. X. Ning, "Analysis of multiplexed-reflection holographic gratings," *J. Opt. Soc. Am. A* **7**, 1436–1440 (1990).

41. D.W. Diehl and N.George, "Analysis of multitone holographic interference filters by use of a sparse Hill matrix method," *Appl. Opt.* **43**, 88–96 (2004).

12

Diffraction Efficiency: An Alternative Approach Using the PSM Model

12.1 Introduction

An alternative model to Kogelnik's coupled wave theory [1] of the volume holographic grating can be developed through a more refined version of the parallel stacked mirror (PSM) model [2,3], which we presented in the last chapter. Indeed, in 1982, Ludman [4] published a simple analysis of the volume transmission phase grating, which used the Fresnel reflection formulae to provide an estimation of the diffraction efficiency at Bragg resonance. In 2009, Heifetz et al. [5] extended this idea somewhat. However, we shall see that the PSM model can be taken *much* further and that it in fact provides a truly alternative picture to the coupled wave theory. It also sheds some useful light on the assumptions made in Kogelnik's work. This more rigorous treatment of the PSM model allows a proper consideration of multiple-colour reflection gratings. We will see that it is also capable of analytically describing spatially multiplexed polychromatic reflection gratings [3] and in providing useful formulae for the scaling of the diffractive efficiency of full-colour reflection holograms.

The PSM model is based on a particular mathematical description of the permittivity distribution of the unslanted volume holographic grating. The distribution is broken up into an infinite number of infinitesimal discontinuities or step functions. The Fresnel reflections of an incident plane wave from each of these discontinuities are then summed up in a consistent way. The resulting first-order coupled partial differential equations are solved in a rotated frame of reference to arrive at analytical expressions for the diffraction efficiency of the general slanted grating. The PSM model can be viewed as a differential representation of the chain matrix model [6,7]. Both the chain matrix model and the basic differential equations of the PSM model rigorously describe the phenomenon of diffraction from unslanted planar reflection gratings.

12.2 Formulation of the Simplest Model—The Unslanted Reflection Grating at Normal Incidence

We will consider a reflection holographic grating with the following index profile

$$n = n_0 + n_1 \cos\left(\frac{4\pi n_0}{\lambda_r} y\right) = n_0 + \frac{n_1}{2}\left\{ e^{\frac{4i\pi n_0}{\lambda_r} y} + e^{-\frac{4i\pi n_0}{\lambda_r} y} \right\} \tag{12.1}$$

Here, n_0 is the average index and n_1 is generally a small number representing the index modulation. We can imagine that this grating has been created by the interference of two counterpropagating normal-incidence plane waves within a photosensitive material, each with a wavelength of λ_r.*

We wish to understand the response of the grating to a plane reference wave of the form

$$R^{\text{ext}} = e^{i\beta y} \tag{12.2}$$

---

* Both λ_r and λ_c refer to wavelengths in vacuum.

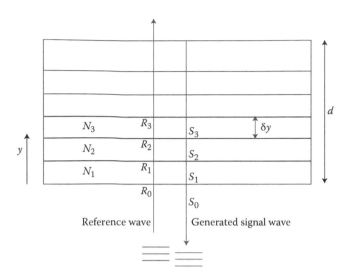

FIGURE 12.1 A simple model of the unslanted normal-incidence reflection grating. The grating is assumed to be made up from discrete layers of different refractive indices. The reference wave illuminates the grating and is reflected and transmitted according to the standard Fresnel rules, generating the signal wave.

where

$$\beta = \frac{2\pi n_0}{\lambda_c} \tag{12.3}$$

We shall assume that the grating is surrounded by a zone of constant index, n_0, to circumvent the complication of refraction and reflection at the grating interface. We start by modelling the grating of Equation 12.1 by a series of many thin constant-index layers, $N_0, N_1, N_2, \ldots N_M$, between each of which there exists an index discontinuity (Figure 12.1). Across each such discontinuity, we may derive the well-known Fresnel formulae [e.g., 8] for the amplitude reflection and transmission coefficients from Maxwell's equations by demanding that the tangential components of the electric and magnetic fields be continuous. An illuminating plane wave will, in general, generate many mutually interfering reflections from each discontinuity. We therefore imagine two plane waves within the grating—the driving reference wave, $R(y)$, and a created signal wave, $S(y)$. Using the Fresnel formulae, we may then write the following relationship for either the σ- or π-polarisations[*]

$$R_J = 2e^{i\beta n\delta y/n_0}\left\{\frac{N_{J-1}}{N_J + N_{J-1}}\right\}R_{J-1} + e^{i\beta n\delta y/n_0}\left\{\frac{N_{J-1} - N_J}{N_J + N_{J-1}}\right\}S_J$$

$$S_J = 2e^{i\beta n\delta y/n_0}\left\{\frac{N_{J+1}}{N_{J+1} + N_J}\right\}S_{J+1} + e^{i\beta n\delta y/n_0}\left\{\frac{N_{J+1} - N_J}{N_{J+1} + N_J}\right\}R_J \tag{12.4}$$

Here, the terms in curly brackets are just the Fresnel amplitude reflection and transmission coefficients and the exponential is a phase propagator which advances the phase of the R and S waves as they travel the distance δy between discontinuities. We now let

$$X_{J-1} = X_J - \frac{dX}{dy}\delta y - \ldots \tag{12.5}$$

---

[*] Note we have used the shorthand here that $n = N_J$.

and consider the limit $\delta y \to 0$. Further expanding the exponential terms as Taylor series and ignoring quadratic terms in δy, we arrive at the differential counterpart to Equation 12.4

$$\frac{dR}{dy} = \frac{R}{2}\left(2i\beta\frac{n}{n_0} - \frac{1}{n}\frac{dn}{dy}\right) - \frac{1}{2n}\frac{dn}{dy}S$$

$$\frac{dS}{dy} = -\frac{S}{2}\left(\frac{1}{n}\frac{dn}{dy} + 2i\beta\frac{n}{n_0}\right) - \frac{1}{2n}\frac{dn}{dy}R \tag{12.6}$$

These equations are an *exact* representation of Maxwell's equations for an arbitrary index profile, $n(y)$—as letting $u(y) = R(y) - S(y)$, we see that they simply reduce to the Helmholtz equation

$$\frac{d^2u}{dy^2} + \frac{\beta^2 n^2}{n_0^2}u = 0 \tag{12.7}$$

and the conservation of energy

$$\frac{d}{dy}(nR^*R - nS^*S) = 0 \tag{12.8}$$

When $dn/dy = 0$, Equations 12.6 describe two counterpropagating and non-interacting plane waves. A finite index gradient couples these waves.

We now make the transformation

$$R \to R'(y)e^{i\beta y}$$

$$S \to S'(y)e^{-i\beta y} \tag{12.9}$$

where the primed quantities are slowly varying compared with $e^{i\beta y}$. Because they are slowly varying, we can write

$$\langle R' \rangle \sim R$$

$$\langle S' \rangle \sim S \tag{12.10}$$

where the operator $\langle\ \rangle$ takes an average over several cycles of $e^{i\beta y}$. Substituting Equation 12.9 in Equation 12.6 and using Equation 12.10, we then arrive at the following differential equations

$$\frac{dR}{dy} = -i\alpha\kappa S e^{2i\beta y(\alpha-1)}$$

$$\frac{dS}{dy} = i\alpha\kappa R e^{-2i\beta y(\alpha-1)} \tag{12.11}$$

where we have defined

$$\alpha = \frac{\lambda_c}{\lambda_r} \tag{12.12}$$

which is just the ratio of the replay wavelength to the recording wavelength. Introducing the pseudo-field,

$$\hat{S} = S e^{2\beta i y(\alpha-1)} \tag{12.13}$$

and defining Kogelnik's constant,

$$\kappa = \frac{\pi n_1}{\lambda_c} \tag{12.14}$$

these equations may now be written in the form of Kogelnik's equations for the normal-incidence unslanted sinusoidal grating:

$$c_R \frac{dR}{dy} = -i\kappa\hat{S}$$

$$c_S \frac{d\hat{S}}{dy} = -i\vartheta\hat{S} - i\kappa R$$

(12.15)

where

$$c_R = \frac{1}{\alpha}; \; c_S = -\frac{1}{\alpha}; \; \vartheta = 2\frac{\beta}{\alpha}(1-\alpha)$$

(12.16)

For comparison, Kogelnik's coefficients are

$$c_R = 1; \; c_S = (2\alpha - 1); \; \vartheta = 2\alpha\beta(1-\alpha)$$

(12.17)

By imposing boundary conditions appropriate for the reflection hologram,

$$R(y=0) = 1$$

$$\hat{S}(y=d) = 0$$

(12.18)

where d is the grating thickness, Equations 12.15 may be solved analytically. We can then define the diffraction efficiency for both the PSM and Kogelnik models as

$$\eta = \left|\frac{c_S}{c_R}\right| \hat{S}(0)\hat{S}^*(0)$$

$$= \left\{1 - \frac{c_R c_S}{\kappa^2} \Upsilon^2 csh^2(d\Upsilon)\right\}^{-1}$$

(12.19)

where

$$\Upsilon^2 = -\frac{\vartheta^2}{4c_S^2} - \frac{\kappa^2}{c_R c_S}$$

(12.20)

Note that we should ensure that

$$d = m\left(\frac{\pi}{2\alpha\beta}\right)$$

(12.21)

where m is a non-zero integer to prevent a discontinuity in index at $y = d$ (see Moharam and Gaylord [9] for a detailed discussion of the starting and ending conditions of a grating).

For cases of practical interest for display and optical element holography, substitution of Equation 12.16 (the PSM coefficients) or Equation 12.17 (Kogelnik's coefficients) into Equations 12.19 and 12.20 yields very similar results. However, one should note that the only approximation made in deriving the PSM equations (Equations 12.15, 12.16, 12.19 and 12.20) has been that of Equation 12.10. This is an assumption that one would reasonably expect to hold in most gratings of interest. At Bragg resonance, when $\alpha = 1$, both the Kogelnik and PSM models reduce to

$$\eta = \tanh^2(\kappa d)$$

(12.22)

The PSM model provides a useful insight into what is happening within the grating: multiple reflections of the reference wave simply synthesise the signal wave by classical Fresnel reflection and transmission at each infinitesimal discontinuity. This is a rigorous picture for the normal-incidence unslanted reflection grating because Equation 12.6 is an exact representation of Maxwell's equations. The fact that we explicitly need to introduce a "pseudo-field," $\hat{S}$, to get the PSM equations into the same form as Kogelnik's equations reminds us that indeed Kogelnik's signal wave is not the physical electric field of the signal wave for $\alpha \neq 1$. Kogelnik's theory models the dephasing away from Bragg resonance by letting the non-physical wave propagate differently from the physical signal wave. In the PSM analytical theory (Equations 12.15 and 12.16), the pseudo-field is also not the real electric field—but here, the transformations in Equations 12.10 and 12.13 make the relationship between the real and the pseudo-field perfectly clear.

12.2.1 Comparison with a Numerical Solution of the Helmholtz Equation

The Helmholtz equation (see Equation 12.7) for the normal-incidence reflection grating of Equation 12.1 can be written as

$$\frac{d^2 u}{dy^2} - \gamma^2 u = 0 \tag{12.23}$$

where

$$\gamma^2 = -\beta^2 \left\{ 1 + \frac{n_1}{n_0} \cos\left(\frac{4\pi n_0}{\lambda_r} y \right) \right\}^2 \tag{12.24}$$

and where u is the tangential electric field. The corresponding boundary conditions to Equation 12.18 can be shown to be

$$u'(0) + i\beta u(0) = 2i\beta$$

$$\beta u(d) = -iu'(d) \tag{12.25}$$

These equations, which constitute a rigorous solution to the diffraction problem at hand, can be conveniently solved on a personal computer (PC) using standard Runge–Kutta integration. Comparison of such numerical results offers a simple way to rate the analytical PSM and coupled-wave predictions of diffractive efficiency for gratings of interest.

In Figures 12.2 and 12.3, the diffraction efficiency is plotted versus replay wavelength for two cases with values typical to display holography. Figure 12.2 pertains to a grating recorded at 532 nm and Figure 12.3 pertains to a grating recorded at 660 nm. These graphs show that the PSM analytical model agrees closely with the numerical Helmholtz calculation for all replay wavelengths. On the other hand,

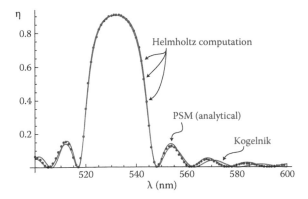

FIGURE 12.2 Comparison of the predictions of the PSM analytical model, Kogelnik's coupled wave theory and a Runge–Kutta numerical solution of the Helmholtz equation for a typical normal-incidence unslanted reflection grating with a thickness of 7 µm and with $n_0 = 1.5$, $n_1 = 0.045$. The grating was recorded at 532 nm.

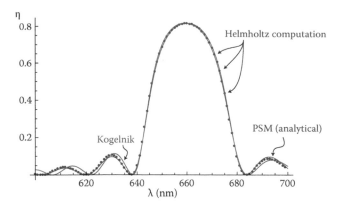

FIGURE 12.3 Comparison of the predictions of the PSM analytical model, Kogelnik's coupled wave theory and a numerical Runge–Kutta solution of the Helmholtz equation for identical parameters to Figure 12.2 with the exception that the grating was recorded here at 660 nm.

Kogelnik's coupled wave model appears to agree closely with the numerical calculation around Bragg resonance but does not predict an accurate behaviour of the sideband structure. In fact, this is broadly the behaviour seen over a wide range of parameters. As a general rule, Kogelnik's off-resonance solutions seem to become more accurate with lower modulations as one might expect. As would also be expected, a direct Runge–Kutta integration of the Helmholtz equation agrees within computational rounding errors to a similar numerical integration of Equation 12.6.

12.3 Unslanted Multiple-Colour Gratings at Normal Incidence

One of the advantages of the PSM model is that it does not limit the grating to a sinusoidal form. This is a great advantage over the simplest variants of standard coupled wave theories, including Kogelnik's.

We start by assuming a general (polychromatic unslanted) index profile

$$n = n_0 + n_1 \cos(2\alpha_1 \beta y) + n_2 \cos(2\alpha_2 \beta y) + \ldots$$

$$= n_0 + \frac{n_1}{2}\left\{e^{2i\beta\alpha_1 y} + e^{-2i\beta\alpha_1 y}\right\} + \frac{n_2}{2}\left\{e^{2i\beta\alpha_2 y} + e^{-2i\beta\alpha_2 y}\right\} + \ldots \tag{12.26}$$

Equations 12.6 then reduce to the following form

$$\frac{dR}{dy} = -S\sum_{j=1}^{N} i\kappa_j \alpha_j e^{2i\beta y(\alpha_j - 1)} \tag{12.27}$$

$$\frac{dS}{dy} = R\sum_{j=1}^{N} i\kappa_j \alpha_j e^{-2i\beta y(\alpha_j - 1)}$$

Assuming that the individual gratings have very different spatial frequencies, these equations lead to a simple expression for the diffractive efficiency when the reference wave is in Bragg resonance with one or another of the multiplexed gratings:

$$\eta_j = \tanh^2(\kappa_j d) \tag{12.28}$$

where

$$\kappa_j = \frac{n_j \pi}{\lambda_c} \tag{12.29}$$

In addition, in the region of the jth Bragg resonance, Equation 12.27 leads to the approximate analytical form

$$\eta_j = \frac{\alpha_j^2 \kappa_j^2}{\beta^2(1-\alpha_j)^2 + (\alpha_j^2 \kappa_j^2 - \beta^2(1-\alpha_j)^2)\coth^2\left\{d\sqrt{\alpha_j^2 \kappa_j^2 - \beta^2(1-\alpha_j)^2}\right\}} \tag{12.30}$$

When the spatial frequencies of the different gratings are too close to one another, these relations break down. For many cases of interest, however, Equations 12.28 through 12.30 provide a rather accurate picture of the normal-incidence polychromatic reflection phase grating. Indeed the following form can often be used to accurately describe an N-chromatic grating:

$$\eta = \sum_{j=1}^{N} \frac{\alpha_j^2 \kappa_j^2}{\beta^2(1-\alpha_j)^2 + (\alpha_j^2 \kappa_j^2 - \beta^2(1-\alpha_j)^2)\coth^2\left\{d\sqrt{\alpha_j^2 \kappa_j^2 - \beta^2(1-\alpha_j)^2}\right\}} \tag{12.31}$$

For example, Diehl and George [10] have used a sparse Hill's matrix technique to computationally calculate the diffraction efficiency of a lossless trichromatic phase reflection grating at normal incidence. They used free-space recording wavelengths of 400, 500 and 700 nm. The grating thickness was 25 μm and the index parameters were taken as $n_0 = 1.5$, $n_1 = n_2 = n_3 = 0.040533$. Comparison of Equation 12.31 with Diehl and George's published graphical results shows very good agreement.

In cases where the gratings are too close to one another in wavelength, Equation 12.6 or Equation 12.27 must be solved numerically.

12.3.1 Numerical Solution for Two-Colour Normal-Incidence Reflection Gratings

Equation 12.6 can be solved numerically using a standard Runge–Kutta method and the results compared with a similar numerical integration of the Helmholtz equation (Equation 12.23). If we compare these computational integrations with the PSM analytical expressions for diffractive efficiency, we see extremely good agreement.

This is illustrated in Figure 12.4, where diffraction efficiency versus replay wavelength is plotted for a two-colour grating recorded at 532 and 660 nm in a 7 μm emulsion. A peak index modulation of 0.045 was chosen as in Figures 12.2 and 12.3, and equal modulation was used for the two wavelengths. Comparison of Figures 12.2 and 12.3 with Figure 12.4 shows that the two-colour grating exhibits a somewhat lower diffractive efficiency than separate monochromatic gratings of an equivalent index modulation depth.

Figure 12.4 shows excellent agreement between the numerical PSM model and the direct Helmholtz integration as expected. In addition, the analytical expressions (Equations 12.28–12.31) are seen to

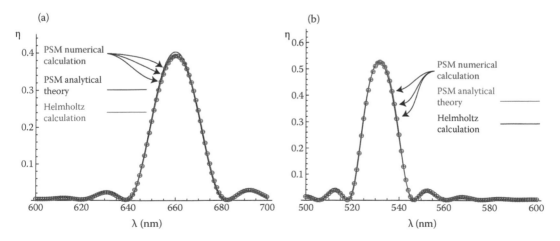

FIGURE 12.4 A normal-incidence unslanted reflection grating recorded using two wavelengths: 532 and 660 nm. A value of $n_1 = n_2 = 0.045/2$ was used giving a total index modulation of 0.045 ($n_0 = 1.5$). The grating thickness was 7 μm. (a) Replay near 660 nm and (b) replay near 532 nm.

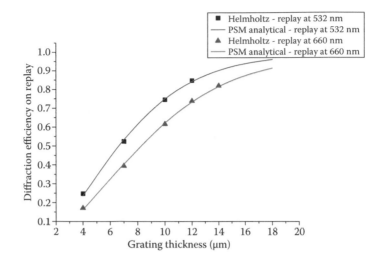

FIGURE 12.5 Diffraction efficiency versus grating thickness for a normal-incidence two-colour unslanted reflection grating. The grating was recorded at 660 and at 532 nm with equal modulations: $n_1 = n_2 = 0.045/2$ ($n_0 = 1.5$). The red and green curves show the analytical PSM prediction of Equation 12.28 for replay at 532 and 660 nm. The markers show numerical simulation results.

give very similar results. Calculations in which the higher 660 nm wavelength is lowered progressively towards the lower 532 nm wavelength show that the analytical expressions seem usefully accurate until the two diffractive peaks start to coalesce at approximately 555 nm.

To investigate whether the diffractive response of a two-colour reflective grating is always less than that of the corresponding one-colour grating with equivalent index modulation, we plot in Figure 12.5 a graph of the diffraction efficiencies at 532 and 660 nm against grating thickness. Clearly, at high thicknesses, two-colour gratings suffer rather less diffractive impairment. Of course, this does assume that the grating is lossless.

Finally, we should make the observation that the choice of the second recording wavelength affects the diffractive efficiency of a two-colour holographic grating quite significantly. Figure 12.6 shows a numerical calculation of the peak diffraction efficiency of a two-colour grating versus the second recording wavelength. This seems to imply that a proper choice of wavelengths may possibly optimise the diffractive response of colour holograms.

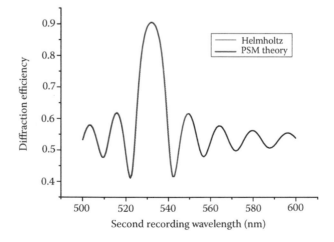

FIGURE 12.6 Two-colour normal-incidence unslanted reflection gratings have different diffractive efficiencies according to the exact choice of the wavelength pair. Here, a grating is recorded at 532 nm, and the peak 532 nm diffractive efficiency is plotted for different choices of the second recording wavelength. Equal modulation is used for the two recording wavelengths with $n_1 = n_2 = 0.045/2$ ($n_0 = 1.5$). Grating thickness is 7 μm.

12.3.2 Numerical Solution for Three-Colour Normal-Incidence Reflection Gratings

Figure 12.7 shows the case of an equal-modulation, three-colour normal-incidence unslanted reflection grating for the same total index modulation used in the previous graphs and for a grating thickness of 7 μm. Recording wavelengths of 440, 532 and 660 nm have been used. Clearly, the agreement between the numerical PSM model and the direct Helmholtz integration is excellent. Once again, the results are also described very closely by Equations 12.30 and 12.31.

With three colours, the diffractive efficiency falls substantially—particularly in the red spectral region—as compared with the monochromatic cases. Figure 12.8 shows a graph of the peak diffractive efficiency versus

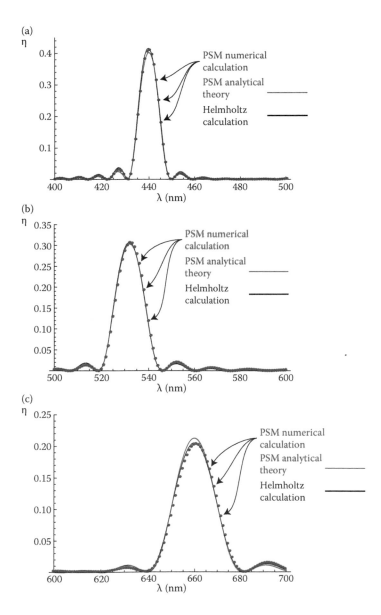

FIGURE 12.7 A normal-incidence unslanted reflection grating recorded using three wavelengths: 440, 532 and 660 nm. Values of $n_1 = n_2 = n_3 = 0.045/3$ were used, giving a total index modulation of 0.045 ($n_0 = 1.5$). The grating thickness is 7 μm. (a) Replay near 440 nm, (b) replay near 532 nm and (c) replay near 660 nm. The markers indicate a solution of the numerical PSM equations. The black lines represent a Runge–Kutta numerical integration of the Helmholtz equation. The red lines indicate the predictions of the PSM analytical theory.

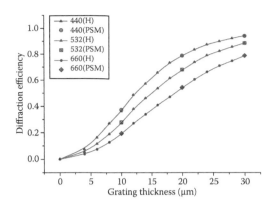

FIGURE 12.8 Peak diffraction efficiency for a three-colour normal-incidence unslanted reflection grating. The grating was recorded at 440, 532 and 660 nm with equal modulations: $n_1 = n_2 = n_3 = 0.03/3$ ($n_0 = 1.5$). The red, green and blue curves show the analytical PSM prediction of Equation 12.28 for replay at 440, 532 and 660 nm. The markers show numerical Runge–Kutta simulation results of the Helmholtz equation.

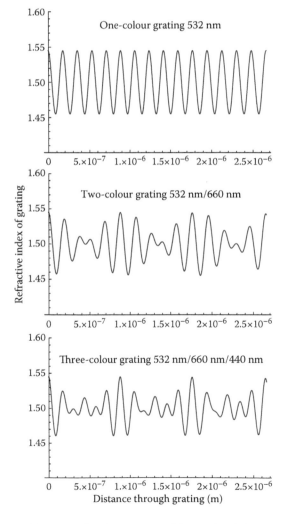

FIGURE 12.9 Index profiles used for the studies of one-, two- and three-colour holographic gratings with total modulation of 0.045.

grating thickness for a total index modulation of 0.03. This graph underlines the fact that thicker emulsions are required for bright, multicolour gratings. Figure 12.9 shows the index profiles used for the one-, two- and three-colour cases.

12.4 Unslanted Reflection Grating at Oblique Incidence

To treat the case of reference wave incidence at finite angle to the grating planes, we must redraw Figure 12.1 using two-dimensional fields, R and S, which we now endow with two indices instead of the previous single index (Figure 12.10). We shall make the approximation that the index modulation is small enough such that the rays of both the R and S waves are not deviated in angle. We shall, however, retain the proper Fresnel amplitude coefficients.

12.4.1 σ-Polarisation

The Fresnel amplitude coefficients for this polarisation may be written as

$$r_{k,k+1} = \frac{N_{k+1}\sqrt{1 - \dfrac{n_0^2}{N_{k+1}^2}\sin^2\theta_c} - N_k\sqrt{1 - \dfrac{n_0^2}{N_k^2}\sin^2\theta_c}}{N_{k+1}\sqrt{1 - \dfrac{n_0^2}{N_{k+1}^2}\sin^2\theta_c} + N_k\sqrt{1 - \dfrac{n_0^2}{N_k^2}\sin^2\theta_c}}$$

$$t_{k,k+1} = \frac{2N_k\sqrt{1 - \dfrac{n_0^2}{N_k^2}\sin^2\theta_c}}{N_{k+1}\sqrt{1 - \dfrac{n_0^2}{N_{k+1}^2}\sin^2\theta_c} + N_k\sqrt{1 - \dfrac{n_0^2}{N_k^2}\sin^2\theta_c}}$$

(12.32)

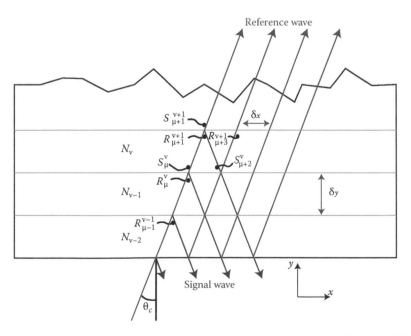

FIGURE 12.10 PSM model of the two-dimensional slanted reflection grating composed of layers of different refractive index. A plane reference wave is incident to the bottom boundary at angle θ_c. Fresnel transmission and reflection at the layer boundaries synthesise the signal or image wave.

where r and t pertain, respectively, to reflection and transmission occurring at the index discontinuity between layers k and $k + 1$ for a wave travelling from k to $k + 1$. The R and S waves in the exterior medium of index n_0 are assumed to be plane waves of the form

$$R = e^{i(k_{cx}x + k_{cy}y)}$$

$$S = S_0 e^{i(k_{ix}x + k_{iy}y)}$$

(12.33)

where S_0 is a constant. Within the grating, we shall assume that R and S are functions of x and y. Using the normal rules of Fresnel reflection, the wave vectors can be written explicitly as

$$\mathbf{k_c} = \beta \begin{pmatrix} \sin\theta_c \\ \cos\theta_c \end{pmatrix}$$

$$\mathbf{k_i} = \beta \begin{pmatrix} \sin\theta_c \\ -\cos\theta_c \end{pmatrix}$$

(12.34)

where the angle θ_c is the angle of incidence of the R wave.

We can now use Figure 12.10 to write down an expression relating the discrete values of R and S within the grating:

$$R_{\mu+1}^{\nu+1} = e^{i\beta n(\sin\theta_c \delta x + \cos\theta_c \delta y)/n_0} R_\mu^\nu \left\{ \frac{2N_{\nu-1}\sqrt{1 - \dfrac{n_0^2}{N_{\nu-1}^2}\sin^2\theta_c}}{N_{\nu-1}\sqrt{1 - \dfrac{n_0^2}{N_{\nu-1}^2}\sin^2\theta_c} + N_\nu\sqrt{1 - \dfrac{n_0^2}{N_\nu^2}\sin^2\theta_c}} \right\}$$

$$+ e^{i\beta n(\sin\theta_c \delta x + \cos\theta_c \delta y)/n_0} S_\mu^\nu \left\{ \frac{N_{\nu-1}\sqrt{1 - \dfrac{n_0^2}{N_{\nu-1}^2}\sin^2\theta_c} - N_\nu\sqrt{1 - \dfrac{n_0^2}{N_\nu^2}\sin^2\theta_c}}{N_{\nu-1}\sqrt{1 - \dfrac{n_0^2}{N_{\nu-1}^2}\sin^2\theta_c} + N_\nu\sqrt{1 - \dfrac{n_0^2}{N_\nu^2}\sin^2\theta_c}} \right\}$$

(12.35)

$$S_{\mu+1}^{\nu-1} = e^{i\beta n(\sin\theta_c \delta x + \cos\theta_c \delta y)/n_0} S_\mu^\nu \left\{ \frac{2N_\nu\sqrt{1 - \dfrac{n_0^2}{N_\nu^2}\sin^2\theta_c}}{N_{\nu-1}\sqrt{1 - \dfrac{n_0^2}{N_{\nu-1}^2}\sin^2\theta_c} + N_\nu\sqrt{1 - \dfrac{n_0^2}{N_\nu^2}\sin^2\theta_c}} \right\}$$

$$+ e^{i\beta n(\sin\theta_c \delta x + \cos\theta_c \delta y)/n_0} R_\mu^\nu \left\{ \frac{N_\nu\sqrt{1 - \dfrac{n_0^2}{N_\nu^2}\sin^2\theta_c} - N_{\nu-1}\sqrt{1 - \dfrac{n_0^2}{N_{\nu-1}^2}\sin^2\theta_c}}{N_{\nu-1}\sqrt{1 - \dfrac{n_0^2}{N_{\nu-1}^2}\sin^2\theta_c} + N_\nu\sqrt{1 - \dfrac{n_0^2}{N_\nu^2}\sin^2\theta_c}} \right\}$$

(12.36)

Because we are assuming that δx and δy are small, we can use Taylor expansions for the fields and index profile

$$R_{\mu+1}^{\nu+1} = R_\mu^\nu + \frac{\partial R_\mu^\nu}{\partial x}\delta x + \frac{\partial R_\mu^\nu}{\partial y}\delta y + \ldots$$

$$S_{\mu+1}^{\nu-1} = S_\mu^\nu + \frac{\partial S_\mu^\nu}{\partial x}\delta x - \frac{\partial S_\mu^\nu}{\partial y}\delta y + \ldots \qquad (12.37)$$

$$N_{\nu-1} = N_\nu - \frac{\partial N_\nu}{\partial y}\delta y + \ldots$$

The exponentials are also written using a Taylor expansion. Before substituting these expressions in Equations 12.35 and 12.36, we make the following additional approximations

$$\sqrt{1 - \frac{n_0^2}{N_\nu^2}\sin^2\theta_c} = \sqrt{1 - \frac{a}{N_\nu^2}} \sim \cos\theta_c = b \qquad (12.38)$$

$$\sqrt{1 - \frac{n_0^2}{N_{\nu-1}^2}\sin^2\theta_c} \sim b - \frac{\partial N_\nu}{\partial y}\frac{a\delta y}{bN_\nu^3} + O(\delta y^2) \qquad (12.39)$$

Then, letting

$$R_\mu^\nu \to R;\ S_\mu^\nu \to S;\ N_\nu \to n \qquad (12.40)$$

Equation 12.35 becomes

$$\left\{R + \frac{\partial R}{\partial x}\delta x + \frac{\partial R}{\partial y}\delta y\right\}\left\{\left(n - \frac{\partial n}{\partial y}\delta y\right)\left(b - \frac{\partial n}{\partial y}\frac{a\delta y}{bn^3}\right) + nb\right\}$$

$$= (1 + i\beta\sin\theta_c\delta x + i\beta\cos\theta_c\delta y)R\left\{2\left(n - \frac{\partial n}{\partial y}\delta y\right)\left(b - \frac{\partial n}{\partial y}\frac{a\delta y}{bn^3}\right)\right\} \qquad (12.41)$$

$$+ (1 + i\beta\sin\theta_c\delta x + i\beta\cos\theta_c\delta y)S\left\{\left(n - \frac{\partial n}{\partial y}\delta y\right)\left(b - \frac{\partial n}{\partial y}\frac{a\delta y}{bn^3}\right) - nb\right\}$$

Ignoring quadratic terms such as δx^2, δy^2 and $\delta x \delta y$ and letting $n\beta/n_o \to \beta(n) = \beta$, this expression simplifies to

$$\frac{\partial R}{\partial x}\delta x + \frac{\partial R}{\partial y}\delta y =$$

$$-R\frac{1}{2n}\frac{\partial n}{\partial y}\delta y\left(1 + \frac{a}{b^2 n^2}\right) + i\beta R(\sin\theta_c\delta x + \cos\theta_c\delta y) - S\delta y\frac{1}{2n}\frac{\partial n}{\partial y}\left(1 + \frac{a}{b^2 n^2}\right) \qquad (12.42)$$

Letting δx, $\delta y \to 0$, we arrive at a partial differential equation for the R field:

$$\frac{\mathbf{k_c}}{\beta}\cdot\nabla R = \sin\theta_c\frac{\partial R}{\partial x} + \cos\theta_c\frac{\partial R}{\partial y} = \frac{R}{2}\left\{2i\beta - \frac{1}{n\cos\theta_c}\frac{\partial n}{\partial y}\right\} - \frac{S}{2n\cos\theta_c}\frac{\partial n}{\partial y} \qquad (12.43)$$

Equation 12.36 may be treated in exactly the same manner, yielding a corresponding partial differential equation for the S field:

$$\frac{\mathbf{k}_i}{\beta} \cdot \nabla S = \sin\theta_c \frac{\partial S}{\partial x} - \cos\theta_c \frac{\partial S}{\partial y} = \frac{S}{2}\left\{2i\beta + \frac{1}{n\cos\theta_c}\frac{\partial n}{\partial y}\right\} + \frac{R}{2n\cos\theta_c}\frac{\partial n}{\partial y} \qquad (12.44)$$

Note the similarity of Equations 12.43 and 12.44 to Equation 12.6. Note also that if we set $\theta_c = 0$, then we retrieve Equation 12.6 exactly. Equations 12.43 and 12.44 are the PSM equations for an unslanted reflection grating at oblique incidence with the σ-polarisation. In the following sections, we shall use these equations to provide useful analytical expressions for both unslanted and slanted gratings. However, before proceeding, we shall briefly derive the PSM equations appropriate for the π-polarisation.

12.4.2 π-Polarisation

The Fresnel amplitude coefficients for this polarisation may be written as

$$r_{k,k+1} = \frac{\dfrac{N_{k+1}}{\sqrt{1 - \dfrac{n_0^2}{N_{k+1}^2}\sin^2\theta_c}} - \dfrac{N_k}{\sqrt{1 - \dfrac{n_0^2}{N_k^2}\sin^2\theta_c}}}{\dfrac{N_{k+1}}{\sqrt{1 - \dfrac{n_0^2}{N_{k+1}^2}\sin^2\theta_c}} + \dfrac{N_k}{\sqrt{1 - \dfrac{n_0^2}{N_k^2}\sin^2\theta_c}}}$$

$$\qquad (12.45)$$

$$t_{k,k+1} = \frac{\dfrac{2N_k}{\sqrt{1 - \dfrac{n_0^2}{N_k^2}\sin^2\theta_c}}}{\dfrac{N_{k+1}}{\sqrt{1 - \dfrac{n_0^2}{N_{k+1}^2}\sin^2\theta_c}} + \dfrac{N_k}{\sqrt{1 - \dfrac{n_0^2}{N_k^2}\sin^2\theta_c}}}$$

The analogues of Equations 12.35 and 12.36 then become

$$\left\{N_{v-1} + N_v\left(1 - \frac{\partial N_v}{\partial y}\frac{a\delta y}{b^2 N_v^3}\right)\right\}R_{\mu+1}^{v+1}$$

$$= 2N_{v-1}e^{i\beta\sin\theta_c\delta x + i\beta\cos\theta_c\delta y}R_\mu^v + e^{i\beta\sin\theta_c\delta x + i\beta\cos\theta_c\delta y}S_\mu^v\left\{N_{v-1} - N_v\left(1 - \frac{\partial N_v}{\partial y}\frac{a\delta y}{b^2 N_v^3}\right)\right\} \qquad (12.46)$$

$$\left\{N_{v-1} + N_v\left(1 - \frac{\partial N_v}{\partial y}\frac{a\delta y}{b^2 N_v^3}\right)\right\}S_{\mu+1}^{v-1} = e^{i\beta\sin\theta_c\delta x + i\beta\cos\theta_c\delta y}S_\mu^v\left\{2N_v\left(1 - \frac{\partial N_v}{\partial y}\frac{a\delta y}{b^2 N_v^3}\right)\right\}$$

$$+ e^{i\beta\sin\theta_c\delta x + i\beta\cos\theta_c\delta y}R_\mu^v\left\{N_v\left(1 - \frac{\partial N_v}{\partial y}\frac{a\delta y}{b^2 N_v^3}\right) - N_{v-1}\right\} \qquad (12.47)$$

Following our previous analysis, we can then derive the corresponding partial differential equations for the π-polarisation:

$$\frac{\mathbf{k_c}}{\beta} \cdot \nabla R = \sin\theta_c \frac{\partial R}{\partial x} + \cos\theta_c \frac{\partial R}{\partial y} = \frac{R}{2}\left\{2i\beta - \frac{1}{n}\frac{\cos 2\theta_c}{\cos\theta_c}\frac{\partial n}{\partial y}\right\} - \frac{S}{2n}\frac{\cos 2\theta_c}{\cos\theta_c}\frac{\partial n}{\partial y} \tag{12.48}$$

$$\frac{\mathbf{k_i}}{\beta} \cdot \nabla S = \sin\theta_c \frac{\partial S}{\partial x} - \cos\theta_c \frac{\partial S}{\partial y} = \frac{S}{2}\left\{2i\beta + \frac{1}{n}\frac{\cos 2\theta_c}{\cos\theta_c}\frac{\partial n}{\partial y}\right\} + \frac{R}{2n}\frac{\cos 2\theta_c}{\cos\theta_c}\frac{\partial n}{\partial y} \tag{12.49}$$

12.4.3 Simplification of PSM Equations to Ordinary Differential Equations

The PSM equations may be simplified under boundary conditions corresponding to monochromatic illumination of the grating.

Let

$$R \to R(y)e^{i\beta\sin\theta_c x}$$
$$S \to S(y)e^{i\beta\sin\theta_c x} \tag{12.50}$$

Under this transformation, Equations 12.43 and 12.44 yield the following pair of ordinary differential equations

$$\cos\theta_c \frac{dR}{dy} = \frac{R}{2}\left\{2i\beta\cos^2\theta_c - \frac{1}{n\cos\theta_c}\frac{dn}{dy}\right\} - \frac{S}{2}\left\{\frac{1}{n\cos\theta_c}\frac{dn}{dy}\right\}$$
$$-\cos\theta_c \frac{dS}{dy} = \frac{S}{2}\left\{2i\beta\cos^2\theta_c + \frac{1}{n\cos\theta_c}\frac{dn}{dy}\right\} + \frac{R}{2}\left\{\frac{1}{n\cos\theta_c}\frac{dn}{dy}\right\} \tag{12.51}$$

Similarly, the π-polarisation equations yield

$$\cos\theta_c \frac{dR}{dy} = \frac{R}{2}\left\{2i\beta\cos^2\theta_c - \frac{\cos 2\theta_c}{n\cos\theta_c}\frac{dn}{dy}\right\} - \frac{S}{2}\left\{\frac{\cos 2\theta_c}{n\cos\theta_c}\frac{dn}{dy}\right\}$$
$$-\cos\theta_c \frac{dS}{dy} = \frac{S}{2}\left\{2i\beta\cos^2\theta_c + \frac{\cos 2\theta_c}{n\cos\theta_c}\frac{dn}{dy}\right\} + \frac{R}{2}\left\{\frac{\cos 2\theta_c}{n\cos\theta_c}\frac{dn}{dy}\right\} \tag{12.52}$$

Equations 12.51 and 12.52 are approximate only because we have assumed an approximate form for the direction vector of the waves within the grating. We may, however, approach the problem differently and derive exact equations directly from Equation 12.6. For example, in the case of the σ-polarisation, we use the optical invariant

$$\tilde{\beta}(y) \to \tilde{\beta}(y)\cos\theta(y) \tag{12.53}$$

where

$$\tilde{\beta}(y) = \frac{\beta n(y)}{n_0} \tag{12.54}$$

Then using Snell's law

$$\frac{d\tilde{\beta}(y)}{dy}\sin\theta(y) + \tilde{\beta}(y)\frac{d\theta(y)}{dy}\cos\theta(y) = 0 \tag{12.55}$$

it is simple to see that Equation 12.6 reduces to

$$\cos\theta \frac{dR}{dy} = \frac{R}{2}\left\{ 2i\tilde{\beta}\cos^2\theta - \frac{1}{\tilde{\beta}\cos\theta}\frac{d\tilde{\beta}}{dy} \right\} - \frac{S}{2}\left\{ \frac{1}{\tilde{\beta}\cos\theta}\frac{d\tilde{\beta}}{dy} \right\}$$

$$-\cos\theta \frac{dS}{dy} = \frac{S}{2}\left\{ 2i\tilde{\beta}\cos^2\theta + \frac{1}{\tilde{\beta}\cos\theta}\frac{d\tilde{\beta}}{dy} \right\} + \frac{R}{2}\left\{ \frac{1}{\tilde{\beta}\cos\theta}\frac{d\tilde{\beta}}{dy} \right\}$$

(12.56)

where θ is now a function of y throughout the grating. If we now replace Equation 12.50 with the more general behaviour

$$R \rightarrow R(y)e^{i\tilde{\beta}(y)\sin\theta(y)x}$$

$$S \rightarrow S(y)e^{i\tilde{\beta}(y)\sin\theta(y)x}$$

(12.57)

then Equation 12.56 can be seen to be an exact solution of the Helmholtz equation. It follows therefore that solution of Equations 12.55 and 12.56, subject to the boundary conditions of Equation 12.18 and $\theta(0) = \theta_c$, constitutes a rigorous solution of the Helmholtz equation. Note that this is independent of periodicity required by a Floquet solution. Because these equations are none other than a differential representation of the chain matrix method of thin films [6,7], it is simple to show that this implies that the chain matrix method is itself rigorous.

Equations 12.55 and 12.56, or alternatively the approximate Equations 12.51 and 12.52, are simple to solve on a PC for arbitrary index profiles. However, Equations 12.51 and 12.52 also possess analytic solutions of interest.

12.4.4 Analytic Solutions for Sinusoidal Gratings

We start by defining an unslanted grating with the following index profile

$$n = n_0 + n_1\cos(2\alpha\beta\cos\theta_r y) = n_0 + \frac{n_1}{2}\left\{ e^{2i\alpha\beta\cos\theta_r y} + e^{-2i\alpha\beta\cos\theta_r y} \right\}$$

(12.58)

where we imagine θ_r to be the recording angle of this grating. Then letting

$$R \rightarrow R(y)e^{i\beta\cos\theta_c y}$$

$$S \rightarrow S(y)e^{-i\beta\cos\theta_c y}$$

(12.59)

and using Equation 12.10, Equation 12.51 reduces to

$$\cos\theta_c \frac{dR}{dy} = -\frac{1}{2n_0}n_1 i\beta\alpha\frac{\cos\theta_r}{\cos\theta_c}\left\langle \left\{ e^{2i\beta\alpha\cos\theta_r y} + ... \right\}Se^{-2i\beta y\cos\theta_c} \right\rangle$$

$$= -\frac{n_1 i\beta(\alpha\cos\theta_r)}{2n_0\cos\theta_c}Se^{2i\beta y(\alpha\cos\theta_r - \cos\theta_c)}$$

$$\cos\theta_c \frac{dS}{dy} = \frac{1}{2n_0}n_1 i\beta\alpha\frac{\cos\theta_r}{\cos\theta_c}\left\langle \left\{ e^{-2i\beta\alpha\cos\theta_r y} + ... \right\}Re^{2i\beta y\cos\theta_c} \right\rangle$$

$$= \frac{n_1 i\beta(\alpha\cos\theta_r)}{2n_0\cos\theta_c}Re^{-2i\beta y(\alpha\cos\theta_r - \cos\theta_c)}$$

(12.60)

As before, we now define the pseudo-field

$$\hat{S} = Se^{2i\beta y(\alpha\cos\theta_r - \cos\theta_c)} \tag{12.61}$$

at which point Equations 12.60 reduce to the standard form of Kogelnik's equations

$$c_R\frac{dR}{dy} = -i\kappa\hat{S}$$

$$c_S\frac{d\hat{S}}{dy} = -i\vartheta\hat{S} - i\kappa R \tag{12.62}$$

The coefficients for the PSM model and for Kogelnik's model are as follows:

$$c_{R(PSM)} = \frac{\cos^2\theta_c}{\alpha\cos\theta_r} \qquad\qquad c_{R(KOG)} = \cos\theta_c$$

$$c_{S(PSM)} = -\frac{\cos^2\theta_c}{\alpha\cos\theta_r} \qquad\qquad c_{S(KOG)} = \cos\theta_c - 2\alpha\cos\theta_r \tag{12.63}$$

$$\vartheta_{PSM} = 2\beta\left(1 - \frac{\cos\theta_c}{\alpha\cos\theta_r}\right)\cos^2\theta_c \qquad \vartheta_{KOG} = 2\alpha\beta\cos\theta_r(\cos\theta_c - \alpha\cos\theta_r)$$

Equation 12.62, in conjunction with the boundary conditions in Equation 12.18, then leads to the general analytic expression for the diffractive efficiency of the unslanted reflection grating:

$$\eta_\sigma = \frac{|c_S|}{c_R}\hat{S}(0)\hat{S}*(0) = \frac{\kappa^2\sinh^2(d\Upsilon)}{\kappa^2\sinh^2(d\Upsilon) - c_Rc_S\Upsilon^2} \tag{12.64}$$

where

$$\Upsilon^2 = -\frac{\vartheta^2}{4c_S^2} - \frac{\kappa^2}{c_Rc_S} \tag{12.65}$$

Note that at Bragg resonance, both the PSM theory and Kogelnik's theory reduce to the well-known formula

$$\eta_\sigma = \tanh^2(\kappa d\sec\theta_c) \tag{12.66}$$

The π-polarisation may be treated in an exactly analogous way, leading to the following pair of ordinary differential equations for R and $\hat{S}$:

$$c_R\frac{dR}{dy} = -i\kappa\cos 2\theta_c\hat{S}$$

$$c_S\frac{d\hat{S}}{dy} = -i\vartheta\hat{S} - i\kappa\cos 2\theta_c R \tag{12.67}$$

These are just Kogelnik's equations with a modified κ parameter. The PSM model therefore distinguishes the π- and σ-polarisations in exactly the same manner as Kogelnik's theory does! In both theories, in the case of the unslanted grating, Kogelnik's constant is simply transformed according to the rule

$$|\kappa| \rightarrow |\kappa\cos 2\theta_c| \tag{12.68}$$

We shall see shortly that the practical predictions of Kogelnik's model and the PSM model are very close for gratings of interest to display and optical element holography. This is partly due to the effect of Snell's law, which acts to steepen the angle of incidence in most situations. At very high angles of incidence within the grating, larger differences appear.

12.4.5 Multiple Colour Gratings

Let

$$
n = n_0 + n_1 \cos(2\alpha_1\beta\cos\theta_{r1}y) + n_2\cos(2\alpha_2\beta\cos\theta_{r2}y) + ...
$$

$$
= n_0 + \frac{n_1}{2}\left\{e^{2i\alpha_1\beta\cos\theta_{r1}y} + e^{-2i\alpha_1\beta\cos\theta_{r1}y}\right\} + \frac{n_2}{2}\left\{e^{2i\alpha_2\beta\cos\theta_{r2}y} + e^{-2i\alpha_2\beta\cos\theta_{r2}y}\right\} + ... \tag{12.69}
$$

$$
= n_0 + \frac{1}{2}\sum_{j=1}^{N} n_j\left\{e^{2i\alpha_j\beta\cos\theta_{rj}y} + e^{-2i\alpha_j\beta\cos\theta_{rj}y}\right\}
$$

In this case, the PSM σ-polarisation equations yield

$$
\cos\theta_c\frac{dR}{dy} = -S\sum_{j=1}^{N} i\kappa_j\alpha_j\frac{\cos\theta_{rj}}{\cos\theta_c}e^{2i\beta y(\alpha_j\cos\theta_{rj}-\cos\theta_c)}
$$

$$
\tag{12.70}
$$

$$
\cos\theta_c\frac{dS}{dy} = R\sum_{j=1}^{N} i\kappa_j\alpha_j\frac{\cos\theta_{rj}}{\cos\theta_c}e^{-2i\beta y(\alpha_j\cos\theta_{rj}-\cos\theta_c)}
$$

where

$$
\kappa_j = \frac{n_j\pi}{\lambda_c} \tag{12.71}
$$

Once again, as in Section 12.3, if we assume that the individual gratings have very different spatial frequencies, then these equations lead to a simple expression for the diffractive efficiency when the reference wave is in Bragg resonance with one or another of the multiplexed gratings:

$$
\eta_{\text{PSM}/\sigma_j} = \tanh^2(\kappa_j d\sec\theta_c) \tag{12.72}
$$

The corresponding result for the π-polarisation is

$$
\eta_{\text{PSM}/\pi_j} = \tanh^2(\kappa_j d\sec\theta_c\cos 2\theta_c) \tag{12.73}
$$

If we recalculate Figures 12.5 and 12.8 using Equation 12.72, for typical values of θ_c, we see very little difference in the plots. The effect for the σ-polarisation is usually to increase the diffractive response of the grating by a small amount. The π-polarisation generally produces an inverse behaviour, at least at small angles of incidence.

As in Section 12.3 in the region of the jth Bragg resonance, Equations 12.70 lead to the approximate analytical form

$$
\left.\begin{array}{l}
\eta_{\sigma j} = \dfrac{\kappa_j^2\sinh^2(d\Upsilon_{\sigma j})}{\kappa_j^2\sinh^2(d\Upsilon_{\sigma j}) - c_Rc_S\Upsilon_{\sigma j}^2} \\[4mm]
\Upsilon_{\sigma j}^2 = -\dfrac{\vartheta_{\sigma j}^2}{4c_S^2} - \dfrac{\kappa_j^2}{c_Rc_S} \\[4mm]
\vartheta_{\sigma j} = 2\beta\left(1 - \dfrac{\cos\theta_c}{\alpha_j\cos\theta_r}\right)\cos^2\theta_c
\end{array}\right\} \tag{12.74}
$$

where c_R and c_S are given by $c_{R(\text{PSM})}$ and $c_{S(\text{PSM})}$ of Equation 12.63.

Again, as long as there is a sufficient difference in the spatial frequencies of each grating, we can add each response to give a convenient analytical expression for the total diffraction efficiency:

$$\eta_\sigma = \sum_{j=1}^{N} \frac{\kappa_j^2 \sinh^2(d\Upsilon_{\sigma j})}{\kappa_j^2 \sinh^2(d\Upsilon_{\sigma j}) - c_R c_S \Upsilon_{\sigma j}^2} \tag{12.75}$$

In cases in which the individual gratings are too close to one another in wavelength or in which small amplitude interaction effects between gratings are to be described, Equation 12.70 must be solved numerically.

12.4.6 Comparison of Kogelnik's Theory with the PSM Theory for Unslanted Gratings at Oblique Incidence

In Figures 12.11 and 12.12, we plot the diffractive efficiency (σ-polarisation) against replay wavelength and replay angle as predicted by both Kogelnik's theory and by the PSM analytical theory for typical sinusoidal gratings of interest to display and optical element holography. It is immediately clear that the two theories produce very similar predictions.

In Figure 12.13, results are presented from a rigorous chain matrix computation (see Section 11.12.2) for the π-polarisation case at finite incidence angle. These are compared with predictions of the analytical Kogelnik and PSM models. The chain matrix algorithm used is based on the technique described by Heavens [7], and models the grating as a stack of 100,000 layers. As can be seen, in this case, the chain matrix computation is somewhat closer to the PSM model than Kogelnik's, although the differences

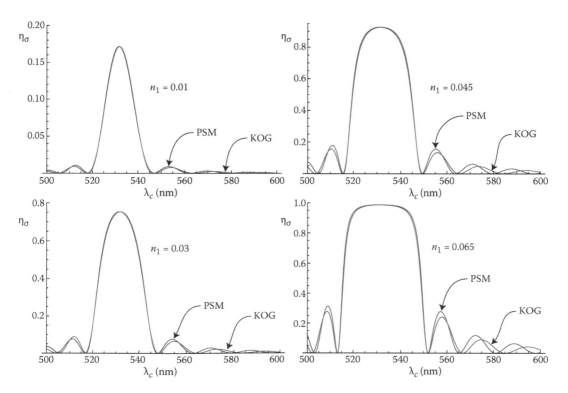

FIGURE 12.11 Four graphs showing the predicted diffractive response (σ-polarisation) versus replay wavelength of typical unslanted single-colour reflection gratings using the PSM analytical model (blue lines) and Kogelnik's theory (red lines). Recording wavelength, $\lambda_r = 532$ nm; recording angle, $\theta_r = 20°$; replay angle, $\theta_c = 20°$; grating thickness, $d = 7$ μm; $n_0 = 1.5$; index modulations shown on graphs. Note that all angles are interior angles.

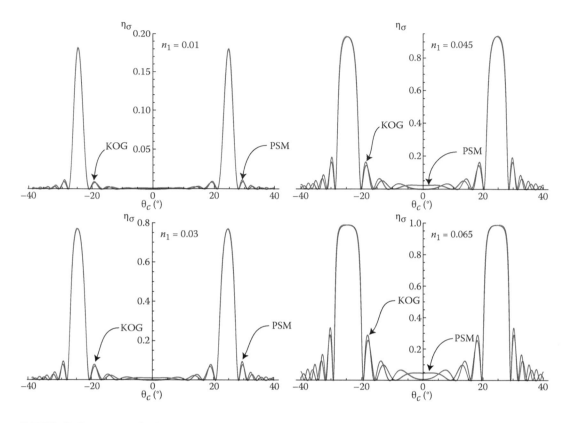

FIGURE 12.12 Four graphs showing the predicted diffractive response (σ-polarisation) versus reconstruction angle of typical unslanted single-colour reflection gratings using the PSM analytical model (blue lines) and Kogelnik's theory (red lines). Recording wavelength, $\lambda_r = 550$ nm; reconstruction wavelength, $\lambda_c = 532$ nm; recording angle, $\theta_r = 20°$; grating thickness, $d = 7$ μm; $n_0 = 1.5$; index modulations shown on graphs. Note that all angles are interior angles.

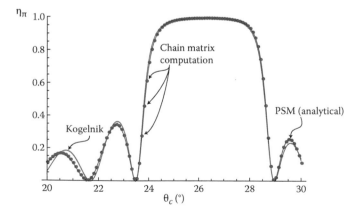

FIGURE 12.13 Comparison of the PSM (analytic) and Kogelnik models with a rigorous chain matrix computation for the π-polarisation diffractive efficiency versus replay angle, θ_c. For the purpose of the numerical calculation, the grating index profile has been modelled using 100,000 discrete thin films. The 12 μm unslanted reflection grating ($n_0 = 1.5$; $n_1 = 0.065$) has been recorded at $\lambda_r = 532$ nm with a recording angle of $\theta_r = 30°$. The replay wavelength is $\lambda_r = 550$ nm. Note that all angles are interior angles.

are small. However, this is not always the case. In particular, as the angle of incidence (within the grating) becomes very high, the balance can shift to favouring Kogelnik's model. Indeed, the PSM model has been derived as an approximate differential representation of the rigorous chain matrix model. In the limit of zero incidence angle, this representation is exact, but as the incidence angle increases, the model should be expected to become somewhat less accurate as can be appreciated from the discussion in Section 12.4.3.

12.5 Slanted Reflection Gratings

We may use the PSM equations for the unslanted grating to derive corresponding equations for the general slanted grating. To do this, we define rotated Cartesian coordinates (x', y') that are related to the unprimed Cartesian system by

$$
\begin{pmatrix} x' \\ y' \end{pmatrix} = \begin{pmatrix} \cos\psi & -\sin\psi \\ \sin\psi & \cos\psi \end{pmatrix} \begin{pmatrix} x \\ y \end{pmatrix}
\tag{12.76}
$$

This is illustrated in Figure 12.14. In the unprimed frame, we have

$$
\mathbf{k_c} = \beta \begin{pmatrix} \sin\theta_c \\ \cos\theta_c \end{pmatrix}; \quad \mathbf{k_i} = \beta \begin{pmatrix} \sin\theta_c \\ -\cos\theta_c \end{pmatrix}
\tag{12.77}
$$

whereas in the primed frame, we have

$$
\mathbf{k'_c} = \beta \begin{pmatrix} \sin(\theta_c - \psi) \\ \cos(\theta_c - \psi) \end{pmatrix}; \quad \mathbf{k'_i} = \beta \begin{pmatrix} \sin(\theta_c + \psi) \\ -\cos(\theta_c + \psi) \end{pmatrix}
\tag{12.78}
$$

Derivatives in the primed system are related to those in the unprimed system by Leibnitz's chain rule

$$
\frac{\partial}{\partial x} = \frac{\partial x'}{\partial x}\frac{\partial}{\partial x'} + \frac{\partial y'}{\partial x}\frac{\partial}{\partial y'} = \cos\psi\frac{\partial}{\partial x'} + \sin\psi\frac{\partial}{\partial y'}
$$

$$
\frac{\partial}{\partial y} = \frac{\partial y'}{\partial y}\frac{\partial}{\partial y'} + \frac{\partial x'}{\partial y}\frac{\partial}{\partial x'} = -\sin\psi\frac{\partial}{\partial x'} + \cos\psi\frac{\partial}{\partial y'}
\tag{12.79}
$$

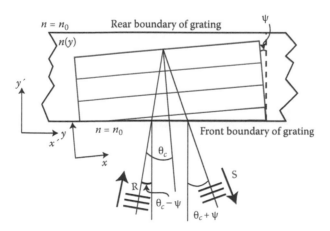

FIGURE 12.14 Geometry for rotating the unslanted reflection grating in the (x,y) system to a slanted grating in the primed system.

The PSM equations for the σ-polarisation may therefore be written as

$$\frac{\mathbf{k}_c'}{\beta} \cdot \nabla' R = \frac{\partial R}{\partial x'}\sin(\theta_c - \psi) + \frac{\partial R}{\partial y'}\cos(\theta_c - \psi)$$

$$= \frac{R}{2}\left\{2i\beta - \frac{1}{n\cos\theta_c}\frac{\partial n}{\partial y}\right\} - \frac{S}{2n\cos\theta_c}\frac{\partial n}{\partial y} \tag{12.80}$$

and

$$\frac{\mathbf{k}_i'}{\beta} \cdot \nabla' S = \frac{\partial S}{\partial x'}\sin(\theta_c + \psi) - \frac{\partial S}{\partial y'}\cos(\theta_c + \psi)$$

$$= \frac{S}{2}\left\{2i\beta + \frac{1}{n\cos\theta_c}\frac{\partial n}{\partial y}\right\} + \frac{R}{2n\cos\theta_c}\frac{\partial n}{\partial y} \tag{12.81}$$

Note that we have kept the unprimed frame on the right hand side (RHS) on purpose because, in this system, the index profile is one-dimensional and as such, much easier to evaluate.

12.5.1 Analytical Solutions for Single-Colour Gratings

To study the single-colour grating, we shall use the unslanted index profile (Equation 12.58) in the unprimed frame leading to the following profile in the primed frame

$$n = n_0 + n_1\cos(2\alpha\beta\cos\theta_r\{\sin\psi x' - \cos\psi y'\}) = n_0 + \frac{n_1}{2}\left\{e^{2i\beta\alpha\cos\theta_r\hat{\mathbf{K}}.\mathbf{r}'} + e^{-2i\beta\alpha\cos\theta_r\hat{\mathbf{K}}.\mathbf{r}'}\right\} \tag{12.82}$$

where $\hat{\mathbf{K}}$ is the unit grating vector in the primed frame.

We now let

$$R \to Re^{i\beta\left\{\sin(\theta_c - \psi)x' + \cos(\theta_c - \psi)y'\right\}}$$

$$S \to Se^{i\beta\left\{\sin(\theta_c + \psi)x' - \cos(\theta_c + \psi)y'\right\}} \tag{12.83}$$

Equations 12.80 and 12.81 then become

$$\sin(\theta_c - \psi)\frac{\partial R}{\partial x'} + \cos(\theta_c - \psi)\frac{\partial R}{\partial y'}$$

$$= -\frac{S}{2}\left\{\frac{1}{n\cos\theta_c}\frac{\partial n}{\partial y}\right\}e^{i\beta\left\{[\sin(\theta_c + \psi) - \sin(\theta_c - \psi)]x' - [\cos(\theta_c - \psi) + \cos(\theta_c + \psi)]y'\right\}}$$

$$= -\frac{S}{2}\left\langle\left\{\frac{1}{n_0\cos\theta_c}\frac{\partial}{\partial y}\frac{n_1}{2}\left\{e^{2i\beta\alpha(\cos\theta_r)y} + e^{-2i\beta\alpha(\cos\theta_r)y}\right\}\right\}e^{2i\beta\cos\theta_c\left\{[\sin\psi]x' - [\cos\psi]y'\right\}}\right\rangle \tag{12.84}$$

$$= -\frac{S}{2}\left\langle\left\{\frac{n_1 i\beta\alpha(\cos\theta_r)}{n_0\cos\theta_c}\left\{e^{2i\beta\alpha(\cos\theta_r)y} + ...\right\}\right\}e^{-2i\beta\cos\theta_c y}\right\rangle$$

$$= -\frac{i\beta n_1}{2n_0}\alpha\frac{\cos\theta_r}{\cos\theta_c}Se^{2i\beta(\alpha\cos\theta_r - \cos\theta_c)(y'\cos\psi - x'\sin\psi)}$$

and

$$\sin(\theta_c + \psi)\frac{\partial S}{\partial x'} - \cos(\theta_c + \psi)\frac{\partial S}{\partial y'} = -\frac{i\beta n_1}{2n_0}\alpha\frac{\cos\theta_r}{\cos\theta_c}Re^{-2i\beta(\alpha\cos\theta_r - \cos\theta_c)(y'\cos\psi - x'\sin\psi)} \tag{12.85}$$

Next, we make the transformation

$$\hat{S} = S(y')e^{2i\beta(\alpha\cos\theta_r - \cos\theta_c)(y'\cos\psi - x'\sin\psi)}$$

$$\hat{R} = R(y')$$

(12.86)

at which point the PSM equations once again reduce to a simple pair of ordinary differential equations in the form of Kogelnik's equations:

$$c_R \frac{dR}{dy'} = -i\kappa\hat{S}$$

$$c_S \frac{d\hat{S}}{dy'} = -i\vartheta\hat{S} - i\kappa R$$

(12.87)

with coefficients

$$c_{R(\text{PSM})} = \frac{\cos\theta_c \cos(\theta_c - \psi)}{\alpha\cos\theta_r}$$

$$c_{S(\text{PSM})} = -\frac{\cos\theta_c \cos(\theta_c + \psi)}{\alpha\cos\theta_r}$$

$$\vartheta_{\text{PSM}} = 2\beta\left(1 - \frac{\cos\theta_c}{\alpha\cos\theta_r}\right)\cos^2\theta_c$$

(12.88)

For comparison, Kogelnik's coefficients are

$$c_{R(\text{KOG})} = \cos(\theta_c - \psi)$$

$$c_{S(\text{KOG})} = \cos(\theta_c - \psi) - 2\alpha\cos\theta_r\cos\psi$$

$$\vartheta_{\text{KOG}} = 2\alpha\beta\cos\theta_r(\cos\theta_c - \alpha\cos\theta_r)$$

(12.89)

With the usual reflective boundary conditions, $\hat{R}(0) = 1$ and $\hat{S}(d) = 0$, we can then use the standard formula to describe the diffraction efficiency of the slanted reflection grating:

$$\eta_\sigma = \frac{|c_S|}{c_R}\hat{S}(0)\hat{S}*(0) = \frac{\kappa^2 \sinh^2(d\Upsilon)}{\kappa^2 \sinh^2(d\Upsilon) - c_R c_S \Upsilon^2}$$

(12.90)

where

$$\Upsilon^2 = -\frac{\vartheta^2}{4c_S^2} - \frac{\kappa^2}{c_R c_S}$$

(12.91)

Substitution of either Equation 12.88 or Equation 12.89 into Equation 12.90 gives the required expression for the diffractive efficiency in either the Kogelnik or PSM model.

When $\psi = 0$, $\eta_{\text{PSM}/\sigma}$ reduces to the unslanted formula that was derived in Section 12.4.4. In the case of finite slant and Bragg resonance (where $\cos\theta_c = \alpha\cos\theta_r$), we have

$$\eta_{PSM/\sigma} = \tanh^2(d\kappa\sqrt{\sec(\theta_c - \psi)\sec(\theta_c + \psi)}) \tag{12.92}$$

which is identical to Kogelnik's solution. Note that the behaviour of the π-polarisation is simply described by making the transformation (Equation 12.68) in all formulae of interest.

The PSM model for the slanted grating under either the σ- or π-polarisations gives expressions very similar to Kogelnik's theory. We shall see shortly that for most gratings of practical interest to display and optical element holography, the two theories produce predictions that are extremely close.

12.6 Slanted Reflection Gratings in Three Dimensions

The PSM model for the slanted grating at oblique incidence can easily be extended to the case of oblique incidence in three dimensional space.

We start by recognising that the PSM equations for an unslanted grating in three dimensions are no different from their two-dimensional counterparts. We use the same index profile as before (Equation 12.58) and simply generalise the coordinates from the (x,y) Cartesian system to an (x,y,z) Cartesian system. The reference and image rays are therefore, as before, in the (x,y) plane. We now introduce a primed system of Cartesian coordinates which has been rotated first about the z-direction (as before) and then about the x-direction. Accordingly, we have

$$
\begin{pmatrix} x' \\ y' \\ z' \end{pmatrix} = R_x R_z \mathbf{r} = \begin{pmatrix} 1 & 0 & 0 \\ 0 & \cos\varphi & -\sin\varphi \\ 0 & \sin\varphi & \cos\varphi \end{pmatrix} \begin{pmatrix} \cos\psi & -\sin\psi & 0 \\ \sin\psi & \cos\psi & 0 \\ 0 & 0 & 1 \end{pmatrix} \begin{pmatrix} x \\ y \\ z \end{pmatrix}
$$
$$
= \begin{pmatrix} \cos\psi & -\sin\psi & 0 \\ \cos\varphi\sin\psi & \cos\varphi\cos\psi & -\sin\varphi \\ \sin\varphi\sin\psi & \sin\varphi\cos\psi & \cos\varphi \end{pmatrix} \begin{pmatrix} x \\ y \\ z \end{pmatrix} \tag{12.93}
$$

The primed system of coordinates is then, as before, the real-world coordinates of the grating. In these coordinates, the grating has a slant in two directions defined by the angles ψ and φ.

The PSM equations (Equations 12.43 and 12.44) can then be written in vector notation as

$$\frac{\mathbf{k}'_c}{\beta} \cdot \nabla' R = \frac{R}{2}\left\{ 2i\beta - \frac{1}{2n\cos\theta_c}\frac{\partial n}{\partial y} \right\} - \frac{S}{2n\cos\theta_c}\frac{\partial n}{\partial y} \tag{12.94}$$

and

$$\frac{\mathbf{k}'_i}{\beta} \cdot \nabla' S = \frac{S}{2}\left\{ 2i\beta + \frac{1}{n\cos\theta_c}\frac{\partial n}{\partial y} \right\} + \frac{R}{2n\cos\theta_c}\frac{\partial n}{\partial y} \tag{12.95}$$

where the reference and image k-vectors in the primed frame are defined as

$$\mathbf{k}'_c = R_x R_z \mathbf{k}_c = \beta\begin{pmatrix} \sin(\theta_c - \psi) \\ \cos\varphi\cos(\theta_c - \psi) \\ \sin\varphi\cos(\theta_c - \psi) \end{pmatrix}; \quad \mathbf{k}'_i = R_x R_z \mathbf{k}_i = \beta\begin{pmatrix} \sin(\theta_c + \psi) \\ -\cos\varphi\cos(\theta_c + \psi) \\ -\sin\varphi\cos(\theta_c + \psi) \end{pmatrix} \tag{12.96}$$

These equations may now be solved in exactly the same way as the two-dimensional equations using

$$R \rightarrow R e^{i\mathbf{k}'_c \cdot \mathbf{r}'}$$
$$S \rightarrow S e^{i\mathbf{k}'_i \cdot \mathbf{r}'}$$

(12.97)

giving

$$\frac{\mathbf{k}'_c}{\beta} \cdot \nabla' R = -\frac{S}{2n \cos \theta_c} \frac{\partial n}{\partial y} e^{i(\mathbf{k}'_i - \mathbf{k}'_c) \cdot \mathbf{r}'}$$

$$= -\frac{S}{2} \left\langle \left\{ \frac{n_1 i \beta \alpha (\cos \theta_r)}{n_0 \cos \theta_c} \left\{ e^{2i\beta \alpha (\cos \theta_r) y} + \dots \right\} \right\} e^{-2i\beta \cos \theta_c y} \right\rangle$$

$$= -i \kappa \alpha \frac{\cos \theta_r}{\cos \theta_c} S e^{2i\beta(\alpha \cos \theta_r - \cos \theta_c) \left\{ (R_x R_z)^{-1} \cdot \begin{pmatrix} x' \\ y' \\ z' \end{pmatrix} \right\} \begin{pmatrix} 0 \\ 1 \\ 0 \end{pmatrix}}$$

$$= -i \kappa \alpha \frac{\cos \theta_r}{\cos \theta_c} S e^{2i\beta(\alpha \cos \theta_r - \cos \theta_c)(-\sin \psi x' + \cos \psi \cos \varphi y' + \cos \psi \sin \varphi z')}$$

(12.98)

and

$$\frac{\mathbf{k}'_i}{\beta} \cdot \nabla' S = -\frac{i \beta n_1}{2n_0} \alpha \frac{\cos \theta_r}{\cos \theta_c} R e^{-2i\beta(\alpha \cos \theta_r - \cos \theta_c)(-\sin \psi x' + \cos \psi \cos \varphi y' + \cos \psi \sin \varphi z')}$$

(12.99)

We then choose

$$\hat{S} = S(y') e^{2i\beta(\alpha \cos \theta_r - \cos \theta_c)(-\sin \psi x' + \cos \psi \cos \varphi y' + \cos \psi \sin \varphi z')}$$
$$\hat{R} = R(y')$$

(12.100)

at which point the three-dimensional PSM equations reduce to the standard form of Kogelnik's equations (e.g., Equation 12.87) with coefficients

$$c_{R(\mathrm{PSM})} = \frac{\cos \varphi \cos \theta_c \cos(\theta_c - \psi)}{\alpha \cos \theta_r}$$

$$c_{S(\mathrm{PSM})} = -\frac{\cos \varphi \cos \theta_c \cos(\theta_c + \psi)}{\alpha \cos \theta_r}$$

(12.101)

$$\vartheta_{\mathrm{PSM}} = 2\beta \left(1 - \frac{\cos \theta_c}{\alpha \cos \theta_r} \right) \cos^2 \theta_c$$

We therefore see that the effect of tilting the grating in the second dimension is simply to multiply the obliquity constants, c_R and c_S, by a factor of $\cos \varphi$. Note that we have treated the σ-polarisation above with respect to the grating planes. The polarisation with respect to the grating surface is, in general, mixed for finite φ.

At Bragg resonance, the diffraction efficiency for the three-dimensional case reduces to

$$\eta_{\mathrm{PSM}/\sigma} = \tanh^2 \left(d \kappa \sec \varphi \sqrt{\sec(\theta_c - \psi) \sec(\theta_c + \psi)} \right)$$

(12.102)

Likewise,

$$\eta_{\mathrm{PSM}/\pi} = \tanh^2 \left(d \kappa \cos 2\theta_c \sec \varphi \sqrt{\sec(\theta_c - \psi) \sec(\theta_c + \psi)} \right)$$

(12.103)

12.7 Transmission Gratings with Slanted Fringes

In this chapter we have concentrated on the reflection hologram. However, the PSM model can be applied to the transmission hologram by simply using the appropriate boundary conditions to solve the PSM equations in a rotated frame. Alternatively, the analysis of Section 12.4 can be repeated for Figure 12.10 rotated by 90°. The first strategy is very simple and works in two or three dimensions. For example, in two dimensions and with a single slant angle, ψ, we use the transmission boundary conditions

$$R(0) = 1$$
$$S(0) = 0 \tag{12.104}$$

to solve Equation 12.87, which, at Bragg resonance, results in the standard formula given by Kogelnik's theory.

$$\eta_{\sigma T/\text{PSM}} = \sin^2\left(\kappa d / \sqrt{c_R c_S}\right) = \sin^2\left(\kappa d / \sqrt{-\cos(\theta_c - \psi)\cos(\theta_c + \psi)}\right) \tag{12.105}$$

12.8 Comparison of the PSM Theory with Kogelnik's Theory for Slanted Gratings

Figure 12.15 shows typical plots comparing diffraction efficiency versus replay angle according to the PSM model and Kogelnik's model for the two-dimensional slanted reflection grating. The plots refer

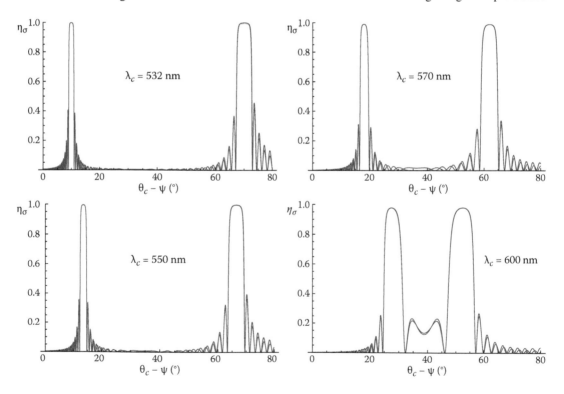

FIGURE 12.15 Diffraction efficiency (σ-polarisation) against replay angle in degrees for a 12 μm grating recorded at 532 nm with an index modulation of $n_1 = 0.03$ ($n_0 = 1.5$) at a recording reference angle of $\theta_r = 30°$ and with a fringe slant of $\psi = -40°$. The grating is replayed at four different wavelengths. Note that, as expected, when the replay and recording wavelengths are the same, reconstruction is observed at 70° and 10° (PSM, blue lines; Kogelnik, red lines). Note that all angles are interior angles.

to a 12 μm grating recorded at 532 nm, with an index modulation of $n_1 = 0.03$ ($n_0 = 1.5$) at a recording reference angle of $\theta_r = 30°$ and with a fringe slant of $\psi = 40°$. Clearly, the two models give *extremely* similar predictions. This includes an almost exact replication of the asymmetry between the two reconstruction maxima. A survey of a wide range of typical scenarios reveals that there are some differences between the two models, but these are at larger interior grating angles than permitted by Snell refraction when gratings are used in air or at very high index modulation ($n_1 > 0.25$)—the overall behaviour of the two models is thus very similar for practical purposes. In Appendix 8 rigorous coupled wave theory is reviewed and its predictions compared to those of both Kogelink's theory and the PSM theory.

12.9 Polychromatic Slanted Reflection Gratings

The discussion in Section 12.4.5 applies identically to the case of the polychromatic slanted reflection grating. The PSM model yields simple ordinary differential equations for both the two- and three-dimensional cases and these can be solved very easily on a PC. Alternatively, if the primary diffractive peaks in a polychromatic grating do not overlap, approximate formulae such as Equations 12.74 and 12.75 may be used very effectively with the coefficients in Equations 12.88 or 12.101.

12.10 Extending PSM to Describe Spatially Multiplexed Monochromatic Gratings

The fundamental idea of the PSM model is to describe the permittivity distribution of a grating as an infinite sum of discontinuous step functions. The Fresnel reflections of an incident plane wave can then be summed to calculate the properties of the grating. We started our discussion of the PSM model at finite incidence angle by considering the following simple index distribution:

$$n = n_0\left(1 + \frac{\kappa}{\beta}\left\{e^{2i\alpha\beta\cos\theta_r y} + e^{-2i\alpha\beta\cos\theta_r y}\right\}\right) \tag{12.106}$$

We shall now extend our discussion to consider the case of a spatially multiplexed monochromatic grating of the form

$$n = n_0\left(1 + \sum_{j=1}^{N}\frac{\kappa_j}{\beta}\left\{e^{2i\alpha_j\beta\cos\theta_{rj}(\sin\psi_j x' - \cos\psi_j y')} + e^{-2i\alpha_j\beta\cos\theta_{rj}(\sin\psi_j x' - \cos\psi_j y')}\right\}\right) \tag{12.107}$$

where we have used the same notation as in Sections 12.5 and 12.6.

We can imagine that this grating has been created by the holographic interference of a single plane reference wave with many plane object waves, each incident at a different angle $\Phi_{oj} \equiv -(\theta_{rj} + \psi_j)$ to the physical substrate normal. The reference wave is incident at an angle $\Phi_r \equiv \theta_{rj} - \psi_j$ for all j. In this way, the grating can be regarded as being composed of N unslanted gratings each tilted by a different angle ψ_j and each having a common angle of incidence with respect to the substrate normal. We assume that interference only occurs between the reference and each of the object waves and not between the object waves themselves. This is equivalent to the assumption of sequential recording or to the assumption that the reference wave is of a much larger amplitude than each of the object waves.

The PSM model can now be used to treat Equation 12.107 by summing the Fresnel reflections of a single incident plane wave from N infinite series of stacked mirrors. Each of these infinite mirror sets is characterised by a different tilt. To see how this works, we need to imagine a single plane reference wave, R, which is incident onto the multiplexed grating. This wave interacts with each grating, generating a separate signal wave, which we can label S_j.

The analysis [3] follows Section 12.5 except that now Equations 12.84 and 12.85 become

$$\cos\Phi_c \frac{\partial R}{\partial y'} = -i\alpha \sum_{j=1}^{N} \kappa_j \frac{\cos(\Phi_r + \psi_j)}{\cos(\Phi_c + \psi_j)} S_j \left\langle e^{2i\beta\alpha\zeta_j(\Phi_r,\psi_j)} e^{i(\mathbf{k}'_{ij}-\mathbf{k}'_c)\cdot r'} \right\rangle + \left\langle \ldots \right\rangle$$

$$= -i\alpha \sum_{j=1}^{N} \kappa_j \frac{\cos(\Phi_r + \psi_j)}{\cos(\Phi_c + \psi_\mu)} S_j \, e^{2i\beta\alpha\zeta_j(\Phi_r,\psi_j)} e^{i(\mathbf{k}'_{ij}-\mathbf{k}'_c)\cdot r'}$$

$$-\sin\Phi_{ij}\frac{\partial S_j}{\partial x'} - \cos\Phi_{ij}\frac{\partial S_j}{\partial y'} = -i\alpha\kappa_j \frac{\cos(\Phi_r + \psi_j)}{\cos(\Phi_c + \psi_j)} R \left\langle e^{-2i\beta\alpha\zeta_j(\Phi_r,\psi_j)} e^{-i(\mathbf{k}'_{ij}-\mathbf{k}'_c)\cdot r'} \right\rangle + \left\langle \ldots \right\rangle$$

$$= -i\alpha\kappa_j \frac{\cos(\Phi_r + \psi_j)}{\cos(\Phi_c + \psi_j)} R e^{-2i\beta\alpha\zeta_j(\Phi_r,\psi_j)} e^{-i(\mathbf{k}'_{ij}-\mathbf{k}'_c)\cdot r'}$$

$$(12.108)$$

where

$$\left. \begin{array}{l} \theta_{cj} - \psi_j = \Phi_c \\ \theta_{rj} - \psi_j = \Phi_r \\ \theta_{cj} + \psi_j = -\Phi_{ij} \\ \theta_{rj} + \psi_j = -\Phi_{oj} \end{array} \right\} \quad \forall j \le N \qquad (12.109)$$

and where

$$\zeta_j(\theta_{rj}, \psi_j) \equiv \cos\theta_{rj}(y'\cos\psi_j - x'\sin\psi_j) \qquad (12.110)$$

We then define different pseudo-fields for each signal wave according to the following rule:

$$S_j = \hat{S}_j(y')e^{-2i\beta\alpha\zeta_j(\theta_{rj},\psi_j)-i(\mathbf{k}'_{ij}-\mathbf{k}'_c)\cdot r'} \qquad (12.111)$$

This leads directly to the N-PSM equations for the spatially multiplexed monochromatic grating

$$\frac{\partial R}{\partial y'} = -i \sum_{j=1}^{N} \frac{\kappa_j}{c_{Rj}} \hat{S}_j$$

$$c_{Sj}\frac{\partial \hat{S}_j}{\partial y'} = -i\vartheta_j \hat{S}_j - i\kappa_j R \qquad (12.112)$$

where for the σ-polarisation

$$c_{Rj} = \frac{\cos\theta_{cj}\cos(\theta_{cj} - \psi_j)}{\alpha_j \cos\theta_{rj}}$$

$$c_{Sj} = -\frac{\cos\theta_{cj}\cos(\theta_{cj} + \psi_j)}{\alpha_j \cos\theta_{rj}} \qquad (12.113)$$

$$\vartheta_j = 2\beta\left(1 - \frac{\cos\theta_{cj}}{\alpha_j\cos\theta_{rj}}\right)\cos^2\theta_{cj}$$

Equations 12.112 may now be solved using the boundary conditions appropriate for a reflection multiplexed grating, that is,

$$R(0) = 1$$

$$\hat{S}_j(d) = 0 \ \forall j \le N \tag{12.114}$$

At Bragg resonance, c_R becomes a constant

$$c_{Rj} = \frac{\cos\theta_{cj}\cos(\theta_{cj} - \psi_j)}{\alpha_j \cos\theta_{rj}} = \cos\Phi_c \tag{12.115}$$

and Equation 12.112 then gives the following expression for the diffractive efficiency of the *j*th grating:

$$\eta_j \equiv \frac{1}{c_R}\left|c_{Sj}\right|\hat{S}_j(0)\hat{S}_j^*(0) = \frac{1}{c_{Sj}}\frac{\kappa_j^2}{\displaystyle\sum_{k=1}^{N}\frac{\kappa_k^2}{c_{Sk}}}\tanh^2\left\{d\sqrt{-\frac{1}{c_R}\sum_{k=1}^{N}\frac{\kappa_k^2}{c_{Sk}}}\right\} \tag{12.116}$$

The total diffraction efficiency of the entire multiplexed grating is likewise found by summing the diffractive response from each grating:

$$\eta \equiv \sum_{j=1}^{N}\eta_j = \tanh^2\left\{d\sqrt{\frac{1}{\cos\Phi_c}\sum_{k=1}^{N}\frac{\kappa_k^2}{\cos\Phi_{ik}}}\right\} \tag{12.117}$$

Here, Φ_c is the incidence angle of the replay reference wave and Φ_{ik} is the incidence angle of the *k*th signal wave. These results are of course none other than the expressions that we obtained in Chapter 11 using the *N*-coupled wave theory (see Equation 11.165)! At Bragg resonance, the PSM model of the multiplexed grating gives an identical description to the corresponding *N*-coupled wave theory just as the simple PSM theory gave an identical description at Bragg resonance to Kogelnik's theory. Here again, however, the advantage of the PSM model over the *N*-coupled wave theory is that we can immediately extend the result to the polychromatic spatially multiplexed grating in the limit that the recording wavelengths are sufficiently separated. Here, the diffraction efficiency at the Bragg resonance associated with the *j*th replay angle at the *m*th recording wavelength is given by

$$\eta_{mj} \equiv \frac{1}{c_{Sj}}\frac{\kappa_{mj}^2}{\displaystyle\sum_{k=1}^{N}\frac{\kappa_{mk}^2}{c_{Sk}}}\tanh^2\left\{d\sqrt{-\frac{1}{c_R}\sum_{k=1}^{N}\frac{\kappa_{mk}^2}{c_{Sk}}}\right\} \tag{12.118}$$

Again, the total diffractive response at the *m*th wavelength is then simply

$$\eta_m \equiv \sum_{j=1}^{N}\eta_{mj} = \tanh^2\left\{d\sqrt{\frac{1}{\cos\Phi_c}\sum_{k=1}^{N}\frac{\kappa_{mk}^2}{\cos\Phi_{ik}}}\right\} \tag{12.119}$$

In the limit that $N \to \infty$ this then leads to formulae for the diffractive efficiency of the polychromatic hologram

$$\eta_m(\Phi_c,\Phi_i) = \frac{\kappa_m^2(\Phi_i)}{L_m\cos\Phi_i}\tanh^2\left\{d\sqrt{\frac{L_m}{\cos\Phi_c}}\right\}$$

$$\eta_m = \frac{1}{\Delta\Phi}\int\frac{\kappa_m^2(\Phi')}{L_m\cos\Phi'}\tanh^2\left\{d\sqrt{\frac{L_m}{\cos\Phi_c}}\right\}d\Phi' = \tanh^2\left\{d\sqrt{\frac{L_m}{\cos\Phi_c}}\right\} \tag{12.120}$$

where

$$L_m = \frac{1}{\Delta\Phi}\int\frac{\kappa_m^2(\Phi)}{\cos\Phi}d\Phi \tag{12.121}$$

and where Φ is the replay image angle and $\Delta\Phi$ is the total reconstructed image angle range. If we assume that we record the hologram such that

$$\kappa_m^2(\Phi) \rightarrow \kappa_m^2\cos\Phi \tag{12.122}$$

then this reduces to the simpler form

$$\eta_m = \tanh^2\left\{\frac{d\kappa_m}{\sqrt{\cos\Phi_c}}\right\} \tag{12.123}$$

This tells us that a lossless polychromatic hologram is therefore theoretically capable of perfect diffractive replay at each and every wavelength. In other words, if it is co-illuminated by P co-propagating reference plane waves, each of a different wavelength, then a hologram is capable of producing a perfect diffractive response to each of these P waves simultaneously. However, the all-important parameter here is κ_m—the effective index modulation achievable with a given material at the mth wavelength. If this is low, then a larger grating thickness will be needed to achieve a bright hologram. The PSM model can be extended to model lossy holograms by consideration of a complex index. In this case, the strategy of simply making a hologram thicker to produce a brighter diffractive response does not work because the further the reference wave penetrates into the hologram, the more it is absorbed. This leads to the lossy polychromatic hologram being characterised by a maximum attainable diffractive response, which is innately dependent on the achievable modulation in a given material.

Another insight to take away from Equation 12.120 is that if we record a spatially multiplexed grating that has a diffuse object beam spanning a finite but small angle range, the usual formula for a single polychromatic grating applies but now with an effective modulation and an average replay angle. To see this, we make the approximation

$$L_m = \frac{1}{\Delta\Phi}\int\frac{\kappa_m^2(\Phi_i)}{\cos\Phi_i}d\Phi_i \sim \frac{1}{\langle\cos\Phi_i\rangle\Delta\Phi}\int\kappa_m^2(\Phi)d\Phi \sim \frac{1}{\cos\langle\Phi_i\rangle}\langle\kappa_m^2(\Phi)\rangle \tag{12.124}$$

which then leads to a total diffraction efficiency at the mth wavelength of

$$\eta_m = \tanh^2\left\{d\langle\kappa_m^2(\Phi)\rangle^{1/2}\sqrt{\sec\Phi_c\sec\langle\Phi_i\rangle}\right\} \tag{12.125}$$

Figure 12.16 shows graphically the results of Equation 12.123 for several typical three-colour recording parameter sets. In Figure 12.16a we assume an index modulation of $n_1 = 0.01$ for each wavelength. This is typical of a Silver Halide emulsion such as PFG-03CN which has a thickness of between 9 and 10 μm. We can therefore expect a diffraction efficiency for each colour in the region of 25% for reflection holograms made in this type of material. In Figure 12.16b we use an index modulation of 0.02 for each chromatic component. This is more characteristic of a modern photopolymer. For a 12 μm emulsion we would therefore expect a diffraction efficiency for each colour component of around 70%. Clearly these numbers are for the lossless case and loss can be expected to diminish the actual efficiencies somewhat.

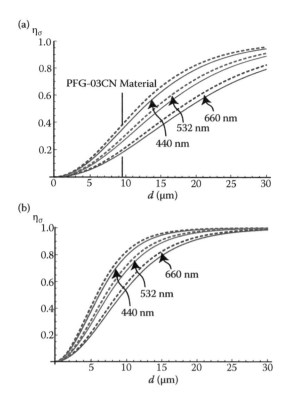

FIGURE 12.16 Diffractive efficiency versus grating thickness for typical lossless three-colour reflection volume phase holograms according to the N-PSM theory (σ-polarisation). All holograms are recorded at 660 nm, 532 nm and 440 nm. Dashed lines indicate a 45 degree reference illumination beam (in air). Solid lines indicate the case of normal incidence illumination. (a) $n_1(660) = n_1(532) = n_1(440) = 0.01$ and (b) $n_1(660) = n_1(532) = n_1(440) = 0.02$. Average index of grating assumed to be $n_0 = 1.5$.

12.11 Coupled Wave Theory, PSM and the Rigorous Coupled Wave Theory

In Section 12.10 we have generalized the PSM model to describe the spatially multiplexed volume phase grating. Just as the simple PSM model provides a simple and intuitive method of understanding and calculating diffraction in simple volume phase gratings, so the N-PSM model builds on this simplicity to construct an intuitive model of diffraction capable of accurately describing spatially multiplexed volume phase gratings and colour reflection holograms.

The basic idea behind PSM is that all volume phase gratings, and *in particular* reflection phase gratings, can be described very well by the process of Fresnel reflection of the illuminating wave from the fringe planes. To understand why this should be so is simple. The rigorous coupled wave theory of Moharam and Gaylord [11], which is reviewed in Appendix 8, demonstrates that almost always reflection gratings are dominated by the first order "+1" diffractive mode. But this is precisely the mode which the process of Fresnel reflection describes. The basic PSM equations are essentially rigorous for the un-slanted reflection grating and here Fresnel reflection describes the grating perfectly. PSM simply assumes that Fresnel reflection continues to operate as the sole diffractive process even when the fringe planes are tilted. Since only the "+1" mode actually is of any importance this is a rather good approximation. Snell's law also helps in practice as this acts to steepen all incidence angles and lessen grating slants. In a tilted grating at Bragg resonance the Fresnel description is most accurate at Bragg resonance. Here it is hardly surprising that Kogelnik's theory [1] agrees identically to PSM. In fact Kogelnik assumes that the signal ray propagates according to the formula

$$\mathbf{k}_i = \mathbf{k}_c + \mathbf{K} \tag{12.126}$$

where $\mathbf{k}_i$ represents the wavevector of the diffracted signal ray, $\mathbf{k}_c$ is the illuminating wavevector and $\mathbf{K}$ is the grating vector.[*] PSM, on the other hand assumes that

$$\mathbf{k}_i = \mathbf{k}_c - \frac{2\mathbf{K} \cdot \mathbf{k}_c}{|\mathbf{K}|^2}\mathbf{K} \tag{12.127}$$

Finally we show in Appendix 8 that rigorous coupled wave theory implies that actually the "+1" mode must propagate according to the law

$$\mathbf{k}_i = (k_x + K_x)\hat{\mathbf{x}} - \sqrt{\beta^2 - (k_x + K_x)^2}\,\hat{\mathbf{y}} \tag{12.128}$$

All these expressions are the same at Bragg resonance but all three differ away from resonance.

In Appendix 8 we compare the N-PSM theory with rigorous coupled wave theory for a variety of simple monochromatic and polychromatic spatially multiplexed gratings. Here we see that N-PSM provides a good description of the spatially multiplexed grating—and for precisely the same reasons we have discussed above. When rigorous coupled wave theory is applied to the spatially multiplexed grating once again only first order "+1" type modes turn out to be of importance. Cross-modes which are driven by several gratings at once are nearly always extremely small unless index modulations are extremely high. This then means that grating cross-coupling can be effectively ignored and a single signal wave can be attributed to a single grating in the multiplex.

12.12 Lippmann Photography

The PSM theory can be used to analyse Lippmann photography in the limit of a large focal length Lippmann camera. Here, we can approximate the Lippmann photograph as a normal-incidence broadband polychromatic reflection grating. As such, we can use Equation 12.27 to model playback. Of course, Lippmann photographs in reality are rather different from holograms and gratings because each of the many object rays in a finite focal length Lippmann camera act as reference beams to all other object beams. As such, the grating structure is rather more complex than the standard hologram, where it is usual to make the approximation that the individual object rays only interfere with the reference wave. This evident complexity is neglected in the present analysis. Nevertheless, the analysis is still instructive as it serves to clearly demonstrate the concept of interferometric spectral recording.

We start by assuming that the Lippmann photograph in the limit of infinite focal length recording can be described by the following index distribution:

$$n = n_0 + \frac{1}{2(\hat{\beta}_1 - \hat{\beta}_0)}\int_{\hat{\beta}_0}^{\hat{\beta}_1} \tilde{n}(\hat{\beta})\left\{e^{2i\hat{\beta}y} + e^{-2i\hat{\beta}y}\right\}d\hat{\beta} \tag{12.129}$$

where

$$\hat{\beta} \equiv \frac{2\pi n_0}{\lambda_r} = \alpha\beta \tag{12.130}$$

We imagine that this distribution has been created by a polychromatic wave of the form

$$E(y) = \int_{\hat{\beta}_0}^{\hat{\beta}_1} E_0(\hat{\beta})e^{i\hat{\beta}y}d\hat{\beta} \tag{12.131}$$

[*] Note that we have used a different sign convention for the grating vector, $\mathbf{K}$ and the grating vector, $\mathbf{G}$ used elsewhere.

The spectral information in the wave $E_o(\hat{\beta})$ has therefore been transferred to the polychromatic grating in the form of the index spectrum $\tilde{n}(\hat{\beta})$. We can now generalise Equation 12.27 to write

$$\frac{dR}{dy} = -\frac{iS}{\beta(\hat{\beta}_1 - \hat{\beta}_0)} \int_{\hat{\beta}_0}^{\hat{\beta}_1} \hat{\beta}\kappa(\hat{\beta}) e^{2i(\hat{\beta}-\beta)y} \, d\hat{\beta}$$

$$\frac{dS}{dy} = \frac{iR}{\beta(\hat{\beta}_1 - \hat{\beta}_0)} \int_{\hat{\beta}_0}^{\hat{\beta}_1} \hat{\beta}\kappa(\hat{\beta}) e^{-2i(\hat{\beta}-\beta)y} \, d\hat{\beta}$$

(12.132)

where

$$\kappa(\hat{\beta}) = \frac{\beta\tilde{n}(\hat{\beta})}{2n_0}$$

(12.133)

If the index modulation spectrum is flat between the start and end wavelength, that is, $\kappa(\alpha) = \kappa_1$, then these equations can be solved numerically using a Runge–Kutta integration subject to the boundary conditions

$$R(0) = 1$$
$$S(d) = 0$$

(12.134)

We shall also consider an index profile which possesses m peaks between the two limits, $\hat{\beta}_0$ and $\hat{\beta}_1$, where

$$\kappa(\hat{\beta}) = \kappa_1 \sin^2\left\{ m\pi \frac{(\hat{\beta} - \hat{\beta}_0)}{\hat{\beta}_1 - \hat{\beta}_0} \right\}$$

(12.135)

The integrals in Equation 12.129 can then be calculated analytically and the resulting equations solved using a numerical Runge–Kutta method. Figure 12.17 shows four plots of the index distribution, $\kappa(\lambda)$, and the associated spectrum on replay, $\eta(\lambda) \equiv S(y = 0, \lambda) \, S^*(y = 0, \lambda)$. If we assume that the index modulation on recording is proportional to the square of the electric field, then $\kappa(\lambda)$ represents the recording spectrum and $\eta(\lambda)$, the reconstruction spectrum. The first diagram (a) illustrates the case of a boxcar spectrum. Here, the response is roughly flat in the middle of the spectrum, but there are two peaks at each extremity. The blue peak is rather bigger as might be anticipated by realising that the blue radiation will "feel" a deeper grating due to its smaller wavelength. The remaining three diagrams (b–d) show the cases of $m = 5$, 6 and 7 in Equation 12.135. These cases pertain to a 2 µm-thick grating with a value of $\kappa_1 = n_1\pi/\lambda = 0.03\pi/\lambda$. As can be seen, the spectrum is reproduced well until the case of $m = 7$. Here, the grating becomes too thin to reconstruct properly and the spectral peaks are averaged. If one increases the thickness to 5 µm, then the case of Figure 12.17d becomes almost identical to that of Figure 12.17c with the spectral peaks being well reproduced. If even thicker emulsions are used, cases of $m = 30$ and beyond lead to good reconstruction spectra.

There are two other points to note from Figure 12.17. The first is that all the graphs show a larger diffractive response in the blue spectrum than in the red spectrum. Again, this can be understood by the ratio of the wavelength to grating thickness, which favours diffraction at shorter wavelengths. The second point is that some peaks are shifted slightly on reconstruction with respect to their maximum on recording. This phenomenon is due to interactions between the different fundamental gratings.

The PSM equations (Equation 12.132) are an approximation to the exact equations (Equation 12.6). However, numerical integration of Equation 12.6 using the index profile in Equation 12.129 produces substantially identical results to the above analysis. This is because, even in the fairly extreme case of a Lippmann photograph, the approximation (Equation 12.10) is still valid. As we have seen, Equation 12.6

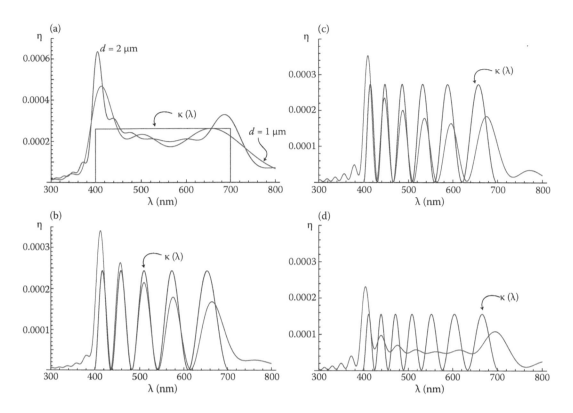

FIGURE 12.17 PSM spectral calculations of the Lippmann photograph in the limit of long focal length. (a) A box-car recording spectrum and the diffractive response of 1 μm-thick and 2 μm-thick Lippmann photographs. (b) Different recording and (c) replay spectra for a 2-μm-thick Lippmann emulsion ($n_1 = 0.03$, $n_0 = 1.5$ for all graphs).

constitutes a differential representation of the chain matrix method [6,7]. In 1991, Nareid and Pedersen [12] used the chain matrix method to analyse Lippmann photographs and compared their numerical solutions to a solution of the Helmholtz equation in the first Born approximation. In a recent publication on Lippmann photography, Kellerbauer [13] explains that when viewing a Lippmann photograph, there is a slight distortion of the recorded spectrum in which the most strongly pronounced parts of the original spectrum are additionally amplified.

12.13 Discussion

The main result of this chapter is that a general volume holographic grating may be conveniently and accurately described as an infinite set of infinitesimal parallel-stacked mirrors. At each of these mirrors, the classical laws of Fresnel reflection and transmission can be applied. In this way, a reference wave illuminating the grating gives rise to an infinite sum of secondary waves. By adding the waves together in a consistent manner, the electric field distribution within the grating and the diffractive response of the grating can be simply calculated.

Summing the waves from finite dielectric stacks is of course nothing new. For instance, Rouard [14] described this in 1937. In addition, the numerical chain matrix method [6,7] has been used with great success to calculate the optical properties of stratified media. What is important here, however, is that by using a *differential formulism*, the picture of parallel-stacked mirrors operating under Fresnel's laws offers a completely alternative description of the general holographic grating. Not only do we find that the PSM model is capable of largely reproducing, with very good accuracy, the results of Kogelnik's coupled wave theory through an abundance of analytical expressions for the diffractive efficiency, but the model is also naturally capable of treating multicolour holographic gratings, spatially multiplexed gratings and even full-colour holograms!

REFERENCES

1. H. Kogelnik, "Coupled wave theory for thick hologram gratings," *Bell Syst. Tech. J.* **48**, 2909–2947 (1969).
2. D. Brotherton-Ratcliffe, "A treatment of the general volume holographic grating as an array of parallel stacked mirrors," *J. Mod. Optic.* **59**, 1113–1132 (2012).
3. D. Brotherton-Ratcliffe, "Analytical treatment of the polychromatic spatially multiplexed volume holographic grating," *Appl. Opt.* **51**, 7188–7199 (2012).
4. J. E. Ludman, "Approximate bandwidth and diffraction efficiency in thick holograms," *Am. J. Phys.* **50**, 244 (1982).
5. A. Heifetz, J. T. Shen and M. S. Shariar, "A simple method for Bragg diffraction in volume holographic gratings," *Am. J. Phys.* **77**, 623–628 (2009).
6. F. Abeles, "Recherches sur la propagation des ondes électromagnétiques sinusoïdales dans les milieux stratifiés. Application aux couches minces," *Ann. Phys. (Paris)* **5**, 596–640 (1950).
7. O. S. Heavens, "Optical properties of thin films," Reports on Progress in Physics, Vol XXIII, (1960) p. 1.
8. R. Guenther, *Modern Optics*, John Wiley and Sons, Hoboken, NJ (1990).
9. M. G. Moharam and T. K. Gaylord, "Chain-matrix analysis of arbitrary-thickness dielectric reflection gratings," *J. Opt. Soc. Am.* **72**, 187–190 (1982).
10. D. W. Diehl and N. George, "Analysis of multitone holographic interference filters by use of a sparse Hill matrix method," *Appl. Opt.* **43**, 88–96 (2004).
11. M. G. Moharam and T. K. Gaylord, "Rigorous coupled wave analysis of planar grating diffraction," *J. Opt. Soc. Am.* **71**, 811–818 (1981).
12. H. Nareid and H. M. Pedersen "Modeling of the Lippmann color process," *J. Opt. Soc. Am.* **8**, 257–265 (1991).
13. A. Kellerbauer, "Farbbilder aus gefrorenem Licht," *Phys. Unserer Zeit* **41**, 16–22 (2010).
14. M. P. Rouard, "Etudes des propriétés optiques des lames métalliques très minces," *Ann. Phys. (Paris)* Ser. II **7**, 291–384 (1937).

13

Illumination of Colour Holograms

13.1 Introduction

In this chapter, we discuss how holograms can be displayed. We shall pay particular attention to colour holograms. Often, the term *reconstruction* is used to describe the display of a hologram. This refers to the fact that if the reference beam, which is used to record the hologram, is used to also illuminate the processed hologram plate, the wavefront emitted from the object during recording will be faithfully *reconstructed*, generating the holographic image. This underlines the critical fact that to obtain a faithful holographic reconstruction, the properties of the reference recording and reconstruction beams must usually be identical. Indeed, it is well established in the literature that proper illuminating sources must be used to display holograms; this was already stressed as early as 1971 by Collier et al. [1].

The illumination of holograms is nevertheless a topic that is seldom included in books on holography; but it is as important as the techniques to record holograms. Remember that holography is a *two-part process* in which the holographic plate, with its recorded interference pattern, constitutes only one of the two parts needed to create the holographic image. The second part is the reference light, which is used to both record and replay the hologram. The characteristics of this reference light fundamentally control the display process. Only if the properties of the reference and replay lights are identical (spatial coherence, ray divergence, angle of incidence and wavelength) will a distortion-free, correct holographic image be generated.

13.1.1 Chromatic and Source-Size Blurring

In most practical situations, it is rare that the replay light is *absolutely* identical to the recording light. In particular, conventional broadband white-light sources are frequently used to illuminate reflection holograms. Here, one uses the wavelength-selective properties of the reflection hologram to filter out light having a different wavelength from that of the recording light. However, this comes at a price, as we have studied in Chapter 11. Illumination by a broadband source will bring with it chromatic blurring and, in general, the clear image depth of the hologram will be compromised.

One of the most important properties of any light source used to illuminate a hologram is spatial coherence. When a hologram is recorded, the reference laser light emanates from a spatial filter having a diameter of between 10 and 25 µm. This is several orders of magnitude smaller than the source size of current popular illumination sources. For example, halogen spotlights commonly used to display reflection holograms have source sizes from one to several centimetres. As we discussed in Chapter 11, a finite spatial coherence* in the reconstruction source leads to source-size blurring (Figures 13.1 and 13.2), which limits the clear image depth of the hologram. If our aim is to be free of such source-size blurring, a rough rule that can be applied is that the reconstruction source size must be less than 1 mm in diameter for every 1 m of diagonal distance that separates the illumination source and any point on the hologram. This ensures that any residual source-size blurring will be below the perception level of the standard human observer—as the average human eye can resolve image details of approximately 1 mm at a distance of 1 m.

---

* Strictly speaking, if the illumination source is temporally coherent, one should talk about the étendue of the source rather than its spatial coherence.

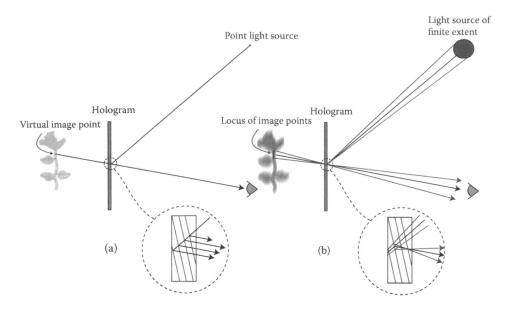

FIGURE 13.1 Source-size blurring in a reflection hologram causes the reflection of rays from the grating planes emanating from different parts of the source. (a) The case for a point source. Here, a single ray from the source is reflected from the grating to create a unique virtual image point. (b) The corresponding case for a source of finite size. Now, there exist various rays connecting the original point on the hologram to the light source. Each of these rays is reflected by the same grating at slightly different angles (angle of incidence = angle of reflection). This causes a blurred image.

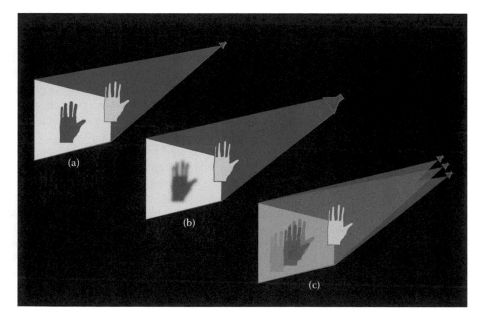

FIGURE 13.2 Layperson's diagram showing the importance of using a point source light to illuminate a hologram. Here, we make an analogy to the shadow cast by a spotlight on a screen. (a) The spotlight emits light from a small point, resulting in a sharp image. (b) A large-area diffuse spotlight results in a diffuse image. (c) Many spotlights illuminate the hologram, resulting in multiple images.

It is interesting to note that Lippmann photographs present a very different behaviour from holograms with regard to the illuminating source size. Here, one needs a fundamentally large and diffuse white-light source for their illumination (Figure 13.3).

The bandwidth of the light source also exerts a fundamental influence on the displayed image quality. Reflection holograms act as discriminating filters, but this intrinsic Bragg filter is usually rather broad. Certainly, for popular photosensitive materials, it is narrow enough to effectively stop any cross-talk between the primary colours of a three-colour reflection hologram, but it is very unusual for it to be narrow enough to annul chromatic blurring. For most commercial silver halide materials, the full-width half-maximum (FWHM) of the Bragg filter is approximately 15 to 25 nm. When a broadband source such as a halogen spotlight is used to illuminate a hologram, it is this filter that determines the clear depth of the hologram if source-size blurring is less than chromatic blurring. As we discussed in Chapter 11, a rough rule is that to be visually free of chromatic blurring, the bandwidth of the illuminating source must be less than 1 to 2 nm. As we shall see in later sections, a new paradigm is likely to operate in the future concerning the illumination of ultra-realistic reflection holograms such as high virtual volume

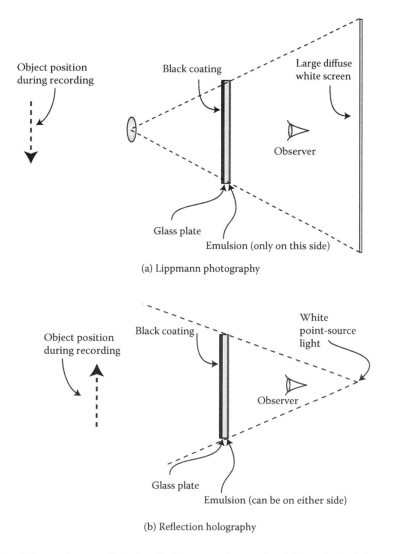

(a) Lippmann photography

(b) Reflection holography

FIGURE 13.3 The difference between displaying (a) a Lippmann photograph and (b) a reflection hologram, showing that a large-area diffuse light is needed to display the Lippmann photograph, but the light source for a hologram has to come from a point source which can be located at infinity.

(HVV) displays. Currently, if source-size blurring is controlled, it is the hologram, through its wavelength discriminating Bragg grating, which determines the clear image depth. However, with new laser diode illumination sources, which have a bandwidth of 1 to 2 nm, it is now the intrinsic properties of the source and not the hologram that are likely to determine the clear image depth.

13.1.2 Geometry Matching

The illumination reference source may be different from the recording reference source in its ray geometry. For example, it is often the case that holograms are recorded with collimated reference beams, but it is rather rare that a hologram is illuminated with such a beam. More likely, an approximation is sought to the original recording geometry and a point illumination source is used at some distance from the final hologram. For small holograms, this is usually a workable solution because the difference in ray angles is rather small if the light is placed at a distance of, for instance, five to ten times the diameter of the hologram. However, for larger displays, it clearly becomes difficult to place the light at a great distance from the hologram. Take, for example, a 1 m × 1 m reflection hologram recorded with a collimated reference beam. Ideally, one would like to situate the light at a diagonal distance from the hologram centre of at least 5 m, but this is often just not practical. As we have discussed in Chapter 7, the issue is with variable reference beam systems in digital printers. Here, it is not much more difficult to implement the proper recording reference beam geometry such that the hologram can be replayed accurately with a close point source. This is also the *great advantage* of the direct-write digital technique. On the other hand, for certain analogue holograms, this is definitely not the case: for example a 1 m × 1 m analogue H_2 reflection hologram would require a very expensive off-axis parabolic mirror to produce the correct reference beam.

If the illuminating reference beam differs substantially from the recording reference beam, aberration is introduced into the hologram. Such aberration is both chromatic and geometric in that it distorts the colours and the geometry of the recorded image. The human observer can usually tolerate a fair amount of geometric aberration without this creating too much of a problem. However, this does depend on the image in question. For example, an average observer is fairly resilient to aberration in holographic por-

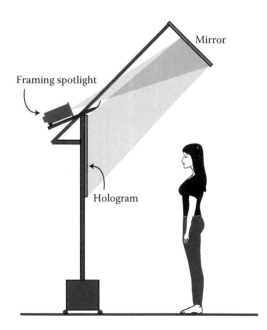

FIGURE 13.4 For large holograms which have been recorded with a collimated reference beam, there is often no other choice than to use a folded illumination beam path to attain a greater distance from the light source to the hologram. The diagram shows a large reflection hologram illuminated by a high-power framing spotlight.

traits, but where a hologram contains the image of a geometrical shape, such as a sphere, the eye will very quickly perceive the replayed image as an ellipsoid.

It is important to realise that induced aberration is always worse at the sides of a hologram. It is usually a fairly simple matter to match the ray directions in the centre by simply locating the spotlight at the correct position and angle. The induced aberration due to reference beam disparity can nonetheless have serious consequences regarding the entire image. If the light source is too close on replay, the image projection will be pushed out and magnified and the image behind the plate will be squashed. Large-depth holograms will therefore exhibit unnaturally small depths, and large-projection holograms will exhibit anomalous projections that can potentially make the image undecipherable. Colour holograms are particularly susceptible to reference beam disparity because the induced chromatic aberration means that colours will be altered with the observation location. Although this may be regarded as acceptable in a monochromatic reflection hologram, this is almost always not the case with a full-colour hologram.

Matching the ray geometry on recording and replay should therefore be taken extremely seriously; even more so for full-colour reflection holograms. For digital holograms, the best solution is to use a printer fitted with a variable reference beam system. For small holograms, either analogue or digital, a distant light provides a workable solution. For large analogue holograms, there are two solutions. The first is to use relay mirrors on replay to attain a larger distance from the light to the hologram (Figures 13.4 and 14.27). The second is to record the holograms in smaller panels (each with its own reference beam) and to use individual framing spotlights to illuminate each panel. We shall discuss this second option a little later in the context of HVV displays.

13.1.3 Illumination of HPO Holograms

Horizontal-parallax only (HPO) holograms may be illuminated by more than one spotlight as illustrated in Figures 13.5 and 13.6. Because these displays do not encode parallax in the vertical dimension, arrays of vertically stacked spotlights may be used to increase both the brightness of the holographic image and its vertical field of view without the introduction of blurring. In such an arrangement, each of the

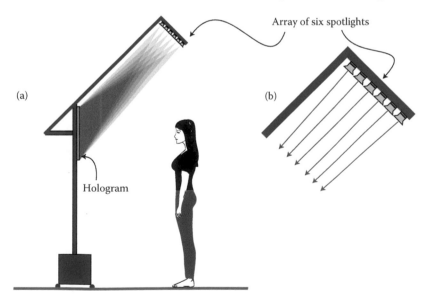

FIGURE 13.5 HPO reflection and rainbow transmission holograms can be illuminated by vertical stacks of broadband spotlights. Each of the lamps is directed towards the centre of the hologram. The centre lamp is usually arranged such that it corresponds to the reference beam geometry at recording. The lamps to either side therefore have an angle of incidence either above or below the recording angle. The Bragg condition is fulfilled for these lamps at a slightly different wavelength leading to a desaturation of the image. There is no induced geometrical distortion as the parallax information is exclusively in the horizontal dimension. (a) Display of HPO hologram with vertically stacked spotlights. (b) Detail drawing of the spotlights.

FIGURE 13.6 DWDH hologram and mounting stand made by Geola illuminated by a spotlight array comprising six individual 75 W MR16 halogen lamps.

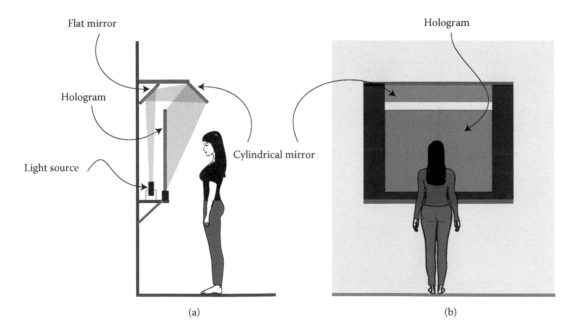

FIGURE 13.7 Compact illumination of an HPO hologram using a cylindrical mirror. Often, the angle of incidence of recording and illumination is a little larger than usual such that the mirror width can be minimised. Seen from (a) side and from (b) behind the viewer.

spotlights is directed towards the centre of the hologram. The central lamp is usually arranged such that it corresponds to the reference beam geometry at recording. The lamps to either side therefore possess an angle of incidence either above or below the recording angle. The Bragg condition is fulfilled for these lamps at a slightly different wavelength than the recording wavelength leading to a desaturation of the image and a greater vertical range of angles in which the hologram is visible. There is no induced geometrical distortion, as the parallax information exists exclusively in the horizontal dimension. In addition to arrays of lights, one-dimensional extended sources may also be used. The technique is applicable to all HPO holograms, whether of the transmission or of the reflective variety, and is extremely effective. One must be careful, however, not to use light of too small a bandwidth. Arrays of laser diodes, for example, will not work. Arrays of light-emitting diodes (LEDs) can be used, but the array length will be limited by their smaller inherent bandwidth.

Another technique that can be successfully applied to HPO holograms is the use of a cylindrical mirror. This is a version of the relay mirror technique in which the beam is collimated only in the horizontal direction, thus perfectly preserving the parallax information. The technique can be used to produce relatively compact displays (Figure 13.7).

13.1.4 Illumination of Large Rainbow and Rainbow-Achromatic Displays

Large rainbow-type holograms are particularly sensitive to the illumination geometry. As far as source-size blurring goes, they are no different. However, chromatic blurring operates in a rather different way in these holograms. The rainbow technique, if applied properly, allows large depth images to be attained by encoding the parallax information in the horizontal dimension and simply letting the vertical dimension chromatically disperse. This strategy works very well if there is a collimated reference beam on replay; however, if this is not present, then chromatic dispersion also occurs in the horizontal dimension, as we underlined in Chapter 11. Of course, large-depth HPO reflection holograms also suffer from chromatic blurring in the horizontal dimension under such circumstances—but there is very little Bragg wavelength discrimination in the transmission hologram, and this can make it more sensitive than the reflection hologram. However, more importantly, with the reflection hologram, there is always the option (at least in principle) to match the recording and replay reference beam geometries. The clear image depth (assuming a point source) is then determined by either Bragg wavelength discrimination or the bandwidth of the illuminating source. With rainbow-type holograms, however, one is forced to replay with a collimated beam—as only with a collimated beam will chromatic dispersion operate exclusively in the vertical direction. Transmission rainbow holograms are then, by their nature, rather more sensitive to the constraint of a properly collimated replay reference beam.

13.2 Illumination of Holograms by Laser Sources

In the early days of display holography, lasers were often used to illuminate monochrome transmission holograms. This resulted in spectacular images characterised by large clear image depths; in no uncertain way, this illumination technique led in large part to the creation of the public fascination for holography. The deep and detailed images available with laser illumination are due to the high temporal and spatial coherence characteristic of the laser source. The helium–neon laser was often used to display small or mid-size holograms. In early public exhibitions, other more powerful continuous wave (CW) lasers, such as argon–ion lasers, krypton–ion lasers, metal–vapour lasers and even tunable dye lasers were also employed. Soon enough, however, laser safety regulations were introduced, making it difficult for high-power lasers to be used for the display of holograms to the public.

Today, there are many new types of small, powerful and rather inexpensive semiconductor and solid-state lasers in the market that can be used to illuminate holograms. Provided that laser safety regulations can be observed, these sources promise to quite simply create a revolution in the illumination of display holograms. The main problem with laser safety is not the output power level itself but the spatial coherence (i.e., the fact that the light is emitted from a point source)—as it is this property of the light that determines the size of the focussed spot at the retina. Of course, in one sense, a very high spatial coherence is desirable because one eliminates the most ubiquitous source of blurring in any hologram—source-size blurring. However, one only needs to get below a source size of approximately 1 mm per 1 m distance from lamp to hologram for source-size blurring to become imperceptible to the average human observer. As such, most lasers have an effective source size of several orders of magnitude smaller than required. This leads to an irradiance at the human eye, in the case of direct ocular exposure, of some hundreds of thousands of times greater than that needed.

Using lasers to display reflection holograms can completely circumvent chromatic blurring if the bandwidth of the laser is chosen to be less than approximately 1 to 2 nm. However, if the laser radiation is too (temporally) coherent, the problem of laser speckle arises. This is caused by the intrinsically coherent nature of laser light and results from interference at the surface of the retina. Generally, this becomes a problem at a bandwidth smaller than several nanometres [2]. Lee et al. [3] have reported that the appearance of speckle also depends on the brightness of the image, the illumination level of the room, the viewing distance and whether the image is moving or stationary. Furthermore, the perception of speckle depends on the wavelength, which can be explained by the sensitivity and the resolution of the eye: speckle is less visible in blue, than in the green and red spectra. The contrast in the image also influences the perception of speckle: in contrast-rich images, a given speckle contrast is less annoying than in a uniform image.

In addition to wavelength diversity, there are additional techniques that may be used to efficiently reduce speckle noise. The introduction into the laser beam of a moving diffuser is perhaps the best-known solution. This creates a dynamic ray angle diversity that averages out the speckle noise. For example, the company Optotune, Dietikon, Switzerland [4] now offers a unique commercial product based on this principle. Electroactive polymers (so-called artificial muscles) are used to laterally oscillate a small diffuser at high frequency, averaging out the speckle. The device is extremely compact, completely free of mechanics and has low power consumption. Optotune's laser speckle reducer (LSR) can be customised in terms of size, frequency, coatings and diffuser structure (Figures 13.8 and 13.9). The LSR consists of a diffuser bonded onto a polymer membrane that includes four independent dielectric elastomer actuators. When activated, the surface area of the electrodes increases and causes a motion of the rigid diffuser in the membrane plane. Four independent electrodes are used to obtain displacement of the diffuser in both lateral directions. The moving frequency is optimal when the mechanical resonance frequency of the system is attained and this provides the highest speckle reduction.

As well as reducing speckle, LSRs such as Optotune's reduce the spatial coherence of the source. For example, the LSR-3005 increases the effective source size at a 10° divergence angle to just under 5 mm. This corresponds to a fairly ideal value of étendue for a hologram illumination source, given the dual concerns of source-size blurring and laser eye safety.

FIGURE 13.8   A Laser speckle reducer from Optotune. (Photo courtesy of Optotune.)

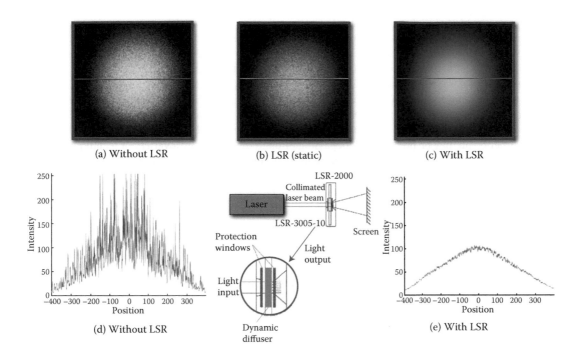

FIGURE 13.9 Optotune laser speckle reduction system. (a) Typical laser speckle patterns without LSR, (b) with LSR in passive mode and (c) with LSR switched on. Intensity profiles from an LSR-3005-10 device (5 mm aperture, 10° diffusion angle, 20 μm speckle; d and e). The schematic in the centre of the diagram illustrates how the LSR is used.

13.2.1 Importance of Wavelength Matching

An inconvenience of illuminating a reflection hologram with a laser source is that the laser wavelengths on recording and replay must usually be matched quite precisely. The precision of matching depends on the level of induced aberration that can be tolerated in a given image and also on the brightness reduction due to non-compliance with Bragg resonance. With some of the more modern photosensitive materials, very high index modulation can be achieved (see Chapter 4) and, as a result, high diffraction efficiency is possible from a thin emulsion with a very broad Bragg filter. With digital holography techniques such as DWDH, this opens up the possibility of actually accepting a small mismatch between the wavelengths at recording and replay but correcting for the induced aberration by image predistortion. Because the Bragg grating is broad in this case, its only function is then to avoid chromatic cross-talk, and as long as the difference in wavelengths is not so great, the diffraction efficiency of the hologram will be largely unaffected. Analogue holography does not allow for such a possibility and here one really is obliged to rigorously match the wavelengths at recording and replay.

Another point to mention about laser illumination is the deformation of the emulsion and the change of its refractive index on processing. Again, to maintain freedom from aberration and a maximum diffractive response, both the refractive index and the emulsion thickness should be invariant between the phases of recording and playback. Once again, digital holography does have the innate flexibility of image predistortion, which can be used to compensate for unavoidable changes in index and thickness. (See Appendix 4.)

13.2.2 Illumination of Full-Colour Transmission Holograms

Full-colour transmission holograms may be illuminated effectively by several lasers if each laser illuminates the hologram at a substantially different angle. Transmission holograms are generally more discriminating in reference beam angle than reflection holograms as we saw in Chapter 11. This makes them more suitable for angle multiplexing just as the reflection hologram, having greater wavelength discrimination, is more suitable for wavelength multiplexing. The three different reference beams required for the illumination of a trichromatic transmission hologram can be organised at different altitudinal angles or different azimuthal angles depending on the amount of Bragg angle discrimination available from a given photosensitive material. The trichromatic or polychromatic transmission hologram is generally more difficult to illuminate than the corresponding reflection hologram. However, in certain circumstances, these holograms may offer several advantages. For example, the illuminating laser sources are always situated behind the hologram and so they are essentially hidden from view. In addition, the zeroth order beam may be prevented from passing through the display by a variety of techniques such as the inclusion of an additional reflective grating in front of the transmission grating. Trichromatic digital HVV displays have the advantage that they can be integrated into false walls, giving the impression that the display is completely self-illuminating.

Finally, we should mention that laser illumination is not suitable for rainbow-type displays. These types of displays require a broadband illumination.

13.2.3 Gas Lasers

In Chapter 3, different types of gas lasers, such as helium–neon lasers, argon–ion, krypton–ion lasers and metal-vapour lasers were described. With the possible exception of the He–Ne laser, and the copper-vapour laser these lasers are almost never used today to display holograms. This is because there are now new laser sources available that are much smaller and much cheaper. These are the semiconductor and solid-state lasers which we shall discuss below.

13.2.4 Semiconductor Laser Diodes and Solid-State CW Lasers

High-power semiconductor laser diodes are now becoming available at many wavelengths. The typical bandwidths of these ultracompact sources are in the order of several nanometres, making such diodes ideal for the illumination of holograms. In addition, ultracompact diode-pumped solid-state (DPSS) lasers, typically based on Nd:YVO$_4$ crystals with frequency doubling using potassium dihydrogen phosphate, lithium triborate or beta barium borate (see, for example, Figure 13.10) are now available, providing watt level TEM$_{00}$ emissions in red, green and blue. Light from these miniature lasers may be combined into a compact red, green, and blue (RGB) source far more easily than the lambertian light produced by LEDs. With prices at the time of writing ranging from $50 to several hundred dollars per unit (depending on output power), compact laser sources are now a definite option for hologram installations requiring the best quality of illumination.

FIGURE 13.10   Example of a DPSS laser pointer producing watt level TEM$_{00}$ emission at 457 nm from an ultracompact low-power package (beam diameter, 5 mm; divergence, 1.5 mRad). (Photograph courtesy of Wicked Lasers.)

Solid-state lighting based on lasers offers significant advantages for illuminating holograms—most notably, the complete absence of source-size and chromatic blurring, leading to the real possibility of essentially infinite clear-image depths. However, it has also been suggested by Neumann et al. [5] that white light produced by a set of lasers of different colours could well be suitable for general illumination. In this recent study, an RYGB (red–yellow–green–blue) white laser light source was used to illuminate various objects and an experimental survey was conducted which compared people's perception of the laser-illuminated objects with that of the same objects illuminated by other sources. The source was composed of the following four discrete laser lines with the following specifications:

- Red: 635 nm, 800 mW maximum power
- Yellow: 589 nm, 500 mW maximum power
- Green: 532 nm, 300 mW maximum power
- Blue: 457 nm, 300 mW maximum power

The study concluded that the four-colour white laser illuminant studied was *virtually indistinguishable* from high-quality state-of-the-art white reference illuminants. The major perceived advantage of a laser source is, of course, that it is likely to be rather more efficient than current LED sources. This is good news for holography, as we may realistically expect that a strong research effort directed at laser illumination sources for general illumination will also inevitably generate a much larger range of laser light sources suitable for holographic illumination. Another area in which RGB laser sources are being developed and which may be expected to drive the laser lighting industry is the field of laser projection for large-screen television and cinema [2].

13.3 Non-Laser Light Sources Used for Hologram Illumination

Today, most holograms are displayed by non-laser sources. This is both because of cost—the non-laser sources are usually substantially cheaper—but also because there are still relatively few groups capable today of making the type of large-depth holograms that can really benefit from laser illumination. The narrow-beam halogen lamp has historically been the most popular source for display holography and is still commonly selected for exhibitions. However, today there are new types of non-laser light sources that are more suitable for illuminating holograms. First among these sources and by far the most promising is the LED. It is interesting to note that modern LEDs can use as little as 15% of the energy required by incandescent lights (see Table 13.1). At this point in time, solid-state lighting based on blue-emitting InGaN LEDs and phosphors has demonstrated the highest luminous efficacy of any white-light source (265 lm/W), although this is only for small current densities. One of the key features of LED lighting devices is that many may last 10 to 20 years before needing replacement. The long LED lifetime is very important when illuminating holograms—remember that a hologram without illumination means no visible image!

Issues that are all too-often overlooked when selecting a light source to illuminate a hologram are

- Electrical efficiency of the source
- Generation of heat and infrared radiation

TABLE 13.1

Efficacy and Lifetime of Different Light Sources

Light Source	Efficacy (lm/W)	Lifetime (h)
Incandescent lamp	10–18	1500
Halogen lamp	12–24	2000
Compact fluorescent lamp (CLP)	40–50	5000
LED	70–90	40,000

- Temperature of source and associated fire risk
- Requirement for ventilation
- Generation of UV radiation

New light sources such as LEDs often have a positive influence on many of these aspects.

13.3.1 Halogen Lights

Halogen lights have, for a long time, been the light source of choice used to illuminate both monochrome and colour reflection holograms. The halogen lamp contains a small quantity of an active halogen gas such as bromine. The inert gas suppresses the evaporation of the tungsten filament, while the halogen gas acts to reduce the amount of tungsten that plates the interior wall of the lamp. The halogen gas reacts with the tungsten that has evaporated, migrated outward and has been deposited on the lamp wall. When the lamp wall temperature is sufficient, the halogen reacts with this tungsten to form tungsten bromide, which is then freed from the wall of the lamp and migrates back to the filament. The tungsten bromide compound reacts at the filament of the lamp where temperatures close to 2500°C cause the tungsten and halogen to dissociate. The tungsten is deposited on the filament and the cycle repeats. Unfortunately, the tungsten is not always deposited in the same zone at which evaporation initially took place and so the filament becomes thinner in places and eventually fails.

The light output of a halogen lamp is more stable than a non-halogen gas lamp due to the cleaning action of the halogen gas on the lamp envelope. This feature, coupled with the high colour temperature of the light and long-life, makes these lamps very desirable for many industrial and scientific applications.

Halogen lamps with aluminium reflectors, dichroic reflectors or focussed ellipsoidal dichroic reflectors are suitable for illuminating holograms. Most often narrow-beam 20, 50 and 75 W, 12 V MR11 or MR16 lamps are used. A typical 50 W lamp is shown in Figure 13.11a, and its spectrum is illustrated in Figure 13.11b.

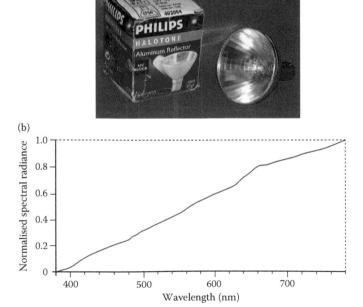

FIGURE 13.11   (a) A typical 50 W halogen lamp suitable for hologram illumination. (b) Spectrum of the emitted light.

13.3.2 Special Lamps

International Light Technologies, Peabody, MA [6] have many special types of lamps suitable for hologram illumination. Miniature lamps have advantages and should be considered for those applications requiring higher light output and better spatial coherence. Reflectors and lenses may be incorporated to shape the emitted light distribution.

For the illumination of larger holograms, the Electronic Theatre Controls Source Four Junior Zoom (also known unofficially as Source 4 or S4), which is an ellipsoidal reflector spotlight used in stage lighting, can be used quite effectively. This is an excellent, if bulky, source suitable for square-metre plus colour reflection holograms. It is a "framing projector" in the sense that the incorporated baffles and zoom may be used to arrange exclusive illumination of the rectangle of the hologram, eliminating any overspill of light. First released in 1992, the Source 4 features an improved lamp and reflector compared with previous ellipsoidal reflector spotlight designs, tool-free lamp adjustment, and a rotating, interchangeable shutter barrel. It also uses a faceted dichroic borosilicate reflector behind the lamp. The proprietary high-performance lamp uses four compact filaments and an ellipsoidal reflector system. It is rated at 575 W but produces light equivalent to a 1000 W spotlight using older technology. High-performance lamps are available at two colour temperatures: 3250 K (300–400 h lifetime) and 3050 K (1500–2000 h lifetime).

13.3.3 Mercury Lamps

As an alternative to lasers, in the early days of holography, short-arc mercury-vapour lamps were used to display off-axis monochrome transmission holograms. Combined with corresponding interference filters, the green (546.1 nm) or yellow–orange (578.2 nm) lines could be used (Figure 1.11, which shows a mercury lamp display cabinet). There are other similar lamps, such as metal-halide lamps, which use various compounds in an amalgam with the mercury in addition to lamps based on sodium iodide and scandium iodide. All these lamps can potentially be used for the illumination of holograms, but all suffer from severe problems. In addition, the use of mercury-vapour lamps for lighting purposes will be banned in the European Union in 2015 (the United States banned them in 2008).

13.3.4 Arc Lamps

Miniature arc lamps such as the Sōlarc lamps (Figure 13.12) available from Welch Allyn, Skaneateles Falls, NY [7] constitute an alternative to the common halogen lamp. The Sōlarc lamp is a metal halide light source in the class of high-pressure, high-intensity-discharge lights, which differs in a fundamental way from halogen, incandescent, fluorescent or LED illumination sources. Light is emitted from an arc discharge between two closely spaced electrodes, hermetically sealed inside a small quartz glass envelope. During operation, small amounts of metal are heated to a liquid state that provides the needed vapour to create a desired light colour. The light emitted from this arc tube is intense and generates more

FIGURE 13.12 The Sōlarc lamp from Welch Allyn. (a) Packaging, (b) lamp and (c) close-up showing arc gap.

than 60 lm/W. Sōlarc lamps feature a small, typically 1.2 mm arc gap, the smallest gap available in a metal halide arc lamp at the time of writing. Combined with elliptical reflectors, this arc gap allows one to focus illumination with laser-like precision into very small areas. The low-voltage (9–16 V) miniature arc lamp emits more blue light than halogen lamps. The colour temperature is between 6500 and 13,700 K. It has a very short arc and is equipped with a parabolic reflector, providing a narrow beam. The small intrinsic source size and a well-designed reflector translate into a somewhat sharper holographic image. The drawback is the rather short lamp life (350–700 h).

13.3.5 Plasma Lamps

Continuous full-spectrum lighting can be generated by a sulphur plasma light source. A typical emission spectrum from this type of illuminant is shown in Figure 13.13. The spectrum is very similar to the light from the sun. Plasma lamps use an electrode-less bulb design. Externally generated high-frequency electromagnetic energy is used to generate plasma within the bulb, converting the high-frequency energy into light in an efficient manner. The high-frequency energy (microwaves) is generated by a magnetron, which is powered from an adjustable stabilised power source. A prototype sulphur plasma lamp is shown in Figure 13.14. This unit, although bulky, can emit light with better spatial coherence than is normally

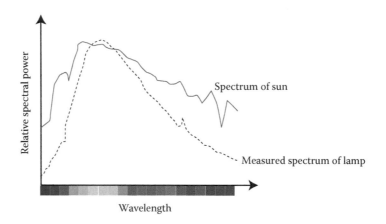

FIGURE 13.13 Typical sulphur plasma lamp spectrum.

FIGURE 13.14 A prototype sulphur plasma lamp.

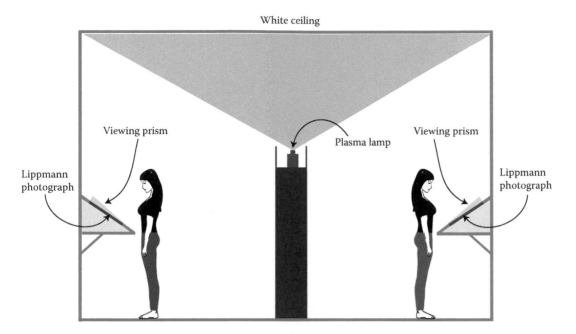

FIGURE 13.15 Display of Lippmann photographs using a sulphur plasma lamp. Most Lippmann photographs are recorded in sunlight and, as such, sulphur lamps, with their solar-like spectrum, provide an ideal illumination source, assuring optimal reconstruction of colours.

the case from many other commercial lighting sources. The colour rendering index (CRI) of the source is between 98 (60 lm/W) and 80 (140 lm/W).

One manufacturer of such light systems is Plasma International Lighting Systems [8] in Germany which markets two variations of the plasma lamp: the Standard Sulphur lamp (SS0) and the Triple A class Solar Simulator lamp (AAA). The lamp power can be set by software to a desired level between 500 and 1300 W, and the light spectrum remains almost unchanged between 600 and 1300 W power setting.

This type of light is not really merited (principally due it its high cost, bulky size and its broadband spectrum) for displays of reflection holograms, but it is an excellent (if expensive) source for large-format rainbow holograms. Its most interesting application is, however, for displaying Lippmann photographs. If the light source is made to illuminate the white ceiling in a gallery, many Lippmann photographs can be displayed in a truly excellent fashion according to the arrangement shown in Figure 13.15.

13.3.6 LEDs

The past few decades have seen a continuing and rapidly developing race among manufacturers of LEDs to produce ever cheaper and ever more efficient illuminants. This process is now producing a fundamental transformation in the field of general lighting. These miniature semiconductor devices will undoubtedly lead to the obsolescence of the common incandescent light bulb in the near future.

The rapid progress in solid-state LED lighting has opened up new possibilities to illuminate colour reflection holograms. A significant advantage of LEDs is that they possess a much smaller bandwidth than broadband white-light sources. Although typical bandwidths are much larger than those commonly associated with lasers and laser diodes, LED light sources should nevertheless be matched to the recording laser wavelengths (or vice versa). This guarantees that only the white light from the LED source (which is a mixture of the primary LED wavelengths) contributes to creating the holographic image. Using a halogen spotlight, a large part of the light spectrum emitted illuminates the surface of the plate without having any effect on the intensity of the image. Instead, this light is scattered, lowering the image contrast. The lack of this scattered light in LED illumination can lead to significantly higher image fidelity. In addition, LED light sources have considerable advantages over halogen and other traditional lighting sources, such as

- Long life (20,000 to 100,000+ h)
- Small size
- Small étendue
- High durability and robustness to thermal and vibration shocks
- Low energy usage/high energy efficiency
- No infrared or UV in beam output
- Directional light output
- Digital dynamic colour control—white point tunable
- Relatively low cost

13.3.6.1 Theory of LED Operation

LEDs are semiconductors that convert electrical energy into light energy. The colour of the emitted light depends on the semiconductor material and on its composition. LEDs are formed from various doped semiconductor materials in the form of a PN junction. When an electrical current passes through the junction in the forward direction, the electrical carriers give up energy in the form of photons at a level proportional to the forward voltage drop across the diode junction. The amount of energy is relatively low for infrared or red LEDs. Although LEDs are semiconductors and need a minimum voltage to operate, they are still diodes and need to be operated in a current mode. The band-gap of the semiconductor determines the wavelength of the emitted light. Shorter wavelengths equate to greater energy and therefore higher band-gap materials emit at shorter wavelengths. High-efficiency LEDs can be produced in many wavelength ranges. Each material technology has a peak efficiency within the operational wavelength range, but it is difficult to make high-efficiency LEDs that operate at the edge of the material technology.

The semiconductor material typically takes the form of a very small chip or die, which is mounted onto a lead frame and encapsulated in a clear or diffused epoxy. The shape of the epoxy and the amount of diffusing material in the epoxy control the light output angle of emission. The construction of a common LED package is illustrated in Figure 13.16.

The output of high-power LEDs is typically expressed either in terms of luminous flux, which is measured in lumens, or in terms of radiant flux in watts. Alternatively, it is sometimes expressed as luminous

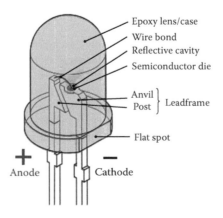

FIGURE 13.16   Common LED package.

intensity (in candela) or radiant intensity (in watts per steradian). If specified as an intensity, the value is usually measured along the projection axis of the device. The viewing angle for LEDs is specified as the included angle between the half-intensity points on either side of the output beam (Figure 13.17).

The wavelength of light emitted is determined by the difference in energy between the recombining electron-hole pair of the valence and conduction bands. The approximate energies of the carriers correspond to the upper energy level of the valence band and the lowest energy of the conduction band because of the tendency of the electrons and holes to equilibrate at these levels. Consequently, the wavelength, λ, of an emitted photon can be approximated by the following expression:

$$\lambda = \frac{c}{h} E_{bg} \qquad (13.1)$$

where h represents Planck's constant, c is the velocity of light, and E_{bg} is the band-gap energy. To change the wavelength of emitted radiation, the band-gap of the semiconducting material used to fabricate the LED must be changed. Gallium arsenide is a common diode material and may be used as an example, illustrating the manner in which a semiconductor's band structure can be altered to vary the emission wavelength of the device. Gallium arsenide has a band-gap of approximately 1.4 eV, and emits in the infrared at a wavelength of 900 nm. To increase the frequency of emission into the visible red region (~650 nm), the band-gap must be increased to approximately 1.9 eV. This can be achieved by mixing gallium arsenide with a compatible material having a larger band-gap. Gallium phosphide, having a band-gap of 2.3 eV, is a candidate for this mixture. LEDs produced with the compound GaAsP (gallium arsenide phosphide) can be customised to produce band-gaps of any value between 1.4 and 2.3 eV, through adjustment of the content of arsenic to phosphorus. A major development occurred in the late 1980s, when LED designers borrowed techniques from the rapidly progressing laser diode industry, leading to the production of high-brightness visible light diodes based on the indium–gallium–aluminium–phosphide (AlGaInP) system. Again, this material allows changes in the emission colour by adjustment of the band-gap. The same production techniques can be used to produce red, orange, yellow, and green LEDs. More recently, blue LEDs have been developed based on gallium nitride and silicon carbide materials. One of the most important aspects of a blue LED is that it completes the RGB primary colour family to provide an additional mechanism of producing solid-state white light, through the mixing of these component colours. In particular, the indium gallium nitride (InGaN) system is the leading candidate for the production of blue LEDs and is also a primary material in the development of white LED lights.

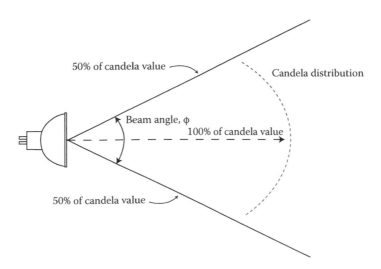

FIGURE 13.17 When luminous or radiant intensity is used to specify the emission characteristic of an illuminant, the candela value on-axis is quoted. The beams angle is then twice the angle from the axis to the half-intensity line.

13.3.6.2 Blue to Green LEDs: 450 to 530 nm

The material for this wavelength range of products is InGaN. Most large suppliers concentrate on creating blue (450–475 nm) LEDs for making white light with phosphors, and green LEDs that fall into the 520 to 530 nm range for green traffic signal lighting. Rapid advancements and improvements in efficiency are noted in the blue wavelength range especially as the race to create brighter white illumination LEDs continues.

13.3.6.3 Yellow–Green to Red LEDs: 565 to 645 nm

Aluminium indium gallium phosphide (AlInGaP) is the semiconductor material used for this wavelength range. It is predominately used for amber traffic signals (590 nm) and red (625 nm) lighting. The lime green (or yellowish–green 565 nm) and orange (605 nm) are also available from this technology, but they are somewhat limited. The technology is rapidly advancing, for the red wavelength in particular, because of the growing commercial interest in making red–green–blue white LED lights.

13.3.6.4 LEDs and Hologram Illumination—Wavelength Matching

In applications in which LEDs are used for illuminating colour holograms, the peak wavelengths can be as important—or even more important—than the output in lumens or candela. Many of the current photosensitive materials have a relatively important Bragg wavelength discrimination, meaning that if the wavelengths of the LED and the recording lasers are not matched fairly well, then the hologram will suffer from poor diffraction efficiency. The matching problem is certainly much less severe than for laser sources (the comments we made in Section 13.2.1 are valid here too), but nonetheless, it is still an important constraint that is just not present when a hologram is illuminated by a broadband source.

13.3.6.5 White Phosphor LEDs

Most white-light diodes use a semiconductor chip emitting at a short wavelength (blue, violet or ultraviolet) and a wavelength converter (usually a yellow phosphor) that absorbs light from the diode and produces secondary emission at a longer wavelength. Such diodes, therefore, emit light at two or more wavelengths; combined, these appear as white light. The phosphors typically used are composed of an inorganic host substance containing an optically active dopant. Yttrium aluminium garnet is a common host material, and for diode applications, it is usually doped with one of the rare earth elements or a rare earth compound. Cerium is a common dopant element in yttrium aluminium garnet phosphors designed for white LEDs.

Viewed directly, the LED will appear to be white as the blue and yellow wavelengths are mixed together. This product is ideal for general lighting. However, for illumination of holograms in which colour rendering is important, this type of LED is not the most suitable. Visible LED wavelength bands are listed in Table 13.2 with their corresponding general applications.

TABLE 13.2

Visible LED Wavelength Bands and Applications

Wavelengths (nm)	Main Application
430–470 (Blue)	White LEDs using phosphor, blue for RGB white lights
520–530 (Green)	Green traffic signal lights, green for RGB white
580–590 (Amber)	Amber traffic signal lights, amber for RGBA white lights
630–640 (Red)	Red signal lights, red for RGB white lights

13.3.6.6 Multicolour LEDs

By mixing together a variety of semiconductor, metal and gas compounds, the following list of LEDs can be produced.

- Gallium arsenide (GaAs)—infrared
- Gallium arsenide phosphide (GaAsP)—red to infrared, orange
- Aluminium gallium arsenide phosphide (AlGaAsP)—high-brightness red, orange–red, orange, and yellow
- Gallium phosphide (GaP)—red, yellow and green
- Aluminium gallium phosphide (AlGaP)—green
- Gallium nitride (GaN)—green, emerald green
- Gallium indium nitride (GaInN)—near ultraviolet, bluish-green and blue
- Silicon carbide (SiC)—blue as a substrate
- Zinc selenide (ZnSe)—blue
- Aluminium gallium nitride (AlGaN)—ultraviolet

The most popular type of tricolour LED comprises a single red and a single green LED combined in one package with their cathode terminals connected together, producing a three-terminal device. They are called tricolour LEDs because they can give out a single red or a green colour by turning "on" only one LED at a time. They can also generate additional shades of colours (the third colour) such as orange or yellow by turning "on" the two LEDs in different ratios of forward current.

13.3.6.7 LED Colour Rendering

Colour rendering is very important when LED lights are used in museum illumination of paintings and other works of art. Poor colour rendering is often evident when a hologram is illuminated by white LEDs. This is usually because the green and red components are too weak. With phosphor-pumped white LEDs, there is no smooth output-versus-wavelength behaviour, and this leads to errors in the colours perceived by the eye. Here, one uses the colour rendering index (CRI) as a quantitative measure of the ability of a light source to reproduce the colours of various objects in comparison with a natural light source. Getting the correct CRI is very important for such applications as photography, cinema, television and holography. The CRI on many high-quality lamps is between 85 and 90. This translates into the human eye actually seeing the true colour of the object being lit. The CRI of most traditional light sources can be as low as 25 to 50, with high-pressure sodium lamps being the worst with a CRI of approximately 25. In Table 13.3, the CRI for different LEDs show the high value of the trichromatic LED.

People who spend a lot of time in museums and galleries have become very accustomed to viewing art under halogen lights. Halogens have a colour temperature somewhere between 2700 and 3000 K, but this technology definitely adds warmth (yellow) to the light. With LEDs, control over the colour temperature of the light becomes much more precise. LEDs hold their colour temperature even though the lamps can be dimmed to approximately 5% of their light output. Traditional lighting in museums and galleries has often appeared rather dim because of the need to reduce damage to the art over time

TABLE 13.3

LED Efficiency and CRI

LED type	Luminous Efficiency (lm/W)	General CRI
Dichromatic LED	33.6	10
Broadened output dichromatic LED	30.6	26
Trichromatic LED	28.3	60
Phosphor-based LED	28.0	57

owing to the UV and heat generated by halogen lamps. With LEDs, it is now possible to light such fragile art a bit brighter owing to the fact that LEDs do not emit UV and will not cause heat damage to the art. This is also important when illuminating holograms. UV light can cause printout in silver halide holograms. And heat, over an extended period, can create emulsion shrinkage affecting the colour of the holographic image.

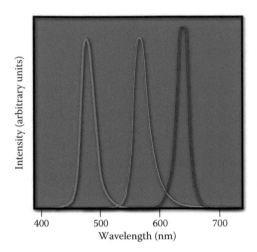

FIGURE 13.18 RGB LEDs combined technology: combined spectral curves for blue, yellow–green and high-brightness red solid-state semiconductor LEDs. The FWHM spectral bandwidth is approximately 24 to 27 nm for all three colours.

TABLE 13.4

Current LEDs Suitable for Holographic Illumination

	Power	Bandwidth (FWHM)	Dominant Wavelength	Emission Area
Red Diode				
Luxeon Rebel	720 mW	30 nm	655 nm	1.4 mm × 1.4 mm
Luxeon Rebel	102 lm	30 nm	627 nm	1.4 mm × 1.4 mm
Luminous SBT 16	155–300 lm	18 nm	623 nm	1.6 mm × 1.0 mm
Luminous PT39	390–750 lm	18 nm	623 nm	2.09 mm × 1.87 mm
Luminous PT54	600–1050 lm	19 nm	623 nm	2.7 mm × 2.0 mm
Luminous PT120	1360–2225 lm	19 nm	623 nm	4.6 mm × 2.6 mm
Green Diode				
Luxeon Rebel	161 lm	20 nm	530 nm	1.4 mm × 1.4 mm
Luminous SBT 16	315–600 lm	40 nm	525 nm	1.6 mm × 1.0 mm
Luminous PT39	815–1650 lm	36 nm	525 nm	2.09 mm × 1.87 mm
Luminous PT54	1275–2150 lm	36 nm	525 nm	2.7 mm × 2.0 mm
Luminous PT120	2700–4300 lm	36 nm	525 nm	4.6 mm × 2.6 mm
Blue Diode				
Luxeon Rebel	70 lm	20 nm	470 nm	1.4 mm × 1.4 mm
Luxeon Rebel	1120 mW	20 nm	448 nm	1.4 mm × 1.4 mm
Luminous SBT 16	60–110 lm	20 nm	460 nm	1.6 mm × 1.0 mm
Luminous PT39	1.5–2.75 W	20/25 nm	460/462 nm	2.09 mm × 1.87 mm
Luminous PT54	275–480 lm	20/25 nm	460/462 nm	2.7 mm × 2.0 mm
Luminous PT120	550–970 lm	20 nm	460/462 nm	4.6 mm × 2.6 mm

13.3.6.8 Preferred LEDs for Colour Holography

An alternative white LED technology to the phosphor-pumped LEDs is offered by RGB or RGBA LEDs. These combine red, green and blue or red, green, blue and amber chips onto one discrete package allowing the generation of white light or any of 256 colours by utilising circuitry that drives the three diodes independently. In applications requiring a full spectrum of colours from a single point source, this type of RGB diode format is the preferred type and is the most suitable for illuminating colour holograms. Currently, one of the best methods for the generation of white light is the "RGB LEDs combined" technology—that is, the generation of white light using a combination of red, green and blue. Note, however, that this form of white light relies on the electrical control of three LED chips. A spectrum from a typical RGB combined LED device is illustrated in Figure 13.18. Typical FWHM spectral widths are 24 to 27 nm.

When different-colour LED lights were first marketed in 2003, the OptiLED chip, a commercial LED spotlight, was introduced to illuminate monochrome reflection holograms. The 4° red (627 nm, linewidth 20 nm), green (530 nm, linewidth 35 nm) and amber (590 nm, linewidth 14 nm) versions provided suitable illumination for such holograms. The 2.5 W LED spotlight had an electric circuit allowing them to be operated at 90 to 240 V AC. Different interface types were offered: E27, E14, B22 and MR16, for example. These new LED lights were promoted and distributed by Laser Reflections in San Francisco, CA [9,10]. A list of some current LEDs suitable for holographic illumination is given in Table 13.4.

13.3.6.9 Lenses for LED Spotlights

It is important when illuminating holograms to be able to adjust the divergence of the illuminating source such that the illuminated area is best matched to the hologram size. Optotune has developed a new type of adaptive lens for LED lights [4], which is shown in Figure 13.19. The adaptive lens is a liquid polymer in a flexible container. The focal length of the lens can be controlled by electrically changing the thickness of the lens. The design includes an LED, a secondary optic and an adaptive condenser lens for the tuning of the spotlight. The secondary optic (usually a total internal reflection or TIR lens) defines the maximum angle of the spotlight. The secondary optic may have a diffuse layer on top, preventing the display of the lens or LED structure, but this may not always be warranted because it can increase the étendue of the source, thereby increasing source-size blurring in the hologram. The adaptive condenser lens is added as a tertiary optic. By tuning the adaptive condenser lens, the divergence of the spotlight can be changed.

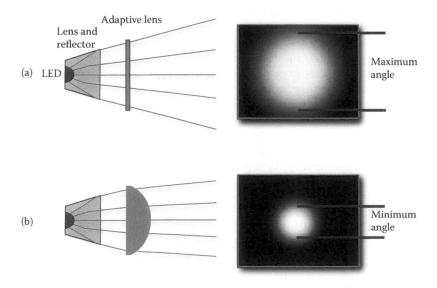

FIGURE 13.19 Adaptive lighting principle using a tunable condenser lens.

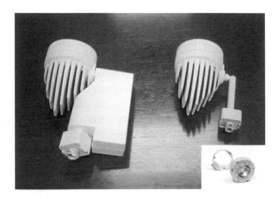

FIGURE 13.20   The Westport track lights based on Bridgelux/Molex Helieon modules.

Another company that offers special lenses for LED lights is Aether Systems Inc., Taipei, Taiwan [11]. Aether Systems designed and produced a digital lens for LEDs. The Aether Digital Lens is a high-performance lens for LEDs and produces excellent light shape.

The concept of focussing the light from one or more LEDs with a lens can actually be taken a few steps further. In Appendix 6, we summarise a design study of an RGB LED illuminant of high spatial coherence that uses a specially moulded aspheric lens of high index to achieve an almost perfect irradiance distribution at the hologram surface (see, for example, Figure A6.11). The technique allows an effectively top hat distribution of irradiance to be created at a defined distance from the illuminant with a defined boundary. For example, an LED illuminant of high spatial coherence can be constructed to accurately illuminate a tilted rectangular hologram with very little overspill of the light. To design such lens systems, one must use a computational procedure that optimises a desired boundary, a flat irradiance distribution and the final spatial coherence of the source. The problem is complicated by its non-linear nature, but the results are quite impressive. Clearly, this is a technique that can be applied to laser illumination as well. In fact, as will be clear from Appendix 6, rather more accurate framing of the hologram may be expected in this case.

13.3.6.10 White LED Spotlights

Not all white LEDs are suitable for the illumination of holograms. In particular, we have already discussed white phosphor-type LED spotlights. These are not really suitable for illuminating colour holograms. The spectrum of these devices is broadband with normally a low-intensity red region. The manufacture of these white LEDs involves coating the LEDs of one colour (mostly blue LEDs made of InGaN) with phosphor of different colours to form white light. If several phosphor layers of distinct colours are applied, the emitted spectrum is broadened. Due to the simplicity of manufacturing, the phosphor method is currently still the most popular method for making high-intensity white LEDs.

As we have already mentioned, LED spotlights suitable for illuminating colour reflection holograms should be constructed from separate RGB LED chips. This way of making white LED lamps is particularly interesting because of the inherent flexibility of colour mixing; the mechanism also has a higher quantum efficiency in producing white light. We shall review commercial spotlights based on this concept in Section 13.3.6.11.

There are, however, many white phosphor-type LED spotlights on the market that can be used for the illumination of monochromatic and rainbow holograms.* For example, the Bridgelux/Molex Helieon modules, Livermore, CA [12], which are compatible with Westport tracks, are shown in Figure 13.20. There are also modules already on the market that are based on the Zhaga standards [13].† One example

---

* One should nevertheless be mindful of the potential low red intensity of many of these illuminants.

† Zhaga is an industry-wide cooperation aimed at the development of standard specifications for the interfaces of LED light engines.

FIGURE 13.21   Lytespan Mini LED. (Photo courtesy of Philips Lightolier.)

FIGURE 13.22   Definity PAR38 LED lamp. (Photo courtesy of Lighting Science Group.)

is the socket-able LED light engine, which fits GE's Infusion module and Phillips' Fortimo spotlight module. Another example of a white phosphor-type LED spotlight is the Lytespan Mini LED manufactured by Philips Lightolier [14] (Figure 13.21), which offers excellent beam control. Interchangeable optics also provide flexibility to adjust the lighting. Finally, Lighting Science Group, Satellite Beach, FL [15] manufactures a new family of 18 to 24 W LED lamps, the Definity lamps, which are intensely bright (lumen output, 840–1460 lm) all in one package shown in Figure 13.22.

13.3.6.11 Special LED Spotlights for Colour Holograms

In Greece, the Hellenic Institute of Holography has developed a special LED spotlight to illuminate colour holograms. The HoLoFoS LED spotlight, based on Cree LEDs, is manufactured at AutoTech, Athens, Greece [16] and is commercialised by TAURUS SecureSolutions Ltd., Athens, Greece [17]. Through proper choice of the component LEDs in terms of bandwidth and wavelength, the HoLoFos LED spotlight is capable of achieving high-quality reproduction of deep full-colour reflection holograms.

The device consists of an illuminating head, extending arms and a mounting base. The illuminating head contains the RGB LEDs, mixing optics, lenses and heat sinks. The system has an embedded microcontroller for intensity control of each LED with DMX protocol decoding and a miniature wireless receiver. Remote adjustment of the colour mixing by DMX protocol communication is achieved by a handheld wireless remote control. A small switching power supply provides the power needed for EU or US mains.

The optics incorporated in the unit provide for an axial mixing of the LED beams resulting in a homogeneous colour mixing over the full extent of the projected beam. The small footprint of the LED die (~2 mm) is small enough to produce clear and deep holograms even at small illuminating distances.

The illuminating head can be fitted with a variety of LEDs at selected wavelengths and more than three different LEDs can be fitted to match various recording wavelengths. For example, an RRGGBB configuration can be achieved. This is important for colour holograms that will be recorded with four or five laser wavelengths to obtain more or less perfect colour rendering.

The current prototype unit uses three LEDs with the following spectral characteristics for the red, green and blue LEDs, which correspond to the lasers of the Z3 Holographic Camera (see Chapter 14).

- Red 620–630 nm
- Green 520–535 nm
- Blue 450–465 nm

The LED spectrum of the HoLoFoS illuminating system is shown in Figure 13.23 and the LED spotlight with its colour control box is shown in Figure 13.24.

To demonstrate the advantage of using this LED light for displaying colour holograms, the same hologram was illuminated with a conventional halogen spotlight and the new HoLoFoS LED light; the results are shown in Figure 13.25a and b. Note the increased contrast obtainable with the new LED light. This is also a good illustration of the vital importance of the illumination source in holography. Pulsed digital holograms, which are often printed using a red wavelength of 660 nm, require a version of the HoLoFoS light with a deeper red diode.

LED spotlights such as the HoLoFoS light can also be mounted in stacks for the illumination of HPO reflection holograms as described in Section 13.1.3.

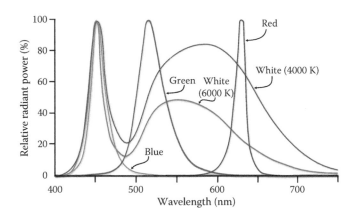

FIGURE 13.23   The LED spectrum of the HoLoFoS illuminating system.

FIGURE 13.24   The HoLoFoS LED spotlight for the illumination of colour holograms.

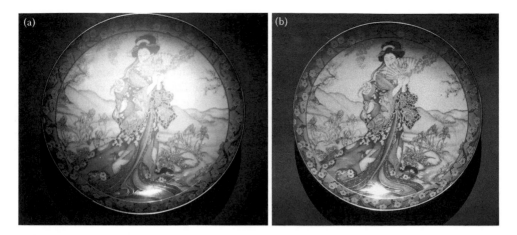

FIGURE 13.25 The same colour hologram illuminated (a) with halogen light and (b) LED light.

13.3.6.12 LED and Lamp Safety

It is important to be aware of the safety aspects of operating powerful light sources such as lasers and LEDs. If these sources are used for the public display of holograms, it is even more necessary to be aware of existing safety regulations. Laser safety regulations have existed for a long time, but because the new very bright LED lights also pose a risk hazard to people, new safety regulations for these products have been introduced. Initially, these standards were intended to be incorporated with the safety regulations for lasers, but the International Electrotechnical Commission (IEC) committee finally decided to remove LEDs from the scope of the IEC 60825-1:2007 laser safety standard. This was, in fact, an acknowledgement of the existing standard CIE S 009:2002, which addresses the photobiological safety of lamps and lamp systems, including LED sources. This standard was prepared by the International Commission on Illumination (CIE). The safe exposure limits in the CIE lamp safety standard are, like those in the IEC laser safety standard, based on the underpinning International Commission on Non-Ionizing Radiation Protection (ICNIRP) safety data and guidelines. Also, the measurement methods in CIE S 009:2002 are similar in format to those contained in the IEC laser safety standard. Unfortunately, the CIE standard does not provide detailed information on how to measure a source configured within an array, nor does it include the hazards posed by aided optical viewing, and there is also no information on product safety labelling requirements unlike the case for lasers.

Furthermore, the exposure measurement is undertaken at a fixed distance of 200 mm from the (apparent) source location and this approach may not be universally applicable to the evaluation of LED and LED array sources. The standard does, however, provide a clear delineation of the various ocular hazard bands and its methodology is directly applicable to broadband sources (such as a white HB-LED). To accommodate the removal of LED sources from IEC 60825-1, the IEC published a lamp and LED standard (IEC 62471:2006) in 2006, which was harmonised with (and indeed directly based on) the CIE S 009:2002 lamp safety standard. This relatively new IEC standard, which was released in 2008, is already undergoing revision to better reflect the needs of LED source hazard assessment and labelling. Although there is no specific information on labelling requirements, the lamp safety standard does contain its own risk classification scheme for potentially hazardous lamps, namely, exempt, low risk, moderate risk and high risk, which follows a similar reasoning with the IEC laser product classification. In summary, for the purposes of international trade of LED products, manufacturers and vendors of LED products should consider IEC 62471:2006 to be the currently applicable product standard for LED safety. Again, this is notwithstanding any specific national regulatory requirements or directives that refer to an alternative standard or assessment method such as the IEC 60825-1:2001 laser (and LED) safety standard.

Caution needs to be advised that a fixed assessment distance of 200 mm may not be sufficient to fully ascertain the maximum optical radiation hazard posed by the source and consideration might also need to be given to the effect of aided viewing (e.g., magnifiers and telescopes) on the hazard assessment.

13.4 Exhibition Facilities and Galleries Suitable for Displaying Holograms

A disturbing fact often associated with hologram exhibitions is the presence of many bright spotlights, as illustrated in Figure 13.26. If possible, one should try to avoid this situation when arranging hologram exhibitions. In addition, one should try to arrange exhibition facilities for holograms in such a way that it is easy to change and mount different holograms. The first task is to obtain a suitable type of spotlight, preferably with a lens system for the control of beam divergence. In addition, the spotlight tracks should be positioned perpendicularly to the walls (not parallel with) so that the illuminating angle of each spotlight can be easily adjusted. The tracks in existing galleries are often too close to the walls to be used for illuminating holograms properly.

In a specialised hologram gallery, the following design makes it possible to hide all the spotlights so the visitor cannot see them. The spotlights are arranged in the ceiling with a track system as mentioned previously. A suspended ceiling, underneath the ceiling on which the spotlights are mounted, should be

FIGURE 13.26   Many bright spotlights in the ceiling at a hologram exhibition can distract the viewer and detract from his or her viewing experience.

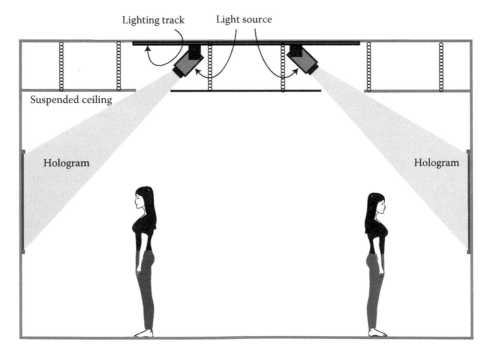

FIGURE 13.27   Layout of a hologram exhibition with hidden spotlights.

constructed with openings for the light to pass through it. This means that the light from the spotlights can pass through openings in the ceiling (a grid system) to illuminate the holograms without the spotlights being visible. The visitor is therefore not disturbed by many bright light sources in the ceiling and, in the best case, is not even aware that spotlights are used to illuminate the holograms on display. This makes a radical improvement to the exhibition (Figure 13.27).

Another way of hiding the spotlights is to arrange the holograms on 45° shelves along the walls with shielded spotlights mounted overhead on the ceiling (Figure 13.28). This type of installation is very suitable for the display of artefact holograms in museums. A further advantage of this arrangement is that both tall and short people (as well as children) can easily see the holograms by moving closer or further away from the hologram. A Swedish museum used this method of displaying an early artefact hologram, which was mentioned in Chapter 1 (Figure 1.31).

Horizontally mounted, overhead-illuminated holograms can be displayed on a table with a spotlight directly above them. This technique is used to display holographic maps, which we will describe in Chapter 14. An example of a horizontally mounted, overhead-illuminated hologram mounted as a tabletop and made by Laser Reflections [9] is shown in Figure 13.29.

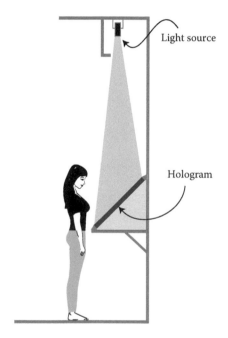

Light source

Hologram

FIGURE 13.28 Hologram on 45° shelf with spotlight mounted overhead.

FIGURE 13.29 Tabletop hologram by Laser Reflections.

When arranging exhibitions of display holograms, it is recommended that spotlights of the type used in theatres be utilised—so-called framing projectors. Such spotlights can project an adjustable frame around the hologram, which means that only the hologram area is illuminated with no illumination outside the hologram frame, eliminating any disturbing light on the wall. Figure 13.30 illustrates typical illumination of holograms using spotlights in which the walls are illuminated around the hologram frames. This is both visually disturbing and also reduces the contrast of the holographic image. Illumination of a large-format hologram using an Electronic Theatre Control Source 4 Junior Zoom is shown in Figure 13.31. As we mention in Appendix 6, LEDs in conjunction with high-index aspheric lenses can be expected to dramatically decrease the size and efficiency of framing projectors.

FIGURE 13.30   Spotlights illuminating the wall surrounding the holograms.

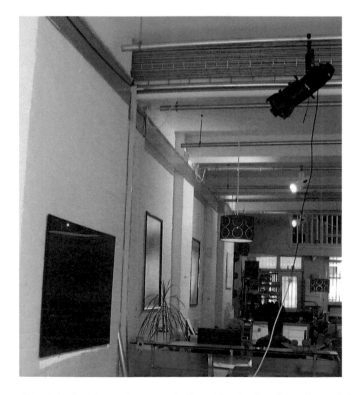

FIGURE 13.31   A spotlight of the framing projector-type is shown illuminating a large-format hologram.

13.5 Edge-Lit Holograms

Considering all the problems associated with the correct illumination of holograms, it would be rather nice if it were possible to integrate the light with the hologram to create an essentially self-luminous display. One way to do this is by introducing the illuminating light into the edge of the hologram plate. Such holograms are referred to as edge-illuminated or edge-lit holograms.

One type of edge-lit hologram is based on total internal reflection (TIR) and was described in 1968 by Nassenstein [18]. In the recording arrangement for this type of hologram, the incidence angle of the reference wave must be greater than the critical angle. In fact, the critical angle is not strictly defined here as there is weak light absorption by the photosensitive medium—this is the case of attenuated TIR. In 1969, Bryngdahl [19] described how TIR holographic recordings could be understood in terms of evanescent waves. Stetson [20] published the first experimental investigations of TIR gratings. Edge-lit TIR gratings and holograms can only be used for the recording of small-size plates—for example, holograms used in fingerprint detectors.

A lot of research has been undertaken to investigate how edge-lit holograms can be successfully recorded. Various solutions have been suggested, but so far, it has only been possible to apply these to displays for small holograms. Lin [21] was the first to introduce the concept of the edge-lit technique to display holography.

Edge-lit holograms not using the TIR principle are illuminated at a very steep angle by a light source positioned nearby. One solution here is to attach the holographic plate to a glass or plastic block. Light is introduced through the edge of the block and directed onto the holographic emulsion. There are two types of such holograms: one is the reflection edge-lit hologram and another is the transmission edge-lit hologram, as illustrated in Figure 13.32. Several variants are possible. For example, in a colour transmission display, the red, green and blue primary reference beams can be brought in from three sides of the block. Reflection holograms can also be treated in this way, although it is more usual to use a single white reference beam.

Holograms somewhat larger than TIR holograms have been displayed successfully in this fashion. However, no medium- or large-format displays have thus far been reported. From an absolute point of view, it should be possible to create large displays using this principle. However, the larger the display, the thicker the glass or plastic block must be. Because the edge of this block must be optically polished, the larger blocks are much more expensive. They are also heavy. This is certainly an area that needs more work. We should, in addition, point out that digital holograms may have some advantages here, as distortions in the reference beam (such as non-collimation and even astigmatism) may be corrected when the hologram is written using a variable reference beam system.

Unlike the standard types of transmission and reflection holograms, edge-lit holograms are not susceptible to inteference and image blurring from extraneous light in the viewing environment. This allows such holograms to be displayed in well-lit areas with little regard for other light sources that might otherwise affect the holographic image. However, the main advantage is the fact that the hologram forms an effectively compact self-luminous display with no zeroth order reference beam reaching the observer's side of the hologram. Over many years, research on edge-lit holograms has been undertaken by several groups [18–33]. The ability to create full colour edge-lit holograms has also been considered and this has been described by Ueda et al. [27].

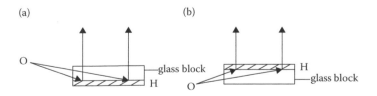

FIGURE 13.32 (a) Reflection and (b) transmission edge-lit holograms.

13.6 Illumination of Large Displays

Large displays pose particular problems for illumination. With digital holographic printing technologies improving all the time, it is now becoming realistically possible to print high-definition HVV displays. These types of displays require illumination sources of a small bandwidth and high spatial coherence if chromatic and source-size blurring are to be avoided. From a practical point of view, this means that HVV displays must be illuminated using laser technology.

Full-colour HVV displays can be made using reflection holograms or transmission holograms. Transmission holograms have the advantage that the zeroth order illumination beam may be suppressed by an additional grating and an in-wall installation arranged. The inconvenience with colour transmission holograms is that angle multiplexing is required and this does represent a significant complication. Reflection-type HVV displays are therefore likely to be the first type to appear commercially. However, unless upscaled edge-lit technologies are used, it is not really possible to suppress the zeroth order illumination beam here. This places a constraint on the maximum size that a single HVV reflection display can have due to laser safety concerns.

Essentially, an observer must be able to stand next to a laser-illuminated hologram and look directly into the laser illumination source without incurring eye damage. Key elements here will be limiting the spatial coherence of the RGB laser source (no doubt using an LSR) to no greater than that required to abolish source-size blurring, maximising the diffractive efficiency of the hologram and, if possible, restricting the field of view of the display. With these measures, quite large panels could be lit using compact RGB laser sources. Nevertheless, above a certain size (which will depend on the above points), too powerful a laser will be required for ocular safety. One particularly good solution is therefore to break-up the display into rectangular panels, each with their own illumination source (Figure 13.33). Of course, there can be very little overspill of light from one panel to the next as multiple images will then result. However, the spatially coherent light from an RGB laser source can be easily focussed into an almost perfect achromatic rectangle on the wall using only simple compact optics.

We are fundamentally used to making windows, mirrors and even pictures out of panels. There is therefore little problem in including thin borders between the holographic panels and this may conveniently cater to some overspill from the individual illuminating light sources if necessary. The one disadvantage of splitting an HVV display into panels is the relatively high precision (~0.1 mm per metre distance between light and hologram) that will be required in the alignment of the light sources such that the individual images all align!

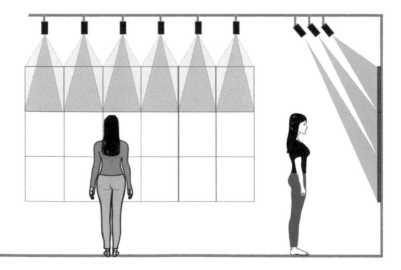

FIGURE 13.33 Illumination of HVV reflection hologram panels using individual RGB laser framing projectors for each panel.

REFERENCES

1. R. J. Collier, C. B. Burckhardt and L. H. Lin, *Optical Holography*, Academic Press, New York (1971) pp. 170–174.
2. P. Janssens and K. Malfait, "Future prospects of high-end laser projectors," in *Emerging Liquid Crystal Technologies IV*, L.-C. Chien and M. H. Wu eds., Proc. SPIE **7232**, OY-1–12 (2009).
3. Y. M. Lee, J. M. Park, S. Y. Park and S. G. Lee, "A study of the relationships between human perception and the physical phenomenon of speckle," SID Int. Symp. Digest of Tech. Papers P-45, 1347–1350 (2008).
4. Optotune AG, Switzerland; www.optotune.com (Sept. 2012).
5. A. Neumann, J. J. Wierer, Jr., W. Davis, Y. Ohno, S. R. J. Brueck and J. Y. Tsao, "Four-color laser white illuminant demonstrating high color-rendering quality," *Opt. Expr.* **19**, A982–990 (2011).
6. International Light Technologies, USA; www.intl-lighttech.com (Sept. 2012).
7. Welch Allyn, USA; www.welchallyn.com
8. Plasma International Lighting Systems, Germany; www.plasma-i.com (Sept. 2012).
9. Laser Reflections, USA; www.laserreflections.com (Sept. 2012).
10. H. A. Jones-Bey, "LEDs illuminate holographic displays," *Laser Focus World*, February 1, (2003).
11. Aether Systems Inc., Taiwan; www.aether-systems.com (Sept. 2012).
12. Molex Corporation, USA; www.molex.com (Sept. 2012).
13. The Zhaga Consortium, USA; www.zhagastandard.org (Sept. 2012).
14. Philips Lightolier, USA; www.lightolier.com (Sept. 2012).
15. Light Sciences Group, USA; www.lsgc.com (Sept. 2012).
16. AutoTech Wireless Automation, Greece; www.autotech.gr (Sept. 2012).
17. Taurus SecureSolutions Ltd. Greece; www.taurus.com.gr (Sept. 2012).
18. H. Nassenstein, "Holographie und Interferenzversuche mit inhomogenen Oberflächenwellen," *Phys. Lett. A* **28**, 249–251 (1968).
19. O. Bryngdahl, "Holography with evanescent waves," *J. Opt. Soc. Am.* **59**, 1645–1650 (1969).
20. K. A. Stetson, "Holography with total internally reflected light," *Appl. Phys. Lett.* **11**, 225–227 (1967).
21. L. H. Lin, "Edge-illuminated holograms," *J. Opt. Soc. Am.* **60**, 714A (1970).
22. S. Benton, S. Birner and A. Shirakura, "Edge-Lit Rainbow Holograms," in *Practical Holography IV*, T. H. Jeong, ed., Proc. SPIE **1212**, 149–157 (1990).
23. N. J. Phillips, C. Wang and T. E. Yeo, "Edge-illuminated holograms, evanescent waves and related optical phenomena," in *Int'l Symp on Display Holography*, T. H. Jeong, eds., Proc. SPIE **1600**, 18–25 (1991).
24. Q. Huang and H. J. Caulfield, "Edge-lit reflection holograms," in *Int'l Symp on Display Holography*, edited by Jeong, T. H., Proc. SPIE **1600**, 183–186 (1991).
25. J. Upatnieks, "Edge-illuminated holograms," *Appl. Opt.* **31**, 1048–1052 (1992).
26. N. J. Phillips, C. Wang and Z. Coleman, "Holograms in the edge-illuminated geometry—new materials developments," in *Practical Holography VII: Imaging Materials*, S. A. Benton, ed., Proc. SPIE **1914**, 75–81 (1993).
27. H. Ueda, K. Taima, and T. Kubota, "Edge-illuminated color holograms," in *Holographic Imaging and Materials*, T. H. Jeong, ed., Proc. SPIE **2043**, 278–286 (1993).
28. H. Ueda, E. Shimizu, and T. Kubota, "Image blur of edge-illuminated holograms," *Opt. Eng.* **37**, 241–246 (1998).
29. S. Sainov and R. Stoycheva-Topalova, "Total internal reflection holographic recording in very thin films," *J. Opt. A: Pure Appl. Opt.* **2**, 117–120 (2000).
30. Y. J. Wang, M. A. Fiddy and Y. Y. Teng, "Preshaping of reference beam in making edge-illuminated holograms," in *Diffractive and Holographic Device Technologies and Applications VI.*, I. Cindrich and S. H. Lee eds., Proc. SPIE **3291**, 190–198 (1998).
31. W.-C. Su, C.-C. Sun and N. Kukhtarev, "Multiplexed edge-lit holograms," *Opt. Eng.* **42**, 1871–1872 (2003).
32. M. Metz, "Edge-lit holograms," in *Holography for the New Millennium*, Chapter III:3, J. Ludman, H. J. Caulfield and J. Riccobono eds., Springer-Verlag (2002) pp. 59–78.
33. W. Farmer, "Edge-lit holography," in *Holographic Imaging*, Chapter 18, S. A. Benton and V. M. Bove Jr., Wiley-Interscience, John Wiley & Sons Inc. (2008) pp. 193–205.

14

Applications of Ultra-Realistic Holographic Imaging

14.1 Introduction

This book is predominantly targeted at visual imaging applications that have arisen because of new and improved holographic methods. In particular, during the course of this book, we have described how ultra-realistic full-colour three-dimensional (3D) images may now be generated using analogue and digital holographic techniques. In this final chapter, using several examples, we illustrate how such ultra-realistic 3D images can be used today. Along these lines, we discuss how high-quality holographic reproductions of priceless artefacts and items of interest to cultural heritage are gaining interest with museums, how new holographic techniques offer a very effective method of reproducing oil paintings to a level of detail which is impossible using any other technique, and how digital holographic printing has already produced stunning large-format 3D posters that are miles ahead of other technologies, such as lenticular displays. We conclude the chapter with a brief discussion of such future applications as holographic windows and real-time holographic displays.

Before embarking on a presentation of the direct visual imaging applications of ultra-realistic holography, we will review several applications of scientific interest that fall somewhat to the side of our main theme. These are bubble chamber holography, holographic endoscopy and holographic microscopy. All are based on the capacity of holography to generate ultra-realistic structural images.

General applications of holography, conventional holographic optical elements, volume holographic gratings, holographic memory and holographic interferometry are covered in many books [1–6]. Here, we limit ourselves simply to making the comment that, today, holography is being adopted more and more in various "non-imaging" applications. For example, thick low-loss volume holographic gratings are leading to better and more readily available single-frequency laser sources [7–9]. Here, the hologram acts as a frequency-selective mirror. The applications of such single-frequency lasers are extremely wide. Another example is holographic data storage [2–4]. Finally, improved holographic optical elements are finding applications in an increasing number of areas. For example, high-quality gratings are now useful in visual astronomy for eliminating certain parasitic wavelengths [10,11], and very large holographic gratings are being made for chirped pulse amplification systems in the petawatt lasers that are required for inertial confinement nuclear fusion [12].

14.2 Some Scientific Applications of Holographic Imaging

One considerable advantage that holography has over photography is its ability to offer a high-resolution recording over a large volume—in particular, when using very short exposure times. In photography, one needs to know in advance where to focus one's camera to record a high-resolution image.* Three applications that make use of this principle are bubble chamber holography, holographic endoscopy and holographic microscopy.

---

* One can of course use deconvolution techniques to deblur a photograph but only to a certain point—holography is intrinsically more stable than deconvolution as the hologram contains far more information than the photograph.

14.2.1 Bubble Chamber Holography

A bubble chamber holography project [13] took place in the mid-1980s at the Fermi National Accelerator Laboratory in Batavia, Illinois.* Nowadays, bubble chambers have largely been replaced by faster and more accurate solid-state electronic detectors. Nevertheless, the same principles used here may be applied to similar detectors. For example, large cloud chambers are currently being used to measure the effect of cosmic rays on precipitation levels in clouds for climate models [14].

For efficient detection of short-lived particles with lifetimes of the order of 10^{-13} s, bubble chambers must be equipped with high-resolution imaging systems. High-energy physics research interest in the 1980s was centred on particles containing heavy-flavour quarks (charm, bottom, etc.). These particles, as well as the heavy lepton (τ), have lifetimes of the order of 10^{-12} to 10^{-13} s, which was much shorter than those previously observed. The experimental difficulty in decay-vertex detection of short-lived particles with mean lifetimes of approximately 10^{-13} s is that their flight paths before decay are of the order of several hundred microns up to a few centimetres. This requires a vertex detector with a spatial resolution better than several microns. The most ideal vertex detector is one that tags short-lifetime particles by precisely measuring the impact parameter of their decay products† and efficiently assigning the bubble chamber tracks to their correct parents. Bubble chambers equipped with high-resolution imaging systems were the most prominent "visual" vertex detectors in the 1980s.

As a charged particle traverses the superheated metastable liquid in a bubble chamber, it produces bubbles along its path through ionisation. By improving the resolution in conventional photography, a rapid decrease in the depth of field occurs, thereby resulting in a smaller useful volume photographed. Using holography to supplement conventional photography remedies this problem. Holography separates resolution from depth of field and achieves a high resolution over a large depth of field.

The E-632 Physics Experiment used a dark-field holographic recording technique in the 15 ft. (4.6 m) bubble chamber at Fermilab [13,15]. A high-power pulsed ruby laser was used as the light source. The laser, which we reviewed in Chapter 6 along with the electronics required to lengthen its pulses, produced light pulses of 2 to 100 µs duration with energies of more than 8 J. The Fermilab bubble chamber with the holographic recording system installed is shown in Figure 14.1.

The laser beam entered the 15 ft. bubble chamber through a small window at its bottom after traversing 30 m in an underground vacuum pipe. The laser beam was expanded using a special dispersing lens. In addition to the holographic recording camera, the bubble chamber was equipped with conventional cameras. All the cameras had concentric hemispherical lenses called *fisheye* windows. The centre of the expanding lens was on an axis passing through the centre of the holographic camera; the film was placed only 10 cm from the inner surface of the fisheye window. The laser pulses used were mostly of 40 µs duration with approximately 600 mJ energy in a 2.5 cm-diameter beam entering the bubble chamber. The beam had a flat light intensity distribution before hitting the lens. The expanding lens distributed the laser light inside the chamber with increased intensity at large angles to compensate for the reduced light diffraction off bubbles illuminated at large angles. Rolls of 70 mm Agfa-Gevaert Holotest 10E75 film were used for recording the holograms. During the experiment, approximately 100,000 holograms were recorded. Conventional bubble chamber photographs were recorded 10 ms after each hologram. An example of what could be achieved with the holography system is shown in Figure 14.2. In the upper part of the figure is a conventional photograph of a neutrino event taken 12 ms after the particle beam entered the bubble chamber (the diameter of the bubbles is <1 mm). The lower part of the figure depicts the same event photographed from its reconstructed holographic image showing much smaller bubbles (190 µm diameter) as the hologram was recorded only 2 ms after the neutrino beam injection. Whereas the large bubbles in the conventional photographs obstruct the details of a close-in interaction at a distance of approximately 2 cm from the primary vertex, the hologram clearly allows a detailed study of the secondary interaction.

---

* One of the authors (HB) was involved in this project.
† This is the distance by which the trajectory of a charged particle coming from a secondary misses the production vertex.

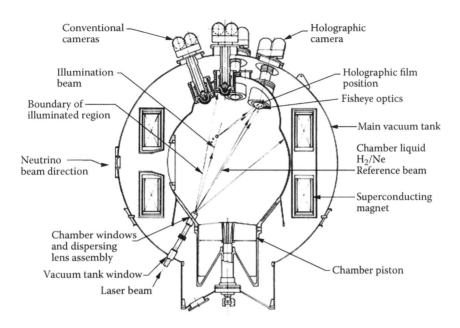

FIGURE 14.1 The 15 ft. Fermilab bubble chamber with holographic components shown (side view).

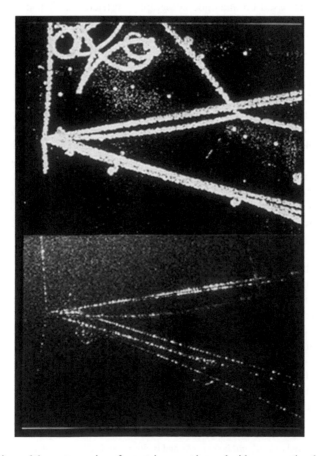

FIGURE 14.2 Comparison of the vertex region of a neutrino event imaged with a conventional photograph (top picture) and with the hologram of the same event, having a resolution of 150 μm (bottom picture).

14.2.2 Holographic Microscopy

Holography can be used to provide high image resolution over large microscopic volumes. There are many examples of early holographic microscopy applications that used holographic photosensitive plates or film. Today, it is possible to record holograms on electronic detector arrays such as charged coupled devices (CCDs) and to replay them numerically using fast Fourier transform (FFT) algorithms.* Digital holographic microscopy (DHM) has many important applications that were either not possible or not convenient with analogue recording. Digital holographic recording (DHR) eliminates the need for wet chemical processing and other time-consuming procedures, so recording and numerical reconstruction can be done in almost real time. The basic principles of DHR, holographic microscopy and state-of-the-art applications have been reported by Sang et al. [16].

An example of the application of DHM to real-time process control has been reported by Osanlou [17]. Here, holographic microscopy is used to monitor the production of various emulsions in specialised production vessels. This study is particularly relevant to online production monitoring and control in continuous manufacturing processes. The technique reported is capable of quantitative 3D mapping of moving fluids in a snapshot. It is non-invasive, high-resolution and precise. Important features of the methodology include quality of image capture and reconstruction without the use of a pulsed laser. Examples of reconstructed image quality are shown in Figure 14.3.

An ambitious and difficult application of DHR for recording marine particles has been studied by Watson [18]. Watson has constructed several holographic cameras for underwater imaging of plankton and other marine particles. These devices use both analogue and digital holography to record water volumes of up to a cubic metre. For in-water deployment, however, the weight and size of the early instruments, which were based on analogue holography, restricted their use on advanced observation platforms such as remotely operated vehicles and limited operational depth to a few hundred metres. Recent advances made in DHR on electronic sensors coupled with fast numerical reconstruction has led to the development of smaller underwater holographic recording equipment (Watson's eHoloCam). This has freed holography from many of its constraints, allowing rapid capture and storage of images including holographic video of moving objects. Digital holography's ability to record true 3D, full-field, high-resolution, distortion-free images *in situ* from which particle dimensions, distribution and dynamics can be extracted is hard to match. Figure 14.4 shows some examples of images of marine creatures calculated by the numerical reconstruction of the recorded interference patterns at 100 m depth.

DHM is very suitable for use in the fields of biological and medical research. Here, the in-line recording technique is preferentially used (see Xu et al. [19] for the properties of such digitally recorded holograms). For many years, DHM has been applied by the University of Münster, Germany, for studying living cells. An off-axis configuration has been introduced by Carl et al. [20] for studying tumourous human hepatocytes *in vivo*. It should be mentioned that DHM has significant advantages over standard phase-contrast microscopy in the very high resolution it can offer.

Another recent biological application of DHM was presented by Jourdain et al. [21], in which neuronal activity was observed in real time and in three dimensions, with a resolution of up to 50 times better than previously available. The new technique accurately visualises the electrical activity of hundreds of neurons simultaneously, at up to 500 images per second without damage by electrodes. In comparison other techniques are only able to record activity from a few neurons at a time.

Traditional microscopes are limited to a resolution of approximately 500 nm. However, DHM allows a resolution down to as low as 10 nm in some cases. DHM constitutes a fundamentally novel application for studying biological objects such as neurons, and the technique has many advantages over traditional microscopes. It is non-invasive, allowing extended observation of neural processes without the need for electrodes or dyes that inevitably damage cells. By inducing an electric charge in a culture of neurons using glutamate (the main neurotransmitter in the brain), charge transfer carries water inside the neurons and changes their optical properties in a way that can be detected only by DHM. A computer is then able

---

* This process is often referred to as *digital holography* and is closely related to computer-generated holography. Note that digital holographic printing techniques such as DWDH and MWDH, as described in Chapter 7, are often also referred to as digital holography. However, the two fields of digital holography are rather different.

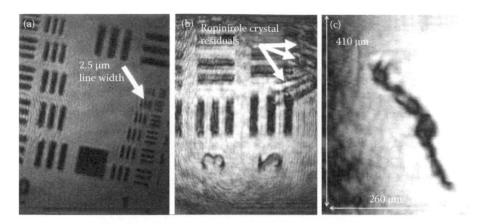

FIGURE 14.3 Reconstructed images (real-time holographic process control) showing (a) US Air Force test target (b) ropinirole crystal residuals (width, 6.2 μm) in front of the US Air Force test target and (c) benzoic crystal chains floating in the gently stirred fluid within the reactor vessel. (Photos courtesy of A. Osanlou.)

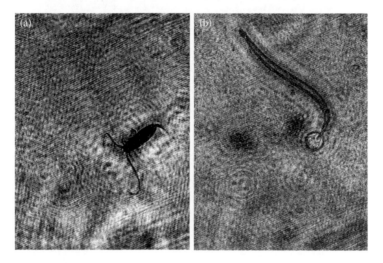

FIGURE 14.4 Reconstructed DHR images of (a) a calenoid copepod (3 mm body length) and (b) an arrow worm (1.75 mm length). (Photos courtesy of J. Watson.)

to numerically reconstruct a real-time 3D image of the neurons. DHM is the first imaging technique that has been able to monitor the activity of these cotransporters dynamically and *in situ* during physiological or pathological neuronal conditions.

There are many other applications of digitally recorded holograms and we can expect the field to expand considerably in the coming years. Today, the possibility to record digital holograms is limited to small angles between the object and reference beams. This is because current electronic recording devices only possess a certain finite resolution. This is why today's applications are limited primarily to microscopic recording techniques in which in-line holograms can provide the required image resolution.

14.2.3 Holographic Endoscopy

One of the authors (HB) has been involved in the 2006 *Holoendoscope* project [22] to develop high-resolution colour holographic endoscopy for cellular analysis. Endoscopic holography or *endohography* combines the features of endoscopy and holography. The purpose of endoholographic imaging is to provide the physician with a unique means of extending diagnosis by providing a lifelike record of tissue. Endoholographic recording provides a means for microscopic examination of tissue and, in some

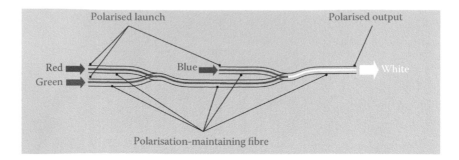

FIGURE 14.5   Fibre-coupler used to generate the "white" laser light for the endoscope. (Reproduced with permission from Gooch & Housego.)

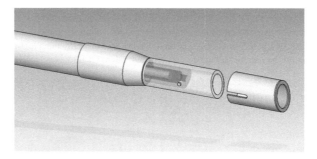

FIGURE 14.6   Prototype holoendoscope. The holographic film is inserted at the tip of the endoscope. (Reproduced with permission from Gooch & Housego.)

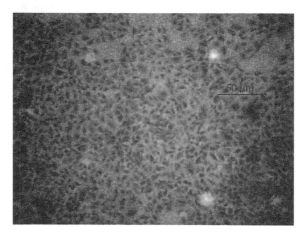

FIGURE 14.7   *In vitro* endoscopic colour hologram of human endothelial cells.

cases, may obviate the need to excise specimens for biopsy. In the project, *in vitro* colour holograms were recorded to demonstrate the feasibility of recording high-resolution tissue images. The prototype endoscope* used small RGB lasers with the laser light being combined using single-mode optical fibres to provide illumination through the endoscope. Denisyuk single-beam colour reflection holograms were recorded in close contact with the tissue at the distal end of the endoscope [23]. The recorded holograms were viewed and photographically documented under a microscope. Earlier, in 1988, a demonstration of a *monochrome* hologram recording using an earlier prototype holoendoscope (holographic

---

* The *holoendoscope* project partner *Sifam-Gooch & Housego* was responsible for the RGB lasers, the fibre-optic coupler, and the prototype holoendoscope.

sigmoidoscope) was performed to obtain *in vivo* holograms of the colon of an anaesthetised dog [24]. The principle of the fibre coupler colour laser combiner is shown in Figure 14.5, the prototype holoendoscope is illustrated in Figure 14.6, and a microscopic photograph derived from an *in vitro* endoscopic colour hologram of human endothelial cells is shown in Figure 14.7.

14.3 Visual Applications of Full-Colour Holographic Imaging

In this section, we focus on current applications of the technologies of full-colour analogue hologram recording (as described in Chapter 5) and digital holographic printing (Chapters 7–10). Despite holography now being more than 60 years old, many of these applications are in an early stage of market development. This is really because the technology improvements that we have presented in this book have, in many cases, only recently led to a positive reassessment of old ideas.

14.3.1 Holographic Copies of Museum Artefacts

In the days of the British Empire, museums filled their collections with exotic items from around the world. British museums today still possess large collections from foreign countries, but there is increasing pressure to repatriate priceless artefacts to their respective homes. Holography now offers the possibility to essentially duplicate such artefacts—and to a point where observers practically cannot tell whether they are looking at the real exhibit or at a holographic copy. Although such holographic reproduction can never match the value of actually possessing the real artefact, it can allow the museum to fulfil one of its most important functions—to maintain display of the exhibits. Of course, the real exhibit potentially allows future scientific tests to be performed on an artefact to verify scientific theories—such as CAT scans, material testing and non-visual spectral analysis—but analogue holography does offer a means to preserving a faithful visual recording of unprecedented microscopic detail.

We should mention that digital holographic printing can also be usefully applied to museum recordings. By using a two-dimensional tracking camera system,* high-resolution digital image data from a museum exhibit can be recorded from over a million different angles. This data may then be written onto a very high-resolution digital hologram producing a digital holographic copy. Although the resolution of this type of hologram is less than that of an analogue hologram, at the smallest hogel sizes now available (~250 μm), it can nevertheless be very difficult for an observer to tell the difference between an analogue and a digital hologram with the unaided eye. The digital hologram also offers several sizeable advantages. For instance, new holograms can be generated from the digital data whenever required. This means that as long as the original data is stored securely, there are no image lifetime problems. In addition, the same image data can be used to produce holograms of small and very large sizes. Thus, museum exhibits can be displayed in any format. Finally, such digital images are not constrained to be behind a glass plate.

Holography can also help museums with travelling exhibitions, as we shall see in later sections. It is difficult for some people to travel to museums, and, as a result, there is pressure on museums to take exhibits to the people. Transporting priceless artefacts, however, is both hazardous and expensive. Transporting holographic copies on the other hand is not.

A final reason why ultra-realistic full-colour holograms are useful to museums is related to insurance costs of exhibiting within a museum. Most museums have large collections "downstairs" which they do not exhibit. The reason is that it generally costs more to exhibit something than to securely store it—as the risk of damage or loss is greater when an exhibit is on display. Again, holography can help solve this problem. If the hologram is indistinguishable from the real exhibit, why not just securely store the real item and display the hologram?

14.3.1.1 Virtual Museum Exhibitions

One interesting colour holography project,† which was recently carried out by the Centre for Modern Optics in North Wales, was a project funded by the Esmee Fairbairn Foundation entitled *Bringing the*

---

* See Chapter 10 for a discussion of holocam systems.

† One of the authors (HB) was involved in this project.

Artefacts Back to the People. The project involved collaborations with a number of major museums including the National Museum of Wales, the British Museum, the Maritime Museum in Liverpool, as well as the Royal Commission for Ancient and Historical Monuments in Wales. Full-colour holograms of various artefacts were recorded using the analogue techniques described in Chapter 5. The holograms were completed by the end of 2009, after which they were displayed as a travelling exhibition that toured North Wales and its borders. The exhibition first opened at Llangollen Museum in June 2010 and later at the museums of Grosvenor (Chester), Wrexham, Llandudno, Bangor and many others [25].

One of the recorded artefacts, supplied by the British Museum, was a 14,000-year-old decorated horse jaw bone from the Ice Age, or late glacial period of Britain [26]. The recording setup is shown in Figure 14.8 and the hologram in Figure 14.9. Another hologram recorded was the *Tudor Owl Jug* and *Sergeant at Arms Ring* shown in Figure 14.10. These artefacts were from the Grosvenor Museum in Chester, United Kingdom. In total, ten full-colour holograms of different artefacts were included in the touring museum exhibition (Figure 14.11).

FIGURE 14.8 Horse jaw recording setup.

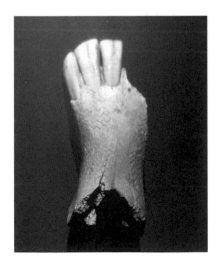

FIGURE 14.9 Decorated horse jaw hologram.

FIGURE 14.10 Hologram of the *Tudor Owl Jug* and *Sergeant at Arms Ring.*

FIGURE 14.11 Exhibition of the artefact colour holograms at the Llangollen Museum in Wales, 2010.

14.3.1.2 Museum Holography in Greece

In Greece, the Hellenic Institute of Holography has recently invested in a full-colour recording facility and is currently promoting *Realistic Colour 3D Holography*, according to its director Alkis Lembessis [27]. The primary goal of the Hellenic Institute of Holography is to record Greek cultural artefacts through the "*HoloCultura: Applied Holography in Cultural Heritage*" project. The project consists of three parts:

- Phase A: Study necessary for implementation of the colour holography programme
- Phase B: Recording of experimental colour holograms
- Phase C: Pilot project involving the recording of cultural artefacts

The institute is active in a country with a unique cultural tradition of worldwide influence extending from classical ancient Greece to orthodox Byzantium Christianity. The use of display holography in the preservation, recording and public visual dissemination of artefacts from this cultural heritage is at the core of the activities of the Hellenic Institute of Holography.

In 2011, the institute built a small, portable three-colour analogue holographic camera, the Z3 RGB Holography YSB1 prototype camera, which it is now using to record holograms of museum artefacts (Figure 14.12). The camera is a computer-controlled optomechanical device capable of exposing selected, commercially available or experimental, panchromatic silver halide emulsions to combined red, green and blue CW laser beams at appropriate energy levels. The device consists of a main camera unit (MCU) and a control electronics unit (CEU). The MCU is built on top of a lightweight aluminium honeycomb optical board to ensure portability, minimum spatial deformation under nominal temperature variation and fast damping of induced vibrations.

There are three lasers housed in the MCU with wavelengths selected to cover a broad triangle of hues in the Commission Internationale de l'Eclairage (CIE) chromaticity diagram.

- Red laser: 638 nm at an output power of 80 mW (*CrystaLaser* laser)
- Green laser: 532 nm at an output power of 100 mW (*Cobolt Samba* laser)
- Blue laser: 457 nm at an output power of 50 mW (*Cobolt Twist* laser)

The lasers produce TEM_{00} emissions with coherence lengths of more than 5 m each, and the MCU contains suitable optics to generate a clean collinear mixed RGB beam.

The CEU houses all power supplies, A/D and D/A subsystems plus a specially designed F/P scanning interferometer for beam monitoring. The CEU also connects to an external PC running custom software to control all aspects of the holographic exposure plus full monitoring and tuning of each laser's stability and beam quality.

The Z3 has been successfully tested under wide ambient temperature and humidity ranges. The system becomes stable after 30 min to 1 h, depending on the ambient temperature. At 24°C, the system stabilises after 20 to 30 min. The Z3 camera is accompanied by auxiliary equipment for beam orientation and a flexible vibration-absorbent setup for the positioning of the object. One example of a hologram recorded with the camera is shown in Figure 14.13. The Hellenic Institute of Holography has also produced a portable darkroom for on-site processing (Figure 14.14).

In addition to recording their own holograms with the Z3, the Hellenic Institute of Holography has produced museum holograms in collaboration with Yves Gentet (Figure 14.15) and the Colour Holographic Company in London.

FIGURE 14.12   Z3 prototype portable RGB camera (model YSB1) made by the Hellenic Institute of Holography.

FIGURE 14.13 Full-colour Denisyuk reflection hologram made using the Z3 prototype portable RGB holography camera produced by the Hellenic Institute of Holography.

FIGURE 14.14 Andreas Sarakinos and Alkis Lembessis in front of the portable darkroom produced by the Hellenic Institute of Holography.

FIGURE 14.15 Full-colour Denisyuk reflection hologram of a Greek ceramic vase made by Yves Gentet. This analogue hologram was made in collaboration with the Hellenic Institute of Holography. The right-hand photograph shows an enlarged detail.

14.3.1.3 Museum Holography in England

In 2010, the Colour Holographic Company in London started to record colour holograms on their own ultrafine-grain panchromatic material. Some examples of the full-colour museum holograms that they have produced on this material are shown in Figures 14.16 and 14.17. A new product that the company has introduced is a nice wooden box for the display of colour holograms with an integrated LED light source built into the lid. When the box is opened, correct illumination of the hologram is provided. It is

FIGURE 14.16 Full-colour Denisyuk reflection test hologram made by Colour Holographic Ltd.

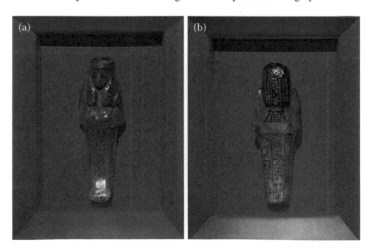

FIGURE 14.17 Full-colour Denisyuk reflection hologram made by Colour Holographic Ltd. The hologram shows (a) the front and (b) back of an Ushabti figure from the Theban cache of royal mummies found in 1881.

FIGURE 14.18   (a) Colour Holographic hologram box and (b) colour hologram.

possible to switch holograms and the company offers the box with different holograms for sale. The box and a colour hologram are shown in Figure 14.18.

14.3.1.4 Holographic Reproduction of Oil Paintings

Full-colour holographic copies of oil paintings offer another interesting application. Because the depth of the recorded image is essentially dictated by the thickness of the brushstrokes, this type of quasi two-dimensional recording can be rather easier to illuminate than conventional holograms. For example, source-size blurring is much less of an issue here. Holographic copying provides a method of producing copies of valuable or priceless paintings, which is unlike any other technique. An oil painting does not look the same from every angle and a photographic reproduction only records the view from straight ahead. A holographic reproduction, on the other hand, faithfully records how the light reflects at all angles, as well as accurately reproducing the relief of the brushstrokes.

One of the authors (HB), working with Dalibor Vukičević, introduced this potential application in 2000 [28]. An example of a *still life* oil painting (20 cm × 25 cm), which was copied using this technique, is shown in Figure 14.19a. The painting was selected mainly because it was painted on

FIGURE 14.19   (a) Oil painting and (b) full-colour analogue Denisyuk reflection holographic reproduction.

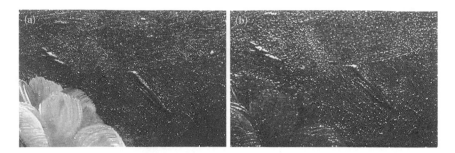

FIGURE 14.20   (a) Detail of brush strokes in the oil painting of Figure 14.19a and (b) in the holographic reproduction of Figure 14.19b.

FIGURE 14.21   Higher magnification of brush strokes in the hologram of Figure 14.19b.

wood* with a pronounced surface texture. A Denisyuk single-beam three-colour holographic setup (described in Chapter 5) was used to record the colour hologram onto a Slavich panchromatic PFG-03C silver halide plate (Figure 14.19b). A detail of the painting texture, visible when observed at a certain angle, is shown in Figure 14.20a. The same surface texture visible in the hologram is shown in Figure 14.20b. The holographic reproduction can be studied in more or less the same way as the real painting can be investigated. Brush strokes, visible in the hologram under high magnification, are illustrated in Figure 14.21.

Colour holograms of paintings have also been demonstrated by Yves Gentet, who recently took recording equipment to the Louvre in Paris to demonstrate the potential of this new reproduction technique. A transportable holographic camera is often required in applications regarding the recording of items of significant cultural heritage, including oil paintings—often because the item in question simply cannot be moved due to insurance or security reasons. Gentet and Shevtsov [29] were the first to develop a small, mobile full-colour analogue holographic camera system especially for this purpose. Much like the system currently used by the Hellenic Institute of Holography, Gentet and Shevtsov's system is based on three solid-state continuous wave lasers: a semiconductor red laser at 639 nm giving 25 mW, a diode-pumped solid-state laser at 532 nm giving 120 mW and another diode-pumped solid-state (Cobolt) laser at 473 nm giving 70 mW. The overall dimensions of the camera system, which allows a hologram format of up to 30 cm × 40 cm to be recorded, are 30 × 40 × 50 cm with a weight of 12 kg.

In copying oil paintings and indeed other cultural heritage items, the reproduction of spectral information is of particular importance. As we mentioned in Chapter 5, three-colour analogue holograms produce a good, but certainly not perfect, spectral representation. Future work will hopefully extend holographic copying to four or five wavelengths.

---

* With the available lasers at the time, wood provided better interferometric stability.

14.3.2 Digital Display Holograms for Advertising and Product Promotion

14.3.2.1 Introduction

Large-format full-colour 3D reflective digital holographic displays can be extremely impressive. Over the years, this application has had a number of false starts for a variety of reasons—mostly connected with colour issues, reliability, speed of fabrication, illumination and price point. From time to time, great interest has been shown by advertising agencies, printing companies and by organisations wishing to promote their products. A number of studies by reputable market research companies in the last 10 years have estimated potential yearly returns for this sector approaching the billion dollar level.

Many of the issues that have frustrated the penetration of holographic displays in the advertising display sector are now being resolved. Advances in laser technology are making the process of writing the holograms easier. Better materials, such as a new photopolymer material from Bayer, allow higher quality and cheaper images to be produced. Advances in illumination technology make the final hologram brighter, deeper and easier to light. As such, the recent proliferation of 3D films and the increasing popularity of 3D televisions can reasonably be expected to drive a reappraisal and renaissance of this application.

14.3.2.2 Key Organisations

The main companies currently active in the area of digital holographic printing are Zebra Imaging Inc., Austin, TX, and Geola Digital UAB, Vilnius, Lithuania. Zebra was the first group to produce digital full-colour holograms using continuous wave lasers, hogel by hogel. Geola was the first group to do this using RGB-pulsed lasers. Until recently, XYZ Imaging Inc., Montreal, Canada originally a spin-off from Geola and subsequently operating under the name of RabbitHoles Media Inc.,* also operated a digital holographic printing facility in Canada. Ultimate Holography in France, the Dutch Holographic Laboratory in Holland as well as Ceres Imaging Ltd., Spatial Imaging Ltd. and View Holographics Ltd. in the United Kingdom have also developed equipment for the digital printing of colour reflection holograms.

14.3.2.3 Zebra Imaging

Zebra was formed in 1996 by Michael Klug and fellow graduate students from the Massachusetts Institute of Technology's (MIT) Media Lab [30]. In 1999, Zebra produced the world's largest single-colour hologram, which is shown in Figure 14.22. It was developed in collaboration with the Ford Motor Company and was exhibited at the 1999 North American International Auto Show in Detroit. The hologram showed Ford's concept vehicle, the *P2000 Prodigy*, at a 50% scale. The 3D image bisected the 40 ft.^{2} panel, which was composed of individually printed smaller tiles. The exterior of the *P2000 Prodigy* hologram was translucent red, allowing the viewer to see inside the vehicle's futuristic hydrogen fuel cell power train. The hologram contained 900,000 individual exposures taken directly from computer design data supplied by the Ford development team. Ford's Advanced Design Studio was, at the time, working on "replacement reality" techniques—and full parallax large computer-generated colour holograms fit very well into this description. Four terabytes of data storage was needed for the *Prodigy* hologram.

In December 2001, three large Zebra holograms—*Austin Dimensions; Music, Nature and Technology*—were installed at the Austin-Bergstrom International Airport in Texas. The holograms were presented to the airport by Zebra Imaging, Samsung Austin Semiconductor and Frog Design. Measuring 30 ft. long and 4 ft. high, the three panels comprised the world's largest holographic display at the time of its completion. A photograph of the holograms is shown in Figure 14.23.

14.3.2.4 XYZ, Geola and Sfera-S

In 2001, Geola teamed up with Québecois entrepreneur, Eric Bosco, to launch the company, XYZ Imaging (Imagerie XYZ Inc.) in Montreal. This was the first company to target commercial mass

---

* At the time of going to press, we understand that XYZ Imaging Inc. were in the process of restarting their activities.

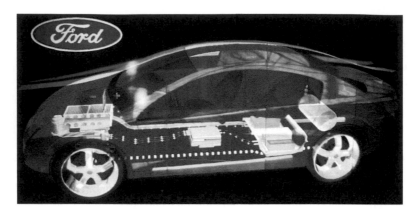

FIGURE 14.22   The 40 ft^2 full-colour digital reflection hologram of Ford's *P2000 Prodigy* produced by Zebra Imaging in 1999.

FIGURE 14.23   *Austin Dimensions* Zebra airport holograms (2001).

production of full-colour 3D holographic printers rather than simply supplying holograms to clients. At the time, the Lithuanian operation of Geola numbered around 30 people. A new company, comprising around 10 people, was created in the United Kingdom to manufacture the RGB lasers for the printers. This was Geola Technologies Ltd., which was based on-campus at the University of Sussex. XYZ itself hired an additional 25 people. The Russian company Sfera-S AO, formed by ex-Micron director at Slavich AO, Yuri Sazonov, was also a critical partner to the XYZ–Geola group. Sazonov worked with Geola and XYZ to produce a vital silver halide film compatible with RGB-pulsed lasers. Zebra used DuPont's panchromatic photopolymer material due to its special contacts, but this material was not available on the general market, and it also had uncertain compatibility with Geola's lasers.

From 2001 to 2006, XYZ, Geola and Sfera-S worked together as a team of around 60 people towards the goal of establishing digital holographic printing as a viable industry for advertising display applications. The initial plan was to develop a true commercial digital holographic printer based on an RGB pulsed laser. Initially, an in-house service bureau in Montreal was to be set up to test the printer under real operational conditions and also to identify any problems with the holograms. However, investor concern over the project timescale led to the service bureau being farmed out to a third-party printing company, and sales of the printers to end-users being brought forward. Despite this, the printer, original equipment manufacturer (OEM) laser and film development were rather successful, and by 2003 to 2004, excellent quality holograms on a new commercial film were being produced regularly on the new printers. XYZ started to win prestigious awards within the advertising industry for its holograms and, under pressure for sales revenue and because of the optimism created by the awards, commenced commercial sales of its printers. Unfortunately, in hindsight, the company rushed into this process. In addition, its first sales were international and required numerous international service visits. The lead investors, seeing these problems, put the company up for sale, and despite offers from its founding shareholders, new management was brought in and took the decision to sell to a third party. At this point, the synergy of the greater group was broken and the new company, which was again later sold, became an independent 3D printing company known under the name RabbitHoles Media.

FIGURE 14.24 President and Founder of XYZ Imaging, Eric Bosco, in front of one of his company's first digital full-colour holograms. Note that the display has been tiled from smaller panels.

FIGURE 14.25 Digital display hologram made by XYZ Imaging advertising Puma shoes in downtown Montreal (2005).

FIGURE 14.26   Another Puma digital display hologram made by XYZ (2005).

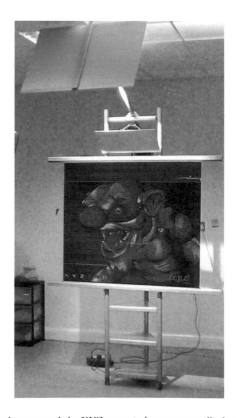

FIGURE 14.27   Digital display hologram made by XYZ, mounted on a custom display unit manufactured by a UK Geola subsidiary, Power Imaging Ltd. (2006).

FIGURE 14.28 *Avatar* digital hologram (16 in. × 16 in.) by RabbitHoles Media (2011). The company produced limited edition series in three formats: 32 in. × 32 in. holograms (edition of 10), 16 in. × 16 in. holograms (edition of 50) and 8 in. × 8 in. holograms (edition of 200).

During its heyday in the mid-2000s, XYZ produced many interesting holograms. Some of these are shown in Figures 14.24 through 14.27. Under the name of RabbitHoles, XYZ developed close connections with the movie industry in Hollywood and produced many excellent promotional display holograms for movies [31]. Most notable was their hologram of the 3D movie *Avatar* (Figure 14.28), which won the RealD award for Innovation in 3D. The company also produced holograms for trade show promotion. Notable clients included Bombardier, Audi, Toyota, Puma and Zeiss Meditec.

14.3.3 Digital Display Holograms for Mapping and Architectural Design

14.3.3.1 Horizontally Mounted Overhead-Illuminated Holograms

Horizontally mounted digital full-colour, full-parallax reflection holograms with overhead illumination are useful for a variety of applications in mapping and architectural design. Large full-parallax holograms may be laid out on a table and viewed by a group of people. 3D terrain can project up out of the hologram and viewers can easily perceive the 3D structure of mountains and valleys. Such holograms can potentially be rolled up into tubes and taken out for display when required. As such, they are of potential interest to military organisations.

At the time of writing, the US Army had purchased more than 10,000 horizontally mounted overhead-illuminated (HMOI) holographic displays for military mapping applications from Zebra Imaging. Zebra has also now started producing holographic printers for the US Army. The company has developed a new generation of high-speed colour and enhanced monochromatic digital hologram printers, which it calls *Imagers*. These allow the rapid production of digital holograms and, in particular, HMOI holographic displays. Like the printers made by Geola and XYZ, a wide variety of data sources can be used for the creation of the 3D images including computer-aided design/engineering/manufacturing, aerial photos, radar and laser scans such as synthetic aperture radar (SAR) and light detection and ranging.

The Hellenic Institute of Holography has also recorded map holograms for geographical services of various national armies, including the Greek Army. The institute recently cooperated with the NATO Fast Deployment Unit for the Balkan areas based in the city of Thessaloniki (NDC-GR) in the making of one trial map. In this case, geographical data, which were freely available on the web, were used (colour satellite terrain picture of an area near the Evros River in Northern Greece). Figure 14.29 shows a topographic reflection hologram, 50 cm × 41 cm, created by General Command of Mapping, Ankara, Turkey and printed by the Geola organisation [32].

(a)   (b)

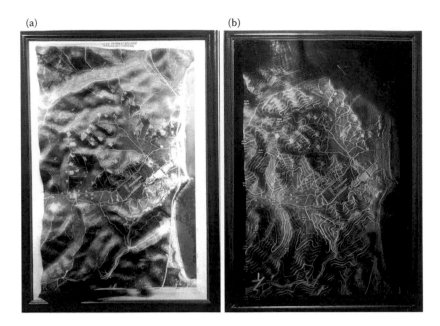

FIGURE 14.29   (a) Normal relief map, and (b) topographic holomap or Digital Carto–hologram of the same area as in (a). Photos from "Ihsan Seref Dura Exhibition Room" at General Command of Mapping, Ankara, Turkey. (Reproduced with permission from H. Dalkiran, GCM, printed at Geola 2010.)

HMOI holographic displays can also be used for architectural and urban planning applications (Figure 14.30). Urban planners in the United Kingdom have long sought detailed physical models from developers for large residential or commercial developments. Such models allow an effective visualisation of a major project but can be very costly to make and extremely costly to modify when the planners request design changes. More recently, 3D virtual models have tended to replace such physical models. Such models are cheaper, can be modified easily and portray truly photorealistic detail, but usually, for convenience, the planner will see just a programmed "fly-through". Writing such computer models to an HMOI display produces the best of both worlds. The HMOI hologram is cheap, easily transportable (unlike the model) and can be laid out quickly for viewing. Once lit, the display acts and feels just like a physical model except that the rendering is usually completely photorealistic. By replacing one display with another (an operation of several seconds), an alternative model can be displayed. In addition, design changes can easily be implemented in the computer model and another display printed—rather easier than changing the model!

14.3.4 Digital Holographic Colour Portraits

Despite early enthusiasm, pulsed-laser monochrome portraits never became terribly popular. Their single-colour waxy appearances seemed, on the one hand, strangely devoid of life and, on the other hand, appeared just too realistic. Digital holograms, however, can now offer a far more palatable solution. In the simplest variant, a series of photographs are taken using a holocam device. As we described in Chapter 10, this is just an automated camera on a horizontal rail. As the camera moves along the rail (typically from one to several metres long), digital photographs (typically several hundred to several thousand) are taken of a subject from different angles (Figure 14.31). The recorded images are then processed and a full-colour horizontal parallax-only (HPO), direct-write digital holography (DWDH)* portrait is printed. An example of such a hologram is the 30 cm × 40 cm 2006 colour holographic portrait of Nick Phillips shown in Figure 14.32. There are several advantages inherent to this technique. First and

---

* Full-colour reflection portraits can also be written using MWDH.

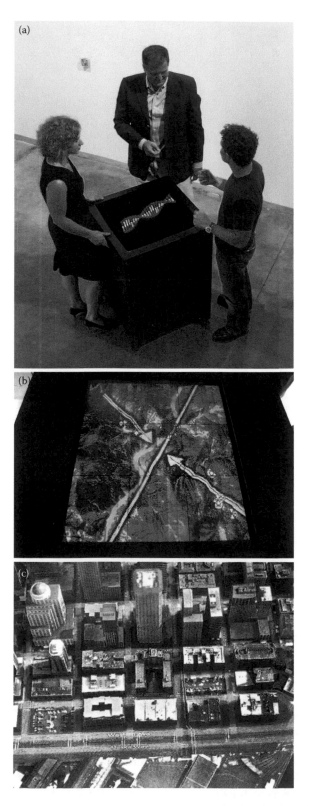

FIGURE 14.30 Full-parallax HMOI hologram displays by Zebra Imaging. (a) Double helix model, (b) military map and (c) city map. (Photos courtesy of M. Klug, Zebra Imaging).

FIGURE 14.31   Large-format 2005 studio holocam system manufactured by Geola.

FIGURE 14.32   Full-colour digital hologram portrait (30 cm × 40 cm) of Nick Phillips shot at the 2006 *International Symposium on Display Holography* held in Wales and produced by Geola.

foremost, the hologram is a true-colour portrait, but almost as important as this is how the image data are derived. The subject is lit, not by highly coherent monochromatic laser beams, but by normal wide-band white illumination. This bypasses the old problems associated with the skin's reflective properties at certain laser frequencies. Because the image data is collected in digital form, the data may be easily manipulated and retouched. Backgrounds may be inserted and other features such as text inserted into the hologram. Finally, limited animation may be included, which makes the hologram feel more alive.

Full-parallax digital holographic portraits are also possible. In Chapter 10, we reviewed how structured-light techniques have led to the ability to acquire complete 3D computer image data from a subject in real-time and in Appendix 9 we review a revolutionary new technique which allows a single device such as iPad from Apple Computers to acquire complete 3D data-sets. Such data sets allow the printing of full-parallax colour portraits using the DWDH technique.

14.3.5 Digital Art Holograms

Artists have been attracted to the new digital full-colour holograms. Paula Dawson in Australia, who has always been interested in large-format holograms, has recently created several large digital colour holograms which were printed by Geola. The *Luminous Presence* hologram (0.95 m × 1.5 m), shown in Figure 14.33, was created in 2007. Martin Richardson at DeMontfort University in the United Kingdom has also created art portraits with several composite image sets. One such hologram is *Psychedelic Amy*, which is shown in Figure 14.34. At first sight, this is a digitally retouched portrait of a young woman, but a ghost-like image can be seen to reflect in the subject's eyes as one moves from left to right, giving the impression of sixties-like psychedelic imagery. Many other artists have started to work with this new and exciting medium.

14.3.6 Smaller Full-Colour Holograms

Small photopolymer holograms tend to be the easiest type of hologram to mass replicate. They also exhibit a high diffractive efficiency, and if image depth is small, they can be less sensitive to lighting conditions. As such, they are sometimes used on book or magazine covers. In 1999, a DAI Nippon

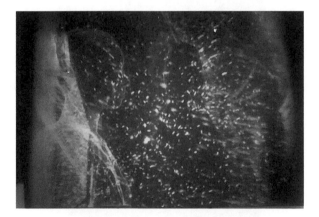

FIGURE 14.33 Full-colour digital hologram entitled *Luminous Presence* by Paula Dawson (printed by Geola, 2007).

FIGURE 14.34 Full-colour digital hologram entitled *Psychedelic Amy* by Martin Richardson (printed by Geola, 2010).

colour hologram was used on the cover of the Eighth Edition of *Holography Marketplace*. The mass-produced hologram shown in Figure 14.35 was recorded in DuPont's panchromatic photopolymer material. Although the DuPont material is not yet available to general customers, Utimate Holography North America (UHN) in Canada has recorded similar colour holograms on the new panchromatic photopolymer material from Bayer—an example is shown in Figure 14.36.

FIGURE 14.35   Small, full-colour analogue reflection hologram by Dai Nippon (DNP) recorded using the DuPont panchromatic photopolymer.

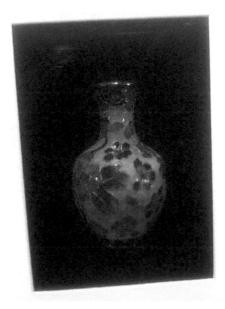

FIGURE 14.36   Small, full-colour analogue reflection hologram by UHN recorded using the recent panchromatic photopolymer material from Bayer.

FIGURE 14.37   Model 840 Classica Brass Fire from Valor incorporating an analogue reflection hologram of the fire bed.

14.3.7 Holographic Fuel-Effect Electric Fires

To date, monochromatic holograms have only found display applications of significant sales volume in the document security field. One exception to this rule, however, is an application based on the use of a monochromatic reflection hologram in the mass-produced Valor *Electric Fireplace*. Valor has recently introduced a new type of household electric fireplace (Figure 14.37), in which a hologram is used to create a realistic-looking fuel bed of burning coal. Traditionally, fireplaces mimicking burning coal have made use of a bed of cut-glass "pebbles" illuminated by red light. However, such technology requires a bulky device that all too often simply does not fit into modern homes which frequently have no chimney. The great advantage of using a hologram to create the illusion of a fire is that it gives the illusion that the fireplace is much deeper than it really is. According to Valor, the design "changes the way you view fuel effect fires forever, offering you an unrivalled degree of realism".

Valor's fire bed hologram is relatively small—20 cm × 30 cm. The master reflection holograms were made by Colour Holographic Ltd. in the United Kingdom. Mass production of the holograms was then carried out by DuPont Holographics, Inc. in Logan, Utah using a monochromatic DuPont photopolymer film. The laminated transparent reflection hologram is mounted at an angle in the fireplace and is illuminated via a mirror by a 50 W, 12 V halogen lamp mounted at the rear of the device. Behind the hologram is a diffuser through which a flame effect is generated using material strips that move in the flowing air generated by a silent fan. These material strips are illuminated by zeroth-order light, which the hologram does not diffract. The diffuser makes the flame effect more realistic by making the moving material strips appear slightly diffuse through the hologram plate. Overall, the result is rather pleasing and the untrained eye does not immediately suspect that the fire is a hologram.

The reason this application works with a monochromatic hologram is that burning coals are often very red. As such, a red monochromatic hologram works satisfactorily. The company is nonetheless now working on an improved version of the fireplace, which will use a three-colour hologram.

14.4 Future Applications

14.4.1 Holographic Windows and Super-Realistic 3D Static Displays

One of the most intriguing applications of holography is the use of large-format digital holograms to mimic windows. This is perhaps the ultimate test of what is possible to extract from static holography.

The basic idea is to create glass holographic panels that, on appropriate illumination, reveal landscape-type images with virtual volumes of up to many square kilometres. At first sight, this might seem impossibly ambitious. After all, the standard credit card hologram might have, under ideal illumination, an image volume of 1 cm^3, a modern panchromatic hologram of a typical museum artefact might possess an image volume of 0.3 m^3, and even a large modern digital hologram might only have an image volume of say, 20 m^3. Despite these figures, however, calculations clearly show that much larger image volumes should be possible.

Put simply, a holographic window is a large-format panchromatic digital reflection hologram that emulates the view outside a typical window. The characteristics of the hologram are chosen such that the intrinsic image blurring properties of the hologram, together with its illumination source, fall below the level of human discrimination. As such, an observer standing in front of this hologram interprets it, not as a hologram, but as a window. If properly implemented, such holograms have the ability to transform basements or small windowless rooms into top-floor penthouses with spectacular views over city centres or mountain ranges. Equally well, architects and interior designers could use this technology in commercial projects.

14.4.1.1 Creation of Space—High Virtual Volume Displays

Closely related to the concept of the holographic window is holography's ability to create space. A super-realistic hologram can effectively create virtual space behind its glass surface. Take for example, a small corridor in a typical London office suite. Often, such corridors are windowless and dark. Developers spend many millions of pounds opening up such small dark corridors with glass walls looking onto a central light well. Such developments produce nicer spaces in which to work, as the eye can see further and the darkness is replaced by light; but the central light well takes space and is extremely expensive. Holography can potentially achieve the same effect with glass holographic panels. Here, the extra space is all virtual and costs nothing. All that needs to be done to create the illusion of the brightly lit central light well is the installation of the panels and the illumination sources.

Holographic windows and super-realistic holographic displays will undoubtedly find a future market in architecture. Such displays may also be expected to find applications in museums. Imagine for example, going to the British Museum and seeing the pyramids in full-scale through a giant window display. Because such displays are intrinsically digital, one is not constrained to modern landscapes of course. Therefore, real-scale super-realistic images may be created of historical events. Even limited animation of the image may be included in such displays. The larger the display, the easier such animation is to encode and the longer it can be.

14.4.1.2 Constraints on Digital Holograms

Clearly, holographic windows or super-realistic high virtual volume (HVV) displays require digital full-parallax writing techniques such as DWDH. Analogue holograms could not, in any form or manner, be expected to generate such HVVs. Today's digital printers have not yet targeted HVV displays; these printers are capable only of producing digital holograms of rather limited depth. The actual depth possible depends on the field of view that the printer is required to create in the printed hologram. When the field of view is maximised, the capacity of the hologram to reproduce large depths decreases. Up until now, this has not been important as typical illumination sources have introduced greater blurring into the hologram than any restriction due to the printer's optical construction. In the following sections, we list the basic requirements for printing and displaying HVV and holographic window displays.

14.4.1.2.1 Optimal Hogel Size

The hogel size of a digital HVV display needs to be chosen in an optimal fashion. As we have already mentioned in Chapter 11, digital diffractive blurring becomes larger than the average human eye perception level at hogel diameters smaller than approximately 0.5 mm. This means that the virtual volume of a HVV display will decrease dramatically if the hogel size is reduced much beyond this point. Nevertheless, a hogel size of 0.5 mm is tolerable for a large-format display. Generally, the eye will interpret this as a fine mesh covering the window when the display is viewed at a distance of less than 0.5 m.* At an observation distance of more than 0.5 m, the display will appear effectively non-pixelated.

14.4.1.2.2 Field of View

The field of view of a digital DWDH hologram is defined by the numerical aperture of the primary-colour writing objectives within the digital printer. A (horizontal) field of view for a DWDH hologram of between 100° and 120° is typical nowadays. Although HVV displays can be created with a field of view of only 120°, a holographic window ideally requires a field of view that is nearer to 140°. Values greater than these are usually not necessary as windows start to reflect the interior view as the angle becomes much flatter. In addition, an installation can be designed in certain circumstances such that panels may be mounted so that glancing angle views are physically obstructed. This reduces the requirement on field of view. One must also remember that field of view and hologram brightness are conjugate variables—the greater one is, the smaller the other is. As such, the vertical field of view may often be chosen to be rather smaller, thus increasing the effective brightness of the display.

14.4.1.2.3 Pixel Dimensions of Printer SLM

The three primary-colour spatial light modulators (SLMs) in a DWDH printer, together with the numerical aperture of the optical objectives, dictate the maximum possible angular resolution of the hologram at replay. The required horizontal numerical dimension† of the SLM is related to the field of view by the formula (Figure 14.38)

$$N \sim \cot\frac{\theta_{Eye}}{2} \times \tan\frac{\Phi}{2} \sim 2000 \times \tan\frac{\Phi}{2} \tag{14.1}$$

This means that for a horizontal field of view, $\Phi = 120°$, one needs an SLM with a horizontal pixel dimension of 3464. Assuming a slightly smaller vertical field of view of $\Phi = 100°$, the vertical dimension must be 2384. This is nearly double the 1080p HDTV standard of 1900 × 1200. Accordingly, today, even for a 100° × 120° hologram, one must use four 1080p SLM panels per primary colour. Combining the optical images of these panels is of course possible, but it considerably complicates the design of the printer. For a true 140° × 140° field of view, an SLM of dimensions around 5500 × 5500 is required. No doubt, such products will become available within the not-so-distant future and, at this point, digital printer design for HVV displays will become much simpler. Until then, one has to either use multiple panels or accept a smaller field of view if one is not to compromise virtual image volume. For instance, a 1080p panel equates to a maximum field of view of 87° × 62° (H × V), if one is not to incur blurring due to SLM insufficiency. If one does build a digital printer using a single 1080p panel (per primary colour channel) with an optical system having a higher numerical aperture, then one introduces a very significant blurring. The maximum image depth is given by the formula

$$x_{CRIT} = \frac{h}{\dfrac{1}{(N-1)}\cot\dfrac{\theta_{Eye}}{2}\tan\dfrac{\Phi}{2} - 1} \sim \frac{h}{\dfrac{2000}{N}\tan\dfrac{\Phi}{2} - 1} \tag{14.2}$$

If we choose a horizontal field of view of even 120°, then, at an observation distance of 0.5 m, we can expect to start to observe blurring due to the insufficiency of the printer SLM at an image depth of only 40 cm! The equations are unfortunately highly non-linear, and there is no way around this. A digital printer capable of printing HVV displays requires sufficient SLM resolution. And for the lowest possible standard of 100° × 120°, this equates to requiring four 1080p panels per primary colour channel.

---

* Of course DWDH itself has limitations in HVV display. For example very fine detail cannot be encoded "close-up". For this computer generated holography (CGH) techniques must be used.
† Also called the horizontal pixel count.

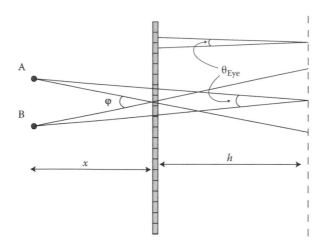

FIGURE 14.38 Diagram showing a side view (or overhead view) of a DWDH hologram. The two points (A and B) are just resolvable when θ_{Eye} is equal to the angular resolution of the human eye. However, the digital hologram can only replay the required rays so that A and B are actually resolved if (a) the SLM which records these rays has sufficient resolution and (b) the hogel size is below a critical value for a given observation distance.

14.4.1.2.4 Aberrations of Printer Optical System

A digital printer capable of printing HVV or window-type displays must have a very high numerical aperture optical system with extremely low aberration. Basically, the optical system must produce an image of the SLM at the given primary wavelength that is capable of resolving the SLM pixels. Holographic optical elements promise to provide a better solution to attaining this goal than current compound objectives.

14.4.1.2.5 Printer Variable Reference Beam System

To avoid introducing aberrations into the final hologram, it is important that the printer possesses a variable reference beam system that is able to properly synthesise a macroscopic point source. Current designs tend to use compound optical systems that work well for normal holographic displays but which may induce too much aberration for HVV displays. This is a technological issue only.

14.4.1.2.6 Constraints on Digital Data

Digital data must be available for each SLM pixel at each hogel coordinate. If part or all of the data originates from a Holocam device, then the Holocam must have the capability of generating the required image resolution. In the case of a final hologram field of view of 140° × 140°, a full-parallax Holocam might realistically need to generate approximately 25 million high-resolution images. If the image data originates from a computer model, then a similar number of render views will be required.

An HVV display differs fundamentally from a holographic window display, which has no foreground data. Such window holograms might portray distant landscapes. An example from the real world would be the view out of the window from an extremely high cliff-face. Only very distant features are visible. To produce this type of hologram, only one high-resolution photograph is required.

14.4.1.2.7 Constraints on Illumination Source

The illumination source is vital for all large-depth holograms. In Chapter 11, we reviewed the important role of source-size and chromatic blurring on the hologram. Most full-parallax reflection holograms nowadays are depth-limited primarily by the illumination source. In fact, it is common to use a broadband source and rely on the Bragg grating to filter the light; this gives the hologram a finite depth. Generally, the size of the source is, however, what finally determines the depth unless the hologram is particularly thick. Thick holograms become dim if lit by a broadband source; therefore, one usually restricts the thickness and tries to increase the permittivity modulation to optimise diffractive efficiency.

As we have seen in Chapter 13, new narrow-band semiconductor illumination sources of much better spatial coherence are now becoming available. This breaks the old equation. One no longer needs to rely on the Bragg grating to assure hologram depth. Of course, a certain frequency selection is still needed so that cross-talk between the three primary illuminating colours does not become an issue. However, chromatic blurring can now be determined solely by the bandwidth of the illumination source. As we have seen in Chapter 11, for chromatic blurring in the reflection hologram to fall below the level of human discrimination, the source needs to have a bandwidth of less than 1 to 2 nm. Speckle noise becomes a problem at bandwidths below 1 nm. Therefore, the ideal bandwidth for an HVV display source is 1 to 2 nm. Source-size blurring falls below the human perception level when the size of the source decreases to less than 1 mm per metre of distance from the hologram. To control image aberration, the wavelength of each illumination source must correspond exactly to a primary-colour laser recording wavelength.

The illumination constraints of HVV or holographic window displays are therefore identical to laser TV. This will undoubtedly mean that the price point of the best illumination sources will fall in due course.

14.4.1.2.8 Constraints on Photographic Material

The photosensitive material is an integral part of any HVV display. Because the angle of view is usually very large for these displays, the efficiency of the hologram must also be high. In addition, high image fidelity can only be attained if high spatial frequencies are recordable. As such, an extremely high-resolution material is required.

14.4.1.2.9 Summary

HVV and holographic window displays primarily require improvements in the current digital holographic printing technology. Currently, no printer has the required resolution. However, the modifications necessary are certainly feasible, even with today's technology. The illumination and materials technologies are also here. This is a real application of ultra-realistic imaging that appears to be well within reach but remains unrealised today.

14.4.2 Updateable 3D Holographic Displays

An updateable holographic display is a 3D display system in which the image data can be erased and reloaded. Following Geola's demonstration [33–36] in 1999 that pulsed lasers could be used effectively in a DWDH digital holographic printer to significantly increase the hogel-write rate and eliminate the interferometric stability problems of digital printers, Zebra filed a US patent in 2003 [37], which described an updateable holographic display using a pulsed laser. At the most basic level, all that was required to transform a pulsed laser DWDH printer into an updateable display system was a suitable material that could temporarily store a hologram before being erased. An experimental device based on a 4 in. × 4 in. photorefractive polymer screen* was then demonstrated in 2010 by workers at the University of Arizona in the United States [38]. Using a 50 Hz single-colour pulsed laser, the small display, capable of a 0.5% diffractive efficiency, could be updated every 2 s using a hogel size of 1 mm². Both single and full-parallax images were used. To attain full colour, the group used a transmission geometry with angle multiplexing.

The updateable holographic display is still in a very early stage of development; as such, it is still currently far from what one might describe as constituting an ultra-realistic display technology. Nevertheless, it is very possible that progress will be made here. Two important areas are the screen technology and the hogel write-rate. Clearly, the technology faces some significant challenges; a typical-size commercial display screen is unlikely to be smaller than 20 cm × 30 cm, and the required hogel size is unlikely to be greater than 0.5 mm. This leads to 240,000 hogels in the best of cases, which one assumes will need to be updated within some seconds to be useful. This in turn leads to a write-rate of approximately 100,000 hogels per second. Using optical multiplexing of a factor of 100, this leads to a required basic write-rate of 1 kHz. Pulsed lasers are certainly available at this repetition rate, and even though current twisted nematic liquid crystal displays cannot operate so fast, there are clearly ways around this. Perhaps more

---

* Produced by Nitto Denko Technical.

important, however, is the data processing required to create the high-resolution image data. Arrays of computers would be needed today to cope with the full-parallax data processing required to update a high-resolution screen of useful size. One answer could be to sacrifice the vertical parallax and to print only HPO holograms. Here, the data-processing issues are trivial, but there are far fewer applications for such restricted parallax displays. Given time, however, the cost of the required computational resources required for full-parallax calculation should fall dramatically.

Updateable holographic displays, unfortunately, suffer from the problem that current devices are based on fast macroscopic movement of an optical head over a suitable photosensitive material. This is a fundamental limitation; unless it can be addressed and a fully solid-state system is identified, the market for updateable displays is likely to be relatively specific. Progress in a number of areas will, in any case, be required to bring such displays to the level of a commercial product. Their first application is likely to be in the military domain, where somewhat larger updateable displays could be useful in mapping and tactical visualisation.

14.4.3 Real-Time 3D Display Technologies Based on Holography

14.4.3.1 Simple Autostereoscopic Systems Based on Holographic Screens

Glasses-free autostereoscopic screens for 3D visualisation systems may be produced using both reflection and transmission holograms [39,40]. Such systems offer some significant advantages over the better-known lenticular or barrier solutions: generally, these types of screens sacrifice two-dimensional image resolution to attain 3D, which is not the case for the holographic screen.

Generally, real-time holographic autostereoscopic visualisation systems work by projecting both a left-eye and a right-eye image, using standard digital projector technology, onto a special holographic screen. The screen then directs the respective images to each eye. By using ultrafine-grain holographic materials to record special gratings, high-resolution screens may be constructed. The difficulty with all standard simple glasses-free autostereoscopic visualisation systems (including those with holographic screens) is that the observer must hold his or her head in a unique horizontal position; this is the unique location where the two images fall into the correct eyes.

One way around this *observer location* problem is to use head tracking. Here, either the entire system, including the screen, continuously moves, tracking the observer's head or, alternatively, optics within the device move in order to move the left and right "eye-boxes". Both solutions are workable in some cases, although moving the optics has quite severe limitations. One of the authors (HB) has recently been involved in developing such a system for 3D visualisation by surgeons of medical endoscopy images with the US company Absolute Imaging Inc. Both transmissive and reflective systems are being developed. Analogue holographic screens have been recorded in cooperation with the Centre of Modern Optics in North Wales. Digitally written screens have been supplied by the Geola organisation. Figure 14.39 shows a prototype autostereoscopic system based on a transmission holographic screen.

14.4.3.2 Multiprojector Autostereoscopic Systems Based on Holographic Screens

Another way around the observer location problem is to use a holographic screen in conjunction with many projectors. Sang et al. [41] have recently produced such a system using a transmission holographic screen of 1.8 m × 1.3 m using 64 digital projectors (Figure 14.40). The system is capable of displaying real-time full-colour HPO 3D images over a total horizontal viewing angle of 45° and can be viewed by many observers at once. Depths of up to 1 m are possible. The group have also developed an array of 64 digital cameras (640 × 480 pixels) so that real-time camera data may be displayed on the screen. A similar technology is being developed in Hungary [42–45]. These types of systems show some promise as extremely small and cheap pico-projectors are becoming available. This will potentially allow the incorporation of many more projectors, thereby improving depth and image resolution. Nevertheless, the fundamental architecture of such a system imposes certain limits and it is therefore difficult to see how useful full-parallax compact systems might be produced. The most likely scenario is that these systems will be adopted first for applications such as military engagement simulation in which large and bulky real-time HPO 3D glasses-free display systems could reasonably be employed for multiperson use.

FIGURE 14.39 *Absolute* medical autostereoscopic 3D display system being demonstrated to Dalibor Vukičević of the University of Strasbourg by Absolute's president, William Pinkerton.

FIGURE 14.40 Professor Tung H. Jeong standing in front of the multiprojector autostereoscopic display system of Sang et al. [41]. The system is based on a large holographic screen, showing here the 3D portrait of Professor Jeong.

14.4.4 Future Real-Time True Holographic Displays

The ultimate application of holography is perhaps the true real-time display of full-parallax 3D images. Potentially, one can easily imagine a type of high-resolution two-dimensional transmission phase-modulated holographic screen (essentially a high-resolution SLM) that could be electrically updated in real-time. RGB laser illumination of such a screen could then be used to generate full-colour real-time images using temporal multiplexing to avoid chromatic cross-talk. Presumably, thin reflective screens could also be used, thereby producing a model for true 3D TV.

However, what appears in various science fiction films to be ostensibly simple is, in fact, rather difficult to achieve. The best way to understand why this is so is to consider the information rate required to write the 3D images. As we have seen in Chapter 9, even a discretised DWDH full-parallax hologram contains a lot of information. Supposing we are able to limit the optical information required to that pertaining to 1 mm² hogels,* then for a display with a reasonable field of view and a reasonable angular resolution, we can calculate that we will require approximately 300 MB of uncompressed memory storage per square centimetre of display. Because for real-time operation, the screen must be updated at a rate of at least 30 Hz, one can see that for a reasonably sized screen of say 40 cm × 50 cm, an information rate of nearly 20 TB/s will be required. In the simplest case of writing the real-time optical interference pattern digitally to a single high-resolution screen, the actual raw digital information required will be rather higher. For example, if one requires a field of view of 100° in both the vertical and horizontal directions, then any transmission grating capable of replaying this will require, using the results of Chapter 11, a maximum pixel spacing of $\lambda/2\sin(50°)$ or roughly 0.35 μm. For a 40 cm × 50 cm display, this then equates to 2×10^{12} pixels. Using temporal multiplexing for three-colour display, this then gives an information rate of approximately 200 TB/s, but even the lower rate of 20 TB/s, which corresponds to producing a real-time hologram from a reduced data set, is extremely high for today's technology and would require massive parallelisation strategies for the image data calculations.

Another major problem is the lack of high-resolution electrically addressable displays. Twisted nematic liquid crystal displays are available today in 1080p panels down to pixel sizes of somewhat less than 10 μm. Recently, Boulder Nonlinear Systems, Lafayette, CO, released a 2 cm × 2 cm one-dimensional panel (composed of 12,288 liquid crystal columns rather than square pixels) with a pixel spacing of 1.2 μm and a gap between pixels of 0.6 μm. The panel is intended for solid-state beam-steering applications and can deflect a beam over a range of around ±5°. Although extremely encouraging, this is still some way from the submicron resolution required for holographic displays with a good viewing angle. Perhaps more important, however, is the fact that today's micropanels would need to be scaled up in area by approximately 1000 times to be useful as a simple holographic screen.

The lack of large ultrahigh-resolution electrically addressable spatial light modulators has led research groups in this field to adopt demagnification and stacking strategies. Several groups have been active in the field for some years. The first group to start serious work was Stephen Benton's group at MIT in the United States [46,47]. Benton used the HPO approximation to simplify these initial experiments, employing computational calculation of the fringe patterns, which were then written in sequence to a one-dimensional tellurium dioxide acousto-optic modulator. A demagnified image of this modulator was raster scanned to produce a composite HPO holographic image of 75 mm × 150 mm × 160 mm with an angle of view of 30°. The work was subsequently extended to full-colour. The most recent MIT work uses a new light modulator technology that uses surface acoustic waves in a slab of lithium niobate [48].

Work at QinetiQ in the United Kingdom [49] has focussed on full-parallax displays using optically addressable spatial light modulators—this type of SLM possesses a significantly better spatial resolution than current electrically addressable spatial light modulators. Computational calculation of the fringe patterns of a full-parallax hologram are performed using clusters of computers and the resulting data displayed on a 1024 × 1024 ferroelectric LCD running at several kilohertz. A demagnified image of this LCD is then (optically) tiled onto an optically addressable spatial light modulator using time multiplexing. A number of such modules can be stacked together to produce a larger full-colour, full-parallax display. Tiling from electrical to optical SLMs can currently produce an effective linear increase in SLM resolution of up to 10 times while maintaining video speeds. Although extremely impressive, these systems are still rather small and have a limited field of view. QinetiQ have so far demonstrated an effective 24 billion pixel display (Figure 14.41)—which really does prove that the technology can work—but the pixel count for a typical-size commercial screen will need to increase by nearly two orders of magnitude before such displays are optically equivalent to today's static full-colour digital holograms.

Stephen Benton was a firm believer that, given enough time, the computational task of calculating the fringe patterns required for full-colour, wide-angle real-time holographic display would cease to

---

* In fact, the quantisation of the hologram into hogels forms the basis of diffraction-specific algorithms for calculating the fringe structure.

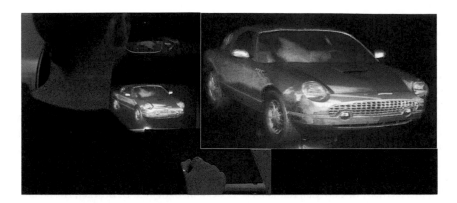

FIGURE 14.41 Real-time digital, full-parallax holographic display (3 × 8 billion pixels) produced by QinetiQ.

become an issue. In 2004, the QinetiQ system used a cluster of 102 Pentium III CPUs each with 1 GB of memory to calculate the required real-time fringe patterns. Since then, as one might expect, the cost and physical size of computing power have decreased substantially. Indeed, it would seem that despite the seemingly enormous task of calculating real-time fringe patterns, if current trends continue, Benton will almost certainly be proved correct. When this stage is reached, ultra-realistic holographic imaging will truly have come of age.

REFERENCES

1. L. Solymar and D. J. Cooke, *Volume Holography and Volume Gratings,* Academic Press, New York (1981).
2. H. J. Coufal, D. Psaltis and G. T. Sincerbox (eds), *Holographic Data Storage*, Springer, New York (2000).
3. Ting-Chung Poon (ed.) *Digital Holography and Three-Dimensional Display—Principles and Applications,* Springer, New York (2010).
4. K. Curtis, L. Dhar, A. J. Hill, W. L. Wilson and M. R. Ayres, *Holographic Data Storage : From Theory to Practical Systems*, Wiley-Interscience, John Wiley & Sons Inc. (2010).
5. S. Benton and V. M. Bove Jr., *Holographic Imaging*, Wiley-Interscience, John Wiley & Sons Inc. (2008).
6. J. Robillard and H. J. Caulfield, *Industrial Applications of Holography,* Oxford University Press, New York (1990).
7. F. Havermeyer, L. Ho and C. Moser, "Compact single mode tunable laser using a digital micro-mirror device ," *Opt. Expr.*, **19**, 14642–14652 (2011).
8. P. Leisher, K. Price, S. Karlsen, D. Balsey, D. Newman, R. Martinsen and S. Patterson, "High-performance wavelength-locked diode lasers," in *High-Power Diode Laser Technology and Applications VII*, M. A. Zediker, ed., Proc. SPIE **7198**, 7198–38 (2009).
9. S. L. Rudder, J. C. Connolly, G. J. Steckman, "Hybrid ECL/DBR wavelength & spectrum stabilized lasers demonstrate high power & narrow spectral linewidth," in *Laser Beam Control and Applications*, A. V. Kudryashov, A. H. Paxton, V. S. Ilchenko, A. Giesen, D. Nickel, S. J. Davis, M. C. Heaven and J. T. Schriempf, eds., Proc. SPIE **6101**, 61010I 1–8 (2006).
10. S. J. Barden, J. A. Arns and W. S. C. Willis, "VPH gratings and their potential for astronomical applications," in *Optical Astronomical Instrumentation*, S. D'Odorico, ed., Proc. SPIE **3355**, 866–876 (1998).
11. J. G. Robertson, K. Taylor, I. K. Baldry, P. R. Gillingham and S. C. Barden, "ATLAS: A Cassegrain spectrograph based on VPH gratings," in *Optical and IR Telescope Instrumentation and Detectors*, M. Iye and A. F. Moorwood, eds., Proc. SPIE **4008**, 194–202, (2000).
12. P. K. Rambo, J. Schwarz, I. C. Smith, C. S. Ashley, E. D. Branson, D. R. Dunphy, A. W. Cook, S. T. Reed and W.A. Johnson, "Development of an efficient large-aperture high damage-threshold sol-gel diffraction grating," *SANDIA REPORT* , SAND2004-5496 (2004).

13. H. Akbari and H. I. Bjelkhagen, "Pulsed holography for particle detection in bubble chambers," *Opt. Laser Technol.* **19**, 249–255 (1987).

14. J. Duplissy, M. B. Enghoff, K. L. Aplin, F. Arnold, H. Aufmhoff, M. Avngaard, U. Baltensperger, T. Bondo, R. Bingham, K. Carslaw, J. Curtius, A. David, B. Fastrup, S. Gagné, F. Hahn, R. G. Harrison, B. Kellett, J. Kirkby, M. Kulmala, L. Laakso, A. Laaksonen, E. Lillestol, M. Lockwood, J. Mäkelä, V. Makhmutov, N. D. Marsh, T. Nieminen, A. Onnela, E. Pedersen, J. O. P. Pedersen, J. Polny, U. Reich, J. H. Seinfeld, M. Sipilä, Y. Stozhkov, F. Stratmann, H. Svensmark, J. Svensmark, R. Veenhof, B. Verheggen, Y. Viisanen, P. E. Wagner, G. Wehrle, E. Weingartner, H. Wex, M. Wilhelmsson and P. M. Winkler, "Results from the CERN pilot CLOUD experiment," *Atmos. Chem. Phys.* **10**, 1635–1647 (2010).

15. H. Bingham, J. Lys, L. Verluyten, S. Willocq, J. Moreels, K. Geissler, G. Harigel, D. R. O. Morrison, F. Bellinger, H. I. Bjelkhagen, H. Carter, J. Ellermeier, J. Foglesong, J. Hawkins, J. Kilmer, T. Kovarik, W. Smart, J. Urbin, L. Voyvodic, E. Wesly, W. Williams, R. Cence, M. Peters, R. Burnstein, R. Naon, P. Nailor, M. Aderholz, G. Corrigan, R. Plano, R. L. Sekulin, S. Sewell, B. Brucker, H. Akbari, R. Milburn, D. Passmore and J. Schneps, "E-632 Collaboration: Holography of particle tracks in the Fermilab 15-Foot Bubble Chamber," *Nucl. Instr. and Meth.* **A297**, 364–389 (1990).

16. X. Sang, C. Yu, M. Yu and D. Hsu, "Applications of digital holography to measurements and optical characterization," *Opt. Eng.* **50**, 091311-1-8 (2011).

17. A. Osanlou, "Holographic digital microscopy in on-line process control," *Opt. Eng.* **50**, 091312-1-4 (2011).

18. J. Watson, "Submersible digital holographic cameras and their application to marine science," *Opt. Eng.* **50**, 091313-1-5 (2011).

19. L. Xu, J. Miao and A. Asundi, "Properties of digital holography based on in-line holography," *Opt. Eng.* **39**, 3214–3219 (2000).

20. D. Carl, B. Kemper, G. Wernicke and G. Von Bally, "Parameter-optimized digital holographic microscope for high-resolution living cell analysis," *Appl. Opt.* **43**, 6536–6544 (2004).

21. P. Jourdain, N. Pavillon, C. Moratal, D. Boss, B. Rappaz, C. Depeursinge, P. Marquet and P. J. Magistretti, "Determination of transmembrane water fluxes in neurons elicited by glutamate ionotropic receptors and by the cotransporters KCC2 and NKCC1: A digital holographic microscopy study." *Journal of Neuroscience* **31**, (No. 33)11846–11854 (2011).

22. H. Bjelkhagen, D. Boggett, P. Crosby, P. Henderson, S. Lilington, E. Mirlis, B. Napier, A. Osanlou, S. Rider, A. Robertson and A. Shore, "Full colour holographic endoscope (Holoendoscope) project" UK-DTI TP/3/IMG/6/I/15727 (2006).

23. H. I. Bjelkhagen, J. Chang and K. Moneke, "High-resolution contact Denisyuk holography," *Appl. Opt.* **31**, 1041–1047 (1992).

24. M. Friedman, H. I. Bjelkhagen and M. Epstein, "Endoholography in medicine," *J. Laser Appl.* **1**, 40–44 (1988).

25. H. I. Bjelkhagen, "Color holography for museums: bringing the artifacts back to the people," in *Practical Holography XXV: Materials and Applications*, H. I. Bjelkhagen, ed., Proc. SPIE **7957**, 0B-1 – 7 (2011).

26. H. I. Bjelkhagen and J. Cook, "Colour holography of the oldest known work of art from Wales," *The British Museum Technical Research Bulletin* **4**, 87–94 (2010).

27. A. Lembessis, "Realistic colour 3D holography: is it here?" *HOLO-PACK HOLO-PRINT 2011*, Las Vegas (2011).

28. H. I. Bjelkhagen and D. Vukičević, "Color holography: a new technique for reproduction of paintings," in *Practical Holography XVI and Holographic Materials VIII*, S. A. Benton, S. H. Stevenson and T. J. Trout, eds., Proc. SPIE **4659**, 83–90 (2002).

29. Y. Gentet and M. K. Shevtsov, "Mobile holographic camera for recording color holograms," *J. Opt. Technol.* **76**, 399–401 (2009).

30. M. Klug, "Display applications of large scale digital holography," in *Holography: A Tribute to Yuri Denisyuk and Emmett Leith*, H. J. Caulfield, ed., Proc. SPIE **4737**, 142–149 (2002).

31. V. Bates, "Down the Rabbithole – Avatar Neytiri with Sprite," *HOLO-PACK HOLO-PRINT 2010*, Malaysia (2010).

32. H. P. Dalkiran, "Hologram – The future of cartographic publishing (Holocartography)," *HOLO-PACK HOLO-PRINT 2010*, Malaysia (2010).

33. D. Brotherton-Ratcliffe, F. M. Vergnes, A. Rodin and M. Grichine, *Method and apparatus to print holograms,* Lithuanian Patent, LT4842, (1999).

34. D. Brotherton-Ratcliffe, F. M. Vergnes, A. Rodin and M. Grichine, *Holographic printer*, US Patent 7,800,803 (filed 1999, granted 2010).
35. D. Brotherton-Ratcliffe and A. Rodin, *Holographic printer*, US Patent 7,161,722 (filed 2002, granted 2007).
36. A. Rodin, F. M. Vergnes and D. Brotherton-Ratcliffe, *Pulsed multiple colour laser*, EU Patent, EPO 1236073 (2001).
37. M. A. Klug, C. Newswanger, Q. Huang and E. Holzbach, *Active digital hologram display*, US Patent 6,859,293 (filed 2003, granted 2005).
38. P.-A. Blanche, A. Bablumian, R. Voorakaranam, C. Christenson, W. Lin, T. Gu, D. Flores, P. Wang, W.-Y. Hsieh, M. Kathaperumal, B. Rachwal, O. Siddiqui, J. Thomas, R. A. Norwood, M. Yamamoto and N. Peyghambarian, "Holographic three-dimensional telepresence using large-area photorefractive polymer," *Nature* **468**, 80-83, (4 Nov. 2010).
39. C. Newswanger, *Real time autostereoscopic displays using holographic diffusers*, US Patent 4,799,739 (filed 1987, granted 1989).
40. D. Brotherton-Ratcliffe, H. Bjelkhagen and J. Fischbach, *Holography apparatus and system,* US Patent Application, Patent Serial Number 11/870,442 (2007).
41. X. Sang, F. C. Fan, C. C. Jiang, S. Choi, W. Dou, C. Yu and D. Xu, "Demonstration of a large-size real-time full-color three-dimensional display," *Opt. Lett.* **34**, 3803–3805 (2009).
42. T. Balogh, *Method and apparatus for displaying three dimensional images*, US Patent 6,201,565 (filed 1998, granted 2001).
43. T. Balogh, *Method and apparatus for generating 3D images*, PCT Application WO2005/117458A2 (2004).
44. T. Balogh, *Method and apparatus for Displaying 3D images*, US Patent 6,999,071 (filed 2000, granted 2006).
45. T. Balogh, *Method and apparatus for producing 3D pictures*, PCT Application, WO9423541A1 (1993).
46. M. Lucente, S. A. Benton and P. St.-Hilaire, "Electronic holography: the newest," in *Int'l Symposium on 3D Imaging and Holography*, Osaka, Japan, November (1994).
47. P. St-Hilaire, S. A. Benton, M. Lucente, J. Underkoffer and H. Yoshikawa, "Realtime holographic display: improvements using multichannel acousto-optic modulator and holographic optical elements," in *Practical Holography V,* S.A. Benton, ed., Proc. SPIE **1461**, 254–261 (1991).
48. D. E. Smalley, *High-Resolution Spatial Light Modulation for Holographic Video*, Thesis submitted for Master of Science in Media Technology, MIT (2008).
49. C. Slinger, C. Cameron, S. Coomber, R. Miller, D. Payne, A. Smith, M. Smith, M. Stanley and P. Watson, "Recent developments in computer-generated holography: toward a practical electroholography system for interactive 3D visualisation," in *Practical Holography XVIII: Materials and Applications*, T. H. Jeong and H. I. Bjelkhagen, eds., Proc. SPIE **5290**, 27–41 (2004).

15

Acronyms

A/D	Analogue to digital
AgX	Silver halide
ANSI	American National Standards Institute
AR	Antireflective (coating)
CAD	Computer-aided design
CAE	Computer-aided engineering
CAM	Computer-aided manufacturing
CCD	Charge-coupled device
CEU	Control electronics unit
CFL	Compact fluorescent lamp
CFSL	Continuous full-spectrum lighting
CGH	Computer-generated hologram
CIE	Commission Internationale de l'Eclairage
CLP	Centre for Laser Photonics (located in OpTIC, Wales, UK)
CMO	Centre for Modern Optics (located in OpTIC, Wales, UK)
CMOS	Complementary metal oxide semiconductor
CRI	Colour rendering index
CT	Computed tomography
CW	Continuous wave (laser)
CW	Coupled wave (theory)
CWC2	Cook and Ward developer
CWT	Coupled wave theory
D/A	Digital to analogue
DCG	Dichromated gelatin
DE	Diffraction efficiency
DEA	Dielectric elastomer actuator
DHM	Digital holographic microscopy
DHR	Digital holographic recording
DI-HO	Digital input – holographic output
DMX	DMX512-A, the ESTA standard for controlling lighting equipment and related accessories
DPSS	Diode-pumped solid-state
DPSSL	Diode-pumped solid-state laser
DWDH	Direct-write digital holography
ECDL	External cavity diode laser
EDTA	Ethylenediaminetetraacetic acid
EM	Electro-magnetic
ERS	Ellipsoidal reflector spotlight
ESTA	Entertainment Services Technology Association
FFT	Fast Fourier transform

FOV	Field of view
F/P	Fabry-Perot (interferometer)
FWHM	Full-width half-maximum
GCM	General Command of Mapping (Turkish military organisation)
GEOLA	General Optics Laboratory (Anglo-Lithuanian group of Companies from 1995 now comprising Geola Digital UAB in Lithuania and Geola Technologies Ltd. in the UK)
GUI	Graphical user interface
H_1	Master hologram to be used for copying
H_2	Transmission or reflection copy of a master hologram (H_1)
H_3	Hologram copied from an H_2 (e.g., an embossed hologram)
HDTV	High-definition television display
HeNe	Helium–neon laser
HeCd	Helium–cadmium laser
HF	High-frequency (electromagnetic radiation)
HiH	Hellenic Institute of Holography
HIRF	High-intensity reciprocity failure
HiPER	High power laser energy research facility
HMOI	Horizontally mounted, overhead-illuminated (type of hologram)
HOE	Holographic optical element
Hogel	Holographic element (the component hologram of a DWDH hologram)
Holocam	Holographic (image data) camera (camera on a motorised rail for 3D image acquisition)
Holopixel	Holographic pixel (same as hogel)
HPIV	Holographic particle image velocimetry
HPO	Horizontal parallax–only (type of hologram without vertical parallax)
HRLF	Holographic reciprocity law failure
HUD	Heads-up display
HVV	High virtual volume (type of holographic display)
HWHM	Half-width half-maximum
ICNIRP	International Commission on Non-Ionizing Radiation Protection
IEC	International Electrotechnical Commission
IMS	Industrial methylated spirits
IPS	Isopropyl alcohol
IR	Infrared
I-to-**S**	Transformation from image data to SLM data in a digital DWDH or MWDH printer
ISIS	Interference security image structures
Laser	Light amplification by stimulated emission of radiation
LCD	Liquid crystal display
LCOS	Liquid crystal on silicon (reflective SLM)
LED	Light-emitting diode

LIDAR	Light detection and ranging
LIRF	Low-intensity reciprocity failure
LSR	Laser speckle reducer
MAX	Abbreviation for 3D StudioMax™ (3D modelling program)
MaxSCRIPT™	Max Scripting language (for 3D modelling)
MB	Methylene blue (dye)
MCU	Main camera unit
MMA	Methyl methacrylate
MOPA	Master oscillator power amplifier
MPGH	Multiple-photo generated holography
MRI	Magnetic resonance imaging
MTF	Modulation transfer function
MWDH	Master-write digital holography
N-CWT	N-Coupled wave theory
N-PSM	N-Parallel Stacked Mirror (theory)
NIKFI	Cinema and Photographic Research Institute in Moscow
NIP	National ignition facility
OEM	Original equipment manufacturer
OPO	Optical parametric oscillator
OTF	Optical transfer function
OVD	Optical variable device
PCB	Printed circuit board
PBU	Phillips Bjelkhagen ultimate bleach
PC	Personal Computer
PET	Polyethylenterephtalate
PMC	Printer and monitoring controller (software package used by Geola UAB in its 2001 DWDH printing systems)
PMMA	Polymethyl methacrylate
PSM	Parallel stacked mirror (theoretical model of a colour holographic grating)
PTP	Peak-to-peak
PVA	Polyvinyl alcohol
R6G	Rhodamine 6G (dye)
RCA	Royal College of Art (London, UK)
RCW	Rigorous coupled wave (theory)
RCWT	Rigorous coupled wave theory
RFM	Ripping, filing and monitoring software package (used by Geola UAB in its 2001 DWDH printing systems)
RGB	Red, green, blue
RMS	Root mean square
SAR	Synthetic aperture radar
SBS	Stimulated Brillouin scattering
SCR	Silicon controlled rectifier
SHG	Second harmonic generation
SHSG	Silver halide–sensitised gelatin
SLM	Single longitudinal mode
SLM	Spatial light modulator

SM-6	Popular developer for pulsed AgX materials (Salim's mistake)
SRS	Stimulated Raman scattering
SSDL	Solid-state dye laser
TEA	Triethanolamine
TEM_{00}	Transverse electromagnetic mode 00
TFT	Thin-film transistor
THG	Third harmonic generation
TIR	Total internal reflection
TMG	1,1,3,3-Tetramethylguanidine (chemical promoter)
TTL	Transistor-transistor logic
UV	Ultraviolet
VCSEL	Vertical surface cavity emitting laser

Laser Crystals

Nd:YAG	Neodymium-doped yttrium aluminium garnate
Nd:YALO	Neodymium-doped yttrium aluminium oxide
Nd:YAP	Neodymium-doped yttrium aluminium perovskite
Nd:YLF	Neodymium-doped yttrium lithium fluoride
$Nd:YVO_4$	Neodymium-doped yttrium orthovanadate
Yb:YAG	Ytterbium yttrium aluminium garnate

Non-linear Crystals

BBO	Beta barium borate
DKDP	Deuterated potassium dihydrogen phosphate
KDP	Potassium dihydrogen phosphate
$KNbO_3$	Potassium niobate
KTP	Potassium titanyl phosphate
LBO	Lithium triborate
PPKTP	Periodically poled KTP

Passive Q-Switches

Co:MALO	Cobalt-doped MALO—$Co:MgAl_2O_4$
Cr:YAG	Chromium-doped yttrium aluminium garnate
V:YAG	Vanadium-doped yttrium aluminium garnate

Semiconductor Materials

AlGaN	Aluminium gallium nitride
AlGaP	Aluminium gallium phosphide
GaAs	Gallium arsenide
GaAsP	Gallium arsenide phosphide
AlGaAsP	Gallium arsenide phosphide
GaInN	Gallium indium nitride
GaN	Gallium nitride
GaP	Gallium phosphide
SiC	Silicon carbide
ZnSe	Zinc selenide

Chemicals

| Acetic acid | CH_3COOH |
| Amidol | (see diaminophenol dihydrochloride) |

Ammonium dichromate	$(NH_4)_2Cr_2O_7$
Ammonium rhodanide	(see Ammonium thiocyanate)
Ammonium thiocyanate	NH_4SCN
Ammonium thiosulphate	$(NH_4)_2S_2O_3$
L-Ascorbic acid	$CH_2OHCHOH(CHCOH:COHCOO)$
p-Benzoquinone	$C_6H_4O_2$
Benzotriazole	$C_6H_4NHN_2$
Borax	(see Sodium tetraborate, decahydrate)
Bromine	Br
Calgon	(see Sodium hexametaphosphate)
Carbon tetrachloride	CCl_4
Catechol	C_6H_4-1,2-$(OH)_2$
Chlorohydroquinone	ClC_6H_3-1,4-$(OH)_2$
Chrom (III) acetate hydroxide	$(CH_3CO_2)7Cr_3OH_2$
Chromium III chloride	$CrCl_3 \cdot 6H_2O$
Chromium III nitrate	$Cr(NO)_3 \cdot 9H_2O$
Chromium III potassium sulphate	$CrK(SO_4)_2 \cdot 12H_2O$
Citric acid, monohydrate	$HOC(COOH)(CH_2COOH)_2 \cdot H_2O$
Cupric bromide	$CuBr_2$
Cupric sulphate	$CuSO_4$
Cupric sulphate, pentahydrate	$CuSO_4 \cdot 5H_2O$
Decahydronaphthalene	$C_{10}H_{18}$
Decalin	(see Decahydronaphthalene)
2,4-Diaminophenol dihydrochloride	$(NH_2)_2C_6H_3OH \cdot 2HCl$
Dimethyl phalate	$C_8H_4(COOH CH_3)_2$
EDTA	(see Ethylenediaminetetraacetic acid)
Elon	(see *p*-Methylaminophenol sulphate)
Ethanol	(see Ethyl alcohol)
Ethylenediaminetetraacetic acid	$(HOOCCH_2)_2NCH_2CH_2N(CH_2COOH)_2$
Ethyl methyl ketone	$CH_3:C(CH_3)COO CH_3$
Ferric chloride	$FeCl_2 \cdot 4H_2O$
Ferric nitrate, nonahydrate	$Fe(NO_3)_3 \cdot 9H_2O$
Ferric sulphate	$Fe_2(SO_4)_3$
Formaldehyde	HCHO
Formalin	(see Formaldehyde)
Glutaraldehyde	$CH_2(CH_2CHO)_2$
Glycerin	(see Glycerol)
Glycerol	$CH_2OHCHOHCH_2OH$
Hydroquinone	C_6H_4-1,4-$(OH)_2$
Hypo	(see Sodium thiosulphate, pentahydrate)
Iodine	I_2
Isopropyl alcohol	$(CH_3)_2CHOH$
Kodalk	(see Sodium metaborate)
Methanol	(see Methyl alcohol)
Methyl alcohol	CH_3OH
p-Methylaminophenol sulphate	$(HOC_6H_4NHCH_3)_2 \cdot H_2SO_4$
4-Methylaminosulphate	$CH_3NH(C_6H_4)_2H_2SO_4$
Methylphenidone	$C_{10}H_{12}N_2O$

Methyl methacrylate	$CH_2:C(CH_3)\ COO\ CH_3$
Metol	(see *p*-Methylaminophenol sulphate)
Paraformaldehyde	$(CH_2O)_x$
Phenidone (A)	(see 1-Phenyl-3-pyrazolidone)
Phenidone (B)	(see Methylphenidone)
Phenosafranine	$C_{18}H_{15}ClN_4$
p-Phenylendiamine	$C_6H_4(NH_2)_2$
1-Phenyl-3-pyrazolidone	$C_6H_5\text{-}C_3H_5N_2O$
Potassium biborate	(see Potassium tetraborate)
Potassium bisulphate	(see Potassium hydrogen sulphate)
Potassium bromide	KBr
Potassium carbonate	K_2CO_3
Potassium citrate	$K_3C_6H_5O_7 \cdot H_2O$
Potassium dichromate	$K_2Cr_2O_7$
Potassium dihydrogen orthosulphate	$K_2H_2PO_4$
Potassium ferricyanide	$K_3Fe(CN)_6$
Potassium hydrogen sulphate	$KHSO_4$
Potassium hydroxide	KOH
Potassium iodide	KI
Potassium metabisulphite	$K_2S_2O_5$
Potassium nitrate	KNO_3
Potassium permanganate	$KMnO_4$
Potassium persulphate	$K_2S_2O_8$
Potassium phosphate, dibasic	K_2HPO_4
Potassium pyrosulphite	(see Potassium metabisulphite)
Potassium rhodanide	(see Potassium thiocyanate)
Potassium thiocyanate	KSCN
Propanol	(see Propyl alcohol)
Propyl alcohol	$CH_3CH_2CH_2OH$
Pyrocatechol	(see Catechol)
Pyrogallol	$C_6H_3\text{-}1,2,3\text{-}(OH)_3$
Quinol	(see Hydroquinone)
Quinone	(see *p*-Benzoquinone)
Rhodamine 6G	$C_{28}H_{31}N_2O_3Cl$
Rhodamine B	$C_{28}H_{31}ClN_2O_3$
Silver nitrate	$AgNO_3$
Sodium acetate	CH_3COONa
Sodium bisulphate, monohydrate	$NaHSO_4 \cdot H_2O$
Sodium borohydride	$NaBH_4$
Sodium carbonate, anhydrous	Na_2CO_3
Sodium hexametaphosphate	$(NaPO_3)_6$
Sodium hydrogen carbonate	$NaHCO_3$
Sodium hydrogen sulphate	(see Sodium bisulphate)
Sodium hydroxide	NaOH
Sodium metabisulphite	$Na_2S_2O_5$
Sodium metaborate tetrahydrate	$NaBO_2 \cdot 4H_2O$
Sodium metasilicate	$Na_2SiO_3 \cdot 5H_2O$
Sodium nitrite	$NaNO_2$

Sodium sulphate	Na_2SO_4
Sodium sulphite, anhydrous	Na_2SO_3
Sodium tetraborate, decahydrate	$Na_2B_4O_7 \cdot 10H_2O$
Sodium thiocyanate	NaSCN
Sodium thiosulphate, anhydrous	$Na_2S_2O_3$
Sodium thiosulphate, crystal, pentahydrate	$Na_2S_2O_3 \cdot 5H_2O$
Sulphuric acid	H_2SO_4
Tetraethylammonium bromide	$(C_2H_5)_4NBr$
Thiocarbamide	(see Thiourea)
Thiourea	H_2NCSNH_2
Triethanolamine	$(HOCH_2CH_2)_3N$
Urea	NH_2CONH_2
Vitamin C	(see Ascorbic acid)
o-Xylene	$C_6H_4(CH_3)_2$

Appendix 1: Historical Origins of Display Holography: Spreading Awareness

A1.1 Hologram Exhibitions

It has always been important to make people aware of holography, and one of the best ways to do this has been through the organisation of hologram exhibitions. The very first hologram exhibition took place in 1968 at Cranbrook Academy in Michigan. Margaret Benyon exhibited her first holograms in 1969 at the Nottingham University Art Gallery in England. In 1975, a large exhibition, *Holography '75: The First Decade*, was held at the International Center of Photography in New York. It was organised by Joseph (Jody) Burns, Jr. and Rosemary (Posy) Jackson and represented the work of artists and scientists from the United States and six other countries. Artists and scientists met for the first time to acknowledge each other's presence in the field—this resulted in the first use of the word *holographer*.

A sense of community grew. In March 1976, Burns and Jackson arranged the first exhibition outside the US at the Cultural Centre in Stockholm, Sweden: *Holografi: Det 3-Dimensionella Mediet*. The exhibition was extremely successful, with 60,000 visitors during the 2 weeks (12–28 March) over which it was held, breaking all previous attendance records for the most people per day, per week and per event. The second exhibition in New York, *Through the Looking Glass*, opened in December 1976, this time at the newly opened Museum of Holography. In 1977, the exhibition moved to Toronto in Canada and later to other places.

In 1977, the first French hologram exhibition, *Sculptures de Lumière*, took place in Strasbourg. Nick Phillips of Loughborough University was part of the HOLOCO team (with Anton Furst and John Wolff) and was involved in recording large-format display holograms. HOLOCO was supported by the rock group, the WHO. Phillip's holograms were on display at two large exhibitions, both arranged at the Royal Academy of Arts in London: *Light Fantastic I* in 1977 and *Light Fantastic II* in 1978. The first exhibition was very successful, with 96,000 people in 28 days and created a lot of interest in holography. The second exhibition attracted 250,000 visitors.

Many successful hologram exhibitions were arranged in many countries in the late 1970s and early 1980s. *Alice in the Light World* was one such exhibition held in August 1978 at the Isetan Department Store in Tokyo, Japan. Many of Nicholson's portraits from Hawaii were part of this show. In 1979, the number of exhibitions increased; shows occurred, for example, in Eindhoven, Rotterdam, Berlin, Milan and Liverpool in Europe. In 1980, there were many international exhibitions and also one of interest in London—*Light Years Ahead*—at the Photographer's Gallery; this show broke all previous attendance records.

Another UK exhibition was *Light Dimensions*, which took place in Bath in 1983 at the National Centre of Photography. This was the world's biggest hologram exhibition thus far and was organised by Eve Ritscher Associates in London. Her Royal Highness Princess Margaret, Countess Snowdon, opened the exhibition on 21 June 1983. The exhibition moved to The Science Museum in London in December that year and continued until the end of April 1984.

The Art and Science of the Soviet Union was a 1985 exhibition of holograms from the former USSR organised at the Trocadero in London. Mostly, it contained monochrome Denisyuk holograms of artefacts from the Hermitage and other museums (Figure A1.1). A total of 132 holograms—many, very large rainbow plates—were on display at the exhibition *Decouvrez l'Holographie*, which was organised by Ap-Holographie in France, and supported by Philip Morris, France and Paris Match. It opened in May 1985 and took place at the Palais de la Decouverte in Paris.

FIGURE A1.1   Trocadero exhibition entrance.

Teit Ritzau in Copenhagen, Denmark, arranged a permanent hologram exhibition *Holography or Reality* at the *Holographic World* in the Tivoli Gardens, H. C. Andersen's Castle, in 1986. It had many large Denisyuk holograms from the Cinema and Photographic Research Institute (NIKFI) in Moscow. The holograms were integrated in specially designed rooms. It continued for many years.

The travelling exhibition, *Images in Time and Space*, was referred to as "the most significant international holography exhibition of the decade." It was organised by the Associates of Science and Technology Inc. in Ottawa, Canada. It opened at Montreal's EXPOTEC in 1987. With more than 170 historical, scientific and artistic works, it drew large crowds, estimated at 250,000, in its first three month period. In 1988, it moved to San Jose, California where the exhibition ran into financial problems. This marked the end of large international holographic exhibitions.

Smaller exhibitions, however, continued to be organised in association with holographic conferences such as those at Lake Forest and the Millennium Conference on Holography in St Pölten, Austria. After the fall of the Berlin Wall, there were also a number of significant exhibitions of holography in the countries of Eastern Europe. One such exhibition was organised by one of the authors (DBR) in Bucharest, Romania in 1996. The exhibition, which featured many large-format holograms from Australian Holographics and Geola, was held at the Bucharest Polytechnical University and attracted many visitors; the opening ceremony was attended by the vice president of the house of deputies, Ion Ratiu. Sizeable exhibitions were also held in Hungary, Lithuania and Estonia.

In 2009, Jonathan Ross in London started to arrange touring hologram exhibitions in the United Kingdom, mainly with holograms in his collection. These exhibitions attracted many visitors, most of whom were aware only of credit card holograms at best, and who now had the opportunity to see large art and display holograms for the first time. A large Belarusian exhibition, *HOLOEXPO 2011*, with holograms from around the world, opened in Minsk in October 2011. In addition to various historic holograms, it included many new analogue and large digital colour holograms.

A1.2 Commercial and Educational Entities Involved in Holography

A1.2.1 Companies Producing Holograms

Over the years, holographic companies and individual holographers around the world have produced holograms and hologram products. One of the first companies to offer holograms for sale was *The Holex*

Corporation in Norristown, Pennsylvania. The company's president, Larry Goldberg, claimed that they were "creating a new reality". The holograms were bleached transmission film holograms delivered with a converter filter, peaked at a wavelength of 580 nm, designed to be used with a slide projector for illumination of the hologram with monochromatic light. The company also offered the "solar powered" *Spectrol hologram*, which was a rainbow hologram that could be viewed in sunlight or illuminated with an ordinary tungsten light bulb. The holograms were distributed through the Edmund Scientific company in the United States.

In the early 1970s, companies started to offer custom-made display holograms or limited edition holograms produced by artists and commercial holographers around the world. Only a few of these companies remain today and, if they are still active, they are mainly producing holograms for the document security market. *Hologram Industries* run by Hughes Souparis in France is an example of this.

A1.2.2 Hologram Galleries

In addition to advertising or exhibition holograms, holographic products and holograms were also manufactured for shops and galleries. Outlets were established in many shopping malls and city centres during the 1980s. There were many hologram galleries established in both Europe and the United States. One of the first to open a gallery was Gary Zellerbach with his *HOLOS Gallery* in San Francisco. In 1981, Peter Woodd opened Britain's first permanent gallery of holography named *Light Fantastic*, which was located in Covent Garden in London. However, most of these galleries closed in the early to mid-1990s, including the *Light Wave Gallery* in North Pier, Chicago shown in Figure A1.2.

FIGURE A1.2   Part of the *Light Wave* hologram gallery in Chicago.

A1.2.3 Holographic Museums

Various holographic museums opened during the 1970s, but most of them have closed now. One such museum was Posy Jackson's *Museum of Holography* located in New York, which opened in 1976. The important historic hologram collection of this museum is now preserved as part of the *MIT Museum*. One of the earliest and the longest running museums, which opened in 1977, was Loren Billing's *Gallery 1134: Fine Arts Research and Holographic Center*. Later, it was renamed as the *Holographic Museum in Chicago*. It closed in 2010 when Loren Billings was not able to continue running it because of age and poor health. In 1977, Denisyuk opened the *Museum of Holography* in St. Petersburg; now closed, the holograms are kept in the Vavilov Institute.

In Europe, Matthias Lauk's *Museum für Holographie & Neu Visuelle Medien* in Pulheim near Cologne, Germany, opened in 1979 with a large collection of holograms, including several of the McDonnell Douglas holograms, but it is no longer open. In 1980, Anne Marie Christakis opened the *Musée Française de l'Holographie* in Paris, but again, this museum closed some years ago. Christakis retained the collection of holograms and arranges exhibitions now and then. Another German museum, *Holowood*, was opened in the 1990s by Matthias Frieb in Bamburg and included several of the large-format rainbow holograms from Australian Holographics. Frieb's museum closed when the Bamburg city council undertook extended public works and tourists could no longer easily reach the museum's location.

A1.2.4 Educational Institutes

In the beginning, university physics and electrical engineering departments around the world were the main places where students could learn about holography. Many of the museums also opened educational centres aimed at display and art holography with artist-in-residence programmes. Tung H. Jeong (TJ) at Lake Forest College, located outside Chicago, started in 1972 to offer week-long courses on holography at the college and, later, he arranged the first *International Symposium on Display Holography* with associated workshops in 1982. Jeong's conferences focussed on art and display holography and attracted both scientists and artists from around the world. Jeong was also behind the *School of the Art Institute of Chicago*; he started giving lectures on holography there in 1975. Ten years later, holography became a full-time instructional area. Specialised schools of holography for artists opened as well—for example, Lloyd Cross' school in San Francisco, which opened in 1971. In the beginning, the San Francisco area became a main centre for artists to learn about holography. The *Holography Institute* opened in 1980 and, for many years, was run by Jeffrey Murray.

In 1973, Jody Burns, Jr. opened *The New York School of Holography*. In 1978, it was taken over by Daniel Schweitzer (1946–2001) and Samuel Moree and is now run by Moree as *Holographic Ocean Laboratories*. In 1974, David Hlynsky and Michael Sowdon opened the *Fringe Research* educational centre in Toronto, Canada. In Europe, in 1980, Michael Wenyon started to run workshops at the *Goldsmiths' College of Art* in London; he continued this activity until 1983. London became the United Kingdom's centre for educational art and display holography through the *Royal College of Art* courses and Edwina Orr's private *Richmond Holographic Studios*. Both places offered pulsed holography courses. Dieter Jung's *Academia of Media Arts* in Cologne, Germany also attracted many art holography students over the years. In Japan, the *Holography School of the Tamara Art College* in Dawasaki opened in 1980.

Appendix 2: History of the Geola Organisation

A2.1 The Beginning

The chain of events that led to the creation of Geola can probably be traced back to a trip made by one of us, David Brotherton-Ratcliffe, to Soviet Russia in the late 1980s just before perestroika. David returned from the three-week trip, famished but deeply impressed by what he had seen in Moscow and Leningrad. At the time, he had been working as a postdoctoral fellow at the Flinders University of South Australia on theoretical physics problems connected with controlled nuclear fusion. In 1986, he had met another Flinders postdoctoral fellow, Igor Bray, now director of the Institute of Theoretical Physics at Curtin University in Western Australia. It was with Igor that David made this first trip to Russia.

David left Flinders University in 1989 to form his own company, Australian Holographics Pty. Ltd., with the aim of manufacturing large-format holograms for advertising and display. While searching for laboratory optics and lasers for this company, David met Dmitry Konovalov, a newly arrived Russian postdoc at Flinders. Dmitry, who now worked with Igor, told David that it might be possible to source the items that Australian Holographics needed from the Soviet Union. This was now the time of perestroika and the borders of the Eastern Bloc were starting to open. The two men guessed that if Soviet laser and optical equipment turned out to be useful to Australian Holographics, then perhaps universities and other companies in Australia would also be interested. With this idea in mind, a new company, General Optics Pty. Ltd. was formed and Dmitry started to look for contacts in the Soviet Union.

On 11 March 1990, Lithuania had declared independence from Soviet Russia and it became possible to think seriously about exporting from Russia through Lithuania to the West. By mid-1991, Dmitry had made contact with Lev (Leon) Isacenkov, who was an early postindependence Lithuanian entrepreneur—Leon spoke fluent Russian, Lithuanian and English—and could also be contacted, albeit with great difficulty, by telephone. By the end of 1991, Dmitry had come up with a list of possible photonic products, all with price tags of around a hundred times cheaper than equivalent western products. The one problem was that there was no serious possibility of transporting the goods out of Lithuania. David, remembering the fascination of his earlier trip to Soviet Russia, therefore, decided to drive a lorry himself to Lithuania to pick up the first shipment for General Optics (Figures A2.1, A2.2 and A2.3).

David set off from London in early February 1992. He invited John Fenton, a London music producer to accompany him on the trip for company and also for greater security. The trip took them through France, Belgium, West Germany, East Germany, Poland and finally into Lithuania through the Lazdijai border crossing. The roads at this time were often rather poor and weather conditions in Poland were difficult. The border crossings to both Poland and Lithuania had immense queues of traffic. The Lithuanian queue was easily several miles long; here, David and John understood through somewhat awkward attempts to communicate with other drivers that the wait could last many days. With temperatures going down to −20°C, they decided to take their chances and made their way to the front of the enormous queue. After being stopped by armed Polish guards with submachine guns, they were forced to stay in the lorry for several hours. Eventually John decided to brave it and walked to the soldier's cabin with the idea of explaining that he was a western journalist. Somehow, this worked and 10 minutes later, the lorry was waived through to the Lithuanian side. Here, John distributed copies of western music magazines to the soldiers who became amazingly friendly and welcomed the two Englishmen to Lithuania. As the lorry left the Lazdijai border crossing, the Lithuanian border guards, music magazines in hand, all gathered to wave them off.

Unfortunately, with the bad roads, bad weather and the problems with the border crossings, the lorry was 30 hours late for its scheduled meeting with Leon Isacenkov. They had agreed to all meet up at the border crossing, but after 12 hours of waiting, Leon had returned to Vilnius, disappointed. David and

FIGURE A2.1 David Brotherton-Ratcliffe, in early 1992, seen here driving General Optics' first lorry back from Lithuania.

FIGURE A2.2 Leon Isacenkov (early 1992) in the Vilnius offices of Infortechnika UAB during the first meeting with David Brotherton-Ratcliffe.

John had barely enough petrol left in the lorry to reach Vilnius. In those days, there were no satellite navigation systems and all the road signs in Lithuania were still written in Cyrillic. David had been awake for probably 45 hours by now, as John did not drive. The travellers therefore decided to park the lorry and sleep.

In the morning, David and John awoke to find the lorry surrounded by people who had apparently never seen a western lorry before. This would become a familiar pattern that would follow the two men pretty much wherever they went in the coming weeks. In the centre of Vilnius, with the petrol gauge firmly on empty, David parked the lorry, and once again, the lorry was surrounded, this time by a group of children. By chance, one of the children, a boy who was around 11 years old, spoke a little English; David asked him if he could somehow make a telephone call to the telephone number he had for Leon Isacenkov. The child took David back to his parents' apartment where he dialled Leon's number using the family telephone. The child's parents were both out. Within 30 minutes Leon arrived.

FIGURE A2.3 Large building on Naugarduko Street (seen here in 2003), which has been home to Geola since 1994. Geola occupies around 1000 m² of laboratories and offices over three floors. The entrance is almost exactly in the centre of the picture, just in front of the bus stop.

A2.2 First Meeting in Vilnius

Leon had worked in a well-known electronics institute in Vilnius. After Lithuania's declaration of independence, however, he decided to leave and set up his own business, Infortechnika UAB, assembling and selling computers. Over the next few weeks, David accompanied Leon on numerous visits to factories and universities in Lithuania and in Moscow. At the time, the train connection between Vilnius and Moscow was still uncontrolled and it was easy to go between the two cities. By the end of the trip, an official cooperation had been signed by General Optics and Infortechnika for the export of scientific equipment. David returned to the United Kingdom with a lorry full of holographic plates, large-frame helium–neon lasers, optical tables, holograms and optical components (Figure A2.4). This was the first of what would be many such journeys.

The equipment purchased in Lithuania from Infortechnika proved extremely interesting. General Optics started to make sales, most notably to university departments. Australian Holographics opened a shop called Rainbow Bridge in South Australia, which sold, amongst other things, the Russian holograms from General Optics. Australian Holographics also started to use the Soviet equipment itself in its own laboratories.

As their joint business started to produce results, Leon and David decided that they needed a laboratory in Vilnius to test the equipment being exported. Orders soon became quite complex, with university customers ordering custom-pulsed lasers. These lasers needed verification before shipping. With this in mind, Leon asked his friend and colleague, Viktor Karaganov to head up a small team responsible for this task in Vilnius. Within the year, however, Viktor had decided to take up an offer of employment at Flinders University in Australia (invited by Igor and Dmitry's group) and with his wife, Sveta, left for Australia in 1993. In Australia, Sveta became office manager and accountant for Australian Holographics. Previously, she had been an accountant and lawyer for Infortechnika. Viktor also became

FIGURE A2.4 Large-frame He–Ne lasers brought back from Lithuania on the first lorry in 1992 being unloaded at Paxton Automation Ltd. in the United Kingdom. Note the contract marking visible in (a) on one of the cases: 12/91/ AU-001—this was the first of many contracts between General Optics and Infortechnika.

a part-time scientific consultant for the company. In Vilnius, Mikhail Grichine left his job at a large scientific laser laboratory in the Shatura region outside Moscow and moved to Vilnius with his wife Galia to replace Viktor. Mikhail was an expert in Soviet laser technology and saw the possibility not just to export lasers but also to build them.

In fact, as the sales of General Optics got better and better, a general problem appeared with the manufacturers in Russia. Their electricity was being turned off, they were complaining about unserviceable equipment and salaries were not being paid. The vast Soviet scientific infrastructure was in effect collapsing. In certain cases, laboratories and companies managed to rearrange themselves in time and avoid being closed. In other cases, activities ceased. In any case, it was slowly becoming more difficult to source the cheap optical items Dmitry Konovalov had identified in 1991. The decision was therefore taken to follow Mikhail's suggestion and to design and actually produce pulsed lasers in Vilnius.

In parallel to the events in Vilnius and Australia, in 1993, David took part in several artistic exhibitions of holography in Paris. In one such exhibition, he met Florian Vergnes, then a young Parisian entrepreneur just out of university. Florian was fascinated by the large holograms that David displayed at the exhibition. When David showed him examples of Soviet optical components several days later, Florian proposed that they open a joint company together. David accepted and, with several other associates, they formed LMC France Instruments SARL. Within several months, LMC had signed a rental contract and the company started business in St. Denis. For the next few years, David lived mostly in Paris, travelling frequently to Adelaide and Vilnius. In the United Kingdom, Ralph Cullen from UK Optical Supplies Ltd. had become a distributor for General Optics and was helping with UK sales. LMC developed a customer base across mainland Europe. Simon Edhouse had joined Australian Holographics and was now also sales director for General Optics. Simon started to look after sales in Australasia and David concentrated more on Europe.

A2.3 Incorporation of Geola UAB

By some time in early 1994, problems with the supply end of the import/export business had reached a level where it was decided that more resources had to be directed at producing a stable western-type office in Vilnius. More effort needed to be spent forging relationships with newly emerging Soviet enterprises and in-house manufacturing had to be accelerated. With this in mind, David moved from Paris to Vilnius, leaving Simon Edhouse to run General Optics in Australia and Florian Vergnes to run LMC in France. With Mikhail Grichine and Leon Isacenkov, David identified two new engineers who joined the Vilnius team. These were Gleb Skokov (Figure A2.5) and Aleksej Rodin. Gleb had just finished technical university in optical physics and Aleksej was in the middle of his PhD in quantum electronics. A new company called Geola UAB was formed. It was Mikhail who thought up the name—an acronym for General Optics Laboratory.

Serendipitously, just at this time, the large Soviet electronics institute on Naugarduko Street, where Leon had worked during the communist period, was partially privatised and it became possible, for a minimal price, to buy rooms in this enormous building. Both David and Leon took the opportunity and brought in builders to renovate the spaces. Within months, Geola had a small modern office in the heart of Vilnius (Figure A2.3).

Geola worked closely with General Optics, LMC, but also with Australian Holographics. In Vilnius, they built a pulsed holographics facility based on a 5 J neodymium laser for the creation of large-depth transmission holograms (Figure A2.6). Several of Australian Holographics' large stock holograms were produced in this studio as well as holograms for well-known artists such as H. R. Giger (Figure A2.6).

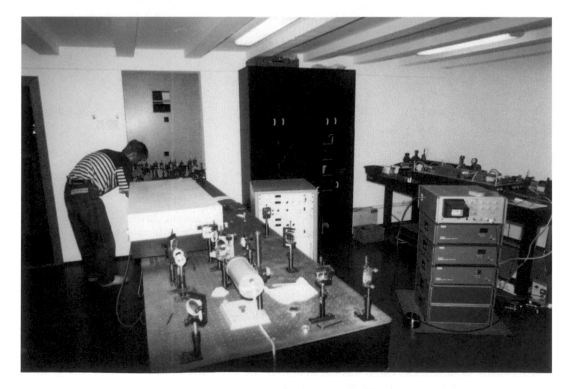

FIGURE A2.5 Gleb Skokov working in 1994 at Geola UAB in the main laser room.

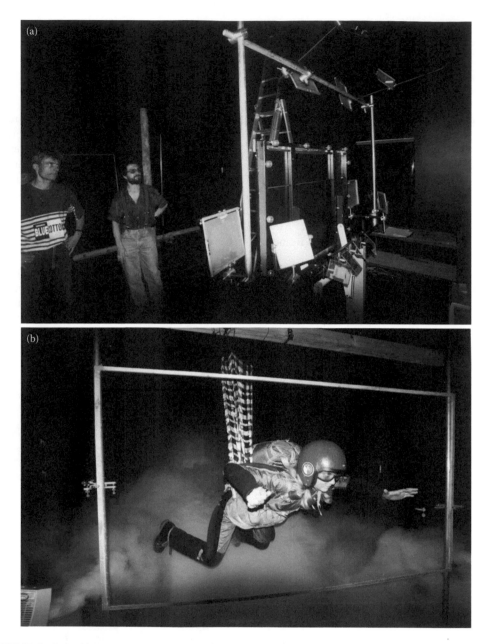

FIGURE A2.6   Pulsed hologram room at Geola UAB in 1994. (a) Gleb Skokov and Mikhail Grichine looking at the large plate holder and laser diffuser system. (b) A large-format rainbow master hologram being shot for Australian Holographics.

Sergei Vorobiov visited and taught Mikhail and David the basics of human pulsed laser portraiture. Aleksej Rodin worked on building a Raman laser to make colour pulsed holograms.

In the winter of 1994, David unfortunately contracted pneumonia—there was often no heating in the average Vilnius apartment then and if you wanted to remain clean, one was obliged to take cold showers. Of course, in the winter, the gas pressure fell, so you could forget about heating the water. Most people showered rarely to cope with this phenomenon, but westerners like David found this habit unacceptable. The inevitable occurred, and David was forced to travel back to Paris, having undergone an old Soviet treatment that would leave him weak for some years to come.

A2.4 First Romanian Exhibition of Large-Format Holography

Some months later, partially recovered from the pneumonia, David was walking near his apartment in the centre of Paris, when he noticed that he was in a street called "*rue de Bucarest*". Having nothing better to do that weekend and having never visited Romania, he decided to buy an airline ticket to Bucharest. In Bucharest, he tried to identify people who could be interested in holography. During this process, he met several people by chance whom he became close friends with. Over the next few months, David visited Bucharest and Transylvania many times, after which he decided that it would be a good idea to spend some more time there. In mid-1995, he rented a flat in Petru Poni Street next to the north railway station and began organising the first Romanian exhibition of large-format holography with Professor Chisleag from the Polytechnic University of Bucharest.

The Romanian exhibition was a great success, but afterwards, David fell ill again and remained in Romania for more than a year convalescing. During this time, the Australian Holographics business was reorganised; Simon Edhouse formed his own company to market the holograms (this became Australian Holographics Pty. Ltd.), and David retained the manufacturing facility (which became Australian Holographic Studios Pty. Ltd.). Geoff Fox remained chief holographer for some time at AH Studios until his apprentice, Mark Trinne, took over. Florian Vergnes, delegating his duties at LMC, moved to Vilnius and became general manager of Geola UAB. By 1997, Geola had grown to around 10 people and the company had already constructed and delivered its first neodymium-pulsed laser. It had also produced its first holographic portraiture system, and by 1998, it was selling such systems. The company now occupied a space of around 1000 m^2. Florian visited the Slavich company in Moscow many times and signed an exclusive distribution agreement for the international sale of Russian holographic materials.

In late 1998, David, now completely recovered from his earlier illness, returned to live in Vilnius. He and Mikhail decided to look at making digital holograms with pulsed lasers. Soon, they had a small pulsed laser printer assembled that could print master H$_1$ type holograms using a Sony LCD display. At this time, Eric Bosco, then working at Pixel Systems Inc. in Montreal, contacted Geola and ordered an 8 J pulsed neodymium laser. Subsequently, it seemed that Eric's company was interested in a larger collaboration to develop digital pulsed holography. Geola and Pixel started negotiations for an extensive partnership whereby Geola was to undertake a funded scientific research programme with a view to creating a commercial full-colour digital holographic printer. Pixel wanted two types of systems investigated—single-step and two-step printers. Geola had already demonstrated a monochromatic two-step system but lacked appropriate RGB pulsed lasers for a colour system. During the negotiations, David worked out how to build both types of printing system. Mikhail Grichine and Aleksej Rodin also thought they knew how to build the pulsed lasers. In 1999, David and Florian flew to Montreal to sign the multi-million dollar contract between Pixel and Geola. On the aeroplane back, the two men celebrated with a bottle of wine and David, joking, mentioned that the only thing that could go wrong was if the Canadian company was put into liquidation.

Unfortunately, Pixel was indeed inexplicably wound up the next day and Geola was forced to tear up the contract. Nevertheless, Geola had advanced during the negotiations with Pixel to such a point that it was now capable of producing both one- and two-step printers. Within several months, it had demonstrated the technology in its laboratories and David had filed a Lithuanian patent for the first RGB pulsed digital holographic printer.

A2.5 Start of XYZ

Eric Bosco was, however, still on the case and sought to remedy the mess at Pixel by reorganising the company's digital holography team into a new company, XYZ Imaging Inc. Within a year, Eric had reached an agreement with Geola, Pixel's main shareholder and with several Canadian venture capital organisations to go ahead with the project. Contracts were signed in Montreal and XYZ and Geola joined forces to develop digital colour pulsed holography. Geola became the major shareholder—a situation that lasted for quite some time—and had two directors on the board of XYZ. Nonetheless, it was the institutional shareholders who really controlled XYZ, as it was their money. XYZ was to develop one-step printing, but Geola retained rights to its laser and two-step printing technology. The first task for Geola under the new agreements was to construct a digital RGB pulsed laser printer, which was installed in Montreal in 2001 (Figure A2.7). Mikhail Grichine left the company to work at Ekspla UAB, another well-known laser manufacturer in Vilnius, and Stanislovas Zacharovas, who had already been working with Geola for several years as head of sales and marketing, became Geola's director. Aleksej Rodin became chief laser engineer and played a pivotal role in producing a stable pulsed RGB laser. Marcin Lesniewski designed the critical optical writing objectives.

A new UK company, Geola Technologies Ltd., was formed in 2000. Some Canadian investors were nervous about building a manufacturing organisation in an ex-Soviet country like Lithuania. The new UK company would manufacture the critical RGB pulsed lasers required by the XYZ printers. It would also do research for new types of more compact lasers for the future. David moved back to the United Kingdom to become managing director of Geola Technologies Ltd. Aleksej Rodin also moved to the United Kingdom, as did Nataly Vidmer, one of Geola's highly talented precision mechanical engineers. Over the next few years, the team in Vilnius grew to around 30 people whereas the UK team increased to around 10. In Montreal, XYZ hired around 25 people, some of whom came from Geola or were trained in Geola.

Every so often, David and Eric would meet somewhere (Paris, London, Vilnius, Montreal, etc.) to discuss the way forward. Both men were adamant that their respective teams should cooperate well. All too often, sensitivities between the Lithuanian, English and Canadian teams developed which could, without intervention, erupt into disputes. There were also sensitivities on the side of XYZ's institutional investors—some found it difficult to work with a company in an ex-Soviet country and felt intrinsically uncomfortable with this. Eric managed both sides of the equation with great mastery. Although he now had some brilliant engineers in-house—Eugene Kosenko, Jean-Jacques Cotteverte, Vladimir Fedorenchik, Roman Rus and many others—he realised that XYZ and Geola still had much to gain through close cooperation.

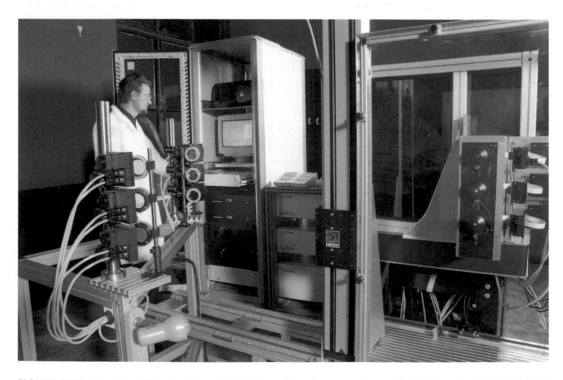

FIGURE A2.7   Very first RGB pulsed laser digital holographic printer manufactured in Vilnius at Geola UAB for XYZ Imaging Inc., seen here in 2001 with Geola's first electronics engineer, Sergei Cimliakov.

A2.6 Panchromatic Film and Sfera-S

One of the critical areas of concern to XYZ was the availability of a suitable panchromatic material. The DuPont material that Zebra Imaging was using was not available to XYZ or Geola and anyhow there were doubts as to its sensitivity to nanosecond pulses. Eric worked with many people to try and produce a film product. However, only one produced a real commercial product that XYZ could really use for mass production. This was a silver halide emulsion developed by Geola's partner, Yuri Sazonov. Yuri was the ex-Micron Director at Slavich and had worked closely with Geola for some years. Geola had after all been appointed the international Slavich sales office some years before. Yuri formed a Russian company, Sfera-S AO, to commercialise his new material. He then signed an exclusive agreement for the distribution of this product through Geola. Eric decided to order a holographic plate plant through Geola and to license this new silver halide product so that XYZ could have an in-house facility for a suitable photosensitive material. Eric also travelled to Moscow with David to meet Yuri at Sfera-S (Figure A2.8). Later, Yuri and Stanislovas installed the equipment in Montreal and Eric hired chemists to start producing holographic plates for his digital printers. But despite now having a facility to make large holographic plates suitable for pulsed colour holograms, XYZ could not make a film product. For this, a tripartite agreement between Geola, XYZ and Sfera-S was signed.

XYZ's first printers (Figures A2.9) to be installed outside Montreal were in Paris. By 1995, investors were starting to lose patience with the business and there was increasing pressure on Eric to start commercial sales. The company had started to win prestigious awards for its full-colour digital holograms and, spurred on by this success, XYZ decided to accept the first commercial orders it received. In retrospect, this was premature and the company rushed into this process. However, new US investors had recently come on board and were not in the mood to hang around. The company probably would have been shut down by these investors had another decision been taken. Unfortunately, these first printers seemed to have had various flaws, which stopped them from functioning perfectly. XYZ was called on to make many service visits to Paris and eventually this became extremely expensive. At about this time, it became clear that XYZ would have to slow down its programme for printer sales and as such, smaller

FIGURE A2.8 Visit of groups from Geola and XYZ to Slavich and Sfera-S near Moscow in 2003. From left to right, Yuri Sazonov (director of Sfera-S), Eric Bosco (president of XYZ), Olga Gradova (Sfera-S), Lynne Hrynkiw (XYZ), Stanislovas Zacharovas (director of Geola UAB) and Aleksej Rodin (scientific manager, Geola Technologies Ltd.).

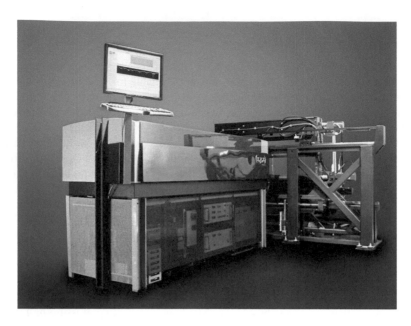

FIGURE A2.9 First commercial RGB pulsed laser digital holographic printer produced by XYZ (2005)—examples were installed in Montreal, Paris and Vilnius.

quantities of film would be required from Sfera-S. Unfortunately, Sfera-S had just scaled up operations for film production to meet its commitments under the tripartite film agreement it had signed with Geola and XYZ. When Eric announced that XYZ would have to renegotiate the contract, there was therefore a sizeable problem. Geola Technologies had also scaled up operations for XYZ in the United Kingdom, and it too now had problems with overhead expenses. Eric, David, Yuri and Stanislovas met up in Sussex, UK, to try and find a solution.

A2.7 Vilnius Digital Printer

The solution the four men came up with was that a new XYZ printer would be made available to Geola Technologies Ltd. This printer would be paid for by a percentage discount of film supplied by Sfera-S though Geola UAB. In return, a new film supply contract would be made showing revised minimum yearly film-order quantities. The printer would then be installed in Lithuania on loan to a new company, 3D Print UAB, which would be owned jointly by Sfera-S and Geola Technologies Ltd. To compensate the Russian side of the business, Sfera-S would commercialise the digital one-step holograms in Russia. Yuri believed that this would represent a sizeable market and so could compensate his losses caused by XYZ's change in orders. David too believed that the XYZ holograms should sell well in the United Kingdom. Geola Technologies therefore formed a subsidiary under the name of Power Imaging Ltd. to commercialise the holograms made by 3D print UAB (Figures A2.10 and A2.11).

This sounded like a good solution. XYZ had a spare printer available. Contracts were exchanged and the printer was sent to the United Kingdom and then on to Vilnius. However, problems arose almost immediately after the XYZ team had installed the printer. Like the problems encountered in Paris, the printer seemed to have various faults. Geola's chief engineer, Ramunas Bakanas (Figure A2.12), immediately contacted XYZ, but the problems could not be resolved easily. Orders started to come in from Power Imaging Ltd., but they could not be fulfilled. XYZ printed some of the holograms—some occasionally were of good enough quality from Vilnius. Slowly, however, the business became poisoned. Clients became disenchanted and ended up disappearing. To make matters worse, a new manager was appointed over Eric. Geola asked XYZ for circuit and mechanical diagrams as they felt they could repair the problems, but XYZ, under the new leadership, was adamant—Geola could only carry out service

FIGURE A2.10   Marcin Lesniewski and David Brotherton-Ratcliffe at Power Imaging Ltd. in 2005 with an XYZ hologram.

procedures according to the instructions of XYZ. This process went on for some months until exasperation amongst Geola's engineers reached boiling point. Ramunas and his team had worked nights and weekends for months with continual pressure for results. They started to understand what was really wrong with the printer but were not allowed to implement their conclusions. Finally, acting CEO, Bill Meder, wrote to Geola, giving the company permission to fix the printer as it thought necessary. Because XYZ would still give no details of its electronics or software for fear of intellectual property loss, Geola was basically forced to replace system after system with its own devices until finally the company had

FIGURE A2.11   Ardie Osanlou, Stanislovas Zacharovas, Hans Bjelkhagen and David Brotherton-Ratcliffe at the 2006 International Symposium on Display Holography in Wales, where some of the first holograms produced by Power Imaging Ltd. were exhibited.

FIGURE A2.12   Ramunas Baranas, chief laser engineer and technical director of Geola Digital UAB in the lab at Geola's Vilnius facilities in 2010.

changed a large part of the printer. At this stage, roughly a year after installation, the printer worked well and predictably. It was too late for Power Imaging Ltd., but nevertheless, the Vilnius printer was now printing and selling digital holograms on a regular basis. In retrospect, despite the problems that appeared, XYZ's first printer was basically well designed. Had the company had access to just a little extra time, there is little doubt that these flaws could have been detected before proper commercial sales commenced. Geola had ended up changing most of the printer systems, not because they were intrinsically bad, but because of the lack of information available during the crisis created by the technical issues in Paris and Vilnius. David and Stanislovas precipitated in the ISDH 2006 Welsh conference, chaired by Hans Bjelkhagen, to promote the large colour holograms produced by Geola (Figure A2.11). The Geola chief laser engineer and technical director Ramunas Baranas is responsible for the lasers produced in the Vilnius lab (Figure A2.12).

A2.8 Sale of XYZ

The French company that had purchased the two digital printers from XYZ was not so lucky as Geola. They did not have engineers capable of replacing the critical printer systems with their own. Instead, the owners decided to bring the case to court in Montreal. This was the final straw for the institutional shareholders backing XYZ. A plan was made to sell the company and despite ostensibly valid inside offers from various shareholders pursuant to the shareholders' agreement, the company was sold to another Canadian company.

The new owners held XYZ for only a short period, after which it was sold and its name was changed to RabbitHoles Media Inc. The new director contacted Geola and asked to meet with David and Yuri in Vilnius to discuss collaboration several months after XYZ's acquisition. Both David and Yuri travelled to Vilnius to meet with him to see if they could work together. RabbitHoles had inherited the old XYZ contracts and was in serious default of some with Sfera-S and Geola UAB. Geola wished to normalise these defaults and was willing to wipe out all debts owed to them. Sadly, however, no agreement could be found. Basically, Geola had come away from the breakup with XYZ with the ability to print and sell one-step and two-step digital holograms and this business, for the first time, was going quite well. RabbitHoles seemed to want to position Geola as their distributor—the two groups obviously had rather different ideas. Despite their differences, however, workers within XYZ and Geola still kept in touch. Roman Rus, one of the original optical engineers at Geola who worked on the first generation printer project in Vilnius, became chief engineer at RabbitHoles and over the coming years, produced some wonderful digital holograms using the XYZ printer technology.

Appendix 3: Active Cavity Length Stabilisation in Pulsed Neodymium Lasers

A3.1 Introduction

The application of digital holographic printing currently requires low-repetition rate (10–100 Hz) pulsed nanosecond laser sources having a very narrow linewidth spectral emission in the visible region, a super-Gaussian or TEM_{00} spatial profile and an exceptionally stable pulse-to-pulse energy and frequency stability.

In the context of lamp-pumped lasers, which are today the most common laser source for digital holographic printing, current technology can only provide the required stability properties using the technique of injection seeding. This technique essentially employs a special thermostabilised small cavity (preferably monolithic) SLM TEM_{00} CW laser to provide a seeding signal for the main lamp-pumped master oscillator. For injection seeding to work, the cavity length of the master oscillator must be actively controlled using a piezo-mounted rear mirror and matched to the longitudinal mode of the seeding signal. This matching process is usually performed by a minimisation of the time for a laser pulse to appear after an electro-optic Q-switch in the master oscillator cavity is opened.

Such injection-seeded systems are quite complex and costly as they require a seeding laser, an expensive and complex electro-optical Q-switching system requiring nanosecond switching of multi-KV signals, a piezo feedback system and complex fast feedback and optimisation electronics.

Here, we describe a much simpler technique [1,2] whereby excellent energy and frequency stability of a lamp-pumped pulsed neodymium laser may be obtained. In its simplest form, thermal heating of either the cavity rear mirror mount, the output coupler or an intracavity glass wedge is used to control the cavity length according to a simple algorithm that either maximises output energy or minimises the standard deviation of output energy. In a slightly more sophisticated version, a piezo element is used to introduce a small amplitude harmonic oscillation into the cavity length and the computed cross-correlation between output laser energy and the test signal allows the cavity length to be maintained at an optimal point.

A3.2 Example of Cavity Stabilisation System Using Heated Rear Mirror Holder

We illustrate the active cavity length stabilisation scheme on the 1064 nm Nd:YAG laser oscillator, which forms the green channel of the RGB laser described in Figure 6.24. This laser has been described in Chapter 6 in detail.

Figure A3.1 shows a diagram of the laser oscillator with the stabilisation system. In the simplest version, the rear cavity mirror (M1g) is attached to a 1.5 cm-long aluminium tube (PZTg) that is in turn attached to a mount through an insulating layer. Heaters embedded in the aluminium tube act to control its temperature. This in turn creates a proportional change in the laser cavity length. The temperature of the aluminium tube is monitored by a thermistor sensor that sends an electric current to a small printed circuit board (PCB). The PCB controls the heating of the heaters in such a way that the temperature reported by the thermistor remains at the given set point temperature to an accuracy of ±0.01°C. The set point temperature is defined by a computer that communicates the desired temperature via an RS-485 link to the heating system controller.

The cavity length can also be controlled by mounting the rear cavity mirror on a piezo tube. In this case, the computer then communicates via RS-485 the desired mirror displacement to a piezo controller.

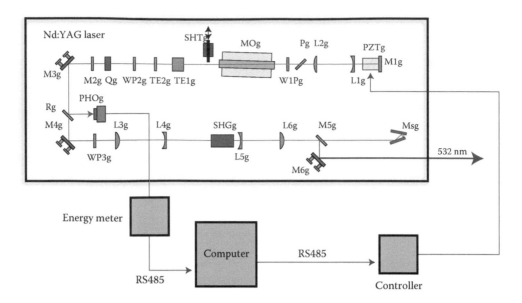

FIGURE A3.1   Schematic diagram of a pulsed Nd:YAG laser oscillator and frequency-doubling optics with active stabilisation scheme. The silicon photodiode PHOg detects the pulse energy of each laser pulse. The signal is processed, calibrated and transferred to a computer in real time by an RS-485 link. The computer then uses an algorithm to control the cavity length through the element PZTg (a heated mirror mount or a piezo mount) to optimise the lasing stability.

To monitor the energy of each laser pulse, a small uncoated glass wedge (Rg) is used to divert some of the output beam towards a diffuser and a silicon photodiode (PHOg). The signal from this photodiode is used to calculate the energy of the output laser pulse and this information is then transferred to the computer via an RS-485 link.

The laser resonator described should be constructed using a temperature-stabilised super-invar mechanical structure and should be enclosed in a temperature-controlled environment to ensure that the physical cavity length remains as stable as possible. If this is not done, fluctuations in environmental conditions may make it impossible for the active stabilisation system to work. Despite the use of a super-invar mechanical resonator and an active temperature control system, without active stabilisation of the cavity length, the output energy of the laser will inevitably drift from one longitudinal mode to another as the optical path length (rather than the physical length) varies. This is caused by several processes, but the most dominant stems from the fact that the xenon lamp wears out. As this happens, the thermal energy deposited into the Nd:YAG crystal changes, and this causes a change in the cavity optical path.

Lamp wear and other processes cause mode drifting, which leads to bad energy stability at 1064 nm and even worse stability at 532 nm. At 532 nm, in the transition region between stability for one longitudinal mode and another, two modes may oscillate and this causes mode beating and increased frequency conversion efficiency.

In Figure A3.2, we show what happens when the rear mirror mount temperature is changed (thus changing the cavity length)—we plot energy per pulse averaged over 1000 points. Figure A3.3 shows the corresponding graph for the standard deviation of energy calculated over 1000 points.

Clearly, both the energy and standard deviation of energy vary with rear mirror mount temperature. What is interesting is that as one increases the temperature of the rear mirror mount by an amount that gives a thermal expansion equivalent to approximately 1 μm, the graph undergoes a complete cycle. Hence, one can conclude that the high-energy, low standard deviation regions correspond to an optical cavity length that is matched to a given longitudinal mode. As the temperature is increased away from this region, one falls into a region in which one particular mode by itself is not optimised. At the peak standard deviation, one falls into a zone in which the longitudinal mode may change frequently from shot to shot.

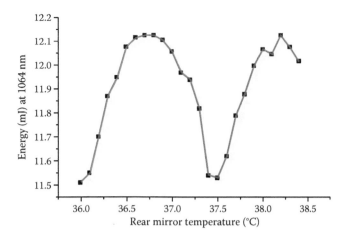

FIGURE A3.2 Graph showing laser output energy at 1064 nm versus rear mirror mount temperature.

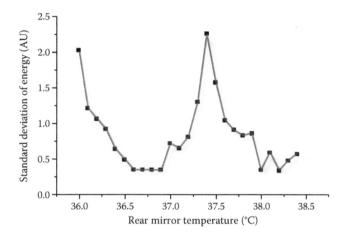

FIGURE A3.3 Graph showing standard deviation of laser output energy at 1064 nm versus rear mirror mount temperature (corresponding to Figure A3.2).

Given that the laser cavity described above exhibits such a clear behaviour concerning energy stability versus rear mirror mount temperature, we use a computer to continually calculate and instruct the thermo-controller to set the optimum temperature of the rear mirror mount such that the averaged energy (more than 1000 points) is a maximum and standard deviation (more than 1000 points) of energy is a minimum. This is the simplest type of algorithm for active cavity stabilisation.

Figure A3.4 shows a typical flow chart of the computer program that we use to iterate the rear mirror mount temperature. Typically, more than 1000 pulses are used for averaging; N_1 is approximately 2000 pulses and N_2 is approximately 1600 pulses.

The function f is usually chosen such that the jump in temperature is smaller as the standard deviation of energy gets smaller. Usually, below a certain standard deviation (typically 0.5%), one wants to put $f = 0$.

Figure A3.4 shows a simple algorithm based on optimisation of the averaged energy.* Figures A3.5 through A3.7 show how the laser responds to this scheme. Figure A3.5 shows a typical plot of output energy at the second harmonic (532 nm) versus time for the case of active stabilisation applied (Zone A)

---

* One could, of course, optimise the reciprocal of the standard deviation of energy. Alternatively, one could optimise the average energy times a coefficient + the reciprocal of the standard deviation of energy times another coefficient.

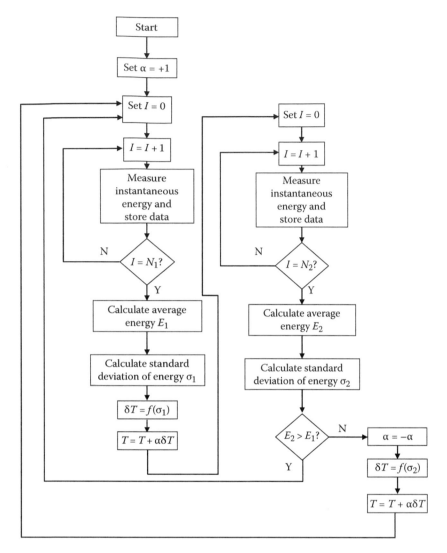

FIGURE A3.4 Flow chart of a simple optimisation algorithm.

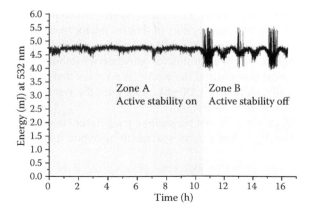

FIGURE A3.5 Typical plot of the output energy at the second harmonic (532 nm) versus time for the case of active stabilisation applied (Zone A) and no active stabilisation applied (Zone B).

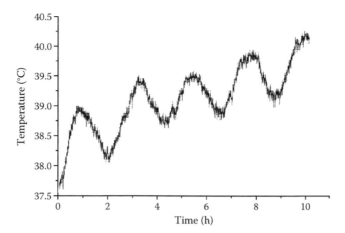

FIGURE A3.6 Plot of temperature of rear mirror mount versus time corresponding to Figure A3.5.

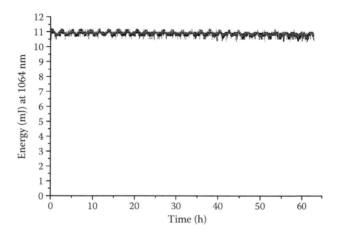

FIGURE A3.7 Typical longer duration plot of output energy (1064 nm) versus time for the case of active stabilisation applied.

and no active stabilisation applied (Zone B). Figure A3.6 shows a plot of the temperature of the rear mirror mount versus time corresponding to Figure A3.5. Figure A3.7 shows a longer plot of output energy (1064 nm) versus time for the case of the active stabilisation applied. Note that the periodic oscillations in temperature that can be observed in all the traces presented (Figures A3.5 through A3.7; period = approximately 2 h) are due to ambient temperature cycling.

Clearly, active stabilisation of the rear mirror mount temperature produces a long-term laser emission of greatly increased energy stability. This is also true of the emission at the second harmonic, where ±4% peak-to-peak energy stabilities and less than ±1% RMS energy stabilities have routinely been obtained over several days. Because the stable energy regime is produced by optimising conditions for a single unique longitudinal mode to resonate, inevitably one also observes a marked improvement in frequency stability. Such stabilities are vital in the applications of writing dot-matrix holographic optical elements, holographic screens and digital holograms.

A3.3 Active Cavity Length Stabilisation by Piezo Element

A piezo element may be used (Figure A3.8) to alter the cavity length of the laser. The principal advantage of this approach over the heated element approach is the faster time response. This is particularly

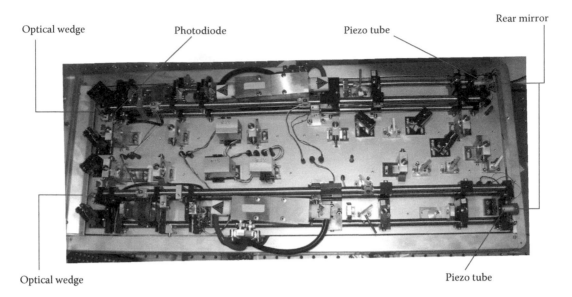

Optical wedge   Photodiode   Piezo tube   Rear mirror

Optical wedge   Piezo tube

FIGURE A3.8   Photograph of a lamp-pumped pulsed RGB Nd:YAG holography laser with piezo-mounted rear mirrors on both the 1064 nm cavity (upper) and the 1319 nm cavity (lower). Laser manufactured by Geola Technologies Ltd.

important at 1.3 μm, as here one needs greater displacements due to the greater wavelength. These greater displacements must be produced either by a greater temperature swing or by lengthening the temperature-sensitive aluminium tube. Greater temperature swings can produce problems themselves and greater tube lengths lead to a poorer temporal response. If the laser system is not of a certain fundamental stability, it may then occur that the heated element approach cannot keep up with the mode drifts and the system is unable to "lock".

A piezo system, albeit a little more complex, can therefore achieve a better stability. It can also cope with worse environmental conditions (i.e., fluctuations in laboratory temperature) than the simpler heated element approach.

A3.3.1 Statistical Optimisation Algorithm

The fact that the piezo element technique has a much better temporal response allows one to use a statistical method to optimise the cavity length. Specifically, one applies a small test signal to the piezo element:

$$V = V_0 + V_1 \sin \omega t \tag{A3.1}$$

Here, V_0 determines an appropriate midpoint cavity length and $V_1 \ll V_0$. Typically, ω must be chosen to be significantly smaller than the laser repetition rate but large enough to ensure many cycles over the desired cavity length updating time, T. For a 30 Hz laser, T should typically be somewhat more than 10 s.

Every T seconds, the voltage to the piezo element is updated according to the following rule:

$$V^{(n+1)} = V^{(n)} + \alpha \wp \tag{A3.2}$$

where α is a small stepping constant and $\wp$ is an error signal. If α is too small, the iteration will be very stable, but convergence will be slow. On the other hand, if α is too large, the iteration may go unstable.

The error signal is calculated as

$$\wp = \frac{1}{T} \int\limits_{nT}^{(n+1)T} E(t) V_1 \sin \omega t \, dt \tag{A3.3}$$

which may be approximated to

$$\wp \sim \frac{V_i \delta T}{T} \sum_{i=i_0}^{i_f} E_i \sin \omega t_i \qquad (A3.4)$$

Here, $E(t)$ is the pulse energy of the laser and E_i is its measurement at the ith pulse. The parameter δT is the time between laser pulses. The sum in Equation A3.4 is taken over every laser pulse between updating intervals. This simple algorithm is very effective. By using an error signal that is based on a first-order cross-correlation between the test signal and the laser energy, spurious effects not induced by the test signal are eliminated. In this way, a very clear error signal is attained, resulting in quick cavity locking.

A3.4 Extension to Other Lasers

Either of the two stabilisation systems we have described above can be easily applied to ring cavity laser oscillator designs. Here an intracavity heated glass wedge may be used instead of a heated rear mirror mount which is usually used for the linear oscillator. The optical path length of the laser cavity is then changed actively by varying the temperature of the wedge. Active cavity length stabilisation can be used effectively on lasers oscillating at many of the 1 µm and 1.3 µm lines of neodymium lines of Nd:YAG. As such, the technique is extremely useful for stabilisation of both channels (1064 nm and 1319 nm) of the RGB-pulsed laser of Figure 6.24.

REFERENCES

1. D. Brotherton-Ratcliffe, *Laser*, US Patent 7,852,887 (filed 2005, granted 2010).
2. C. Cotteverte, J.-C. Joseph, J. Kosenro and R. Rus, *Laser*, US Patent 7,596,153 (filed 2005, 2009).

Appendix 4: Aberration Correction by Image Predistortion in Digital Holograms

A4.1 Introduction

In Chapter 11, we derived formulae relating how the light rays recorded in a hologram replay when the hologram is illuminated. We saw that when the holographic emulsion swells or shrinks, when the refractive index changes during processing, or when the illumination geometry and wavelength does not exactly match the recording geometry and wavelength, then aberration in the resulting image is present. This aberration can take the form of chromatic aberration or geometric aberration (or both).

In digital holography, it is possible, to some extent at least, to correct for such aberration by predistorting the digital data. In this appendix, we shall give a very brief outline of how exactly this may be done in one specific and very simple case. In particular, we shall consider the case of a slight emulsion shrinkage on chemical processing in the three-colour reflection hologram and how this may be corrected by predistortion and chromatic rebalancing of the digital data.

A4.2 Mathematical Model

In almost all cases in reflection display holography, the Bragg condition provides a good description of how the hologram replays. According to our discussions in Chapter 11, the grating vector of a recorded hologram is given by

$$\mathbf{G} = \mathbf{k}_r - \mathbf{k}_o \tag{A4.1}$$

Here, the $\mathbf{k}$ parameters are the wave vectors of the rays incident at a given hogel of the hologram being recorded. The subscripts "o" and "r" stand for the object and reference ray. When there is no change to the emulsion thickness, we can expect that the hologram will replay most efficiently when

$$\mathbf{k}_c - \mathbf{k}_i = \mathbf{G} \tag{A4.2}$$

Here, the subscripts "c" and "i" denote the reference illumination ray and the image ray at replay. When there is a change in the thickness of the emulsion, the grating vector is changed

$$\mathbf{G} \to \underline{\alpha} \cdot \mathbf{G} \tag{A4.3}$$

with the emulsion deformation matrix usually given by the simple relation

$$\underline{\alpha} = \begin{pmatrix} \tau^{-1} & 0 & 0 \\ 0 & 1 & 0 \\ 0 & 0 & 1 \end{pmatrix} \tag{A4.4}$$

where τ is the swelling constant with $\tau < 1$ indicating shrinkage and $\tau > 1$ indicating swelling.

The hologram can then be expected to replay according to the formula

$$\mathbf{k}_c - \mathbf{k}_i = \underline{\alpha} \cdot \mathbf{G} \tag{A4.5}$$

We can therefore calculate that a hogel recorded with an object wave vector, $\mathbf{k}_o$, and a reference wave vector, $\mathbf{k}_r$, and then suffering an emulsion deformation given by $\underline{\alpha}$ will replay most efficiently with a wave vector, $\mathbf{k}_i$, when illuminated by a wave vector, $\mathbf{k}_c$, where

$$\mathbf{k}_i = \mathbf{k}_c - \underline{\alpha} \cdot \mathbf{G} = \mathbf{k}_c - \underline{\alpha} \cdot (\mathbf{k}_r - \mathbf{k}_o) \tag{A4.6}$$

The optimal free-space wavelength of replay, λ_c, will then be given simply by taking the magnitude of this expression:

$$\frac{2\pi n_c}{\lambda_c} = \left| \mathbf{k}_c - \underline{\alpha} \cdot (\mathbf{k}_r - \mathbf{k}_o) \right| \tag{A4.7}$$

where n_c is the refractive index of the emulsion after processing, which, in general, will be different from that before processing, n_r.

A4.3 Calculation of Optimal Reference Replay Angle

When the emulsion shrinks (or swells) on processing, the hologram may be illuminated from different angles at different wavelengths. However, there will be an optimal illumination angle at which each hogel is associated with an axially projecting image bundle. Basically, we don't want to illuminate our hologram in such a way that the image is directed downwards or upwards. Rather, we wish the image to project straight out in front.

We will limit our discussion here to considering the case of a digital reflection hologram recorded and replayed using a collimated reference beam. This has two important effects. First, every hogel is the same and, second, we only need to consider angles in the vertical plane. We shall also only consider the case of a paraxial writing objective.

We start by considering the central pixel of the printer spatial light modulator (SLM) (corresponding to one primary colour channel in a triple beam DWDH printer). The k-vector associated with this pixel, $\mathbf{k}_o^o$ is known: the unit vector is given by simply subtracting the Cartesian world coordinates of the pixel location on the projected SLM plane from the hogel coordinates and dividing by the length of this vector—then the actual k-vector can be calculated by multiplying the unit vector by $2\pi n/\lambda_r$. This vector can then be written as (see Figure A4.1)

$$\mathbf{k}_o^o = \frac{2\pi n_r}{\lambda_r} \hat{\mathbf{x}} \tag{A4.8}$$

If we wish the hogel to project axially, then $\hat{\mathbf{k}}_i^o = \hat{\mathbf{k}}_o^o$ and this means that

$$\mathbf{k}_i^o = \frac{2\pi n_c}{\lambda_c} \hat{\mathbf{x}} \tag{A4.9}$$

If we now take

$$\mathbf{k}_r = -\frac{2\pi n_r}{\lambda_r} \left\{ \cos\theta_r \hat{\mathbf{x}} + \sin\theta_r \hat{\mathbf{y}} \right\}$$

$$\mathbf{k}_c = -\frac{2\pi n_c}{\lambda_c} \left\{ \cos\theta_c \hat{\mathbf{x}} + \sin\theta_c \hat{\mathbf{y}} \right\} \tag{A4.10}$$

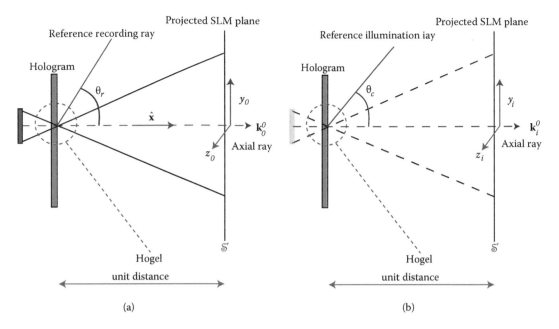

FIGURE A4.1   (a) Recording and (b) playback of the reflection hologram considered in the text. The reference illumination angle is chosen so that the reconstructed image of the axial object ray is axial itself.

the x and y components of Equation A4.6 then tell us that

$$\frac{2\pi n_c}{\lambda_c}(1+\cos\theta_c) = \frac{2\pi n_r}{\lambda_r}\tau^{-1}(1+\cos\theta_r)$$

$$\frac{2\pi n_c}{\lambda_c}\sin\theta_c = \frac{2\pi n_r}{\lambda_r}\sin\theta_r$$

(A4.11)

Note that these are internal angles. These equations may be solved to give:

$$\lambda_c = 2\lambda_r \frac{n_c}{n_r\tau}\left\{\frac{1}{(\tau^{-2}+1)+(\tau^{-2}-1)\cos\theta_r}\right\}$$

$$\cos\theta_c = \frac{(\tau^{-2}-1)+(\tau^{-2}+1)\cos\theta_r}{(\tau^{-2}+1)+(\tau^{-2}-1)\cos\theta_r}$$

(A4.12)

In terms of external angles they become

$$\lambda_c = \frac{2\lambda_r n_c}{n_r\tau\left[\left\{1+\tau^{-2}\right\}+\left\{\tau^{-2}-1\right\}\sqrt{1-\frac{\sin^2\theta_r}{n_r^2}}\right]}$$

$$\sin\theta = \frac{2n_c\tau\sin\theta_r}{n_r\left\{1+\sqrt{1-\frac{\sin^2\theta_r}{n_r^2}}+\tau^2\left(1-\sqrt{1-\frac{\sin^2\theta_r}{n_r^2}}\right)\right\}}$$

(A4.13)

This then determines the optimal replay illumination angle for the hologram. Note that this angle of illumination is the same for each primary colour. One should also observe that the optimal replay wavelength for

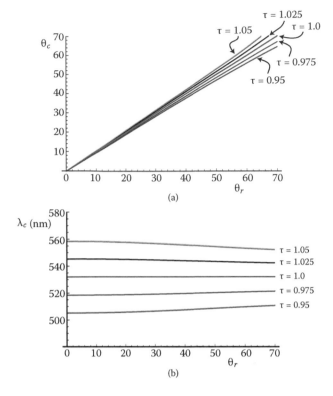

FIGURE A4.2 Optimal replay (assuring an axial projection of the reconstructed image of the axial object ray) of a reflection hologram which has suffered emulsion deformation. (a) Graph of the optimal (external) illumination angle of incidence, θ_c in degrees as defined by Equation 4.13 versus (external) reference recording angle of incidence, θ_r for various values of the emulsion swelling parameter, τ. The refractive indices, n_r and n_c before and after chemical processing of the reflection hologram have both been set to 1.5. The recording wavelength is 532 nm. (b) Corresponding graphs of replay wavelength of the axial ray versus (external) reference recording angle. Note that for more than around 5% of shrinkage or swelling the magnitude of the induced chromatic aberration can make it difficult to rebalance the colours successfully.

the axial ray (given by Equation A4.12) is now different from the recording wavelength (see Figure A4.2). We can anticipate that this optimal replay wavelength will be different for each and every ray in the hogel.

A4.4 Compensation for Geometrical Distortion

Each object ray, characterised by its wave vector, $\mathbf{k}_o$, will replay under the optimal illumination angle (Equation A4.13) to produce an image ray defined by

$$\mathbf{k}_i = \mathbf{k}_c - \underline{\underline{\alpha}} \cdot (\mathbf{k}_r - \mathbf{k}_o) \tag{A4.14}$$

We can write the components of this equation as follows:

$$\frac{2\pi n_c}{\lambda_c \sqrt{1 + y_i^2 + z_i^2}} \begin{pmatrix} 1 \\ y_i \\ z_i \end{pmatrix} = -\frac{2\pi n_c}{\lambda_c} \begin{pmatrix} \cos\theta_c \\ \sin\theta_c \\ 0 \end{pmatrix} + \frac{2\pi n_r}{\lambda_r} \begin{pmatrix} \tau^{-1}\cos\theta_r \\ \sin\theta_r \\ 0 \end{pmatrix} + \frac{2\pi n_r}{\lambda_r \sqrt{\tau^{-2} + y_o^2 + z_o^2}} \begin{pmatrix} \tau^{-1} \\ y_o \\ z_o \end{pmatrix} \tag{A4.15}$$

Note that we have used the k-vectors of Equation A4.10. We have also used a unit x distance from the hogel to the projected SLM plane for the calculation of the object and image k-vector. Note finally that angles are internal!

The three equations (Equation A4.15) determine the intersection of image rays with the unit x-distance projection of the SLM, (y_i, z_i) and the optimal replay wavelength, λ_c of each image ray in terms of the corresponding intersection of object rays (y_o, z_o) and the recording wavelength, λ_r. They may be written in the simplified form

$$\frac{1}{\sqrt{1+y_i^2+z_i^2}}\begin{pmatrix}1\\y_i\\z_i\end{pmatrix} = \varepsilon\left\{\begin{pmatrix}\tau^{-1}\cos\theta_r\\\sin\theta_r\\0\end{pmatrix} + \frac{1}{\sqrt{\tau^{-2}+y_o^2+z_o^2}}\begin{pmatrix}\tau^{-1}\\y_o\\z_o\end{pmatrix}\right\} - \begin{pmatrix}\cos\theta_c\\\sin\theta_c\\0\end{pmatrix} \tag{A4.16}$$

where

$$\varepsilon = \frac{\lambda_c n_r}{\lambda_r n_c} \tag{A4.17}$$

If we simplify this further, we can write

$$1 = (1+y_i^2+z_i^2)(a_x\varepsilon+b_x)^2$$

$$y_i^2 = (1+y_i^2+z_i^2)(a_y\varepsilon+b_y)^2$$

$$z_i^2 = (1+y_i^2+z_i^2)a_z^2\varepsilon^2 \tag{A4.18}$$

Adding all three equations together now gives

$$(a_x\varepsilon+b_x)^2 + (a_y\varepsilon+b_y)^2 + a_z^2\varepsilon^2 = 0 \tag{A4.19}$$

or

$$A\varepsilon^2 + B\varepsilon + C = 0 \tag{A4.20}$$

with coefficients

$$A = a_x^2 + a_y^2 + a_z^2$$

$$B = 2(a_yb_y + a_xb_x)$$

$$C = b_x^2 + b_y^2 \tag{A4.21}$$

This may then be solved by the standard formula

$$\varepsilon = \frac{-B \pm \sqrt{B^2 - 4AC}}{2A} \tag{A4.22}$$

Practically, there will only be one sensible root to this equation, which will be the nearest one to real unity.

With ε known, we know the replay wavelength of each ray. This will be vital for chromatic rebalancing, which we will treat in the next section. However, now that we know ε, we can easily solve Equation A4.18 to define (y_i, z_i):

$$y_i = \frac{a_y\varepsilon+b_y}{a_x\varepsilon+b_x}$$

$$z_i = \frac{a_z\varepsilon}{a_x\varepsilon+b_x} \tag{A4.23}$$

Equation A4.23 can be used to formulate a mapping of the SLM data to correct for induced aberration. We start this process by dividing up the SLM into two coordinate systems, (α,β) and (α',β'). We defined the (α,β) system in Chapter 8. It basically just counts the pixels on the SLM. In Chapter 8, we derived **I**-to-**S** transformations, which defined the data $\mathbf{S}_{\alpha\beta}$ at the SLM pixel (α,β) in terms of the raw camera or image data. We can therefore imagine the (α,β) coordinate system corresponding to the (y_o,z_o) coordinate system. Specifically, we can write, following the results of Chapter 8[*],

$$z_o = \left\{ \frac{2(\alpha-1)}{(N_A-1)} - 1 \right\} \tan\left(\frac{\Psi_{PH}}{2} \right)$$

$$y_o = -\left\{ \frac{2(\beta-1)}{(N_B-1)} - 1 \right\} \tan\left(\frac{\Psi_{PV}}{2} \right) \tag{A4.24}$$

where Ψ_{PH} and Ψ_{PV} are the printer field of views (horizontal and vertical) as apodised by the SLM.

If we write data, $\mathbf{S}_{\alpha\beta}$ to the SLM at the pixel (α,β) corresponding to the coordinates (y_o,z_o) on the projected SLM plane, $\mathscr{S}'$ at unit distance from the hologram then we know that on replay the ray will effectively move to a new position (y_i,z_i) on $\mathscr{S}'$. We can now define the system (α',β') to define the effective pixel position this image ray corresponds to

$$z_i = \left\{ \frac{2(\alpha'-1)}{(N_A-1)} - 1 \right\} \tan\left(\frac{\Psi_{PH}}{2} \right)$$

$$y_i = -\left\{ \frac{2(\beta'-1)}{(N_B-1)} - 1 \right\} \tan\left(\frac{\Psi_{PV}}{2} \right) \tag{A4.25}$$

So what we actually want to do is not to write the data $\mathbf{S}_{\alpha\beta}$ to the pixel (α,β). Rather, we want to write the data $\mathbf{S}_{\alpha'\beta'}$ to the pixel (α,β). This way, on replay, the data $\mathbf{S}_{\alpha'\beta'}$ ends up in the right place. Another way of saying this is that we should redefine the SLM data so that

$$\mathbf{S}'_{\alpha\beta} = \mathbf{S}_{\alpha'\beta'} \tag{A4.26}$$

where

$$\alpha' = 1 + \left(\frac{N_A-1}{2} \right) \left\{ 1 + z_i(\alpha,\beta)\cot\left(\frac{\Psi_{PH}}{2} \right) \right\}$$

$$\beta' = 1 + \left(\frac{N_B-1}{2} \right) \left\{ 1 - y_i(\alpha,\beta)\cot\left(\frac{\Psi_{PV}}{2} \right) \right\} \tag{A4.27}$$

In this way, we end up writing an image data byte to a different pixel location on the SLM than we would have done if there were no emulsion thickness change. However, this shift in position is just what is required such that the distortion induced by the emulsion thickness change effectively moves the ray back to the position where it should have been.

The index transformation (Equation A4.27) may be calculated just one time and then applied to all hogels for a given primary colour channel. Typically, bilinear or bicubic interpolation is used (see Appendix 7). Note that the index transformation depends on illumination wavelength, so there will be different transformations necessary for each primary colour.

A4.5 Compensation for Chromatic Aberration

Our reflection hologram is assumed to be illuminated by a broadband white-light source. As such, each ray will be associated with an optimal replay wavelength that is, in general, different from λ_r. This

---

[*] Note that we are using a non-conjugate SLM position here as per Figure A4.1.

optimal wavelength is determined by the parameter, ε, in the ray equation (Equation A4.22). For the axial ray of each hogel

$$\varepsilon = \frac{\lambda_c n_r}{\lambda_r n_c} = \frac{2\tau}{1 + \tau^2 + (\tau^2 - 1)\cos\theta_r} \tag{A4.28}$$

If we consider the case of a three colour reflection hologram λ_r will be replaced by the three laser wavelengths used to record the hologram-namely λ_R, λ_G, and λ_B.

With no emulsion change, the tristimulus values associated with a given object/image ray from a given hogel are given by

$$X = k\left\{\mathbf{S}_G\bar{x}(\lambda_G) + \mathbf{S}_R\bar{x}(\lambda_R) + \mathbf{S}_B\bar{x}(\lambda_B)\right\}$$

$$Y = k\left\{\mathbf{S}_G\bar{y}(\lambda_G) + \mathbf{S}_R\bar{y}(\lambda_R) + \mathbf{S}_B\bar{y}(\lambda_B)\right\}$$

$$Z = k\left\{\mathbf{S}_G\bar{z}(\lambda_G) + \mathbf{S}_R\bar{z}(\lambda_R) + \mathbf{S}_B\bar{z}(\lambda_B)\right\} \tag{A4.29}$$

Here, $\bar{x}$, $\bar{y}$ and $\bar{z}$ are the colour-matching functions of the CIE Standard Colorimetric Observer (see, for example, Giorgianni and Madden [1]) and k is a normalising factor. The parameters $\mathbf{S}_G$, $\mathbf{S}_R$ and $\mathbf{S}_B$ are, respectively, the green, red and blue brightness data written to the three primary-colour SLMs for the case of zero emulsion shrinkage.

When the primary wavelengths change according to Equation A4.22, the tristimulus values will also change:

$$X' = k\left\{\mathbf{S}_G\bar{x}(\lambda'_G) + \mathbf{S}_R\bar{x}(\lambda'_R) + \mathbf{S}_B\bar{x}(\lambda'_B)\right\}$$

$$Y' = k\left\{\mathbf{S}_G\bar{y}(\lambda'_G) + \mathbf{S}_R\bar{y}(\lambda'_R) + \mathbf{S}_B\bar{y}(\lambda'_B)\right\}$$

$$Z' = k\left\{\mathbf{S}_G\bar{z}(\lambda'_G) + \mathbf{S}_R\bar{z}(\lambda'_R) + \mathbf{S}_B\bar{z}(\lambda'_B)\right\} \tag{A4.30}$$

This then describes the chromatic aberration of the ray in question. To ensure that there is zero chromatic aberration, we must change the parameters $\mathbf{S}_G$, $\mathbf{S}_R$ and $\mathbf{S}_B$ and ensure that the tristimulus values are equal to their primed values. Or in other words,

$$k\left\{\mathbf{S}'_G\bar{x}(\lambda'_G) + \mathbf{S}'_R\bar{x}(\lambda'_R) + \mathbf{S}'_B\bar{x}(\lambda'_B)\right\} = X$$

$$k\left\{\mathbf{S}'_G\bar{y}(\lambda'_G) + \mathbf{S}'_R\bar{y}(\lambda'_R) + \mathbf{S}'_B\bar{y}(\lambda'_B)\right\} = Y$$

$$k\left\{\mathbf{S}'_G\bar{z}(\lambda'_G) + \mathbf{S}'_R\bar{z}(\lambda'_R) + \mathbf{S}'_B\bar{z}(\lambda'_B)\right\} = Z \tag{A4.31}$$

where X, Y and Z are given by Equation 4.29.

Equation 4.31 can be written in matrix form:

$$
\begin{pmatrix} \mathbf{S}'_G \\ \mathbf{S}'_R \\ \mathbf{S}'_B \end{pmatrix} = \begin{pmatrix} \bar{x}(\lambda'_G) & \bar{x}(\lambda'_R) & \bar{x}(\lambda'_B) \\ \bar{y}(\lambda'_G) & \bar{y}(\lambda'_R) & \bar{y}(\lambda'_B) \\ \bar{z}(\lambda'_G) & \bar{z}(\lambda'_R) & \bar{z}(\lambda'_B) \end{pmatrix}^{-1} \times \begin{pmatrix} \bar{x}(\lambda_G) & \bar{x}(\lambda_R) & \bar{x}(\lambda_B) \\ \bar{y}(\lambda_G) & \bar{y}(\lambda_R) & \bar{y}(\lambda_B) \\ \bar{z}(\lambda_G) & \bar{z}(\lambda_R) & \bar{z}(\lambda_B) \end{pmatrix} \begin{pmatrix} \mathbf{S}_G \\ \mathbf{S}_R \\ \mathbf{S}_B \end{pmatrix}
$$

$$
= \begin{pmatrix} a_{11} & a_{12} & a_{13} \\ a_{21} & a_{22} & a_{23} \\ a_{31} & a_{32} & a_{33} \end{pmatrix} \begin{pmatrix} \mathbf{S}_G \\ \mathbf{S}_R \\ \mathbf{S}_B \end{pmatrix} \tag{A4.32}
$$

The matrix coefficients a_{ij} need only be calculated for each ray one time, as they are identical (for a particular ray) for all hogels. Application of the transformation (Equation A4.32) to all the SLM brightness data will effectively rebalance the chromatic equation and ensure that each ray from each hogel has the correct tristimulus values.

The above analysis assumes that the hologram is rather thick and that the colour-matching functions of the CIE Standard Colorimetric Observer are only sampled at one exact wavelength. In the case that the hologram is thinner, we can calculate, the exact form of the spectral function for each ray using the results of the PSM theory presented in Chapter 12. In addition, one can include the spectral power distribution of the illumination source. In this case, each of the matrix elements in Equation A4.32 are transformed in a similar fashion to

$$\bar{x}(\lambda_G) \to \int F(\lambda)\eta_G(\lambda,\theta_c,\psi,\varphi)\bar{x}(\lambda)d\lambda \tag{A4.33}$$

Here, $F(\lambda)$ is the spectral power distribution of the illumination source and η_G is the diffractive efficiency in the green channel as given, for example, by Equation 12.90, with the coefficients in Equation 12.101, in the case of illumination by σ-polarised light.

A4.6   Other Corrections

We have considered here only the case of a simple change in emulsion thickness with different refractive indices on recording and processing. This is really the simplest case and serves to illustrate in the most simple and straightforward manner how geometric and chromatic predistortion works. In practice, however, rather more complicated scenarios arise. In particular, often one wants to write a hologram with a collimated reference beam and then replay it with a spot lamp at a certain distance. The mathematics used above can be generalised rather easily to this situation—all that needs to be done is for the vector $\mathbf{k}_c$ in Equation A4.15 to be written in Cartesian form, thus describing a point source at the desired distance—this simply changes b_x and b_y and introduces a new b_z parameter in Equation A4.18. Note, however, that both the geometric and chromatic predistortions are now different for all rays *and* all hogels.

It should also be underlined that predistortion has its limits. If too great a change in emulsion thickness or index occurs or if too great a disparity in the illumination/recording geometry exists, then it may just not be possible to compensate for the induced aberrations. We should also state that it is possible to compensate for slightly larger chromatic aberration if one relaxes somewhat the condition of an axially propagating image ray bundle. In this case one trades chromatic correction for geometric correction. However this scheme can very quickly reduce the vertical field of view.

REFERENCE

1. E. J. Giorgianni and T. E. Madden, *Digital Colour Management—Encoding Solutions*, Addison-Wesley, Reading, MA (1998).

Appendix 5: MAXScript Holocam Program

The following is a full listing of the MAXScript virtual holocam program described in Chapter 10.

```
rollout DoubleParallaxRender "DoubleParallaxRender" width:687 height:361
(
-control definitions
    button BRen "Start Render" pos:[462,13] width:175 height:36
    edittext RenderDir "Render Directory" pos:[248,266] width:373 height:21
    label hOutput "h (mm)" pos:[386,98] width:108 height:18
    spinner Win "W (mm)" pos:[59,288] width:74 height:16 range:[0,5000,600] type:#integer
    spinner Hin "H (mm)" pos:[59,319] width:74 height:16 range:[10,5000,400] type:#integer
    spinner NA "NA" pos:[229,100] width:74 height:16 range:[100,3000,600] type:#integer
    spinner NB "NB" pos:[230,131] width:74 height:16 range:[100,3000,400] type:#integer
    GroupBox grp1 "Calculated Parameters" pos:[371,74] width:268 height:155
    GroupBox grp2 "Hologram Definition" pos:[195,73] width:163 height:155
    GroupBox grp3 "Window Parameters" pos:[3,265] width:173 height:92
    GroupBox grp4 "SLM Definition" pos:[2,73] width:175 height:83
    spinner NM "NM" pos:[56,99] width:74 height:16 range:[100,5000,1280] type:#integer
    spinner NV "NV" pos:[57,129] width:74 height:16 range:[100,5000,1024] type:#integer
    label NKOutput "NKOutput" pos:[385,118] width:66 height:18
    label NGOutput "NGOutput" pos:[386,139] width:74 height:18
    GroupBox grp5 "Printer Optics Definition" pos:[2,173] width:174 height:77
    spinner PsiPH "PsiPH" pos:[56,203] width:72 height:16 range:[10,120,89] \
    type:#integer scale:0.1
    spinner NStopLine "Render: Line Number End" pos:[352,326] width:192 height:16 \
    range:[1,5000,1] type:#integer
    GroupBox grp6 "Job Definition" pos:[195,237] width:444 height:120
    spinner NStartLine "Render: Line Number Start" pos:[350,303] width:196 height:16 \
    range:[1,5000,1] type:#integer
    button BMD "Make Directories" pos:[5,13] width:173 height:34
    button BCal "Calculate" pos:[195,12] width:104 height:34
    label PsiPVOutput "PsiPVOutput" pos:[385,158] width:82 height:18
    label DXOutput "DX (mm)" pos:[385,179] width:87 height:18
    label DYOutput "DY (mm)" pos:[386,198] width:86 height:18
    label W_Output "W Used (mm)" pos:[506,98] width:118 height:18
    label H_Output "H Used (mm)" pos:[508,119] width:124 height:18
    label CompleteColOutput "Completed Column" pos:[491,184] width:135 height:18
    GroupBox grp19 "Render Progress" pos:[487,154] width:144 height:63
    spinner Delta "Hogel Size (microns)" pos:[279,162] width:70 height:16 \
    range:[100,3000,800] type:#integer
    button BCam "Make Camera" pos:[330,13] width:104 height:34
    global W,h1,H,DX,DY,NK,NG,TopDirName-global variable definitions
    on BRen pressed do
    (
        -EVENT HANDLER FOR BRen Button Pressed
        for g=NStartLine.value to NStopLine.value do
            (
            LineText=gas string
            LineDirName=TopDirName+"/Line"+LineText
            for k=1 to NK do
                (
                ColText=k as string        -make filename
                FrameFileName=LineDirName+"/Col"+ColText+".jpg"

                $Cam1.pos=[-W/2+(k-1)*Delta.value/1000.0,-h1,-H/2+ \
                (g-1)*Delta.value/1000.0]    -move camera
                $Tar1.pos=[-W/2+(k-1)*Delta.value/1000.0,0,-H/2+(g-1)*Delta.value/1000.0]

                NI=NK+NA.value-1            -prepare frame parameters
                NJ=NG+NB.value-1
                N1x=1+NK-k
                N2x=N1x+NA.value-1
                N1y=g
                N2y=N1y+NB.value-1
```

```
            Render\
                  OUTPUTFILE: FrameFileName\
                  OUTPUTWIDTH:NI              \
                  OUTPUTHEIGHT:NJ \
                  RENDERTYPE: #RegionCrop \
                  REGION: #(N1x,N1y,N2x,N2y) \
                  VFB:OFF                     —render frame

            gc()                              —empty garbage

            OutText=kas string
            CompleteColOutput.Caption=OutText
            )
        )
)
on NA changed val do
(
NA.value=2.0*floor(NA.value/2.0)         —ensure NA is even
)
on NB changed val do
(
NB.value=2.0*floor(NB.value/2.0)         —ensure NB is even
)

on NM changed val do
(
NM.value=2.0*floor(NM.value/2.0)         —ensure NM is even
)
on NV changed val do
(
NV.value=2.0*floor(NV.value/2.0)         —ensure NV is even

)
on BMD pressed do
(
    —EVENT HANDLER FOR BMD Button Pressed
    TopDirName="C:/"+RenderDir.Text
    makedir TopDirName
    fori=NStartLine.value to NStopLine.value do
        (
        LineText=ias string
        DirName=TopDirName+"/Line"+LineText
        makedir DirName
        )
)

on BCal pressed do
(
    —EVENT HANDLER FOR BCal Button Pressed
    h1=Delta.value/2.0/1000.0*(NM.value-1)/tan(PsiPH.value/2.0)
    hText=h1 as string
    hOutput.caption="h(mm) = "+hText

    NK=1+2*floor(Win.value*1000.0/2.0/Delta.value) as integer
    NG=1+2*floor(Hin.value*1000.0/2.0/Delta.value) as integer
    NKText=NK as string
    NGText=NG as string
    NKOutput.caption="NK = "+NKText
    NGOutput.caption="NG = "+NGText

    W=(NK-1)*Delta.value/1000.0
    H=(NG-1)*Delta.value/1000.0
    WText=Was string
    Htext=Has string
    W_Output.caption="W(mm) = "+WText
    H_Output.caption="H(mm) = "+Htext

    DX=(NA.value-1)*Delta.value/1000.0
    DY=(NB.value-1)*Delta.value/1000.0
    DXText=DX as string
```

```
        DYText=DY as string
        DXOutput.caption="DX(mm) = "+DXText
        DYOutput.caption="DY(mm) = "+DYtext

        PsiPV=2.0*atan(float(NV.value-1)/float(NM.value-1)*tan(PsiPH.value/2.0))
        PsiPVText=PsiPV as string
        PsiPVOutput.caption="PsiPV = "+PsiPVText
    )
    on BCam pressed do
    (
        —EVENT HANDLER FOR BCam Button Pressed
        PsiCH=2.0*atan((W+DX)/2.0/h1)
        TargetCamera  NAME:"Cam1" \
                      SHOWCONE: true \
                      FOV:PsiCH \
                      POS:[-W/2,-h1,-H/2.0] \
                      TARGET: (targetObject NAME:"Tar1" POS: [-W/2,0.0,-H/2.0])
        Viewport.setCamera $Cam1
    )
)
CreateDialog DoubleParallaxRender
```

Appendix 6: Design Study of Compact RGB LED Hologram Illumination Source

A6.1 Introduction

We present here a design study of a new light source that is suitable for holography.* This light source is bright, efficient, small, lightweight and has a long life with low heat output, low power consumption and low étendue. The design allows for a wide range of customised performance in terms of beam footprint, beam irradiance distribution at the target, spectral composition of the emitted radiation, spatial coherence and beam divergence. In particular, the light source can be designed with a highly non-rectangular footprint[†] and with a radiant intensity distribution suitable for even illumination of holograms in which the illumination angle is as large as 70°, and in which the light source is positioned at close proximity. Geometrical illumination efficiency, as defined by the radiant energy used by the hologram (in either the reconstruction of the image or absorption) divided by the total radiant energy emitted by the source, can be as high as 70% for such highly tilted cases. This should be compared with typically less than 10% illumination efficiency which is available from previous light sources. In addition to providing higher geometric efficiency, the new source generally also provides better spatial coherence; in addition, the source can be optimised to produce exceptionally good spatial coherence at the cost of a lesser choice in the definition of the profile of radiant intensity. Finally, the source is characterised by high spectral radiance at one or more wavelengths. These wavelengths may be matched to the recording wavelengths used to write the hologram, leading to a reconstructed image of high brightness and low noise.

The new light source is based on recent developments in semiconductor diodes. These sources (e.g., Philips Lumileds) are available in appropriate wavelengths for display holography and general illumination applications (e.g., 455, 530 and 627 nm), have small emitter sizes (typically 1 mm², emitting within a solid angle of just under 2π rad), are narrow band (typically full-width half-maximum ~30 nm) and emit approximately 145 lm. Due to the Bragg selectivity of a typical reflection hologram made with modern silver halide materials, such diodes produce significantly superior brightness when compared with illumination by a 700 lm, 50 W MR16 halogen lamp. The electrical power consumed by each diode is typically only 1 W, and the operating life is of the order of 100,000 hours. In general illumination applications, approximately five diodes are required to produce the same perception of brightness as a single 50 W MR16 halogen lamp.

Using commercial ray-tracing packages, such as TracePro and Zemax, it is possible to conclude that the light emitted from modern diodes may be very conveniently injected into square light guides of approximately 1mm² sectional area made from glasses with refractive indices in the region of 1.8. [‡]Such guides may be somewhat curved and are typically a few centimetres long. The guides serve to transport light from typically nine diodes (three of each colour) to an optical element that combines and redistributes the light into a beam of defined form. The guides also serve to limit and homogenise the angular and spatial distribution of radiation.

The light-emitting diodes (LED) used in the new light source do not require as much input energy as previous light sources, and this makes it possible, when illuminating a self-animating hologram that requires movement of the light source, to supply the power to the light source without the use of wires

---

* This design study was carried out in 2008 by one of the authors (DBR) in collaboration with John Fleming, whose kind permission to reproduce the study here is gratefully acknowledged by the authors.

† This means that rectangular hologram panels can be illuminated by an exactly matching rectangular illumination.

‡ Plastic light-guides may also be used. In later work these were found to be preferable to glass.

connecting the power source to the light. In addition, the heat output of the new light source is much lower than that from previous devices and this aids in preventing thermally caused chromatic distortions of the image where the light source is mounted close to the hologram being displayed.

A6.2 Monochromatic Light Source

Figure A6.1 shows a diagram of a monochromatic light source comprising three diode sources. The diodes used are LMXL-PM01-0080 Luxeon Rebels emitting at a centre wavelength of 530 nm with a full-width half-maximum bandwidth of 30 nm. The three diodes are optically cemented to square-section light guides 1.27 mm × 1.27 mm made from Schott SF57HHT glass or Ohara S-LAF52. The ends of each light guide comprise a linearly tapered section that increases the collection efficiency and improves the angular distribution of radiation. The LMXL-PM01-0080 diodes are sold with an attached silica gel lens assuring an approximately lambertian emission. This lens is easily removed and the diode is then optically attached directly to the light guide. The size of the light guide is chosen by making it just a little larger than the active area of the diode emitter. The maximum angle of output radiation is then fixed by the angle of total internal reflection of the guide glass, which, in this case, is approximately 33.75°. Having fixed the light-guide size in this way, the overall collection efficiency may be increased by including a tapered section to the smaller diode emission area. This tapered section collects rays that would normally pass outside the 33.75° angle, and so would not normally be collected. The angle of the tapered section redirects such rays to fall within the 33.75° limit. The length and angle of this section can be chosen such that both the collection efficiency is optimal and the output angular distribution of radiation from the guide is closest to a "top hat" distribution. In addition, there exists an optimum refractive index that minimises the size of the light guide for this collection geometry.

Typical collection efficiencies of the light guides (power collected to total power emitted) are approximately 95%. An alternative collection scheme would be to taper and dielectrically coat the guide in the form of a parabolic reflector. However, this method is more expensive and does not produce better

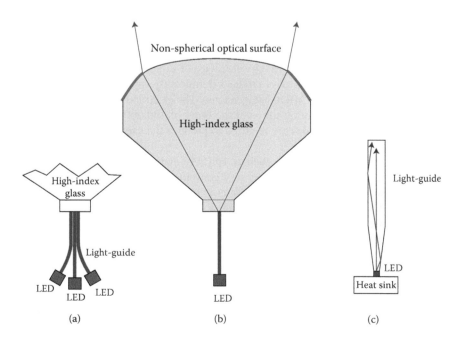

FIGURE A6.1   Monochromatic light source showing details of (a) the coupling of radiation from three diodes to three light guides and (b) subsequent coupling of the light guides to a main high-index optic with non-spherical surface. (c) Details of the tapered light guide matched to the LED.

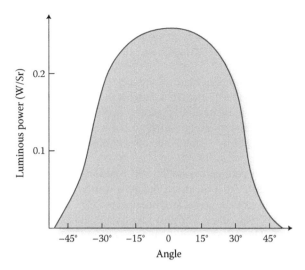

FIGURE A6.2 Radiant intensity distribution emitted from a light guide fed by a LMXL-PM01-0080 diode and coupled into Sumita K-PSFFn202 glass.

results. Each of the light guides may be bent somewhat to transport the radiation from each diode to a desired location on the main optic.

Figure A6.2 shows a typical plot of the radiant intensity distribution emitted at the end of a light guide as calculated by diffractive ray tracing. The light guide is optically attached to the main optic element, which is made from Sumita K-PSFn202 glass having a refractive index of 2.035 at 530 nm. The radiation emission into the Sumita glass is therefore somewhat reduced in angular extent by refraction at the interface. The overall appearance of the distribution is quite close to a top hat with a width of ±32°.

Returning to Figure A6.1, the radiation from the light guides is injected into the main optic. This constitutes a cylindrical block of Sumita K-PSFn202 glass, 41.5 mm long, with a specially designed non-spherical front surface incorporating a graded aperture.

The function of the non-spherical surface is to redirect the radiation coming from each of the three light guides such that the irradiance distribution (usually defined at a target plane) downstream of the lens matches a given target distribution (which, for example, would be the hologram in applications of a light source for display holography). Typically, this target distribution will be a top hat intensity distribution over a rectangular area and the target will usually be tilted. Another function of the refracting surface is to ensure that each point on the target plane "sees" a source of good spatial coherence.

The function of the graded aperture is to eliminate rays impinging on the front surface of the lens at too great an angle without causing a secondary diffractive source. In situations where the light guides are designed to produce a good top hat intensity distribution, a hard aperture is also suitable.

The reason that a single non-spherical refracting surface is used is that it economises the cost of fabrication of the light source (the Sumita glass in question is mouldable) and it allows an efficient and compact design. The geometry of the refracting surface is, however, non-trivial to calculate and this subject will be covered in detail in the following sections.

A6.3 Polychromatic Light Source

The positions at which the three light guides are attached to the main optic in Figure A6.1 must be chosen carefully to optimise the characteristics of the light source. This is because the light guides are of finite size and the lens surface always has optimum performance only for one unique source position. In the case of a bichromatic source, it is sometimes possible to find acceptable positions for two or more light guides carrying radiation of different wavelengths. In this case, a bichromatic light source is extremely similar to the monochromatic source already described. However, for three colours, this is in practice rarely possible (particularly in the case of hologram illumination for realistic hologram illumination geometries).

The light source that we describe here requires the use of high refractive index glasses. However, it is known that such glasses always have low Abbe numbers and, as a result, the behaviour of the main refracting surface changes with wavelength. In practice, we find that as long as the Abbe number is not too low, then a lens surface designed for one wavelength can be used at another if the position of the light guide on the main optic is changed by a small amount. Nevertheless, this displacement is rarely larger than the size of the light guide.

We are therefore almost always obliged to employ dichroic reflectors to deliver the red, green and blue radiation to the main optic. Figure A6.3 illustrates how this works. The main optic is now split into three Sumita K-PSFn202 glass sections that are optically connected. The top section comprises the main lens surface with graded neutral density filter. The next section contains a 45° dichroic mirror embedded in a cylindrical block of glass. Three light guides carrying red radiation from three LMXL-PD01-0040 diodes are optically attached at a location in point A (side by side). The red radiation is reflected by the dichroic mirror, which is transparent to blue and green light. An additional cylindrical block contains a dichroic mirror that reflects green and transmits blue light. Three light guides carrying green radiation from LMXL-PM01-080 diodes are mounted at location B. Finally, three blue guides are mounted at location C.

This design allows the exact positions of the red, green and blue light guides to be adjusted precisely and independently for the best performance. Often, for highly tilted holographic displays, the lens design implies a much greater sensitivity to the positioning of the light guides on the main optic's surface in one dimension rather than the other. This allows similar-coloured light guides to be stacked in that dimension when higher power is required.

The Sumita K-PSFn202 glass is a particularly suitable glass for the present application because it has a high index (2.035 at 530 nm), relatively high Abbe number ($V_d = 21.5$) and melts at a relatively low temperature (yielding point = 486°C) making it easier to mould and polish than other glasses which typically

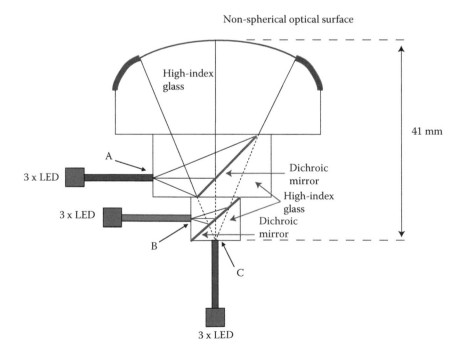

FIGURE A6.3   Trichromatic light source coupled to nine diodes.

have melting points 100°C higher than this. It also has excellent transparency in all visible wavelengths including the blue spectrum.

Other glasses, such as the higher index Sumita K-PSFn214, are unsuitable for polychromatic applications because their transmittance in the blue spectral region is poor and their Abbe number is rather lower. However, their performance is superior for monochromatic applications in the green and red spectral region. Lower index glasses such as SF57 ($n = 1.846$), which have good transparency and Abbe characteristics, can be used for some applications with available diode sources. However, for the majority of applications, glasses with a refractive index of greater than 2 are to be preferred.

A6.4 Design of the Main Lens Surface

The function of the main lens surface is to refract the rays emanating from a light pipe and to redistribute them such that the radiation distribution on a tilted target plane downstream of the lens matches, as closely as possible, a chosen distribution, while the spatial coherence "seen" by individual points on the target plane within a proscribed area remains optimal. Using a geometric optics approach, this may be cast in the form of a standard optimisation problem and may be solved computationally for all cases of interest.

A6.4.1 Monochromatic Light Source

Figure A6.4 (side view) depicts a lighting source optically coupled to a thick lens illuminating a target plane inclined at angle Φ. Figure A6.5 shows the corresponding view from behind the light. The light source emits radiation over a certain solid angle. We assume that the source is point-like to start with; this will be generalised later to a source of finite size. Likewise, we shall assume that the radiant intensity distribution of the source is constant per unit of solid angle and that the emission is in the form of a cone of maximum angle ϕ_m. Later, we will check our results for the real radiant intensity distributions that we calculate explicitly for each light guide.

The source emits rays A, B and C. These propagate through the thick substrate of the lens, hitting the lens surface at points D, E and F, respectively. At the lens surface, the rays are refracted and intersect the target at points H, I and J, respectively. Both Figures A6.4 and A6.5 have a distorted scale for clarity. In practice, the light source and lens are typically some centimetres in size and the target may have dimensions of some tens of centimetres to some tens of metres.

By choosing a general lens surface, all rays emitted from the source can be made to illuminate a target of a chosen shape. Furthermore, by constraining the lens surface more stringently, it is possible to

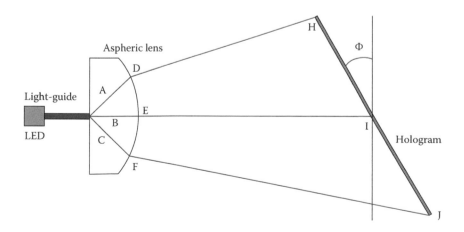

FIGURE A6.4   Side view of a simplified light source comprising a light guide plus the main optic and a tilted target plane.

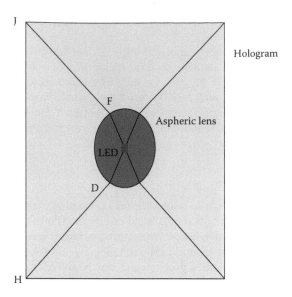

FIGURE A6.5 View of a simplified light source and target plane as seen from behind the light guide.

produce a wide range of power density distributions over the target. Most importantly, it is possible to produce almost exactly rectangular distributions of constant power density on a tilted target as required by the problem of hologram illumination. At extreme parameter values, particularly with hologram tilt, limitations do appear and it may not be possible to then attain exactly flat power distributions. However, numerical studies indicate that there is an extremely wide useful parameter space in which useful light sources may be designed.

A6.4.2 Mathematical Formulation

With reference to Figure A6.6, let the lens surface height, $z(x,y)$, be described by the following polynomial spectral form:

$$z(x,y) = \sum_{\alpha=0}^{N} \sum_{\beta=0}^{\alpha} A_{\alpha\beta} r^{\alpha} \cos(\beta\theta) \tag{A6.1}$$

where $r \geq 0$, $2\pi > \theta \geq 0$, $y = r \sin\theta$ and $x = r \cos\theta$.

We assume the plane $y = 0$ is a plane of symmetry and, accordingly, we omit the sine terms in the expansion. In addition, the fact that z must be single-valued at $r = 0$ implies that $\alpha \geq \beta$ and that $A_{0\beta} = 0 \; \forall \; \beta \geq 1$.

Assuming a point source at $(0,0,0)$ that emits over a cone angle of $\phi \leq \phi_m$, the intersection (x,y) of a given ray at angle θ with the lens surface may be traced using geometric optics through the lens to its intersection with the target plane. By considering the projection of the outer rays as defined by $\phi = \phi_m$ on the target plane, we may choose the coefficients $A_{\alpha\beta}$ such that the intersection of outer rays most closely matches a given target contour.

More specifically, if we define

$$\Xi \equiv \sum_{\phi=\phi_m} (x_{IO} - x_T)^2 + (y_{IO} - y_T)^2 \tag{A6.2}$$

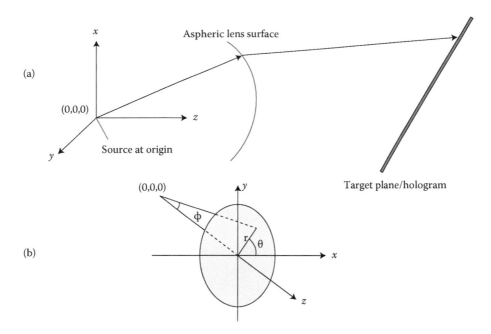

FIGURE A6.6 Geometrical coordinate systems used. (a) Side view and (b) view looking back through the lens towards the LED source.

where x and y represent Cartesian coordinates of the target plane, the subscripts IO and T, respectively, denote outer ray intersections and target contour coordinates, and the sum is carried out over all outer rays, then the equation

$$\frac{\partial \Xi}{\partial A_{\alpha\beta}} = 0 \tag{A6.3}$$

defines a lens that will act to deviate the outer rays in such a manner that they intersect most closely with a given target contour on a given target plane.

The manner in which the target contour coordinates are specified is important. One could, for instance, insist that a given outer ray passes exactly through a certain point on the target contour. However, this is rather restrictive and it is more convenient to define a given target contour coordinate pair as being that coordinate pair whose distance squared to (x_{IO}, y_{IO}) is smallest.

In general, there are too many solutions to Equations A6.2 and A6.3 for such a system to be of use in calculating a real lens. This is because rays "cross over" too easily and most solutions consist of lenses that therefore result in a light source of low spatial coherence. In addition, most solutions are characterised by inappropriate power density distributions on the target plane.

To resolve the inadequacies of the above solution, we introduce a modified form for Ξ:

$$\Xi \equiv \sum_{\varphi=\varphi_m} (x_{IO} - x_T)^2 + (y_{IO} - y_T)^2 + \gamma \sum_{S_m} (\varepsilon_{IO} - \varepsilon_T)^2 \tag{A6.4}$$

where γ is a Lagrange multiplier, ε_{IO} is the calculated power density at a given point on the target plane, ε_T is a chosen target power density for that same location, the additional sum is taken over all solid angle elements of the unit sphere surrounding the source and

$$\sum_{S_m} (\varepsilon_T) = E \tag{A6.5}$$

where E is the total power emitted by the source.

By introducing this second "energy" term in the error function, solutions in which no rays cross may be preferentially generated and, in addition, tailored power density distributions may be catered for.

It is an interesting empirical fact that the degree of freedom implicit in the problem of the design of a general lens for a point light source is such that many power density distributions and illumination contours may be generated without rays crossing. This situation remains broadly the same with the introduction of a small but finite size to the source.

A6.4.3 Geometrical Ray Tracing in the Point-Source Approximation

Assuming a point source at (0,0,0) that emits over a cone angle of $\phi \leq \phi_m$, the intersection (x,y) of a given ray at angle θ with the lens surface is given by the non-linear equation:

$$\sum_{\alpha=0}^{N} \sum_{\beta=0}^{\alpha} A_{\alpha\beta} r^{\alpha} \cos(\beta\theta) \tan\phi = r$$

$$x = r\cos\theta$$

$$y = r\sin\theta \tag{A6.6}$$

This can be effectively solved using a Newton–Raphson method:

$$r_{n+1} = r_n - \frac{g(r_n)}{\left.\dfrac{\partial g}{\partial r}\right|_n}$$

$$g = \sum_{\alpha=0}^{N} \sum_{\beta=0}^{\alpha} A_{\alpha\beta} r^{\alpha} \cos(\beta\theta) \tan\phi - r \tag{A6.7}$$

The normalised propagation vector, $\hat{k}$, of a given ray emitted from the source and intersecting the lens surface at (x,y) is given by (Figure A6.7):

$$\hat{k} = \frac{1}{(x^2 + y^2 + z^2)^{1/2}} \begin{pmatrix} x \\ y \\ z \end{pmatrix} = \begin{pmatrix} \sin\phi\cos\theta \\ \sin\phi\sin\theta \\ \cos\phi \end{pmatrix} \tag{A6.8}$$

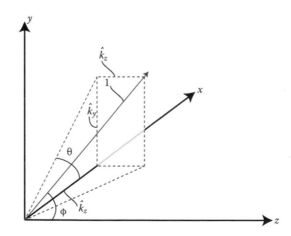

FIGURE A6.7 Diagram showing the unit ray vector $\hat{k}$ in relation to the various coordinates.

At the lens surface, this ray refracts. Denoting all quantities to the right of the surface by starred variables, the vector form of Snell's law states:

$$\hat{\mathbf{k}}^* = \mu\hat{\mathbf{k}} - \hat{\mathbf{N}}\left\{\mu\left(\hat{\mathbf{k}}\cdot\hat{\mathbf{N}}\right) - \sqrt{\left(\left[1-\mu^2\right]+\mu^2\left(\hat{\mathbf{k}}\cdot\hat{\mathbf{N}}\right)\right)^2}\right\} \tag{A6.9}$$

where $\hat{\mathbf{N}}$ is the unit normal vector to the surface at (x,y) and

$$\mu = \frac{n}{n^*} \tag{A6.10}$$

is the ratio of the refractive indices to the left and right of the lens surface, respectively.

Defining the lens function,

$$\Theta(x,y,z) = \Theta(r,\theta,z) \equiv f(r,\theta) - z = 0 \tag{A6.11}$$

where

$$f(r,\theta) = \sum_{\alpha=0}^{N}\sum_{\beta=0}^{\alpha} A_{\alpha\beta}r^{\alpha}\cos(\beta\theta), \tag{A6.12}$$

the surface normal vector of the lens may be defined as

$$\mathbf{N} \equiv -\nabla\Theta = \begin{pmatrix} -\nabla\Theta_x \\ -\nabla\Theta_y \\ -\nabla\Theta_z \end{pmatrix} = \begin{pmatrix} -\cos\theta\,\partial f\big/\partial r + \dfrac{\sin\theta}{r}\,\partial f\big/\partial\theta \\ -\sin\theta\,\partial f\big/\partial r - \dfrac{\cos\theta}{r}\,\partial f\big/\partial\theta \\ 1 \end{pmatrix} \tag{A6.13}$$

and the unit normal as

$$\hat{\mathbf{N}} = \frac{\mathbf{N}}{\sqrt{\mathbf{N}\cdot\mathbf{N}}}. \tag{A6.14}$$

A6.4.4 Ray Intersection with Target Plane

Figure A6.8 shows a diagram of a target plane inclined at an angle ψ to the (x,y) plane. The intersection (x_H, y_H, z_H) of a ray propagating from the lens at point (x_0, y_0, z_0) having k vector $\hat{\mathbf{k}}^*$ with the target plane is given by

$$\begin{pmatrix} x_0 \\ y_0 \\ z_0 \end{pmatrix} + \lambda\hat{\mathbf{k}}^* = \begin{pmatrix} x_H \\ y_H \\ z_H \end{pmatrix} = \begin{pmatrix} z_H\cot\psi - d\cot\psi \\ y_H \\ z_H \end{pmatrix} \tag{A6.15}$$

where we have used the defining equation of the target plane:

$$x_H = (\cot\psi)(z_H - d) \tag{A6.16}$$

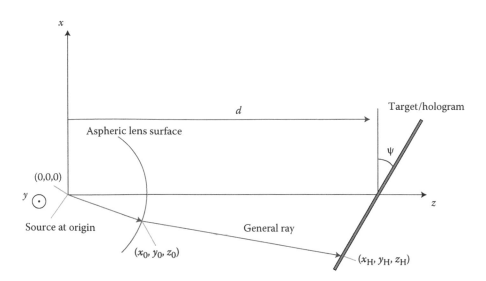

FIGURE A6.8   Diagram showing intersection coordinates on the lens surface and on the tilted target plane.

Solving the above equation yields

$$\begin{pmatrix} x_H \\ y_H \\ z_H \end{pmatrix} = \begin{pmatrix} x_0 \\ y_0 \\ z_0 \end{pmatrix} + \frac{z_0 - d - x_0 \tan\psi}{\hat{k}_x^* \tan\psi - \hat{k}_z^*} \hat{\mathbf{k}}^* \tag{A6.17}$$

In terms of a right-handed coordinate system (x',y') with origin $(0,0,d)$ on the target plane itself, we may write

$$x' = +\sqrt{x_H^2 + (z_H - d)^2} \qquad\qquad z_H \geq d$$

$$x' = -\sqrt{x_H^2 + (z_H - d)^2} \qquad\qquad z_H < d$$

$$y' = y_H \tag{A6.18}$$

This equation defines the intersection point of a given ray with the target plane in target coordinates. Given the spectral polynomial coefficients of a lens and letting θ vary from 0 to 2π and ϕ from a small number, ε to ϕ_m, we are therefore able to calculate the complete pattern of ray intersections on the target plane.

A6.4.5 Calculation of Power Density Distribution at Target Plane

We now assume that the point source emits a constant power, κ, per unit of solid angle within its emission cone. Figure A6.9 shows that portion of the unit sphere surrounding the source through which energy from the source passes. In particular, a general element of solid angle is identified as having a solid angle

$$\delta\Omega = (\cos\phi_N - \cos\phi_{N-1})\delta\theta \tag{A6.19}$$

The four vertices of this element are ray-traced using the equations in the previous section to calculate their intersections with the target plane. These four intersections define a corresponding area element, δA, on the target plane, which may be evaluated using Heron's formula. The energy passing though the element of solid angle, as defined by Equation A6.19, is therefore brought to bear on this elemental area, δA. As $\delta\Omega$ and δA tend to zero, so

(a)   (b)

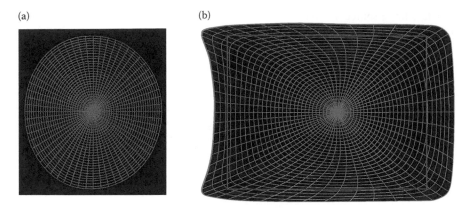

FIGURE A6.9   (a) Lines of equal θ and equal φ on the unit sphere and (b) on the target plane b; target contour shown in red.

$$\lim_{\delta\Omega,\delta A \to 0} \kappa \frac{\delta\Omega}{\delta A} = \frac{dE}{dA} \equiv \varepsilon \tag{A6.20}$$

where ε is the power density at a point on the target plane due to radiation passing through a given point on the unit sphere.

A6.4.6 Numerical Solution

Equations A6.3, A6.4 and A6.5 are solved numerically using the expansion of Equation A6.1. Ray intersections with the general lens surface are solved using the Newton–Raphson solution presented in Equations A6.6 and A6.7. Ray intersections with the target plane are calculated using Equations A6.8 through A6.18. Power density is calculated using Equation A6.20. For reasons of increased computational speed, the energy summation in Equation A6.4 is usually replaced by a summation along several (typically 10) lines of constant θ. Typically, a value of $N = 8$ to 12 is used. Different (θ,φ) meshes from 30×30 to 200×200 are employed.

In general, Equations A6.3 through A6.5 represent a highly non-linear multidimensional system with local minima. It is therefore important to prime the solution with a sensible guess. For many cases, we can use a perfect sphere for this purpose. However, the best initial guess may well depend on each individual case. Where highly tilted target planes are used, solutions should be computed initially for the case with no tilt and then the tilt gradually increased.

Practical experience shows that the best numerical solution is offered by the combination of a quasi-Newton method with a polytrope algorithm. The polytrope algorithm is useful for advancing to the vicinity of a solution, but the quasi-Newton method is often required to find the exact solution. The techniques of Levenberg–Marquardt, conjugate gradients and steepest descents are of little use in the context of the current problem.

A6.4.7 Ray Tracing with Finite Source Size

The best way to calculate useful lenses is by initially using the point-source approximation and then, by checking the results, using a finite-source simulation. Because, in many applications, a lens must be designed for use simultaneously at three component colours, finite-source ray tracing allows a proper characterisation of the lens's behaviour with a change of refractive index with illuminating wavelength. In addition, finite-source ray tracing allows us to compensate for the various effects introduced by the finite source itself or by a different refractive index by varying the lateral position of the source. Indeed, whereas lenses are best calculated using the point-source approximation at a chromatically median refractive index, finite-source simulation must be used to calculate the exact lateral position of each source of a given wavelength.

A6.4.7.1 Method

We assume that the source is composed of a regular grid of elemental point emitters. Typically, we may use a grid of 10×10 such emitters for a square source. Each point emitter at $(x_i, y_j, 0)$ is assumed to emit over a cone angle of $\phi \leq \phi_m$. The intersection (x_{li}, y_{lj}) of a given ray at angle θ with the lens surface from a given elemental emitter is now given by the following non-linear simultaneous equations:

$$g \equiv \sum_{\alpha=0}^{N} \sum_{\beta=0}^{\alpha} A_{\alpha\beta} r^\alpha \cos(\beta\theta) \tan\phi - R = 0$$

$$R^2 + 2R(x_i \cos\theta_s + y_j \sin\theta_s) + x_i^2 + y_j^2 - r^2 = 0$$

$$\tan\theta = \frac{R\sin\theta_s + y_j}{R\cos\theta_s + x_i}$$

$$x_{li} = r\cos\theta$$

$$y_{lj} = r\sin\theta \tag{A6.21}$$

The coordinates (r,θ) refer to the usual polar coordinate system of Figure A6.6 whose origin is at $(x,y) = (0,0)$. The coordinates (R,θ_s) refer to a shifted polar coordinate system whose origin is at the emitter in question, $(x,y) = (x_i, y_j)$. At the boundary $\phi = \phi_m$, these equations are solved for each i and j using a Newton–Raphson method similar to the one introduced in the previous section:

$$r_{n+1} = r_n - \frac{g(r_n)}{\left.\dfrac{\partial g}{\partial r}\right|_n} \tag{A6.22}$$

For each emitter (i,j) this leads to a locus of intersection points $[x_{li}(\theta), y_{lj}(\theta)]$ forming a contour at the lens surface. From this locus of points, a non-orthogonal coordinate system (ξ,ς) is generated to discretise the space interior to the contour. The exact choice of coordinate system is not particularly important. Here we shall choose

$$\xi \equiv r/r_B(\theta)$$

$$\varsigma \equiv \theta \tag{A6.23}$$

where

$$r_B(\theta) = \sqrt{x_{li}(\theta)^2 + y_{lj}(\theta)^2} \tag{A6.24}$$

This coordinate system is illustrated in Figure A6.9a for the case of a 30×60 mesh. For calculations, a 200×200 mesh should be used.

For each elemental emitter, the x, y and z coordinates of each node of the (ξ,ς) system are defined as

$$x = \xi r_B(\theta)\cos(\varsigma)$$

$$y = \xi r_B(\theta)\sin(\varsigma)$$

$$z = \sum_{\alpha=0}^{N} \sum_{\beta=0}^{\alpha} A_{\alpha\beta} (\xi r_B(\theta))^\alpha \cos(\beta\theta) \tag{A6.25}$$

These nodes define a specific (but complete) set of rays emitted from a given elemental emitter, all of which intersect the lens surface. It is a trivial matter, given the nodes and the position of a given elemental source to calculate the set of the normalised k vectors that describe the rays. Once the k vectors are known, then generally each four neighbouring k vectors define an element of area on the lens. The projection of such elemental areas onto the plane of $z = 0$ is of course shown in Figure A6.9a. Using the vector form of Snell's law that we have given in Equation A6.9, we may trace all rays to the target plane where each elemental area will be projected as shown in Figure A6.9b for the case of a 30×60 mesh.

The solid angle of each elemental area on the lens surface is defined as

$$\delta\Omega(\xi,\varsigma) = \iint \frac{\hat{\mathbf{k}} \cdot d\mathbf{A}}{|\mathbf{k}|^2} = \hat{\mathbf{k}} \cdot \hat{\mathbf{N}} \frac{\delta A}{|\mathbf{k}|^2} \tag{A6.26}$$

where $\hat{\mathbf{N}}$ is given by Equations A6.13 and A6.14, and the area element, δA, is calculated by Heron's formula after decomposing the (ξ,ς) element on the lens surface into a pair of triangles.

Because the power emitted per steradian by each elemental source is known, we may easily calculate the power density passing through the lens within this solid angle. In addition, because we know how each elemental area of Figure A6.9a transforms to its corresponding area in Figure A6.9b on the target plane, we may easily work out the power density distribution on the target plane for each elemental emitter. Again, we can use the technique of dividing each elemental area on the target plane into two triangles and applying Heron's formula.

Because each elemental emitter, in general, illuminates the lens in a slightly different way, both the (ξ,ς) coordinate system and its projection onto the target plane will be different for each emitter. For each elemental emitter, we may therefore define a regular Cartesian system at the target plane and interpolate the calculated power density distribution onto this system. In this way, the power density, $P_{ij}(X,Y)$, may be calculated for each elemental emitter (i,j) on the regular Cartesian grid (X,Y), covering the target plane. Then, the total power density distribution at the target plane due to the finite source is simply

$$P = \sum_i \sum_j P_{ij}(X,Y) \tag{A6.27}$$

A6.4.7.2 Spatial Coherence

In the case of the point-source approximation and if the lens in question has also been designed specifically not to allow rays to cross over on their journey to the target plane, then each point on the target plane will effectively "see" only one illuminating point. This corresponds to the case of a light source of perfect spatial coherence. When we replace the point source, however, with a source of finite extent, each point on the target plane will now "see" a variety of rays impinging at different angles from the source. This corresponds to the case of a source of finite spatial coherence. In applications such as holography, the spatial coherence parameter of a source must usually be as large as possible.

To calculate a measure of the spatial coherence of our source, we should first estimate the average value of the k vector illuminating a given point on the target plane. We do this by defining

$$\langle \hat{\mathbf{k}} \rangle \equiv \frac{\sum_i \sum_j \hat{k} P_{ij}}{\sum_i \sum_j P_{ij}} \tag{A6.28}$$

The angle that a given ray from a given elemental source makes to this average value is then given by

$$\psi_{ij} = \cos^{-1}\left(\langle \hat{\mathbf{k}} \rangle \cdot \hat{\mathbf{k}}_{ij}\right) \tag{A6.29}$$

And a measure of the spread in illumination angles for the point (X,Y) is given by

$$\langle \delta\psi \rangle \equiv 2 \left\{ \sum_i \sum_j (\psi_{ij}^2 P_{ij}) \middle/ \sum_i \sum_j P_{ij} \right\}^{1/2} \tag{A6.30}$$

The parameter $\langle \delta\psi \rangle$ effectively measures the angular illuminating source size "seen" by a point (X,Y) on the target plane. It is related to the spatial coherence, l_S, of the source by

$$l_S = \frac{\kappa\lambda}{\langle \delta\psi \rangle} \tag{A6.31}$$

where κ is a geometrical form factor that is 1.22 for circular sources.

A6.5 Selected Cases: Computational Results

A6.5.1 Case 1

We study here the case of a rectangular reflection hologram 500 mm × 600 mm in size (in landscape format), illuminated using an overhead light source positioned 500 mm at a diagonal distance from the centre of the hologram. The light source is arranged such that rays strike the centre of the hologram at 45°. This geometry allows an extremely compact lighting arrangement.

We use the design presented in Figure A6.3 with nine Luxeon Rebel diodes comprising three LMXL-PD01-0040 red diodes (assumed 630 nm), three LMXL-PM01-0080 green diodes (assumed 530 nm) and three LMXL-PR01-0275 blue diodes (assumed 455 nm). The light guides employed are 1.27 mm × 1.27 mm square sections made from Schott SF57HHT glass. The main optic is made from Sumita K-PSFn202 glass. The radiant intensity exiting the light guides at the Schott SF57HHT–Sumita K-PSFn202 interface is calculated by diffractive ray tracing employing the published emission data for the diodes. This is shown in Figure A6.2 for the case of the green diodes for the azimuthal angles of 0, $\pi/2$, π and $3\pi/2$. This data is averaged over the four azimuthal angles, truncated at 34° to model the effect of a hard aperture of this transmission angle and, finally, a top hat distribution independent of azimuthal angle is fitted to the resultant data.

To calculate the shape of the lens surface, we assume initially that the light guide is a point source located at $(0,0,0)$ according to Figure A6.6. We then set up the target geometry to be a rectangle 500 mm × 600 mm tilted at 45°, again according to Figure A6.6. Using an appropriate Lagrange multiplier in Equation A6.4, assuming an eighth order expansion in Equation A6.6 and employing the fitted top hat emission distribution, Equations A6.3 and A6.4 are solved numerically on a PC for $n = 2.035$. A contour plot of the calculated lens is shown in Figure A6.10a. The lens centre is at 41.5 mm from the point source and contours are spaced at 0.485 mm. The green outer line denotes the 34° boundary.

Figure A6.10b shows a plot of the target (hologram) plane and intersections of the set of rays of constant θ and constant ϕ. The red outline shows half of the target rectangle. As the plot is symmetric about the centre horizontal line, the bottom half of the target rectangle is omitted to show the intersections more clearly. The outer contour corresponds to the contour of $\phi = 34°$. Clearly, it can therefore be seen that the lens is acting to shape the light such that it falls nicely within the target contour. In addition, we can see that no contours cross.

Figure A6.10c shows a plot of the irradiance distribution on the target plane. Different colours indicate a difference in irradiance of 12.5%. Clearly, the radiation power is extremely constant over the target, falling rapidly to zero at the edges. The black hole at the centre of the plot indicates the start of the polar grid over which the equations are solved.

Having determined the form of the lens surface by solving the point-source case for 530 nm, we must verify that this lens can work acceptably with the real size of the light guides and with the red and blue emissions. To do this, we use the equations derived in Section A6.4.7 to compute the irradiance

FIGURE A6.10 (a) Calculated height contours of a Sumita K-PSFFn202 glass lens; (b) calculated lines of constant θ and constant φ on the target plane with target contour shown in red; and (c) false-colour plot of the calculated irradiance distribution (c). Case of 500 mm × 600 mm hologram (landscape format) illuminated at 45° from overhead at a distance of 500 mm. Length scales in relative units. Light source positioned to left of diagrams (b) and (c). Power scale: change in colour represents a change in irradiance by 12.5%. Calculations assume a single point source at (0,0,0) and a wavelength of 530 nm.

distribution at the target plane due to the nine actual red, green and blue light guides. Each light guide is divided up for this purpose into 100 equally distributed elemental emitters. Light guides carrying the same colour are stacked symmetrically next to each other in the y direction (Figure A6.8), with the centre diode being located at $y = 0$. For each colour, an optimum offset position in the x direction is calculated for the three guides by solving the equations for the irradiance for a given offset and then re-solving until the best distributions are found for a given offset. Figure A6.11a through c shows graded intensity plots of the optimised irradiance distributions for the green, red and blue emissions, respectively. Differences in shade indicate a 10% difference in irradiance. Clearly, the light source performs extremely well, giving a variation of irradiance over the target of only ±10%. Energetic efficiencies as measured by power within the target area divided by total power falling within and outside are 94%, 80% and 92% for the red, green and blue emissions, respectively. Calculations of the spatial coherence, as measured from multiple points on the hologram surface, indicate that the effective size of the source is approximately 5% of its diameter.

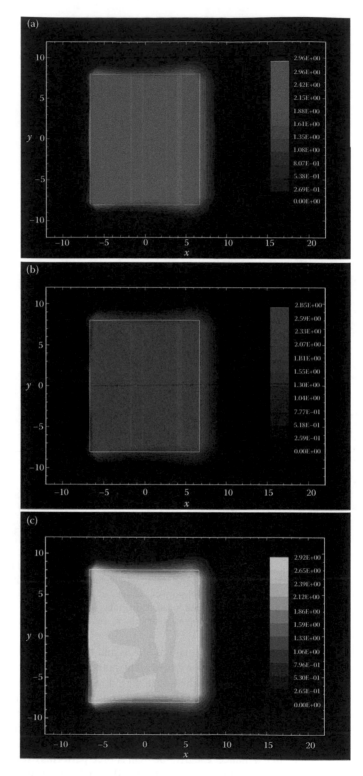

FIGURE A6.11 Graded colour plots of irradiance distribution with hologram outline for (a) green, 530 nm; (b) red, 630 nm; and (c) blue, 455 nm. Case of 500 mm × 600 mm hologram (landscape format) illuminated at 45° from overhead at a distance of 500 mm. Finite-source calculations for the case of nine diodes. A change in graded colour is equivalent to a 10% change in irradiance. Length scales in relative units.

A6.5.2 Case 2

The next case that we shall study is the case of a side-illuminated hologram measuring 808 mm (H) × 888 mm (W). The central angle of incidence is 70° to the vertical and the distance of the source to the centre of the hologram is 1306.246 mm. This geometry allows a large reflection hologram to be mounted in a relatively thin box using a flat side mirror to extend the ray paths (Figure A6.12).

Again, we use the exact same diodes and light guides as described in the previous section. We calculate a new lens surface and new optimal offset positions for the current geometry using exactly the same process as in the previous case, again using an eighth order expansion, except for one detail: we set the target contour to be 10% larger than the actual hologram size. Because the angle of incidence is extreme in the present case, this allows us to accept some decrease in irradiance towards the edges. Figure A6.13

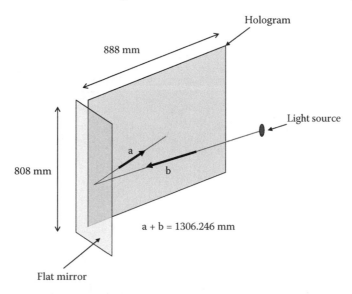

FIGURE A6.12   Hologram (888 mm × 808 mm) side-illuminated by close light source and plane mirror at large angle of incidence allowing a small depth enclosed display.

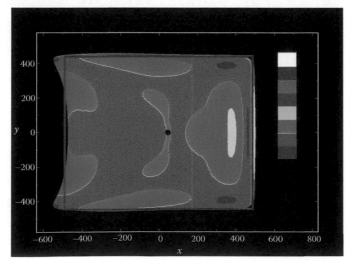

FIGURE A6.13   False-colour plot of the (far field) irradiance distribution on the target plane with target contour shown. Case of 808 mm × 888 mm hologram (landscape format) illuminated at 70° from the side at a distance of 1306.246 mm. Length scales in relative units. Power scale: change in colour represents a change in irradiance by 12.5%. Calculations assume a single point source at (0,0,0) and a wavelength of 530 nm.

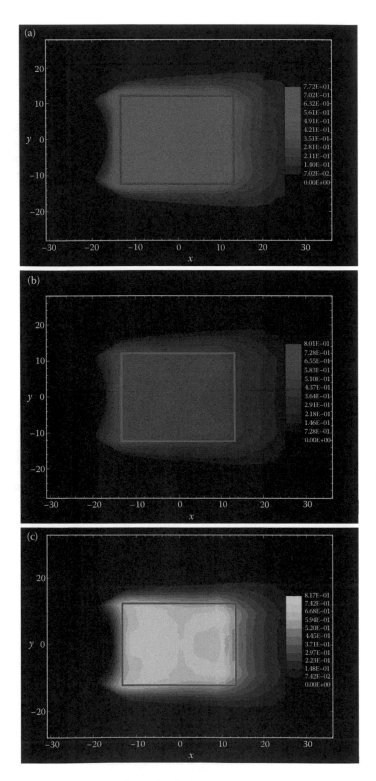

FIGURE A6.14 Graded colour plots of irradiance distribution with hologram outline for (a) green, 530 nm (b); red, 630 nm; and (c) blue, 455 nm. Case of 808 mm × 888 mm hologram (landscape format) illuminated at 70° from the side at a distance of 1306.246 mm. Finite-source calculations for the case of nine diodes. A change in graded colour is equivalent to a 10% change in irradiance. Length scales in relative units.

shows a false-colour map of the calculated irradiance distribution at the target plane, together with an outline of the target contour (note that this is now 10% larger than the hologram size). Different colours indicate a difference in irradiance of 12.5%. Clearly, the result is still good in that the irradiance is relatively flat within the contour and the contour is still well fitted. However, it is clear that the results are not as good as we saw in case 1. This simply shows the limitation of the physics; the 70° angle of incidence is indeed an extreme angle.

Figure A6.14a through c show graded colour plots of the irradiance at the target plane for the three colours using finite-source ray tracing according to Section A6.4.7. The hologram size is also drawn. Differences in shade indicate a 10% difference in irradiance. Clearly, the result is still extremely good for a practical light source. A difference of approximately ±20% in irradiance is apparent within the hologram contour and this falls well into what would be acceptable for a commercial source. Energetic efficiencies for the green, red and blue spectra are, respectively, 72%, 69% and 71%. For the geometry in question, these are excellent figures. Calculations of the spatial coherence, as measured from multiple points on the hologram surface, indicate that the effective size of the source is approximately 10% to 25% of its diameter. By running many cases on the computer, we see that we may accept a reduction in energetic efficiency in return for a flatter irradiance distribution.

A6.5.3 Case 3

The final case that we shall study is one in which a non-rectangular distribution of light is required on the target plane. In particular, we consider a variation of the geometry of case 2, in which the light source is moved up and down while always pointing towards the centre of the hologram. Such a variable geometry can be used to replay repetitive motion in a special type of digital hologram. As such, a reflection hologram may be mounted in a relatively thin box and a light source, which illuminates the hologram from the side and moves up and down continuously, may be incorporated therein: as the light moves, different holographic images are then replayed. This is shown in Figure A6.15. The geometry is the same as for case 2 except that the light source now travels up 47.5 cm and down 47.5 cm.

The geometry of the present case corresponds identically with the geometry of case 2 when the light source is in position A (Figure A6.15). However, if we calculate a lens surface that illuminates the hologram exactly at this position, then when the light source moves to the upper position (position B), clearly not all of the hologram will now be illuminated. The same will be the case at the bottom position (C). To

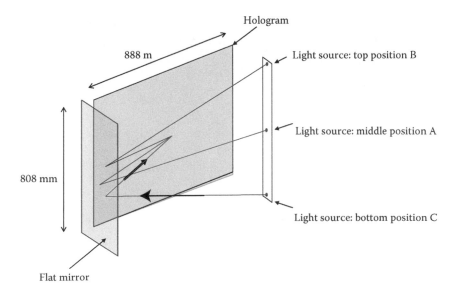

FIGURE A6.15	Hologram (888 mm × 808 mm) side-illuminated by a close light source on a motorised rail allowing motion of the light source in the vertical direction while the light source always points to the centre of the hologram.

choose a target contour that will generally fit the hologram at all light source positions, we use the target contour shown in Figure A6.16.

We now apply the process already described to this case. Figure A6.17 shows the calculated point-source irradiance distribution together with the target contour. Again, different colours indicate a difference in irradiance of 12.5%.

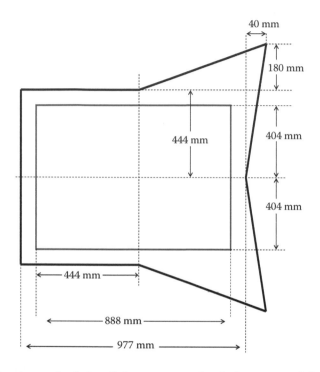

FIGURE A6.16   Target contour used to design a light source appropriate for the geometry of Figure A6.15.

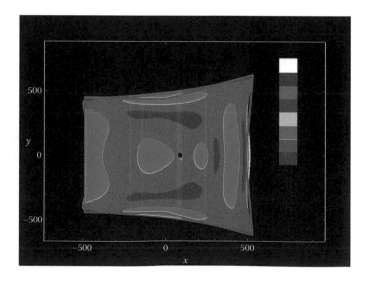

FIGURE A6.17   False-colour plot of the (far field) irradiance distribution on the target plane with target contour of Figure A6.16 shown. Case of 808 mm × 888 mm hologram (landscape format) illuminated at 70° from the side at a distance of 1306.246 mm. Length scales in relative units. Power scale: change in colour represents a change in irradiance by 12.5%. Calculations assume a single point source at (0,0,0) and a wavelength of 530 nm.

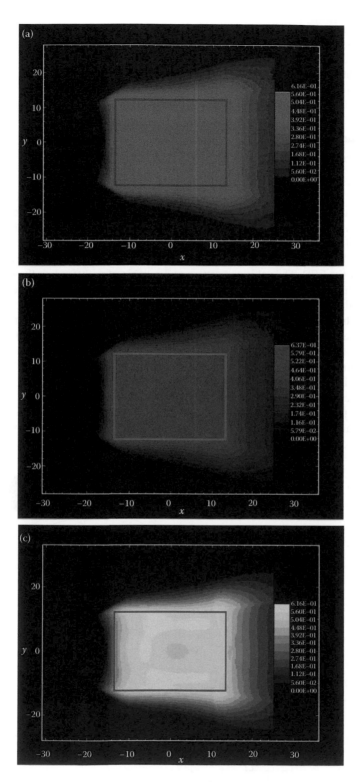

FIGURE A6.18 Graded colour plots of irradiance distribution with hologram outline for (a) green, 530 nm; (b) red, 630 nm; (c) and blue, 455 nm. Case of 808 mm × 888 mm hologram (landscape format) illuminated at 70° from the side at a distance of 1306.246 mm with target contour of Figure A6.16. Finite-source calculations for the case of nine diodes. A change in graded colour is equivalent to a 10% change in irradiance. Length scales in relative units.

Figure A6.18a through c show graded colour plots of the irradiance at the target plane for the three colours using finite-source ray tracing according to Section A6.4.7. The hologram size is also drawn. Differences in shade indicate a 10% difference in irradiance. Clearly, the result is again extremely good for a practical light source. As in case 2, a difference of approximately ±20% in irradiance is apparent within the hologram. Energetic efficiencies for the green, red and blue are, respectively, 61%, 58% and 59%. These are calculated using the real hologram size for a central light source position.

A6.6 Commercial Design of Nine-Diode Framing Light Source

Figure A6.19 shows a commercial design of the light source described conceptually in Figure A6.3. The light pipes and the LEDs are held together in the correct place by a mirrorised plastic holder. The holder touches the light pipes only at sufficient locations to hold and orient it securely, and the mirrored surface ensures that very little light is lost by doing this. The LEDs clip in from the back of the plastic holder and there is optical gel between the light pipes and the surface of the LEDs. The optical gel is matched in refractive index to be the square root of the sum of the squares of the indices of the glass in the light pipe and the surface of the LED. The gel is non-setting and capable of withstanding the surface temperature of the LED dice. In this way, no stress is transmitted from the light pipes to the dice even when there is significant thermal expansion of the dice. The plastic holders in turn push into the lower and upper glass lens parts and are held in place by optical cement, which again is index-matched as described previously. The refractive index of the light pipe glass is chosen so as to provide the minimum light pipe size. The neutral filter fits into the case, and the top glass lens, the dichroic plate and the lower lens are located in the neutral filter and the case that aligns them. Again, optical gel or cement is used between the parts to ensure correct optical performance. The plastic light pipe holders mutually support each other and are in contact with the top lens so that they may be simply pushed down into the case. Heat sink grease is used to make sure there is good thermal contact between the LEDs and the case. The active terminals on the LEDs, which are on the same side as the thermal pad, are arranged to be clear of the case so that an electrical connection can be made.

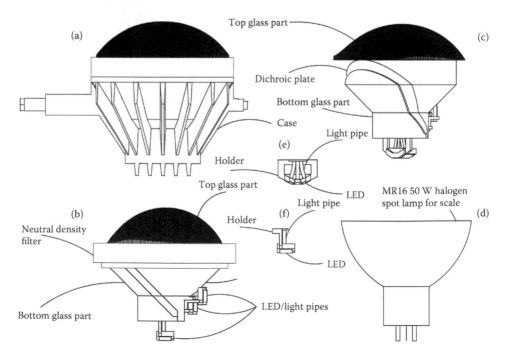

FIGURE A6.19 Elevations of a commercial compact RGB light source (a–c) based on nine LEDs. Details of the LED and light-guide mounting are shown in (e) and (f). A light source (MR16) from previous art is shown to scale for comparison (d).

If appropriate, the top lens can be made into two nesting parts so that a simple change of the top lens will allow field reconfiguration of the beam pattern for many common situations. This feature is something that is impossible with fixed-beam lights such as MR16s.

An important feature of the present design is that the fall-off in the light distribution outside the desired coverage area can be made to be much sharper than that of traditional lamps. These typically have a distribution that falls off in a Gaussian manner, whereas it is easy with the present design to produce a top hat distribution. This means that the light pattern is much more even within the desired area and much less light falls outside that area. It also means that the present design may be used to illuminate abutting holographic panels where different lights illuminate different panels. Previous light sources usually cannot do this.

A scale comparison between the present light and an MR16 lamp is also shown in Figure A6.19d. From this, it can been seen that the physical size of this device is very comparable with an MR16, while at the same time providing all the advantages of power, efficiency, longevity and light distribution pattern described above.

The design lends itself easily to scaling up in size. For instance, 12 diodes may be used instead of 9 in a package broadly the same size as presented here. A version with 30 diodes producing an emission of approximately 3600 lm would be just under twice the size of a standard MR16 when optimised for the display of highly tilted holograms. A version with 90 diodes producing approximately 10,875 lm and optimised for general illumination with an approximately circular footprint would again only be two to three times the size of a standard MR16. This latter example could well be used, when married with a telescopic lens and optional aperture, to replace traditional theatre projectors, which currently use 575-W compact filament halogen lamps giving averages of approximately 12,000 lm (e.g., Source Four Junior Zoom Stage and Studio projector with HPL 575-W CN/240-V bulb made by Ushio) with lifetimes of just 400 h.

A6.7 Additional Considerations

Although the LEDs described are currently state-of-the-art in terms of light output, one of the characteristics of such LEDs is that their lifetime and light output decline significantly with an increase in junction temperature. In the example shown in Figure A6.19, the back of each LED is therefore held firmly against the outer case by the support box. This allows for a good conduction path to the outer case, which is typically die-cast black anodised aluminium and has fins and holes to allow for improved heat dissipation from both the interior and exterior surfaces. This results in a higher light output and a longer life than would otherwise be the case. These particular LEDs have an operating life of up to 100,000 h, which is more than an order of magnitude better than the previous light sources. This is especially important where the light sources are relatively inaccessible for replacement as, for example, in the case where they are used in consumer appliances or in widely dispersed products such as point-of-sale or advertising uses.

Where the light source illuminating a hologram is required to move, it may become difficult to arrange for interconnecting wires. An advantage of the present invention is that because of the low wattages required by the LEDs, it is practical and possible to use a high-frequency electromagnetic coupling between the power source and the LEDs. This removes the necessity for interconnecting wires.

A6.8 Reflective Light-Source

The present light source can also be designed to use a reflector rather than a non-spherical refracting lens. Radiation from one or more diodes may still be delivered to a main reflecting optic via light guides, exactly as we have discussed in previous sections. The shape of the reflector may be parameterised as a polynomial spectral expansion exactly as in the case of a lens. The problem of reflector design may then be cast as a constrained optimisation using one or more Lagrange multipliers exactly as with the refracting lens case; the equations we derived earlier are not changed much except that the vector form of Snell's law must be replaced by the vector reflection law. The numerical solution is then essentially identical.

One advantage of using a reflective geometry is that the vector reflection law does not depend on wavelength. This solution is then better suited to applications in which a truly broad spectrum emission is required and "white" diodes may be employed. As with the design of the lens surface described above, the shape of the reflecting surface is best established by the use of a point-source approximation. Then, the best feeder positions for the light guides are established by ray tracing the case in which all finite-sized light guides are included.

Generally, there exist a number of obvious choices of how to inject light from one or more light guides into a reflective collector. Certain choices (such as where the light guides point themselves in the direction of the target) lead to relatively compact configurations but with the disadvantage of poorer spatial coherence. Other configurations (such as where the light pipes emerge from the centre base of the lamp and are then bent symmetrically to emit radiation at a significant angle to the direction of the target) lead to slightly less compact configurations but with a better spatial coherence.

Appendix 7: Bilinear and Bicubic Interpolation

A7.1 Introduction

As we have seen in Chapters 8 and 9 and also in Appendix 4, digital data are often not available at the exact pixel locations of the image data spatial light modulators (SLMs) in a direct-write digital holography (DWDH) printer. Rather, due to aberration of one kind or another, or non-compliance with the various quantisation rules, exact data are available only at varying distances between the actual pixels. One solution here is simply to use a "nearest integer" interpolation.* Suppose we have a function, $S(x_i, y_i)$, and we want to estimate this function at integer values of x and y. If the data set (x_i, y_i) we are given is dense (in the sense that we always have data near the integer values, i.e., there are no big gaps) but does not contain these integer values themselves, then in nearest integer interpolation, we simply use the estimate $S(\|x_i\|, \|y_i\|)$. So if we know $S(156.1, 293.3)$, then we use this as an approximation to $S(156, 293)$. The problem with this technique is that it creates sudden changes of intensity in S, which leads to interpolation noise in digital holograms. Another way to deal with the situation is to use a single $\mathbf{I}$-to-$\mathbf{S}$ transformation that implicitly incorporates all sources of aberration. This avoids the use of sequential nearest integer operations.

In this appendix, we briefly mention two alternative techniques to the nearest integer interpolation. These are bilinear interpolation and bicubic interpolation. Both seek to minimise abrupt changes in intensity by creating a two-dimensional analytical model of intensity surrounding each pixel. Both techniques are used with great success in conventional image processing and form a vital part of the digital holographer's toolkit.

A7.2 Bilinear Interpolation

Bilinear interpolation is an extension of ordinary linear interpolation which is used to interpolate functions of two variables on a regular grid. The interpolated function uses the bilinear product xy. Bilinear interpolation can be broken down into two operations. The first is a standard linear interpolation in one direction, and the second is another standard linear interpolation but in the orthogonal direction. Although each of these steps is linear in the sampled values and in the positions, the composite bilinear interpolation is not linear but quadratic in the sample location. Bilinear interpolation can be trivially extended to trilinear interpolation in three dimensions.

Suppose that we wish to find the value of an unknown function g at a point (x, y). Suppose further that the value of g is known at the four points surrounding this point. Let us label these points as

$$(x_1, y_1)$$
$$(x_1, y_2)$$
$$(x_2, y_1)$$
$$(x_2, y_2) \tag{A7.1}$$

where

$$x_2 > x > x_1$$
$$y_2 > y > y_1 \tag{A7.2}$$

---

* Also commonly referred to as nearest neighbour interpolation.

The first step is a linear interpolation in the x direction. This gives

$$g(x, y_1) \sim \frac{x_2 - x}{x_2 - x_1} g(x_1, y_1) + \frac{x - x_1}{x_2 - x_1} g(x_2, y_1) \tag{A7.3}$$

and

$$g(x, y_2) \sim \frac{x_2 - x}{x_2 - x_1} g(x_1, y_2) + \frac{x - x_1}{x_2 - x_1} g(x_2, y_2) \tag{A7.4}$$

Next, we use these estimates to interpolate in the y direction:

$$g(x, y) \sim \frac{y_2 - y}{y_2 - y_1} g(x, y_1) + \frac{y - y_1}{y_2 - y_1} g(x, y_2) \tag{A7.5}$$

We can therefore write a single expression for an estimate of the function g at the point (x,y) as

$$g(x, y) \sim \frac{y_2 - y}{y_2 - y_1} \frac{x_2 - x}{x_2 - x_1} g(x_1, y_1) + \frac{y_2 - y}{y_2 - y_1} \frac{x - x_1}{x_2 - x_1} g(x_2, y_1)$$

$$+ \frac{y - y_1}{y_2 - y_1} \frac{x_2 - x}{x_2 - x_1} g(x_1, y_2) + \frac{y - y_1}{y_2 - y_1} \frac{x - x_1}{x_2 - x_1} g(x_2, y_2) \tag{A7.6}$$

In the context of digital holography, we are usually concerned with estimating the SLM data $\mathbf{S}_{\alpha\beta} = \mathbf{S}(x_\alpha, y_\beta)$ in terms of image data $^{kg}\mathbf{I}_{ij}$ through index laws of the form

$$i = X(\alpha, \beta)$$

$$j = Y(\alpha, \beta) \tag{A7.7}$$

where X and Y are real functions. We therefore require an estimation of the image data $^{kg}\mathbf{I}_{ij}$ at non-integer values of i and j. Equation 7.6 can then be used to provide this:

$$I(i, j) \sim \frac{\lceil j \rceil - j}{\lceil j \rceil - \lfloor j \rfloor} \frac{\lceil i \rceil - i}{\lceil i \rceil - \lfloor i \rfloor} I(\lfloor i \rfloor, \lfloor j \rfloor) + \frac{\lceil j \rceil - j}{\lceil j \rceil - \lfloor j \rfloor} \frac{i - \lfloor i \rfloor}{\lceil i \rceil - \lfloor i \rfloor} I(\lceil i \rceil, \lfloor j \rfloor)$$

$$+ \frac{j - \lfloor j \rfloor}{\lceil j \rceil - \lfloor j \rfloor} \frac{\lceil i \rceil - i}{\lceil i \rceil - \lfloor i \rfloor} I(\lfloor i \rfloor, \lceil j \rceil) + \frac{j - \lfloor j \rfloor}{\lceil j \rceil - \lfloor j \rfloor} \frac{i - \lfloor i \rfloor}{\lceil i \rceil - \lfloor i \rfloor} I(\lceil i \rceil, \lceil j \rceil) \tag{A7.8}$$

A7.3 Bicubic Interpolation

Bicubic interpolation is a more sophisticated version of bilinear interpolation. Instead of taking into account only four points around the point where an interpolation is desired, sixteen points are now effectively considered.

Let us assume that we know the value of a function, f, and its derivatives at the four points of the unit square surrounding the point of interest. The derivatives are typically calculated by finite differences using the points immediately surrounding these four points (making sixteen in total). Then, within the unit square, a smooth interpolated surface may be defined:

$$g(x, y) = \sum_{i=0}^{3} \sum_{j=0}^{3} a_{ij} x^i y^j \tag{A7.9}$$

The task is to find the coefficient matrix a_{ij}. Matching function values at the four principal points yields

$$f(0,0) = g(0,0) = a_{00}$$

$$f(1,0) = g(1,0) = \sum_{i=0}^{3} a_{i0}$$

$$f(0,1) = g(0,1) = \sum_{j=0}^{3} a_{0j}$$

$$f(1,1) = g(1,1) = \sum_{i=0}^{3} \sum_{j=0}^{3} a_{ij} \tag{A7.10}$$

In addition, matching the derivatives at the four points yields a further eight conditions, which are

$$f_x(0,0) = g_x(0,0) = a_{10}$$

$$f_x(1,0) = g_x(1,0) = a_{10} + 2a_{20} + 3a_{30}$$

$$f_x(0,1) = g_x(0,1) = \sum_{j=0}^{3} a_{1j}$$

$$f_x(1,1) = g_x(1,1) = \sum_{i=0}^{3} \sum_{j=0}^{3} i a_{ij} \tag{A7.11}$$

and

$$f_y(0,0) = g_y(0,0) = a_{01}$$

$$f_y(1,0) = g_y(1,0) = \sum_{i=0}^{3} a_{i1}$$

$$f_y(0,1) = g_y(0,1) = a_{01} + 2a_{02} + 3a_{03}$$

$$f_y(1,1) = g_y(1,1) = \sum_{i=0}^{3} \sum_{j=0}^{3} j a_{ij} \tag{A7.12}$$

Finally, matching cross-derivatives yields the final four necessary conditions:

$$f_{xy}(0,0) = g_{xy}(0,0) = a_{11}$$

$$f_{xy}(1,0) = g_{xy}(1,0) = a_{11} + 2a_{21} + 3a_{31}$$

$$f_{xy}(0,1) = g_{xy}(0,1) = a_{11} + 2a_{12} + 3a_{13}$$

$$f_{xy}(1,1) = g_{xy}(1,1) = \sum_{i=0}^{3} \sum_{j=0}^{3} i j a_{ij} \tag{A7.13}$$

These relations form a system of linear equations that enable the matrix a_{ij} to be calculated by Gaussian elimination.

Appendix 8: Rigorous Coupled Wave Theory of Simple and Multiplexed Gratings

A8.1 Introduction

Moharam and Gaylord [1] were the first to show how coupled wave (CW) theory could be formulated without approximation. This led to a simple computational algorithm that could be used to solve the wave equation exactly. Although earlier approaches such as the Modal method [2] were also rigorous, they involved the solution of a transcendental equation for which a general unique algorithm could not be defined. This came in contrast to the simple Eigen formulation proposed by Moharam and Gaylord. Here we provide a derivation of the rigorous coupled wave (RCW) theory for the more general spatially multiplexed grating. We shall then show how the resulting equations simplify to Moharam and Gaylord's equations for the simple phase grating. Rather than solving these equations using an Eigen method, we employ an alternative approach using Runge–Kutta integration.[*]

A8.2 Derivation of RCW Equations

For brevity, we shall limit discussions to the lossless case with isotropic permittivity[†] and we shall employ the σ-polarisation for which the Helmholtz equation may be written:

$$\frac{\partial^2 u}{\partial x^2} + \frac{\partial^2 u}{\partial y^2} - \gamma^2 u = 0 \tag{A8.1}$$

where u is the transverse (z) electric field and the parameter

$$\gamma^2 = -\beta^2 - 2\beta \sum_{\mu=1}^{N} \kappa_\mu \left\{ e^{i\mathbf{K}_\mu \cdot \mathbf{r}} + e^{-i\mathbf{K}_\mu \cdot \mathbf{r}} \right\} \tag{A8.2}$$

defines the multiplexed grating.[‡] Following the notation established in Chapter 12, we now consider the case of illumination of the grating by a wave of the form

$$u(y < 0) = e^{i(k_x x + k_y y)} \tag{A8.3}$$

where

$$\begin{aligned} k_x &= \beta \sin(\theta_{c\mu} - \psi_\mu) = \beta \sin(\Phi_c) \\ k_y &= \beta \cos(\theta_{c\mu} - \psi_\mu) = \beta \cos(\Phi_c) \end{aligned} \quad \forall \mu \tag{A8.4}$$

---

[*] This may be programmed with exceptional ease using Mathematica from Wolfram Research Inc using the NDSolve function.

[†] In 1989, Glytis and Gaylord extended the RCW theory to anisotropic media and spatially multiplexed gratings.

[‡] This is just the polychromatic version of Equation 12.107.

In both the front region ($y < 0$) and the rear region ($y < d$), the average index is assumed to be n_0.

Now, the Helmholtz field $u(x, y)$ may be consistently expanded as follows:

$$u(x,y) = \sum_{l_1=-\infty}^{\infty} \sum_{l_2=-\infty}^{\infty} \sum_{l_3=-\infty}^{\infty} ... u_{l_1 l_2 l_3...}(y) e^{i(k_x + l_1 K_{1x} + l_2 K_{2x} + ...)x}$$

$$= \sum_{l_1=-\infty}^{\infty} \sum_{l_2=-\infty}^{\infty} \sum_{l_3=-\infty}^{\infty} ... u_{l_1 l_2 l_3...}(y) e^{ik_x x} \prod_{\sigma=1}^{N} e^{il_\sigma K_{\sigma x} x}$$

(A8.5)

This expression may be substituted into Equations A8.1 and A8.2. On taking the Fourier transform and applying orthogonality, we then arrive at the following RCW equations:

$$\left\{ \left(k_x + \sum_{\sigma=1}^{N} l_\sigma K_{\sigma x} \right)^2 - \beta^2 \right\} u_{l_1 l_2 l_3 ... l_N}(y) - \frac{\partial^2 u_{l_1 l_2 l_3 ... l_N}}{\partial y^2}(y)$$

$$= 2\beta \sum_{\sigma=1}^{N} \kappa_\sigma \left\{ u_{l_1 l_2 l_3 ...(l_\sigma - 1)... l_N}(y) e^{iK_{\sigma y} y} + u_{l_1 l_2 l_3 ...(l_\sigma + 1)... l_N}(y) e^{-iK_{\sigma y} y} \right\}$$

(A8.6)

A8.3 Simplification in the Case of Simple Non-Multiplexed Grating

For the case of the simple sinusoidal grating, the transformation

$$u_l(y) = \hat{u}_l(y) e^{i(k_y + lK_y)y}$$

(A8.7)

reduces Equation A8.6 to the more usual form

$$\frac{\partial^2 \hat{u}_l(y)}{\partial y^2} + 2i(k_y + lK_y) \frac{\partial \hat{u}_l(y)}{\partial y}$$

$$= \left\{ (k_x + lK_x)^2 + (k_y + lK_y)^2 - \beta^2 \right\} \hat{u}_l(y) - 2\beta \kappa \left\{ \hat{u}_{l-1}(y) + \hat{u}_{l+1}(y) \right\}$$

(A8.8)

used by Moharam and Gaylord [1].

A8.4 Derivation of Boundary Conditions

In the zones in front of and behind the grating where $\kappa_\sigma = 0$, Equations A8.6 is reduced to the simpler constant index equations:

$$\left\{ \sum_{\sigma=1}^{N} (k_x + l_\sigma K_{\sigma x})^2 - \beta^2 \right\} u_{l_1 l_2 l_3 ... l_N}(y) - \frac{\partial^2 u_{l_1 l_2 l_3 ... l_N}}{\partial y^2}(y) = 0$$

(A8.9)

These equations define which l modes can propagate in the exterior regions. They have simple solutions of the form

$$u_{l_1 l_2 l_3 ... l_N} = A_{l_1 l_2 l_3 ... l_N} e^{i \sqrt{\left\{ \beta^2 - \sum_{\sigma=1}^{N} (k_x + l_\sigma K_{\sigma x})^2 \right\}} y} + B_{l_1 l_2 l_3 ... l_N} e^{-i \sqrt{\left\{ \beta^2 - \sum_{\sigma=1}^{N} (k_x + l_\sigma K_{\sigma x})^2 \right\}} y}$$

(A8.10)

where the square roots are real for undamped propagation.* Accordingly, we may deduce that the front solution comprising the illumination wave and any reflected modes must be of the form

---

* Note that there are modes that propagate inside the grating but which show damped propagation outside.

$$u(x,y) = e^{ik_x x} e^{i\sqrt{\beta^2 - k_x^2} y} + \sum_{l_1=-\infty}^{\infty} \sum_{l_2=-\infty}^{\infty} \sum_{l_3=-\infty}^{\infty} \cdots \sum_{l_N=-\infty}^{\infty} u_{l_1 l_2 l_3 \ldots l_N} e^{-i\sqrt{\left\{\beta^2 - \sum_{\sigma=1}^{N}(k_x + l_\sigma K_{\sigma x})^2\right\}} y} e^{i\left(k_x + \sum_{\sigma=1}^{N} l_\sigma K_{\sigma x}\right) x} \tag{A8.11}$$

Likewise, the rear solution comprising all transmitted modes must be of the form

$$u(x,y) = \sum_{l_1=-\infty}^{\infty} \sum_{l_2=-\infty}^{\infty} \sum_{l_3=-\infty}^{\infty} \cdots \sum_{l_N=-\infty}^{\infty} u_{l_1 l_2 l_3 \ldots l_N} e^{i\sqrt{\left\{\beta^2 - \sum_{\sigma=1}^{N}(k_x + l_\sigma K_{\sigma x})^2\right\}} y} e^{i\left(k_x + \sum_{\sigma=1}^{N} l_\sigma K_{\sigma x}\right) x} \tag{A8.12}$$

By demanding continuity of the tangential electric field and the tangential magnetic field at the boundaries $y = 0$ and $y = d$, we may now use these expressions to define the boundary conditions required for a solution of Equation A8.1 within the multiplexed grating. At the front surface, these are

$$i\sqrt{\beta^2 - k_x^2}\,(2 - u_{000\ldots}(0)) = \left.\frac{du_{000\ldots}}{dy}\right|_{y=0}$$

$$-i\sqrt{\beta^2 - (k_x + l_1 K_{1x} + l_2 K_{2x} + \ldots)^2}\, u_{l_1 l_2 l_3 \ldots}(0) = \left.\frac{du_{l_1 l_2 l_3 \ldots}}{dy}\right|_{y=0} \tag{A8.13}$$

And at the rear surface, they take the form

$$i\sqrt{\beta^2 - (k_x + l_1 K_{1x} + l_2 K_{2x} +)^2}\, u_{l_1 l_2 l_3 \ldots}(d) = \left.\frac{du_{l_1 l_2 l_3 \ldots}}{dy}\right|_{y=d} \tag{A8.14}$$

The modes available for external (undamped) propagation are calculated using the condition

$$\beta^2 > (k_x + l_1 K_{1x} + l_2 K_{2x} + \ldots)^2 \tag{A8.15}$$

Note, however, that there are internal propagating modes that nevertheless do not propagate outside the grating, and these must be retained.

A8.5 Numerical Solution of RCW Equations

Moharam and Gaylord [1] solved the single grating equations (Equation A8.8) using a state-space formulation in which solutions are obtainable through the eigenvalues and eigenvectors of an easily defined coefficient matrix. However, as mentioned above, we can also solve the more general equations (Equation A8.6), subject to the boundary conditions in Equations A8.13 and A8.14, using simple Runge–Kutta integration. This is a practical method as long as the number of component gratings within the multiplexed grating is relatively small.

Diffraction efficiencies of the various modes are defined as

$$\eta_{l_1 l_2 l_3 \ldots} = \frac{\sqrt{\beta^2 - (k_x + l_1 K_{1x} + l_2 K_{2x} + l_3 K_{3x} + \ldots)^2}}{k_y} u_{l_1 l_2 l_3 \ldots} u_{l_1 l_2 l_3 \ldots}^* \tag{A8.16}$$

where the fields in this equation are defined either at the front boundary in the case of reflected modes or at the rear boundary in the case of transmitted modes. Note that we are treating the lossless case here, so the sum of all transmitted and reflected efficiencies totals to unity.*

---

* In the case of the front reflected 000... mode, one uses $\eta_{000\ldots} = \dfrac{\sqrt{\beta^2 - k_x^2}}{k_y}(u_{000\ldots} - 1)(u_{000\ldots} - 1)^*$.

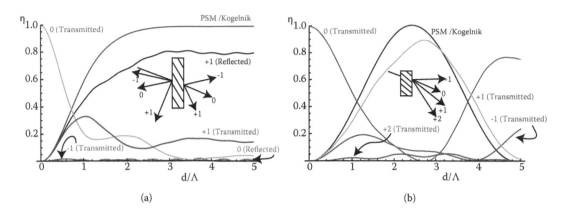

FIGURE A8.1   Diffraction efficiency (η_o) versus normalised grating thickness according to RCW theory and compared to the PSM and Kogelnik theories at Bragg resonance for (a) the simple reflection grating ($n_0 = 1.5$, $n_1/n_0 = 0.331/2$, $\theta_c = \theta_r = 50°$, $\psi = 30°$, $\lambda_c = \lambda_r = 532$ nm) and (b) the simple transmission grating ($n_0 = 1.5$, $n_1/n_0 = 0.121/2$, $\theta_c = \theta_r = 80°$, $\psi = 60°$, $\lambda_c = \lambda_r = 532$ nm).

A8.5.1 Comparison of Kogelnik's Theory and PSM Theory with RCW Theory

Equation A8.6, subject to the boundary conditions in Equations A8.13 and A8.14, is solved by Runge–Kutta integration. This permits the rigorous calculation of the diffraction efficiencies of all modes, which are produced by a general grating. Figure A8.1 shows an example for a simple reflection grating and a simple transmission grating at Bragg resonance.* In the case of the reflection grating, a very high index modulation has been assumed. Nevertheless, the Kogelnik/parallel stacked mirror (PSM) estimation is still only 20% out, and it is clear that most of the "dynamics" of the grating are associated with the +1 reflected mode as both PSM and Kogelnik's CW theories assume. In the case of the transmission hologram, a relatively high index modulation is assumed and also a large incidence angle with respect to the grating planes. Here we see again only a small departure from the Kogelnik/PSM estimation but also the presence of the +2 mode.

A8.5.2 Comparison of N-PSM Theory with RCW Theory

In order to compare the N-PSM model [4] developed in Chapter 12 with the RCW theory, we investigate the typical spatially multiplexed grating, which is illustrated in Figure A8.2. This grating is composed of two simple gratings that have been sequentially recorded in the same material using the same laser wavelength of 532 nm and the same incidence angle of 30°. The component gratings have differing slants and different grating constants and give rise to the multiplexed grating structure shown in Figure A8.2c.

In Figure A8.3, the diffraction efficiency at Bragg resonance as determined by the N-PSM model[†] is compared for different grating thicknesses and different index modulations to an RCW calculation. Typically, 14 modes are retained in the Runge–Kutta integration, the higher order modes being many orders of magnitude smaller than the lower ones. Only the reflected modes are plotted. Figure A8.3a shows an extreme case where an index modulation of $n_1 = 0.3$ for each of the two component gratings of Figure A8.2c is assumed. It is clear that in this case, there are quite important differences between the N-PSM model and the RCW model. In addition, quite a few of the higher order modes allowed under the RCW theory start to oscillate at non-negligible amplitudes, including modes that require both gratings present to propagate. However, this is indeed an extreme case, and one might have anticipated this from the results of Moharam and Gaylord [1]. As the index modulation of each grating drops in Figure A8.3b through d, we see a better and better agreement between the two models. Figure A8.3d represents

---

* Note that at Bragg resonance, the PSM and Kogelnik models give the same predictions.
† Conventional N-CW theory [5] gives an identical prediction to N-PSM at Bragg resonance here.

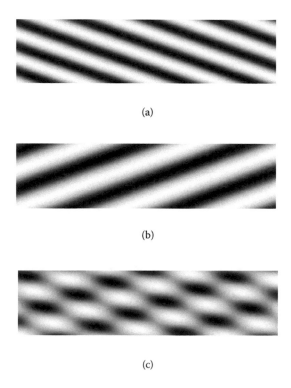

(a)

(b)

(c)

FIGURE A8.2 An example of a spatially multiplexed phase reflection grating. The grating, whose (x,y) index distribution is shown in (c) is formed by the sequential recording of the two simple gratings shown in (a) and (b). Each diagram shows a section of size 0.5 µm by 2 µm. Each simple grating has been recorded with a reference beam angle of $\Phi_c = 30°$ and with a wavelength of 532 nm. One grating has a slope of $\psi_1 = 20°$ and the other has a slope of $\psi_2 = -20°$. Note that the form of the multiplexed grating in (c) is fundamentally different from the characteristic linear shape of its component simple gratings of (a) and (b). Note also that identical index modulations for each of the two component gratings have been assumed in (c).

a typical multiplexed grating made using a modern material such as photopolymer or dichromated gelatine. The N-PSM and RCW models therefore produce extremely good agreement here, and higher order modes only account for less than 1% of the total diffraction.

In Figure A8.4, a comparison of the N-PSM model and the RCW theory for the off-Bragg case is presented. In particular, the diffractive efficiency of the multiplexed grating of Figure A8.2c is investigated as the replay wavelength is changed whilst keeping the replay angle of incidence fixed at 30°. Although a complex analytic solution of the N-PSM equations is available for this problem, for pure convenience, one can solve the N-PSM equations numerically using Runge–Kutta integration. In Figure A8.4a, an index modulation for each of the component gratings of $n_1 = 0.03$ is used; this would be rather typical for diffractive elements made from photopolymer or dichromated gelatin. Agreement between the two theories is clearly excellent, particularly in the primary diffractive band. In Figure A8.4b and c, we plot for comparison the N-PSM diffractive efficiencies when one or another of the two component gratings is removed from the multiplexed element.

It is interesting to note that the diffractive efficiency does not peak in Figure A8.4a for both components at the Bragg angle. This is because the dominant grating within the multiplexed element depletes the reference wave disproportionately around resonance. As the wavelength is changed away from resonance, there rapidly becomes more reference wave available within the grating for signal generation by the second component grating. Even though the intrinsic efficiency of this second grating drops away from Bragg resonance, the increased reference wave left by the first grating more than compensates for this, leading to the characteristic hollow curve with symmetric peaks away from resonance.

Figure A8.4d repeats the case of Figure A8.4a but with individual index modulations of $n_1 = 0.15$. Here the differences between the N-PSM and RCW solutions are somewhat larger as might be expected.

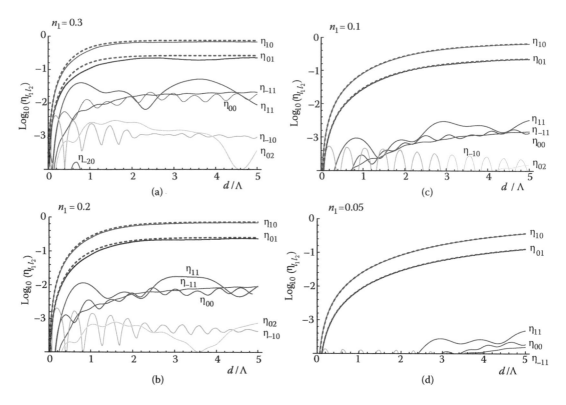

FIGURE A8.3   Diffractive efficiency, η_σ, versus normalised grating thickness, d/Λ, as predicted by the PSM model and by an RCW calculation for the case of the twin multiplexed reflection grating of Figure A8.2c at Bragg resonance. The grating is replayed using light of 532 nm at an incidence angle of $\Phi_c = 30°$. The grating index modulation of each of the component twin gratings has been taken to be $n_1 = 0.3$ in (a), $n_1 = 0.2$ in (b), $n_1 = 0.1$ in (c), and $n_1 = 0.05$ in (d). In each case, the average index inside and outside the grating has been set to $n_0 = 1.5$. The dotted lines indicate the S_1 and S_2 modes of the PSM model and the full lines indicate the modes of the RCW calculation. The most prominent RCW modes are the 01 and 10 modes, which correspond to the S_1 and S_2 modes in PSM. Note that Λ refers to the larger of the two grating periods—i.e. to that of the component grating shown in Figure A8.2b.

A8.5.3   Comparison of N-PSM with RCW Theory for Multicolour Gratings

In Chapter 12, we derived simple equations describing the diffractive response of a multicolour spatially multiplexed grating using the N-PSM model. Here we compare this theory with the RCW theory derived in this appendix for the case of a complex grating composed of three component subgratings as shown in Figure A8.5a through c. All gratings are recorded using sequential exposure and a reference incidence angle of 30°. The grating slants are, respectively, −10°, 0°, and 15°. The first two gratings are recorded at 532 nm whilst the last grating is recorded at 660 nm. Index modulations for the three gratings are chosen to be 0.03, 0.02, and 0.035.

In order to reduce computational time for the multidimensional Runge–Kutta integration, we restrict the modes available in the RCW calculation to only, −1, 0, and 1 for each of the three l integers. Selective introduction of higher order modes is then used to check convergence. Figure A8.6a shows a graph of the results for the dominant 100 and 010 RCW modes and the corresponding S_1, S_2 N-PSM modes for replay at 532 nm. Figure A8.6b shows the corresponding graph for 660 nm where the dominant RCW mode is the 001 mode corresponding to the S_3 mode of N-PSM. Clearly, agreement between the two theories is excellent. The higher order modes of the rigorous theory are typically less than a few percent at these index modulations. At higher modulation, however, we would again expect to see differences between the two theories, as per Figure A8.3. What is important though is that the index modulations of Figure A8.6 are typical for modern holographic techniques and that we can therefore be rather optimistic that

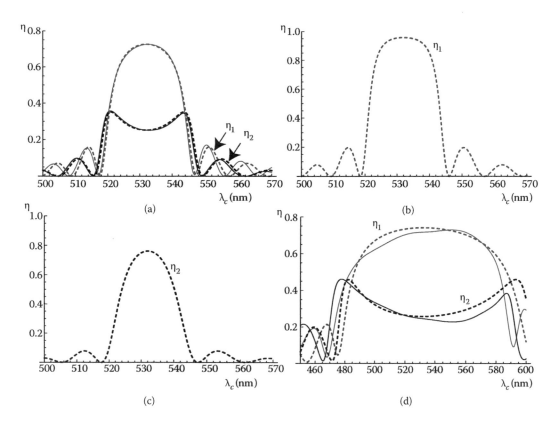

FIGURE A8.4 Diffractive efficiency, η_o, versus replay wavelength, λ_c, as predicted by the PSM model and by an RCW calculation for the case of the twin multiplexed reflection grating of Figure A8.3c at and away from Bragg resonance. The grating is illuminated at its recording angle of $\Phi_c = 30°$. The average index inside and outside the grating has been set to $n_0 = 1.5$. In (a), a grating index modulation of $n_1 = 0.03$ for each component grating is assumed and a grating thickness of $d = 7$ μm is used. The dotted lines indicate the S_1 and S_2 modes of the PSM model, and the full lines indicate the corresponding 10 and 01 modes of the RCW calculation. Panels (b) and (c) show the PSM diffractive response expected when one or the other of the component gratings is deleted from the diffractive element. Panel (d) shows a similar case to (a) but with higher index modulations ($n_1 = 0.15$ for each grating) and with smaller thickness, $d = 2$ μm.

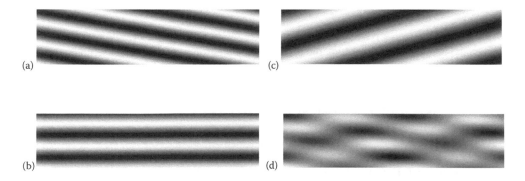

FIGURE A8.5 Example of a panchromatic spatially multiplexed phase reflection grating. The grating, whose (x,y) index distribution is shown in (d), is formed by the sequential recording of the three simple gratings shown in (a) and (b). Each diagram shows a section of size 0.5 μm by 2 μm. Each simple grating has been recorded with a reference beam angle of $\Phi_c = 30°$. Gratings 1 and 2 have been recorded at 532 nm whereas grating 3 has been recorded at 660 nm. The gratings have slopes of $\psi_1 = -10°$, $\psi_2 = 0°$, and $\psi_3 = 15°$ and index modulations of 0.03, 0.02, and 0.035, respectively.

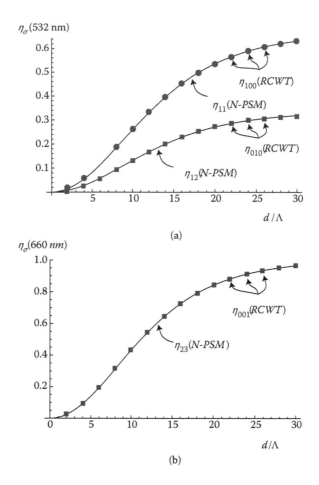

FIGURE A8.6 Diffractive efficiency, η_σ, at 532 nm and 660 nm versus normalised grating thickness, d/Λ, as predicted by the N-PSM model and by an RCW calculation for the case of the triple multiplexed bi-colour reflection grating of Figure A8.5d at Bragg resonance. The grating is replayed using laser light at 532 nm and 660 nm at an incidence angle of $\Phi_c = 30°$. In each case, the average index inside and outside the grating has been set to $n_0 = 1.5$. The full lines indicate the modes of the N-PSM model, and the markers indicate the modes of the RCW calculation. By far, the most prominent RCW modes are the 100 and 010 modes at 532 nm and the 001 mode at 660 nm. Note that Λ refers to the larger of the three grating periods—i.e. to that of the component grating shown in Figure A8.5c.

the N-PSM model at Bragg resonance (or equivalently the N-CW theory of Solymar [5]) is capable of accurately and conveniently describing the multiplexed panchromatic volume phase reflection grating.

REFERENCES

1. M. G. Moharam and T. K. Gaylord, "Rigorous coupled wave analysis of planar grating diffraction," *J. Opt. Soc. Am.* **71**, 811–818 (1981).
2. R. S. Chu and J. A. Kong, "Modal theory of spatially periodic media," *IEEE T. Microw. Theory,* MTT-25, (1977) 18–24.
3. E. N. Glytis and T. K. Gaylord, "Rigorous 3D coupled wave diffraction analysis of multiple superposed gratings in anisotropic media," *Appl. Opt.* **28**, 2401–2421 (1989).
4. D. Brotherton-Ratcliffe, "Analytical treatment of the polychromatic spatially multiplexed volume holographic grating," *Appl. Opt.* **51**, 7188–7199 (2012).
5. L. Solymar, "Two-dimensional N-coupled-wave theory for volume holograms," *Opt. Commun.* **23**, 199–202 (1977).

Appendix 9: Recent Developments

A9.1 New Equipment, Materials, Techniques and Applications

After the manuscript of this book was submitted for typesetting, inevitably new advances in equipment, techniques, and applications came to the authors' attention. Before publication, we were pleased to have the opportunity to include this final appendix, which we believe brings the book properly up to date as of October 2012.

In Chapter 2, we described Lippmann photography and discussed interferential structures in nature. In September 2012, an interesting paper was published revealing that such structures exist not only in insects and butterflies but also in plants. Vignolini et al. [1] present a striking example of multilayer-based strong iridescent colouration in plants, in the fruit of *Pollia condensata* (Figure A9.1). The fruit contains helicoidally stacked cellulose microfibrils that form multilayers in the cell walls of the epicarp. Because the multilayers form with both helicoidicities, optical characterisation reveals that the reflected light from every epidermal cell is polarised circularly either to the left or to the right, a feature that has never been previously observed.

Important papers were presented at the *9th International Symposium on Display Holography* (ISDH 2012) in June 2012 at MIT, which also need to be mentioned here. The reader is referred to the conference website [2] where video presentations [3] and proceedings can be accessed. Zebra Imaging Inc. described a new simple way of capturing 3D information intended for digital colour holograms. Zebra is now offering a printing service based on the technique for people interested in making their own digital colour holograms.

In association with the ISDH conference, there was a hologram exhibition arranged at the MIT museum. One of the exhibited holograms that drew particular attention was a digital achromatic reflection hologram, printed by Geola. The hologram was a new version of the holographic portrait of Queen Elizabeth II, created by the artist Chris Levine and holographer Rob Munday, and demonstrated clearly that high-quality black-and-white holograms can be made using the DWDH technique described in this book. The Queen's portrait was also featured on a recent postage stamp from Jersey Post.

Progress in solid-state CW and pulsed lasers has been rapid, and accordingly, there are now various improved laser systems on the market that are suitable for recording colour holograms.

Progress in recording materials and recording techniques has been made on two fronts. Firstly, a new recording principle for colour holograms based on *surface plasmon waves* has been reported, although more development work will be needed before the technique can become a practical recording method for colour holograms. Additional progress in recording materials has come from Geola, where high efficiency DWDH holograms have recently been written onto photoresist using single nanosecond pulses at energies 10 times smaller than those usually required using conventional CW exposure.

Holografika introduced a new improved HoloVizio™ 3D holographic system at the *Siggraph 2012* Las Vegas exhibition in August 2012. The technique behind Holografika's products has been mentioned in Chapter 14, but the recent products and improvements are worth including here. The company SeeReal in Germany has also reported progress in their VISIO 20 real-time 3D holographic display system.

FIGURE A9.1   Fruits of *Pollia condensata* conserved in the Herbarium collection at the Royal Botanic Gardens, Kew, United Kingdom. Material collected in Ethiopia in 1974 and preserved in alcohol-based fixative. (Vignolini, S. et al. 2012. Pointillist structural color in *Pollia* fruit. *Proceedings of the National Academy of Sciences of the United States*, *PNAS* 109 (39), 15712–15715. Copyright [2012] National Academy of Sciences, U.S.A.)

A9.2 Progress at Zebra Imaging

At ISDH 2012, Craig Newswanger of Zebra Imaging Inc. [4] presented a new simplified technique to capture and print digital colour holograms. Zebra's paper [5] outlined several methods based on *photogrammetry* for producing 3D content for holograms, software applications for editing, positioning and lighting. Photogrammetry is the practice of determining the geometric properties of objects from photographic images. A number of free software tools have become available recently that can extract 3D data from collections of photographs using automated photogrammetry. Examples include commercially available tools such as Autodesk 123D Catch [6]. These types of tools can provide surprisingly accurate 3D models from a variety of cameras, including cell phones, tablets, and point and shoot cameras, and promise to simply revolutionise image acquisition for full-parallax digital holography.

Zebra Imaging has developed the Zscape™ print technology based on the hogel concept. The combination of the easy capture method and a free authoring tool plus an easy web interface for ordering a Zscape™ print finally brings holographic technology "within the reach of the masses" according to Newswanger. 123D Catch (originally named *PhotoFly*) is offered for free from Autodesk. It takes in a set of photos of an object or scene and develops a polygonal model from automatic photogrammetric analysis. It is very simple to learn and builds a model starting from a few to a hundred photographs. More photographs do not necessarily result in a better model. The lion example shown below was based on only 17 photos, sufficient to produce a high-quality model.

The desktop version of 123D Catch is intended for photographs taken with a typical digital camera. By employing the software, it is possible to capture photographs and process the 3D model in the same tool. The desktop tool can export an ".obj" file, which is a ubiquitous file format that is compatible with many 3D tools. In Figure A9.2, 17 captured images of a lion sculpture (from Asolo, Italy) are shown.

By employing Zebra's Zscape™ Preview software, it is possible to display a realistic preview of a potential hologram, including viewing angles, blur from the image plane, lighting, and resolution. Preview is not a modeller but a hologram composition tool that allows the user to scale and position multiple polygonal and point cloud data sets in the same hologram. It is possible to import multiple 3D data models and scale and position them prior to printing, for example, to choose the hologram size, colour, and format that best fit the data. At Zebra's website, it is easy to send the 3D information and to order a hologram. Figure A9.3 shows the hologram preview at the Zebra website. A photograph of the final hologram of the lion sculpture is shown in Figure A9.4.

P1010184.jpg	P1010185.jpg	P1010186.jpg

P1010187.jpg   P1010188.jpg   P1010189.jpg

P1010190.jpg   P1010191.jpg   P1010192.jpg   P1010193.jpg   P1010194.jpg   P1010195.jpg

P1010196.jpg   P1010197.jpg   P1010198.jpg   P1010199.jpg   P1010200.jpg

FIGURE A9.2 Collection of 17 photographs of a sculpted stone lion. Two passes were made at different heights. (Courtesy of Zebra Imaging Inc.)

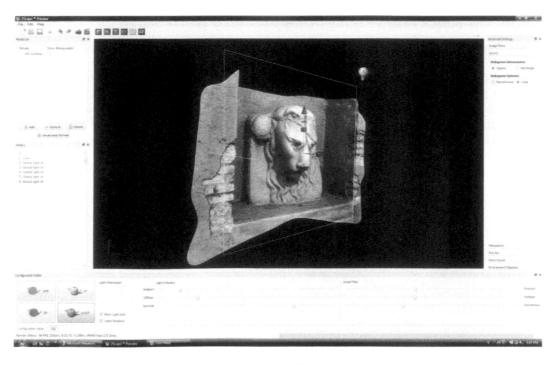

FIGURE A9.3 The "obj" file export from 123D Catch is dropped into Zscape Preview. Note the white frame that indicates the image plane of the hologram. (Courtesy of Zebra Imaging Inc.)

FIGURE A9.4   Photograph of a 60 cm × 60 cm DWDH hologram of the lion sculpture. (Courtesy of Zebra Imaging Inc.)

A9.3 The *Queen Elizabeth II* Portrait and the Jersey Postage Stamp

A9.3.1 Achromatic Portrait of the Queen

The hologram portrait of Queen Elizabeth II, mentioned in Chapter 1, was commissioned by Jersey Heritage Trust. The first hologram portrait *Equanimity* was created by the artist Chris Levine and holographer Rob Munday. The recording of the portrait took place on 24 March 2004 in the Yellow Drawing Room at Buckingham Palace in London. Rob Munday installed the 3D recording equipment with a moving camera that travels along a horizontal rail for recording a sequence of photographic images to be used for the holographic stereogram. In total, 38 sequences were shot, each consisting of 205 frames and each recording taking 8 seconds. The Queen adopted her inimitable regal pose and remained perfectly still for each of the 38 recordings.

A new version of the portrait, *The Diamond Queen*, by Rob Munday was on display at the ISDH 2012 hologram exhibition at the MIT museum. It was a large format (72 cm × 90 cm) one-step HPO digital achromatic reflection hologram, which was printed at the Geola facility in Vilnius using a pulsed RGB laser and the DWDH technique with a hogel size of 0.8 mm. A photograph of the portrait hologram is shown in Figure A9.5.

A9.3.2 Hologram Postage Stamp of the Queen

Jersey Post issued on 1 June 2012 an embossed version of the Queen hologram portrait on a £10 postage stamp. The value and lettering on the stamp was produced in flat simili silver foiling. This is the first time that a 3D holographic portrait has been used on a postage stamp. The embossed hologram for the stamp was created especially for Jersey Post by Rob Munday, and the stamps were printed by Cartor Security Printing in France. A photograph of the stamp is shown in Figure A9.6.

FIGURE A9.5 Photograph of the 72 cm × 90 cm achromatic DWDH portrait hologram, *The Diamond Queen*, on display at the ISDH 2012 MIT hologram exhibition. (Hologram: Joint creative collaboration between C. Levine, R. Munday, and the Jersey Heritage Trust Commemoration. Photo: Courtesy of R. Munday. Printed at Geola.)

FIGURE A9.6 Jersey Post *Queen Elizabeth II* hologram postage stamp. (Hologram: Joint creative collaboration between C. Levine, R. Munday, and the Jersey Heritage Trust Commemoration.)

A9.4 Lasers for Colour Holography

The progress in solid-state CW and pulsed RGB lasers has been fast—in particular regarding output power.

A9.4.1 New Lasers from Cobolt

The range of ultralow-noise DPSS lasers manufactured by Cobolt AB, Solna, Sweden, has been extended to cover higher power red, green, and blue CW lasers. The lasers, which are suitable for colour holography, are based on the company's 05-01 Series platform. The new Flamenco™ laser operates at a wavelength of 660 nm with an output power of 0.5 W. The new Samba™ laser operates at a wavelength of 532 nm with an output power up to 1.5 W. Cobolt has also a blue DPSS laser, the Calypso™ laser, which operates at a wavelength of 491 nm with an output power of 200 mW. The 491 nm wavelength is, however, not suitable for recording colour holograms.

The Cobolt proprietary laser cavity design provides ultralow-noise performance of typically <0.1% rms and a spectral linewidth of <1 MHz. This extremely narrow linewidth can correspond to a coherence length of more than 10,000 m. The lasers provide a TEM_{00} spatial mode and are manufactured using the proprietary HTCure™ technology in a compact and hermetically sealed package, which provides a very high level of immunity to varying environmental conditions and exceptional reliability. A compact controller (CDRH or OEM) is supplied with all the lasers, allowing remote operation and monitoring over RS-232 or USB interfaces.

A9.4.2 New Lasers from Coherent

CW lasers from Coherent (including the high-power Verdi laser) were described in Chapter 3. Here we mention the new range of Genesis MX SLM-Series high-power optically pumped semiconductor lasers (OPSL) with single longitudinal mode and TEM_{00} spatial mode. The two Genesis lasers made by the company are a blue-wavelength laser—the MX 488 nm—with an output power of 1 W and a green-wavelength laser—the MX 532 nm—also with an output power of 1 W.

In addition to the Genesis lasers, Coherent can now also provide lasers with the following wavelengths: 460, 514, 532, 561 and 577 nm with output powers of 0.5 W and, in some cases, up to 1 W.

A9.4.3 New Lasers from Laser Quantum

The Torus SLM lasers from Laser Quantum UK, Stockport, UK, are suitable for colour holography. These lasers have a long coherence length (>100 m) and excellent mode stability provided by active locking. The red-wavelength laser—the Torus 660 nm—has an output power of 200 mW. The green-wavelength laser—the Torus 532 nm—has an output power of 750 mW.

Torus lasers are single-frequency lasers that use intelligent electronics to continually track the longitudinal mode position and to ensure that there are no mode-hops. The patented cavity guarantees single-frequency operation. The photons resonant in the cavity form a travelling wave, removing mode competition, which results in the laser operating at just one frequency. The bandwidths are <5 MHz (the red laser) and <1 MHz (the green laser), and both lasers have a TEM_{00} spatial mode.

The Torus laser power supply is an integral part of each laser system. It has an LCD screen and monitors component temperatures in the laser head, automatically maintains laser output power, and provides diagnostic analysis.

A9.4.4 Progress in RGB Laser Technology at Geola

Geola Digital UAB has very recently extended its range of single-frequency lamp-pumped pulsed RGB lasers for holography. The RGB-α series (see Table 6.4 in Chapter 6) is now commercially available with both amplification and with the option of a mixing unit for applications that require a single white beam (Figure A9.7). Previously, the company had only offered its RGB lasers unamplified and with three

(a)

(b)

(c)

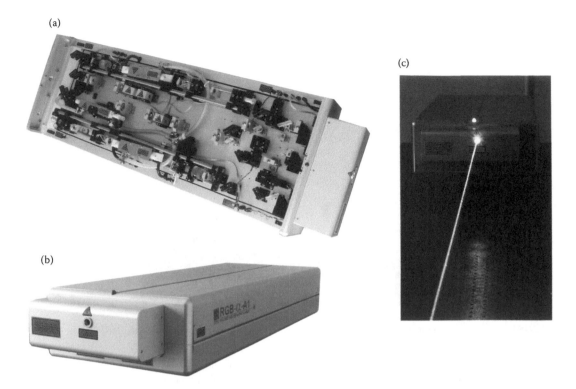

FIGURE A9.7 Several photographs of the new amplified RGB SLM lasers from Geola for holography applications: (a) view of interior optics, (b) external appearance, and (c) laser in action showing output white beam. The laser shown is the RGB-α-A1, which produces just over 0.5 W of white emission at 660 nm, 532 nm, and 440 nm. (Courtesy of Geola Digital UAB.)

separate red, green, and blue output beams. Currently, Geola is able to offer the highest energy output with Nd:YAG. Output wavelengths are 660, 532 and 440 nm. Pulse energies in each colour are up to 15 mJ at a repetition rate of 35 Hz, giving a white beam output of just over 1.5 W. All members of the RGB-α series are available with automatic cavity length stabilisation systems, making these lasers suitable for both digital (single exposure) and analogue (multiexposure) holography. Geola has also released single-colour versions of the technology with power outputs of up to 900 mW at 660 nm (red) or 700 mW at 440 nm (blue). The blue lasers have been extensively tested for the application of writing holograms onto photoresist (see Section A9.5.2).

A9.5 Progress in Recording Materials and Recording Techniques

A9.5.1 Surface Plasmon Waves

On the surface of a metal, there exists a form of slow light in the form of an evanescent wave that is associated with the collective oscillation of free electrons; this is known as a *surface plasmon polariton* (SPP). Such slow light can be used in a variety of applications in nanophotonic materials and devices. Holograms displayed using interactions between light and the collective oscillations of electrons on the surface of a metal are set to transform 3D imaging technology according to Satoshi Kawata and his team at the RIKEN Institute in Japan. They have constructed colour holographic images in a new type of recording material [7]. Plasmons are observed when electrons in a metal collectively oscillate at light wave frequencies. If light falls on a metal with a lower frequency than that characteristic of surface plasmons, it is reflected, while higher-frequency light is transmitted. The characteristic frequencies of surface plasmons propagating in gold and silver are in the visible range and are responsible for the metals'

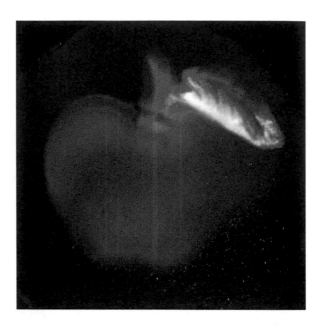

FIGURE A9.8   Surface plasmon hologram of a red apple with a green leaf. (Courtesy S. Kawata, RIKEN.)

distinctive colours. To record a colour hologram in such a recording material, an object is illuminated with red, green, and blue laser light. The light beams are made to diffract onto a glass sheet covered with typically a 150 nm-thick layer of photoresist in which the image is recorded. To make a "plasmon hologram", it is necessary to cover the photoresist with 55 nm silver and 25 nm glass layers. The hologram is reconstructed using white light by illuminating it through a prism from three different angles, one for each colour, in order to create the 3D colour image. A plasmon colour hologram is shown in Figure A9.8.

A9.5.2 Anomalously High Pulsed Sensitivity of Photoresist

Traditionally, photoresist has not been considered a material suitable for ultrarealistic holographic imaging, and indeed, we have not mentioned this material in Chapter 4. With a typical sensitivity only to blue radiation of approximately 60–90 mJ/cm^2, it has really only been used extensively by the embossed hologram industry. Now Zacharovas et al. [8] working at the Geola organisation have demonstrated anomalously high sensitivity of Shipley S1813 photoresist to single pulses of blue radiation (440 nm, 50 ns, pulse energy density 6–8 mJ/cm^2). DWDH transmission holograms were written using a hogel size of 250 μm and the electroplated gratings studied using an NT-206 atomic force microscope from MTM. Diffractive efficiencies obtained were completely equivalent to typical values obtained on the S1813 material with CW exposure but at total energies 10 times lower. Although photoresist is only sensitive to blue radiation, these latest results may well provoke a reconsideration of this material, which has high diffractive efficiency and excellent stability characteristics, for use in high-quality transmission rainbow and achromatic holograms—particularly when the DWDH technique is used.

A9.6 Printers to Record Digital Colour Holograms

A9.6.1 CGH Composite Reflection Hologram Printer

In 2012, Miyamoto et al. [9] reported a new type of digital holographic printer capable of printing full-colour composite digital reflection holograms using computer-generated holography (CGH). Physically, the printer appears almost identical to a standard DWDH printer. It differs principally in the data displayed on the printer's spatial light modulator, in the absence of any optics designed to increase the

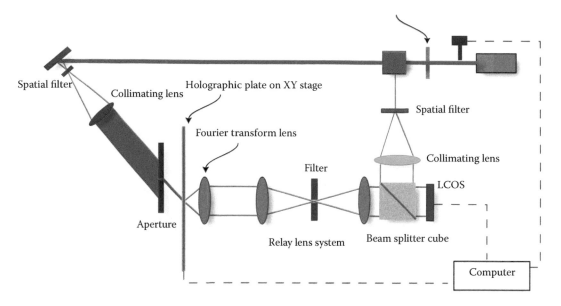

FIGURE A9.9 Simplified optical scheme of the monochrome prototype 2012 CGH reflection hologram printer of Miyamoto et al. [7]. Note that unlike DWDH, first-order diffraction at the SLM is used to create the object beam.

effective étendue of the object beam, which are needed to control hogel size in DWDH, and in its use of first-order diffraction as opposed to the zeroth order that is used in DWDH. In its simplest form (Figure A9.9), a CGH lensless Fourier transform hologram is calculated and displayed on the printer's SLM. This is then illuminated by collimated laser light of high étendue. The first-order diffracted beam produced by the SLM is optically relayed to a Fourier transform lens where it is focused down to the surface of a suitably positioned photosensitive recording material. Here a mutually coherent reference beam co-illuminates the small area illuminated by the diffracted radiation. This results in a small (usually square) elemental reflection hologram of some millimetres in linear dimension being recorded. Further such holograms are then recorded next to this hologram using a step and repeat process. By building up a square matrix of such elemental holograms, a macroscopic reflection hologram is synthesised in a process completely analogous to DWDH.

The main advantage of this new CGH reflection hologram printer over standard DWDH is that each elemental hologram written constitutes effectively a true hologram that contains detailed spatial information exactly as in an analogue hologram. The hogel in DWDH does not have an (x,y) spatial structure, as this is averaged out to increase the fidelity of its main function, which is as a monolithic transmitter of angular information. As such, CGH printers do have the potential to produce better displays than DWDH in applications where close-up viewing of the hologram is paramount. The present state of the technology is, however, far from that of DWDH, which is currently able to produce excellent quality full-parallax full-colour reflection holograms of great depth[*]—but this may well change.

A9.6.2 Pioneer's Compact Holographic Printer

The Pioneer Corporation in Japan has introduced a compact, full-colour DWDH printer for 3 inch × 2 inch (about 75 mm × 50 mm) reflection holograms. The briefcase-shaped consumer electronics product shown in Figure A9.10 has a footprint of 370 mm × 580 mm and is 130 mm high. The printer recording

---

[*] It is often stated that only CGH can produce the correct accommodation cues required in 3D images. This is a common misunderstanding and is misleading. Although CGH is theoretically the technique that is best able to record faithfully in 3D any object or scene with the least optical aberration, from a purely practical point of view, full-parallax DWDH is able to produce full-colour reflection holograms with excellent accommodation cues when hogel sizes are greater than around 0.5 mm.

FIGURE A9.10   New DWDH CW-Laser Colour hologram printer from Pioneer.

wavelengths are 473, 532 and 633 nm. The recording material is the panchromatic photopolymer from Bayer. The holograms are created from graphics images as shown on the unit's monitor.

A9.7 3D Display Systems

A9.7.1 HoloVizio from Holografika

Holografika Kft, Budapest, Hungary [10], demonstrated a new improved HoloVizio system at the *Siggraph 2012* exhibition in Las Vegas. The Hungarian company's 3D holographic technique was mentioned in Chapter 14. A prototype system, the world's first front-projection 140 inch Light Field™ 3D cinema system, was on display. HoloVizio is not a purely holographic system; rather it uses a special holographic screen to redirect light from many small projectors. The pixels of the holographic screen— or, more accurately, voxels as they are intrinsically three-dimensional—emit light beams of different intensity and colour in various directions (Figure A9.11). A light-emitting surface composed of these voxels will act as a digital window or "hologram" and will be able to display 3D information. Many projection engines based on compact LED modules are needed, which are driven by a cluster of nine high-end PCs and sophisticated software. The new large HoloVizio C80 uses the Light Field™ technology, which means that multiple viewers can "look around" objects on the 3.2 m × 1.8 m reflective screen (Figure A9.12). Holografika is working in the framework of various EU projects. The work has been developed in collaboration with Fraunhofer HHI under the MUSCADE European FP7 project. The 3D resolution is 63 megapixels, and the maximum viewing angle is 40°. Recently, Holografika introduced the large HoloVizio 721RC with a 1560 mm × 880 mm 3D display screen with 73 megapixels resolution.

A9.7.2 SeeReal Technologies' Holographic 3D Display

The ultimate 3D television is most likely to be based on the holographic principle; this implies 3D colour images displayed in real time by computer-generated interference fringes generated on an electronic display device of ultrahigh resolution. As we saw in Chapter 14, the great problem here, however, is the insufficient resolution of the available screens.

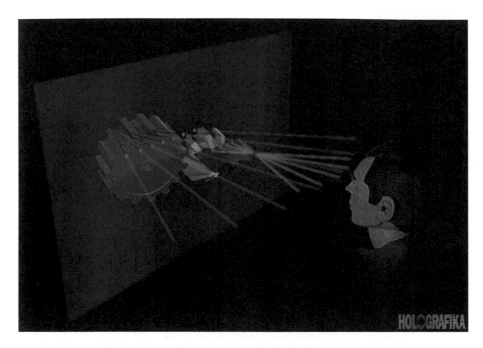

FIGURE A9.11   Principle of the HoloVizio system. (Courtesy of Holografika Kft.)

SeeReal Technologies GmbH, Dresden, Germany, has nonetheless a prototype 3D system up and running that is based on the holographic principle. Eventually, SLMs may well have a resolution high enough to make high-quality full colour real-time holography a reality. The VISIO 20 prototype is, however, based on a real-time computer-generated transmission hologram display with a much lower resolution screen [11–13]. SeeReal's primary goal is to reconstruct the wavefront that is generated by

FIGURE A9.12   HoloVizio 3D cinema system. (Courtesy of Holografika Kft.)

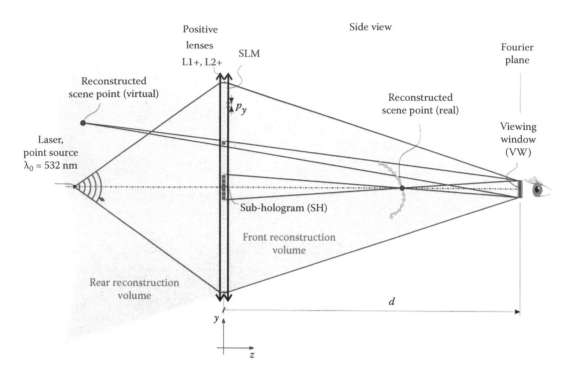

FIGURE A9.13 SeeReal reconstruction geometry. When the hologram (SLM) is illuminated by collimated coherent light (L1+), the 3D scene is reconstructed in the frustum formed by the field lens (L2+). (From Miyamoto, O. et al. *Proceedings of the SPIE*, 8281, paper 82810N, 1–10, 2012. With permission.)

FIGURE A9.14 Display of a colour 3D image. (Courtesy of SeeReal.)

a real existing object in a viewing window at the eye position. In the company's device, the viewing window is the Fourier transform of the hologram and may be as small as the eye pupil. To get over the lack of resolution available, the viewing window is tracked to the eye position (Figure A9.13). The object size is therefore only limited by the hologram size and not by the hologram resolution. 3D scenes extend from in front of the display to a great depth behind the display. In the colour version of the display, RGB lasers are used with sequential multiplexing so that one laser wavelength at a time illuminates the display screen with its corresponding interference pattern. Figure A9.14 shows an example of a displayed colour image.

A9.8 New Holography Camera from the Hellenic Institute of Holography

The Hellenic Institute of Holography (HiH) in Athens has completed the latest version of its commercial transportable camera Z3^{RGB} (model: ZZZyclops, shown in Figure A9.15). The details of the camera design have been described in Chapter 14. The camera is now ready for use in museums around the world. The ultrarealistic Denisyuk colour holograms recorded by the camera are marketed by HiH as OptoClones©. HiH also reported that they have originated a full-colour DWDH holographic map of the island of Kos in Dodecanese (printed by Geola, size 50 cm × 130 cm) on commission from the Greek Infantry Geographical Services using real GIS data and on permanent exhibition at their museum in Athens. More recent information was provided by Lembessis at the *2012 Holo-pack-Holo-print* conference in Austria [14].

A9.9 HoloKit from Liti Holographics

Finally, we should mention that Liti Holographics Inc., Newport News, VA [15], is offering a digital hologram printing service for application within the retail signing, merchandising and point of purchase markets. Liti Holographics is also a manufacturer of HoloKits (small complete kits for making holograms), and in 2012, Liti introduced a new kit for making colour reflection holograms. The kit contains three small (RGB) lasers to record 2 inch × 3 inch holograms on a panchromatic photopolymer material.

FIGURE A9.15 HiH transportable camera Z3^{RGB}. (Courtesy of Hellenic Institute of Holography.)

The price of the kit is relatively affordable, which means that artists, students, and amateur holographers may now be able to experience some of the excitement of recording ultrarealistic 3D holographic images.

REFERENCES

1. S. Vignolini, P. J. Rudall, A. V. Rowland, A. Reed, E. Moyroud, R. B. Faden, J. J. Baumberg, B. J. Glover and U. Steiner, "Pointillist structural color in *Pollia* fruit," in *Proc. National Academy of Sciences of the United States*, *PNAS* **109** (39), 15712–15715 (2012).
2. The 9th International Symposium on Display Holography website: http://isdh2012.media.mit.edu (Oct. 2012).
3. The 9th International Symposium on Display Holography video presentations: http://river-valley.tv/conferences/isdh2012 (Oct. 2012).
4. Zebra Imaging, Inc. USA. www.zebraimaging.com (Oct. 2012).
5. C. Newswanger and M. Klug, "Holograms for the masses," in *Proc. 9th Int'l Symposium on Display Holography*, V. M. Bove Jr. ed., IOP: Conference Series (2013) (to be published).
6. Autodesk®123D® Catch are registered trademarks of Autodesk, Inc. www.123dapp.com/catch (Oct. 2012).
7. M. Ozaki, J.-I. Kato and S. Kawata, "Surface-plasmon holography with white-light illumination," *Science* **332**, 218–220 (April 2011).
8. S. Zacharovas, "Refreshing the concept of security holograms: 3D achromatic masters for manufacturing Embossers," in *Proc. Holo-pack-Holo-print*, www.holopack-holoprint.com (Oct. 2012).
9. O. Miyamoto, T. Yamaguchi and H. Yoshikawa, "The volume hologram printer to record the wavefront of 3D object, in *Practical Holography XXVI: Materials and Applications.* H. I. Bjelkhagen, and V. M. Bove, Jr. eds., *Proc. SPIE*, 8281 (2011) paper 82810N, 1–10.
10. Holografika Kft, Hungary. www.holografika.com (Oct. 2012).
11. S. Reichelt, R. Häussler, G. Fütterer, N. Leicester, H. Kato, N. Usukura and Y. Kanbayashi, "Fullrange complex spatial light modulator for real-time holography," *Opt. Lett.* **37**, 1955–1957 (2012).
12. S. Reichelt and N. Leister, "Computational hologram synthesis and representation on spatial light modulators for real-time 3D holographic imaging," in *Proc. 9th Int'l Symposium on Display Holography*, V. M. Bove Jr. ed., IOP: Conference Series (2013) (to be published).
13. SeeReal Technologies GmbH, Germany. www.seereal.com (Oct. 2012).
14. Liti Holographics Inc, USA. www.litiholo.com (Oct. 2012).
15. A. Lembessis, "Holography for accreditation of works of art," in *Proc. Holo-pack-Holo-print*, www.holopack-holoprint.com (Oct. 2012).

Index

Page numbers followed by f and t denote figures and tables, respectively.

T - #0153 - 111024 - C664 - 254/178/31 - PB - 9780367576431 - Gloss Lamination